PHYSICS OF ATOMIC COLLISIONS

PHYSICS OF
ATOMIC COLLISIONS

J. B. Hasted, M.A., D.Phil.

Professor of Experimental Physics and Head of Department,
Birkbeck College, University of London

Second Edition

LONDON   BUTTERWORTHS

THE BUTTERWORTH GROUP

ENGLAND: BUTTERWORTH & CO (PUBLISHERS) LTD
             LONDON: 88 Kingsway, WC2B 6AB

AUSTRALIA: BUTTERWORTH & CO (AUSTRALIA) LTD
             SYDNEY: 586 Pacific Highway Chatswood, NSW 2067
             MELBOURNE: 473 Little Collins Street, 3000
             BRISBANE: 240 Queen Street, 4000

CANADA: BUTTERWORTH & CO (CANADA) LTD
             TORONTO: 14 Curity Avenue, 374

NEW ZEALAND: BUTTERWORTH & CO (NEW ZEALAND) LTD
             WELLINGTON: 26–28 Waring Taylor Street, 1
             AUCKLAND: 35 High Street, 1

SOUTH AFRICA: BUTTERWORTH & CO (SOUTH AFRICA) (Pty) LTD
             DURBAN: 152–154 Gale Street

First published 1964
Second edition 1972

Suggested U.D.C. numbers 539.184
Suggested additional numbers: 539.186

ISBN 0 408 70148 X

Printed in Hungary

PREFACE TO THE SECOND EDITION

The first edition of this book appeared in 1964, at a time when Atomic Collisions was on the threshold of becoming a growth point in Physics. The growth is now steady, but it has been so considerable that a specialist who had lost contact with the subject over this period would not now recognize it. I have therefore made a complete revision of the book in order to cover all the new concepts and data which have become available. However, I am still conscious that the rate of advance of the subject will inevitably mean that the text will be slightly outdated at the time of publication.

The preparation of this second edition owes much to the invaluable secretarial assistance of Celia Sansom; it is a pleasure to acknowledge this, and also the capable editiorial activities of the staff of Butterworths.

London   J. B. H.

CONTENTS

Chapter 1

COLLISIONS, POPULATIONS, ENERGY DISTRIBUTIONS

1.1 INTRODUCTION. TYPES OF COLLISION PROCESS. NOTATION

Ionized gas phenomena are governed by electric and magnetic fields, by mass, energy and radiation transfer processes, and by the time variation of the populations of different species. This book discusses the study and behaviour of the collision processes which are concerned in these phenomena. These collisions can take place between any of the species present—neutral atoms and molecules in ground and excited states, positive ions, negative ions (the former sometimes multiply charged), and electrons and photons of a wide range of energies. The discussion of the processes which govern the production, the interconversion and the kinetic energies of formation of these species make up the greater part of the book. But no attempt is made to treat collisions in any but the gas or plasma phase, although similar processes in metals, in surfaces and in liquids are familiar.

An important feature is the experimental technique involved in studying the collision processes. Chapter 3 is devoted to technique and includes discussion of the methods of monitoring the species populations in ionized gases. The kinetics of species interconversion are treated in the introductory Chapter 1.

The revival of interest in electronic and ionic collisions has resulted in the multiplication of experimental data, to an extent which makes impossible their detailed tabulation in a work of this size; nevertheless there are such tabulations, compiled by a number of scientists working in 'data collection centres'. Quotations from these tabulations may encourage the reader to seek out and study the originals[1]. There are also two important books covering the experimental side of atomic collision physics[2, 3].

In recent years there has been an increasing emphasis on particle spectroscopy (electron spectroscopy, photoelectron spectroscopy, and heavy-particle spectroscopy). Whereas the purpose of most collision studies has been the elucidation of the nature and probability of a particular collision process, the main purpose of particle spectroscopy is to determine the energies and oscillator strengths of actual energy levels of atoms and molecules, thus adding to the information available from optical spectroscopy. Obviously the different purpose implies differences in the experimental methods, and it is necessary in this edition to lay some emphasis on these differences.

A two-body collision is said to have taken place when any physical change can be detected after the distance between two particles has first decreased and then increased. Such physical changes include angular deflexion or change in magnitude of momentum vectors (within the limits of the uncertainty principle), changes in kinetic or internal energy, chemical changes, and the gain or loss of electrons. Collisions involving only the exchange of kinetic energy are termed 'elastic'; collisions in which kinetic energy is converted into internal energy are termed 'inelastic', and when internal energy is converted into kinetic energy, the collision is termed 'superelastic' or a 'collision of the second kind'.

It is usually necessary both in theory and in experiment to isolate the single collision process from the ensemble of particles in an ionized gas. But there exist situations, such as the diffusion of ions in their own gas, in which elastic collisions cannot be distinguished from inelastic collisions macroscopically; the symmetrical resonance charge transfer process (Chapter 12) produces identical collision products, although the energy distribution between particles is not as in an elastic collision.

In a three-body inelastic collision, a third particle, sometimes termed the 'chaperone', influences the collision process, usually by the removal of excess internal energy as kinetic energy; the physical change taking place in the two-body process is affected by the temporary presence of the third body.

Classification of Inelastic Collisions

A classification of two-body inelastic collision processes is given in Table 1.1 and in Table 1.3 (see page 4).

A concise and occasionally illuminating description of inelastic collision processes may be made with the aid of the notation shown in Table 1.2.

Table 1.1. CHAPTERS IN WHICH COLLISION PROCESSES ARE DISCUSSED

	atoms and molecules	excited atoms and molecules	ionized atoms and molecules	negative ions
Absorption and radiation of photons by	9	9, 13	9	8
Electron collisions with	4, 5, 6	13	5, 6, 7	8
Excited atom and molecule collisions with	13	13	11, 12, 14	8
Ionized atom and molecule collisions with	10, 11, 12, 14	11, 12, 14	11, 12	7, 12
Negative ion collisions with	8, 12	8	7	—
Atom and molecule collisions with	10, 11	13	10, 11, 12, 14	8, 12

Table 1.2. NOTATION FOR THE DESCRIPTION OF INELASTIC COLLISION PROCESSES

Collidant	Symbol
Single photon of any energy	ϕ
Single electron	e
n photons, n electrons	ϕ^n, e^n
Neutral atom or molecule	0
Singly ionized atom or molecule	1
Multiply-charged ion or molecule	2, 3, etc.
Atomic or molecular negative ion	$\bar{1}$
Electronic excitation (no distinction between different states)	Superscript $'$
Two-electron excitation	Superscript $''$
Excitation to a metastable state	Superscript ^{m}
Vibrational excitation	Superscript $'$
Rotational excitation	Superscript ^{r}
Distinction between members of the same species	Subscripts $_{1,\ 2}$

A process in which a projectile of species A collides with a target of species B, producing products of species P and Q ($A+B \to P+Q$) is written for brevity as AB/PQ; thus the single ionization of an atom by an electron $e+X \to e+e+X^+$ would be written $e0/e^21$.

In this notation it is sometimes important that the collidants before and after the collision are written in the same order; thus the single electron capture process (charge transfer) $A^+ +B \to A+B^+$ is written as 10/01 and not as 10/10. It is in the field of heavy-particle collisions that this symbolism is most valuable, even necessary; for example, the distinction between the ionization of the projectile and the target is easily made; ionization of atoms by positive ions is written $10/11e$ and stripping of the projectile as $10/20e$. For electron collisions, the symbolism is of less importance and should not be pedantically followed.

In this notation the collision cross-section σ for a given process, such as charge transfer 10/01, may be written $_{10}\sigma_{01}$. Three-body processes are best represented with the chaperone written after the other reactants, so that, for example, three-body recombination is $e10/00$.

Although no distinction of notation has yet been made between atomic and molecular species, it is valuable in some circumstances to notate each atom separately, using brackets to separate the molecular species from the atoms. Thus the ion-atom interchange process $A^+ +BC \to A^+B+C$ would be written rather clumsily as $1(00)/(10)0$. An advantage of this convention is that, for diatomic positive ions, states AB^+ which separate to $A+B^+$ are distinguished from states A^+B which separate to $A^+ +B$. In general, molecules with more than two atoms are not treated extensively in this book. The notation of molecular species by single figures (0, 1, etc.) is often sufficient for the present symbolism.

Sequences of processes may be written as $AB/PQ/XY$.

Table 1.3 summarizes and notates, without attempting completeness, the important inelastic collision processes encountered in ionized gas studies, and lists the chapters in which they are considered.

Table 1.3. CLASSIFICATION OF INELASTIC COLLISION PROCESSES

Reactants	Collision	Description	Chapter
$\phi0$	$\phi0/0'$	Photoexcitation	9
	$\phi0/0^v$	Vibrational photoexcitation	9
	$\phi0/e1$	Photoionization	9
	$\phi(00)/01e$	Dissociative photoionization	9
	$\phi0/0''/1e$	Autoionization	9
	$\phi(00)/00$	Photodissociation	9
	$\phi_1(00)/0'0/00\phi_2$	Photodissociation into excited atoms	9
	$\phi_10/0'/0\phi_2$	Line fluorescence, chemiluminescence	9
	$\phi^20/0'$	Two-photon excitation	9
	$\phi^n0/1e$	Multi-photon ionization	9
$\phi0', \phi0^v$	$\phi0'/0'$	Photoexcitation	9
	$\phi0'/0'^v$	Vibrational photoexcitation	9
	$\phi0'/1e, \phi0^v/1e$	Photoionization	9
	$\phi(00)'/00$	Photodissociation	9
	$\phi(00)'/0'1e$	Dissociative photoionization	9
	$\phi_10'/0'/0'\phi_2$	Stepwise radiation	
$\phi1$	$\phi1/1'$	Photoexcitation	
	$\phi1/1^v$	Vibrational photoexcitation	
	$\phi1/2e$	Photoionization	
	$\phi(01)/01$	Photodissociation	
$\phi\bar{1}$	$\phi\bar{1}/0e$	Photodetachment	8
	$\phi^2\bar{1}/0e$	Two-photon detachment	8
	$\phi(0\bar{1})/00e$	Dissociative photodetachment	8
	$\phi(0\bar{1})/0\bar{1}$	Photodissociation	8
$\phi e0$	$\phi e0/e0$	Stimulated bremsstrahlung	
$e0$	$e0/e0'$	Electron impact excitation	5
	$e0/e0^v$	Electron impact vibrational excitation	5
	$e0/e^21$	Electron impact ionization	6
	$e(00)/e00$	Electron impact dissociation	5
	$e(00)/10e^2$	Electron impact dissociative ionization	6
	$e(00)/e1\bar{1}$	Simultaneous negative and positive ion production	6
	$e0/\phi\bar{1}$	Radiative attachment	8
	$e0/e0''/1e^2$	Autoionization	6
	$e0/\bar{1}'/\phi\bar{1}$	Dielectronic attachment	8
	$e0/0\bar{1}$	Dissociative attachment	8
	$e00/\bar{1}^v0/\bar{1}0$	Attachment with collisional stabilization	8
	$e0/e0\phi$	Bremsstrahlung	
	$e_10/0e_2$	Electron exchange	4
$e0', e0^v$	$e0'/e0'$	Electron impact excitation	13
	$e0'/e0'^v$	Vibrational excitation by electrons	
	$e0'/e^21$	Electron impact ionization	
	$e(00)'/e00'$	Electron impact dissociation	
	$e(00)^v/e(01)$	Electron impact ionization	
	$e0'/e0$	Superelastic collision	13
$e1$	$e1/e1'$	Electron impact excitation	5
	$e1/e1^v$	Electron impact vibrational excitation	5
	$e1/e^22$	Electron impact ionization	6

Table 1.3. — *continued*

Reactants	Collision	Description	Chapter
	$e(01)/e01$	Electron impact dissociation	6
	$e(01)/e^211$	Electron impact dissociative ionization	
	$e1/\phi0'$	Radiative recombination	7
	$e1/0''/\phi0$	Dielectronic recombination	7
	$e(01)/0'0'$	Dissociative recombination	7
	$e10/00$	Three-body recombination	7
$e\bar{1}$	$e\bar{1}/e^20$	Electron detachment	
	$e(0\bar{1})/e0\bar{1}$	Electron impact dissociation	
$0'0, 0^v0, 0^m0$	$0'0/0'0'$	Excitation	
	$0'0/00'$	Excitation exchange	13
	$0'_10_2/0_10'_2$	Sensitized fluorescence	13
	$0^v0/00^v$	Vibrational excitation exchange	13
	$0'0/0'0^v$	Mixed-gas fluorescence	13
	$0^v0/00$	Vibrational deactivation	13
	$0'(00)/000$	Sensitized chemical reaction; dissociation	
	$0'0/00\phi$	Band resonance radiation	
	$0^m0/01e$	Penning ionization	13
	$0'0/0^m0^v$	Quenching of resonance radiation	13
	$0^m0/(01)e$	Hornbeck–Molnar process	13
	$0'0/(00)'/(01)e$	Associative ionization and band fluorescence	13
	$0'0/10e$	Collision-induced ionization	13
$0'0', 0^m0^m, 0^v0^v,$ etc.	$0'0'/01e$	Ionization	
	$0'0^v/0'0$	Deactivation	
	$0'0'/0'0'$	Excitation exchange	
	$0'(00)'/00'0'$	Dissociation	
$0'1, 0^v1, 0^m1$	$0'1/10$	Charge transfer	12
	$0'1/02e$	Ionization	
	$0^v1/01$	Deactivation	
	$(00)'1/001$	Dissociation	12
$0'\bar{1}, 0^v\bar{1}, 0^m\bar{1}$	$0'\bar{1}/00e$	Detachment (ionization, deactivation, dissociation as above)	
10, 20, etc.	$10/11e$	Ionization	11
	$10/10''/11e$	Autoionization	11
	$10/20e$	Ionization	11
	$10/mne^{(m+n-1)}$	Multiple ionization to charge states m, n	11
	$10/01$	Charge transfer, single electron capture	12
	$20/11$	Partial charge transfer, single electron capture	12, 12
	$10/\bar{1}2$	Two-electron capture	12
	$10/10', 10/1'0$	Excitation	11
	$1(00)(00)/(100)(00)$	Clustering	10
	$1(00)/100$	Dissociation	11
	$1(00)/(10)0$, etc.	Ion–atom interchange	14
	$10/01'/01\phi$	Spark line enhancement	
11	$11/12e$	Ionization	11
	$11/20$	Charge transfer	12

Table 1.3. — *continued*

Reactants	Collision	Description	Chapter
$1\bar{1}$	$11/1'1$	Excitation	11
	$1(01)/101$	Dissociation	11
	$1\bar{1}/0'0'$	Mutual neutralization (ion–ion recombination)	7
	$1\bar{1}/1'\bar{1}$	Excitation	11
	$1\bar{1}/10e$	Detachment	8
	$(01)\bar{1}/01\bar{1}$	Dissociation	11
$\bar{1}0$	$\bar{1}0/0e0$	Collisional detachment	8
	$\bar{1}0/0\bar{1}$	Charge transfer	12
	$\bar{1}0/(00)e$	Associative detachment (also dissociation, ionization, excitation, ion–atom interchange, etc.)	8
$\bar{1}\bar{1}$	$\bar{1}\bar{1}/\bar{1}0e$	Detachment	11
00	$00/10e$	Electron loss (ionization)	11
	$00/1\bar{1}$	Electron capture	8
	$0(00)/(00)0$	Chemical reaction; atom–atom interchange	10
	$00/00^e$	Vibrational excitation	12
	$(00)0/1(00)e$	Ionization by rearrangement	11
	$00/mne^{(m+n)}$	Multiple ionization to charge states m and n	11

The symbols which are used in this book for mathematical discussion of collision processes are listed in Table 1.4.

Table 1.4. NOTATION

Symbol	Representation
a	Adiabatic parameter; radius of curved surface; acceleration
a_0	Radius of first Bohr orbit of hydrogen atom
A	Einstein A coefficient; atomic weight; area
b	Impact parameter; radius of curved surface; constant in van der Waals' equation of state
B	Einstein B coefficient
B_0	Rotational constant
c	Velocity of light
C	Van der Waals' interaction constant
d	Distance
D	Diffusion coefficient; dissociation energy
D_a	Ambipolar diffusion coefficient
D_e	Dissociation energy
e	Electronic charge
E	Energy
E_a	Electron affinity
E_i	Ionization energy
f	Oscillator strength; focal length; distribution function; direct scattering amplitude for electrons; frequency

Table 1.4. — *continued*

Symbol	Representation
f'	Generalized oscillator strength
F	Fraction of beam in a particular state
g	Statistical weight; Kramers–Gaunt g factor; exchange scattering amplitude for electrons
G	Constant in Blanc's law
h	Planck's constant; probability of attachment per collision
H	Hamiltonian operator; heat energy; magnetic field
i	Designation of an excited state; probe current
I	Current or flux of particles; integral; flux of electromagnetic radiation
j	Designation of an excited state
J	Angular momentum; inner quantum number; rotational quantum number; current or flux density
k	Boltzmann constant; rate constant; wave number of particle; damping constant
K	Ionic mobility; momentum exchange in electron scattering; equilibrium constant
l	Collision path length; orbital angular momentum quantum number
l_f	Mean free path
L	Orbital angular momentum
m	Mass; magnetic quantum number
n	Number density; principal quantum number; degree of ionization; refractive index
n_0	Loschmidt's number
N	Flux of particles
N_0	Avogadro's number
p	Gas pressure; linear momentum; radial frequency
P	Polarization factor; probability of collision or transition
q	Line profile; quadrupole moment
Q	Cross-section; energy transferred from kinetic to internal
r	Radius variable; nuclear separation
r_m	Nuclear separation at minimum interaction energy
R	Gas constant; Rydberg constant; radius
R^*	Characteristic excitation in heavy-particle collision
s	Width of slit; spin quantum number
S	Line strength; pumping speed
t	Time
T	Temperature
u	Velocity; energy density of radiation
U	Reduced interaction potential; energy
v	Velocity; vibrational quantum number
V	Interaction potential; electric potential
W	Depth of conduction band of metal; heat energy
x	Distance variable; collision path length
X	Electric field strength
y	Distance variable
z	Distance variable
Z	Nuclear charge; number of electrons
α	Particle absorption coefficient; atomic polarizability; divergence angle; fine structure constant; recombination coefficient; Townsend ionization coefficient

Table 1.4. — *continued*

Symbol	Representation
β	Divergence angle; ratio of negative ions to electrons; afterglow parameter; screening constant; reduced impact parameter
γ	Ratio of mean velocities
Γ	Line width
δ	Hard sphere diameter; Dirac function; phase shift
Δ	Increment of energy or of frequency
ε	Well depth parameter; permittivity; quantum efficiency of surface
η	Phase shift; attachment coefficient
θ	Polar scattering angle
$\varkappa$	Electromagnetic wave absorption coefficient
λ	Wavelength; electron energy loss parameter; Debye length
Λ	Diffusion length; orbital angular momentum quantum number of molecule
μ	Reduced mass; electron mobility; photon absorption coefficient; dipole moment
ν	Frequency
ν'	Wave number
ν_m	Electron collision frequency
π	Target parameter
ϱ	Reduced nuclear separation; density
σ	Cross-section; conductivity
σ_d	Diffusion cross-section
ς	Collision radius of interaction potential
τ	Time constant of exponential decay; reduced temperature; volume
ϕ	Azimuthal scattering angle; wave function; screening function
ψ	Wave function
ω	Radial frequency; vibrational constant
Ω	Solid angle of scattering

Posterior subscripts

0	Initial (zero time or distance); spherically symmetrical; spiralling orbits
1, 2, 3	The particle or number of particles taking part in a collision
e	Electron
$+$	Positive ion
$-$	Negative ion
i	Ion or ionization
ex	Excitation
i, j, m, n	Atomic energy levels
r	Relative
p	Projectile
t	Transferred
c	Centre-of-mass; critical
a	Ante-collision
p	Post-collision

Superscripts

$-$	average
$\sim$	root mean square
$=$	most probable
$\rightarrow$	forward
$\leftarrow$	reverse
m, n	degree of ionization

1.2 COLLISION CROSS-SECTIONS AND RATE COEFFICIENTS

Collision Cross-section in Relation to Mean Free Path

For historical reasons it is necessary to approach from two different standpoints the description of the parameter which is a measure of the probability of a collision process, namely the collision cross-section σ. The first derives from the kinetic theory of gases and the kinetics of electron swarms in gases; the second from experiments with beams.

Following Loschmidt, consider a hard spherical gas atom of radius r moving with velocity v among similar atoms which are at rest. In a time dt, it moves a distance v dt and encounters any atoms whose centres lie within a cylinder of radius $2r$ and length v dt; that is, it makes $\pi n(2r)^2/v$ dt collisions, where n is the gas atom density (atoms cm^{-3}). The mean distance l_f between collisions is therefore given by

$$l_f = \frac{1}{4\pi r^2 n} = \frac{1}{\sigma n} \qquad (1.1)$$

where

$$\sigma = 4\pi r^2 = \pi d^2 \qquad (1.2)$$

is the cross-sectional area of the atomic or molecular sphere[4]. A more careful analysis due to Maxwell includes the relative motions of all the atoms, and leads to a factor $\sqrt{2}$ in the denominator. The distribution of velocities has the effect of raising the average relative velocity of impact. The Maxwell mean free path $(\sqrt{2}\pi nd^2)^{-1}$, or mean of all paths traced out by one atom in unit time, differs by 4 per cent from the Tait mean free path $(\pi nd^2/0\cdot677)^{-1}$, which is the mean of all paths being described at a given instant in the gas.

In atomic collisions, the use of the symbol σ for collision cross-section has for some years been gaining ground over the use of the symbol Q (German Querschnitt); here it is adopted throughout. However, σ has also been used for collision diameter, collision radius and in particular for the nuclear separation r_m at which the potential minimum between two atoms or molecules occurs (see Chapters 10 and 13). Where it is necessary to follow such a convention, the symbol is written ς.

The hard-sphere concept, whilst surprisingly effective, is only an approximation to an interatomic potential with a strongly repulsive core and weakly attractive skin. The 'gas-kinetic cross-sections' deduced from critical data via the van der Waals' constant b, from the temperature variation of virial coefficients and from gas viscosity, are only of approximate significance.

In a binary mixture of particles, the mean free path l_{f1} between all collisions made by the first type of particles is

$$l_{f1} = \frac{1}{4\sqrt{2}\pi n_1 r_1^2 + \pi n_2 (r_1 + r_2)^2 (\bar{v}_1^2 + \bar{v}_2^2)^{1/2}/\bar{v}_1} \qquad (1.3)$$

where $\bar{v}$ represents average velocity, and the subscripts 1 and 2 refer to the two types of particles. For particles of type 1, the mean free path between collisions with particles of type 2 is

$$l_{f12} = \frac{\bar{v}_1}{\pi n_2 (r_1 + r_2)^2 (\bar{v}_1^2 + \bar{v}_2^2)^{1/2}} \qquad (1.4)$$

The 1–2 collision frequency v_{12}, or total number of these collisions per second, is thus

$$v_{12} = \frac{n_1 \bar{v}_1}{l_{f12}} = 4 n_1 n_2 \sigma_{12} (\bar{v}_1^2 + \bar{v}_2^2)^{1/2} \qquad (1.5)$$

with cross-section

$$\sigma_{12} = \pi r_{12}^2 \qquad (1.6)$$

where

$$r_{12} = r_1 + r_2 \qquad (1.7)$$

For a swarm of electrons in a gas of particles of type 2, $\bar{v}_e \gg \bar{v}_2$ and $r_e \ll r_2$, so equation 1.4 approximates to

$$l_{fe} \simeq \frac{1}{\pi n_2 r_2^2} = \frac{1}{n_2 \sigma_e} \qquad (1.8)$$

and equation 1.5 reduces to

$$v_e \simeq \sigma_e n_e n_2 \bar{v}_e \qquad (1.9)$$

where the subscript e refers to the electrons.

In reality, the collision cross-section and mean free path are dependent on the velocity of the particles. Only their averaged values are related by equation 1.1 and equation 1.8:

$$\bar{l}_f = \frac{1}{n \bar{\sigma}} \qquad (1.10)$$

although for atom–atom collisions it follows from equation 1.4 that this average is incorrect. The mean collision frequency for electrons contains an averaged product of cross-section and impact velocity:

$$\bar{v}_e = n_e n_2 \overline{\sigma_e v_e} \qquad (1.11)$$

For atom–atom collisions, the mean collision frequency is

$$\bar{v}_{12} = n_1 n_2 \overline{\left(\frac{v_{12}}{l_{f12}} \right)} \qquad (1.12)$$

However, the use of a similar average

$$\bar{v}_{12} = n_1 n_2 \overline{\sigma v_{12}} \qquad (1.13)$$

does not follow from equation 1.4.

Collision Cross-section in Relation to the Absorption of Beams in Gases

A superior approach to the definition of collision cross-section is made from the standpoint of absorption of a beam or flux of particles passing through a gas. Before the absorption event, such beams must be considered to be free from collisions within themselves, and the particles must be appropriately restricted in angular and energy distribution. The production and handling of beams of the various species of particles is considered in Chapter 3, but it will be valuable at this stage to note the definition of (mass) flux density and its relations to momentum and energy flux density, electric current, etc. A particle flux density is defined as the number of particles crossing normally a unit area in unit time, and the total flux is symbolized as N particles crossing a suitably defined volume in unit time. *Figure 1.1* compares the different units of flux and also the different units of gas pressure; these are related by comparing the number of particles falling on a surface in unit time, when exposed either to gas or to a flux of particles.

When a single fast particle moves in a direction x through a gas whose molecular velocities can be neglected in comparison, the probability P of making a collision in a distance dx is proportional to the number density n, to dx, and to a parameter of the system which is called the collision cross-section σ. Thus

$$P = n\sigma \, dx \qquad (1.14)$$

A flux of N projectile particles cm^{-2} moving at a velocity v is reduced in a distance dx by a decrement

$$dN = nN\sigma \, dx \qquad (1.15)$$

So the flux density N after a collision path length x is

$$N = N_0 \exp\left(-n\sigma x\right) \qquad (1.16)$$

For particles carrying a single charge, the initial beam current

$$I_0 = N_0 v \qquad (1.17)$$

so

$$I = I_0 \exp\left(-n\sigma x\right) \qquad (1.18)$$

The absorption of a flux $\mathcal{J}$ of electromagnetic radiation is described by a similar equation, known as Lambert's law:

$$\mathcal{J} = \mathcal{J}_0 \exp\left(-\mu x\right) \qquad (1.19)$$

where μ is the photon absorption coefficient (Chapter 9).

It can now be shown that the cross-section defined in equation 1.14 is the same as that described by equation 1.10. The 'mean free path' l_f of the beam molecules is the distance which, on average, a molecule travels before it

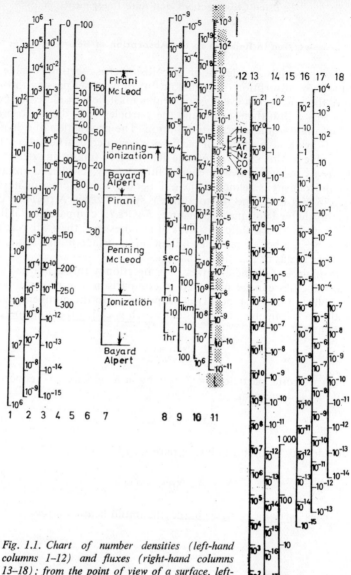

Fig. 1.1. Chart of number densities (left-hand columns 1-12) and fluxes (right-hand columns 13-18); from the point of view of a surface, left-hand and right-hand columns are equivalent, since the surface receives identical numbers of atoms:

column 1, electron plasma resonance frequencies (Hz) corresponding to the number densities of column 10; column 2, pressure (dyne cm^{-2}); column 3, pressure (bar); column 4, height (km) above earth's surface at which atmosphere density is equal to that in column 10; column 5, temperature (°C) of water at which equilibrium vapour pressure is equal to that in column 11; column 6, temperature (°C) of mercury at which equilibrium vapour pressure is equal to that in column 11; column 7, upper and lower limits of pressure gauges corresponding to pressures in column 11; column 8, times (sec) taken for the formation of a monolayer of nitrogen on a surface when the gas pressure corresponds to that of column 11; column 9, mean free path of CO_2 molecules at densities of column 10 and 0°C temperature; column 10, number densities (cm^{-3}); column 11, gas pressure (torr) corresponding at 0°C to number density of column 10; column 12, gases in which the mean free path is 1 cm at densities of column 10 and 0°C temperature; column 13, flux of molecular beam (molecules sec^{-1}) producing identical bombardment of a surface to that produced on the same area by N_2 gas at 0°C; column 14, current (A) of singly charged particles corresponding to flux of column 13; column 15, signal–noise ratio when flux of column 13 is detected by completely efficient detector of time constant equal to 1 sec; column 16, rate of deposition of molecules (g/hr) by flux of column 13; column 17, thrust (dyne) exerted by molecular beam flux of column 13 originated at 300°K; column 18, energy (erg) gained per second by surface receiving flux of column 13 originated at temperature 2000°K higher than that of the surface

undergoes a collision. If the beam is directed along the x-axis,

$$l_f = \frac{\int\limits_0^\infty xI(x)\,\mathrm{d}x}{\int\limits_0^\infty I(x)\,\mathrm{d}x} = \frac{\int\limits_0^\infty xI_0 \exp(-n\sigma x)\,\mathrm{d}x}{\int\limits_0^\infty I_0 \exp(-n\sigma x)\,\mathrm{d}x} \qquad (1.20)$$

Thus $l_f = 1/n\sigma$.

For gas densities sufficiently low that $l_f \gg x$, equation 1.18 approximates to

$$\frac{I}{I_0} \simeq 1 - n\sigma x \qquad (1.21)$$

Such situations are described as satisfying 'single-collision conditions', since each projectile particle is likely to make only one collision.

The quantity $n\sigma$ is sometimes designated by the (particle) absorption coefficient α and is loosely called the 'macroscopic cross-section'. As with the photon and other absorption coefficients, it refers to gas at a standard density; for the particle absorption coefficient, this is the density of gas at pressure 1 torr and temperature 0°C. The absorption coefficient therefore appears in the literature in units of $cm^2\ cm^{-3}\ torr^{-1}$, or for brevity cm^{-1}.

For gas at pressure p torr and temperature $T°K$, the density

$$n = \frac{2 \cdot 6873 \times 10^{19}\, p \times 273}{760T} \qquad (1.22)$$

so

$$\frac{\sigma}{\alpha} = \frac{1}{n} = 2 \cdot 81 \times 10^{-17}\ cm^3\ (\text{atom or molecule})^{-1} \qquad (1.23)$$

An absorption coefficient $\alpha\ cm^2\ cm^{-3}\ torr^{-1}$ must therefore be multiplied by $2 \cdot 81 \times 10^{-17}$ to convert to cross-section $\sigma\ cm^2\ (\text{atom or molecule})^{-1}$. For brevity the (atom or molecule)$^{-1}$ will be omitted hereafter, although this is not always the case in the literature. In problems of the stopping of nuclear radiation, cross-sections are often expressed 'per atom' even for a molecular gas; this practice will not be adopted here.

Cross-sections are often expressed in units of πa_0^2 (sometimes in a_0^2), where $a_0 = 0 \cdot 5292 \times 10^{-8}$ cm is the radius of the first Bohr orbit of the hydrogen atom: $\pi a_0^2 = 8 \cdot 806 \times 10^{-17}\ cm^2$ and $a_0^2 = 2 \cdot 803 \times 10^{-17}\ cm^2$. Nuclear physicists are familiar with cross-sections in 'barns' or units of $10^{-24}\ cm^2$. Megabarns, occasionally used for photon cross-sections, are therefore $10^{-18}\ cm^2$. An additional complication is that photon absorption coefficients μ are customarily referred not to 1 torr pressure but to s.t.p. Therefore $\sigma = \mu/(2 \cdot 6893 \times 10^{19})$.

Frequently, a collision process is studied not as an absorption of projectiles (experiments deriving from an interest in nuclear radiation) but by the measurement of the current I_2 of charged particles produced from the gas molecules (experiments deriving from an interest in ionized gases). For example, in the charge transfer collision 10/01 (Chapter 12), a fraction I_1

of the initial beam current I_0 loses its charge to the gas molecules, so

$$I_0 = I_1 + I_2 \tag{1.24}$$

which with equation 1.21 yields

$$\frac{I_2}{I_0} \simeq n\sigma x \tag{1.25}$$

The current I_2 is, in the low-density limit, linearly dependent on gas density.

The Target Parameter and Fractional Conversion of Beams

A convenient parameter

$$\pi = nx = \frac{px}{kT} \tag{1.26}$$

defines the total gas target through which the projectile beam passes and is termed the 'target parameter'. For laboratory temperature $T = 20°C$, and pressure p expressed in torr,

$$\pi = 3.30 \times 10^{16} \, px \tag{1.27}$$

and equation 1.25 becomes simply

$$\frac{I_2}{I_0} \simeq \pi\sigma \tag{1.28}$$

The second term in the exponential expansion of equation 1.18 is quadratic in $\pi\sigma$:

$$\frac{I_2}{I_0} \simeq \pi\sigma - \frac{1}{2}\pi^2\sigma^2 \tag{1.29}$$

This quadratic part of the pressure dependence may be observed, except at the very lowest pressures. A sequence of two processes with cross-sections σ_A and σ_B, each contributing a charged particle per collision to the current I_2, gives rise to a positive term in π^2:

$$\frac{I_2}{I_0} \simeq \pi\sigma - \frac{1}{2}\pi^2\sigma_A^2 + \frac{1}{2}\pi^2\sigma_A\sigma_B \tag{1.30}$$

A different situation is encountered when both forward and reverse processes are possible: in the forward process, having a cross-section $\vec{\sigma}$, the projectile particle is converted into a different species, whilst in the reverse process, having a cross-section $\bar{\sigma}$, it reverts to its original form. In the low pressure limit, a fraction $1 - F$ of the particles is converted to the second type, and this fraction is linearly dependent on pressure. A fraction F remains of the first type. At sufficiently high pressure, an equilibrium between forward and reverse processes is achieved, and the fraction F_∞ of first type still remaining is pressure independent. The two cross-sections are related to these fractions by Wien's formulae[5], which will be proved and considered in greater detail in Chapter 11:

$$\vec{\sigma} = \frac{1-F_\infty}{\pi} \ln \left(\frac{1-F_\infty}{F-F_\infty} \right) \tag{1.31}$$

$$\vec{\sigma} = \frac{F}{\pi} \ln \left(\frac{1-F_\infty}{F-F_\infty} \right) \tag{1.32}$$

In high-energy heavy-particle collision studies, this simple 'two-component system' is very seldom encountered. It is more usual for the projectile particle to be able to exist in several different charge states and excited states, all of them interconvertible by collision processes. Even the 'three-component system' is only soluble under certain conditions, namely the neglect of three of the six possible charge-changing collisions[6]; the remaining three can be calculated from the equilibrium and low-pressure fractions (see Chapter 11, in which many-component systems are also discussed).

Partial and Differential Cross-sections

The probability of making a collision in unit distance, symbolized as P in equation 1.14, may be divided into P_1, P_2, ..., which are probabilities of various inelastic, elastic or superelastic processes; thus $\sum_n P_n = P$ and $\sum_n \sigma_n = \sigma_t$, where σ_t is the total collision cross-section and σ_n are partial cross-sections for these processes. In order that the reader may get a feeling for the magnitudes of the different partial cross-sections to be encountered in subsequent chapters, a graphical representation is made in *Figure 1.2*. In this chart are also compared the different units in which cross-sections, absorption coefficients, rate constants, etc., are measured. The term 'partial' is usually omitted from discussions of probabilities and cross-sections for inelastic processes.

Consider a projectile particle proceeding along the broken line in *Figure 1.3* and making a collision at O. Suppose that it is scattered into a small solid angle $d\Omega$, at a polar scattering angle θ and azimuthal scattering angle ϕ. Since $r = R \sin \theta$,

$$R^2 \, d\Omega = R^2 \, d\theta \, d\phi \sin \theta \tag{1.33}$$

The differential cross-section $d\sigma$ for scattering into solid angle $d\Omega$ is related to the total cross-section σ by the equation

$$\sigma = \int_0^{2\pi} \int_0^{\pi} \frac{d\sigma}{d\Omega} \sin \theta \, d\theta \, d\phi \tag{1.34}$$

The differential cross-section $d\sigma/d\Omega$, measured in units of cm^2 steradian^{-1}, has often been symbolized as $I(\theta)$ and as $\sigma(\theta)$. The symbol $I(\theta)$ is unsatisfactory, since it should refer to a flux or current and not a cross-section.

Line excit-ation

N_2 Ionization
O_2 Ionization
O_2 Dissociation

CO_2 Dissociation

Thomson

Max dipole

Hg $\frac{kT}{}$ Kr $\frac{kT}{}$ H $\frac{kT}{}$ **Ne**

Ar$^+$Kr 5×10^4 H+H_2 max Ne$^+$Ne H$^+$ H max O_2 K 300eV

Ar Dissociative
Ne Dissociative
I$^-$ Dissociative
Ar$^+$Ne 5×10^4
He Dissociative
H$^-$ Detachment
O$^-$ Ionization Detachment 10^5 H$^+$ H_2

Total
Ionization $n=2$
Elastic $n=3-5$ Total $n=2$ Ionization

Hg **Ar** Ne **He**

Hg 6 $^3P_1$ He 3 ^1P

Cs / K / Na H_2 Xe / Kr / Ar

Xe Ar H_2 He

10^{-3} 10^{-2} kT 10^{-1} 1 Xe Ar H_2 He 10 10^2 10^3

10^{-7} 10^{-8} 10^{-9} 10^{-10} 10^{-11} 10^{-12}

10^{-6} 10^{-7} 10^{-8} 10^{-9} 10^{-10} 10^{-11}

10^{-4} 10^{-5} 10^{-6} 10^{-7} 10^{-8} 10^{-9} 10^{-10}

10^5 10^4 10^3 10^2 10 1

10^5 10^4 10^3 10^2 10 1 10^{-1}

10^5 10^4 10^3 10^2 10 1

10^2 10 1

10^{-11} 10^{-12} 10^{-13} 10^{-14} 10^{-15} 10^{-16} 1Å^2 10^{-17}

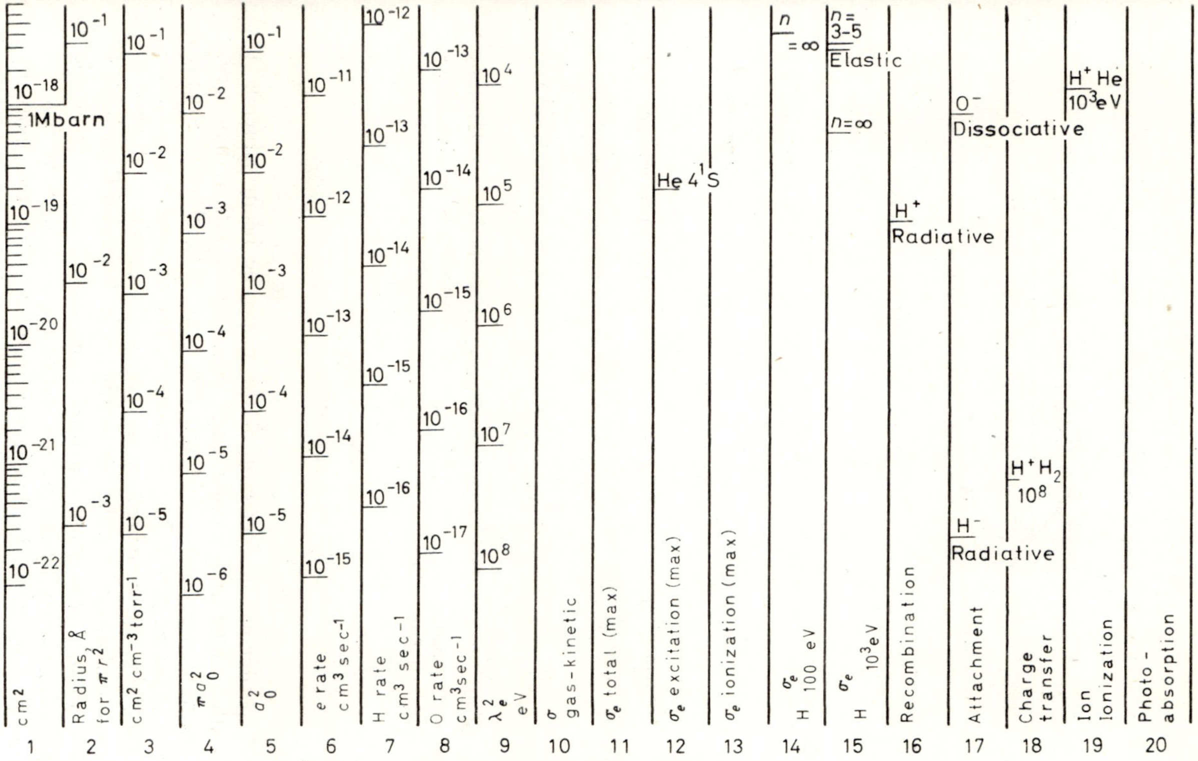

Fig. 1.2. Chart of cross-sections:

columns 1–5, cross-section units; columns 6–8, rate coefficient units; columns 1–5, cross-section units; column 9, electron energies corresponding to the appropriate wavelengths squared; column 10, classical gas-kinetic cross-sections; column 11, total electron collision cross-sections at their maxima; column 12, electron excitation cross-sections at their maxima; column 13, electron ionization cross-section at their maxima; column 14, e–H cross-sections at 100 eV impact energy: total, elastic, ionization and excitation into quantum numbers $n = 1$ to ∞; column 15, e–H cross-sections at 1000 eV impact energy; column 16, thermal recombination rates: Thomson, dissociative and radiative; column 17, thermal attachment ad detachment rates; dissociative and radiative; column 18, charge transfer cross-sections; maximum dipole value and values at specified impact velocities (cm sec^{-1}); column 19, cross-sections for ionization by heavy particles at specified impact energies (eV); column 20, photoabsorption: line, dissociation and ionization cross-sections

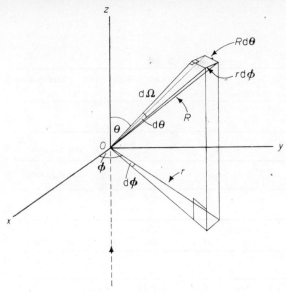

Fig. 1.3. Laboratory spherical polar coordinate system for point projectile approaching, along the z axis, a point target at origin O

The notation $d\sigma/d\Omega$ is preferred to $\sigma(\theta)$, since it is more appropriate to a differential cross-section. For isotropic scattering,

$$\frac{d\sigma}{d\Omega} = \frac{\sigma}{4\pi} \tag{1.35}$$

The integral of equation 1.34 will be convergent provided that $d\sigma/d\Omega$ does not increase with decreasing θ as rapidly as θ^{-2}. Scattering is nearly always isotropic in azimuth, therefore

$$\sigma = 2\pi \int_{0}^{\pi} \frac{d\sigma}{d\Omega} \sin\theta \, d\theta \tag{1.36}$$

In beam experiments, the detector of the projectile beam subtends a known azimuthal angle and a polar angle equal to θ_0 rather than zero. The polar integrals in equations 1.34 and 1.36 cannot therefore be exactly reproduced in experiment. For fast beams travelling through a gas, differential cross-sections are measured by traversing the detector, which subtends a known solid angle, in polar scattering angle. Provided that there is isotropy in azimuthal scattering, the total cross-section can in principle be deduced approximately by integration. Some experiments have reached a higher level of refinement, in that it is possible to cross two mutually perpendicular beams of particles and observe scattering in or out of the plane of the beams. Integration of the two differential functions does not necessarily yield the total cross-section, but a kinematic analysis can be made[7].

Differential cross-sections can be defined for inelastic scattering by anal-ogy with their definition for elastic scattering. For impact energy loss ΔE, the differential cross-section can be denoted by $(d^2\sigma/d\Omega\, dE)\,(\Delta E, \theta)$. Such cross-sections fall off with increasing θ very much faster than the elastic differ-ential cross-section, so in beam–gas experiments the pressure variation of the observed differential currents is complicated. There is a contribution, which may be dominant, from the following double process: a large-angle elastic event followed by a small-angle inelastic effect. This can result in pressure-dependent differential cross-sections, which must be extrapolated to zero pressure to obtain true results: the appropriate relation is

$$I(\Delta E, \theta)\sin\theta\exp\left(\frac{p}{p_0}\right) = I_0 p\,\frac{d^2\sigma}{d\Omega\, dE}\,(\Delta E, \theta)$$

for beam intensity I_0 incident on gas at pressure p and yielding scattered current I (but see equations 3.5 and 3.6). The pressure p_0 is a specific pressure which is inversely proportional to the total cross-section.

The Diffusion Cross-section

In addition to the total collision cross-section σ or σ_t, the *momentum transfer* or *diffusion* cross-section σ_d for electron–atom and electron–molecule colli-sions is of importance. An electron of mass m_e scattered by an atom or a molecule of mass m_2 through a polar scattering angle θ will, on a hard-sphere model (Section 2.5), lose a fraction ΔE_e of its kinetic energy E_e given by

$$\frac{\Delta E_e}{E_e} = 2(1-\cos\theta)\frac{m_e}{m_2} \tag{1.37}$$

The mean fractional energy loss per collision is therefore

$$\overline{\frac{\Delta E_e}{E_e}} = \frac{2m_e}{m_2\sigma_t}\int_0^\pi\int_0^{2\pi}(1-\cos\theta)\sin\theta\,\frac{d\sigma}{d\Omega}\,d\theta\,d\phi = \frac{2m_e}{m_2}\frac{\sigma_d}{\sigma_t} \tag{1.38}$$

The momentum transfer or diffusion cross-section is defined as

$$\sigma_d = \int_0^\pi\int_0^{2\pi}\frac{d\sigma}{d\Omega}(1-\cos\theta)\sin\theta\,d\theta\,d\phi \tag{1.39}$$

It is a measure of the forward momentum loss of electrons drifting through a gas under the action of an electric field. For $d\sigma/d\Omega$ independent of θ (that is, for isotropic scattering in laboratory coordinates), σ_d is equal to the total cross-section σ_t. The cross-sections differ only when there is some concen-tration of forward or back scattering.

Another scattering cross-section sometimes encountered is the viscosity cross-section:

$$\sigma_\eta = 2\pi \int_0^\pi \frac{d\sigma}{d\Omega} \sin^3 \theta \, d\theta \qquad (1.40)$$

in which the scattering perpendicular to the projectile path is weighted most heavily.

Rate Coefficients

Following the tradition of chemical kinetics, inelastic collision probabilities are often defined in terms of 'rate coefficients' rather than cross-sections. In a reaction between two particles, converting them into different particles, the rate of change of number density n_1 of the first is

$$\frac{dn_1}{dt} = -kn_1n_2 \qquad (1.41)$$

where k is the rate coefficient, defined in units of cm³ sec⁻¹ (particle)⁻¹; the bracketed term is usually omitted for brevity. The number density n_2 of the second is occasionally written as a pressure (at $T = 0°C$); the rate constant then being in units of torr sec⁻¹.

The rate coefficient is the integrated product of cross-section, impact velocity and normalized velocity distribution:

$$k = \int_0^\infty v_r \sigma(v_r) f(v_r) \, dv_r \qquad (1.42)$$

where $f(v_r) \, dr$ is the fraction of encounters in which the relative velocity v_r of impact lies between v_r and $v_r + dv_r$. Thus the recombination coefficient α_e (Chapter 7) for positive ions and electrons is defined by the relation

$$\frac{dn_e}{dt} = -\alpha_e n_e n_+ \simeq -\alpha_e n_e^2 \qquad (1.43)$$

where, in a plasma, there is macroscopic balance of charge:

$$n_e \simeq n_+ \qquad (1.44)$$

Under some circumstances, the velocity distribution $f(v_r)$ may be replaced by a mean velocity $\bar{v}_r$, so that

$$k = \sigma \bar{v}_r \qquad (1.45)$$

The calculation of k from equation 1.42 when $\sigma(v_r)$ and $f(v_r)$ are known is straightforward, but will usually require a computer. Equation 1.45 is often an excessively crude approximation, but there exists for Maxwellian distributions a slightly less simple approximation[8], using Gauss–Laguerre quadrature.

The mean relative velocity in a mixture of two gases of masses m_1 and m_2 (reduced mass μ) and atomic weights A_1 and A_2 can be written:

$$\bar{v}_r = \left(\frac{8kT_g}{\pi} \frac{m_1+m_2}{m_1 m_2}\right)^{1/2} \tag{1.46}$$

where T_g is the temperature of the gas mixture. Therefore

$$\bar{v}_r = \left(\frac{8kT_g}{\pi \mu}\right)^{1/2} \tag{1.47}$$

or

$$\bar{v}_r = 1.45 \times 10^4 \left(T_g \frac{A_1+A_2}{A_1 A_2}\right)^{1/2} \quad \text{cm sec}^{-1} \tag{1.48}$$

In collisions between electrons and atoms (and electron–ion collisions in plasmas), the mean electron velocity $\bar{v}_e$ greatly exceeds the mean atomic velocity, so for velocity-independent cross-section, the rate-constant

$$k_e \simeq \bar{v}_e \sigma(v_e) \tag{1.49}$$

For electrons at $T_e = 300°\text{K}$,

$$\bar{v}_e \simeq 6 \times 10^7 \sqrt{0.039} \text{ cm sec}^{-1} \tag{1.50}$$

A typical cross-section for radiative recombination (Chapter 7) is 3×10^{-19} cm^2, and the corresponding room-temperature recombination coefficient is 3.5×10^{-12} cm^3 sec^{-1}. The two-body electron attachment coefficient, corresponding to the attachment of electrons to atoms or molecules to form negative ions (Chapter 8), is similar. Comparisons are made of electron rate coefficients with cross-sections in *Figure 1.2*.

Three-body Collisions

Three-body collisions are at the present stage of development conveniently considered in terms of three-body rate coefficients k_3, defined by the relation

$$\frac{dn_1}{dt} = -k_3 n_1 n_2 n_3 \tag{1.51}$$

and measured in units of cm^6 sec^{-1} (particle)$^{-2}$, or occasionally torr^2 sec^{-1}. Three-body rate coefficients are typically 10^{-27}–10^{-31} cm^6 sec^{-1}, compared with two-body electron rate coefficients of 10^{-7}–10^{-12} cm^3 sec^{-1}.

At gas pressures in excess of 1 torr, three-body collision processes become important, even dominant. The role of the third body, or chaperone, which can be either a gas atom or molecule, or a surface, or a liquid or solid, is usually to remove the excess energy that could otherwise only be liberated as electromagnetic radiation (an inefficient process). Another possible role of the third body is to stabilize collisionally an unstable state, which could otherwise undergo a transition preventing the inelastic collision process under consideration.

Three-body collisions are occasionally described in terms of a pseudo-cross-section σ^*, corresponding to the mean free path for an arbitrary density of third-body particles, usually that corresponding to 1 torr.

The calculation of three-body rate coefficients by classical or quantum methods requires that a suitable three-body 'cross-section' be defined, capable of being related to the scattering matrix (Chapter 2), to the probability of transition, and to the rate coefficient. For a two-body collision, the probability of transition P is calculable in terms of the impact parameter b of the collision (defined in Section 2.4), and the angular momentum

$$L = (2\mu E b^2)^{1/2} \tag{1.52}$$

for kinetic energy of impact E, and reduced mass μ. The collision cross-section is related to the probability by the equation

$$\sigma(E) = 2\pi \int_0^\infty P(E, b) b \, db \tag{1.53}$$

that is,

$$\sigma(E) = \frac{\pi}{\mu E} \int_0^\infty P(E, L) L \, dL \tag{1.54}$$

In this formulation, the collision is considered in the centre of mass co-ordinates, in which the motion of the centre of mass of the collidants is eliminated.

For a three-body collision, one can define similarly a generalized angular momentum Λ and hyperradius r in six-dimensional space, with motion of the centre of mass eliminated[9]. Thus

$$\Lambda^2 = 2\mu_3 E_3 r^2 \tag{1.55}$$

with *total* impact energy E_3 and reduced mass

$$\mu_3 = \left(\frac{m_1 m_2 m_3}{m_1 + m_2 + m_3} \right)^{1/2} \tag{1.56}$$

The three-body hypercross-section is related to the probability of transition by the equation

$$\sigma_3(E_3) = \frac{8\pi^2}{3} (2\mu_3 E_3)^{-5/2} \int_0^\infty P(E_3, \Lambda) \Lambda^3 \, d\Lambda \tag{1.57}$$

For three different types of particle in Maxwell–Boltzmann equilibrium at a temperature T, the rate constant

$$k_3 = (kT)^{-3} (2\mu_3)^{-1/2} \int_0^\infty \exp\left(-\frac{E_3}{kT} \right) E_3^{5/2} \sigma_3(E_3) \, dE_3 \tag{1.58}$$

and for an energy-independent cross-section this becomes

$$k_3 = v_3\sigma_3 \tag{1.59}$$

with

$$v_3 = \frac{15\pi}{8}\left(\frac{kT}{2\pi\mu_3}\right)^{1/2} \tag{1.60}$$

The four-dimensional hypercross-section is the appropriate parameter for a three-body collision process; it corresponds to the cross-section for a two-body collision. Consider the possible mechanism: species $1+$species 2 produce the complex 12, which can either dissociate spontaneously in time t, or collide with species 3 to produce the new species 4. Now

Rate of production of 4 = (concentration of 12)$\times\sigma(3+12 \rightarrow 4)\,v(3+12)$

But

Concentration of $12 = t\sigma(1+2 \rightarrow 12)\,v(1+2)$

The overall rate thus contains the product of two cross-sections, a quantity of dimensions l^4. This hypercross-section is basic to the three-body process and can be calculated from more general considerations. Gerjuoy has shown quantum-mechanically that a term in l^6 is also important, which leads to the question: is a two-body collision fully described by a cross-section? In the case of two-body atom–molecule scattering, the cross-section will often be dependent on the alignment of collision axis and molecular axis. Thus a non-spherical scattering volume is the fundamental collision parameter, and the cross-section only partially describes it.

1.3 KINETICS OF REACTION RATES

Since the publication of the first edition of this book, the study of ionic, electronic, and excited atom processes in the flowing and the time-dependent afterglow has progressed to the stage where complex kinetic situations are encountered. For this reason, it is valuable to consider the possible application of chemical kinetic techniques to ionized gas situations. The treatment given in this section owes much to textbooks of chemical kinetics, such as that of Benson[10].

The time decay of particle concentration is represented generally as

$$\frac{dn_1}{dt} = -k_i n_1^p n_2^q n_3^r \tag{1.61}$$

where a limit of three reacting species is assumed. Such a decay process is said to be of order p with respect to species 1. Thus for a zero order reaction (constant decay rate of species 1),

$$\frac{dn_1}{dt} = -k_0 \tag{1.62}$$

and

$$n_1 = -k_0 t + n_{10} \tag{1.63}$$

where the density at zero time is n_{10}. The rate constant is written by chemists in moles litre^{-1} sec^{-1}, but the units more familiar to physicists are particles cm^{-3} sec^{-1}; conversion is made using the relation

$$10^{-3} N_0 \text{ particles cm}^{-3} \text{ sec}^{-1} = 1 \text{ mole litre}^{-1} \text{ sec}^{-1} \qquad (1.64)$$

with N_0 Avogadro's number.

A zero order rate is seldom encountered in atomic collision physics, since it is neither a gas phase collision process nor a radiative decay; it is found in surface reactions where the particle reacting with species 1 saturates the surface and so constitutes a substance depleting n_1 at a uniform rate.

For a first order reaction (decay of species 1 proportional to its density),

$$\frac{dn_1}{dt} = -k_1 n_1 \qquad (1.65)$$

and

$$\ln\left(\frac{n_1}{n_{10}}\right) = -k_1 t \qquad (1.66)$$

The first order rate constant k_1 is defined in units of sec^{-1}. Equation 1.66 is a simple exponential decay, commonly analysed in terms of the times $t_{1/2}$, $t_{1/10}$ or $t_{1/e}$ in which the density is reduced to $\frac{1}{2}$, $\frac{1}{10}$ or $1/e$ of its original value. It is appropriate to radiative decay (Chapter 9), and the time variation of concentrations under the action of diffusion is also of this type.

A second order rate can belong to one of two types, namely reactions between species 1 and species 1, and reactions between species 1 and species 2. In the first type,

$$\frac{dn_1}{dt} = -k_2 n_1^2 \qquad (1.67)$$

and

$$\frac{1}{n_1} - \frac{1}{n_{10}} = k_2 t \qquad (1.68)$$

This corresponds to the 'inverse first power' decay encountered in electron–ion recombination (Chapter 7). Species 1 is the electron, and since the plasma condition

$$n_e \simeq n_+ \qquad (1.69)$$

applies, the product of electron and ion concentrations is approximately n_1^2. In the second type of second order reaction, as in equation 1.41

$$\frac{dn_2}{dt} = \frac{dn_1}{dt} = -k_2 n_1 n_2 \qquad (1.70)$$

and

$$\ln\left(\frac{n_1}{n_2}\right) = (n_1 - n_2)k_2 t + \ln\left(\frac{n_{10}}{n_{20}}\right) \qquad (1.71)$$

For $n_2 \gg n_1$ and $n_2 \simeq n_{20}$, this reduces to an exponential decay:

$$\ln\left(\frac{n_1}{n_{10}}\right) = -n_2 k_2 t \qquad (1.72)$$

The situation is encountered in the two-body attachment of electrons to molecules (Chapter 8). All second order rate coefficients (both first and second types) are defined in units of cm^3 sec^{-1} $particle^{-1}$ or litres $mole^{-1}$ sec^{-1}.

A third order rate can belong to one of three types, namely reactions between three particles of one species, between two of one species and one of another, and between three different species. In the first case,

$$\frac{dn_1}{dt} = -k_3 n_1^3 \tag{1.73}$$

and

$$\frac{1}{n_1^2} - \frac{1}{n_{10}^2} = 2k_3 t \tag{1.74}$$

In the second case,

$$\frac{dn_1}{dt} = -k_3 n_1^2 n_2 \tag{1.75}$$

and

$$(n_{20} - n_{10}) \left(\frac{1}{n_1} - \frac{1}{n_{10}} \right) + \ln \left(\frac{n_1 n_{20}}{n_{10} n_2} \right) = (n_{20} - n_{10})^2 \, k_3 t \tag{1.76}$$

In the third case,

$$\frac{dn_1}{dt} = -k_3 n_1 n_2 n_3 \tag{1.77}$$

and

$$(n_{30} - n_{20}) \ln \left(\frac{n_{10}}{n_1} \right) - (n_{30} - n_{10}) \ln \left(\frac{n_{20}}{n_{20} - n_{10} + n_1} \right)$$

$$+ (n_{20} - n_{10}) \ln \left(\frac{n_{30}}{n_{30} - n_{10} + n_1} \right) = k_3 (n_{30} - n_{10})(n_{20} - n_{10})(n_{30} - n_{20})t \tag{1.78}$$

For $n_1 \ll n_2$ and $n_1 \ll n_3$, so that $n_{20} \simeq n_2$ and $n_{30} \simeq n_3$, this reduces to an exponential decay:

$$\ln \left(\frac{n_1}{n_{10}} \right) = -k_3 n_2 n_3 t \tag{1.79}$$

The situation is encountered in the three-body attachment of electrons to molecules (Chapter 8). Third order rate coefficients are defined in units of cm^6 sec^{-1} $particle^{-2}$ or $litres^2$ $mole^{-2}$ sec^{-1}.

Since cases are known in chemical kinetics of reactions of fractional order, it is worth mentioning that the general solution for a reaction of order n (type 1) is

$$\frac{dn_1}{dt} = k_3 n_1^n \tag{1.80}$$

and

$$\frac{1}{n_1^{n-1}} - \frac{1}{n_{10}^{n-1}} = (n-1)k_n t \tag{1.81}$$

3*

Opposing, Competing and Consecutive Reactions

Consider now the situation arising when inelastic collision processes can proceed in either direction (opposing reactions). The first order conversion of species 1 into species 2 (without collision with a second particle), proceeding with rate constant $\vec{k}_1$, and in reverse with rate constant $\overleftarrow{k}_1$, can be written

$$\frac{dn_1}{dt} = -\vec{k}_1 n_1 + \overleftarrow{k}_1 n_2 \tag{1.82}$$

and

$$\ln\left(\frac{\vec{k}_1 n_1 - \overleftarrow{k}_1 n_2}{\vec{k}_1 n_{10} - \overleftarrow{k}_1 n_{20}}\right) = -(\vec{k}_1 + \overleftarrow{k}_1)t \tag{1.83}$$

After a sufficient period of time, equilibrium concentrations $n_{1\infty}$ and $n_{2\infty}$ are reached; the equilibrium constant K is defined as

$$K = \frac{\vec{k}_1}{\overleftarrow{k}_1} = \frac{n_{2\infty}}{n_{1\infty}} \tag{1.84}$$

and equation 1.83 can be written as

$$\ln\left(\frac{n_1 K - n_2}{n_{10} K - n_{20}}\right) = -(\vec{k}_1 + \overleftarrow{k}_1)t \tag{1.85}$$

The second order conversion of species 1 and 2 by collision into species 3 and 4, proceeding with rate constant $\vec{k}_2$, whilst the reverse process has rate constant $\overleftarrow{k}_2$, may be described by the equations

$$\frac{dn_1}{dt} = -\vec{k}_2 n_1 n_2 + \overleftarrow{k}_2 n_3 n_4 \tag{1.86}$$

and

$$\ln\left[\frac{n_{10} - n_1 + (\beta - q^{1/2})/2\gamma}{n_{10} - n_1 + (\beta + q^{1/2})/2\gamma}\right] = tq^{1/2} + \ln\left(\frac{\beta - q^{1/2}}{\beta + q^{1/2}}\right) \tag{1.87}$$

where:

$$\alpha = \vec{k}_2 n_{10} n_{20} - \overleftarrow{k}_2 n_{30} n_{40} \tag{1.88}$$

$$\beta = -(\vec{k}_2 n_{10} + \vec{k}_2 n_{20} + \overleftarrow{k}_2 n_{30} + \overleftarrow{k}_2 n_{40}) \tag{1.89}$$

$$\gamma = \vec{k}_2 - \overleftarrow{k}_2 \tag{1.90}$$

$$q = \beta^2 - 4\alpha\gamma \tag{1.91}$$

This situation is encountered in hot plasmas, where the electron capture by multiply-charged ions, of the form $nm/(n-1)(m+1)$, can proceed in either direction.

Another possible situation is the natural decay of species 1 into two products, species 2 and 3, with first order rate constant $\vec{k}_1$; but the reverse process, by which 1 is formed by collision of 2 and 3, has a second order rate constant $\bar{k}_2$. The solution of the equation

$$\frac{dn_1}{dt} = -\vec{k}_1 n_1 + \bar{k}_2 n_2 n_3 \tag{1.92}$$

has the form of equation 1.87; but equations 1.88–1.90 now become:

$$\alpha = \vec{k}_1 n_{10} - \bar{k}_2 n_2 n_3 \tag{1.93}$$

$$\beta = -[\vec{k}_1 + \bar{k}_2(n_{30} + n_{20})] \tag{1.94}$$

$$\gamma = -\bar{k}_2 \tag{1.95}$$

The kinetics of concurrent or competing reactions, such as might be encountered in ionized gases, are comparatively simple. In a mixture of species 1 and 2, the decay of species 1 may be imagined to take place by a first order process (natural decay) with rate coefficient k_1, and by second order processes of both types—that is, two particles of species 1 reacting with rate coefficient k_{21}, and particles of species 1 and 2 reacting with rate coefficient k_{22}.

The solution of the kinetic equation

$$-\frac{dn_1}{dt} = k_1 n_1 + k_{21} n_1^2 + k_{22} n_1 n_2 \tag{1.96}$$

can be written

$$\ln\left(\frac{n_{10}}{n_1} \frac{k_1 + k_{22} n_2 + 2k_{21} n_1}{k_1 + k_{22} n_{20} + 2k_{21} n_{10}}\right) = k_1 + k_{22}(n_{20} - n_{10})t \tag{1.97}$$

In simpler situations, one may simply write k_1, k_{21} or k_{22} equal to zero.

The kinetics of consecutive (sequential) processes are in general much more difficult and may result in situations which cannot be solved analytically. A sequence of first order reactions converting species 1 to species 2, with rate coefficient k_{11}, and thence to species 3, with rate coefficient k_{12}, can be solved using the following equations:

$$\frac{dn_1}{dt} = -k_{11} n_1 \tag{1.98}$$

$$\frac{dn_2}{dt} = k_{11} n_1 - k_{12} n_2 \tag{1.99}$$

$$\frac{dn_3}{dt} = k_{12} n_2 \tag{1.100}$$

$$n_3 = n_{30} + n_{10}[1 - \exp(-k_{11}t)]$$
$$+ n_{20}\left\{1 - \exp(-k_{12}t) - \frac{n_{10}[\exp(-k_{12}t) - \exp(-k_{11}t)]}{n_{20}(1 - k_{12}/k_{11})}\right\} \tag{1.101}$$

Sequences of this type are encountered in radiative decay of excited states. The density of species 2 can increase with time before ultimately decreasing, provided $k_{11}n_{10} < k_{12}n_{20}$. The time t_{max} at which the maximum n_2 is reached is

$$t_{max} = \frac{1}{k_{12}-k_{11}} \ln\left[\frac{k_{12}}{k_{11}}\left(1+\frac{n_{20}}{n_{10}}-\frac{k_{12}}{k_{11}}\frac{n_{20}}{n_{10}}\right)\right] \qquad (1.102)$$

Consider the most general first order sequence possible, namely s components, each capable of being converted to and from all the other components, with rates

$$j \underset{k_{mj}}{\overset{k_{jm}}{\rightleftharpoons}} m$$

The general expression for the number density of each state is given by

$$n_m = \sum_{j=1}^{s} [a_{mj} \exp(-\lambda_j t)] + \theta_m \qquad (1.103)$$

The constants θ_m are related to the initial densities by equations of the type

$$\theta_m = n_{m0} - \sum_{j=1}^{s} a_{mj} \qquad (1.104)$$

The time variation of the number densities is

$$\frac{dn_m}{dt} = \sum_{j=1}^{s} (k_{jm}n_j) \qquad (1.105)$$

with

$$\sum_{j=1}^{s} (k_{jm}\theta_j) = 0 \qquad (1.106)$$

The s^2 coefficients a_{mj} and s exponents λ_j require for their determination $s(s+1)$ equations. First there are s linear sets of s equations:

$$\left.\begin{array}{l} (k_{11}+\lambda_j)a_{1j}+k_{21}a_{2j}+ \ldots + \quad k_{s1}a_{sj} \quad = 0 \\ k_{12}a_{1j}+(k_{22}+\lambda_j)a_{2j}+ \ldots + \quad k_{s2}a_{sj} \quad = 0 \\ \quad\quad\quad \vdots \quad\quad\quad\quad\quad\quad\quad\quad\quad \vdots \\ k_{1s}a_{1j}+k_{2s}a_{2j} \quad + \ldots +(k_{ss}+\lambda_j)a_{sj} = 0 \end{array}\right\} \qquad (1.107)$$

But these s^2 equations are compatible only if the determinants vanish:

$$\begin{vmatrix} k_{11}+\lambda_j & k_{21} & \ldots & k_{s1} \\ k_{12} & k_{22}+\lambda_j & \ldots & k_{s2} \\ \vdots & \vdots & & \vdots \\ k_{1s} & k_{2s} & \ldots & k_{ss}+\lambda_j \end{vmatrix} = 0 \qquad (1.108)$$

These s determinant equations of degree s in λ_j give s pieces of information, namely all λ_j. Further studies have been made by Matsen and Franklin[11].

No exact general method of solution can be given for the situation commonly encountered in afterglows where there are consecutive reactions of order higher than one. But a variety of techniques can be employed, notably: (*i*) the elimination of the time variable; (*ii*) the reduction of second order to first order; (*iii*) the stationary state hypothesis; and (*iv*) the approximation of the rate-determining step.

The elimination of the time variable may be considered in the case of a species 1 capable of being converted by (collisionless) first order reaction to species 2, with rate coefficient k_1; but species 1 can also react with species 2, yielding species 3, with second order rate coefficient k_2. Thus

$$\frac{dn_1}{dt} = -k_1 n_1 - k_2 n_2 n_1 \qquad (1.109)$$

$$\frac{dn_2}{dt} = k_1 n_1 - k_2 n_2 n_1 \qquad (1.110)$$

and

$$\frac{dn_3}{dt} = k_2 n_1 n_2 = -\frac{1}{2}\left(\frac{dn_1}{dt} + \frac{dn_2}{dt}\right) \qquad (1.111)$$

Putting

$$\frac{k_1}{k_2} = K \qquad (1.112)$$

one may deduce and make use of the relation

$$\frac{n_2}{K} + 2\ln\left(1 - \frac{n_2}{K}\right) = -\frac{n_{10}}{K}\left(1 - \frac{n_1}{n_{10}}\right) \qquad (1.113)$$

The reduction of second order to first order rate coefficients enables the general solution given by equations 1.103–1.108 to be applied. Such a reduction can be achieved by the use of a variable z instead of t, where

$$dz = n_1\, dt \qquad (1.114)$$

This method can be applied to the situation where two particles of species 1 are capable of being converted to one of species 2, with a second order rate coefficient k_{21}; species 1 and species 2 can themselves react, with second order rate coefficient k_{22}.

For discussion of the technique of the *stationary state hypothesis*, the following sequence is considered:

$$A \xrightarrow{k} 2M_1$$

$$M_1 + C \xrightarrow{k_{21}} P + M_2$$

$$M_2 + A \xrightarrow{k_{22}} P + M_1$$

$$M_2 + M_1 \xrightarrow{k_{23}} A$$

The overall reaction is thus

$$A + C \to 2P$$

If the concentrations of intermediates M_1 and M_2 are both much smaller than those of A, C and P, then, after an induction period, it is possible to assume that the change of concentrations of intermediates is negligible. The rate of decay A is governed by the relations

$$\arctan\left(\frac{n_A}{n_{C0} - n_{A0}}\right) = -k_{21}\left(\frac{k_{11}}{k_{23}}\right)^{1/2} t + \arctan\left(\frac{n_A}{n_{C0} - n_{A0}}\right)^{1/2} \qquad (1.115)$$

provided $n_{C0} - n_{A0} > 0$,

$$\ln\left[\frac{n_A^{1/2} + (n_{A0} - n_{C0})^{1/2}}{n_A^{1/2} - (n_{A0} - n_{C0})^{1/2}}\right] = k_{21}\left(\frac{k_1}{k_{23}}\right)^{1/2} t + \ln\left[\frac{n_{A0}^{1/2} + (n_{A0} - n_{C0})^{1/2}}{n_{A0}^{1/2} - (n_{A0} - n_{C0})^{1/2}}\right] \quad (1.116)$$

provided $n_{A0} - n_{C0} > 0$, and

$$n_A^{-1/2} - n_{A0}^{-1/2} = k_{21}\left(\frac{k_1}{k_{23}}\right)^{1/2} t \quad (1.117)$$

when $n_{C0} = n_{A0}$. The application of the hypothesis to higher order reactions has been discussed by various workers[12].

The basic idea of the *method of the rate-determining step* is that in a sequence, one rate, if much slower than the others, will dominate. The total time t_p taken to produce unit quantity of product may be regarded as made up of the times taken to produce unit quantity of intermediates. That is,

$$t_p = \sum_n t_1 \quad (1.118)$$

for n stages. If the rate coefficients k_p are inversely proportional to these times, then

$$k_p = t_p^{-1} = \left(\sum_n t_1\right)^{-1} = \left(\sum_n k_1^{-1}\right)^{-1} \quad (1.119)$$

One of these rate coefficients will sometimes be much smaller than the others and will contribute the dominant term. Consider the sequence

$$A \underset{\overleftarrow{k_1}}{\overset{\overrightarrow{k_1}}{\rightleftarrows}} B$$

$$B + C \xrightarrow{k_2} P$$

If $\overleftarrow{k_1} \gg k_2 n_C$

$$\frac{dn_P}{dt} = \frac{\overrightarrow{k_1}}{\overleftarrow{k_1}} + k_2 n_A n_C \quad (1.120)$$

In this case k_2 is the determining rate coefficient. However, if $k_2 n_C \gg \overleftarrow{k_1}$, then

$$\frac{dn_P}{dt} \to \overrightarrow{k_1} n_A$$

and $\overrightarrow{k_1}$ is the determining rate coefficient.

1.4 VELOCITY AND ENERGY DISTRIBUTIONS

Fundamental to the calculation of reaction rates is a knowledge of the distribution of impact velocities. The velocity v achieved by a particle of mass m and charge e, accelerated through a potential difference V, is

$$v = \left(\frac{2\,eV}{m}\right)^{1/2} \quad (1.121)$$

Where V is measured in volts, the electron velocity is

$$v_e = 5\cdot94\times10^7V^{1/2} \tag{1.122}$$

and the velocity v_p of an ion of charge n and atomic weight A is

$$v_p = 1\cdot38\times10^6\left(\frac{n}{A}\right)^{1/2}V^{1/2} \tag{1.123}$$

A feel for the magnitudes of internal energies of atoms, energies of quanta, thermal energy, and of different units of energy and the corresponding velocities may be obtained from *Figure 1.4*. Equation 1.123 is represented graphically in *Figure 1.5* for convenience.

In fluxes and assemblies of particles, there is always a distribution of velocities and of kinetic energies which must be taken into account when relating rate coefficients to cross-sections, as in equation 1.42. The number

$$dN = Nf(v)\,dv \tag{1.124}$$

of particles in a group of N particles whose velocity lies in the interval v to $v+dv$ defines the velocity distribution (function) $f(v)$. The number

$$dN = Nf(E)\,dE \tag{1.125}$$

of particles whose kinetic energy lies in the interval E to $E+dE$ defines the energy distribution (function) $f(E)$. Since

$$mv = \frac{dE}{dv} = \frac{f(v)}{f(E)} \tag{1.126}$$

a velocity distribution must be divided by v to convert it to an energy distribution.

The simplest type of velocity distribution applies to experiments in which some type of momentum analysis of a projectile beam is attempted (see Chapter 3). In deflexion momentum analysers, an input aperture admits a certain angular divergence of beam; this is deflected in such a way that normally incident particles whose momentum lies between certain limits pass through the exit aperture. The ideal velocity distribution is 'Fraunhofer', that is, triangular for equal apertures, and for unequal apertures trapeziodal: for $v_1 \leqslant v \leqslant v_2$

$$f(v) = a(v-v_1) \tag{1.127}$$

for $v_2 \leqslant v \leqslant v_3$

$$f(v) = a(v_2-v_1) \tag{1.128}$$

and for $v_3 \leqslant v \leqslant v_3+v_2-v_1$

$$f(v) = -a(v-v_3) \tag{1.129}$$

Experiment has shown that, in practical momentum analysers, the corners of the distribution are rounded. A functional approximation may be

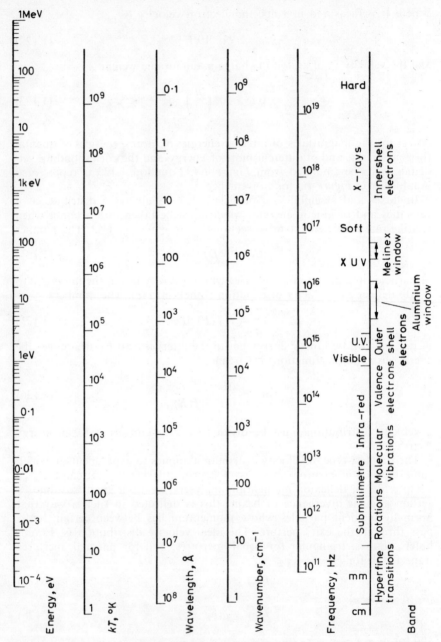

Fig. 1.4. Charts of units of energy, energy levels, wavelengths, wave numbers, frequencies and parameters proportional to energy

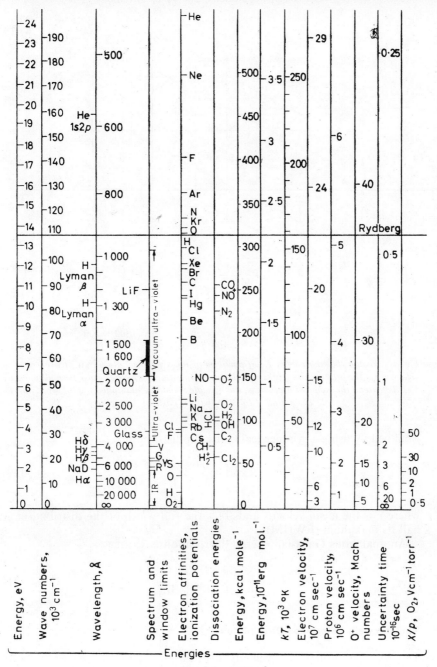

Fig. 1.4. continued

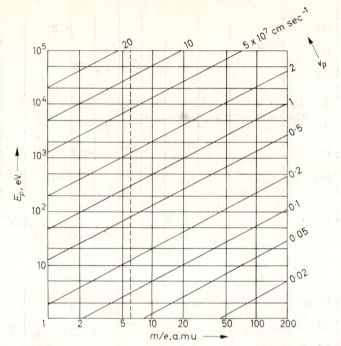

Fig. 1.5. Graph showing energy E_p of particles of mass number m/e a.m.u., travelling with velocity v_p; the broken line represents electrons, for which velocities must be multiplied by 100. To extend the range of usefulness, the v_p scale should be multiplied by 10^n when the E_p scale is multiplied by 10^{2n}, with n positive or negative

obtained by using a symmetrical Gaussian distribution:

$$f(v) = \exp\left[-\frac{0{\cdot}693(v-\bar{v})^2}{\frac{1}{4}\Delta v_{1/2}^2}\right] \tag{1.130}$$

where $\bar{v}$ is the mean velocity and $\Delta v_{1/2}$ the full width of the distribution at half its maximum (FWHM).

An analogous Gaussian energy distribution is written

$$f(E) = \exp\left[-\frac{0{\cdot}693(E-\bar{E})^2}{\frac{1}{4}\Delta E_{1/2}^2}\right] \tag{1.131}$$

The Maxwell–Boltzmann Distribution

The Maxwell–Boltzmann velocity distribution of molecules in a gas at temperature T is

$$f\left(\frac{v}{\bar{v}}\right)d\left(\frac{v}{\bar{v}}\right) = \frac{4N}{\sqrt{\pi}}\left(\frac{v}{\bar{v}}\right)^2 \exp\left(-\frac{v^2}{\bar{v}^2}\right)d\left(\frac{v}{\bar{v}}\right) \tag{1.132}$$

with most probable velocity

$$\bar{v} = \left(\frac{2kT}{m}\right)^{1/2} = \left(\frac{2\bar{E}}{m}\right)^{1/2} \tag{1.133}$$

The root mean square velocity

$$\tilde{v} = \left[\int_0^\infty v^2 f(v)\,\mathrm{d}v\right]^{1/2} = \bar{v}\sqrt{\tfrac{3}{2}} = 1\cdot22\bar{v} \tag{1.134}$$

and the average velocity

$$\langle v \rangle = \int_0^\infty v f(v)\,\mathrm{d}v = \bar{v}\,\frac{2}{\sqrt{\pi}} = 1\cdot13\bar{v} \tag{1.135}$$

The energy distribution follows from equation 1.132:

$$f\!\left(\frac{E}{\bar{E}}\right)\mathrm{d}\!\left(\frac{E}{\bar{E}}\right) = \frac{2N}{\sqrt{\pi}}\left(\frac{E}{\bar{E}}\right)\exp\left(-\frac{E}{\bar{E}}\right)\mathrm{d}\!\left(\frac{E}{\bar{E}}\right) \tag{1.136}$$

Molecular beams effusing from ovens in which thermal equilibrium has been achieved are now supposed[14] to possess Maxwell–Boltzmann velocity distributions. In the first edition of this book, the distribution was taken to be multiplied by a velocity term, since it was incorrectly supposed by many physicists that the probability of effusion of a molecule is proportional to its velocity. Molecular beams emerging from supersonic nozzles, from sputtering sources, from charge transfer neutralization sources and from rotor acceleration do not necessarily possess Maxwell–Boltzmann velocity distribution.

The unsymmetrical nature of the distribution may be seen in *Figure 1.6* where the Maxwellian and Druyvesteyn distributions (equation 1.145) are compared. In Chapter 3, a number of analytical techniques for determining energy distributions are considered.

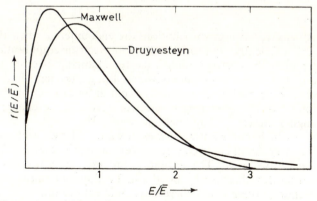

Fig. 1.6. Comparison of Maxwellian and Druyvesteyn energy distributions normalized to the same mean energy $\bar{E}$ and the same number of electrons

Energy Distributions in Ionized Gases

The electron velocity distributions encountered in weakly ionized plasmas ($n_e < 10^{10}$ cm^{-3}) differ from the Maxwellian because thermodynamic equilibrium is not achieved; the 'thermal contact' between low-energy electrons and atoms is minimal because of the large mass difference between them, but relatively large changes of energy occur in inelastic and super-elastic electron–atom collisions; oscillations, waves and a variety of electro-magnetic forces complicate the situation. Some methods of studying the distributions experimentally, using currents to metal probes and walls, radio noise, and spectral line widths, are discussed in Chapter 3. A good example of the electron energy distributions in weakly ionized plasma is to be found in the probe studies of glow discharge striated positive columns made by Boyd and Twiddy[15, 16]. In *Figure 1.7*, it will be noticed that there is a multiple peak structure in their energy distributions. The structure is now partially understood in terms of ionization waves in the column[13].

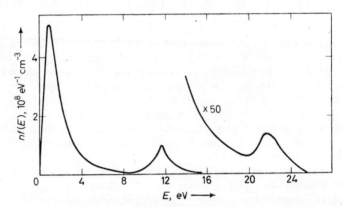

Fig. 1.7. *Measured electron energy distribution nf(e) in a striated positive column* (From Boyd and Twiddy[15], by courtesy of The Royal Society)

Maxwellian electron energy distributions are encountered when the electron density is sufficiently high for the electron–electron equilibration time to be shorter than the corresponding time for changing electron energies by means of inelastic electron–atom collisions. But experiments have shown that Maxwellian distributions can actually occur at lower electron densities; this is attributed to the scattering of electrons from space charge waves, thereby changing the electron momentum vectors[16].

A situation which contributes further towards Maxwellization is the presence of a large number of closely spaced atomic energy levels in the electron temperature range. The excitation processes no longer contribute the peaked structures observed by Boyd and Twiddy, but smear the distribution, even though there is not thermodynamic equilibrium.

One classical method of investigating electron energy distribution in a weakly ionized gas was by inferring excitation and ionization functions and comparing them with known cross-sections. In approximate terms, these experiments support the Maxwellian distribution.

Sufficiently weakly ionized plasma situations are basically similar to the steady diffusion of a swarm of electrons through a gas, under the action of a uniform electric field of strength X. The detailed discussion of swarm electron energy distributions is postponed until Chapter 4. The velocity distributions contain both a spherically symmetrical and a field-directed component, respectively f_0 and f_1. For a field in the direction x, the fraction of electrons $f(\xi, \eta, \zeta)\, d\xi\, d\eta\, d\zeta$ with velocities lying between ξ and $\xi+d\xi$, between η and $\eta+d\eta$ and between ζ and $\zeta+d\zeta$ is given[17] approximately by:

$$f(\xi, \eta, \zeta) = f_0(v) + \left(\frac{\xi}{v}\right) f_1(v) \qquad (1.137)$$

with

$$v^2 = \xi^2 + \eta^2 + \zeta^2 = 2\frac{E_e}{m_e} \qquad (1.138)$$

$$f_0(v) = A \exp\left(-\frac{6m_e}{m}\int_0^\infty \frac{E_e}{E_l^2}\, dE_e\right) \qquad (1.139)$$

$$f_1(v) = -\frac{E_l}{m_e v}\frac{df_0}{dv} \qquad (1.140)$$

and

$$E_l = \frac{Xe}{n_e \sigma_d} = Xel_f \qquad (1.141)$$

In equation 1.139, A is a normalizing constant such that

$$4\pi \int_0^\infty f_0(v)\, v^2\, dv = 1 \qquad (1.142)$$

For an energy-invariant σ_d, the spherically symmetrical part reduces to the 'Druyvesteyn energy distribution'[18]:

$$f\left(\frac{E_e}{E_0}\right) d\left(\frac{E_e}{E_0}\right) = 1\cdot63 N\left(\frac{E_e}{E_0}\right) \exp\left(-\frac{E_e^2}{E_0^2}\right) d\left(\frac{E_e}{E_0}\right) \qquad (1.143)$$

where

$$E_0 = \left(\frac{4m_e}{m}\right)^{1/2} el_{f0}\frac{X}{p} \qquad (1.144)$$

Here m represents the gas molecule mass, l_{f0} the electron mean free path at pressure 1 torr (0°C), and N the total number of electrons.

At sufficiently small Xel_f (that is, X/p), the field-directed component becomes negligible compared with the spherically symmetrical part. The Druyvesteyn distribution differs from the Maxwellian distribution in that the pre-exponential term is E rather than $E^{1/2}$, and the exponential term is $-E^2$ rather than $-E$ ($-v^4$ rather than $-v^2$). The Druyvesteyn distribution has a shorter tail than the Maxwellian, as can be seen from the comparison in *Figure 1.6*. Expressed in terms of velocity, the Druyvesteyn distribution is

$$f\left(\frac{v}{\bar{v}}\right) \mathrm{d}\left(\frac{v}{\bar{v}}\right) = \frac{4v^2}{1 \cdot 23\bar{v}^2} \exp\left(-\frac{v^4}{\bar{v}^4}\right) \mathrm{d}\left(\frac{v}{\bar{v}}\right) \qquad (1.145)$$

Diffusion cross-sections are not usually energy-independent over significantly large energy regions, so the calculation of the exact energy distributions is much more difficult. Certain broad features may be discerned, such as the effect of the Ramsauer–Townsend transparency of gases to low-energy electrons (Chapter 4); this makes the cut-off of the high-energy tail of the distribution even more marked, as is shown in *Figure 1.8*.

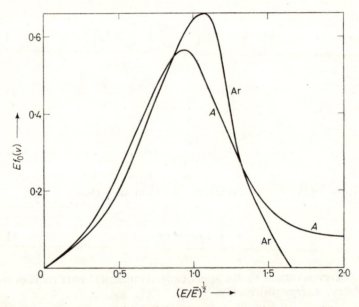

Fig. 1.8. Velocity distributions $E f_0(v)$ of electron swarms in uniform electric field, calculated for energy invariant diffusion cross-section (A), and for real diffusion cross-section derived from swarm experiments in argon (Ar)

Heylen and Lewis[19] have derived other energy distributions which are expected to apply to electrons drifting through rare gases under the action of an electric field. But recent experiments[20] have shown considerable discrepancy with these calculations. Thus for drifting electrons it is difficult to derive an accurate energy distribution. Fortunately, the transport quantities discussed in Chapter 4 are not particularly sensitive to the type of distribution, except when the cross-section varies rapidly with electron energy.

1.5 CONVOLUTION AND DECONVOLUTION OF DISTRIBUTIONS AND FUNCTIONS. APPARENT CROSS-SECTIONS

Deconvolution

The typical experimental situations in which the cross-section function $\sigma(E)$ is measured with a distribution of impact energies $f(E)$ yield mean cross-sections which can be represented in the form:

$$\bar{\sigma}(E_0) = \frac{\int \sigma(E) f(E_0+E)\, dE}{\int f(E_0+E)\, dE} \qquad (1.146)$$

Suppose that a beam of electrons has a thermal energy distribution $f(E)$. It is accelerated through a potential difference which gives all electrons a measured extra energy E_0. This potential difference is varied during the experiment, and the measured cross-section function (obtained, for example, from the current of positive ions formed by ionization $e0/1e^2$) is $\bar{\sigma}(E_0)$. It is usually permissible to neglect the gas atom velocities in comparison with the much greater electron velocities, although striking anomalies can sometimes arise from this practice[21]. The experimental impact energy distribution takes the form $f(E_0+E)$, and the denominator of equation 1.146 is necessary for normalization. The cross-section function $\sigma(E)$ may possess structure, which can be obscured or smoothed out by the existence of the impact energy distribution. The problem is to deduce the structured $\sigma(E)$, knowing $\bar{\sigma}(E_0)$ and $f(E_0+E)$. A further problem which is mathematically almost identical, is to deduce $f(E_0+E)$ knowing $\sigma(E)$ and $\bar{\sigma}(E_0)$.

The calculation of $\bar{\sigma}(E_0)$, when $\sigma(E)$ and $f(E_0+E)$ are known, is straightforward, but will usually require a computer. The inverse problem, derivation of $\sigma(E)$ from known $\bar{\sigma}(E_0)$ and $f(E_0+E)$, is more difficult. (The deduction of energy distributions is considered later in equations 1.172–1.177.) The integral in the numerator of equation 1.146 is known as the convolution of the function f with the function σ and is represented by the notation

$$f*g = f(x)\,g(x)\,dx \qquad (1.147)$$

The solution of the problem of determining $\sigma(E)$ by 'deconvolution' has previously been attempted by trial-and-error techniques, typical of which was the treatment given in the first edition of this book. The problem has been encountered in other fields of physical science[22], and a general solution was first applied to atomic collisions by Morrison[23]. This solution depends upon the 'convolution theorem', which may be stated as

$$T[f*g] = T[f]\,T[g] \qquad (1.148)$$

The operator T represents the Fourier transform

$$T[f(y)] = \frac{1}{\sqrt{2\pi}} \int_{-\infty}^{\infty} f(x) \exp(iyx)\, dx \qquad (1.149)$$

The transform of the experimental function is divided by the transform of the separately determined distribution function, yielding the transform of the cross-section function and hence the function itself.

An improved technique is given by Moore[24]; using the transformations

$$f(x) \rightarrow \mathbf{F} = (a_0, a_1, \ldots, b_0, b_1, \ldots) \tag{1.150}$$

$$g(x) \rightarrow \mathbf{G} = (\alpha_0, \alpha_1, \ldots, \beta_0, \beta_1, \ldots) \tag{1.151}$$

where a_n, b_n, α_n and β_n are the Fourier cosine and sine coefficients of $f(x)$ and $g(x)$ respectively, then

$$\frac{1}{\pi} \int_{-\pi}^{\pi} f(x) g(x) \, \mathrm{d}x = \mathbf{FG} \tag{1.152}$$

which is a special case of Parseval's theorem. Fourier series are better suited than transforms to data functions, which consist of a finite number of observations. The terms of the Fourier series contain a principal and an invisible part; the former is unique, but there can be an infinite number of invisible parts, corresponding to the noise that exists on experimental data. It is also possible for noise, or rather oscillations in the function, to be introduced in the deconvolution process. Both experimental noise and oscillations are removed by truncating the series at a suitable term. If the truncation is carried out at an incorrect term, extra noise may be introduced, therefore trial calculations must be carried out.

The convergence of a Fourier series is not fast, so one may either use the fast-converging Tchebyshev series instead, or alternatively increase the convergence of the Fourier series by multiplying each term by the transform of a narrow ractangular pulse

$$\sigma_K = \frac{\sin K\theta}{K\theta} \tag{1.153}$$

This process of 'sigma smoothing' is applied both to the experimental data and to the deconvoluted cross-section function.

A deconvolution technique particularly applicable to energy distributions which are similar to the Maxwell–Boltzmann distribution (that is, are unsymmetrical and possess an approximately exponential high-energy tail) has been described by Winters, Collins and Courchene[25] and is known as 'electron distribution displacement' (EDD).

Suppose that electrons are accelerated through a potential V after being generated with a Maxwellian energy distribution $f(E)$. An experiment measures a current I proportional to a cross-section $\sigma(E)$:

$$I(V) = C \int f(E - V) \sigma(E) \, \mathrm{d}E \tag{1.154}$$

whilst a further experiment carried out at acceleration V' yields

$$I(V') = C \int f(E - V') \sigma(E) \, \mathrm{d}E \tag{1.155}$$

Multiplying equation 1.155 by constant b and subtracting from equation 1.154,

$$\Delta I(V) = I(V) - bI(V') = C \int [f(E-V) - bf(E-V')] \, \sigma(E) \, dE$$
$$= C \int \Delta f(E-V) \sigma(E) \, dE \qquad (1.156)$$

The 'difference distribution' $\Delta f(E-V)$ is readily shown to be much narrower than $f(V)$ when b and ΔV are suitably chosen. Thus with $\Delta V = 0 \cdot 1$ eV, $b = 0 \cdot 63$ and filament temperature $T = 1730°$K, the FWHM of $\Delta f(E-V)$ is $0 \cdot 14$ eV. The calculation of $\Delta I(V)$ from the experimental data therefore permits a mathematically narrowed distribution to be applied to the experiment.

Convolution Problems with Momentum Analysers

Collision experiments are frequently conducted with beams monochromated and products analysed in momentum. Convolution problems are encountered in the analysis of data from pairs of momentum analysers in tandem. Suppose that these have Gaussian energy distributions (equation 1.131) with FWHM values ΔE_1 and ΔE_2. When the two analysers are ganged together and scanned through a spectral line narrow compared with ΔE_1, and ΔE_2, the observed FWHM can be shown to be:

$$\Delta E_g = \frac{\Delta E_1 \Delta E_2}{(\Delta E_1^2 + \Delta E_2^2)^{1/2}} \qquad (1.157)$$

But when one analyser is scanned against the other, the observed FWHM is that of the convolution of the two distributions:

$$\Delta E_c = (\Delta E_1^2 + \Delta E_2^2)^{1/2} \qquad (1.158)$$

Now when the two analysers are used to investigate a spectral line of Gaussian width ΔE_l, then ganged scansion of both analysers 'yields 'an observed FWHM:

$$\Delta E_s = (\Delta E_g^2 + \Delta E_l^2)^{1/2} \qquad (1.159)$$

But if the first analyser is tuned onto the line and the second is scanned across, then the observed FWHM is:

$$\Delta E_t = (\Delta E_c^2 + \Delta E_l^2)^{1/2} \qquad (1.160)$$

If one finds

$$\Delta E_t^2 - 3\Delta E_g^2 = \Delta E_s^2 \qquad (1.161)$$

then

$$\Delta E_1 = \Delta E_2 \qquad (1.162)$$

which can be used as a check on the internal consistency of particle spectroscopy experiments.

4*

Apparent Cross-sections

Collision experiments are complicated when neither particle velocity can be neglected in comparison with that of the other. The apparent or effective cross-sections measured can be related to the real cross-sections provided that the velocity distributions are known.

For electron beams of energy E_e passing through a gas at temperature $T°K$, the impact energy distribution is broadened by the thermal motion of the gas molecules. This loosely termed 'Doppler broadening' is Gaussian, of FWHM

$$\Delta E_{1/2} = 7{\cdot}23 \times 10^{-4} \left(\frac{E_e T}{m_g}\right)^{1/2} \text{ eV} \tag{1.163}$$

where the gas molecule mass m_g is in a.m.u.

In the scattering of molecular beams by gases and by crossed molecular beams[26], the measured cross-sections are related to the real cross-sections by complicated functional relationships which have recently been calculated by Berkling *et al.*[14] The relative velocity of impact between two molecules of velocities v_1 and v_2 is

$$v_r = v_2 - v_1 \tag{1.164}$$

The effective cross-section is derived from the flux I of molecules unscattered in path l from the flux I_0:

$$\frac{I}{I_0} = \exp\left(-\sigma_{\text{eff}} l\right) \tag{1.165}$$

For a velocity-analysed beam crossed orthogonally with a thermal (un-analysed) beam, a function $Fb_0(x)$ was calculated such that

$$\sigma_{\text{eff}} = \sigma(v_1)\, Fb_0(x) \tag{1.166}$$

where $x = v_1/\bar{v}_2$. For a velocity-analysed beam scattered by a gas, a similar function $Fa_0(x)$ was calculated.

For the crossing of two thermal (un-analysed) beams, a similar function $Gb_0(y)$ was calculated with $y = \bar{v}_1/\bar{v}_2$. For the scattering of a thermal beam by a gas, a similar function $Ga_0(y)$ was calculated.

These functions apply only to an assumed velocity dependence of the scattering cross-section:

$$\sigma \propto v_r^{-2/(s-1)}$$

with $s = 6$ for van der Waals' interaction. In Chapter 4 it will be seen that this expression is sufficiently realistic for the functions to be successful. More complicated velocity dependences have also been treated[27].

In the first edition of this book, the well-known approximate correction function[28] for beam–gas scattering was discussed:

$$\sigma_a = \frac{\bar{\sigma}(\bar{v}_1)}{2\sqrt{\pi}} \left[\mathcal{I}\left(\frac{\bar{\bar{v}}_2}{\bar{v}_1}\right)^2 \right]^{-1} \tag{1.167}$$

where

$$\mathcal{A}(z) = z^2 \int_0^\infty \frac{x^5}{\Psi(x)} \exp(-zx^2) \, dx \qquad (1.168)$$

and

$$x = \frac{v_2}{\bar{\bar{v}}_2} \qquad (1.169)$$

This was derived from the Tait mean free path[29] l_{fT} in which the dependence of collision frequency upon gas atom velocity was taken into account for the first time:

$$l_{fT} = \frac{\sqrt{\pi} \, x^2}{n\bar{\sigma}(\bar{v}_2) \, \Psi(x)} \qquad (1.170)$$

where

$$\Psi(x) = x \exp(-x^2) + (2x^2 + 1) \int_0^x \exp(-y^2) \, dy \qquad (1.171)$$

Although the correct function had already been calculated[30], the approximate function given by equations 1.167–1.169 remained in use until 1962, when the calculations of $F(x)$ and $G(y)$ for both beam–gas and crossed-beam scattering were published. In these calculations, both the gas velocity distributions and the beam velocity distributions are Maxwell–Boltzmann.

Deduction of Energy Distributions from Cross-section Function Data

The problem of deducing an energy distribution $f(E)$ from equation 1.146 when a form of $\sigma(E)$ is inferred from theory and $\bar{\sigma}(E_0)$ is determined by experiment has several simple and important solutions for particular $\sigma(E)$. Such deductions of $f(E)$ can be equal or superior to experimental determinations by momentum analysis.

For a cross-section function $\sigma(E)$ which is an infinitely narrow peak of finite height (delta function), $\bar{\sigma}(E_0)$ is a mirror image of $f(E)$. Such a peak is approximated by the near-zero energy resonance attachment cross-section $e0/\bar{1}'$ to the SF_6 molecule[31], forming SF_6^-, which may be mass-spectrometrically separated from the dissociatively formed SF_5^-.

For a cross-section function with step onset, that is, for $E < E_1$,

$$\sigma = 0 \qquad (1.172)$$

and, for $E \geqslant E_1$,

$$\sigma = \sigma_1 \qquad (1.173)$$

the function $\bar{\sigma}(E_0)$ has a more gradual onset than the sudden step in $\sigma(E)$. It is readily shown that the first differential $d\bar{\sigma}/dE_0$ is proportional to the energy distribution $f(E)$. Step onsets corresponding exactly to equations 1.172 and 1.173 are unknown, but the photoionization functions of atoms $\phi 0/1e$ (Chapter 9) are non-zero at onset, and may without great inaccuracy be considered as step-functions.

For a cross-section function which rises linearly with energy from an onset, that is, for $E < E_1$

$$\sigma = 0 \tag{1.174}$$

and for $E \geqslant E_1$

$$\sigma = a(E - E_1) \tag{1.175}$$

the second differential $d^2\bar{\sigma}/dE_0^2$ is proportional to the energy distribution $f(E)$. The ionization of an atom by an electron $e0/1e^2$ shows *approximately* such a cross-section function, provided that the positive ion is formed only in the ground state (Chapter 6); the deduction of energy distributions from these functions is standard procedure.

In the absence of autoionization, the multiple ionization of atoms by electrons, $e0/ne^{n+1}$ (Chapter 6) follows an nth power law above onset: for $E < E_n$

$$\sigma = 0 \tag{1.176}$$

and for $E \geqslant E_n$

$$\sigma = a(E - E_n)^n \tag{1.177}$$

The nth root of the cross-section rises linearly with E_0, except close to the onset, where the second differential is proportional to the energy distribution $f(E)$. Moreover $d^{n+1}\sigma/dE_n^{n+1}$ is proportional to the energy distribution[23], so there is a choice of performing the $(n+1)$th differential, or of performing the second differential on the nth root plot. Either method becomes increasingly inaccurate with increasing n. The nth root plot is suitable for observing the piecewise structure when excited states contribute (Chapter 6).

These methods supersede those discussed in equations 1.161–1.163 of the first edition, which are limited to Maxwellian distributions.

A further method for deriving energy distributions from scattering data is the 'recoil technique'[31]. The angular distribution of atoms scattered out of the collision of mutually perpendicular crossed beams is measured and converted, using momentum and energy equations, to the velocity distribution of one beam; that of the other must be previously known.

1.6 DIFFUSION OF PARTICLES

The close connection between particle diffusion and atomic collision physics rests upon the relation between the momentum transfer cross-section σ_d (Equation 1.39) and the diffusion coefficient D.

In a gas or plasma contained in a vessel whose dimensions are large compared with the mean free path, a flow of particles takes place in the direction of decreasing particle concentration. The flux density vector J is proportional to this gradient. That is

$$J = -D \nabla n \tag{1.178}$$

The constant of proportionality is known as the diffusion coefficient D cm^2 sec^{-1}; equation 1.178 expresses Fick's law of diffusion.

The analysis leading to the relation between σ_d and D was given by Chapman and by Enskog[32]. For the diffusion of gas 1 into gas 2 in a binary mixture at temperature T,

$$D_{12} = \frac{3}{16(n_1+n_2)} \left(\frac{2\pi kT}{\mu}\right)^{1/2} \frac{1+\varepsilon_0}{\bar{\sigma}_d} \qquad (1.179)$$

with

$$\bar{\sigma}_d = \tfrac{1}{2} \int_0^\infty \tau^2 \sigma_d(\tau) \exp(-\tau) \, d\tau \qquad (1.180)$$

The 'reduced temperature'

$$\tau = \frac{\mu v_r^2}{kT} \qquad (1.181)$$

and the reduced mass is μ. The correction factor ε_0 is usually much smaller than 0·05. The impact velocity to which σ_d is appropriate is denoted by v_r. A plausible functional dependence of σ_d on v_r must be assumed to extract meaningful cross-sections. If σ_d is independent of v_r, then D_{12} is proportional to $T^{1/2}$. At constant temperature, D_{12} is inversely proportional to pressure. There are two contrasting situations in which particle density variation can be studied in its relation to diffusion: (*i*) the time-dependent, and (*ii*) the flowing or drifting system. For plasmas, the decay of ionized species density in the time-dependent and flowing afterglows is discussed in Chapter 3 and elsewhere.

The Time-dependent System

The time-dependent diffusion equation is written as

$$\frac{dn}{dt} = \nabla^2(Dn) \qquad (1.182)$$

where the number density n is both time-dependent and space-dependent. This equation applies to a system of particles diffusing irreversibly across the boundaries of a certain volume (for example, being destroyed by inelastic surface processes). The particle density within the volume decays exponentially with time from an initial value n_0:

$$n = n_0 \exp\left(-\frac{t}{\tau}\right) \qquad (1.183)$$

The decay constant τ is known as the diffusion time. Provided that the diffusion coefficient is constant in space:

$$\nabla^2 n + \frac{n}{D\tau} = 0 \qquad (1.184)$$

This equation may be solved for various geometries as follows.

Infinite parallel plates

If infinite parallel plates are separated by a distance L in the coordinate x,

$$n(x, t) = \sum_{k=1}^{\infty} A_k \cos \left[\frac{x}{(D\tau_k)^{1/2}} \right] \exp \left(-\frac{t}{\tau_k} \right) \qquad (1.185)$$

with

$$\frac{L}{2(D\tau_k)^{1/2}} = (2k-1)\frac{\pi}{2} \qquad (1.186)$$

The particle density decays simultaneously in a number of different modes (different values of k). Since the diffusion time is largest for the lowest mode, this mode will predominate in the end. Nevertheless, any linear combination of A_k values will satisfy the equation.

Parallelepiped

For a parallelepiped of sides a, b and c,

$$n(x, y, z, t) = \sum_{i=1}^{\infty} \sum_{j=1}^{\infty} \sum_{k=1}^{\infty} A_{ijk} \cos \left[\frac{(2i-1)\pi x}{a} \right]$$
$$\times \cos \left[\frac{(2j-1)\pi y}{b} \right] \cos \left[\frac{(2k-1)\pi z}{c} \right] \exp \left(-\frac{t}{\tau_{ijk}} \right) \qquad (1.187)$$

with

$$\frac{1}{\tau_{ijk}} = D\pi^2 \left[\left(\frac{2i-1}{a} \right)^2 + \left(\frac{2j-1}{b} \right)^2 + \left(\frac{2k-1}{c} \right)^2 \right] \qquad (1.188)$$

Higher modes (higher values of i, j and k) persist longer in this geometry than for the parallel plates.

Sphere

For a sphere of radius r_0,

$$n(r, t) = \sum_{k=0}^{\infty} \frac{A_k}{r} \sin \left[\frac{r}{(D\tau_k)^{1/2}} \right] \exp \left(-\frac{t}{\tau_k} \right) \qquad (1.189)$$

with

$$\frac{r_0}{(D\tau_k)^{1/2}} = k\pi \qquad (1.190)$$

Cylinder

For a cylinder of radius r_0 and height H,

$$n(r, z, t) = \sum_{i=1}^{\infty} \sum_{j=1}^{\infty} A_{ij} J_0(\alpha_i r_0) \cos \left[\frac{(2j-1)\pi z}{H} \right] \exp \left(-\frac{t}{\tau_{ij}} \right) \qquad (1.191)$$

with

$$\frac{1}{D\tau_{ij}} = \alpha_i^2 + \left[\frac{(2j-1)\pi}{H}\right]^2 \tag{1.192}$$

The term $\alpha_i r_0$ is the ith root of the Bessel function J_0 of order zero.

Ambipolar Diffusion

In a thermally equilibrated plasma containing only electrons, positive ions and neutral particles, the electrons, which travel the fastest, tend to diffuse outwards and stick to the container walls, where they remain or migrate along the surface until neutralized by an incident positive ion. The excess of positive charge left behind then retards the electrons and their diffusion.

Debye and Hückel[33] showed that the electric field at a distance r from an ion in an electrolyte is reduced by a factor $\exp(-r/\lambda)$ due to shielding by ions of the opposite sign. The theory applies equally well to the positive ions in a plasma, and λ, known as the 'Debye length', is given by the equation[34]

$$\lambda = \left(\frac{D_e}{8\pi n_e e K_e}\right)^{1/2} \tag{1.193}$$

where K_e is the electron mobility, or drift velocity per unit electric field (see equations 1.210–1.212).

If the plasma is contained in a vessel of dimensions greater than λ, the ions will be able to hold their shielding electrons and the 'linked' diffusion is known as 'ambipolar'; but if the dimensions are much smaller than λ, the electrons will diffuse independently of the ions and the diffusion is known as 'free'.

In the ambipolar regime, in a plasma containing electrons and a single species of ions, the time-dependent diffusion equation 1.182 for electrons is

$$\frac{dn_e}{dt} = D_a \nabla^2 n_e \tag{1.194}$$

An analogous equation holds for the positive ions; the ambipolar diffusion coefficient

$$D_a = \frac{D_+ K_e + K_+ D_e}{K_+ + K_e} \tag{1.195}$$

where D_+ and K_+ are the free diffusion coefficient and mobility of the positive ions; D_e and K_e refer to the electrons.

It is shown below (equation 1.213) that

$$\frac{K_e}{D_e} = \frac{e}{kT_e} \tag{1.196}$$

and

$$\frac{K_+}{D_+} = \frac{e}{kT_+} \tag{1.197}$$

where e is the electronic charge and k the Boltzmann constant. Since $K_e \gg K_+$,

$$D_a \simeq D_+ \left(1 + \frac{T_e}{T_+}\right) \tag{1.198}$$

For an isothermal plasma $T_e \simeq T_+$, so

$$D_a \simeq 2D_+ \simeq 2K_+ \frac{kT}{e} \tag{1.199}$$

In this case, measurement of the ambipolar diffusion coefficient yields the positive ion mobility.

In a gas containing more than one type of positive ion, the ambipolar diffusion coefficient decreases with time and approaches the value corresponding to that of the slowest ion.

In an electronegative ionized gas, the presence of negative ions distorts the space charge field. Diffusion of the negative ions is negligible, the distortion of the space field being such as to retard their diffusive motion. The situation[35] is described by the equations:

$$D_{ae} = \frac{(1+\beta)[K_e D_+ + K_+ D_e] + \beta[K_- D_e - K_e D_-]}{K_e + \beta K_- + (1+\beta)K_+} \tag{1.200}$$

$$D_{a-} = \frac{(1+\beta)[K_+ D_- + K_- D_+] + [K_- D_e + K_e D_-]}{K_e + \beta K_- + (1+\beta)K_+} \tag{1.201}$$

$$D_{a+} = \frac{[K_e D_+ + K_+ D_e] + \beta[K_+ D_- + K_- D_+]}{K_e + \beta K_- + (1+\beta)K_+} \tag{1.202}$$

$$\beta = \frac{n_-}{n_e} \tag{1.203}$$

$$1 + \beta = \frac{n_+}{n_e} \tag{1.204}$$

$$n_+ \simeq n_e + n_- \tag{1.205}$$

Assuming that $K_e \gg (1+\beta)K_+ + \beta K_-$ and that, for isothermal plasma,

$$\frac{K_+}{D_+} = \frac{K_-}{D_-} = \frac{K_e}{D_e} = \frac{e}{kT} \tag{1.206}$$

then the following approximate relations are found:

$$D_{ae} \simeq 2D_+(1+\beta) \tag{1.207}$$

$$D_{a-} \simeq \frac{2D_- D_+}{D_e}(1+\beta) \tag{1.208}$$

$$D_{a+} \simeq 2D_+\left(1 + \frac{\beta K_-}{K_e}\right) \simeq 2D_+ \tag{1.209}$$

These diffusion equations are only soluble when some assumption is made concerning the time-dependence of the ratio β. Purely exponential electron density decay in a plasma in which no source of ionization still remains is only possible where β remains small or constant. Oskam[35] has considered the form of electron decay $n_e(t)$ in the presence of inelastic volume processes, of which the simplest is the attachment of electrons to atoms

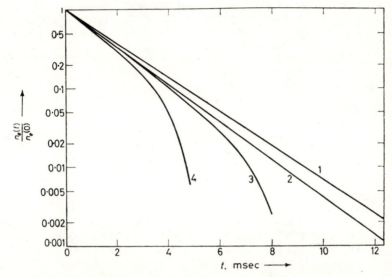

Fig. 1.9. Ambipolar diffusion of electrons in the presence of negative ions
(From Oskam[35], by courtesy of Centrex Publishing Company):

curve 1, exponential decay by diffusion in the absence of volume processes $[\tau^{-1} = 500 \text{ sec}^{-1}, \nu_a = 0, n_-(0) = 0]$;
curve 2, exponential decay by volume processes and by diffusion [attachment frequency $\nu_a = 50 \text{ sec}^{-1}$, $\tau^{-1} = 500 \text{ sec}^{-1}, n_-(t) = n(0) = 0]$; *curve 3, decay under conditions when the negative ion density is constant*
$[\tau^{-1} = 500 \text{ sec}^{-1}, \nu_a = 50 \text{ sec}^{-1}, n_-(t) = n_-(0) = n_e(0)/99]$; *curve 4, decay when electrons are converted into negative ions* $[\nu_a = 50 \text{ sec}^{-1}, n_-(0) = n_e(0)/99, D_{a-} = 0]$

forming negative ions. The results of his calculations are shown in *Figure 1.9*. Afterglow electron decay curves showing such curvature may readily be observed at sufficiently small electron densities ($n_e \sim 10^6$ cm^{-3}).

The Drifting System

Contrasted with the time-dependent systems are the 'flowing afterglow' and also the swarm experiments in drift tubes.

Experiments with charged particles can be conducted by causing them to drift collisionally through gas under the action of a uniform electric field X in direction z (Chapters 4 and 10). The diffusion coefficient of negatively or positively charged particles is related to their velocity of drift v_d:

$$n_{\pm} v_d = -D_{\pm} \frac{dn_{\pm}}{dz} \qquad (1.210)$$

It has been found for ions and, under some circumstances, for electrons that the drift velocity is proportional to the field:

$$v_d = KX \tag{1.211}$$

The constant of proportionality K is known as the mobility and is inversely proportional to particle density. At s.t.p. it assumes the value K_0, known as the 'reduced mobility'.

Across a cylinder of unit cross-section and length dz, a gradient of partial pressure of the charged particles will produce a force on each particle equal to $(1/n_\pm)\, dp_\pm/dz$. This is balanced by the electric force eX, where $\pm e$ is the charge on the particle:

$$\frac{1}{n_\pm}\frac{dp_\pm}{dz} = eX \tag{1.212}$$

Equations 1.210–1.212 lead immediately to the Einstein relation

$$\frac{K}{D_\pm} = \frac{en_\pm}{p_\pm} = \frac{e}{kT_\pm} = \frac{42{\cdot}7\times273}{T_\pm} \tag{1.213}$$

or

$$D_\pm = \frac{2{\cdot}3\times10^{15}}{n_\pm} K_0 T_\pm$$

where k is the Boltzmann constant and $T_\pm$ is the charged particle temperature. Equations 1.213 has been used in the ambipolar regime discussion above. The relations between drift velocities and diffusion cross-sections are discussed further in Chapters 4 and 10.

The time-dependent diffusion equation in cylindrical geometry is applicable to the drift of charged particles through a gas under the action of a uniform electric field. A pulse of charged particles possessing an initial radial distribution $f(r)$ and an initial axial distribution $f(z)$ at $t = 0$ will, after drifting in a uniform electric field in the direction z, possess different $f(r)$ and $f(z)$. In a steady-state system, the axial diffusion will usually be negligible, so the use of pulse electronics for drift velocity measurement, as outlined in Chapters 4 and 10, is an accurate technique. The consequences of taking axial diffusion into account have been considered by Barnes et al.[36] who show that the 'mobility peak' is distorted from a triangular to a skewed Gaussian shape. The radial diffusion also introduces effects.

Neglecting axial diffusion, the radial equation is

$$\frac{dn_\pm}{dt} = \frac{\partial n_\pm}{\partial t} + D_\pm\left(\frac{1}{r}\frac{\partial n_\pm}{\partial r} + \frac{\partial^2 n_\pm}{\partial r^2}\right) \tag{1.214}$$

The term $\partial n_\pm/\partial t$ represents the formation or destruction of the charged particle species by inelastic processes in the drift tube. In the absence of such processes, the radial expansion of $f(r)$ is such that the axial flux $n_{0\pm}$ decreases with time and hence with z. For example, an initial Gaussian

$$n_\pm(r) = n_\pm(0)\exp\left(-\frac{r}{\delta_0}\right)^2 \tag{1.215}$$

of width δ_0 yields a solution of equation 1.214 such that

$$n_{0\pm}(t) = \frac{N_\pm}{\pi\delta_0^2 + 4\pi D_\pm t} \qquad (1.216)$$

where $N_\pm$ is the total particle flux.

At the end of the axis of the 'drift tube' is a particle detector behind an orifice, designed so as to distinguish between different species of charged particle. The calculation of the rates of the inelastic conversion processes depends upon the detailed solution of equation 1.214 for each species, with non-zero $\partial n_\pm / \partial t$. Kaneko, Megill and Hasted[37] have considered these solutions.

1.7 THERMODYNAMIC EQUILIBRIA

Assemblies of particles in which interconversion of species is possible through inelastic processes are studied by determining the time-variation of concentrations of these species. In addition to the time-dependent and drift systems there exist equilibrium systems in which the kinetic energy of the particles, moving in thermal equilibrium at specified temperature or temperatures, supplies the energy necessary for endothermic conversions. Where there is thermodynamic equilibrium, so that the concept of temperature is real, it is often necessary to feed energy into the system by the maintainance of electric fields, by heating solid or liquid surfaces, or by means of electromagnetic radiation. Energy is lost by radiation and conduction of heat. In this section are discussed the temperature dependence of species densities in thermodynamic equilibria, the conditions under which equilibrium is achieved, and the relation of rate constants to equilibrium constants.

The equilibrium constant K was defined in equation 1.84 as the ratio of forward and reverse rate constants. In the general reaction

$$x_1 A_1 + x_2 A_2 + \ldots \rightleftharpoons y_1 B_1 + y_2 B_2 + \ldots$$

the law of mass action states that the equilibrium constant is a function not of the constituent concentrations but of temperature only. That is,

$$\frac{\Pi n_{A_1}^{x_1}}{\Pi n_{B_1}^{y_1}} = \frac{\overleftarrow{k}}{\overrightarrow{k}} = K(T) \qquad (1.217)$$

In chemical kinetics, the form of the function $K(T)$ is exponential, and may be found by applying the van't Hoff isochore:

$$\frac{d[\ln K(T)]}{dT} = \frac{\Delta H}{RT^2} \qquad (1.218)$$

where ΔH is the total heat energy absorbed by the system and R is the gas constant. The temperature dependence of the chemical rate constants k_c

may be represented by the Arrhenius equation:

$$k_c = v_c \exp\left(-\Delta E_{\text{act}}/kT\right) \tag{1.219}$$

with the reduced collision frequency v_c given approximately by

$$v_c = 2\sigma(2\pi kT/\mu)^{1/2} \tag{1.220}$$

for cross-section σ and reduced mass μ. In this formulation, the rate constant for the reaction is taken to rise exponentially with increasing temperature. The activation energy ΔE_{act} of the reaction is the height of the energy barrier that must be surmounted for the reaction to proceed; it must be distinguished from the energy defect ΔE, which is the difference between the initial and final internal energies of the system.

When the collision frequency is temperature independent, the cross-section σ is proportional to $T^{-1/2}$. In chemical reactions and in ion–atom interchange (Chapter 14), this type of behaviour is sometimes applicable; but in many ionic collision processes it is unrealistic.

In general, the thermodynamic equilibrium situation is only found when there is a Maxwell–Boltzmann distribution of energy in discrete states. The equilibrium between positive ions, neutral atoms and electrons, such as might be encountered at a hot metal surface (temperature T_s) exposed to gas, is expressed by the Saha equation[38]. This was deduced by combining the isochore with the Sackur–Tetrode expression for the entropy of a monatomic gas, in order to relate the free energy charge ΔE of the collision process with the total heat energy. For the equilibrium

$$X \rightleftharpoons X^+ + e - \Delta E$$

the Saha equation has the form

$$K = \frac{n_+ n_e}{n_0} = \frac{g_+ g_e}{g_0}\left(\frac{2\pi m_e kT_s}{h^2}\right)^{3/2}\exp\left(-\frac{\Delta E}{kT_s}\right) \tag{1.221}$$

The symbols g represent the statistical weights of the species, which may be made up of degenerate states. The energy ΔE for a metal surface includes the work function ϕ and the ionization energy E_i:

$$\Delta E = E_i - \phi \tag{1.222}$$

For atomic negative ions,

$$\Delta E = \phi - E_a \tag{1.223}$$

where E_a is the electron affinity. The equilibrium constant is related to the flux densities J reflected from the surface by the equations

$$K = \frac{J_+ J_e}{J_0}\left(\frac{kT_s}{2\pi m_e}\right) \tag{1.224}$$

and

$$J_0 = p_0(2m_0 kT_0) \tag{1.225}$$

where J_0, p_0, m_0, and T_0 all refer to the gas atoms. Note that

$$\frac{2\pi m_e kT}{h^2} = 3{\cdot}9 \times 10^{-9} T^{\circ}\text{K} \qquad (1.226)$$

The similar situation for negative ions is discussed in Chapter 8.

An ionized gas contains more species than just the three mentioned in equation 1.221: radiation, free and bound electrons (the latter in many energy levels), negative and positive ions, including ionized atoms in states all the way up to the stripped nucleus, in sufficiently hot plasmas. A number of different regimes may be distinguished.

Under the most complete thermodynamic equilibrium, the absorption and emission of radiation take part in maintaining the equilibrium. A plasma in this condition is described as 'optically thick' rather than 'optically thin', and the diffusion of resonance radiation is further discussed in Chapter 9. For the interpretation of the emitted spectrum, no cross-sections for inelastic processes need to be known. The radiation density B_ν in the frequency interval ν to $\nu + d\nu$ is described by the Planck black-body equation:

$$B_\nu \, d\nu = \frac{2\pi h\nu^3}{c^2} \frac{d\nu}{\exp(-h\nu/kT) - 1} \qquad (1.227)$$

The regime is described as 'full thermodynamic equilibrium'. However, it is seldom encountered in gases, since more frequently the plasma is optically thin and so the equilibrium includes only the free and bound electront and the ions. Under these circumstances, radiation is not taken into accouns and the regime is described as in 'local thermodynamic equilibrium' (LTE); only the radiative process probabilities need to be known for an interpretation of the emitted spectrum to be possible.

In LTE the radiation density is given by the Kirchhoff equation as the ratio of emission coefficient to absorption coefficient

$$\frac{\varepsilon_\nu}{\varkappa_\nu} = B_\nu(T) \qquad (1.228)$$

The energy distribution of the free electrons is described by the Maxwell equation 1.132. But the density of electrons bound in state p is given by the Boltzmann equation for the excitation temperature:

$$\frac{n_p}{n_e} = \frac{g_p}{2} \exp\left(-\frac{E_i - E_p}{kT_e}\right) \qquad (1.229)$$

where E_i is the ionization energy and E_p the energy of state p. For an optically thick plasma in LTE[39], the excitation temperature is non-uniform, decreasing as the plasma boundary is approached.

The density of ions in ionization state m is given by the Saha equation in the form

$$\frac{n_{m+1}}{n_m} = \left(\frac{2\pi m_e kT_e}{h^2}\right)^{3/2} \frac{2g_{m+1}}{g_m} \frac{T_e}{n_e} \exp\left(-\frac{E_{im}}{kT_e}\right) \qquad (1.230)$$

A discussion of LTE and the criterion necessary for its establishment (that is, collision processes dominate radiation processes) has been given by Wilson[40]. The criterion was shown to be

$$n_e > 6 \times 10^{13} E_i^3 (kT_e)^{1/2} \qquad (1.231)$$

with E_i and kT_e both in electron volts. But it is also necessary that the frequency of electron–electron collisions should be much smaller than the frequency of heat exchange and containment collisions.

When condition 1.231 is not satisfied, as in fact it rarely can be, there can still exist non-thermal equilibria in which there is Maxwellian distribution of energy among the free electrons, but with equations 1.229 and 1.230 holding only as inequalities ($\leqslant$).

There is also a regime in which the collisional processes, so far from dominating the radiative processes, are negligible in comparison with them. This is the case in the solar corona; the regime is known as the 'coronal regime'. The criteria[39] for its establishment are that the plasma shall be optically thin, and that

$$n_e < 1 \cdot 5 \times 10^{10} E_i^{-1/2} (kT_e)^4 \qquad (1.232)$$

The excitation and ionization are governed not by equations 1.229 and 1.230 but by the equations

$$\frac{n_p}{n_0} = \frac{n_e k_{0p}}{A_p} \qquad (1.233)$$

and

$$\frac{n_{m+1}}{n_m} = \frac{k_m}{\alpha_{m+1}} \qquad (1.234)$$

where k_{0p} and k_m are the collisional rate functions for excitation and ionization, A_p is the transition probability (Einstein A coefficient) and α_{m+1} is the recombination coefficient for $m+1$ charged ions.

The high-temperature plasmas which satisfy completely neither the LTE nor the coronal criteria (conditions 1.231 and 1.232), but which are optically thin, require for their description a detailed knowledge of excitation, ionization and recombination rates.

As the electron density is increased above condition 1.231, the recombination and ionization rates are enhanced above those for a coronal distribution, because of the gradual diffusion of electrons up and down through states of high principal quantum number n. For the higher levels, the collisional exchange rate with the continuum dominates over the radiative decay rate, so that the excited states are linked to the continuum through equations 1.229 and 1.230. For lower levels, the radiative decay dominates over the collisional exchange and condition 1.231 applies.

The level t which divides these two domains, called the thermal limit, is given by

$$(E_i - E_t)^7 = 6 \times 10^{-28} \frac{E_i}{kT_e} n_e^2 \qquad (1.235)$$

This regime, in which the upper excited states are in equilibrium with the free electrons whilst the lower states are not, is known as 'partial local thermodynamic equilibrium'.

The temperatures T_{gas}, T_e (diagnosed), T_e (equation 1·231) and T_e (equation 1·232) for an argon arc, do not become equal until electron densities as high as 7×10^{15} cm^{-3} are reached. This equality was at first regarded as experimental proof of local thermodynamic equilibrium, but the calculated density at which LTE should be achieved is much higher ($1·3 \times 10^{17}$ cm^{-3}). Between these two densities, the LTE is partial; the resonance radiation has not been included in the observations on which equation 1.231 is calculated. The observation of LTE requires that all radiation be optically thick, and for impurity resonance radiation this is difficult to achieve.

Fully ionized plasmas of nuclear charge Z scale as n_e/Z^7 and as T_e/Z^2 in these equations.

REFERENCES

1. *Atomic and Molecular Collision Cross-sections*, Oak Ridge National Laboratory Report ORNL-3113 – revised, 1964; *Bibliography of Atomic Transition Probabilities*, National Bureau of Standards Miscellaneous Publication 278, 1966; University of Toulouse Data Collection DRME 312/64, 1965; KIEFFER, L. J. and DUNN, G. H. Joint Institute of Laboratory Astrophysics Report 51; *J. atomic Data* 1–2 (1969–70).
2. MASSEY, H. S. W. and BURHOP, E. H. S. *Electronic and Ionic Impact Phenomena*, 2nd edn, 1969. London; Oxford University Press.
3. McDANIEL, E. W. *Collision Phenomena in Ionized Gases*, 1964. New York; Wiley.
4. LOEB, L. B. *Kinetic Theory of Gases*, 1927. New York; McGraw-Hill.
5. WIEN, W. *Ann. Phys. Lpz.* 39 (1912) 528.
6. ALLISON, S. K. *Rev. mod. Phys.* 30 (1958) 1137.
7. COLWELL, J. and FINEMAN, M. A. *J. chem. Phys.* 42 (1965) 4097.
8. WOLF, F. *5th Int. Conf. Phys. electron. and atom. Collisions*, 1967. Akademii Nauk, Leningrad.
9. SMITH, F. T. *J. chem. Phys.* 36 (1962) 248.
10. BENSON, S. W. *The Foundations of Chemical Kinetics*, 1960. New York; McGraw-Hill.
11. MATSEN, F. A. and FRANKLIN, J. L. *J. Amer. chem. Soc.* 72 (1950) 3337.
12. JEN-YUAN CHIEN, *J. Amer. chem. Soc.* 70 (1948) 2256; ADIROVICH, E. I. *Dokl. Akad. Nauk S.S.S.R.* 61 (1948) 467; FUOSS, R. M. *J. Amer. chem. Soc.* 65 (1943) 2406.
13. RUZICKA, T. *Proc. 9th Int. Conf. Ioniz. Phenom. Gases*, p. 454, 1969. Academy of Sciences, Bucurest.
14. BERKLING, K., HELBING, R., KRAMER, K., PAULY, H., SCHLIER, C. H. and TOSCHEK, P. *Z. Phys.* 166 (1962) 406.
15. BOYD, R. L. F. and TWIDDY, N. D. *Proc. Roy. Soc.* 250 (1959) 53.
16. TWIDDY, N. D. *Proc. Roy. Soc.* A262 (1961) 519; 275 (1963) 338; RAYMENT, S. W. and TWIDDY, N. D. *Proc. 9th Int. Conf. Ioniz. Phenom. Gases*, p. 455, 1969. Academy of Sciences, Bucurest.
17. MORSE, P. M., ALLIS, W. P. and LAMAR, E. S. *Phys. Rev.* 48 (1935) 412; DAVYDOV, B. *Phys. Z. Sowjet.* 8 (1935) 59.
18. DRUYVESTEYN, M. J. *Physica, Eindhoven* 10 (1930) 61; 1 (1935) 1003.
19. LEWIS, T. J. *Proc. Roy. Soc.* A244 (1958) 166; HEYLEN, A. E. D. and LEWIS, T. J. *Proc. Roy. Soc.* A271 (1963) 531.
20. ROBERTS, T. D. and BURCH, D. S. *Phys. Rev.* 142 (1966) 100.
21. CHANTRY, G. W. and SCHULZ, G. J. *Phys. Rev. Lett.* 12 (1964) 449.
22. BRACEWELL, R. N. and ROBERTS, J. A. *Aust. J. Phys.* 7 (1954) 615.
23. MORRISON, J. D. *J. chem. Phys.* 39 (1963) 200.
24. MOORE, L. *Brit. J. appl. Phys.* 1 (1968) 237.
25. WINTERS, R. E., COLLINS, J. H. and COURCHENE, W. L. *J. chem. Phys.* 45 (1966) 1931.

26. DATZ, S., HERSCHBACH, D. R. and TAYLOR, E. H. *J. chem. Phys.* 35 (1961) 1549.
27. DÜRER, R., HELBING, R. and PAULY, H. *Z. Phys.* 188 (1965) 468.
28. ROSIN, S. and RABI, I. I. *Phys. Rev.* 48 (1935) 373.
29. TAIT, P. G. *Trans. Roy. Soc.* 33 (1886) 65.
30. WOOD, J. K. *Phys. Rev.* 64 (1943) 42; FONER, S. N. Dissertation, Carnegie Institute of Technology, 1945.
31. RUBIN, K., PEREL, J. and BEDERSON, B. *Phys. Rev.* 117 (1960) 151.
32. CHAPMAN, S. *Phil. Trans.* 216 (1916) 279; 217 (1916) 115; ENSKOG, A. Inaugural Dissertation, University of Uppsala, 1917.
33. DEBYE, P. and HÜCKEL, E. *Phys. Z.* 24 (1923) 190.
34. LIBOFF, R. L. *Phys. Fluids* 2 (1959) 40.
35. OSKAM, H. J. *Philips Res. Rep.* 13 (1958) 335; THOMPSON, J. B. *Proc. phys. Soc., Lond.* 73 (1959) 818.
36. BARNES, W. S., MARTIN, D. W., HARMER, D. S. and McDANIEL, E. W. Georgia Institute of Technology Report on Grant AF-AFOSR-62-306, 1963.
37. KANEKO, Y., MEGILL, L. R. and HASTED, J. B. *J. chem. Phys.* 45 (1966) 3741.
38. SAHA, M. N. *Phil. Mag.* 40 (1920) 472; 40 (1920) 809; *Proc. Roy. Soc.* A99 (1921) 135.
39. HEARN, A. G. *Proc. phys. Soc., Lond.* 31 (1963) 648.
40. WILSON, R. *J. quantve. Spectros. radiat. Transf.* 2 (1962) 447.

Chapter 2

THEORETICAL BACKGROUND, CLASSICAL AND QUANTUM

2.1 INTRODUCTION

The detailed discussion of the theory of atomic and electronic collisions is not possible in a text devoted to the experimental aspects of the subject.

Such a division into theoretical and experimental aspects is not beneficial to the unity of the subject, but experience has shown that it is possible to study in this way, provided that theoreticians are prepared to gain a general knowledge of experimental techniques and experimentalists are prepared to gain a general knowledge of theoretical methods. Study of textbooks of scattering theory[1-8, 106-108], and in particular that of Mott and Massey[1], is essential to experimentalists.

This chapter is intended only as a guide for the experimental worker on the best interpretation of measurements in the light of classical and quantum theory. His aim should be to concentrate on the experiments which are most critical and important to the understanding of the collision process. The interpretation of results is also his business, but not the detailed development of mathematical theories of scattering. In many modern experiments, theoretical results are used to normalize the data or to predict what is to be expected; in this way the range of possible experiments can be greatly extended.

Certain collision cross-sections can be approximately calculated if only the simplest atomic or molecular parameters are known; the test of such calculations requires the widest variation of atomic species. Certain cross-sections can be calculated with the aid of wave functions of varying degrees of sophistication; self-consistent field wave functions with configuration interaction are now available for atoms from hydrogen to neon. But for the deepest understanding of a collision process, only the simplest systems —hydrogen, and to some extent helium—are permitted. This has led to considerable experimental study of atomic hydrogen during the past 15 years.

There are limits within which quantum theory approximates closely to classical mechanics; within these limits, the latter may be used to calculate cross-sections. One condition is that the projectile wavelength be much smaller than the target force field dimensions. A particle of mass m travelling with a velocity v corresponds to a wave of wavelength λ given by:

$$\lambda = \frac{h}{mv} \qquad (2.1)$$

where h is Planck's constant.

A more stringent condition is

$$\theta y \gg \lambda \tag{2.2}$$

where y is the closest distance to which the colliding particles approach each other with relative velocity v; except for polar scattering angle $\theta = 0$, y differs from the impact parameter b. This condition, whilst unimportant for inelastic collisions, is dominant in the analysis of the elastic scattering of electrons, and of ions. Except for force fields $V(r) = -Cr^{-n}$, where $n < 2$, it can be shown that θy decreases for decreasing θ. In most cases the particles scattered with the smallest deflexions are not disciplined by classical but by quantum mechanics. It will be found (Chapter 10) that much angular scattering data can be satisfactorily interpreted classically, whilst for the observation of quantum effects a finer angular resolution of experiment is necessary. Excitation processes could never be handled purely by classical mechanics, without introducing the allowed eigenvalues of the system.

The field of force which operates between two particles is one starting point of scattering calculations. These interactions are be considered in the sections immediately following.

2.2 ELECTRON ATOM INTERACTION

At sufficiently large distances r between a free electron and the nucleus of an atom, the interaction energy is due to the polarization of the atom by the free electronic charge e. For an atomic polarizability α, the interaction energy is

$$V(r) = -\frac{\alpha e^2}{2r^4} \tag{2.3}$$

A table of polarizabilities is given in the Appendix. Using the atomic system of units, with $\hbar = e = m_e = 1$, α and r are in units of $a_0 = 0\cdot5292$ Å and V is in atomic units of energy and is equal to $27\cdot2$ eV. This interaction also applies to non-polar molecules, but the interaction between an electron and a polar molecule at long ranges contains an electrostatic term and is given by the appropriate part of equation 2.34.

For sufficiently small distances, there is Coulomb attraction between nucleus and electron; this is weakened by the partial screening of the atomic electrons.

For the interaction between an electron and a hydrogen atom, the equation

$$V_H(r) = -e^2 \left(\frac{1}{r} + \frac{1}{a_0} \right) \exp\left(-\frac{2r}{a_0} \right) \tag{2.4}$$

is a good approximation. Here a_0 is the radius of the first Bohr orbit of the hydrogen atom.

The best universally available 'static' interaction between an atom and an electron at fixed separation r from the nucleus is that calculated using the self-consistent field technique of Hartree[9, 109] with allowance made by

Fock[10] for exchange. Tabulations have been published[11] for all elements of nuclear charge Z in the form

$$U(r) = -\frac{xV(r)}{2Z} \tag{2.5}$$

$$x = \frac{r}{\mu} \tag{2.6}$$

and

$$\mu = \frac{1}{2}\left(\frac{3\pi}{4}\right)^{2/3} Z^{-1/2} \tag{2.7}$$

This work has superseded the statistical Thomas–Fermi approximation[12, 13], which is, however, adequate for heavy atoms for which the Hartree–Fock–Slater technique is tedious.

Atoms may be classed as 'electronegative' or not, according as to whether or not they form a stable negative ion. More than two-thirds of atoms in the periodic table are electronegative, and perhaps a similar proportion of molecules. It is not easy to predict the binding energy of the atom and electron without resort to good wave functions and a good quantum-mechanical approximation. However, the binding energy (electron affinity E_a, Chapter 8) may usually be determined experimentally. When it is known, the best interaction may be expressed[14] in a form which includes the Hartree–Fock–Slater interaction, the electron affinity and the polarization inter-action:

$$U(r) = U_{HF}(r) - \frac{2}{r}\left[1 - \exp\left(-\frac{r}{r_0}\right)\right] - \frac{\alpha[1 - \exp(-r/r_p)]}{(r^2 + r_p^2)^2} \tag{2.8}$$

The quantity r_p is taken as $1 \cdot 5 a_0$ for the first row of the periodic table, $2 a_0$ for the second, and $2 \cdot 5 a_0$ for the third. The quantity r_0 is chosen so that the correct electron affinity is obtained, and a computer programme[11] exists for this purpose.

It is now known that the 'static' atomic field calculated by the Hartree–Fock–Slater technique is inadequate to predict the correct phase shifts for the differential scattering of electrons, even by helium. Not only must allowance be made for electron exchange, but also for polarization[15].

2.3 ATOMIC AND MOLECULAR INTERACTIONS

The importance of atomic interactions is not only in their dominance of elastic scattering and macroscopic gas properties, but in the understanding of inelastic heavy-particle collisions where a semi-classical approximation is possible.

Interactions between atomic systems will be considered under the following headings:

1. Interactions theoretically predicted,
2. Semi-empirical potentials,
3. Molecular potential energy curves and transitions,
4. Potential energy surfaces.

Theoretically Predicted Interactions

The interaction between two atoms is a 'central force field', that is, a function only of the distance r between the two nuclei. Molecular interactions, on the other hand, are orientation dependent. The atomic interaction contains two regions: the short-range repulsive field ($\lesssim 2a_0$), and the long-range attractive field.

Short-range interactions

The best-established short-range interaction is written in the exponential form usually known as a Born–Mayer potential[16]:

$$V(r) = \frac{Z_1 Z_2 e^2}{r} \exp\left(-\frac{r}{\beta}\right) \qquad (2.9)$$

The Coulomb interaction between the two nuclei, of charges $Z_1 e$ and $Z_2 e$, is screened by the surrounding electron clouds, the factor β being given by Bohr as:

$$\beta = \frac{a_0}{(Z_1^{2/3} + Z_2^{2/3})^{1/2}} \qquad (2.10)$$

A similar expression was given by Zener[110]:

$$V(r) = b \exp\left(-\frac{ar}{a_0}\right) \qquad (2.11)$$

where

$$a = \frac{2a_0}{e^2}\,(E_{i1}^{1/2} + E_{i2}^{1/2}) \qquad (2.12)$$

and E_i are ionization potentials.

It will be seen in Chapter 10 that the analysis of experimental data supports an exponential repulsive potential only for He–He; in other cases, a high-power law is a better approximation, or the Firsov expression given below. The constants of the helium potential do not correspond to the predictions of theory[18].

An empirical repulsion potential of the Born–Mayer type

$$V(r) = A \exp\left(-\frac{r}{b}\right) \qquad (2.13)$$

has been given[18] as

$$A = 2 \cdot 58 \times 10^{-5}\,Z^{11/2}\,\text{eV} \qquad (2.14)$$

and

$$b = \frac{1 \cdot 5 a_0}{Z^{1/3}} \qquad (2.15)$$

for collisions between like systems of nuclear charge Z.

The Firsov expression[19] is

$$V(r) = \frac{Z_1 Z_2 e^2}{r} \phi(x) \tag{2.16}$$

where the screening function $\phi(x)$ has been calculated[12, 13] in connection with the Fermi–Thomas approximation, and

$$x = \frac{(Z_1^{1/2} + Z_2^{1/2})^{2/3}}{0 \cdot 89 a_0} r \tag{2.17}$$

Although satisfactory for the interpretation of high-energy scattering, it cannot be said that this expression is well supported by the low-energy ion–atom data, to which the screening constant β is particularly sensitive[20].

Long-range attractive interactions

The longer range attractive interaction between two atoms is dominated either by weak 'dispersion' or van der Waals' interaction or by strong chemical interaction.

Weak dispersion or dynamic polarization (van der Waals' interaction) only dominates if the electron spins of the two atoms are parallel, or if one or both of the atoms are rare gases, or atoms or molecules for which chemical bonding is not possible. The maximum attractive energy of this interaction is between 0·001 eV and 0·05 eV; it extends to distances greater than 10 Å.

Strong chemical attraction force occurs when the spins of the two atoms are antiparallel, so that they can be chemically bound into a diatomic molecule. The interaction also occurs between an atom A and a molecule BC when either ABC or $AB + C$ can be formed. The maximum attractive energy can be several electron volts.

When chemical force exists between two atoms, the repulsive force field is contracted in range. This is apparent from the graphical representations of the H–H and Ar–Ar potentials in *Figure 2.1*. At sufficiently large separations, all interactions are attractive, even that of the $^3\Sigma_u$ state of H$_2$, there being a shallow minimum around 4·5 Å, which is responsible for the existence of the state known as para-hydrogen under suitable conditions at low temperatures.

The van der Waals' interaction is dominated by the inverse sixth power interaction

$$V(r) = -\frac{C}{r^6} \tag{2.18}$$

This interaction arises from dispersion or dynamic polarization; on a classical model, the electrons e_1 and e_2 in the two systems move in their orbits in such phases that their mean separation r_{12}, averaged over many periods, is as large as possible. This phase difference gives rise to the dispersion force. The interaction can be calculated by expanding the charge distributions of the two atoms in terms of electric multipole moments[9].

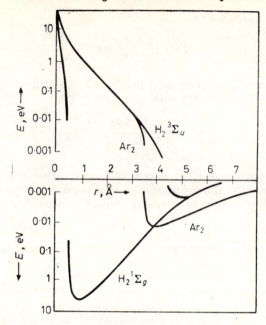

Fig. 2.1. Contrasting interactions between atoms (H–H and Ar–Ar) in the presence and absence of chemical interaction

One obtains a series in even inverse powers of r, of which the first non-vanishing term for two neutral atoms is in r^{-6}. Overlapping charge distributions, such as exist at smaller nuclear separations, require a more complicated expansion[10]. The dispersion constant in equation 2.18 is calculated by Slater and Kirkwood[21] in the form

$$C_{\text{disp}} = \frac{3e\hbar}{2m_e^{1/2}} \frac{\alpha_1\alpha_2}{(\alpha_1/N_1)^{1/2} + (\alpha_2/N_2)^{1/2}} \tag{2.19}$$

where α_1 and α_2 are electric dipole polarizabilities, and N_1 and N_2 are numbers of outer shell electrons. It has been argued[22] that N_1 and N_2 should actually symbolize total numbers of atomic electrons. A good approximate formula[23, 24] is:

$$C_{\text{disp}} = \frac{3}{2}\left(\frac{E_1E_2}{E_1+E_2}\right)\alpha_1\alpha_2 \tag{2.20}$$

The energies E_1 and E_2 are found to be close to the ionization energies E_{i1} and E_{i2} of the atoms. Thus dispersion constants C can be calculated approximately, and there is a theoretical basis for the interpolation formula

$$C_{ab} = (C_{aa}C_{bb})^{1/2} \tag{2.21}$$

Using a classical model[25], the higher order terms are found to be

$$V_{\text{disp}}(r) = -\frac{C}{r^6} - \frac{C'}{r^8} - \frac{C''}{r^{10}} \tag{2.22}$$

where

$$C' = \frac{45}{8} \frac{E_1 E_2 \alpha_1 \alpha_2}{e^2} \left(\frac{E_1 \alpha_1}{2E_1 + E_2} + \frac{E_2 \alpha_2}{E_1 + 2E_2} \right) \tag{2.23}$$

and

$$C'' = \frac{315}{16} \frac{E_1^2 E_2^2 \alpha_1^2 \alpha_2^2}{e^4 (E_1 + E_2)} \tag{2.24}$$

More recent treatments are also available[26]. Anisotropic dispersion contributions to interactions between molecules will be discussed below.

Polarizabilities are normally derived from Stark effect, dielectric constant and molecular beam measurements; a table is given in the Appendix. Polarizabilities may also be calculated from a series of n energy levels E_{mn} and oscillator strengths f_{mn}, applying the sum rule (Chapter 9)

$$\sum_n \frac{f_{mn}}{E_{mn}^2} = \frac{\alpha_m}{4} \tag{2.25}$$

where α_m is the polarizability of the system in level m. Levels in the continuum must be taken into account. A list of calculated van der Waals' constants is available[27].

The existence of van der Waals' attraction gives rise to dimers of Ar, N_2, H_2 and other chemically stable species. These occur only at low temperatures, being produced in the isentropic expansion of gas through nozzles and detected by mass-spectrometric analysis[28]. In alkali metal vapours, and possibly in other metal vapours, molecular clusters are common and, because of their small ionization potentials, might dominate the electrical conductivity of the hot gas[29].

Chemical interactions arise between pairs of atoms possessing electrons with unpaired spins. Between atoms containing a pair of electrons with opposed spins, there is a bonding energy which may be as large as several electron volts. A pair of electrons with parallel spins tends to oppose bonding. Thus the hydrogen molecule[30] can exist in a symmetrical (gerade) $^1\Sigma_g$ state, and also an antibonding, unsymmetrical (ungerade) $^3\Sigma_u$ interaction (see Appendix, and also *Figure 2.1*). Many chemical interactions have not been calculated and are approximated by the semi-empirical potentials described below. However, molecular spectra provide a considerable amount of information, and there are also theoretical techniques of quantum chemistry, in particular the molecular orbital method of Mulliken[31]. Using self-consistent field linear combination of atomic orbital methods, preferably with configuration interaction, the ground state and excited state energies of few-electron molecules (first row up to F_2) can be calculated with reasonable accuracy. The improved wave functions contain both covalent and ionic bonding. For atoms containing many electrons, the situation is much less satisfactory.

There is an important resonance or exchange interaction[14, 24] between two identical atoms of which one is in an excited state n and the other in the ground state 0. This interaction may be either repulsive or attractive, and arises from the tendency of the two atoms to exchange radiative energy:

$$V(r) = \gamma \left(\frac{e^2 h f_{0n}}{8\pi^2 m_e \nu_{0n}} \right) r^{-3} \tag{2.26}$$

with f_{0n} and v_{0n} respectively the oscillator strength and the frequency of the 0–n transition. Where m (the magnetic quantum number of state n) is zero, $\gamma = -2$; but where $m = \pm 1$, $\gamma = +1$. It is not entirely significant to define a static interaction potential $V(r)$ between such identical systems, since the wave functions describing the pseudo-molecule must allow for the interchange of the identical nuclei. There are two simultaneous interactions $V^+(r)$ and $V^-(r)$, corresponding to symmetric and anti-symmetric wave functions. The difference in energy between these two interactions is

$$V^+ - V^- = \frac{C_3}{r^3} \qquad (2.27)$$

where $C_3 \simeq \mu^2$ for an optically allowed transition of temporary dipole moment μ. For a disallowed transition of $s-1$ pole moment,

$$V^+ - V^- \simeq \frac{C_s}{r^s} \qquad (2.28)$$

and for an s–s transition,

$$V^+ - V^- = C_0 \exp(-\lambda r) \qquad (2.29)$$

Adjustable parameters have been taken for this interaction. For example, in the interaction between an ion and an atom approaching with velocity v, the expression

$$V^+ - V^- = \frac{E_i r}{\hbar v a_0} \exp\left[-\left(\frac{E_i}{R}\right)^{1/2} \frac{r}{a_0} \right] \qquad (2.30)$$

has been used (Chapter 12).

A weak interaction arises due to the adiabatic coupling between nuclear and electronic motion[32]. A repulsive r^{-1} term dominates at large separations, but an attractive r^{-2} term takes over at smaller separations. The coefficients are only non-zero when one atom has a non-zero orbital or spin angular momentum, and even in this case they are small. A weak r^{-3} interaction arises from coupling between the two atomic spins[11].

Van der Waals' and other interactions discussed above are termed 'long-range forces'. There are also 'very long range forces' which arise from the retardation caused by the finite time taken by a photon to travel between the two atoms[33]. The existence of these interactions, which can give rise to shallow potential minima, was first suspected by Verwey, Overbeek and van Ness[34] during investigations of the stability of lyophobic colloids, especially thixotropic substances. Relativistic quantum electrodynamics was applied by Casimir and Polder to the problem[17]. The range of the retardation effect is determined by the wavelength $\lambda = hc/E$ of the photon concerned in the interaction (resonance photons). This wavelength is of the order 1000 Å, but the retardation effect becomes noticeable as close as 200 Å. The most general expression of the dispersion energy[35, 36] is in terms of the electric polarizability at imaginary frequency $\omega = i\xi$. This polarizability refers to the dipole moment induced by an electric field with exponential time dependence $\exp(\xi t)$; such an expression allows for retardation effects, and reduces to the dispersion equation at small ranges:

$$V = -\frac{3\hbar}{\pi r^6} \int_0^\infty \alpha_1(i\xi)\, \alpha_2(i\xi)\, d\xi \qquad (2.31)$$

for $r \ll \lambda/2\pi$, and

$$V = -\frac{23\hbar c}{4\pi} \frac{\alpha_1 \alpha_2}{r^7} \qquad (2.32)$$

for $r \gg \lambda/2\pi$.

Electrostatic interactions

The interactions so far discussed are central force fields and therefore are not, in general, good approximations to intermolecular forces. Intermolecular forces are anisotropic and contain terms arising from the permanent electric moments and not found in atomic interactions:

Monopole (ionic charge) e,
Dipole μ,
Quadrupole[111] Q.

The induced dipoles are not isotropic and may be defined according to whether the applied field is parallel to a specific bond ($\alpha_{\parallel}$) or perpendicular to it ($\alpha_{\perp}$). These polarizabilities are bond-additive, so if the three-dimensional molecular configuration is known, the polarizability can be calculated for a given direction of application of field. For application at an angle θ to a bond

$$\alpha_\theta = \alpha_{\parallel} \cos^2 \theta + \alpha_{\perp} \sin^2 \theta \qquad (2.33)$$

The electrostatic and electrically induced forces depend upon the polar angles θ and the azimuthal angles ϕ between the molecular dipole and the line joining the molecule centre of mass to that of the second molecule (or atom). The derivation of the expressions is entirely classical. The electrostatic interaction is

$$V(r, \theta, \phi) = +\frac{e_1 e_2}{r} - \frac{e_1 \mu_2}{r^2} \cos \theta_2 + \frac{e_1 Q_2}{4r^3} (3 \cos^2 \theta_2 - 1)$$

$$-\frac{\mu_1 \mu_2}{r^3} [2 \cos \theta_1 \cos \theta_2 - \sin \theta_1 \sin \theta_2 \cos (\phi_1 - \phi_2)]$$

$$+\frac{3\mu_1 Q_2}{4r^4} [\cos \theta_1 (3 \cos^2 \theta_2 - 1) - 2 \sin \theta_1 \sin \theta_2 \cos \theta_2 \cos (\phi_1 - \phi_2)]$$

$$+\frac{3Q_1 Q_2}{16r^5} \{1 - 5 \cos^2 \theta_1 - 5 \cos^2 \theta_2 - 15 \cos^2 \theta_1 \cos^2 \theta_2$$

$$+ 2[\sin \theta_1 \theta_2 \cos (\phi_1 - \phi_2) - 4 \cos \theta_1 \cos \theta_2]^2\} \qquad (2.34)$$

When two molecules approach sufficiently slowly, their relative orientation assumes the position where the potential energy is minimum. However, at large nuclear separations, where the maximum potential energy variation with angle is much less than thermal energy, it is possible to calculate a spherically symmetrical mean interaction

$$\bar{V}(r) = \frac{\iint V(r, \theta, \phi) \exp\left[-V(r, \theta, \phi)/kT\right] d\omega_1 \, d\omega_2}{\iint \exp\left[-V(r, \theta, \phi)/kT\right] d\omega_1 \, d\omega_2} \tag{2.35}$$

where $d\omega = \sin\theta \, d\theta \, d\phi$. The separation r is held fixed while an average is made over $\theta_{1,2}$, $\phi_{1,2}$. The Boltzmann weighting factor appears because molecules spend more time in those orientations for which the potential energy is small. The electrostatic interaction becomes

$$\bar{V}(r) = +\frac{e_1 e_2}{r} - \frac{1}{3kT}\frac{e_1^2 \mu_2^2}{r^4} - \frac{1}{20kT}\frac{e_1^2 Q_2^2}{r^6} - \frac{2}{3kT}\frac{\mu_1^2 \mu_2^2}{r^6}$$

$$- \frac{1}{kT}\frac{\mu_1^2 Q_2^2}{r^8} - \frac{7}{40kT}\frac{Q_1^2 Q_2^2}{r^{10}} \tag{2.36}$$

The induction interaction is

$$V(r, \theta) = -\frac{e_1^2 \alpha_2}{2r^4} - \frac{\mu_1^2 \alpha_2 (3\cos^2\theta + 1)}{2r^6} \tag{2.37}$$

and the second term yields a mean interaction

$$\bar{V}(r) = -\frac{\mu_1^2 \alpha_2}{r^6} \tag{2.38}$$

Contributions from induced quadrupole moments have not been considered in the above. The terms of longest range arise from charged systems (ion–ion and ion–molecule interactions). It is to be noted that, in ion–molecule interactions, the dispersion contribution is not negligible in comparison with the inverse fourth power polarization contribution, despite the small polarizabilities of positive ions[37, 38].

In calculating interactions, it is necessary to add the interactions of molecule 1 with molecule 2 to those of molecule 2 with molecule 1. Thus the complete interaction arising from equation 2.38 above would be

$$\bar{V}(r) = -\frac{\mu_1^2 \alpha_2}{r^6} - \frac{\mu_2^2 \alpha_1}{r^6} \tag{2.39}$$

The dispersion interaction is also anisotropic for molecules which possess anisotropy of polarizability. At medium ranges, the theory yields rather complicated expressions[39], but at long ranges

$$V(r, \theta, \phi) = -\frac{C(\theta_1 \phi_1 \theta_2 \phi_2)}{r^6} \tag{2.40}$$

and

$$C = (\mathcal{A}-\mathcal{B}-\mathcal{B}'+\mathcal{C})\,[\sin\theta_1\sin\theta_2\cos(\phi_1-\phi_2)-2\cos\theta_1\cos\theta_2]^2$$
$$+3(\mathcal{B}-\mathcal{C})\cos^2\theta_1+3(\mathcal{B}'-\mathcal{C})\cos^2\theta_2+(\mathcal{B}+\mathcal{B}'+4\mathcal{C}) \tag{2.41}$$

where:

$$\mathcal{A} = \frac{h}{4}\,(\alpha_{\|1}+\alpha_{\|2})\,\frac{E_{\|1}E_{\|2}}{E_{\|1}+E_{\|2}} \tag{2.42}$$

$$\mathcal{B} = \frac{h}{4}\,(\alpha_{\|1}+\alpha_{\perp2})\,\frac{E_{\|1}E_{\perp2}}{E_{\|1}+E_{\perp2}} \tag{2.43}$$

$$\mathcal{B}' = \frac{h}{4}\,(\alpha_{\perp1}+\alpha_{\|2})\,\frac{E_{\perp1}E_{\|2}}{E_{\perp1}+E_{\|2}} \tag{2.44}$$

$$\mathcal{C} = \frac{h}{4}\,(\alpha_{\perp1}+\alpha_{\perp2})\,\frac{E_{\perp1}E_{\perp2}}{E_{\perp1}+E_{\perp2}} \tag{2.45}$$

The spherically symmetric mean interaction is

$$\bar{V} = -\frac{2}{3r^6}\,[\mathcal{A}+2(\mathcal{B}+\mathcal{B}')+4\mathcal{C}] \tag{2.46}$$

It is usually possible to assume

$$E_{\|} = E_{\perp} = E_i \tag{2.47}$$

A representation of anisotropic interactions between an atom and a diatomic molecule may be made[40, 41] in terms of Legendre polynomials:

$$V(r,\theta) = V_r(r)\sum_n b_nP_n(\cos\theta)+V_a(r)\sum_n a_nP_n(\cos\theta) \tag{2.48}$$

where V_r is a repulsive and V_a an attractive potential. The terms a_n and b_n are asymmetry coefficients whose sign and magnitude may be determined by scattering experiments. The spherically symmetric average, taking advantage of the orthogonality properties of Legendre polynomials, is

$$\bar{V}(r) = V_r+V_a \tag{2.49}$$

For a homonuclear molecule $b_1 = 0$, and b_1 is also negligible for heteronuclear diatomic molecules in the region of large r, provided the r^{-6} interaction dominates. The anisotropy of polarizability[42] is given by the equation

$$a_2 = \frac{\alpha_{\|}-\alpha_{\perp}}{\alpha_{\|}+2\alpha_{\perp}} \tag{2.50}$$

Semi-empirical Potentials

The quality of a semi-empirical potential may be judged by its mathematical simplicity, the minimization of the number of adjustable parameters used, its accuracy in representing the real interaction, and its theoretical grounding.

All of the potentials to be listed have been used in the calculation of virial coefficients and other gas properties; the more sophisticated are of value in the interpretation of scattering.

The impenetrable elastic sphere interaction contains only a single parameter, the collision radius ς:

$$V(r) = \infty \tag{2.51}$$

for $r < \varsigma$, and

$$V(r) = 0 \tag{2.52}$$

for $r \geqslant \varsigma$. The square-well interaction introduces a second parameter, the well depth ε, and also a second radius $a\varsigma$:

$$V(r) = \infty \tag{2.53}$$

for $0 \leqslant r < \varsigma$;

$$V(r) = -\varepsilon \tag{2.54}$$

for $\varsigma \leqslant r \leqslant a\varsigma$; and

$$V(r) = 0 \tag{2.55}$$

for $a\varsigma < r$. Other simple potentials are: point centres of repulsion, $V(r) = dr^{-\delta}$; and the Sutherland attracted sphere potential $V(r) = \infty$ for $r \leqslant \varsigma$, and $V(r) = Cr^{-\gamma}$ for $r > \varsigma$.

A more realistic potential must combine the long-range inverse sixth power dispersion attraction with the short-range exponential repulsion. The combination used by Buckingham and Corner[43] was

$$V(r) = b \exp\left(-ar\right) - \frac{c}{r^6} - \frac{c'}{r^8} \tag{2.56}$$

which tends to $-\infty$ as $r \to 0$, although correctly representing a strongly repulsive region, provided b is large.

There are four adjustable parameters in this potential, but there is no mention of the well depth ε or the equilibrium separation r_m at which this energy minimum occurs. Since these parameters are often known, the potential is conveniently written as

$$V(r) = b \exp\left(-\alpha' \frac{r}{r_m}\right) - \left(\frac{c}{r^6} + \frac{c'}{r^8}\right) \exp\left[-4\left(\frac{r_m}{r} - 1\right)^3\right] \tag{2.57}$$

for $r \leqslant r_m$, and

$$V(r) = b \exp\left(-\alpha' \frac{r}{r_m}\right) - \left(\frac{c}{r^6} + \frac{c'}{r^8}\right) \tag{2.58}$$

for $r > r_m$. The terms ε and r_m are related to the parameters b, c and c' as follows:

$$b = [-\varepsilon + (1+\beta)cr_m^{-6}] \exp \alpha' \tag{2.59}$$

$$c = \frac{\varepsilon \alpha' r_m^6}{\alpha(1+\beta) - 6 - 8\beta} \tag{2.60}$$

$$c' = \beta r_m^2 c \tag{2.61}$$

A more satisfactory interaction is the 'modified Buckingham 6-exp' potential:

$$V(r) = \frac{\varepsilon}{1-6/\alpha'} \left\{ \frac{6}{\alpha'} \exp \left[\alpha' \left(1 - \frac{r}{r_m} \right) \right] - \left(\frac{r_m}{r} \right)^6 \right\} \tag{2.62}$$

for $r \geqslant r_{max}$, and

$$V(r) = \infty \tag{2.63}$$

for $r < r_{max}$. The spurious maximum occurring at r_{max} is removed, without the introduction of any further adjustable parameters, since

$$\left(\frac{r_{max}}{r_m} \right)^7 \exp \left[\alpha' \left(1 - \frac{r_{max}}{r_m} \right) \right] = 1 \tag{2.64}$$

There are three parameters: ε, ς and the repulsion parameter α'.

A simple form of combined potential for ion–atom interaction has been proposed[20] as

$$V = -\frac{\alpha e^2}{\beta^4 x^4} \left[1 - \left(1 + x + \frac{1}{2} x^2 + \frac{1}{6} x^3 \right) \exp(-x) \right] \tag{2.65}$$

with $x = r/\beta$, the screening constant β being defined in equation 2.10. This avoids a singularity at the origin, and is successful in interpreting differential scattering.

The repulsive and attractive parts of the interaction between two atoms is often represented as a combination of power laws[44]

$$V(r) = \frac{a}{r^m} - \frac{b}{r^n} \tag{2.66}$$

The most usual representation, which is connected historically with the Kammerlingh–Onnes equation of state, is the Lennard–Jones 6–12 potential:

$$V(r) = 4\varepsilon \left[\left(\frac{\varsigma}{r} \right)^{12} - \left(\frac{\varsigma}{r} \right)^6 \right] \tag{2.67}$$

The choice of the power 12 presents certain mathematical advantages and is not severely unrealistic. The potential minimum appears at a separation $r_m = 2^{1/6} \varsigma$. A best fit of the modified Buckingham potential to the repulsive part of the Lennard–Jones 6–12 potential gives $\alpha' = 13\cdot772$.

A modification was introduced by Kihara to displace $V(r) = \infty$ from $r = 0$ to $r = \alpha'' r_m$:

$$V(r) = \varepsilon \left[\left(\frac{1-\alpha''}{r/r_m - \alpha''} \right)^{12} - 2 \left(\frac{1-\alpha''}{r/r_m - \alpha''} \right)^6 \right] \tag{2.68}$$

for $r/r_m \geqslant \alpha''$, and

$$V(r) = \infty \tag{2.69}$$

for $0 \leqslant r/r_m < \alpha''$.

A table of characteristic ε and r_m values is given in Table 2.1.

Table 2.1. CONSTANTS FOR THE LENNARD–JONES POTENTIAL BETWEEN GAS
MOLECULES

Gas molecule	ε (eV)	ς (Å)	$r_m = 1\cdot12\varsigma$ (Å)
He–He	0·00088	2·556	2·863
H_2–H_2	0·00318	2·928	3·279
Ne–Ne	0·00307	2·789	3·124
N_2–N_2	0·00789	3·681	4·123
CO–CO	0·00950	3·590	4·021
Ar–Ar	0·0107	3·418	3·828
O_2–O_2	0·00976	3·433	3·845
CH_4–CH_4	0·0118	3·822	4·281
Kr–Kr	0·0163	3·61	4·043
Xe–Xe	0·0197	4·055	4·542
CO_2–CO_2	0·0163	3·996	4·476

Semi-empirical anisotropic potentials have also been proposed. Apart
from impenetrable ellipsoids and spherocylinders, and impenetrable spheres
containing a point dipole, there is the Stockmayer potential, which consists
of the Lennard–Jones 6–12 plus a dipole–dipole interaction

$$\frac{\mu_1\mu_2}{r^3}[2\cos\theta_1\cos\theta_2-\sin\theta_1\sin\theta_2\cos(\phi_2-\phi_1)] \qquad (2.70)$$

Molecular Potential Energy Curves and Transitions

An important expression that represents the typical potential function of the
diatomic molecule was proposed by Morse[45] in the form

$$V(r-r_m) = D\{1-\exp[-\beta(r-r_m)]\}^2 \qquad (2.71)$$

with

$$\beta = \left(\frac{2\pi^2c\mu}{Dh}\right)^{1/2}\omega_e = 1\cdot2177\times10^7\omega_e\left(\frac{\mu}{D}\right)^{1/2} \qquad (2.72)$$

where D is the dissociation energy and ω_e is the frequency of vibration; the
vibrational term values[46] are given by

$$G(v) = \beta\left(\frac{Dh}{2\pi^2c\mu}\right)^{1/2}\left(v+\frac{1}{2}\right)-\frac{h\beta^2(v+\frac{1}{2})^2}{8\pi^2c\mu} \qquad (2.73)$$

for vibrational quantum number v and reduced mass μ.

The Morse function is consistent with a solution of the wave equation,
but the term values contain no terms in higher powers of $v+\frac{1}{2}$, as they
should do. Klein[47] and Rydberg[48] have given a method for constructing
the potential curve, point by point, from the observed vibrational and rotational levels without assuming an analytical expression for it; the results
are surprisingly close to the Morse function.

Hulburt and Hirschfelder proposed a more complicated analytical expression[49, 112], taking into account the rotational constant B_e and the rotation–vibration interaction parameter α_e, which like D, ω_e and the anharmonicity constant $\omega_e x_e$, are available from spectroscopic data. The expression is

$$V(r-r_m) = D(\{1-\exp\,[-\beta(r-r_m)]\}^2 + K_1\beta^3 \exp\,[-2\beta(r-r_m)]$$
$$\times [1+K_2\beta(r-r_m)]\,(r-r_m)^3) \qquad (2.74)$$

with

$$K_1 = 1 - \frac{1}{\beta r_m}\left(1+\frac{\alpha_e\omega_e}{6B_e^2}\right) \qquad (2.75)$$

and

$$K_2 = 2+\frac{1}{K_1}\left[\frac{7}{12}-\frac{1}{\beta^2 r_m^2}\left(\frac{5}{4}+\frac{5\alpha_e\omega_e}{12B_e^2}+\frac{5\alpha_e^2\omega_e^2}{144B_e^4}-\frac{2\omega_e x_e}{3B_e}\right)\right] \qquad (2.76)$$

The authors tabulate the constants K_1 and K_2 for a considerable number of diatomic molecules.

Some of the most accurate fittings of spectroscopic data have been made with the Lippincott empirical potential[49]:

$$V(r)-V(r_m) = D[1-\exp\,(-x)]\left\{1-abx^{1/2}\exp\left[-b\left(\frac{xr_m}{r}\right)^{1/2}\right]\right\} \qquad (2.77)$$

where

$$x = \frac{hc\omega_e^2(r-r_m)^2}{4B_eDr_mr} = \frac{n(r-r_m)^2}{2r} \qquad (2.78)$$

with a, b and n empirical constants.

Although spectroscopic observations are the principal source of information about molecular potential energy curves, information can be obtained from appearance potentials of ions (Chapter 6), from electron spectroscopy (Chapter 5), from thermochemistry, and from molecular orbital calculations. Of particular importance are the Wigner–Witmer rules (See Appendix), which govern the configurations of the constituent atomic states which make up particular configurations of the molecule and the molecular ion. A molecular positive ion XY^+ can derive from X and Y^+ or from X^+ and Y. The lowest (ground) state of XY^+ will normally derive from the atomic ion formed most easily from its atom (lowest ionization potential). A number of well-established diatomic potential energy diagrams are given in the Appendix.

The equilibrium separation r_m of a diatomic molecule may be calculated empirically, when the vibrational constant k_e is known, using Badger's rule:

$$r_m = \left(\frac{c_{ij}}{k_e}\right)^{1/3} + d_{ij} \qquad (2.79)$$

where

$$k_e = 4\pi^2\mu^2c^2\omega_e^2 \qquad (2.80)$$

thus for the oxygen molecule c_{ij} and d_{ij} are empirical constants characteristic of the molecule; $c_{11} = 1·86 \times 10^5$ dyne cm^{-1} and $d_{11} = 0·680$ Å.

An important part of the understanding of inelastic processes, both atomic and molecular, lies in understanding the transitions which can be made between potential energy curves. A transition can occur between two separated curves when energy is supplied by radiation or by electron impact. A transition can also occur with nuclear separation change, when two curves cross adiabatically in an atom–atom collision; the system can undergo transition when the pseudo-crossing is reached.

The conditions that two energy levels should be degenerate at crossing cannot often be met. If they cannot, then it is not possible for the two curves to cross, and they will appear to have configurations as in *Figure 12.28* (inset). Transitions between them can occur, and for an *s–s* transition (active electron starting and finishing in an *s* orbital), theoretical calculation of the probabilities of transition is possible, and will be discussed in Chapter 12. But when the two levels exist at identical energy and nuclear separation, then a 'diabatic' crossing is said to occur.

A transition brought about by radiation or a colliding electron will normally take place in a time very short compared with the vibrational or other relative motion of the nuclei. This is in effect a statement of the Franck–Condon principle[50]. A transition corresponds to a vertical line drawn between two potential energy curves. In a transition to a higher electronic state, or to the ground state of the singly ionized molecule (*Figure 2.2b*) a line drawn vertically (unchanging *r*) will cross the upper potential energy trough, but not necessarily at a vibrational level; the nearest vibrational states will be formed, with the weakest 'Franck–Condon violation' possible.

In the quantum theory description, the probability P_{ab} of a transition between states a and b is given by

$$P_{ab} \propto \left| \int \Psi_a \mathbf{r} \Psi_b \, d\tau \right|^2 \tag{2.81}$$

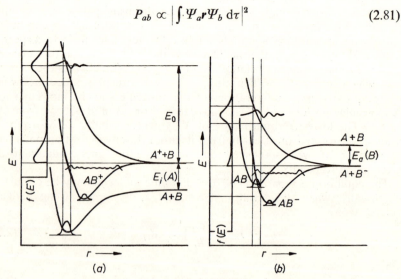

Fig. 2.2. *Vertical transitions in diatomic molecules*

The total wave function Ψ_a can be written in terms of the nuclear separation vector r and the electron-nucleus vector r_e. Thus

$$\Psi_a = \psi_a(r)\psi_e^*(a, r_e) \qquad (2.82)$$

so

$$P_{ab} \propto \left| \int \psi_a(r)\,\psi_e^*(a, r_e)\,r\psi_b(r)\,\psi_e^*(b, r_e)\,\mathrm{d}r\,\mathrm{d}r_e \right|^2 \qquad (2.83)$$

It is often assumed as an approximation that the oscillator strength f_{ab} for the transition $a \rightarrow b$ is independent of r. Since

$$f_{ab}(r) = \int \psi_e(a, r_e)\,r\psi_e(b, r_e)\,\mathrm{d}r_e \qquad (2.84)$$

one has[51]

$$P_{ab} \propto |f_{ab}|^2 \left| \int \psi_a(r)\,\psi_b(r)\,\mathrm{d}r \right|^2 \qquad (2.85)$$

The wave function of the ground state, which may be schematically represented as in *Figure 2.2a*, is of greatest intensity at $r = r_m$. The most probable transitions are likely to reflect this fact, but the upper state wave functions have different forms (*Figure 2.2a*): a decaying oscillatory function for anti-bonding states, and periodic functions for the bonding states. For a transition to an antibonding state, a reasonable approximation to the solution of equation 2.85 would be to reflect ψ_a^2 in the antibonding potential energy curve; a distribution of energies will result, as in *Figures 2.2a* and *2.2b*.

For an upward transition to a bonding state, the simplest situation arises when the two states have identical values of r_m (r_m and r'_m), and the 'vertical transition' is from $v = 0$ to $v' = 0$. When the separations are different, the vertical transitions give rise to a distribution of higher vibrational quantum states and the situation is complicated by the existence of negative parts of the wave functions. As the difference between r_m and r'_m increases, the Frank–Condon integral becomes broader, including more vibrational states; at the same time it becomes asymmetrical. Calculations of the functions for several diatomic molecule transitions are available. There also exist[52] generalized recursion formulae applicable to any pair of states whose spectroscopic data are known.

A procedure which is often adopted, without much better justification than that of simplicity, is to replace the upper vibrational wave functions by delta functions placed at the extrema of vibration, on the potential energy curve. The probabilities of transitions to different vibrational levels are taken as proportional to the ψ_a^2 values at the nuclear separations corresponding to vertical transitions. The most probable upper vibrational level is that nearest to a vertical transition from the lower potential minimum.

A transition to an antibonding state of a molecular positive ion AB^+ produces the atomic ion and atom with a kinetic energy distribution as illustrated in *Figure 2.2a*. The energy is divided in the ratio of the masses

$$E_{A^+} = \frac{m_B E_0}{m_{A^+} + m_B} \qquad (2.86)$$

A transition to a bonding state of a molecular positive ion produces a distribution of vibrationally excited states, corresponding to wave functions of the type shown in *Figure 2.2a*; but the atomic ion and atom may sometimes be produced with near-zero energy distribution, as is also shown in the figure. The equivalent situations for transitions to negative molecular ion states are shown in *Figure 2.2b*.

Potential Energy Surfaces

The energy conditions between three atoms may be expressed in terms of a 'potential energy surface'[53]. This is a significant model provided that the free frictional motion of a particle on the surface, under the influence of gravity, can be made to represent the analogous motion of the three-particle reacting system. The energy of such a system depends upon the relative orientation of the internuclear axes, but for the case when it is a minimum[53] with three particles (m_1, m_2 and m_3) lying on a straight line, it is required that the internal kinetic energy $\mathcal{E}$ shall be expressible in the form

$$\mathcal{E} = \frac{1}{2}\, m\left(\frac{dx}{dt}\right)^2 + \frac{1}{2}\, m\left(\frac{dy}{dt}\right)^2 \tag{2.87}$$

where x and y represent suitably chosen Cartesian coordinates, the vertical coordinate z representing the potential energy, as in the case of the diatomic molecule. Now the internal kinetic energy $\mathcal{E}$ of the system in centre-of-mass coordinates is

$$\mathcal{E} = \frac{1}{2}\, \frac{m_1 m_2}{m_1 + m_2}\left(\frac{dr_{12}}{dt}\right)^2 + \frac{1}{2}\, \frac{(m_1 + m_2)m_3}{m_1 + m_2 + m_3}\left(\frac{dr_{23}}{dt} + \frac{m_1}{m_1 + m_2}\, \frac{dr_{12}}{dt}\right)^2 \tag{2.88}$$

This expression can be reduced to the form of equation 2.87 by writing

$$m = \frac{m_1(m_2 + m_3)}{m_1 + m_2 + m_3} \tag{2.89}$$

and

$$r_{12} = x - y \tan \theta \tag{2.90}$$

$$r_{23} = \left[\frac{m_1(m_2 + m_3)}{m_3(m_1 + m_2)}\right]^{1/2} y \sec \theta \tag{2.91}$$

Thus

$$\sin \theta = \left[\frac{m_1 m_3}{(m_1 + m_2)(m_2 + m_3)}\right]^{1/2} \tag{2.92}$$

The x and y axes of the potential energy surface diagram are therefore skewed through an angle θ so determined; it is possible to eliminate the error caused in considering the motion of the frictionless particle, by measuring the distances not along the actual surface, but in the xy plane.

Potential energy surfaces for several simple three-nuclei systems have been calculated, and are usually represented graphically by contours[113].

A familiar situation is the approach of an atom to a diatomic molecule, along an x-axis 'valley', the passage over a 'saddle', and out along a y-axis valley; this represents an atom–atom interchange or a chemical interchange reaction.

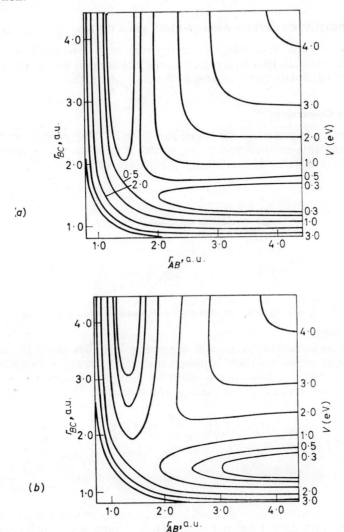

Fig. 2.3. Potential energy surfaces for contrasted triatomic systems: (a) linear; (b) rectangular

The semi-empirical methods[53] of calculating potential energy surfaces have been further developed in recent years[54]. The modifications lead to reasonable values of activation energies, and no longer predict stable species $AB + C \rightarrow ABC$ for which definite experimental evidence is lacking. Previously neglected overlap and three-centre integrals are now included, making possible the calculation of surfaces for non-linear configurations.

A comparison of linear with rectangular triatomic systems is shown in *Figure 2.3*.

A system of n nuclei can be represented in n-dimensional skewed Cartesian coordinates.

2.4 COORDINATE SYSTEMS AND ENERGY RELATIONS

The collision between a particle moving with respect to the laboratory and a particle initially fixed with respect to the laboratory is usually described in terms of either laboratory or centre-of-mass coordinates.

Laboratory Coordinates

Laboratory coordinates are illustrated in *Figure 2.4*. The point particle of mass m_1 and ante-collision velocity v_{1a} would, if there were no interaction, pass within a distance b, the 'impact parameter', of the point particle of

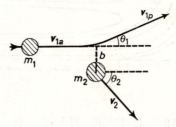

Fig. 2.4. *Elastic scattering in laboratory coordinates, in gravity-free space*

mass m_2, here assumed to be initially at rest in gravity-free space. By virtue of the interaction, m_1 is scattered through an angle θ_1, emerging with a post-collision velocity v_{1p}, whilst m_2 is scattered through an angle θ_2, with a velocity v_2. Conservation of both energy and momentum leads to the equations

$$v_{1p}^2 = v_{1a}^2 \left(1 - \frac{4\mu^2}{m_1 m_2} \cos^2 \theta_2 \right) \tag{2.93}$$

and

$$v_2 = \frac{2v_{1a}}{m_2} \mu \cos \theta_2 \tag{2.94}$$

However, this is not the simplest coordinate system under which the process can be described, since it has not a minimum of variables. A reduction is achieved in centre-of-mass coordinates.

Centre-of-mass (Barycentric) Coordinates

Here the centre of mass of the system is in steady motion with velocity v_c with respect to the laboratory, in a direction defined by the collision path of the projectile. A momentum equation in this direction gives:

$$(m_1 + m_2)v_c = m_1 v_{1a} \tag{2.95}$$

so

$$v_c = \frac{v_{1a}\mu}{m_2} \tag{2.96}$$

for reduced mass

$$\mu = \frac{m_1 m_2}{m_1 + m_2} \tag{2.97}$$

In the centre-of-mass system (*Figure 2.5*), the particle velocities are unchanged by collision, and the two particles are scattered through the same polar scattering angle θ_c. The total kinetic energy E_c in the centre of mass frame is

$$E_c = \tfrac{1}{2}\mu v_{1a}^2 \tag{2.98}$$

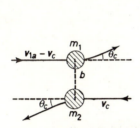

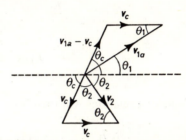

Fig. 2.5. Elastic scattering in centre-of-mass coordinates

Fig. 2.6. Velocity vector diagram of elastic collision in centre-of-mass coordinates

and the total angular momentum is

$$L_c = \mu v_{1a} b \tag{2.99}$$

The relations between the scattering angles in the two systems may be obtained by the vector addition of v_{1a} and v_c, also v_2 and v_c. Referring to *Figure 2.6*,

$$\theta_2 = \tfrac{1}{2}(\pi - \theta_c) \tag{2.100}$$

$$\tan \theta_1 = \frac{\sin \theta_c}{(m_1/m_2) + \cos \theta_c} \tag{2.101}$$

and

$$v_{1p} \sin \theta_1 = \frac{v_{1a}\mu}{m_1} \sin \theta_c \tag{2.102}$$

The total and differential cross-sections remain unchanged in the two frames of reference:

$$\frac{d\sigma}{d\Omega}(\theta, \phi) = \frac{d\sigma}{d\Omega_c}(\theta_c, \phi_c) \tag{2.103}$$

The transformation of solid angle is made using the relation:

$$d\Omega_c = \frac{[(m_1/m_2)^2 + 2(m_1/m_2)\cos \theta_c + 1]^{3/2} \, d\Omega}{1 + (m_1/m_2)\cos \theta_c} \tag{2.104}$$

For elastic collisions certain approximations are useful. When $m_1 \ll m_2$ (electron–atom collisions), $\theta_1 \simeq \theta_c$. When $m_1 = m_2$, $\theta_1 = \theta_c/2$, so there is no back-scattering in laboratory coordinates; when $m_1 \gg m_2$, $\theta_1 \simeq (m_2/m_1) \sin \theta_c$, and only small-angle scattering is possible in laboratory coordinates.

When $m_1 > m_2$, there are two values of θ_c for each θ_1. As θ_1 increases from zero to $\theta_{1\max} = \arcsin(m_2/m_1)$, which is less than $\pi/2$, θ_c increases from zero to $\theta_c = \arccos(-m_2/m_1)$; as θ_c further increases, θ_1 decreases to zero. For the smaller value of θ_c corresponding to a given θ_1, the value of v_{1p} is greater.

For inelastic collisions involving a conversion of a quantity of kinetic energy ΔE to some other form, the energy equation is

$$E_{1a} = E_{1p} + E_2 + \Delta E \qquad (2.105)$$

where:

$$E_{1a} = \tfrac{1}{2} m_1 v_{1a}^2$$
$$E_{1p} = \tfrac{1}{2} m_1 v_{1p}^2$$
$$E_2 = \tfrac{1}{2} m_2 v_2^2 .$$

The following relations will then hold:

$$\tan \theta_1 = \frac{\sin \theta_c}{(m_1/m_2)\,[E_c/(E_c - \Delta E)]^{1/2} + \cos \theta_c}$$

$$\cos \theta_2 = \frac{\Delta E + [1 + (m_2/m_1)]E_2}{2(E_{1a} E_2 m_2/m_1)^{1/2}}$$

$$\cos \theta_1 = \frac{\Delta E + E_{1p} - E_2 m_2/m_1}{2(E_{1a} E_{1p})^{1/2}}$$

$$E_{1p} \frac{\sin^2 \theta_1}{\sin^2 \theta_c} = E_{1p} + E_{1a}\left(\frac{m_1}{m_1 + m_2}\right)^2 - 2\,\frac{m_1}{m_1 + m_2}\,(E_{1a} E_{1p})^{1/2} \cos \theta_1$$

$$E_2 \frac{\sin^2 \theta_2}{\sin^2 \theta_c} = E_2 + E_{1a}\frac{m_1 m_2}{(m_1 + m_2)^2} - 2\,\frac{(m_1 m_2)^{1/2}}{m_1 + m_2}\,(E_{1a} E_2)^{1/2} \cos \theta_2$$

In an elastic collision $v_{cp} = v_{ca}$, but in a chemical collision (reactive scattering) the particles of masses m_1 and m_2 are converted into particles of masses m_3 and m_4, the energy defect of the reaction being ΔE. The azimuthal scattering angles ϕ_1 and ϕ_2 are defined with respect to axes perpendicular to v_{1a} and v_{2a}.

The condition for there to be one–one correspondence between angles in the laboratory and centre-of-mass systems is then

$$(\lambda^2 - 1)\left(\frac{v_{2a}}{v_{1a}}\right)^2 + \lambda^2 - \frac{m_1^2}{m_2} > 2\,\frac{v_{2a}}{v_{1a}}\cos \gamma \left(\lambda^2 + \frac{m_1}{m_2}\right) \qquad (2.106)$$

The development of crossed-beam scattering experiments has made necessary the production of transformations for conversion of these events to the centre-of-mass system[55]. The complete transformation from centre-of-mass to laboratory coordinates has been studied, and the even more valuable reverse transformation is contained in the following equations which apply to the case of beams mutually inclined at an angle γ:

$$\tan \theta = \frac{\lambda k(b-a^2)^{1/2}}{1-\lambda ka} \tag{2.107}$$

for $0 \leqslant \theta \leqslant \pi$, and

$$\tan |\phi| = \frac{\sin \theta_c \sin |\phi_c|}{\cos \delta \sin \theta_c \cos |\phi_c| + c \sin \delta} \tag{2.108}$$

for $|\phi_c| \geqslant 0$ and $\phi \leqslant \pi$. Also,

$$\phi_c > 0 \leftrightarrow \phi > 0$$
$$\phi_c < 0 \leftrightarrow \phi < 0$$

and

$$\frac{\mathrm{d}\Omega}{\mathrm{d}\Omega_c} = \frac{k^2}{\varLambda^3} [\lambda \,|\cos \delta - \lambda a + \lambda kc + \lambda^2 kb\,|] \tag{2.109}$$

where

$$k = \frac{v_{1ca}}{v_{1a}} = \frac{m_2}{m_1 + m_2} \left(1 + \frac{v_{2a}^2}{v_{1a}^2} - \frac{2v_{2a}}{v_{1a}} \cos \gamma \right)^{1/2}$$

$$\lambda = \frac{v_{1cp}}{v_{1ca}} = \left[1 + \frac{\Delta E}{E} f(m) \right]^{1/2}$$

$$f(m) = \frac{m_1 m_4}{m_2 m_3}$$

$$\varLambda = \frac{v_{1p}}{v_{1a}} = 1 - 2\lambda ka + \lambda^2 k^2 b$$

$$a = \sin \theta_c \sin \delta \cos \phi_c - \cos \theta_c \cos \delta + \frac{\cos \delta}{\lambda}$$

$$b = 1 + \frac{1}{\lambda^2} - \frac{2}{\lambda} \cos \theta_c$$

$$c = \cos \theta_c - \lambda^{-1}$$

and where

$$\tan \delta = \frac{v_{2a} \sin \gamma}{v_{1a} - v_{2a} \cos \gamma}$$

for $0 \leqslant \delta \leqslant \pi$. In all these equations θ, ϕ and Ω refer to θ_1, θ_2, ϕ_1, ϕ_2, Ω_1 and Ω_2, and the definitions of k, λ, $\varLambda$, a, b, c and δ must be written accordingly.

These equations apply to an experiment in which both beams of particles are mono-energetic. The experiment including velocity distributions requires a more complicated analysis, for which reference must be made to the original papers. For nearly all problems it is sufficient to study the 'in-plane scattering', that is, the particles scattered in the same plane as that containing the two beams.

Three-body collisions may be described in terms of a six-dimensional normalized centre-of-mass system whose origin represents coincidence of the three particles [56]. The six-space is traversed by three three-dimensional tubes representing the interaction of the particles in pairs. These tubes intersect in a region around the origin where all three particles interact simultaneously. The derivation of hypercross-sections in this system has been attempted both classically and with quantum theory.

Further transformations of two-body and three-body elastic and inelastic collisions between laboratory and barycentric systems have been reported[57-61].

2.5 ENERGY LOSSES IN COLLISIONS BETWEEN IMPENETRABLE ELASTIC SPHERES

The importance of the impenetrable-sphere analysis derives from the fact that the force fields between atom pairs, with their strong exponential short-range repulsion and weak long-range attraction, may be very crudely so approximated.

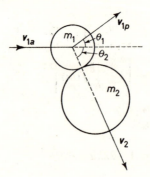

Fig. 2.7. Collision between impenetrable spheres in laboratory coordinates

The fractional kinetic energy loss incurred by the projectile in a collision between two elastic spheres (*Figure 2.7*) is:

$$\frac{\Delta E}{E} = \frac{v_{1a}^2 - v_{1p}^2}{v_{1a}^2} \tag{2.110}$$

for initial kinetic energy of impact E.

It follows that

$$\frac{\Delta E}{E} = \frac{m_2 v_2^2}{m_1 v_{1a}^2} \tag{2.111}$$

with

$$v_2 = \frac{2m_1 v_{1a} \cos \theta_2}{m_1 + m_2} \tag{2.112}$$

Now the probability of impact in which the angle between centres is greater han or equal to θ_2 but less than or equal to $\theta_2 + \delta\theta_2$ is $P(\theta_2) = \sin 2\theta_2 \, d\theta_2$, so by integration of these expressions over all scattering angles

$$\frac{\Delta E}{E} = \frac{2m_1 m_2}{(m_1 + m_2)^2}$$ (2.113)

For collisions between identical atoms, with $m_1 = m_2$,

$$\frac{\Delta E}{E} = \frac{1}{2}$$

a result which is of importance in kinetic theory of gases. Furthermore $\theta_1 + \theta_2 = \pi/2$, that is, the spheres leave each other along paths mutually at right angles.

It is found that in this case the scattering of the target particle is isotropic in the centre-of-mass system, with

$$\frac{d\sigma}{d\Omega}(\theta) = \frac{\pi(r_1 + r_2)^2}{4\pi}$$ (2.114)

For the scattering of electrons by atoms,

$$\frac{\Delta E}{E} \simeq \frac{2m_e}{m_2}$$ (2.115)

which is a very small loss per collision; furthermore

$$\theta_2 = \frac{\pi}{2} - \frac{\theta_1}{2}$$ (2.116)

$$v_2 \simeq \frac{2m_e}{m_2} v_{1a} \cos \theta_2$$ (2.117)

and

$$\frac{\Delta E}{E} \simeq \frac{2m_e}{m_2}(1 - \cos \theta_1)$$ (2.118)

For an inelastic collision of internal energy transfer E_{int}, this equation becomes

$$\frac{\Delta E}{E} \simeq \frac{2m_e}{m_2}\left[1 - \frac{E_{int}}{2E} - \left(1 - \frac{E_{int}}{E}\right)^{1/2}\cos\theta\right]$$ (2.119)

In an electron-atom collision, $d\sigma/d\Omega(\theta)$ is given by equation 2.114, and the classical scattering of electrons by hard-sphere atoms is therefore isotropic in laboratory coordinates.

For a mixture of two gases[62] at temperatures T_1 and T_2, the mean fractional energy loss has been shown to be:

$$\left\langle \frac{\Delta E}{E} \right\rangle = \frac{8}{3}\frac{m_1 m_2}{(m_1 + m_2)^2}\left(1 - \frac{T_2}{T_1}\right)$$ (2.120)

In quantum theory, the scattering of elastic spheres is not isotropic, the differential cross-section rises steeply within a critical angle $\theta_{crit} = \pi/ka$, where $k = \mu v_r/2h$ and a is the radius of either sphere. In *Figure 2.8*, the quantum theory $d\sigma/d\Omega(\theta)$ due to Mott and Massey is represented approximately for the numerical value $ka = 20$. At $\theta = 0$, $d\sigma/d\Omega(\theta)$ rises to $100a^2$. It will be seen that for $\theta > \theta_{crit}$ the scattering is isotropic but the apparent differential cross-section is one-half as great as the classical differential

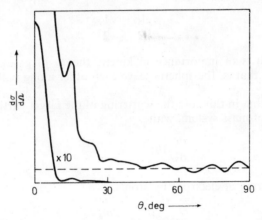

Fig. 2.8. Quantum differential scattering of impenetable spheres in the centre-of-mass system[1]; broken line shows classical cross-section

scattering cross-section. For atoms of atomic weight A and impact energy E electron volts:

$$\theta_{crit} \simeq \frac{8}{(AE)^{1/2}} \text{ rad} \qquad (2.121)$$

taking the typical value $a = 2 \times 10^{-8}$ cm.

Atomic scattering cross-sections correspond to the quantum rather than the classical model. Although there is appreciable large-angle scattering, crudely isotropic, most of the scattering is small angle, with $d\sigma/d\Omega(\theta)$ increasing and oscillating as θ decreases to zero (see Chapter 10).

The hard-sphere classical model has proved successful in the study of ionic mobility[63], and of ionization of atoms by protons[64].

2.6 CLASSICAL TREATMENT OF SOME INELASTIC COLLISION PROCESSES

The simplest type of classical theory of inelastic electron–atom processes is based upon binary encounters between the incident and orbital electrons, between which there is Coulomb interaction. In fact the unexpected success of classical theory is due largely to the fact that Coulomb scattering (e on e) is the same in both classical and quantum theory. In the calculations of Thomson[66] of the ionization of atoms of ionization potential E_i by electrons of kinetic energy E_e, the initial motion of the bound electron is neglected, and the energy transferred E_T in the collision between the bound and the free electron is shown to be:

$$E_T = \frac{E_e}{1 + (b_c E_e/e^2)^2} \tag{2.122}$$

The assumption is made that all collisions are ionizing in which $E_T \geqslant E_i$, so that $\sigma_{\text{class}} = n\pi b_c^2$ where

$$b_c^2 = \left(\frac{e^2}{E_e}\right)^2 \left(\frac{E_e}{E_i} - 1\right) \tag{2.123}$$

and n is the number of outer shell electrons. Close to threshold, the theory over-estimates the cross-section σ by a factor of about five, whilst at large energies it under-estimates the cross-section, with $\sigma_{\text{exp}}/\sigma_{\text{class}}$ increasing with $\ln E_e$. For ionization of atoms in highly excited states, it grossly over-estimates at all energies. Nevertheless, the theory is useful for the estimation of ionization functions for neutral atoms where no experimental data exist; the quantity $E_i^2\sigma/n$ should be a universal function of E_e/E_i, and the best form of this function may be obtained empirically from the assembly of experimental data, and applied with probably not more than ± 25 per cent error in the medium energy-range $(2E_i$–$30E_i)$. The scaling of data for isoelectronic species according to this theory is discussed in Chapter 6.

For excitation of an allowed level E_j similar binary encounter theory may be applied, with the assumption that an atom will be excited into level E_j when the transferred energy lies within limits E_j and E_k, where $k > j$.

When account is taken of the classical motion of the bound electron in orbit before the collision, the model becomes more complex.

The revival of interest in classical theory dates from the contributions of Gryzinski[67, 68]. Gryzinski realized that since Coulomb and gravitational force fields are similar in form, results obtained for the interaction of planetary bodies in orbit are applicable to binary encounters between charged projectile and orbiting electron, neglecting the Coulomb interaction between projectile and nucleus, but not the orbital motion of the bound electron. Such results had been obtained by Chandrasekhar. Gryzinski's original treatment included unnecessary assumptions[69]; the formulae for ionization (which are included in Section 6.1) were not presented in the simplest form because scattering angle rather than momentum transfer was used as a variable. The analysis was repeated in a simpler manner, without using the assumptions, by Stabler[70]. Later, Vriens[71] included an allowance for electron exchange:

$$\sigma = \sigma_{\text{direct}} + \sigma_{\text{exchange}} + (1 + \cos\theta)\sigma_{\text{interference}}$$

and applied the technique to proton collisions[72].

The classical theories yield reasonably correct threshold behaviour for ionization of atoms by electrons, and the absolute cross-sections are realistic in the region of the maximum, as can be seen from the discussion in Chapter 6. But the form of the fall-off of the cross-section at high energies is incorrect; the correct form $(\sigma \propto E^{-1}\ln E)$ can only be obtained by averaging over an incorrect distribution of electron orbit velocities. This failure arises because the long-range electron–electron collisions are incorrectly treated in this theory; the single interaction cannot in this case be isolated from those with other electrons and with the nucleus. The theory is at its best for short-range electron–electron interactions, and is a good semi-empirical formula for total ionization cross-sections; one of the poorest results is the velocity distribution of electrons formed in the ionization process.

Classical theory has been applied to ionization by heavy particles, and to charge transfer, including allowance for tunnel effect[73]; Monte Carlo methods have also been applied successfully[74]. At high energies, the classical approximation can be in error if small-angle scattering predominates, since this is dominated by quantal scattering.

Momentum and energy equations have been applied to the ionization of atoms by ions[65], as follows. An inelastic collision between two atomic systems involving an internal energy defect which is small compared with

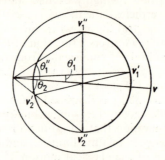

Fig. 2.9. Velocity vector diagram of inelastic scattering of two particles of equal mass

the impact energy E_p may be represented in velocity vector form as in *Figure 2.9*, which is appropriate to the case of equal atomic masses. The difference between the diameters of the two circles represents the energy defect, greatly exaggerated. It is easily seen that for a single value of target scattering angle ϕ there are two values of v_2 (v_2' and v_2''), each corresponding to a projectile velocity (v_1' and v_1'') and scattering angle (θ_1' and θ_1''). For values v_2', v_1' and θ_1' as shown in *Figure 2.9*, where target particles are scattered at small energies while projectile ions are scattered through small angles and do not lose much energy, the process is referred to as 'soft' scattering; by contrast, the 'hard' scattering process involves the scattering of target projectiles with larger energies, while projectile ions are scattered through larger angles. Such splitting of the scattered components has been observed in multiple ionization processes (though not in single ionization), both in the energy spectrum and angular distribution of the target particles, and in the angular distribution of the projectiles. Only a small proportion of the ionization of a gas by fast particles is found to be hard scattering.

The inelastic energy defect, or energy E_T transferred into internal transition energy, can greatly exceed the ionization energy E_i; the excess contributes to the kinetic energy of the liberated electrons. The value of E_T can be deduced from the following equations:

$$E_T(\phi) = 2\left(\frac{m_2 E_2 E_1}{m_1}\right)^{1/2} \cos \phi - \frac{m_1 + m_2}{m_1} E_2 \qquad (2.124)$$

for projectile and target energies E_1 and E_2. In addition

$$\tan\theta = \frac{m_2\left[\cos\phi\pm\cos^2\phi-\dfrac{m_1+m_2}{m_2}\dfrac{E_T}{E_1}\right]^{1/2}\sin\phi}{m_1+m_2-m_2\left[\cos\phi\pm\left(\cos^2\phi-\dfrac{m_1+m_2}{m_2}\dfrac{E_T}{E_1}\right)^{1/2}\right]\cos\phi} \qquad (2.125)$$

An impenetrable sphere treatment of the ionization of atoms by protons[64] is fruitful. It is possible to make certain deductions from the momentum and energy equations for the collisions of an incident particle with an electron in orbit. Neglecting the Coulomb forces exerted by the nucleus and other orbital electrons, the kinetic energy T which is possessed by the orbital electron after collision and escape from orbit is given by the equation

$$T = -E_i + 2\frac{m_e}{m_1}E_p\cos^2\theta - 2(\cos\psi\cos\omega + \sin\psi\cos\phi)\left(\frac{m_1}{m_e}E_iE_p\right)^{1/2}$$

$$+ 2\left(\frac{m_e}{m_1}E_p\cos^2\theta\right)^{1/2}\left[\frac{m_e}{m_1}E_p\cos^2\theta + E_i - 2(\cos\psi\cos\omega + \sin\psi\cos\phi)\right.$$

$$\left.\times\left(\frac{m_e}{m_1}E_iE_p\right)^{1/2}\right]^{1/2} \qquad (2.126)$$

for incident particle of mass m_1 and energy E_p, electron scattering angle θ, electron phase angle ψ, orbit inclination angles ω and ϕ, and ionization energy E_i. For $\psi = \phi = \pi/2$ and $\theta = \omega = 0$, this equation leads to the upper limit of the scattered electron energy distribution[75]. One may also attempt to set limits upon the impact energy at which, for all phase angles ψ, electron emission with a positive kinetic energy $T \geqslant 0$ is possible. Within these limits, the total single ionization cross-section function should reach its maximum. Putting $\partial T/\partial\psi = 0$ (for $\psi \neq 0$) and $T \geqslant 0$ leads eventually to the limits

$$1\cdot46E_i \leqslant \frac{m_e}{m_1}E_p \leqslant 2\cdot25E_i$$

which are found to hold for the ionization functions of H^+, H_2^+ and He^+ in all rare gases and in H_2, N_2 and CO. The similarity of this criterion to the 'adiabatic maximum rule' (Chapter 12) should be noted.

Information regarding the angular distribution of the scattered electrons may also be deduced from equation 2.126.

2.7 CLASSICAL ELASTIC SCATTERING OF ATOMS IN A CENTRAL FORCE FIELD

For the consideration of this problem (see references 76, 77, and also Chapter 10), the two-particle collision is reduced to a single particle problem by transformation into the centre-of-mass system, with reduced mass μ and relative ante-collision velocity v_{ca}. The collision is represented in the plane of the paper in *Figure 2.10*, the assumed force field here being taken as consisting of an attractive plus a short-range repulsive component. The angle χ defines the instantaneous orientation of the nuclear separation vector r.

The effective force field along the r axis is

$$V_{\text{eff}}(r) = V(r) + \frac{\mu v_{ca}^2 b^2}{2r^2} \tag{2.127}$$

or

$$V_{\text{eff}}(r) = V(r) + \frac{L^2}{2\mu r^2} = V(r) + V_{\text{cent}} \tag{2.128}$$

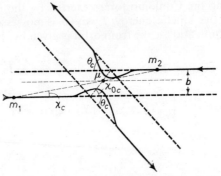

Fig. 2.10. *Planar centre-of-mass representation of classical scattering in an atomic central force field*

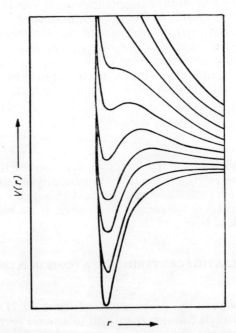

Fig. 2.11. *Characteristic centrifugal interactions of impacting system for different angular momenta*

where b defines the impact parameter of the collision, and $L = b\mu v_{ca}$ is the collisional angular momentum. The term V_{cent} represents the rotational energy of the system, or 'centrifugal potential'. It is positive, and can be regarded as arising from a fictitious repulsive force. The shape of $V_{eff}(r)$ will depend on L; if L is sufficiently large, V_{eff} will decrease monotonically with increasing r, but as L decreases the influence of the potential minimum of $V(r)$ becomes more significant. There is a region of L where $V_{eff}(r)$ will show both a minimum and, at larger r, a maximum; for very small L, the distortion of $V(r)$ does not disturb its general shape. *Figure 2.11* illustrates this situation.

The energy equation is written

$$\tfrac{1}{2}\mu v_{ca}^2 = \tfrac{1}{2}\mu(\dot{r}^2 + r^2\dot{\chi}^2) + V(r) \qquad (2.129)$$

and the angular momentum

$$L = \mu r^2 \dot{\chi}$$

Therefore

$$\frac{dr}{d\chi} = \frac{dr/dt}{d\chi/dt} = -\frac{r^2}{b}\left[1 - \frac{V(r)}{\tfrac{1}{2}\mu v_{ca}^2} - \frac{b^2}{r^2}\right]^{1/2} \qquad (2.130)$$

When

$$\frac{V(r)}{\tfrac{1}{2}\mu v_{ca}^2} + \frac{b^2}{r^2} = 1 \qquad (2.131)$$

the separation reaches its minimum value r_0, and χ reaches a value χ_0. The scattering angle

$$\theta_c = \pi - 2\chi_0 \qquad (2.132)$$

The Langevin equation for the scattering angle follows:

$$\theta_c(v_{ca}, b) = \pi - 2b \int_{r_0}^{\infty} \frac{dr}{r^2\{1 - [V(r)/\tfrac{1}{2}\mu v_{ca}^2] - (b^2/r^2)\}^{1/2}} \qquad (2.133)$$

This equation is fundamental to the classical and semi-classical analysis of atomic scattering; it can be written in terms of the collisional angular momentum and the impact energy $E_c = \tfrac{1}{2}\mu v_{ca}^2$:

$$\theta_c = \pi - 2 \int_{r_0}^{\infty} \frac{\partial}{\partial L}\left[2\mu(E_c - V) - \frac{L^2}{r^2}\right]^{1/2} dr \qquad (2.134)$$

When the interaction energy is written in the usual form, containing well depth parameter ε and separation r_m for minimum energy, then there are advantages in introducing a reduced system of units:

$\beta = b/r_m$ (reduced impact parameter),
$\varrho = r/r_m$ (reduced separation),
$\mathcal{E} = E_c/\varepsilon$ (reduced impact energy),
$U = V/\varepsilon$ (reduced interaction).

The differential cross-section can easily be related to the impact parameter and the angular momentum:

$$\frac{d\sigma}{d\Omega_c} = \frac{r_m^2}{\sin\theta_c}\beta\left|\frac{d\beta}{d\theta_c}\right| = \frac{L}{m^2 v_{ca}^2 \sin\theta_c}\left|\frac{dL}{d\theta_c}\right| \tag{2.135}$$

The typical elastic scattering function $\theta_c(\beta)$ [or $\theta_c(L)$] is of the form shown in *Figure 2.12*. The application of equation 2.135 yields $d\sigma/d\Omega_c(\theta_c)$ functions which exhibit certain special features, as explained in the following sections.

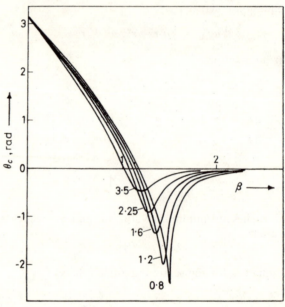

Fig. 2.12. Functions $\theta(\beta_e)$ characteristic of atomic scattering, calculated classically[55] for Lennard–Jones potential, at characteristic values of reduced energy $\mathcal{E}$ from 3·5 to 0·8

Interference Scattering

This is an effect arising only in quantum (semi-classical) theory. In *Figure 2.12* it will be seen that there can be two values of θ_c (θ_{c1} and θ_{c2}) for a single value of β. In purely classical scattering,

$$\left(\frac{d\sigma}{d\Omega_c}\right)(\theta_c) = \left(\frac{d\sigma}{d\Omega_c}\right)_1(\theta_c) + \left(\frac{d\sigma}{d\Omega_c}\right)_2(\theta_e) \tag{2.136}$$

but in quantum (semi-classical) scattering an extra term arises:

$$\left(\frac{d\sigma}{d\Omega_c}\right)(\theta_c) = \left(\frac{d\sigma}{d\Omega_c}\right)_1(\theta_c) + \left(\frac{d\sigma}{d\Omega_c}\right)_2(\theta_c)$$

$$+ 2\left(\frac{d\sigma}{d\Omega_c}\right)_1^{1/2}\left(\frac{d\sigma}{d\Omega_c}\right)_2^{1/2}\cos(\beta_2 - \beta_1) \tag{2.137}$$

where β_1 and β_2 are certain phase angles. These interference effects give rise to undulations in $d\sigma/d\Omega_c(\theta_c)$ (Chapter 10).

Rainbow Scattering[78]

Since the classical cross-section contains the factor $(d\theta_c/d\beta)^{-1}$, it will have a singularity where

$$\frac{d\theta_c}{d\beta} = 0 \tag{2.138}$$

Since the optical analogue of this phenomenon is responsible for the rainbow, the peak in $d\sigma/d\Omega_c(\theta_c)$ occurring at a value $\theta_c = \theta_{crit}$ is termed 'rainbow scattering'. Near θ_{crit} the deflection function may be expanded in the form

$$\theta_c = \theta_{crit} + \frac{1}{2}\left(\frac{d^2\theta_c}{d\beta^2}\right)_{\theta_{crit}} (\beta - \beta_r)^2 \tag{2.139}$$

or

$$\theta_c = \theta_{crit} + q(\beta - \beta_r)^2 \tag{2.140}$$

On the bright side of the rainbow ($\theta_c < \theta_{crit}$), the cross-section is

$$\frac{d\sigma}{d\Omega_c} = \frac{r_m^2 \beta_r}{q(\theta_c - \theta_{crit})^{1/2} \sin\theta_{crit}} \tag{2.141}$$

but on the dark side ($\theta_c > \theta_{crit}$), the cross-section is zero (see Chapter 10). Quantum expressions for rainbow scattering exist, and yield differential cross-sections which do not have an infinite singularity, but only a sharp maximum. For more complicated interaction functions, more than one rainbow can exist.

Scattering at Very Small Angles

The large increase in differential cross-section at small angles arises from large impact parameters for which $d\beta/d\theta_c \to 0$. Only the attractive term in the interaction need be considered, and for

$$V(r) = -\frac{C}{r^s} \tag{2.142}$$

$$\theta_c = \frac{(s-1)}{b^s} f(s) \frac{C}{E_c} \tag{2.143}$$

with

$$f(s) = \frac{\sqrt{\pi}}{2} \frac{\Gamma[\frac{1}{2}(s-1)]}{\Gamma(\frac{1}{2}s)} \tag{2.144}$$

The differential cross-section can be represented[79] as

$$\frac{d\sigma}{d\Omega_c}(\theta_c) = \frac{1}{s}\left[\frac{(s-1)f(s)C}{E_c}\right]^{2/s} \theta_c^{-(2s+2)/s} \tag{2.145}$$

Glory Scattering

When $\theta_c(L)$ passes through zero or an even multiple of π, there is a singularity in the forward direction, or 'forward glory'. When it passes through an odd multiple of π, there is a singularity in the backward direction or 'reverse glory'. The glory effect for elastic head-on collisions corresponds to $\theta_c = \pi$, since $\beta \to 0$. But if inelastic processes such as chemical reactions occur, this need not be the case.

For a reverse glory occurring at

$$\theta_c = \pi + a(L - L_g) \tag{2.146}$$

the classical cross-section is

$$\frac{d\sigma}{d\Omega_c}(\theta_c) = \frac{2L_g}{m^2 v_{ca}^2 |a|(\pi - \theta_c)} \tag{2.147}$$

which is infinite at the reverse glory; this is not so in quantum (semi-classical) theory.

Spiralling Orbits

Consider the situation where the function $V_{eff}(r)$ shows a maximum at r_c, arising from the balance of attractive $V(r)$ by repulsive centrifugal potential (*Figure 2.11*). Near this maximum, but outside it, the radial velocity is very small, but because the angular momentum is not small, the projectile particle makes several revolutions around the target particle before returning to infinite separation. In the limit $E_c = V_{eff}$, the number of revolutions is infinite and classically $\theta_c \to \infty$ and L tends to a critical value L_{orb}. One finds

$$\theta_c = \theta_{c1} + a \ln \frac{L - L_{orb}}{L_{orb}} \tag{2.148}$$

for $L > L_{orb}$, and

$$\theta_c = \theta_{c2} + 2a \ln \frac{L_{orb} - L}{L_{orb}} \tag{2.149}$$

for $L < L_{orb}$. Also

$$a = \frac{2L_{orb}}{\mu r^4} \left| \frac{d^2 V_{eff}}{dr^2} \right|_{r_{orb}} \tag{2.150}$$

and

$$\frac{d\sigma}{d\Omega_c}(\theta_c) \simeq \frac{L_{orb}^2}{\mu^2 v_{ca}^2 a \sin \theta_c} \left[\exp\left(\frac{-\theta_c + \theta_{c1}}{a}\right) + \frac{1}{2} \exp\left(\frac{-\theta_c + \theta_{c2}}{2a}\right) \right] \tag{2.151}$$

For the important inverse fourth power polarization interaction which dominates at large separations between an ion and an atom, the critical impact parameter b_{orb} within which inward spiralling orbits occur is readily shown to be

$$b_{orb} = \left(\frac{4e^2 \alpha}{\mu v_{ca}^2}\right)^{1/4} \tag{2.152}$$

For $b = b_{orb}$ the particles pass classically into an exactly circular orbit of radius $b_{orb}/\sqrt{2}$ (resonance atomic scattering). For $b > b_{orb}$ the scattering is non-spiralling, and for $b < b_{orb}$ the orbits are inward spiralling until there is sufficient repulsive interaction for the system to spiral outwards. During this relatively long collision time, transitions may occur, such as excitation exchange, charge transfer or ion–atom interchange[80], which are discussed in Chapters 12 and 14. Calculated orbits are illustrated in *Figure 2.13*. Further references[81-85] describe orbit calculations for various interactions. Quantum (semi-classical) discussion of the process has been given[78, 114].

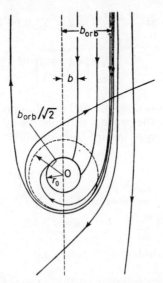

Fig. 2.13. *Classical orbiting in inverse fourth power field of centre O, critical impact parameter* b_{orb}; *orbits within sphere of radius* r_0 *are not shown*

Evidence is beginning to emerge of the formation of spiralling temporary molecules in some thermal collisions between neutral species.

Rotational Predissociation

At the separation r_c at which the effective potential passes through a maximum, the first differential of $V(r)$ is

$$V'(r)_{r_c} = \frac{L^2}{\mu r_c^3} \qquad (2.153)$$

In a molecule, the rotational angular momentum is quantized with rotational quantum number J:

$$L^2 = J(J+1)\hbar^2 \qquad (2.154)$$

Therefore, for a purely attractive interaction

$$V(r) = -Cr^{-s} \qquad (2.155)$$

one has

$$r_c \simeq \left[\frac{s\mu C}{\hbar^2 J(J+1)}\right]^{1/(s-2)} \qquad (2.156)$$

A diatomic molecule for which this equation applies will fly apart when it is rotationally excited with quantum number greater than or equal to J. This effect, which manifests itself as a termination in a rotational spectrum, is known as 'rotational predissociation'. From the 'limiting curve of dissociation' (LCD)[86], C and s may be calculated[87]:

$$E_{\text{LCD}} = D + \frac{V_{\text{eff}}(r_c)}{hc} \simeq D + S_s[J(J+1)]^{s/(s-2)} \qquad (2.157)$$

where

$$S_s = \frac{s-2}{4\pi c}\left[\frac{\hbar^{s+2}}{(s\mu)^s C^2}\right]^{1/(s-2)} \qquad (2.158)$$

The symbol D denotes the dissociation limit, but predissociation takes place at E_{LCD}. The inverse of the process is resonance atomic scattering (Chapter 10), the eigenstate version of inward spiralling orbits.

Rutherford Scattering

For a Coulomb interaction between two nuclei of charges $Z_1 e$ and $Z_2 e$, the differential cross-section is

$$\frac{d\sigma}{d\Omega_c}(\theta_c) = \frac{Z_1^2 Z_2^2 e^4}{4\mu^2 v_{ca}^4 \sin^4 \frac{1}{2}\theta_c} \qquad (2.159)$$

This expression is historically important[88, 89], insofar as it was used by Rutherford in the interpretation of the scattering of α particles by nuclei.

2.8 QUANTUM THEORY OF SCATTERING

Elementary concepts of importance in the quantum theory of collisions derive from the theory of scattering of a plane wave by a central force field, resulting in an outgoing spherical wave. This formalism, developed by Faxen and Holtsmark[90], is essentially the same as that developed by Rayleigh[91] for the scattering of sound waves by spherical obstacles.

The incident particle of mass m is represented as a plane wave $\ell \exp(ikz)$ of wavelength

$$\lambda = \frac{h}{mv} \qquad (2.160)$$

travelling along the z axis with velocity v. The number of particles passing in unit time through unit area of the plane xy is the flux density A, where

$$\ell = \left(\frac{A}{v}\right)^{1/2} \qquad (2.161)$$

The wave number

$$k = \frac{2\pi}{\lambda} = \frac{2\pi m v}{h} \tag{2.162}$$

The flux density is related to the wave function by the equation

$$A = \frac{h}{2m} (\psi \ \text{grad} \ \psi^* + \psi^* \ \text{grad} \ \psi) \tag{2.163}$$

The outgoing spherical wave is of form $r^{-1} \mathcal{C} f(\theta) \exp{(ikr)}$, where

$$\frac{d\sigma}{d\Omega} (\theta) = |f(\theta)|^2 \tag{2.164}$$

The wave equation is

$$\nabla^2 \psi + [k^2 - U(r)]\psi = 0 \tag{2.165}$$

where

$$U(r) = \frac{2m}{\hbar^2} V(r) \tag{2.166}$$

Moreover ψ must have the asymptotic form

$$\psi \sim \exp{(ikz)} + r^{-1} \exp{(ikr)}f(\theta)$$

A differential equation of this type can be solved in spherical polar coordinates, taking

$$\psi = P_l(\cos \theta) f_l(r) \tag{2.167}$$

where $f_l(r)$ is a solution of the equation

$$\frac{1}{r^2} \frac{d}{dr} \left(r^2 \frac{df}{dr} \right) + \left[k^2 - \frac{l(l+1)}{r^2} \right] f = 0 \tag{2.168}$$

and $P_l(\cos \theta)$ are Legendre polynomials of order l; $l = 0$ represents s wave scattering, $l = 1$ represents p wave scattering, $l = 2$ represents d wave scattering, and so on. The solution is

$$f(\theta) = \frac{1}{2ik} \sum_{l=0}^{\infty} \{(2l+1)[\exp{(2i\eta_l)} - 1]P_l(\cos \theta)\} \tag{2.169}$$

and the total cross-section for specified l is

$$\sigma_l = \frac{4\pi}{k^2} (2l+1) \sin^2 \eta_l \tag{2.170}$$

where η_l is the phase shift experienced by the wave of order l. The total cross-section is that summed over all l. The time delay suffered by the particle due to its interaction with the scattering field is $d\eta/dk$. When $\eta_l \ll 1$ for all l, the distortion suffered by the plane wave is negligible, and

$$f(\theta) = \frac{1}{k} \sum_{l=0}^{\infty} (2l+1)\eta_l P_l(\cos \theta)] \tag{2.171}$$

The summation of this expression yields the Born approximation, discussed in Section 2.9.

The application of partial wave expansions[92] enables certain upper bounds to be placed upon scattering cross-sections, as follows:

$$\sigma_{l\,(\text{elast})} \leqslant \frac{4\pi}{k^2}\,(2l+1)$$

Moreover, if the range of the scattering field is small compared with the wavelength, then only the zero-order partial wave is scattered and

$$\sigma_{\text{elast}} \leqslant \frac{4\pi}{k^2}$$

The formulae may be generalized to include inelastic cross-sections:

$$\sigma_{l\,(\text{inelast})} \leqslant \frac{\pi}{k^2}\,(2l+1)$$

and, for $R \ll \lambda$,

$$\sigma_{\text{inelast}} \leqslant \frac{\pi}{k^2}$$

On the other hand, when the wavelength is very short compared with the range R of the scattering field,

$$\sigma_{\text{elast}} \leqslant 4\pi R^2$$

for $\lambda \ll R$, and

$$\sigma_{\text{inelast}} \leqslant \pi R^2$$

The latter holding when $\sigma_{\text{elast}} = \pi R^2$.

A feature of equation 2.169 is that it leads to non-isotropic scattering by an impenetrable sphere[93] of radius a. The wave function vanishes at the boundary of the sphere, giving $\eta_0 = -ka$. The low velocity limit of the total cross-section is $4\pi a^2$, four times the classical value. For $ka \gg l$, the phase shifts are approximately

$$\eta_l = -ka - \tfrac{1}{2}l\pi \tag{2.172}$$

and the summation may be replaced by an integral:

$$\sigma = \frac{4\pi}{k^2} \int_0^{ka} (2x+1) \sin^2\left(ka - \frac{1}{2}x\pi\right) \mathrm{d}x \simeq 2\pi a^2 \tag{2.173}$$

Although the quantum cross-section is thus twice the classical cross-section, one-half of this effect arises from the contribution at very small angles. For $\theta > \pi/ka$, the scattering is more or less isotropic; close to the classical limit,

$$\frac{\mathrm{d}\sigma}{\mathrm{d}\Omega}(\theta) \simeq \frac{a^2}{2} \tag{2.174}$$

but for $\theta \ll \pi/ka$, $d\sigma/d\Omega$ rises with decreasing θ to a value $k^2a^4/4$ at $\theta = 0$. In addition, there are undulations (*Figure 2.8*) which emphasise the wave diffraction arising from destructive interference of the incident and scattered wave[94]. Since atomic interactions, with their strongly repulsive short-range and weak long-range components, may crudely be approximated by an impenetrable sphere, these results are not completely unrealistic.

When the interaction potential falls off as r^{-s} at large r, then certain low velocity limits may be calculated for the phase shifts:

$$\lim_{k \to 0} k^{2l+1} \cot \eta_l = a_l \tag{2.175}$$

for $l < (s-3)/2$, and

$$\lim_{k \to 0} k^{s-2} \cot \eta_l = \text{constant} \tag{2.176}$$

for $l > (s-3)/2$. The parameter a_l is known as the scattering length. For $l = 0$

$$\lim_{k \to 0} \sigma_0 = 4\pi a_l^2 \tag{2.177}$$

and

$$\sigma_0 \simeq \frac{4\pi a_l^2}{\left(1 - \frac{1}{2} k^2 a_l r_0\right)^2 + k^2 a_l^2} \tag{2.178}$$

where r_0 is known as the effective range[115-122].

For $l > 0$, and provided $s > 2l+5$,

$$k^{2l+1} \cot \eta_l \simeq -\frac{1}{a_l} + \frac{1}{2} r_0 k^2 \tag{2.179}$$

But for inverse fourth power interaction, no effective range exists for any l, and a scattering length only exists for $l = 0$ when

$$k \cot \eta_0 = -\frac{1}{a_l} + \frac{\pi\alpha}{3a_l^2} k + \frac{2\alpha}{3a_l} k^2 \ln\left(\frac{\alpha k^2}{16}\right) + \text{terms of order } k^2 \tag{2.180}$$

For $l > 0$,

$$k^2 \cot \eta_l \simeq \frac{8\left(l + \frac{3}{2}\right)\left(l + \frac{1}{2}\right)\left(l - \frac{1}{2}\right)}{\pi\zeta} \tag{2.181}$$

where the interaction energy

$$V(r) = -\frac{\zeta}{r^4} \tag{2.182}$$

Provided that there are no collisions in which particles are absorbed or change their energy, then it can be shown that:

$$\sigma = \frac{4\pi}{k} \text{Im} f(\theta) \tag{2.183}$$

$$f(\theta) = \sum_l c_l P_l(\cos \theta) \tag{2.184}$$

$$c_l = \frac{2l+1}{2ik} [\exp(2i\eta_l) - 1] \tag{2.185}$$

This relation, the 'optical theorem', may be generalized to inelastic scattering processes.

Another method of calculating cross-sections, particularly valuable for collisions between electrons and positive ions, is the quantum defect method[95]. The bound states between an electron and a positive ion can be represented as energy levels of the atom, in the form

$$E_n = -\frac{2\pi^2 m_e (Ze^2)^2}{n'^2 h^2} \tag{2.186}$$

where

$$n' = n - \mu_0(n) \tag{2.187}$$

n being the principal quantum number and Z the nuclear charge. The number $\mu_0(n)$ is the quantum defect for the nth bound state of zero angular momentum. For a pure Coulomb field $\mu_0(n) = 0$. Scattering cross-sections are represented as

$$\frac{\cot \sigma_l(k^2)}{1 - \exp(\pi\beta/k)} = \cot \pi\mu_l(k^2) \tag{2.188}$$

where

$$\frac{\beta^2}{4n'^2} = -\frac{8\pi^2 m_e E_n}{h^2} \tag{2.189}$$

$u_l(k^2)$ is the quantum defect for angular momentum l, considered as a function of energy and extrapolated to the positive energy $k^2 h^2 / 2m_e$.

2.9 THE BORN APPROXIMATION

An exact expression for the scattering of a plane wave of wave number $k = mv/h$ by a centre of force can be written as

$$\psi \sim \exp(ikz) - \frac{r^{-1}\exp(ikr)}{4\pi} \int \exp(ik\mathbf{n}r') V(r')\psi(r') \, d\tau' \tag{2.190}$$

Here a new radius variable r' is introduced for the purpose of making a volume integration over τ'. The vector $\mathbf{n}$ is a unit vector in the direction of $\mathbf{r}$:

$$\mathbf{n} = (\sin\theta\cos\phi, \sin\theta\sin\phi, \cos\theta) \tag{2.191}$$

In the Born approximation[123], the assumption is made that the wave is negligibly diffracted, and $\psi(r')$ is replaced in the integral by the unperturbed wave function $\exp(ikz')$. Then

$$f(\theta) = -\frac{1}{4\pi} \int \exp[ik(\mathbf{n}_0 - \mathbf{n})\mathbf{r}] U(r) \, d\tau \tag{2.192}$$

where n_0 is a unit vector along the z axis, so that $z = n_0 . r$. In spherical polar coordinates

$$f(\theta) = -\frac{2m_e}{\hbar^2} \int_0^\infty \frac{\sin Kr}{Kr} V(r) r^2 \, dr \qquad (2.193)$$

where the collisional momentum exchange $K = k \, | n_0 - n |$.

If $V(r)$ is an atomic interaction, a further simplification can be made, yielding

$$f(\theta) = \frac{e^2}{2m_e v^2} [Z - F(\theta)] \operatorname{cosec}^2 \left(\frac{\theta}{2} \right) \qquad (2.194)$$

where Z is the nuclear charge and

$$F(\theta) = 4\pi \int_0^\infty q(r) \frac{\sin Kr}{Kr} r^2 \, dr \qquad (2.195)$$

and $q(r)$ is the atomic charge density function. The quantity F is known as the atomic scattering or form factor; it has been tabulated over a range of K for all elements[1].

The Born approximation is obtained as a first approximation to the integral equation 2.193. It is possible to calculate the distortion of the incident plane wave using this first Born approximation and to insert the distorted wave into the integral equation, thus obtaining the second Born approximation, and higher approximations by a process of iteration. It is important that the Born series should converge, but for the charge transfer collision at high energies, the Born cross-section is the first term of a different convergent series (Chapter 13). In general, it has been established that for potentials which satisfy the conditions $V(r) = 0, r > 0, V(r)$ and dV/dr finite and bounded, $0 \leqslant r \leqslant a$, and

$$\frac{2m}{\hbar^2} \int_0^a |V(r)| \, r \, dr < 1$$

the Born expansion converges for all wave numbers k.

The conditions necessary for validity of the Born approximation are, in general, the opposite of those necessary for validity of classical calculations. For the scattering of a particle by a force field extending over a distance a, by a potential of order D, the classical approximation requires that the orbit be well-defined, that is:

$$mva \gg \hbar$$

which is the condition that the wavelength should greatly exceed the scattering field range. This corresponds to

$$aK \gg \hbar$$

where K is the momentum transfer. However

$$K \sim \frac{D}{v}$$

therefore classical validity requires

$$\frac{Da}{\hbar v} \gg 1$$

Born approximation validity requires the opposite condition

$$\frac{Da}{\hbar v} \ll 1$$

For calculations of differential cross-section, classical validity requires

$$a\theta \gg \frac{\hbar}{mv}$$

and Born approximation validity requires

$$\left| V\left(\frac{\hbar}{mv\theta}\right) \right| \ll mv^2\theta$$

2.10 THE OSCILLATOR STRENGTH AND THE BORN–BETHE APPROXIMATION

The probability of excitation of an atom to a higher level is governed by the extent to which the wave function can be distorted by an electric vector, with the production of a temporary electric dipole moment. The magnitude of this 'transition moment' is related to a dimensionless quantity known as the optical oscillator strength f, which compares the moment with that of a classical harmonic oscillator and is proportional to the radial overlap integral of the atomic wave functions

$$\int \psi_n \psi_m^* \, dr$$

for the two states m and n between which the excitation process takes place. The oscillator strength is related not only to the absorption but to the emission of electromagnetic radiation from the higher level, and the discussion of these processes is postponed until Chapter 9. However, the oscillator strength plays an important part in collision processes in which an incident charged particle excites or ionizes the atom; it can be inferred from collisional data, using an approximation due to Bethe[96], sometimes called the optical approximation (but distinct from the optical model or potential, discussed in the next section).

The differential cross-section for excitation of an atom from level m to level n by an electron is given by the Born approximation as

$$d\sigma(\theta) = \frac{4\pi^2 m_e}{\hbar^4} \frac{k_p}{k} \left| \int V(r, R) \exp\left[i(k_{mn}n_1 - kn_0)R\right] \psi_n^*(r)\, \psi_m(r) \, dr \, dR \right|^2 \tag{2.196}$$

where the initial and post-collision momentum vectors of the colliding electron are $k\hbar n_0$ and $k_p\hbar n_1$, and the interaction energy V is the Coulomb interaction $e^2/|r-R|$ between incident and atomic electrons. r and R are respectively the radius vectors of the atomic electron and the projectile. A transformation is made into a system using momentum variables, so that, for a collision along the path z,

$$\exp\left[i(k_p \mathbf{n}_1 - k\mathbf{n}_0)\mathbf{R}\right] = \exp\left(iKz\right) \qquad (2.197)$$

where the change of momentum of the incident electron scattered through angle θ is

$$Kh = |k_p \mathbf{n}_1 - k\mathbf{n}_0|\, h = (k_p^2 + k^2 - 2kk_p \cos \theta)^{1/2}\, h \qquad (2.198)$$

If the atom contains N atomic electrons numbered $s = 1, 2, \ldots, N$, then

$$d\sigma(K) = \frac{128\pi^5 m_e^2 e^4}{k^2 h^4 K^3}\, |\varepsilon_{mn}(K)|^2 \qquad (2.199)$$

where

$$\varepsilon_{mn}(K) = \sum_{s=1}^{N} \exp\left(ikz_s\right)\psi_m \psi_n^{*}\, d\mathbf{r} \qquad (2.200)$$

A quantity known as the generalized oscillator strength

$$f_g(K) = \frac{m_e^2 e^4 v_{mn}}{h^4 R K^2}\, |\varepsilon_{mn}(K)|^2 \qquad (2.201)$$

is introduced. Here v_{mn} is the characteristic frequency of the transition and R is the Rydberg constant.

The differential electron excitation cross-section is directly related in the Born approximation to the generalized oscillator strength

$$\frac{d\sigma}{d(Ka_0)} = \frac{4a_0^2 R}{E_{mn}} \left(\frac{E_e - E_{mn}}{E_e}\right)^{1/2} \frac{f_g(K)}{(Ka_0)^2} \qquad (2.202)$$

and the total excitation cross-section

$$\sigma = \frac{8\pi a_0^2 R^2}{E_e E_{mn}} \int_{(Ka_0)_{min}}^{(Ka_0)_{max}} f_g(K)\, \frac{d(Ka_0)}{Ka_0} \qquad (2.203)$$

When

$$K \rightarrow (k_p^2 + k^2)^{1/2}$$

that is, for inelastic scattering of the electron through angle $\theta = 0$, the generalized oscillator strength reduces, in the limit of infinite impact energy, to the optical oscillator strength, and the electron collision cross-section $\Delta\sigma_0$ for $\theta = 0$ and excitation of a level of energy E_{mn} (in atomic units 27·2 eV) is:

$$\Delta\sigma_0 = \frac{2fk_p a_0^2}{E_{mn} K^2 k} \qquad (2.204)$$

For incident electron energy E_e

$$K^2 \simeq 8\left(E_e - \frac{E_{mn}}{2}\right)\left\{\sin^2\frac{\theta}{2} + \left[\frac{E_{mn}}{4\left(E_e - \frac{1}{2}E_{mn}\right)}\right]^2\right\} \tag{2.205}$$

and

$$\Delta\sigma_0 = \frac{4\left(E_e - \frac{1}{2}E_{mn}\right)f\beta a_0^2}{E_{mn}^3} \tag{2.206}$$

where

$$\beta = \frac{\theta^2}{\alpha} + 1 \tag{2.207}$$

and

$$\alpha = \frac{E_{mn}}{2\left(E_e - \frac{1}{2}E_{mn}\right)} \tag{2.208}$$

Therefore, in the limit of high-energy incident electrons and zero-angle scattering, $\beta \to 1$. Since β varies slowly with E_{mn}, a scansion of momentum analysis of post-collision zero-angle scattered electrons scans a spectrum of energy levels with peak heights proportional to oscillator strengths multiplied by the factor $\left(E_e - \frac{1}{2}E_{mn}\right)E_{mn}^{-3}$.

This basic experiment of 'electron spectroscopy' is discussed further in Chapters 5 and 6. It is often conducted for scattering angles close to zero, since only in this way will optical oscillator strengths be obtained. It must be conducted with sufficiently high energy of incident electrons for the Born approximation to hold. Since electrons are accepted into the post-collision momentum analyser over a small range of angles θ, the correction factor β may readily be calculated. Accumulation of oscillator strength data for discrete levels and for the ionization continuum is of fundamental importance to atomic physics and astrophysics. The generalized oscillator strength function $f_g(K^2)$ is also of importance, and relationships between generalized oscillator strengths and total scattering cross-sections were given by Miller and Platzman[97]. For transitions in which the atom is ionized, the Born–Bethe approximation can be applied to continuum transitions in the same way as to discrete levels. The equivalent radiation process is photoionization (Chapter 9). The appropriate equation is

$$\frac{d^2\sigma}{d\Omega\,dE} = \left(\frac{df}{dE}\right)\frac{2k_p a_0^2}{EkK^2} \tag{2.209}$$

At energies lower than that at which the Born approximation is valid, it is valuable to define an 'effective' generalized oscillator strength directly from equation 2.202 just as if it held as an equality.

2.11 MORE EXACT QUANTUM THEORY APPROXIMATIONS

These will only be listed, and described in the briefest outline.

Reference should be made to the original papers, or to the textbooks on scattering theory[1-8].

Distorted Wave Approximation

The wave equation for the system of incident electron (suffix a) plus hydrogen atom (containing electron suffix b) is

$$\left[\frac{\hbar^2}{2m}(\nabla_a^2+\nabla_b^2)+E_a+E_b+\frac{e^2}{r_a}+\frac{e^2}{r_b}-\frac{e^2}{r_{ab}}\right]\Psi = 0 \qquad (2.210)$$

An expansion is made in terms of certain important functions $F_n(r_1)$:

$$\Psi(r_a, r_b) = \left(\sum_n + \int\right)\psi_n(r_b)\,F_n(r_1) \qquad (2.211)$$

where the functions $\psi_n(r)$ are the unperturbed eigenfunctions of the hydrogen atom, satisfying

$$\left(\frac{h^2}{2m}\nabla^2+E_n+\frac{e^2}{r}\right)\psi_n = 0 \qquad (2.212)$$

Consider a collision between two atomic systems a and b capable of existing only in two states 0 and 1. Putting

$$U_{10}(r) = \frac{2m}{\hbar^2}\iint V(r, r_a, r_b)\psi_0\psi_1^*\,dr_a\,dr_b \qquad (2.213)$$

one obtains

$$[\nabla^2+k_0^2]F_0 = U_{01}F_1+U_{00}F_0 \qquad (2.214)$$

and

$$[\nabla^2+k_1^2]F_1 = U_{10}F_0+U_{11}F_1 \qquad (2.215)$$

When the coupling potential U_{01} is small, and U_{00} and U_{11} are large, then $F_1 \ll F_0$, and an approximation may be made by solving the equations

$$(\nabla^2+k_0^2-U_{00})F_0 = 0 \qquad (2.216)$$
$$(\nabla^2+k_1^2-U_{11})F_1 = U_{01}F_0 \qquad (2.217)$$

with suitable boundary conditions.

This is the distorted wave approximation, which is valuable in calculating cross-sections for various electronic and atomic collision processes.

The Optical Model

Another approximation consists of neglecting all but the elastically scattered wave; one writes

$$\left[\nabla^2+k^2-\frac{2m}{\hbar^2}V_{00}(r)\right]F_0(r) = 0 \qquad (2.218)$$

with the volume integration

$$V_{00} = \int V(r, r_a)\psi_0\psi_0^*\,d\tau_a \qquad (2.219)$$

This equation represents the motion of an incident electron in the static field of the atom, which has potential V_{00}. The effect of the terms neglected in the equation can be represented by adding a non-local interaction, approximated by a local but energy-dependent interaction. This combination is termed the 'optical model potential', and is the basis of an approximation valuable in the understanding of low-energy electron–atom scattering. It is in effect an attempt to solve a problem of multiple centre scattering by finding an equivalent single centre interaction which gives the same scattering.

Variational Methods

A variety of 'variational methods' have been applied to collision problems[98, 124–128]. A variational method is normally a technique for obtaining a good wave function using the fact that it must remain correct to first order when a functional variation is applied. In this way, an upper bound can be placed on the energy of the lowest bound state of a system. If no bound state exists, then the lowest allowed energy is zero, and one can obtain either an upper or a lower bound on the scattering length. It is more difficult, but possible, to calculate phase shifts and scattering amplitudes.

Born–Oppenheimer Approximation

So far, only collision processes in which electrons were not interchanged between projectile and target have been considered. Processes such as charge transfer in which there is interchange are termed 'rearrangement collisions'; the simplest of these is the electron exchange process, in which, in an electron–atom collision, the scattered electron is not identical with the incident electron. Only electrons with opposed spins can be distinguished; those with parallel spins are indistinguishable. The exchange must be allowed for in excitation collisions, and the method for making this allowance is due to Oppenheimer[99]. In equation 2.211 above, a function $F_n(r)$ was defined. One defines an analogous function $G_n(r)$ such that

$$\Psi = \left(\sum_n + \int\right) F_n(r_1)\,\psi_n(r_2) \qquad (2.220)$$

and

$$\Psi = \left(\sum_n + \int\right) G_n(r_2)\,\psi_n(r_1) \qquad (2.221)$$

where G_n is taken to have the asymptotic form

$$G_n \sim r^{-1} \exp\,(ik_n r)\, g_n(\theta,\,\phi) \qquad (2.222)$$

The differential cross-section for capture of electron 1 into state n and ejection of electron 2 is

$$\frac{d\sigma}{d\Omega} = \frac{k_n}{k_0}\,|g_n(\theta,\,\phi)|^2 \qquad (2.223)$$

To retain the equivalent of the first Born approximation, one must solve the equation

$$(\triangledown^2 + k_n^2)G_n = \frac{2m}{h^2} \int \left(\frac{e^2}{r_{12}} - \frac{e^2}{r_2} \right) \psi_n^*(\mathbf{r}_1)\,\psi_0(\mathbf{r}_2)\,\exp\,(ik_0\mathbf{n}_0.\mathbf{r}_1)\,\mathrm{d}\tau_1 \qquad (2.224)$$

The solution yields the Born–Oppenheimer approximation. An important improvement to this approximation has been made by Ochkur[100]; the improved approximation can no longer yield worse results than the Born approximation itself.

JWKB Approximation

Another approximation which is frequently encountered is Jeffreys' or the JWKB (Jeffreys Wenzel Kramers Brillouin) approximation[101]. It is of value for the calculation of large phase shifts in atomic elastic scattering.

A solution can be obtained to the one-dimensional wave equation

$$\frac{\mathrm{d}^2\psi}{\mathrm{d}z^2} + f(z)\,\psi = 0 \qquad (2.225)$$

with

$$(z) = \frac{8\pi^2 m}{h^2}\,(W - V) \qquad (2.226)$$

for the interaction of a particle of kinetic energy W in a potential V, which is assumed not to vary appreciably in a distance of the order of the wavelength $h/[2m(W-V)]^{1/2}$. The solution is

$$\psi = f^{-1/4} \sin \left\{ \frac{\pi}{4} + \int_{z_0}^{z} [f(z)]^{1/2}\,\mathrm{d}z \right\} \qquad (2.227)$$

in the range $z > z_0$, where $f(z) > 0$, $z > z_0$ and $f(z) < 0$, $z < z_0$.

In the quantum theory of atomic scattering, a solution is required of the radial wave equation

$$\frac{\mathrm{d}^2 G_l}{\mathrm{d}r^2} + F(r)G_l = 0 \qquad (2.228)$$

with

$$F(r) = \frac{2m}{\hbar^2}\,(E - V) - \frac{l(l+1)}{r^2} \qquad (2.229)$$

and $G(r) = rR(r)$, with $R(r)$ the radial wave function.

The appropriate solution is

$$r^{1/2}G_l \simeq F_1^{-1/4} \sin \left[\frac{\pi}{4} + \int_{r_0}^{r} [F_1(r)]^{1/2}\,\mathrm{d}r \right] \qquad (2.230)$$

where

$$F_1(r) = \frac{2m}{\hbar^2}(E-V) - \frac{(l+\frac{1}{2})^2}{r^2} \tag{2.231}$$

The asymptotic form of the solution is proportional to

$$\sin\left[\frac{\pi}{4} + \int_{r_0}^{\infty} (F_1^{1/2} - k)\,\mathrm{d}r + k(r-r_0)\right]$$

which is equal to

$$\sin\left(kr - \frac{l\pi}{2} + \eta_l\right)$$

Thus, the phase shift η_l is calculable when V is known. Further methods of analysis of phase shifts in atomic scattering are given in Chapter 10.

It is possible to treat the collisional angular momentum as a continuous variable and make use of the relation

$$\theta_c(J) = \frac{2}{\partial l}\frac{\partial \eta_l}{\partial l} \tag{2.232}$$

where $J = (l+\frac{1}{2})\hbar$.

The classical scattering formula

$$|f(\theta)|^2 = \frac{J\,|\partial J/\partial\theta_c|}{m^2 v^2 \sin\theta_c} \tag{2.233}$$

is obtained when the value of l for which

$$\frac{\partial \eta_l}{\partial l} = \frac{\theta_c}{2} \tag{2.234}$$

is large, and η_l is large for this value. The condition for the validity of the Born approximation is η_l small for all l.

The Impulse Approximation

An approximation which follows naturally from the quantum treatment of scattering from a centre of force is the 'impulse' or 'sudden' approximation. This concerns the scattering from a system containing more than one centre of force. The mean separation between the centres must be large compared with the reduced wavelength of relative motion of the particles and the system. The problem is then attacked on the assumption that the collision of the incident particle with one of the particles of the system takes place as if the latter were free except for its possession of a momentum probability distribution determined by the presence of the other particles of the system. The approximation has proved valuable when applied to the production of excited positive ions in electron–atom collisions, which is of importance in the understanding of positive-ion lasers.

Impact Parameter Methods

Many calculations of inelastic collisions between two atomic systems are formulated using the semi-classical conception of 'impact parameter'; this has been introduced in Section 2.2. If it is possible to calculate the probability P of a transition from initial to final energy levels as a function of impact parameter b, then the total cross-section for this process is

$$\sigma = \int_0^\infty P\, 2\pi b\ \mathrm{d}b \tag{2.235}$$

where the maximum possible value of P is 1. This can be expressed in terms of the angular momentum L, given by the equation

$$L = \mu v b \tag{2.236}$$

as

$$\sigma = \frac{\pi}{(\mu v)^2} \int_0^\infty P\, \mathrm{d}(L^2) \tag{2.237}$$

For quantization,

$$L^2 = l(l+1)\hbar^2 \tag{2.238}$$

and

$$\mathrm{d}(L^2) = (2l+1)\hbar^2\, \mathrm{d}l \tag{2.239}$$

Therefore

$$\sigma = \frac{\pi}{k^2} \sum_l (2l+1)P \tag{2.240}$$

for $k = \mu v/\hbar$.

Perturbed Stationary States, and Polarized Orbitals

One of the first quantum theory methods ever applied to atomic collisions was the method of 'perturbed stationary states'. The two atoms are regarded as a quasi-molecule, with fixed nuclei and initial wave function ψ_i. The unperturbed Hamiltonian H_0 satisfies the equation

$$H_0\psi_i = i\hbar\frac{\partial \psi_i}{\partial t} \tag{2.241}$$

The interaction V is treated as a perturbation:

$$H = H_0 + V \tag{2.242}$$

giving

$$H\psi = i\hbar\frac{\partial \psi}{\partial t} \tag{2.243}$$

The function ψ is expanded in terms of the unperturbed functions:

$$\psi = \sum_i c_i(t)\psi_i \qquad (2.244)$$

The probability that the system passes into a state n is found by calculating the value of $|c_n(t = \infty)|^2$, using the boundary condition that before the collision $|c_i(-\infty)|^2 = 1$. However, the expansions do not have the correct asymptotic form unless they are multiplied by additional factors which are difficult to calculate. The only case where this correction is unnecessary is that of symmetrical resonance charge transfer[102], which is discussed further in Chapter 13. Otherwise the method is not now in use[129–131]. However, a method which is similar in its use of a perturbation is that of 'polarized orbitals'[103]. In the scattering of electrons by the hydrogen atom, the wave function is taken as

$$\Psi = (1+P_{12})\,[\psi_0(\boldsymbol{r}_1)+\phi_0(\boldsymbol{r}_1,\,\boldsymbol{r}_2)]\,F_0(\boldsymbol{r}_2) \qquad (2.245)$$

where $\phi_0(\boldsymbol{r}_1,\,\boldsymbol{r}_2)$ represents the change in the hydrogen atom wave function where the atom is perturbed, or polarized, by the second electron fixed at $\boldsymbol{r}_2$. Thus the incident electron must be travelling with a velocity smaller than that of the unperturbed electron. The method is successful in calculating realistic phase shifts, and a simplified approximation to it, the 'exchange adiabatic' approximation, is adequate for many purposes.

2.12 THE PRINCIPLE OF DETAILED BALANCING

The principle of detailed balancing[104] states that the transition probabilities for a process and its reverse are equal:

$$A_{rs} = A_{sr} \qquad (2.246)$$

This principle follows from the quantum theory principle that transition probabilities are proportional to the square of the moduli of the matrix elements of the perturbations which cause the transitions. That is,

$$A_{rs} \propto |rH's|^2 = (rH's)^* (rH's) \qquad (2.247)$$

and

$$A_{sr} \propto |sH'r|^2 = (sH'r)^* (sH'r) \qquad (2.248)$$

where H' is the coupling interaction or perturbation in the Hamiltonian of the collision system. By the Hermitian property of the matrix elements of real operators,

$$(sHr)^* = (rH's) \qquad (2.249)$$

so, when spin-dependent forces are not involved, the principle is easily deduced.

An example of the application of the principle of detailed balancing is provided by the charge transfer process 10/01. As two atomic systems approach, the configuration of the wave functions may be regarded as changing to that of the pseudo-molecule, whilst conforming to the Wigner–Witmer rules[105] (see Appendix). The multiplicity of molecular states is greater than those of the separate atoms, and a pseudo-molecule can be formed in any one of them; furthermore, the collision products may be formed from any one of them provided the Wigner–Witmer rules are not violated. The transition appropriate to a charge transfer process is therefore a transition between two states of the pseudo-molecule; furthermore, it must be between two states of the same symmetry, or angular momentum. According to the principle of detailed balance, the ratio of cross-sections $\vec{\sigma}$ and $\overleftarrow{\sigma}$ for forward and reverse processes from state 1 to state 2 is in proportion to the ratio of statistical weights g_1 and g_2 of the pseudo-molecule states arising from states 1 and 2:

$$\frac{\vec{\sigma}}{\overleftarrow{\sigma}} = \frac{g_2}{g_1} \qquad (2.250)$$

The ratio applies to forward and reverse collision processes taking place at the same impact velocities. The available data are found to conform reasonably well to equation 2.250 (see Chapter 12).

The photoionization process $\phi 0/1e$ and its reverse, radiative recombination, can also be interpreted in terms of a detailed balance. The rate of photoionization per unit volume by radiation of frequency between v and $v+dv$ is:

$$B(v)\, n_0 u(v)\, dv$$

where B is the Einstein B coefficient (Chapter 9) and u is the energy density of the radiation. The rate of spontaneous radiative recombination is

$$n_+ v_e \sigma(E_e)\, n_e(E_e)\, dE_e$$

where σ is the radiative recombination cross-section and E_e is the electron energy.

In thermal equilibrium, these two rates are equal; therefore, for $dE_e = h\, dv$,

$$\sigma(E_e) = \frac{B(v)\, n_0 A(v)}{h n_e(E_e) n_+ v_e} \qquad (2.251)$$

The Rayleigh–Jeans law

$$A(v) \simeq \frac{8\pi h v^3}{c^3} \exp\left(-\frac{hv}{kT}\right) \qquad (2.252)$$

may be substituted in this expression to give

$$\sigma(E_e) \simeq \frac{B(v)\,(E_e + E_i)}{m_e c^3 E_e} \qquad (2.253)$$

where the ionization energy $E_i = hv - E_e$. It will be seen in Chapter 9 that $B(v)$ approaches a finite limit as $hv \to E_i$; thus the radiative recombination cross-section becomes very large as $E_e \to 0$, although it is found in experiments that, even with E_e as small as $kT/2$, the cross-section is small.

REFERENCES

1. MOTT, N. F. and MASSEY, H. S. W. *The Theory of Atomic Collisions*, 3rd edn, 1967. London; Oxford University Press.
2. SCHIFF, L. I. *Quantum Mechanics*, 2nd edn, 1955. New York; McGraw-Hill.
3. LANDAU, L. D. and LIFSCHITZ, E. M. *Course in Theoretical Physics, Vol. 3: Quantum Mechanics (Non-relativistic Theory)*, 1958. London; Pergamon.
4. BATES, D. R. *Atomic and Molecular Processes*, 1962. New York; Academic Press.
5. HIRSCHFELDER, J. O., CURTISS, C. F. and BIRD, R. B. *Molecular Theory of Gases and Liquids*, 1954. New York; Wiley.
6. GOLDSTEIN, H. *Classical Mechanics*, 1950. Reading, Mass.; Addison-Wesley.
7. DRUKHAREV, G. F. *The Theory of Electron–Atom Collisions*, 1965. New York; Academic Press.
8. GELTMAN, S. *Topics in Atomic Collision Physics*, 1969. New York; Academic Press.
9. HARTREE, D. R. *Proc. Camb. phil. Soc.* 24 (1928) 89, 111, 426.
10. FOCK, V. *Z. Phys.* 61 (1930) 126; 62 (1930) 795.
11. HERMAN, F. and SKILLMAN, S. *Atomic Structure Calculations*, 1963. Englewood Cliffs, New Jersey; Prentice-Hall.
12. BUSH, V. and CALDWELL, S. H. *Phys. Rev.* 38 (1931) 1898.
13. FERMI, E. *Z. Phys.* 48 (1928) 73.
14. ROBINSON, E. J. and GELTMAN, S. *Phys. Rev.* 153 (1967) 4.
15. MOTT, N. F. and MASSEY, H. S. W. *Theory of Atomic Collisions*, 3rd edn, p. 577, 1967. London; Oxford University Press.
16. BOHR, N. and DANSKE, K. *Vidensk, Selsk, Mat.-fys. Medd.* 18, No. 8 (1948).
17. CASIMIR, H. B. G. and POLDER, D. *Phys. Rev.* 73 (1938) 360.
18. PAULING, L. *The Nature of the Chemical Bond*, 1960. London; Oxford University Press.
19. FIRSOV, O. B. *J. exp. theor. Phys.* 33 (1957) 696; *Soviet Phys. JETP* 6 (1958) 534.
20. SMITH, F. T., MARCHI, R. P., ABERTH, W., LORENTS, D. C. and HEINZ, O. *Phys. Rev.* 161 (1967) 31.
21. SLATER, J. and KIRKWOOD, J. *Phys. Rev.* 37 (1931) 682; DEBYE, P. *Phys. Z.* 21 (1920) 178.
22. MAVROYANNIS C. and STEPHEN M. J. *Molec. Phys.* 5 (1962) 629; SALEM, L. *Molec. Phys.* 3 (1960) 441.
23. LONDON, F. *Z. Phys.* 63 (1930) 245.
24. EISENSCHITZ, R. and LONDON, F. *Z. Phys.* 60 (1930) 491.
25. MARGENAU, H. *J. chem. Phys.* 6 (1938) 897; HORNIG, J. F. and HIRSCHFELDER, J. O. *J. chem. Phys.* 20 (1952) 1812.
26. HIRSCHFELDER, J. O. and LÖWDIN, P. O. *Molec. Phys.* 2 (1959) 229; 9 (1965) 491.
27. DALGARNO, A. and KINGSTON, A. E. *Proc. phys. Soc., Lond.* 73 (1959) 455.
28. BENTLEY, P. G. *Nature, Lond.* 190 (1961) 432; GREENE, F. T. and MILNE, T. A. *J. chem. Phys.* 39 (1963) 3150.
29. LECKENBY, R. E. and ROBBINS, E. J. *Proc. Roy. Soc.* A291 (1966) 389; *Advanc. Phys.* 16 (1967) 739.
30. HEITLER, W. and LONDON, F. *Z. Phys.* 44 (1927) 455; COOLIDGE, A. S. and JAMES, H. M. *J. chem. Phys.* 6 (1938) 730.
31. MULLIKEN, R. S. *Rev. mod. Phys.* 4 (1932) 1; *Phys. Rev.* 41 (1932) 49.
32. DALGARNO, A. and McCARROLL, R. *Proc. phys. Soc., Lond.* A239 (1957) 412; A237 (1956) 383.
33. POWER, E. A. *Intermolecular Forces*, Ed. J. O. Hirschfelder, Chap. 4, 1967. New York; Interscience.
34. VERWEY, E. J. W., OVERBEEK, J. T. G. and VAN NESS, K. *Theory of Stability of Lyophilic Colloids*, 1948. Amsterdam; Elsevier.
    POWER, E. A. and ZIENAU, S. *Nuovo Cim.* 6 (1957) 7.

36. McLachlan, A. D. *Proc. Roy. Soc.* A271 (1963) 387; A274 (1963) 80; *Disc. Faraday Soc.* 40 (1965) 239.
37. Mason, E. A., Schamp, H. W. and Vanderslice, J. T. *Phys. Rev.* 112 (1958) 445; Weber, G. G. and Bernstein, R. B. *J. chem. Phys.* 42 (1965) 2166.
38. Nakshbandi, M. M. Thesis, University of London, 1967.
39. London, F. *J. phys. Chem.* 46 (1942) 305.
40. London, F. *J. phys. Chem.* 46 (1942) 305.
41. Pitzer, K. S. *Advanc. chem. Phys.* 2 (1959) 59.
42. Kolker, H. J. and Karplus, M. *J. chem. Phys.* 39 (1963) 2011.
43. Buckingham, R. A. and Corner, J. *Proc. Roy. Soc.* A189 (1947) 118.
44. Hirschfelder, J. O., Curtiss, C. F. and Bird, R. B. *Molecular Theory of Gases and Liquids*, 1954. New York; Wiley.
45. Morse, P. M. *Phys. Rev.* 34 (1929) 57.
46. Herzberg, G. *Molecular Spectra and Molecular Structure, Vol. 1: Spectra of Diatomic Molecules*, 2nd edn, 1951. Princeton, New Jersey; Van Nostrand.
47. Klein, O. *Z. Phys.* 76 (1932) 226.
48. Rydberg, R. *Z. Phys.* 73 (1932) 376; 80 (1933) 514.
49. Steele, D., Lippincott, E. R. and Vanderslice, J. T. *Rev. mod. Phys.* 34 (1962) 239.
50. Franck, J. *Trans. Faraday Soc.* 21 (1925) 536; Condon, E. U. *Phys. Rev.* 32 (1928) 858.
51. McDowell, M. R. C. Private communication
52. Coon, J. B., de Warnes, R. E. and Loyd, C. M. *J. molec. Spectrosc.* 8 (1962) 285; Henderson, J. R., Willett, R. A., Muramoto, M. and Richardson, D. C. Douglas Aircraft, Report SM 45807, 1964.
53. Glasstone, S., Laidler, K. J. and Eyring, H. *The Theory of Rate Processes*, 1941. New York; McGraw-Hill; Hirschfelder, J. O. Dissertation, Princeton University, 1935.
54. Sato, S. *J. chem. Phys.* 23 (1955) 592, 2465; Porter, R. N. and Karplus, M. *J. chem. phys.* 40 (1964) 1105.
55. Pauly, H. and Toennies, J. P. *Advanc. atom. molec. Phys.* 1 (1965) 195; Helbing, R. Report of the Institute of Applied Physics of the University of Bonn, 1963; Herschbach, D. R. University of California Radiation Laboratory Report 9379, 1960; Russek, A. *Phys. Rev.* 120 (1960) 1536; Morse, F. A. and Bernstein, R. B. *J. chem. Phys.* 37 (1962) 2019; Datz, S., Minturn, R. E. and Taylor, E. H. *J. chem. Phys.* 35 (1961) 1549.
56. Smith, F. T. *Phys. Rev.* 120 (1960) 1058; *Disc. Faraday Soc.* 33 (1962) 183.
57. Maier, W. B. Los Alamos Laboratory Report LA 3972 UC 34 Physics TID 4500, University of California, 1969.
58. Dalitz, R. H. *Phil. Mag.* 44 (1953) 1068.
59. Whitten, R. C. *J. math. Phys.* 4 (1963) 622.
60. Delves, L. M. *Nucl. Phys.* 9 (1958–59) 391.
61. Several articles in *Rev. mod. Phys.* 37, No. 3 (1965).
62. Cravath, A. M. *Phys. Rev.* 36 (1930) 248.
63. Hornbeck, J. A. and Wannier, G. H. *Phys. Rev.* 82 (1951) 458.
64. Lee, A. R. and Hasted, J. B. *Proc. phys. Soc., Lond.* 79 (1962) 1049.
65. Fedorenko, N. V. *Usp. fiz. Nauk* 68 (1959) 481.
66. Thomson, J. J. *Phil. Mag.* 23 (1912) 419.
67. Gryzinski, M. *Phys. Rev.* 115 (1959) 374.
68. Gryzinski, M. *Phys. Rev.* 138 (1965) A305, A322.
69. Ochkur, V. I. and Petrunkin, A. M. *Optika Spektrosk.* 14 (1963) 245; Burgess, A. *Proc. 3rd Int. Conf. Phys. electron. atom. Collisions*, Ed. M. R .C. McDowell, p. 237, 1964. Amsterdam; North Holland Publishing Company.
70. Stabler, R. C. *Phys. Rev.* 133 (1964) A1268.
71. Vriens, L. A. *Proc., phys. Soc. Lond.* 89 (1966) 13; *Phys. Rev.* 141 (1966) 88; *Phys. Lett.* 9 (1964) 295; 10 (1964) 170.
72. Vriens, L. A. *Proc. phys. Soc., Lond.* 90 (1967) 935.

73. BATES, D. R. and MAPLETON, R. A. *Proc. phys. Soc., Lond.* 87 (1966) 657.
74. ABRINES, R. and PERCIVAL, I. C. *Proc. phys. Soc., Lond.* 88 (1966) 861, 873.
75. BLAUTH, E. *Z. Phys.* 147 (1957) 288.
76. MASSEY, H. S. W. and BURHOP, E. H. S. *Electronic and Ionic Impact Phenomena*, 2nd edn, 1969. London; Oxford University Press.
77. PAULY, H. and TOENNIES, J. P. *Advanc. atom. molec. Phys.* 1 (1965) 195.
78. FORD, K. W., HILL, NAKANO and WHEELER, J. A. *Ann. Phys., Lpz.* 7 (1959) 239.
79. MOTT-SMITH, H. M. *Phys. Fluids* 3 (1960) 721.
80. GIOUMOUSIS, G. and STEVENSON, D. P. *J. chem. Phys.* 29 (1958) 294.
81. LANGEVIN, P. *Ann. Chim. (Phys.)* 5 (1905) 245.
82. HIRSCHFELDER, J. O., BIRD, R. B. and SPOTZ, E. L. *J. chem. Phys.* 16 (1948) 968.
83. MASON, E. A. *J. chem. Phys.* 22 (1954) 169.
84. VOGT, E. and WANNIER, G. H. *Phys. Rev.* 95 (1954) 1190.
85. NAKSHBANDI, M. M. Thesis, University of London, 1967.
86. HERZBERG, G. *Spectra of Diatomic Molecules*, 2nd edn, 1951. Princeton, New Jersey; Van Nostrand.
87. LINNETT, J. W. *Trans. Faraday Soc.* 36 (1940) 1123.
88. RUTHERFORD, S. L., CHADWICK, J. and ELLIS, E. *Radiation from Radioactive Substances*, p. 191, 1930. London; Cambridge University Press.
89. TEMPLE, H. *Proc. Roy. Soc.* A121 (1928) 673.
90. FAXEN, H. and HOLTSMARK, J. *Z. Phys.* 45 (1927) 307.
91. RAYLEIGH, LORD *Theory of Sound, Vol. 2*, 1896. New York and London; Macmillan.
92. MOTT, N. F. *Proc. Roy. Soc.* A133 (1931) 228.
93. MASSEY, H. S. W. and MOHR, C. B. O. *Proc. Roy. Soc.* A141 (1933) 434.
94. MOTT, N. F. and MASSEY, H. S. W. *The Theory of Atomic Collisions*, 3rd edn, p. 40, 1967. London; Oxford University Press.
95. SEATON, M. J. *C. R. Acad. Sci., Paris* 240 (1955) 1317; *Mon. Not. Roy. astr. Soc.* 118 (1958) 504.
96. BETHE, H. A. *Ann. Phys., Lpz.* 5 (1930) 325.
97. MILLER, W. F. and PLATZMAN, R. L. *Proc. phys. Soc., Lond.* A70 (1947) 299.
98. DEMKOV, YU. A. *Variational Principles in the Theory of Collisions*, 1963. London; Pergamon.
99. OPPENHEIMER, J. R. *Phys. Rev.* 32 (1928) 361.
100. OCHKUR, V. I. *Soviet Phys. JETP* 18 (1964) 503.
101. JEFFREYS, H. *Proc. Lond. math. Soc.* 23 (1924) 428.
102. MASSEY, H. S. W. and SMITH, R. A. *Proc. Roy. Soc.* A142 (1933) 142.
103. TEMKIN, A. and LAMKIN, J. C. *Phys. Rev.* 121 (1961) 788.
104. TOLMAN, R. C. *Statistical Mechanics*, 1927. New York; Chemical Catalogue Company.
105. WIGNER, E. P. and WITMER, E. E. *Z. Phys.* 51 (1928) 859.
106. KENNARD, E. H. *Kinetic Theory of Gases*, 1938. New York; McGraw-Hill.
107. MASSEY, H. S. W. *Negative Ions*, 2nd edn, 1950. London; Cambridge University Press.
108. WHITTAKER, E. T. *A Treatise on the Analytical Dynamics of Particles and Rigid Bodies*, 4th edn, 1937. London; Cambridge University Press.
109. HARTREE, D. R. *Rep. Progr. Phys.* 11 (1946) 113.
110. ZENER, C. *Phys. Rev.* 37 (1931) 556.
111. CH'EN, S. and TAKEO, M. *Rev. mod. Phys.* 29 (1957) 1.
112. HULBURT, H. M. and HIRSCHFELDER, J. O. *J. chem. Phys.* 9 (1941) 61.
113. EYRING, H., HIRSCHFELDER, J. O. and TAYLOR, H. S. *J. chem. Phys.* 4 (1936) 479.
114. ELIASON, M. A. and HIRSCHFELDER, J. O. *J. chem. Phys.* 30 (1959) 1426.
115. GERJUOY, E. *Rev. mod. Phys.* 33 (1961) 544.
116. GERJUOY, E. and SAXON, D. S. *Phys. Rev.* 94 (1954) 478.
117. PEKERIS, C. L. *Phys. Rev.* 112 (1958) 1649; 115 (1959) 1216.
118. ROSENBERG, L., SPRUCH, L. and O'MALLEY, T. F. *Phys. Rev.* 118 (1960) 184.
119. SPRUCH, L. *Phys. Rev.* 109 (1958) 2149.

120. SPRUCH, L. and KELLY, M. *Phys. Rev.* 109 (1958) 2144.
121. SPRUCH, L. and ROSENBERG, L. *Phys. Rev.* 116 (1959) 1034; 117 (1960) 143, 1095; 120 (1960) 474.
122. BATES, D. R. and GRIFFING, G. W. *Proc. phys. Soc., Lond.* A66 (1953) 961; A67 (1954) 663; A68 (1955) 90.
123. BORN, M. *Z. Phys.* 37 (1926) 863; 38 (1926) 803.
124. TAMM, I. G. *J. exp. theor. Phys.* 14 (1944) 21.
125. HULTHEN, L. *K. fysiogr. Sällsk. Lund Förh.* 14 (1944) 1.
126. ERSKINE, G. A. and MASSEY, H. S. W. *Proc. Roy. Soc.* A212 (1952) 521.
127. MASSEY, H. S. W. and MOISEIWITSCH, B. L. *Proc. Roy. Soc.* A205 (1951) 483.
128. MOISEIWITSCH, B. L. *Proc. Roy. Soc.* A219 (1953) 102.
129. BATES, D. R. *Proc. Roy. Soc.* A243 (1957) 15.
130. BATES, D. R., MASSEY, H. S. W. and STEWART, A. L. *Proc. Roy. Soc.* A216 (1953) 437
131. BATES, D. R. and McCARROLL, R. *Advanc. Phys.* 11 (1962) 1.

Chapter 3

THE EXPERIMENTAL METHODS EMPLOYED IN COLLISION PHYSICS

3.1 INTRODUCTION

The perspective of experimental atomic collision physics is of continual approach towards the ideal two-body collision experiment in which two particles, identified within the limits of uncertainty in all quantum numbers, in momentum and in energy, collide in such a way that the post-collision identification is also complete. Where photons are absorbed or emitted, their energies and polarization must be identified. The time schedule of the events must be accurately known.

Even a 'single-collision condition' experiment using velocity-analysed crossed beams of ground state atoms can only be said to fulfil this ideal in part. The limitations imposed by the means of production, control and detection of particles make possible only the deduction of incomplete information from present-day experiments; since the first edition of this book, several important advances in technique have brought the ideal experiment much closer to fulfilment.

In this edition, four major experimental situations are discussed critically before detailed consideration is given to the production, control and detection of electrons, photons and atomic species. These four situations are:

1. Crossed and merged beams of particles,
2. Beams and swarms passing through a gas,
3. Experiments using plasmas,
4. Time-dependent and flowing afterglows.

3.2 CROSSED AND MERGED BEAMS OF PARTICLES

There are three particular advantages of crossed-beam techniques over beam–gas techniques: (*i*) collision region confinement; (*ii*) extension of the range of available impact velocities; and (*iii*) extension of the range of types of available target particles. These are now considered individually.

Collision Region Confinement

In a crossed-beam experiment, the collision region is confined to a volume which can not only be reduced to minimal size, but also can be defined, by auxiliary beam flux density observations, with greater accuracy than is usually possible in beam-gas experiments. This is of greatest importance in differential cross-section measurement, but it can also be of value in total scattering and in experiments in which inelastic collision products are required to be collected over a defined collision path length. The difficulties in defining this path in beam–gas experiments are discussed in Section 3.3.

In crossed ribbon beam experiments, the collision region is confined to a parallelepiped when the beams are mutually perpendicular. The region is surrounded only by background gas molecules, and may be studied by observation through either wide or narrow aperture detectors, thus allowing flexibility in collecting inelastic collision products scattered over wide angles.

Extension of Range of Available Impact Velocities

The advantages of two-beam technique for extending the range of available impact velocities are likely to be considerable, but have as yet hardly been exploited. It is well-known that difficulties of beam control, arising mostly from space-charge and surface effects, at present prevent complicated ion-beam experiments from being carried out at energies much below 1 eV. One method of crossing this 'low-energy barrier' can be by 'merging' along the same laboratory path two beams of collidants accelerated to appropriate velocities v_1 and v_2, so that the velocity of impact $v_2 - v_1$ is suitably small. No doubt the concept of merging beams owes much to the familiar 'merging traffic' signs on American expressways. The idea is due to Cook and Ablow[1], among others, and experiments of this type were first reported by Trujillo, Neynaber and Rothe[2], and by Belyaev, Brezhnev and Erastov[3].

The centre-of-mass interaction energy

$$W = \tfrac{1}{2}\mu(v_2 - v_1)^2 \tag{3.1}$$

is the important quantity in collision studies; μ is the reduced mass of the system, and v_1 and v_2 are the laboratory velocities of the two beams. For collidants of equal mass m, moving along identical laboratory paths, it is readily shown that

$$W \simeq \frac{(\Delta E)^2}{8E} \tag{3.2}$$

where ΔE is the absolute value of the difference in beam energies E_1 and E_2 and is assumed to be much smaller than these energies, whose average value is E. The quantity $D = \Delta E / W$, an 'energy deamplification factor', is a useful parameter.

The effect of energy spread or energy fluctuation in the merging beams can be appreciated by considering a perturbation δE in ΔE, which gives rise to a perturbation δW in W:

$$\frac{\delta W}{W} \simeq \frac{2\delta E}{\Delta E} \tag{3.3}$$

An impact energy of order 0·25 eV, with 3 per cent energy resolution, can be achieved.

The mathematics of extracting a cross-section from a merged-beam measurement which includes appreciable transverse beam velocities is tedious. But where the transverse velocities are very small, being limited by the passage of both beams through a pair of collimating apertures of diameter d and separated by a distance l equal to the length of the interaction region, then it can be shown that

$$F = \frac{\overline{W} - W}{W} < 0.54 \left(\frac{Ed}{l\,\Delta E} \right)^2 \tag{3.4}$$

where $\overline{W}$ is the centre-of-mass interaction energy averaged over all possible intersection angles and all points of the interaction region. This figure F of merit of the experiment can readily be kept as low as 14 per cent. For an assumed form of cross-section function $\sigma(W)$, the appropriate correction to the measured cross-section can be calculated.

A merged-beam experiment consists of determining S, the number of collision product particles per second which have been generated in the interaction region. Then

$$\sigma = \frac{S}{\mathscr{I}} \left(\frac{E_1 E_2}{\mu W} \right)^{1/2} \tag{3.5}$$

where the overlap integral

$$\mathscr{I} = \int I_1 I_2 \, \mathrm{d}x \, \mathrm{d}y \, \mathrm{d}z \tag{3.6}$$

is performed over the volume of the interaction region. The symbols I_1 and I_2 denote beam fluxes.

Charge transfer collisions are studied by merging an ion beam with a fast molecular beam, after which the charged particles in the molecular beam are separated from those in the ion beam. The interaction region is surrounded by an electrostatic mesh screen maintained at potential $+\Delta E$, and field penetration through beam orifices is minimized by the use of superposed grids. After the interaction region, the beams pass through a 'demerging magnet'. If they have identical mass numbers, then the ion beam, which is

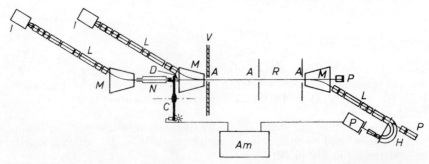

Fig. 3.1. Schematic diagram of merged-beam apparatus for ion–atom inelastic collisions[4]: The vacuum wall V divides the system into two separately pumped chambers; I, ion sources; L, lenses; M, magnets; N, neutralizing chamber; C, beam chopper with lamp and photocell for phase signal; D, deflector plates for removal of charged particles from beam of neutrals; A, collimating apertures; R, region of interaction; P, particle detectors; H, hemispherical electrostatic momentum analyser; Am, phase-sensitive amplifier

arranged to travel faster, is removed by retardation in an electric field. The technique is particularly suitable for the study of ion-molecule reactions, in which collision products are emitted at wide angles. In the laboratory frame, however, the large forward velocity vectors minimize the collection difficulty. The complicated set-up necessary to carry out such an experiment is illustrated schematically in *Figure 3.1*.

It is possible to study ion-neutral collisions in which the reactants are of the same species by using only one source instead of two. The beam is partially neutralized, and the remaining charged component is slightly retarded. The signal-noise ratio is superior to that in the two-source experiment[3].

Extension of Range of Available Target Particles

The use of crossed-beam (not merged-beam) technique extends the range of target particles available to experimenters. The technique has been exploited in experiments with atomic gases[5, 6] and also with collisions between electrons and ions[7]. It is important that the signal arising from beam-target collisions should be clearly distinguished from the noise arising from beam collisions with background gas. Thus the density of target particles in the collision region must be maximized and the background gas density minimized.

Suppose that a flux of projectiles I_2 is defined in a ribbon beam by a slot of width $2h_2$ and indeterminate length. The collimation is such that the entire flux passes through a flux I_1 of target particles, which is defined by a slot of width $2h_1$ and indeterminate length. The target particles undergo an inelastic collision, and a fractional flux I_2' of them is then separated and detected. It may readily be shown that the cross-section for the inelastic collision is

$$\sigma = \frac{I_1'}{I_1 I_2} \left(\frac{v_1 v_2}{v_1^2 + v_2^2} \right) F \tag{3.7}$$

where the factor

$$F = \frac{\displaystyle\int_{-h_1}^{+h_1} I_1(z)\,\mathrm{d}z \int_{-h_2}^{+h_2} I_2(z)\,\mathrm{d}z}{\displaystyle\int_{-h_1}^{+h_1} I_1(z) I_2(z)\,\mathrm{d}z} \tag{3.8}$$

takes into account the inhomogeneities in the flux densities across the slot width (distance variable z). It is usually possible to neglect inhomogeneities along the slot length. When either beam has uniform density, $F = 1$.

The beam profiles are studied by measurements with a movable slotted shutter[8], as in *Figure 3.2*. Profile monitoring can be carried out differentially (collector behind movable slit) or integrally (collector behind movable knife edge), and also in the reverse modes (movable wire in front of collector, and movable knife edge mounted on the opposite side of the beam). All four modes should give consistent results. For collision processes producing electromagnetic radiation, the beam profiles in all three dimensions must be known.

It has been shown that the elimination of errors arising from noise produced in collisions with background gas is the most serious problem in a crossed-beam experiment. Not only are collision products produced by impacts of one or both beams with background gas and surfaces, but also elastic collisions of a charged beam with background gas increase the dimensions of the beam and so its space charge exerts a different force on the second beam (if both carry a charge). A positive ion beam is deflected by the space charge of an electron beam towards its own central axis[8].

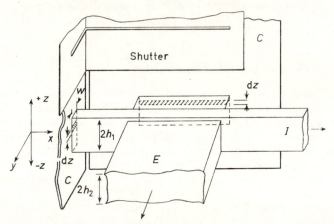

Fig. 3.2. Differential profile monitoring of crossed ion (I) and electron (E) beams, collimated by orifices in plates C

Certain types of measurement, such as that of the photon emission from excitation collisions of electrons with thermal atoms, can be made in a crossed-beam experiment with only minimal noise signal from background gas, since neither beam–gas collision is capable of producing photons of the frequency that is being measured, except where the thermal atom beam and background gas are identical.

In most types of crossed-beam experiment, the background gas noise signal is so serious that beam modulation must be employed to reduce it. One beam is interrupted a few hundred times a second, so that the signal arising from the experiment is no longer steady, but amplitude-modulated at this frequency; it is received by a narrow bandwidth phase-sensitive amplifier tuned to the appropriate frequency. The noise in a phase-sensitive detector is much reduced[9]. The device which interrupts the beam is arranged to provide a reference signal to control the phase of the amplifier, whose bandwidth can be as small as 0·03 Hz, provided that the chosen frequency is not too high; if it is chosen too low, noise can arise from a Fourier component of the time variation of background pressure. A suitable compromise is around 160 Hz. The use of phase control effects a reduction of a factor of about 4 in the noise signal.

The periodic interruption of a neutral beam can be achieved by the rotation of a slotted disc; a phase reference signal is obtained from a photocell illuminated through the slotted disc (not simply a disc mounted on the same axis). An equally useful interrupter can be constructed from a vibrating reed or tuning fork chopper, such as is used in spectrometers[10]; this provides sinusoidal rather than square-wave modulation, and can be arranged to provide its own phase reference signal.

Since the interruption of a neutral beam might also change the background gas pressure in the collision region, it is sometimes argued that it would be better practice instead to deflect the beam a small distance from the collision region by means of a specially designed collimator.

Charged beams can of course be interrupted by the application of pulses of potential to a suitable electrode.

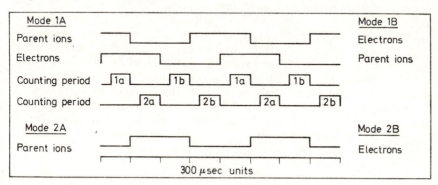

Fig. 3.3. Pulse design suitable for handling two crossed beams, both capable of producing noise in collision with background gas

When both beams are capable of producing noise signal by collisions with background gas, each can be modulated at a different frequency, and the detector amplifier tuned to the heterodyne frequency.

In modern experiments, it is often necessary to gain the greatest sensitivity by counting single particles rather than amplifying currents of charged particles. The counterpart of the tuned phase-sensitive amplifier in this case is the biasing of the signal amplifiers by square pulses. The most complicated situation, in which both beams are capable of producing noise signal in collision with background gas, may be handled[11] by a pulse design shown in *Figure 3.3*. Referring to mode 1A in this figure: scaler 1 counts signal and both backgrounds simultaneously during period 1a, and scaler 2 counts each background separately during periods 2a and 2b. If the mean values of the pulsed currents are $\bar{I}_1$ and $\bar{I}_2$ and the measured count rates are R_1 and R_2, then the collision rate

$$\mathcal{R} = \frac{6(R_1 - R_2)}{4\bar{I}_1\bar{I}_2\mathcal{E}} \tag{3.9}$$

where $\mathcal{E}$ is the detector efficiency. Systematic errors are introduced if the beam pulses are not flat-topped, because scaler 1 counts during the first half of the pulse applied to beam 1 and during the second half of the pulse applied to beam 2; scaler 2 counts vice versa. These errors are eliminated by interchanging at a slow recurrence the waveforms applied to beam 1 and beam 2 (mode 1B). Inequalities in the scalers or gating periods are averaged out by alternating at a slow recurrence the count rate alternately in modes

1 and 2. In mode 2, a suitable delay is applied in order that scaler 2 counts the signal and scaler 1 the background. Thus the complete automatic sequence of the experiment is: mode 1A, mode 1B, mode 2A and mode 2B consecutively.

3.3 BEAMS AND SWARMS PASSING THROUGH A GAS

For setting up a beam-gas experiment under single collision conditions, a collision chamber must be constructed within a vacuum system sufficiently large and efficient to ensure that beam collisions with background gas outside the collision chamber contribute only negligibly to the experiment. The chamber must include an entrance and an exit orifice for the beam; the gas pressure within the chamber must be controllable and be maintained constant at a measured value, whilst the gas temperature is also measurable and constant. The unscattered beam path within the chamber, whether straight or curved by electric or magnetic fields, must be measurable. A suitable electrode or orifice system may be arranged within the collision chamber in order to collect or extract products of inelastic collisions, and sometimes the collection of radiation by an optical system is necessary. Thus each quantity

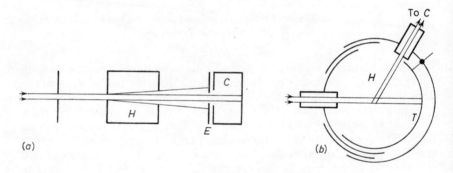

(a)   (b)

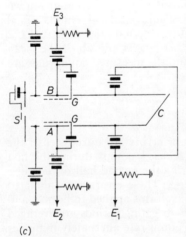

(c)

Fig. 3.4. Collision chambers: (a) collision chamber H suitable for measurement at collector C of fractional absorption of ion beam by scattering through angles defined by the electrode E; (b) collision chamber H suitable for measurement at collector C of differential scattering over a wide range of polar scattering angles, the unscattered beam being collected at electrode T; (c) condenser electrode system for collection of charged collision products (S, collision chamber orifice with electric field to trap secondary electrons; C, primary beam collector with electric field to trap secondary electrons; A, electron collector; B, positive ion collector; E_1, to primary beam electrometer; E_2 and E_3, to electrometers; G, secondary electron suppression grids; guard electrodes are included for the minimization of fringing electric field)

in equation 1.25 is controlled and measured accurately, and the internal consistency of the equation is verified.

Experiments in which electron or ion swarms drift through a gas are less refined than beam experiments, but allow the study of inelastic processes at impact energies lower than those which can be obtained with beam technique. The drift regime is considered in detail in Chapter 4 and Chapter 10.

Certain general problems of beam collision chamber design will now be discussed, followed by questions of gas pressure and temperature. *Figure 3.4* illustrates a number of commonly encountered collision chamber situations.

Figure 3.4a shows the simplest type of electrodeless chamber, suitable for measuring total collision cross-sections, either in absorption or by collecting the flux of particles scattered to the walls. In the absence of elastic scattering, the exit orifice of the chamber must be sufficiently large to allow the entire collimated beam to pass. A total collision cross-section measurement will not be of the true total collision cross-section, but of the cross-section for scattering into all polar angles greater than a certain value θ_0 (see Chapter 10). Even θ_0 is not single-valued, but is an appropriate integration over all points in the collision path. If θ_0 is defined not by the exit orifice but by the electrode E, then the integration is over a smaller range of θ. The absorbed current must be measured only as a difference and not as a current to the collision chamber.

The collision path length in a chamber is not simply the distance between the (infinitely thin) orifices. It would only be so if the gas pressure were to remain constant between the two orifices and if the gas pressure outside were to fall suddenly to zero. But the effusion of gas results in appreciable numbers of collisions outside the orifices. Although the effective collision path can readily be calculated on a beam axis, the computations for a cylindrical beam of finite cross-sectional area are more tedious, and the simplest solution is to experiment with a collision chamber presenting two interchangeable path lengths and subtract the end correction. A further error arises from velocity direction anomalies in the orifice regions, which may be avoided by making both orifices identical[12].

Figure 3.4b illustrates a basic differential scattering situation, designed so that only particles scattered from a small collision region are collected after collimation. By means of a system of cylinders sliding within each other, and also by means of the tilted sliding seal, a wide range of polar scattering angles θ can be investigated[13]. For a narrower range of θ, a bellows construction is often sufficient. A bellows construction capable of covering $-20°$ to $+90°$ has been described[14].

For a rectangular beam monitored by a rectangular slit, the differential cross-section is related to the incident beam flux I_0 and detected flux $I(\theta)$ by the equation

$$\frac{\mathrm{d}\sigma}{\mathrm{d}\Omega} = \frac{I(\theta)}{I_0 n_0 \int\limits_{\Delta x} \mathrm{d}\Omega \, \mathrm{d}x} \tag{3.10}$$

where n_0 is the collision chamber gas density; the integral is the geometrical factor calculated[15] by integrating over Δx (the total beam path length contributing to the scattering signal) the scattering solid angle $\mathrm{d}\Omega$ seen by element dx of the primary beam path. In first order approximation

$$\int_{\Delta x} d\Omega \; dx \simeq \frac{abh}{d(d+l)\sin\theta} \qquad (3.11)$$

where h is the length of the final detector slit; a and b are the widths of the entrance and exit slits; d and l are the distances from the first exit slit to the collision region and the second exit slit respectively. For scattering through very small angles, the effective path length is smaller owing to the finite length of the collision chamber. Discussions have been given of the monitoring of a ribbon-shaped beam by a movable circular orifice system[16], and of a cylindrical beam by the same system[17]. Several other discussions have been published[18-22] (see also Chapter 10). The special problem of measuring particles scattered through 180° has also been treated[23, 24].

Figure 3.4c illustrates the most commonly employed 'condenser plate' electrode system for the collection of charged products of inelastic ion–atom collisions. The momentum transfer in such collisions is usually relatively small, with a few exceptions such as the ion–atom interchange process. Therefore the charged particles formed from the gas molecules may readily be separated from those in the 'primary' beam by means of electric, and sometimes magnetic, fields. In a uniform electric field transverse to the projectile beam, the collision products follow parabolic paths to the appropriate collecting plate; the use of guard electrodes to produce uniform fields ensures that the collision path length is exactly equal to the length of the collecting electrode. There should be no variation of collected current with electric field when the conditions of collection are properly defined, but such 'saturation' of the characteristic is not found for very small and very large fields, particularly when there is more than one species of particle being created in the collision. The variety of causes contributing to these conditions will not be discussed at this stage.

In order to assist saturation and the complete collection of the products, uniform transverse magnetic fields are sometimes applied. The collision products then travel in helical paths with axes transverse to the projectile beam; it is necessary to be certain that the radii of the helices are small compared with the collision path length. Thus the collision products with greatest momenta require the largest fields in order that saturation conditions be achieved. The angular distribution of the momentum vectors of collision products also influences the collection efficiency.

A magnetic field is sometimes applied parallel to the projectile beam, the strength of the field being sufficient to limit electron path radii to values which are small compared with electrode dimensions, whilst allowing ion path radii to remain comparable to or larger than these dimensions. The separation of ions and electrons so achieved is particularly valuable when several different types of inelastic collision occur.

When it is necessary not only to collect all the slow positive ions from a collision region, but also to mass-analyse them, the techniques must be further refined. It is possible to collect ions of a wide range of momenta on a metal plate, but it is much more difficult to cause all of these ions to pass through a slit or to enter a mass-spectrometer. The paths of slow ions are seriously affected by surface charges occurring at insulating films which can form on metal electrodes. One important source of such films is the oil from diffusion pumps; the popular practice of painting metal electrodes with colloidal graphite probably reduces the films by physical absorption. When ion momenta are sufficiently small, as is the case at thermal energies, it is possible to observe saturation conditions for ions collected behind a mass-spectrometer slit which lies parallel to the projectile beam. For collision products whose energies are tens of electron volts, it is possible to reproduce these conditions by elevating the transverse field and maintaining the resolution of the mass-spectrometer by means of a suitably large accelerating potential. But the projectile beam is only unaffected if its energy is sufficiently high; at lower energies, more sophisticated extraction techniques must be employed. One such technique is the use of a strong focusing quadrupole lens to form a stigmatic image of the collision region upon the entrance slit of the mass-spectrometer. The transverse deflexion of the primary beam remains the central problem of such collection systems. Pulsed beams and fields are capable of providing a solution.

Among the problems encountered in controlling and measuring the particle density in a collision chamber are:

1. Minimization of errors arising from the dynamic equilibrium which is set up in most vacuum systems;
2. Minimization of errors in the absolute measurement of pressure[25];
3. Minimization of background gas density.

The problems of maintaining a dynamic equilibrium of pure gas in the collision chamber have not always received the attention they deserve. One must consider the minimization of background gas density, purification techniques, the control of flow input, the dependence of the equilibrium density on the pressure gauge reading, and sometimes the problems of pre-mixed gases.

The minimization of background gas density becomes of greatest importance when the background atom or molecule cross-sections greatly exceed the atom or molecule cross-section under study. In this case one must use ultra-high vacuum techniques, a stainless steel (or aluminium alloy) vacuum system, metal gaskets, high-quality diffusion, getter-ion or sorption pumps, bakeable liquid nitrogen traps (preferably two in series) and optional bake-out. A collision chamber is normally pumped through a valve of suitable admittance which is closed when the gas is passed in for the experiment to take place; its pumping during the experiment is through one or more orifices, and the background gas pressure is enhanced by gas or vapour coming from the collision chamber walls. This increase in background gas pressure is avoided by attention to cleanliness, leakage and bake-out. A monitor can be kept upon the background partial pressure by means of a small mass-spectrometer. Certain vacuum materials are capable of emitting impurities—for example, pyrex glass emits chlorine ions when bombarded in a discharge, whereas kodial glass does not.

Not only must the background gas be minimized in the collision chamber, but also the path traversed by the beam before it reaches the collision chamber must be small compared with the mean free path. At a pressure p torr, the gas kinetic mean free path l_f is approximately $10^{-2}/p$ cm.

for nitrogen at 25°C. When the collision chamber, at pressure p_c, is bounded by infinitely thin orifices of total area A, the pumping speed S litres sec^{-1} necessary to maintain a pressure p_s in the vacuum system is approximately

$$S \simeq \frac{10Ap_c}{p_s} \tag{3.12}$$

The factor 10 is appropriate to air, and is inversely proportional to the square root of the atomic weight; thus it is approximately 30 for helium.

Although the pumping speeds of diffusion pumps should also be inversely proportional to the square root of the atomic weight, exceptionally small pumping speeds are encountered for light particles, such as hydrogen and helium. These small pumping speeds are due to the ability of the gases to backstream through the diffusion pump. This can be countered by the use of exceptionally low backing pump pressure, and by operating the diffusion pump at an elevated temperature. In complicated experiments with these gases, each diffusion pump should be provided with its own backing pump.

Gas and vapour purification techniques are discussed in textbooks of practical chemistry. The important gases in atomic collision physics (rare gases, atmospheric gases, and hydrogen) are usually commercially available with high purity. Hydrogen can be purified by leakage through heated palladium, and helium through a sufficiently thin heated quartz tube. Hydrogen can also be introduced or replenished by heating titanium hydride, and the production of atomic hydrogen by heating lithium hydride has been reported. A purification technique especially suitable for helium is cataphoresis. This consists of passing the gas through an electric discharge. Impurities such as hydrocarbons have lower ionization potentials than rare gases, and so make up the large majority of ions formed; these are extracted by collection at a strongly negative electrode, whilst fewer helium ions are lost—in any case they chemisorb less easily.

In a collision chamber at temperature T_c °K, the gas density n is related to the pressure p_c:

$$n = 3 \cdot 535 \times 10^{16} p_c \left(\frac{273}{T_c} \right) \tag{3.13}$$

It is not always possible to situate a pressure gauge within the collision chamber, but at least it must be connected by a short tube of large diameter. When the collision chamber temperature T_c differs from the gauge temperature T_g,

$$\frac{p_c}{p_g} = \left(\frac{T_c}{T_g} \right)^{1/2} \tag{3.14}$$

for gauge pressure p_g.

Gas is normally leaked by viscous flow into the collision chamber through a needle valve, which must be capable of sustaining a constant flow rate throughout the experiment; both bakeable and non-bakeable types are commercially available. The gas leaks out of the collision chamber by molecular flow through one or more orifices. The molecular flow rate F_{m1} of a gas of molecular weight m_1 is proportional to $m_1^{1/2}$, but for the viscous flow rate F_{v1} there is no such proportionality. The molecular and viscous flow rates are proportional to the partial pressure p_1. Consider a gas mixture of two components of partial pressures p_1 and p_2, which is leaking by viscous flow into a collision chamber, where the partial pressures are p_1', p_2'; out of this they leak by molecular flow. Equating the input and output flow rates, $F_{v1} = F_{m1}$ and $F_{v2} = F_{m2}$, so

$$\frac{p_1}{p_2} = \frac{p_1'}{p_2'} \left(\frac{m_1}{m_2} \right)^{1/2} \tag{3.15}$$

Thus the unexpected conclusion is reached that a pre-mixed gas, such as air at atmospheric pressure, does not maintain its proportions in a collision chamber. Although most experiments are performed with pure gases, pre-mixed gases are encountered occasionally (especially in drift tube and afterglow studies), and must be treated with caution for this reason. Mass-spectrometer experiments can readily be made in support of equation 3.15; furthermore, a mass-spectrometer fed by a truly molecular flow leak will reflect partial pressures in inverse proportion to the square root of the molecular weight ratios.

When it is required to maintain inside a collision chamber a gas mixture of known partial pressures, it is common practice to leak each gas in separately, and measure its partial pressure in the collision chamber. This places great reliance on the time-invariance of needle-valve flow rates, and also on the conditions being such that the addition of one gas does not change the outflow of another from molecular to viscous. For experiments on gas mixtures, it is advisable to monitor the collision chamber partial pressures through an orifice by means of a fast-pumped mass-spectrometer. This also helps to check on certain of the pure gas problems discussed below.

Mass-spectrometric monitoring is a technique much in demand in chemical kinetics, and may be applied with advantage to collision chamber pressure control. The assumption is made that the flow from the sampling orifice into the mass-spectrometer is molecular, and calibrations of the mass spectrometer for different pure gases are made against pressure gauges under conditions when the partial pressure is approximately equal to the total pressure. For orifices about 0·02 mm in diameter, the molecular flow assumption does not hold at pressures much greater than 10 torr, and the resulting complications have been studied by Barber, Farren and Linnett[26]. It was found in their experiments that the Knudsen condition is satisfied at higher pressures than would at first sight be expected. However, these pressures are higher than those encountered even in swarm experiments.

Where a pure gas is under study, the measured values of pressure and temperature must correspond with the actual values in the collision chamber itself. Temperature gradients, such as those arising from liquid nitrogen traps in the feed line, are to be avoided; the gas molecules should make sufficient collisions in the feed line and collision chamber to equilibrate to laboratory temperature. Pressure gradients can develop from temperature gradients as a result of the transpiration effect[27]. If two volumes of gas are connected by a thermally insulating orifice, the two pressures p_1 and p_2 are equal provided that the orifice diameter is much greater than the mean free path. However, when it is much smaller than the mean free path, an equation similar to

3.14 can be used to relate the two temperatures to the two pressures:

$$\frac{p_1}{p_2} = \left(\frac{T_1}{T_2}\right)^{1/2} \tag{3.16}$$

It follows that pressure gradients will exist in a tube across a diameter gradient d_1–d_2–d_1 at the same place as a temperature gradient T_1–T_2–T_1. This is the situation encountered in a cold trap, unless the tube diameter is carefully equalized throughout the trap. The usual type of trap consists of an inner tube of internal diameter d_{ii} and outer diameter d_{oi}, sealed in an outer tube of internal diameter d_{io}. The condition for zero transpiration pressure gradient across the trap is found to be:

$$d_{io} = d_{oi} + d_{ii} \tag{3.17}$$

But it is safer to construct cold traps simply as U-tubes immersed in wide-necked Dewar flasks. Thermal effects can also be introduced by ionization gauges; in order to avoid them, miniature gauges have been designed[28].

For measurement of total pressure in a collision chamber, a gauge must be attached to the chamber. Errors can arise if the gauge is attached in the feed line, with the pumping resistance of the line between gauge and chamber not negligible compared with the pumping resistance of the collision chamber orifice.

Not only is absolute collision chamber pressure measurement required, but also a continuous record during the experiment. Since the absolute manometers do not always provide a continuous record, a relative instrument is necessary. The ionization gauge is often employed, but suffers from possessing both pumping and heating capabilities. It can also operate in more than one mode[29]. Where the pressures are in excess of 10^{-4} torr, the Pirani gauge is a stable and accurate monitor. The capacitance manometer is extremely reliable when operated as a comparison between two equal pressures.

The choice of absolute manometer is between the McLeod gauge, the Knudsen gauge, the absolute capacitance manometer and the expansion technique combined with oil manometer. The most frequently used is the McLeod gauge, which was considered reliable until the discovery by Ishii and Nakayama[30] that, when topped by a cold trap, the gauge acts as a molecular diffusion pump, introducing errors which are dependent upon the temperature of the mercury and on the radius r of the connecting tube. Application of Gaede's theory of the molecular diffusion pump[31] suggests that the real pressure p in the apparatus differs from the experimentally measured pressure p_c as follows:

$$\ln\left(\frac{p}{p_c}\right) = 0{\cdot}0905 r p_v \frac{T^{1/2}}{D_0} \tag{3.18}$$

where p_v is the vapour pressure (in torr) of the mercury, which of course is temperature-dependent[32]:

$$\log_{10}\frac{p_v}{10^3} = 11{\cdot}0372 - \frac{3204}{T} \tag{3.19}$$

D_0 is the diffusion coefficient for the gas molecules in mercury vapour, which is largest for lightest atoms, and may readily be calculated (Chapter 10). Corrections calculated from equation 3.15 can be as large as 50 per cent for xenon, but might be only a few per cent for helium. The theory assumes that the tube joining the gauge to the apparatus is more than $100r$ in length.

Experimental studies[33-36] have largely substantiated the theory[37]. Where it is necessary to estimate corrections to previous experiments in which no account was taken of the 'Ishii effect', equation 3.18 is the best that is available. But in modern work, the error should be avoided by refrigerating the mercury of the McLeod gauge, by restricting the diameter of tube directly above the mercury reservoir, or by abandoning the gauge in favour of another absolute manometer. It is possible for droplets of liquid mercury to become lodged in the bulb or tubing of the gauge, and to emit vapour which counteracts the good effects of restricting the tube diameter above the mercury reservoir, or of refrigerating the mercury reservoir alone. Restriction of the tube above the gauge results in time constants which are inconveniently long for the achievement of equilibrium. Refrigeration (0°C) of the entire gauge in a chamber, with precautions against moisture condensation, is satisfactory[35]. Low vapour pressure oils have been used[38], but they absorb air.

Probably the first substantiated examples of incorrect atomic data arising from McLeod gauge errors are the van der Waals' coefficients discussed in Chapter 10. In using McLeod gauges, it is also necessary to take account of capillary meniscus corrections, and of deviations from Boyle's law. Other references on the McLeod gauge should be consulted[39, 40].

Certain other solutions of the absolute pressure measurement problem are possible. It is unlikely that the Knudsen radiometer gauge will ever prove to be a satisfactory absolute instrument in the range 10^{-3}–10^{-6} torr, because: (*i*) its overall accuracy amounts to little better than ± 10 per cent; (*ii*) it is difficult to construct and operate; (*iii*) the repelling force fails to comply with Knudsen's equation owing to inadequacy of the assumption that the distance between the heater and the movable vane is vanishingly small compared with the mean free path[41]; and (*iv*) the anomalous accommodation coefficient of hydrogen affects the result. Nevertheless, there are laboratories utilizing Knudsen gauges at 10^{-4} torr and below, avoiding the use of hydrogen; gauges constructed of silica exhibit the smallest differences of calibration from gas to gas.

It is not generally realized that the oil manometer operated under carefully controlled conditions is sensitive to $\sim 10^{-3}$ torr. It is necessary to view from underneath the breaking of the surface of the oil by a micrometer point. Since this accuracy is not always sufficient for absolute pressure measurement problems, the expansion technique may have some value.

An expansion technique of low-pressure measurement has been proposed and investigated in several laboratories[42, 43]. It is based upon the fact that the conductance C of an orifice can be calculated from first principles. When two chambers are connected by such an orifice and gas flows through the system at a rate R, then

$$p_1 - p_2 = \frac{R}{C} \qquad (3.20)$$

Since the McLeod gauge error does not apply in the region of diffusive flow, pressures in the 0·1 torr range can be measured absolutely without difficulty; alternatively an oil manometer may be employed. The conductance of the orifice is arranged to be such that p_1/p_2 is large, so that p_1 is $\sim 0·1$ torr, and p_2 can be calculated from equation 3.20. The conductance C is calculated to 0·1 per cent accuracy, including the Clausing correction for orifice thickness and a further correction for diameter of the receiving chamber. The rate R is measured with the aid of a liquid mercury device to measure pressure reduction in a gas reservoir over a long period of time. Use of a cascade of orifices enables absolute pressure measurement to be made in the ultra-high vacuum region, in order to check the limits of linearity of other gauges.

However, the serious competitor of the McLeod gauge is the absolute capacitance manometer. Metal and glass versions of this 'pressure transducer', used as a differential instrument, were described in the first edition of this book, but since then the all-metal instrument has shown its superiority on account of greatly increased sensitivity and its capability of absolute measurement. A metal membrane is supported on sensitive metal bellows, at a short distance from a fixed plate from which it is insulated, so as to form a capacitor whose magnitude can be measured by an electrical bridge or a resonant circuit. One side of the membrane is at high vacuum, the other exposed to the vacuum system under study, so that the capacitance changes with pressure. The capacitance can be restored to its original value by means of a thrust that can be measured absolutely; electrostatic force and the application of weights have been used. The complications in designing these instruments make it advisable to accept the use of commercially available versions, the performance of which can be as good as $\pm 0·5$ per cent at 10^{-3} torr. The instrument is also of great value for comparison of two pressures.

Certain old-fashioned but beautiful methods of measuring membrane strain are still valuable for comparative pressure measurement in the range 0·1–10 torr. It is possible to silver a glass membrane, which becomes a concave mirror of varying focal length. The deflexion of a spiral glass capillary and of a hollow glass spoon form the basis of well-established but comparatively insensitive devices.

Piezo-electric or resistive transducers are often mounted in the walls of shock-tubes to measure pressure transients.

In general, the accuracy of absolute pressure measurement increases with increasing pressure; therefore there is much to be said for using relative pressure gauges with linear scales. The Pirani gauge is suitable down to 10^{-4} torr, and, despite dual mode operation[29], the hot-cathode ionization gauge is satisfactory down to much lower pressures; the background current arising from soft X-rays is reduced to negligible proportions by using the Bayard–Alpert version, and an extension of the upper limit of pressure can be achieved by using a miniaturized version[44]. Nevertheless the hot-cathode ionization gauge operates both as an ion pump and as a pyrolyser. A typical pumping speed is 0·4 litre sec^{-1}; speeds are fastest for CO and H_2, but they can be much reduced if the grid is heated to a suitable temperature[45]. There is also some pumping in the cold-cathode Philips or Penning gauge. With suitable thermal screening and stabilization, accuracy of a few parts per hundred can be achieved at 10^{-4} torr.

There is a need for a very reliable relative gauge designed solely to hold its calibration without drift for long periods. Probably the most reliable ionization gauge would be one based on the apparatus of Tate and Smith described in Chapter 6, and using the weakest electron beams possible in order to minimize pumping and soft X-rays; Klopfer[46] has gone some distance towards this end.

The Bayard–Alpert gauge is linear down to 10^{-10} torr; the soft X-ray background can be partly removed by modulating the field which accelerates the electrons, and observing the modulated ion signal. The lower limit of the gauge is then reduced to 10^{-11} torr. Another gauge, due to Schumann[47], functions as a pentode in which electrons formed by the ionizing collisions are suppressed. A cold-cathode Penning (Philips) gauge, when stimulated into operation by a filament, maintains linearity down to 10^{-12} torr. Expansion techniques are used for these linearity investigations.

In the first edition of this book, certain relevant problems of vacuum technology, such as getter-ion pumping, and gas recovery by recirculation, were briefly considered. Getter-ion pumps of various designs are now commercially available (speeds up to 100 litre sec^{-1}) and need not be made in the laboratory. Efficiencies of pumping rare gases can in a suitable instrument be as high as 30 per cent of the efficiency of pumping active gases.

Background gas can be temporarily reduced in an experimental chamber by evaporating titanium onto a cold surface which then functions (until saturated) as a getter, without the need for ionization. Such a system was until recently difficult to operate owing to the corrosive properties of pure liquid titanium; the titanium could not simply be wound as wire onto a filament. However, the availability of titanium–molybdenum alloy in the form of wire has made this technique relatively simple. The operating time of the getter film is only conveniently long (hours) when the background pressure is $\sim 10^{-8}$ torr.

General references to vacuum technique are of value to the atomic collision physicist[48-50].

Gas recovery by recirculation[51] is of importance in experiments with scarce gases (such as ^3He, which is used for producing the important stripped nucleus He^{2+}); only with the ^3He isotope is this ion readily distinguishable from background H_2^+ in mass analysis. Recirculation of gas pumped by a diffusion pump is only satisfactory if an attempt is made to purify the gas, or at least to freeze out organic vapours introduced in oil diffusion pumps or from stopcock grease. It is necessary to compress and store the recovered gas, either by use of an adjustable column of mercury or by means of a Toepler or diffusion pump or both. Gases which are not easily condensed may be removed as follows: oxygen and nitrogen by a zirconium filament maintained at 1300°C; and hydrogen by cupric oxide at 550°C, the resulting water being frozen out. A recovery circuit was illustrated in the first edition of this book.

Collision chamber experiments on condensable vapours are sometimes necessary, and are conducted in the following way. The entire collision chamber is maintained at a suitable known temperature; it contains sufficient liquid or solid to present a surface area large enough to maintain the saturated vapour pressure p within the collision chamber. Several square centimetres are absolutely necessary. It is sometimes possible to make use of published saturated vapour pressure data[52], so that the number density is deduced from the measured temperature only. When the published vapour pressure is only available at a different temperature, one may use the Clausius–Clapeyron equation, assuming that it is applicable to saturated vapour:

$$\frac{dp}{dT} = \frac{\Delta H}{T\,\Delta V} \tag{3.21}$$

where ΔH is the latent heat of vaporization. Neglecting the condensed phase volume, one may write

$$\frac{d(\ln p)}{dT} = \frac{\Delta H}{N_0 k T^2} \tag{3.22}$$

where N_0 is Avogadro's number. ΔH is temperature-variant and may be expressed as a power series in T:

$$\Delta H = \Delta H_0 + aT + bT^2 + \ldots \tag{3.23}$$

yielding

$$\ln p = -\frac{\Delta H_0}{N_0 k T} + \frac{a}{N_0 k}\ln T + \frac{bT}{N_0 k} + \ldots \tag{3.24}$$

Tables are usually expressed in the form

$$\log_{10} p = A + \frac{B}{T} + CT + D\log_{10} T \tag{3.25}$$

However, condensable vapours are much better suited to crossed-beam than to beam–gas collision chamber experiments.

3.4 EXPERIMENTS USING PLASMAS

Certain collision investigations can be conducted with modulated or even steady state plasmas; in general, these investigations are not so refined or so flexible as the crossed-beam or beam–gas experiments, but they may yield cross-sections for processes not easily accessible by the more refined methods. Therefore plasma diagnostics are sometimes necessary in atomic collision studies, and are discussed in this section. The principal difficulty of experiments with plasma is the presence of noise and instabilities, now fairly well understood.

One class of collision investigation in plasma involves the modulation of the energy or density of one species (usually the electrons), and the study of the resulting modulation in the density of another species, or in the radiation. These investigations include ionization and excitation coefficient measurements, and momentum transfer cross-section deduction from cross-modulation (Chapter 4). In general, such modulation methods are less refined than the afterglow studies considered in the next section; they will not be considered in detail.

An interesting class of experiments is that in which fast neutral particles are directed through the plasma, and their absorption, or their fast collision products, or the electromagnetic radiation, is observed. There are conditions

Table **3.1.** PLASMA DIAGNOSTIC TECHNIQUES

	Density	*Temperature*

Electrons	Single probe
	Double probe
	Pulsed and thermionic
	  probes
	Microwave cavity
	Microwave interferometry
	
	Electrostatic resonances
	Radiofrequency probe
	
	Stark broadening of line
	  emission
	Electron cyclotron resonance
		Single probe
		Double probe
		Thermionic probes
		
		Microwave emission
		Microwave collision-
		  frequency functions
		Ionization rate in gas
		Ionization rate in slow
		  neutral probe
		Thomson scattering of laser
		  radiation
		Relative intensities of
		  emitted lines

Ions	Orifice extraction, mass-
	  spectrometer and detector
	Fast neutral beam probe
	Ion cyclotron resonance
		Doppler broadening of
		  emitted lines

Neutrals	Acoustic absorption
	Alphatron
	Ion trap gauge
	Orifice extraction, ionizer
	  and mass-spectrometer
	Acoustic emission
	Electron spin resonance
		Wall temperatures
		Doppler broadening
		Acoustic emission
		Line reversal
		Orifice extraction with ion-
		  izer and momentum
		  analyser

Excited states	Radiation absorption Radiation emission	Doppler broadening Line reversal

when the production of such radiation from the neutral gas is either negligible or can be minimized by modulation of the plasma or the fast-particle beam and use of phase-sensitive detection. The observation of particles effusing from a plasma, such as ions monitored by means of a mass-spectrometer, also provides valuable information.

The limitations of plasma experiments arise from the fact that very few plasmas are in thermodynamic equilibrium. Therefore the energy distributions of colliding particles are not determined by statistical mechanics, but must be diagnosed by experiment. However, there are systems in thermodynamic equilibrium available for collision experiments. Hot metal surfaces, useful for studying electron attachment, are discussed in Chapter 8. Plasmas produced and heated by shock waves are in thermodynamic equilibrium, and are used for oscillator strength determination (Chapter 9), and also for ionization processes between atoms and for electron–ion recombination (Chapter 7).

There have been considerable advances in the plasma diagnostics of both weak and fully ionized plasmas during the last decade, and several detailed discussions are available[53]. The present outline concentrates mainly on those techniques which are most relevant to collision experiments and collision physics in general. A list of density and temperature diagnostic techniques is given in Table 3.1, and the more important of these are now discussed.

Currents to Metal Probes

A metal probe inserted in the plasma of a glow discharge containing a negligible proportion of negative ions will, when connected electrically via a battery to a reference electrode also exposed to the plasma, receive a current whose value depends upon the potential V_{Pr} to which it is biased, and upon the densities and temperatures of the electrons and ions, as well as its surface area and shape. In its simplest form, the theory of the probe was developed by Langmuir and Mott-Smith[54], who assumed a Maxwellian electron energy distribution. Later, this assumption was shown to be unnecessary, and the foundations were laid for determinations of detailed distribution functions by Druyvesteyn[55]. The techniques for their measurement were developed by Sloane and McGregor[56], and more recently in a series of papers by Boyd and his colleagues[57], which include a treatment of the probe in a plasma containing negative ions. The elementary theory is to be found, for example, in the textbooks of Loeb[58] or of von Engel and Steenbeck[59].

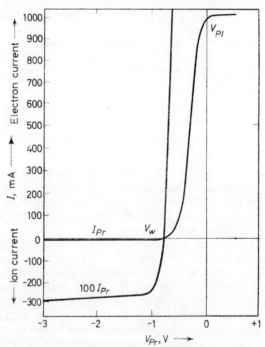

Fig. 3.5. Langmuir probe current–potential characteristic; note the small magnitude of the positive ion current

A plasma, or ionized gas whose charge density is balanced macroscopically such that

$$|n_+ - n_e| \ll n_+$$

$n_+ \simeq n_e$ and $n_- = 0$, will take up macroscopically a potential distribution of its own V_{Pl}, which is a function of position; V_{Pl} will not be the same as the wall potential V_w or the anode potential V_A; its value can be inferred from the probe current–voltage characteristics.

An isolated (floating) metal probe assumes a dynamic equilibrium with the plasma. Positive ions leave the plasma much slower than electrons, and a potential gradient therefore develops between plasma ($+$) and probe ($-$). The floating probe potential V_{Pr} is equal to the wall potential V_w. When V_{Pr} is artifically varied with respect to a reference electrode of large area (e.g. the anode) a current will flow to or from the probe; the area of the reference electrode must be much larger than that of the probe, so that the particle flow to the latter is the rate-determining process.

A plane probe biased so that V_{Pr} is much more negative than V_{Pl} will receive a space-charge-limited current i_+ of positive ions, whose value satisfies the space-charge equation

$$j_+ = \frac{1}{9\pi} \left(\frac{2e}{m_+}\right)^{1/2} \frac{V^{3/2}}{d_+^2} = \frac{en_+ \bar{v}_+}{4} \tag{3.26}$$

with

$$V = |V_{Pr} - V_{Pl}| \tag{3.27}$$

$$\frac{\pi m_+ \bar{v}_+^2}{8} = kT_+ \tag{3.28}$$

and

$$i_+ = j_+ F \tag{3.29}$$

where the current density is j_+, and the average positive ion velocity is $\bar{v}_+$. For a plane probe of area F, the current

$$i_+ = eFn_+ \left(\frac{kT_+}{2\pi m_+}\right)^{1/2} \tag{3.30}$$

In the region of the probe, a large density of positive ions will exist; the thickness of this 'space-charge sheath' is d_+. It increases in magnitude with V, and may at sufficiently large negative potentials reach macroscopic dimensions and produce a visible dark space.

For spherical (subscript s) and cylindrical (subscript c) probes of radius r_p (and length l if cylindrical), the surface area must be replaced by the area of the outer surface of the space-charge sheath, whose thickness is d and radius r_{sh}, so:

$$r_{sh} = r_p + d \tag{3.31}$$

The currents to such probes are

$$i_{s+} = \frac{4}{9} \left(\frac{2e}{m^+}\right)^{1/2} \frac{V^{3/2}}{A} = \pi r_{sh}^2 e \bar{v}_+ n_+ \tag{3.32}$$

and

$$i_{c+} = \frac{2}{9}\left(\frac{2e}{m^+}\right)^{1/2}\frac{lV^{3/2}}{r_p A} = \frac{1}{2}\pi r_{sh}e\bar{v}_+ n_+ l \qquad (3.33)$$

with the correction factor A given by

$$A \simeq \left(\frac{r_{sh}}{r_p} - 1\right)^{3/2} \qquad (3.34)$$

These positive ion currents may be regarded as relatively invariant with V (saturation i_+) for sufficiently large potentials.

However, it has been shown by Allen, Boyd and Reynolds[60], following Bohm[61], that these equations are unrealistic, in that the ion current depends not on the ion temperature but on the electron temperature, because the

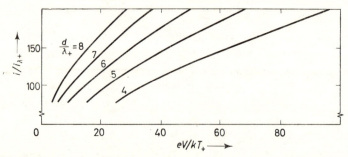

Fig. 3.6. Current to spherical probe at potential V in plasma in electro-negative gas, as a function of positive ion Debye length λ_+ and positive ion temperature T_+ (after Allen, Boyd and Reynolds[60])

latter determines the strength of the electric field which draws the ions towards the sheath. A graphical representation of these authors' calculations of ion current to a spherical probe in terms of electron energy and positive ion Debye length λ_+ is given in *Figure 3.6*. The Debye length

$$\lambda_+ = \left(\frac{kT_+}{8\pi n_+ e^2}\right)^{1/2} \qquad (3.35)$$

is the effective distance through which such an electrical disturbance is propagated in a plasma. The current $i_{\lambda+}$ is defined as $2(\pi m_e/m_+)^{1/2}$ times the random current across a sphere of radius equal to the Debye length. Thus

$$i_{\lambda+} = \frac{(kT_+)^{3/2}}{e(2m_+)^{3/2}} \qquad (3.36)$$

This theory has been verified experimentally, and it is concluded that positive ion currents, whilst almost useless for ion temperature measurement, are still valuable for electron density determinations. As will appear shortly, the classical method of electron density determination is by means of a saturation electron current measurement. However, since the positive ion currents are much smaller than the electron currents, the disturbance of the plasma resulting from the drain of charged particles is greatly reduced, and less heat is dissipated at the probe.

Returning to the simple Langmuir probe theory, consider now the effect of raising V_{Pr}, that is, making it less negative with respect to V_{Pl}. As V_{Pr} is brought close to V_{Fl}, a point is reached where

$$i_{Pr} = 0 = i_+ - i_e \tag{3.37}$$

or

$$i_+ = i_e \tag{3.38}$$

This is when $V_{Pr} = V_w$ the wall potential at which the negative electron diffusion current is equal to the positive ion space-charge-limited current. A much more rapid variation of i_{Pr} with V_{Pr} appears, owing to the penetration of the faster electrons in the distribution through the positive sheath, which is now becoming smaller and contains fewer positive ions. For a Maxwellian energy distribution,

$$i_e = i_{Pr} - i_+ = eFn_e \left(\frac{kT_e}{2\pi m_e} \right)^{1/2} \exp \left(\frac{eV}{kT_+} \right) \tag{3.39}$$

so

$$\frac{\mathrm{d}(\ln i_e)}{\mathrm{d}V} = \frac{e}{kT_e} \tag{3.40}$$

from which T_e can be evaluated by measuring the slope of the probe characteristic (neglecting i_+ variation). Unfortunately, Maxwellian distributions are comparatively rare in running d.c. discharges, as can easily be demonstrated by examining the linearity of $(\ln i_e)$ (V) characteristics in a laboratory positive column.

As V_{Pr} is raised above V_w, it soon reaches the plasma or space potential V_{Pl} at which ions and electrons both diffuse to the probe unaffected by electric fields, and the sheaths have zero thickness. At probe potentials V_{Pr} more positive than V_{Pl}, the electrons are attracted to the probe and a negative space-charge sheath builds up. The positive ion current is rapidly reduced to zero, as T_+ is usually small. The electron current ceases to be dominated by diffusion and becomes space-charge-limited. For a plane probe, the electron current becomes

$$i_e = \left[eFn_e \left(\frac{kT_e}{2\pi m_e} \right)^{1/2} \right. \tag{3.41}$$

and the $(\ln i)$ (V) curve undergoes a comparatively sudden change of slope. The intersection of two straight lines on the logarithmic characteristic serves to fix the point $V_{Pr} = V_{Pl}$, and for previously determined T_e (assuming Maxwellian distribution), n_e is deduced from the value of i_e (any positive ion current contribution can usually be neglected). Further, if $n_e = n_+$ can be assumed, then T_+ might be calculated from the equation 3.30, although this technique has been criticized above.

The identification of the probe current at the knee of the characteristic is the simplest and oldest diagnostic of electron density. Because of the curvature of the characteristic, a number of techniques have been used to improve the identification. A thermionically emitting probe will in fact exhibit much sharper curvature, but even with a cold-metal probe, advantages accrue from maintaining the probe negative and pulsing it up to a potential in the region of the knee; from the dynamic response, measured as a function of probe potential, the plasma potential can be deduced[62]. This arrangement is of particular value when the replenishment problem is serious[63].

With plane probes more positive then the plasma potential, the electron current is space-charge-dominated; for the plane probe

$$i_e = eFn_e\left(\frac{kT_e}{2\pi m_e}\right)^{1/2} = \frac{2}{9}F\left(\frac{2e}{m_e}\right)^{1/2}\frac{V^{3/2}}{d_e^2} \tag{3.42}$$

Thus the negative sheath thickness d_e increases with increasing V, the probe current remaining approximately constant. For spherical and cylindrical probes, there are complications; it may be said that this region is not one of particular importance, as the currents are large and no additional information is made available. In a d.c. glow discharge, a positive probe tends to act as a subsidiary anode. In contrast, an isolated probe, or insulated discharge wall, will take up the wall potential V_w, and owing to the difference between positive ion and electron velocities it is not the same as the plasma potential V_{Pl}, but rather

$$V_{Pl} - V_w = \frac{kT_e}{2e}\ln\left(\frac{T_e m_+}{T_+ m_e}\right) \tag{3.43}$$

The determination of wall potential with the aid of a metal probe is a relatively simple matter with sufficiently high impedance circuitry; if there is no probe available, the potential can still be measured at the outside glass wall painted with colloidal graphite; the resistance of the electrometer used to measure the current must greatly exceed the resistance of the glass wall[64].

Druyvesteyn showed that, in the region between V_w and V_{Pl}, the second differential of a plane probe characteristic yields the complete energy distribution $f(V_e)$, according to the equation[55, 65]

$$n_e f(V_e) = \frac{2}{Fe}\left(\frac{2m_e V}{e}\right)^{1/2}\frac{d^2 i_e}{dV^2} \tag{3.44}$$

This method has been widely employed in studies of the electron energy distribution.

A convenient graphical technique for deriving the electron energy distribution from probe characteristics is as follows[66]. The distribution function is proportional to $V^{-1/2}\,\Delta i_e/\Delta V$, where Δi_e is the distance between the intersections of the two tangents to the probe curve at the ends of the voltage interval ΔV with the current axis, and V is the voltage at the centre of ΔV.

Since graphical differentiation is a low-precision technique, data-logging and computer programming are preferable. An a.c. technique was developed by Sloane and McGregor[56], and a more sophisticated electronic technique by Boyd and Twiddy[67, 68].

If a modulated voltage ε is superimposed upon the probe potential, then by Taylor's theorem the probe current is

$$i = f(V+\varepsilon) = f(V)+\varepsilon f'(V)+\frac{\varepsilon^2}{2!}f''(V)+ \cdots \qquad (3.45)$$

The voltage ε is in the form of a sine wave modulated with square waves, the high harmonics being removed to eliminate the possibility of a sideband lying close to the fundamental modulation frequency

$$\varepsilon = \mathcal{E}\left[\frac{1}{2}+\frac{2}{\pi}\left(\cos pt-\frac{1}{3}\cos 3pt+ \cdots\right)\right]\cos \omega t \qquad (3.46)$$

where $\mathcal{E}$ is the peak value of the modulated signal, and p and ω are the angular frequencies respectively of the modulation and carrier signals.

Even-order derivatives make a contribution to the component i_p of the probe current at angular frequency p:

$$i_p = \left[\frac{\mathcal{E}^2}{2!}\left(\frac{1}{\pi}\right)f''(V)+\frac{\mathcal{E}^4}{4!}\frac{3}{8}\left(\frac{1}{4}+\frac{1}{\pi^2}\right)f'''(V)+ \cdots\right]\cos pt \quad (3.47)$$

Neglecting the contribution of the fourth derivative, the modulation frequency component of the probe current is proportional to $\mathcal{E}^2$, a fact which may be checked experimentally. The electron energy distribution is given by:

$$n_e f(V_e) = \frac{3\cdot 8\times 10^{12}V^{1/2}yv_c}{F\mathcal{E}^2Rx} \qquad (3.48)$$

where y is the deflexion due to the probe signal, x is the deflexion due to a calibration signal v_c applied to the probe current amplifier, and R is the load resistor. Attempts to automate the computation further are unrewarding in noisy discharges, so the apparatus (*Figure 3.7*) consists of a glow discharge probed by a 2 kHz signal modulated 100 per cent by a 300 Hz square wave. The d.c. level of this voltage with respect to earth is variable, so that the probe amplifier may be maintained near earth potential throughout the

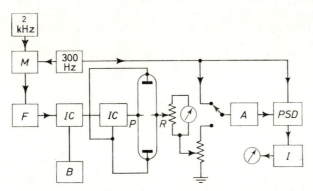

Fig. 3.7. Schematic diagram of apparatus for automatic recording of electron energy distribution: M, modulator; F, filter; B, bias circuit; IC, impedance changer; P, probe; R, reference probe; A, amplifier; PSD, phase-sensitive detector; I, integrator (after Boyd and Twiddy[67])

measurements. For any d.c. voltage V applied, the signal across R contains a 300 Hz component proportional to $Rf''(V)$, from which $n_e f(V_e)$ can be calculated.

A treatment of the metal probe in an electro-negative gas has been given by Boyd and Thompson[68]. It is concluded that the dominating factor in these discharges is the ratio β of negative ion density n_- to electron density n_e:

$$\beta = \frac{n_-}{n_e} \tag{3.49}$$

For $\beta \leqslant 2$, no very serious errors are involved in using the treatment given above. For $\beta > 2$, one must use an entirely new treatment, the results of which are discussed below. The determination of β requires additional techniques, among which may be mentioned:

1. Mass-spectrometric monitoring of negative ions and positive ions;
2. Microwave permittivity measurements of n_e, coupled with probe characteristic change of slope measurements to give the total negative particle current i_- (close correspondence between the two should indicate low β);
3. In afterglows, the ambipolar diffusion coefficient may be measured by $n_e(t)$ time decay studies: the apparent coefficients D_a' are related to the true values[69] by the equation

$$D_a' = D_a(1+\beta) \tag{3.50}$$

and the true coefficients may be determined from ionic mobilities. For large β in the late afterglow, the differences are easily observed[70].

The spherical probe immersed in a strongly electro-negative plasma ($\beta > 2$) collects a positive ion current which is the random current across the sheath edge. The potential of the sheath edge[61] is $\frac{1}{2}ekT_-$, and the field inside the sheath becomes so large that the absorption surface radius is approximately equal to the sheath radius. Following a method similar to that of Allen, Boyd and Reynolds[60], a calculation is made of 'reduced' probe currents $i_+/i_{\lambda+}$ as a function of reduced probe potentials $\eta = V/kT_+$, for different values of T_+ (*Figure 3.7*). Assuming a realistic value of T_+, n_+ may then be calculated from the experimental currents i_+ and the equations

$$i_{\lambda+} = 4\pi\lambda_+^2 \frac{n_+}{4} \frac{8kT_+}{\pi m_+} e \tag{3.51}$$

and

$$\lambda_+ = \left(\frac{kT}{8\pi n_+ e^2}\right)^{1/2} \tag{3.52}$$

It turns out that under the conditions of this theory i_+ is very closely proportional to $n_+^{1/2}$. Strongly biased probes ($V \gg ekT_-$) in electro-negative afterglows can be shown to conform approximately to this condition[71], provided that the electron density is sufficiently high to maintain a sheath.

The equality $n_+ \simeq n_-$ enables negative ion densities to be estimated; an investigation of temperatures in the glow discharge may then be made in the following manner. The radial variation of $n_+ = n_-$ in a cylindrical glow discharge is measured; it is well-known that it follows a Boltzmann distribution[72]

$$n_-(r) = n_-(0) \exp\left(-\frac{V}{kT_-}\right) \tag{3.53}$$

from which T_- may be calculated. Provided that T_- is taken as equal to T_+, the positive ion temperatures so obtained may be substituted back into the calculation of n_+.

The limitations of probe techniques are numerous and only partially understood. One is the fact that the drain of charged particles, particularly under positive probe conditions, may seriously disturb the plasma. Not only does this effect invalidate many of the early rocket probings of the ionosphere, but even in glow discharges the effect can be serious. Replenishment of the electrons by means of a hot cathode is often necessary[75]. Pulsed probe techniques have also been used[76]. A further limitation concerns the surface condition of the probe, which can be changed by chemical and ionic reactions; this is particularly important in afterglow studies, where probe 'fouling-up' can result in false $n_e(t)$ functions. The only remedy is to conduct the experiments in such a short time (single-shot technique) that fouling-up has no chance to take place.

Yet another limitation concerns the orbiting motion of electrons in the sheath region around a cylindrical probe. Plane probes are normally used only to detect anisotropy in electron energy distributions, spherical or cylindrical probes being customary for other purposes. The use of a probe consisting of a short length of very fine wire (0·02 mm diameter; radius ≪ length) minimizes the current collected, and hence minimizes replenishment problems. However, in such a system the sheath radius r_{sh} may greatly exceed the probe radius r_p. When this is so, not all the electrons passing from the plasma within the sheath boundary will be collected at the probe; some will orbit and leave the sheath region without collection. Under such conditions, the probe current is 'orbit limited', and is given[54] by the equation

$$i_e^2 = \frac{2A^2 n_e^2}{\pi^2 m_e^2}(eV_{Pr}+kT_e) \tag{3.54}$$

provided that

$$\frac{eV_{Pr}}{kT} \geqslant 4$$

that

$$\frac{r^2(eV_{Pr}/kT_e)}{r_{sh}^2 - r_p^2} \geqslant 0$$

and that the mean free path $l_f > r_p$ and $l_f > r_{sh} - r_p$. Thus both electron temperatures and electron densities are available under these conditions.

The sheath radius around a cylindrical probe can be related to the Debye length by the expression[79]

$$r_{sh} = 1·66\lambda_+\left(\frac{eV_{Pr}}{kT}\right)^{3/4} + r_p \tag{3.55}$$

The pressure limitations upon the operation of the cylindrical probe[73, 74, 77] are set by the condition $l_f > \lambda_+$ and also by the conditions $l_f > r_p$ and

$l_f > r_{sh} - r_p$ given above. But because of the importance of lower ionosphere studies, much attention is being given to the operation of probes outside this pressure limit[78]. The conditions $l_f > r_p$ and $l_f > r_{sh} - r_p$ also describe an electron density limitation on the cylindrical probe, since the mean free path for electron collisions of any type are included in this definition of l_f.

The floating double probe[80] is under some circumstances a more convenient device for diagnosing electron temperature than is the single probe. Two metal probes of equal area are isolated electrically from the plasma electrodes and walls, but are connected together via a voltage supply. Both remain negative to the plasma, but the current flowing between them is a function of the inter-probe potential V_{12}. As V_{12} is made more positive, the inter-probe current increases gradually, but at a certain potential there comes an inflexion. This is also true as V_{12} is made more negative, so a symmetrical S-shaped curve is seen. The difference of potential ΔV between these two inflexions is related to the electron temperature T_e by the equation

$$T_e = \tfrac{1}{4}\, \Delta V \text{ eV} = 2900\, \Delta V \text{ }^\circ\text{K} \qquad (3.56)$$

The double probe current i_d is related to the positive ion density as follows:

$$n_+ = 1\cdot34 \times 10^{27} \left(\frac{m_+}{T_+}\right)^{1/2} \frac{i_d}{F} \qquad (3.57)$$

where the area of each probe is F.

The operation of the single probe in the presence of a magnetic field has been treated by Bohm[61], and much information can be derived from measurements under these conditions, provided that plasma oscillations do not occur. Plasma oscillations are the most serious drawback to probe studies even in the absence of magnetic fields; they produce a noise background superimposed upon the direct probe currents.

Anisotropies of electron energy can be diagnosed by measuring the electron currents to a rotatable positively biased planar probe[81]. Taking the z axis as normal to the plane of the probe, it can be shown that

$$f(v_z) = \frac{m_e}{e^2} \frac{\partial j}{\partial V}_{Pr} \qquad (3.58)$$

where j is current density. Reorientation of the probe enables the components to be compared with each other.

An advantage which is enjoyed by probe diagnostics over radiofrequency diagnostics is that they are applicable over more orders of magnitude of electron density; this is of importance in afterglow studies.

Probes may be used with applied radiofrequency potentials for special applications. The real part ε' of the dielectric constant for a plasma may be represented by the equation:

$$\varepsilon' = 1 - \frac{4\pi n_e e^2}{m_e \omega^2} \qquad (3.59)$$

for angular applied frequency ω. The contributions from ionic motion and collisions are neglected. The term ε' vanishes at the plasma electron resonance frequency ω_{Pl} given by the equation

$$\omega_{Pl} = \left(\frac{4\pi n_e e^2}{m_e}\right)^{1/2} \tag{3.60}$$

Although it is well-known that the radiation field strength from a dipole antenna vanishes under these conditions, there remains the induction field:

$$E_r = 2E_\theta = \frac{\mathscr{M} \sin \theta}{4\pi\varepsilon' r^3} \tag{3.61}$$

$$E_\phi = 0$$

This field exceeds the radiation field at short distances r from the dipole, whose electric moment is $\mathscr{M}$; for $\omega = \omega_{Pl}$ it rises to large intensities, limited only by electron collision damping and radiation from the plasma. The plasma resonance frequency is measured by observing the maximum reception at a small high-frequency probe placed opposite a small transmitting probe. At electron densities $n_e \simeq 10^8$ cm^{-3}, a frequency of 300 MHz is appropriate; thus it is possible to make the distance between antenna probes much smaller than one wavelength. The theory of the radiofrequency probe in a fully ionized plasma has been considered by Derfler[82], and further discussion of electromagnetic waves in plasmas is given below.

Plasma Diagnosis by Microwaves and Millimetre Waves

Both electron density and electron temperature can be inferred from microwave measurements; the former from the real permittivity, the latter from the noise emitted by the plasma. In addition, it is possible to deduce the electron collision frequency from the unreal part of the permittivity (absorption).

The real part ε' of the permittivity of a plasma at radial frequency of the electromagnetic wave is given by equation 3.59 above. The value of the plasma electron radial frequency ω_{Pl} at which ε' vanishes is given by equation 3.60. Conveniently, one may remember that the plasma frequency f_{Pl} (in Hz) is related to the electron density (in cm^{-3}) (see Table 3.1) by the equation

$$f_{Pl} \simeq 9000\, n_e^{1/2} \tag{3.62}$$

Neglecting the dielectric loss due to electron collisions, the refractive index n is given by

$$n^2 = \varepsilon' = 1 - \frac{\omega_{Pl}^2}{\omega^2} = 1 - \frac{n_e}{n_{ec}} \tag{3.63}$$

where

$$n_{ec} = \frac{\omega^2 m_e}{4\pi e^2} \tag{3.64}$$

For an electromagnetic wave passing through a plasma contained within $x = 0$ and $x = L$, the phase shift $\Delta\eta$ is

$$\Delta\eta = \frac{2\pi}{\lambda} \int_0^L [n(x) - 1]\, dn \tag{3.65}$$

or

$$\Delta\eta = \frac{2\pi L}{\lambda}(n-1) \tag{3.66}$$

for uniform plasma.

When a steady magnetic field H is applied parallel to the E vector, n is unchanged. But for the E vector and direction of propagation perpendicular to H,

$$n_+^2 = \frac{[1-(\omega_{pl}^2/\omega^2)]^2-(\omega_c^2/\omega^2)}{1-(\omega_{pl}^2/\omega^2)-(\omega_c^2/\omega^2)} \tag{3.67}$$

where the electron cyclotron frequency ω_c is given by the equation

$$\omega_c = \frac{eH}{m_e c}$$

There is one further regime which is of great importance, namely the weakly ionized plasma (no magnetic field) probed by radiation whose frequency greatly exceeds the plasma frequency. Equation 3.63 is not applicable because the neglect of electron collisions with gas molecules is not justified. The general equation for the complex conductivity σ is

$$\sigma = -\frac{4\pi e^2}{3m_e}\int \frac{\nu_e-j\omega}{\nu_e^2+\omega^2}\,v^3\,\frac{df_0}{dv}\,dv \tag{3.68}$$

where ν_e is the electron collision frequency (inelastic collisions neglected), and f_0 is the spherically symmetrical part of the electron velocity distribution. Margenau[83] treated the steady state with uniform field intensity and electron density. In the high-frequency case ($\nu_e^2 \ll \omega^2$), the solution reduces to

$$\varepsilon' = \frac{e^2 n_e}{m_e \pi f^2} \tag{3.69}$$

Thus the real part of the permittivity is directly proportional to the electron density in a weakly ionized plasma subjected to microwave radiation of frequency f.

Electron density diagnostic techniques can be summarized under the following headings.

Direct transmission

Transmitting and receiving probe antennae are inserted into the plasma-containing vessel, and the absorption of radiation of frequency, say, 300 MHz is monitored. At the plasma frequency, the transmission is maximum; although equation 3.59 implies that there is then no transmission, the local field

$$E_r = 2E_\theta = \frac{\mathcal{M}\sin\theta}{4\pi\varepsilon' r^3} \tag{3.70}$$

maximizes (with $E_\phi = 0$).

The direct transmission technique is of limited application, since the situation must be arranged in such a way that this maximization is observable.

However, other types of aerial, notably a cage of parallel rods, may be used. The method can also be used in the centimetre wavebands, with horn antennae, and in the form of a radiofrequency probe, as follows.

Radiofrequency probe[66, 84]

As the frequency of a low-voltage radiofrequency signal superposed on a metal probe is swept within a suitable frequency range, the d.c. component of the electron current due to the non-linearity of the sheath impedance is measured. A resonant increase appears in the current at a frequency near to the plasma electron frequency f_{Pl} (equation 3.60). For a spherical probe of radius R, the frequency f_r at which resonance occurs is related to the Debye length λ_+ (equation 3.34)

$$f_r = \frac{f_{Pl}}{1 + (R/k\lambda_+)} \tag{3.71}$$

where k is a numerical factor experimentally found to be ~ 5.

By using two spherical probes of radii R_1 and R_2, k and λ can be eliminated to give

$$f_{Pl} = \frac{f_{r1}[(R_1/R_2) - 1]}{(f_{r1}R_1/f_{r2}R_2) - 1} \tag{3.72}$$

Phase shift interferometry[85]

Figure 3.8 shows a schematic diagram of an interferometer suitable for displaying time-varying phase shifts of a millimetre wave passing through a plasma. The deflexion of the visible 'fringes' from the horizontal is directly

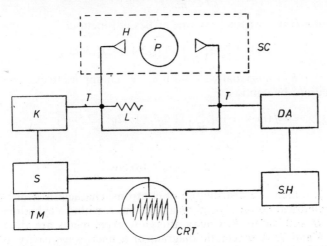

Fig. 3.8. Schematic diagram of millimetre wave interferometer for diagnostics of electron density in plasma: H, horns; P, plasma; SC, screened room; T, magic T; K, klystron; L, load; S, sawtooth generator for frequency modulation; TM, timebase; CRT, intensity-modulated cathode ray tube; SH, shaping circuit; DA, differential amplifier

proportional to the phase shift, provided that reflection at the plasma boundary is neglected.

Double polarization technique

In the double polarization technique, a steady magnetic field is applied parallel to E for radiation applied at 35 GHz, and perpendicular to E for radiation simultaneously applied at 70 GHz. Provided that the cyclotron frequency is comparable to the plasma frequency and is known in space and time, the method is capable of giving information about the space variation of electron density.

Oblique incidence interferometry

In the oblique incidence interferometry technique, the electron density profile can to some extent be determined by an analysis of the phase shift as a function of angle of incidence to the plasma boundary.

Microwave cavity afterglow

The diagnostic of time-dependence of electron density in the afterglow, build-up, or modulation of a weak plasma is conveniently carried out in a resonant TM$_{010}$ microwave cavity. The shift Δf in the resonance frequency of the cavity can be shown (using equation 3.69) to be

$$\frac{\Delta f}{f} = \frac{4\pi e^2}{m_e \omega^2} f(\tilde{r})\, n_e(t) \tag{3.73}$$

The function $f(\tilde{r})$ is written in terms of Bessel functions

$$f(\tilde{r}) = 1\cdot85\left(\frac{h}{H}\right)\varrho^2[J_0^2(2\cdot4\varrho)+J_1^2(2\cdot4\varrho)] \tag{3.74}$$

(where $\varrho = r/R$) for spatially uniform n_e, which applies to electron loss by recombination; by contrast

$$f(\tilde{r}) = 1\cdot16\left(\frac{h}{H}\right)[1\cdot6\varrho^2 - 1\cdot42\varrho^4 + 2\cdot58\varrho^6] \tag{3.75}$$

for

$$n_e = 0\cdot274 n_{e0} J_0(2\cdot4\varrho)\cos\frac{\pi z}{h} \tag{3.76}$$

and $\varrho = \frac{1}{2}$, which is a distribution of electrons, characteristic of diffusion, inside a cylindrical bottle of height h and radius r, within a cylindrical cavity of height H and radius R. Cavity afterglow experiments are discussed in Chapters 4 and 7. A schematic diagram of a microwave cavity afterglow experiment, including high-power pulse to break down the gas, is shown in *Figure 3.9*.

Electron temperature diagnostic techniques include the following.

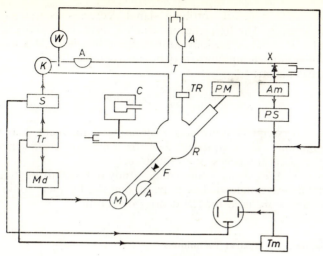

Fig. 3.9. Schematic diagram of microwave cavity afterglow apparatus: C, cavity; M, magnetron; K, klystron; S, sawtooth frequency modulator; Tr, trigger pulse generator; Md, magnetron pulse modulator; A, flap attenuator; F, ferrite isolator; R, rat race; TR, protective transmit–receive switch; PM, power monitor; T, magic T; X, crystal detector; Am, amplifier; PS, pulse sharpener; Tm, timebase

The millimetre wave radiometer

Hot plasma emits sufficient thermal noise radiation in the millimetre waveband for its measurement by superheterodyne techniques, etc., to be possible. The thermal radiation per unit frequency band

$$P(f) = kT_e \, df \qquad (3.77)$$

So in a bandwidth of 1 MHz, a plasma whose electrons have mean thermal energy corresponding to 10 eV emits 1.60×10^{-12} W.

It is important to the black-body assumption of equation 3.77 that the dimensions of the plasma are much greater than the wavelength corresponding to f. Hence radiometers are commonly designed in the 70 GHz and 140 GHz bands. The noise characteristics of the gallium arsenide crystal are at present unequalled in these bands.

The superheterodyne receiver is designed not to make absolute noise power measurements, since this requires knowledge of bandwidth and gain, but to compare the unknown radiation with a standard reference source. The wide operating range of the indium antimonide photoconductive detector is particularly valuable; the detector is competitive in sensitivity with the superheterodyne receiver. Various other radiometers are available, but they are less sensitive.

Ionization in a modulated plasma

Electron temperature can be inferred from microwave measurements of the ionization rate in a modulated plasma[86-88]. The depth of modulation is a function of modulation frequency, and the modulated component of the microwave electron density probing signal decreases as the modulation frequency increases. The frequency at which it reaches $1/\sqrt{2}$ of its low-frequency value is equal to the ionization collision frequency. A knowledge of the electron-gas ionization function (Chapter 5) enables the electron temperature to be calculated:

$$|\tilde{n}_e| = \frac{\alpha}{(\omega^2 + \alpha^2)^{1/2}} \tilde{n}_{e0} \tag{3.78}$$

where the superscript $\sim$ represents alternating component, and the subscript 0 represents zero frequency; α is the ionization coefficient or rate, and ω is the radial frequency of modulation.

Conveniently, the plasma column under study is passed through the broad faces of a TE_0 waveguide at S or X band, and the Dattner–Tonks electrostatic resonances are used for the density diagnostic.

Plasma Diagnosis by Optical Techniques

The emission and absorption of optical radiation by plasmas constitute diagnostic techniques. In Section 9.1, the relations between excited state densities and intensities of emitted radiation are discussed. A measurement of intensity at a certain frequency is, under optically thin conditions, capable of yielding the density of particles in a certain excited state. This is of special importance in afterglows in which the excited states are formed by dissociative recombination (Chapter 7).

From certain excited state populations it is possible, making use of thermodynamic equilibria discussed in Chapter 1, to calculate electron temperatures. In the coronal regime the populations depend primarily on the electron excitation functions (Chapter 5); when these are known, the electron temperature can be computed from the ratio of two suitable intensities[89]. For diagnosis of temperatures encountered in hot plasmas containing stripped hydrogen-like ions (emitting spectra of C V and N VI), suitable electron–ion excitation functions have been calculated (Chapter 5). The temperature diagnostic is highly sensitive to inaccuracy in these functions.

The Doppler broadening of a spectral line emitted from a plasma is a diagnostic of the gas (or appropriate positive ion) temperature. The appropriate equations can be found in Chapter 13.

An optical technique capable of diagnosing densities greater than $10^{15} \, cm^{-3}$ and without apparent upper limit is that of laser interferometry. A laser is mounted together with the plasma in an optical cavity bounded at one end by a movable mirror. The radiation intensity is monitored from one end of the cavity as the mirror is moved, and the fringe distance is related to the refractive index n of the plasma. The latter contains contributions from electrons, positive ions and excited species. For electrons

$$n - 1 = 4.36 \times 10^{14} \lambda^2 n_e \tag{3.79}$$

and for gas

$$n - 1 = A\left(1 + \frac{B}{\lambda^2}\right) \tag{3.80}$$

with A and B constants. The number N of fringes in distance L is

$$N = \frac{2(n-1)L}{\lambda} \tag{3.81}$$

With two wavelengths λ_1 and λ_2, the electron contribution can be separated out from the two refractive indices n_1 and n_2:

$$n_2\lambda_1 - n_1\lambda_2 \simeq 4 \cdot 49 \times 10^{-14}(\lambda_1^2 - \lambda_2^2)n_e \tag{3.82}$$

In plasmas containing a sufficiently high density of electrons, the Doppler broadening of emitted radiation is dominated by the Stark broadening, which has been used as an electron density diagnostic. The determination of electron densities by means of microwave techniques is limited by the frequency bands available. Even with the shortest millimetre wave generators, one cannot diagnose densities in excess of $n_e \sim 10^{15}$ cm^{-3}. In the range $n_e \sim 10^{15}$–10^{18} cm^{-3} the most widely used diagnostic is the Stark broadening of spectral lines; this range of densities is found in hot plasma machines and in shock tubes. Following the theory of Inglis and Teller, Griem and his collaborators[90, 91] have worked out the technique for hydrogen, ionized helium, hydrogen-like states of neutral helium and isolated visible lines of neutral and singly ionized atoms of helium to calcium and also of caesium.

The absorption of optical radiation by a plasma is a diagnostic of the density of those particular states of an atom or molecule which are capable of being excited by this radiation. The line absorptions discussed in Section 9.1 are suitable for this technique. It has been most used in the determination of concentrations of metastables, but if the appropriate transition probabilities are known there is no reason why some other states, excited and ground, should not be monitored in this way. The limit of sensitivity of this technique would be similar to that of the emission technique, were it not for the fact that the plasma under study emits spontaneously precisely that frequency which is being used for the absorption measurement. The two radiation intensities are separated by modulating that which is used for absorption. It is necessary that the modulation frequency be high compared with the characteristic time of variation of particle density. A suitable modulated light source is described in Section 3.9.

In the absence of suitable line absorption, radiation is still scattered by a plasma. The Thomson scattering of a laser beam is used as a diagnostic of electron temperature[92].

Diagnostics Using Particle Detection

Information can be obtained both by detecting particles emitted by the plasma and by measuring the absorption and inelastic changes undergone by particle beams projected through the plasma.

Positive ions emerging from a plasma through an orifice can be accelerated, mass-analysed and detected conventionally. Single particle counting is desirable, because the large difference of pressure between typical weakly ionized plasmas and the mass-spectrometer requires that the extraction orifice be very small, (for example, 0·01 mm). Since the orifice behaves similarly to the wall of a plasma, it is covered by a sheath. The potential gradient through which the ions pass depends upon the potential of the orifice with respect to the rest of the wall. The ion current to the orifice behaves similarly to the ion current collected at a metal probe. The situation has been discussed[93], and investigated experimentally with the aid of a miniature insulated orifice of variable potential. If the orifice is strongly biased, then the ions will be accelerated across the sheath, and deductions may be made concerning epithermal ion–molecule reactions[94].

Neutral particles from a plasma can also be monitored through an orifice[95, 96] and may be ionized by electron impact or by stripping, after which the ions are accelerated, mass-analysed or momentum analysed, and detected. As with ion monitoring, single particle counting is often desirable. The momentum analysis yields information from which the temperature can be deduced. In a hot plasma machine, the neutral particle temperature is usually determined by the ion temperature, since the neutral particles are formed by charge transfer.

Whilst monitoring experiments of this type lead to a knowledge of slow ion and neutral densities, beam probing experiments are also valuable. The fast neutral probe[97] is primarily an ion density diagnostic for hot plasmas. An ion beam accelerated to kilovolt energies is neutralized by charge transfer (unneutralized particles removed by electrostatic field) and passed through the plasma; at these energies the elastic collisions are only through small angles, so a well-defined beam emerges from the plasma. This beam is partially ionized, the dominant collision process being charge transfer with the plasma ions. The proportions of fast ions in different charge states is a diagnostic of the nature of the ions in the plasma.

A neutral beam probe operates in a different manner. The ionized part of an alkali metal beam directed through the plasma is monitored and used as a diagnostic of combined electron density and temperature. Charge transfer with positive ions is unlikely compared with the ionization of the alkali metal beam by hot electrons. It is important that the ionization functions be accurately known, and it is convenient to use this technique simultaneously with another electron temperature or density diagnostic. Alternatively, two different alkali metal beams can be passed through the plasma simultaneously; from the different ionized proportions, both temperature and density might be extracted.

The passage of an electron beam through plasma is important mainly for the information it can yield concerning time-varying electric fields, arising from an oscillating sheath in the plasma[98].

Diagnostic Techniques for Neutral Particles

The problems of measuring gas density and temperature in a plasma differ from the corresponding problems in an un-ionized gas, and a miscellany of methods exists to help solve them. In flowing plasmas, plasma heating machines and shock-generated plasma, the most satisfactory methods depend on the absorption of radiation which does not disturb the plasma, but whose absorption coefficient as a function of static gas density is independently known. Thus the absorption of α-radiation has been used to measure gas density in shock-tubes[99]. Commercial instruments using this principle are known as 'alphatrons'.

The absorption of ultrasonic acoustic waves can be used to derive gas densities in steady plasmas, but the situation is complicated by the fact that pulsed or modulated discharges themselves emit acoustic waves[100, 101]. These are detected with a piezoelectric transducer or with a mobility-limited thermionic diode. They can modulate the plasma luminosity and the micro-wave absorption coefficient. From the velocity of propagation of the acoustic waves (deduced from wavelength and modulation frequency), the gas temperature can be derived; in an experiment with mixed gases, the proportions of mixture can also be found.

Electron spin resonance in absorption may be used to monitor paramagnetic atom densities, provided that the region of study can be contained within a microwave cavity in a steady magnetic field. Techniques have been worked out for nitrogen, hydrogen and oxygen atoms[102]. Atomic hydrogen may be monitored by absorption of the 1400 MHz line familiar in radio-astronomy.

3.5 TIME-DEPENDENT AND FLOWING AFTERGLOWS

The afterglow of an electric discharge or other type of plasma excitation event is a fruitful field for experimental study. During the time immediately after the switch-off of applied energy (that is, during the afterglow) various volume and wall processes contribute to the diminishment and sometimes replenishment of populations of electrons, ions, excited species and radicles; measurement of the time-dependence of these populations provides data from which a whole variety of inelastic collision rates may be calculated, with the aid of the kinetics discussed in Chapter 1. Usually, the assembly of particles is contained in a vessel whose geometry is conducive to simple solutions of the diffusion equation. This is termed a time-dependent afterglow, although sometimes the term 'static afterglow' is used, to distinguish it from the 'flowing afterglow'. A flowing afterglow is one in which the assembly of particles is carried along in a gas flowing along a tube; the flow is not molecular but viscous and usually laminar. The populations of the species under study are monitored along the axis, so that the time variation is converted into a spatial variation along one dimension; pulsed operation is often used.

The advantages of the flowing afterglow derive from the flexibility with which the rate processes can be controlled. Unwanted excited species are difficult to avoid in a time-dependent afterglow, because it is not possible to introduce the reactant gas after the excitation energy has been switched off; in a flowing afterglow, a reactant gas is easily introduced downstream. Many complications of the time-dependent afterglow are now well known, and its limitations are well defined. Some of the most serious can be avoided by substituting a flowing afterglow. However, the complications and limitations of the flowing afterglow are less well understood, although experiments on ion–molecule reactions are extensive[103]. It is therefore desirable that a number of well-understood collision rates (such as two-body electron–ion recombination) be investigated in flowing afterglows, so that the two techniques can be critically compared and all their complications understood.

The relevant collision processes involve ions, electrons, and excited species; they can be produced in the flowing afterglow by the methods outlined in other sections of this chapter. Usually, but not always, a plasma must be produced. The total gas pressure is in the range 0·1–10 torr, and the methods of setting up a plasma include the following:

1. High-frequency discharge,
2. High voltage between electrodes in the gas,
3. Ultra-violet radiation,
4. Flux of electrons,
5. Flux of fast heavy particles.

All of these methods must be capable of being pulsed when employed in the time-dependent afterglow; the trailing edge of the pulse should not exceed a few tens of nanoseconds. It is advisable to employ only the minimum power necessary to set up an observable afterglow, otherwise unwanted species will be multiplied and could interfere with the process under observation. Hence the minimum period of excitation is advisable, a pulse of 0·5 μsec duration being typical. If necessary to the purity of the gas in the afterglow chamber, a 'single shot' technique can be used; otherwise, the pulses are repeated at a rate sufficiently slow for the populations to have subsided well below the limit of detection sensitivity.

The high-frequency discharge is a commonly used source of plasma for afterglow experiments, particularly decimetre and centimetre waves. Reference should be made to the brief discussion of cavity resonator ionization sources given in Section 3.7. In a high-frequency discharge at 1 torr pressure, the electron temperature is a few electron volts; because this is the energy region in which molecules frequently possess compound negative ion states, the cross-sections for vibrational excitation of molecules by electrons are important (Chapter 5). The vibrationally excited molecules, possessing long lifetimes unless they are polar, can complicate the afterglow kinetics; therefore the high-frequency discharge is inferior to the more refined plasma sources which do not heat the electrons.

A d.c. glow discharge, struck by the application of a potential difference between two electrodes, need not heat electrons as much as a high-frequency discharge. The cathode glow region is suitable, and in particular the properties of the brush cathode have been studied by Persson and found to be particularly stable. The brush cathode consists of pointed tungsten wires, several centimetres long, imbedded in a metal disc. The action is like that of many hollow cathodes together. It has been found that the electrons fall into two groups as far as their energy is concerned, one group having approximately the energy of the applied voltage, and the other a temperature slightly superior to gas temperature. A carbon hollow cathode also acts in this way.

Radiation in the vacuum ultra-violet can break down a gas (without appreciably accelerating the electrons formed in ionization), provided that the radiation is sufficiently energetic and the ionization potential of the gas sufficiently low. Thus, Lyman α-radiation is capable of ionizing nitric oxide. The problem in such an experiment is one of vacuum-isolating the ultra-violet source from the gas which is to be ionized. It is usually sufficient to connect the source to the plasma chamber by means of a pair of capillaries between which is a fast-pumped region; but aluminium windows can now be made sufficiently thin (10^3 Å) to answer this purpose.

Such windows will also pass pulses of electrons of energy $\gtrsim 40$ keV, which will ionize gases, although not very efficiently. Linac relativistic electron pulses have been used to strike afterglows and have the advantage that the windows need not be so thin. Fluxes of slow electrons can be produced in a gas by photoemission, or by thermionic emission from a barium zirconate cathode.

Once the time-dependent or flowing afterglow has been set up by one of these methods, the observations of species population are carried out by methods discussed in Section 3.4 under the heading of plasma diagnostics. A modern time-dependent afterglow, measuring ion, metastable and emitted photon densities simultaneously, is described by Sauter, Gerber and Oskam[104].

The decay of a species proceeds by volume processes, usually obeying kinetics of first or second order, as well as by wall processes, preceded by diffusion. Hence the kinetic equations can become complicated and the problem of the experiment is that of setting up conditions under which a single process can be isolated. An example is the behaviour of electrons in the afterglow of a weakly ionized plasma. The decay is through three processes:

1. Ambipolar diffusion (Chapter 1), depending essentially on the mobility of positive ions (Chapter 10), but complicated by the presence of negative ions; the diffusion of electrons and positive ions to the wall is followed by fast wall recombination;
2. Recombination of electrons and positive ions (Chapter 7);
3. Attachment of electrons to neutral particles (Chapter 8).

These may be summarized in the time-dependent equation

$$\frac{dn_e}{dt} = D_a \nabla^2 n_e - \alpha_e n_e^2 - k_a n_e n_0 \qquad (3.83)$$

where n_e and n_0 are respectively electron density and gas density, D_a is the ambipolar diffusion coefficient, α_e is the electron–ion recombination coefficient, and k_a is the rate coefficient for electron attachment to the gas. It is assumed that the electron and ion densities are equal and that both are very much smaller than gas density.

Equation 3.83 is difficult to solve in the general case, but neglect of one of the three terms enables solutions to be obtained.

For decay purely by recombination

$$\frac{dn_e}{dt} = -\alpha_e n_e^2 \qquad (3.84)$$

the solution of which is

$$\frac{1}{n_e} - \frac{1}{n_{e0}} = \alpha_e t \qquad (3.85)$$

where n_{e0} is the electron concentration at $t = 0$, the end of the discharge pulse. The electron density is thus inversely proportional to time.

For the purely diffusive process, an exponential decay of electron density with time constant τ is observed:

$$n_e = n_{e0} \exp\left(-\frac{t}{\tau}\right) \qquad (3.86)$$

This is appropriate to a diffusion equation

$$\nabla^2 n_e + \frac{n_e}{\tau D_a} = 0 \qquad (3.87)$$

This equation can be solved for the conditions of n_e finite, but vanishing at the walls of the container. Solutions are discussed in Chapter 1 and may be expressed in the form

$$\frac{1}{\tau} = \frac{D_a}{\Lambda^2} \qquad (3.88)$$

where Λ is the diffusion length, which depends on the container dimensions and geometry.

For decay solely by two-body electron attachment of rate k_a the solution is

$$n_e = n_{e0} \exp\left(-k_a n_0 t\right) \qquad (3.89)$$

Again an exponential decay is to be expected, in this case independent of container dimensions and directly proportional to gas pressure. Since, from arguments given in Chapter 1, $n_0 D_a$ is constant, different behaviour is observed for diffusive and attachment decay with respect to gas density. The three pure decay processes behave with respect to experimental variables according to the summary in Table 3.2.

Table 3.2.   LOSS RATE DEPENDENCES ON AFTERGLOW PARAMETERS

Process	Electron density	Gas pressure	Container dimension
Recombi-nation	Quadratic	Independent	Independent
Diffusion	Linear	Inverse	Inverse
Attachment	Linear	Linear (2-body) or quadratic (3-body)	Independent

Although the single-process afterglows considered above are instructive, these conditions do not correspond exactly to real situations. Much better approximations are obtained by treating the simultaneous action of two processes. In a gas which can form no negative ions, the simultaneous action of recombination and diffusion is realistic, and an approximate solution of equation 3.83 is

$$\frac{n_e}{1+\alpha_e\tau n_e} = \frac{n_{e0}\exp\left(-t/\tau\right)}{1+\alpha_e\tau n_{e0}} \qquad (3.90)$$

with τ given by equation 3.88. In physical terms, recombination requires the collision of two charged particles and will therefore be most important at high charge densities; in the late afterglow (perhaps $t > 5$ msec) the diffusion process dominates in that $\alpha_e \tau n_e \ll 1$, so the decay is purely exponential. In a graphical analysis, $n_e(t)$ is plotted semi-logarithmically as a straight line in the sufficiently late afterglow. In the early afterglow, the experimental values of n_e exceed the values that correspond to an extrapolation of this line. The differences $n_e = n_{e(exper)} - n_{e(extrap)}$ might be regarded as arising from recombination and be plotted against t^{-1}. The resulting straight line yields α_e, following equation 3.85.

However, it was shown by Gray and Kerr[105] that the recombination coefficients so deduced by analysing data over only one order of magnitude in Δn_e, will often be artificially high. These authors demonstrated that inverse time-dependence of n_e does not necessarily imply that the correct value of α_e will be deduced by analysing the data according to equation 3.90, unless also the condition is fulfilled:

$$\beta = \frac{\alpha_e \Lambda^2 n_{e0}}{D_a} \geqslant 5 \qquad (3.91)$$

The larger the parameter β, the finer the limits of accuracy that can be placed upon experimental values of α_e. In early measurements upon helium, β was scarcely greater than unity, but in more recent studies[104, 106–108] careful attention has been paid to the achievement of high β, using an intense discharge pulse and large diffusion length. Computed solutions of the exact two-process differential equation are to be found in Gray and Kerr's original paper. The most complete series of numerical solutions has been reported by Frommhold and Biondi[109]. The solutions apply to a finite cylinder, a rectangular parallelepiped and some one-dimensional examples. The electron densities are space averaged, assuming weighting functions appropriate to microwave cavity resonator field distributions, since this diagnostic technique is most frequently used.

For measurement of electron attachment in an electro-negative gas, it is necessary to wait sufficiently late in the afterglow for the recombination to be negligible. Calculations of combined attachment and recombination have been made, but the purely exponential regime is always achieved and the real complication arises from the fact that the presence of negative ions affects the ambipolar regime of diffusion. The ambipolar diffusion coefficient D_{ae}, normally measured in the probe or microwave diagnostics of an afterglow, is sensitive to the ratio β' of negative ions to electrons. Thus an exponential decay of n_e, corresponding to a time-dependent D_{ae}, implies either a time-dependent β' ($n_- \gg n_e$ throughout the experiment), or a very small β' ($n_e \ll n_-$ throughout the experiment). When β' is much smaller than unity, it is only at the lowest electron densities ($n_e \sim 10^6$ cm^{-3}) that departure from exponential decay is observed. Calculations illustrating these effects are illustrated in *Figure 1.9* (page 49).

The processes governing electron decay in the time-dependent afterglow apply also in a flowing afterglow system.

The measurements of two-body conversion rates may be carried out by observing either: (*i*) the exponential dependence of a species density upon the density of an added species; (*ii*) the exponential dependence of a species density upon the distance down the flow tube axis; or (*iii*) the exponential dependence of a species density upon the time after the ion-forming pulse has been struck.

It is first necessary to obtain a feeling for the rates of flow which are sufficient to convert the time dependences of the species densities to convenient space dependences. Millisecond times imply flow rates of dekametres per second, which are most conveniently obtained with pumping speeds of several hundreds of litres per second. However, the diffusion speeds of reactants must not exceed their rate of flow down the tube; therefore pressures of the order of 1 torr are mandatory. Diffusion pumps do not operate at these pressures, so fast mechanical pumps (Roots blowers) or booster pumps must be used.

The rates of radial diffusion should be sufficient for good gas mixing to take place in a distance short compared with the length of the afterglow. At the same time, the flow down the tube should be laminar, and the diagnostic equipment should be so shaped that the shock fronts produced by the impacting gas are minimized. One method of adding a gas to the buffer is by means of a narrow tube terminated upon the axis of the afterglow tube. Alternatively, the narrow tube may be terminated in a sprinkler tube containing many holes and bent into a flat spiral.

The flow velocity on the afterglow tube axis can be measured by pulse techniques, using excitation and photodetection, or ionization and charged particle detection. The flow rate can be measured by reduction of pressure in a gas reservoir, or by pressure differential across a flow resistance. Correspondence of axial flow velocity with twice the velocity calculated from average flow rate is proof that the flow is laminar (parabolic). The adequacy of gas mixing may be tested by optical monitoring, but in this case the gas should be optically thin. Mass spectrometric tests of mixing efficiency may also be made.

A schematic illustration of a flowing afterglow system is provided in *Figure 3.10*. The gas temperature can be raised by electric discharge, and pre-heating of an entire apparatus has been achieved (80°–600° K). Shock-heating of afterglow recombination equipment has been reported[110] and could possibly be extended to fully flowing systems. A shock driven through a monatomic gas provides a temperature bath accurately calculable and operating for periods of about 1 msec at temperatures up to 10^4 °K. Some movement of the gas molecules takes place as the shock moves through them; this can be measured and taken into account.

The ionization production technique nearly always produces metastable species, particularly those of helium, the most popular carrier gas in a flowing afterglow system. Helium metastables can be destroyed by the addition of hydrogen or argon gas. Oscilloscope monitoring of ionic species is most helpful in setting up the optimum conditions of study in the flowing afterglow.

Metal vapours can be introduced into a flowing system by evaporation from a furnace, and ozone can be introduced from a commercial ozonizer, from which flows a mixture of O_2 and O_3. A microwave discharge applied to nitrogen (or other molecular species) before introduction will produce both excited and atomic species. Vibrationally excited nitrogen can be removed by a glass wool plug, which will still transmit atomic nitrogen. Chemical titration of added gases is possible, with optical emission as the end-point.

This chapter continues with discussions of the sources, methods of analysis and methods of detection of the possible species needed in atomic collision experiments.

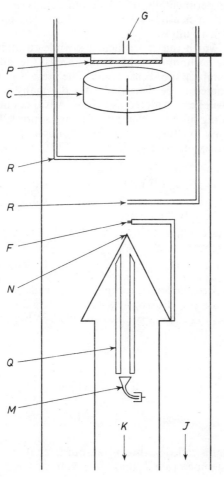

Fig. 3.10. Schematic diagram of flowing afterglow system: G, main gas inlet; P, porous plug; C, cylindrical cathode and wire anode; R, reactant gas inlets; F, floating double probe; N, nose-cone (movable) with ion sampling orifice; Q, quadrupole mass-spectrometer; M, channel multiplier and Faraday cup; J, to Roots mechanical pump; K, to high-vacuum system

3.6 SOURCES OF ATOMIC AND MOLECULAR BEAMS

Since the appearance of the first edition of this book, the technique of producing molecular beams has undergone change. For almost 50 years, the effusive flow of gas from a chamber (oven) into a vacuum had been exploited as the

11*

source of the beam of atoms or molecules. But it now appears possible to exploit flow through many-channel arrays, and also supersonic flow from a nozzle; this latter method produces beams over two orders of magnitude more intense, of narrower velocity distribution, and under certain circumstances much faster than before. Nevertheless the nozzle technique, whilst reliable in that experimentally tested design details are available, is at the time of writing insufficiently tested in the school of experience; it also requires expensive fast pumps.

It is probably worth the experimenter's while to attempt to exploit the supersonic nozzle technique if he really needs the advantages and can afford the high pumping speeds necessary. The present state of the supersonic nozzle art will therefore be discussed, and the many-channel and effusive flow techniques will be considered as before. Advances have also been made in the understanding of many-channel capillary sources, but these, although superior in intensity to effusive sources, do not compare with the supersonic nozzle. Typical (maximum) axial intensities $\mathscr{J}(0)$ are as follows:

1. Effusive source (source mean free path $l_f >$ source orifice dimension d), $H_2 : \mathscr{J}(0) = 2\cdot5\times10^{17}$ mol. steradian^{-1} sec^{-1};
2. Effusive source (source mean free path $l_f <$ source orifice dimension d), $H_2 : \mathscr{J}(0) = 2\cdot5\times10^{18}$ mol. steradian^{-1} sec^{-1};
3. Many-channel capillary source (crinkly foil), $CO_2 : \mathscr{J}(0) = 5\times10^{18}$ mol. steradian^{-1} sec^{-1};
4. Supersonic nozzle, $H_2 : \mathscr{J}(0) = 2\times10^{19}$ mol. steradian^{-1} sec^{-1}.

Reviews of the supersonic nozzle technique are available[111, 112]. A number of monographs discuss the effusive flow source[113-115]. Data for the intensities and mean energies of different sources are given in Figure 3.14.

Effusive Flow Source

An atomic or molecular beam of mean velocity

$$\bar{v}_r = \left(\frac{8kT_s}{\pi m}\right)^{1/2} \qquad (3.92)$$

effuses from a thin-walled orifice in a source chamber containing gas at pressure p_s torr, temperature T_s and molecular mass m, in accordance with the relation

$$\mathscr{J}(\theta) = \frac{n_s\bar{v}_r\mathscr{A}\cos\theta}{4\pi l^2} \qquad (3.93)$$

so that on the axis normal to the orifice

$$\mathscr{J}(\theta) = \frac{3\cdot513\times10^{22}p_s\mathscr{A}}{(AT_s)^{1/2}} \text{ mol. steradian}^{-1} \text{ sec}^{-1} \qquad (3.94)$$

Here equation 3.93 represents the flux falling on unit area, at distance l from the orifice and at polar angle θ to the axis perpendicular to the orifice plane; $\mathscr{A}$ is the orifice area, n_s the source gas density, and A is the atomic or molecular weight. Note that the number density *in the beam* ($\mathcal{J}/v$), is proportional to $p_s/vA^{1/2}$. Since v is proportional to $A^{-1/2}$ the number density depends only on p_s. Therefore, in a crossed-beam experiment, the signals arising from two different gases depend only on the source pressures. This fact may be used in the measurement of relative cross-sections, which are calibrated against a previously established cross-section.

The intensity cannot be raised without limit simply by increasing p_s, since two limiting conditions apply. Firstly, the source pressure must not exceed the value at which the source gas mean free path l_f is equal to the small dimension d of the orifice. The equality

$$l_f = d \tag{3.95}$$

is known as the Knudsen condition or limit. The theory and practice of source operation above the Knudsen limit have been discussed[116-119]. Above the Knudsen limit, the effusing beam has usually been considered to be no longer free from collisions; its intensity is reduced by cloud formation in front of the orifice. However, gas-dynamical considerations predict that the effect should not produce such serious intensity limitation as was previously anticipated, and in fact a nitrogen beam intensity has been increased thirty-fold by raising d/l_f from one to 70. Fluxes as intense as one-sixth of those obtained with supersonic nozzles have been achieved.

For the application of the Knudsen condition, it is usual to take the 'hard sphere' molecular diameters ς or cross-sections (*Figure 1.2*, pages 16–17) calculated from viscosity measurements. Then

$$l_f^{-1} = \sqrt{2}\,\pi n_s \varsigma^2 \tag{3.96}$$

For nitrogen at 300°K, $\varsigma = 3\cdot6\times10^{-8}$ cm.

The Knudsen condition concerns only the small dimension of the orifice, that is, the diameter of a circular aperture or the width of a slit. Therefore a slit, having greater area $\mathcal{A}$, will, according to equation 3.94, emit a more intense flux than an orifice, since the source can in either case be maintained at the same pressure without violating the Knudsen condition. A slit is 25 per cent superior to a one-dimensional array of holes. An array of slits, producing a rectangular orifice, maximizes the intensity without geometrical sacrifice, but the permissible minimum distance between slits is not known.

For mercury atoms at 100°C, with $p_s = 0\cdot3$ torr, $l_f \sim 0\cdot1$ mm; for a slit of length 5 mm and width 0·1 mm, $\mathcal{J}(\theta)$ at $l = 10$ cm is $1\cdot5\times10^{14}$ atoms cm^{-2} sec^{-1}. This corresponds to the bombardment of the area in question by the molecules of residual gas at a pressure around 10^{-6} torr (*Figure 1.1*). Fast ultra-high vacuum pumps and beam modulation techniques make experiments with such effusion sources possible, but a greatly improved intensity would be highly desirable.

The second limit upon p_s in equation 3.94, which therefore limits the effusion source intensity, is imposed by the pumping speed available in the region of experiment. This limits the length, or area, of the slits which can be used, but the limit is supplementary to the Knudsen limit, not alternative to it. One may calculate the maximum intensity available in a system with pumping speed C litres sec^{-1}, such that the gas pressure in the experimental region is held down to a value p at temperature T:

$$\mathscr{I}(0) \simeq \frac{10^{22} pC}{\pi l^2 T} \tag{3.97}$$

It is usually necessary to separate the region of collision experiment from the region of the effusive source by an intermediate pumped chamber. Typically, the chamber containing the source is at 10^{-4}–10^{-5} torr, the intermediate chamber at 10^{-6}–10^{-7} torr, and the collision region chamber at 10^{-7}–10^{-8} torr.

The pumping-speed limitation on molecular beam intensity is interesting because it emphasizes the fact that, owing to the widely divergent effusion from a thin orifice, by far the greatest part of the effusing gas is wasted.

A further limit on oven gas pressure is imposed when there is unwanted dimerization of condensable species, such as lithium. The dimer fraction is given[120] by the equation

$$\gamma = \frac{4\pi^{1/2} h R^2}{(mkT)^{1/2}} \frac{n_0}{T} \frac{273}{760} p \exp\left(\frac{D_e}{kT}\right) \tag{3.98}$$

where R is the radius of gyration of the dimer molecule and D_e its energy of dissociation; n_0 is Loschmidts' number, and p is the pressure in torr. For atmospheric pressure in the oven source, the proportion of Na_2 in the beams is of the same order as that of Na[121]. The presence of molecular clusters of larger size has been observed, and the effect is particularly serious because their ionization potential is intermediate between that of the single atom and the work function of the metal[121, 122].

Ovens for the vaporization of alkali metals and other commonly used easily detectable atomic beams must incorporate various design features[123].

Many-channel Capillary Sources

One can utilize a larger fraction of the effusing molecules, and gain in intensity whilst saving in vacuum pumps, by allowing the molecules to effuse through an array of n_c (hypodermic) capillary tubes instead of a thin orifice[115, 124]; angular resolution (FWHM) $\Delta\theta_{1/2}$ as small as 0·05 rad can readily be obtained with 'crinkly foil', made by serrating metal foil between gears, rolling it up and packing it into a tube. The method is particularly important when only small quantities of gas are available. The behaviour of these devices has been shown[125] to follow the equations

$$\mathscr{I}(0) = \frac{3^{1/2}}{8 \times 2^{1/4}} \frac{\bar{v}_r^{1/2} a^{1/2} \mathscr{N}^{1/2}}{\pi^{1/2} \zeta} \quad \text{mol. steradian}^{-1} \text{ sec}^{-1} \tag{3.99}$$

and

$$\theta_{1/2} = \frac{2^{7/4} \times 3^{1/2} \zeta \mathscr{N}^{1/2}}{1 \cdot 78 a^{1/2} \bar{v}_r^{1/2}} \tag{3.100}$$

for sufficiently long tubes and high flow rates $\mathscr{N}$; that is, for the condition

$$\frac{3^{1/2} L \zeta \mathscr{N}^{1/2}}{2^{3/4} \bar{v}_r^{1/2} a^{3/2}} \gg 1$$

Here it is assumed that the gas is emerging from a single capillary of length L and radius a (satisfying the Knudsen condition); ς is the gas-kinetic diameter and $\mathcal{N}$ is the rate of gas flow (mol. sec^{-1}). To apply these equations it is necessary to measure the flow rate. Experiments have shown that equation 3.99 is correct absolutely, the dependence of both $\mathcal{I}(0)$ on $\mathcal{N}^{1/2}$ and of $\theta_{1/2}$ on $\mathcal{N}^{1/2}$ are also correct. Angular resolution obtainable by this technique is contrasted with orifice effusion angular resolution in *Figure 3.11*.

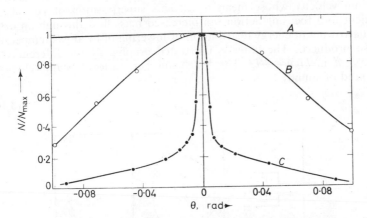

Fig. 3.11. *Angular flux distribution (in radians) in molecular beams, normalized to unity at* $\theta = 0$: *A, cosine flux distribution from thin-walled orifice; B, flux distribution from cylindrical array of 19 hypodermic needles of internal diameter 0·25 mm and length 13·5 mm*[124]; *C, flux distribution from crinkly foil channel*[115]

The ratio $\mathcal{R}$ of intensities obtainable from capillary and orifice sources, with the same pumping speed, is

$$\mathcal{R} \simeq \frac{0 \cdot 32}{\pi^{1/4}\varsigma} \left(\frac{\bar{v}_r}{\mathcal{N}} \right)^{1/2} (\mathcal{A}\tau n_c)^{1/4} \qquad (3.101)$$

where $\mathcal{A}$ is the total area of the orifice or capillary source, τ is the transparency of the capillary source; in approximate terms

$$\theta_{1/2} \simeq \frac{1 \cdot 05}{\mathcal{R}} \qquad (3.102)$$

The parameter $(\tau n_c/\mathcal{A})^{1/4}$ may be regarded as a figure of merit for a capillary source. For an array of hypodermic needles it scarcely exceeds unity, but for crinkly foil it can be as high as 3·7. A porous plastic matrix[126] can be made superior to crinkly foil, up to 4·3. Conflicting results have also been reported[127]. Perforated glass discs of nearly 90 per cent transparency are commercially available.

Since low-velocity molecules are preferentially scattered as the gas passes through a capillary source, beams from these sources have mean velocities higher than those from effusion sources.

Supersonic Nozzle Sources

The supersonic nozzle technique possesses advantages of intensity, acceleration, and narrowness of velocity distribution. When a source nozzle is
designed after the fashion of *Figure 3.12* with orifice and source pressure
such that the Knudsen limit is well exceeded, the gas issuing from it is travelling with mean velocity v relative to the laboratory; there is in addition
a random velocity whose mean is $\bar{v}_r$. This superposition of velocities is
equivalent to a focusing action; an increased mass flow through an orifice
and a narrowing of energy distribution or lowering of local gas temperature
are also produced. The latter is illustrated graphically for different Mach
numbers $\mathcal{M}$ in *Figure 3.13*. The Mach number (ratio of beam velocity to
local speed of sound)

$$\mathcal{M}^2 = \frac{v^2}{\gamma R T_{\text{local}}} \tag{3.103}$$

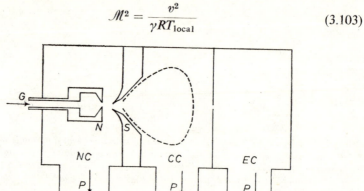

*Fig. 3.12. Schematic diagram of supersonic nozzle atomic beam source: N, nozzle (internal
diameter 0·075 cm); S, skimmer (internal diameter 0·075 cm); NC, nozzle chamber; CC,
collimating chamber; EC, experiment chamber; G, gas inlet; P, pump*

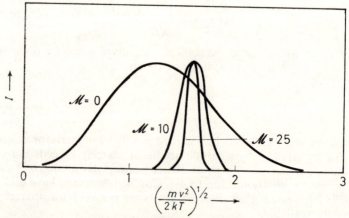

*Fig. 3.13. Calculated axial velocity distributions for monatomic gas nozzle source beams
of Mach number $\mathcal{M}$ at skimmer; the distribution for thin-walled orifice beam ($\mathcal{M} = 0$) is
shown for comparison*

is conveniently used to describe supersonic beams because the local speed of sound is a measure of temperature. With speed ratio w defined as

$$w = \left(\frac{mv^2}{2kT_{\text{local}}}\right)^{1/2} \tag{3.104}$$

then

$$\mathcal{M} = w\left(\frac{2}{\gamma}\right)^{1/2} \tag{3.105}$$

where γ is the specific heat ratio. After isentropic expansion,

$$\frac{T_{\text{local}}}{T_{\text{nozzle}}} = \left(1 + \frac{\gamma+1}{2}\mathcal{M}^2\right)^{-1} \tag{3.106}$$

As the gas moves further from the nozzle, it approaches molecular (as opposed to viscous) flow, eventually reaching this condition after a distance l_1. At a certain distance l_s downstream from the nozzle, a 'skimmer' is introduced to remove the off-axis portion of the flowing gas. A collimator may be introduced still further downstream.

At a distance x downstream from the skimmer entrance, the beam flux (provided $\mathcal{M} \geqslant 4$) is[111]

$$\mathcal{J} \simeq n_s\left(\frac{l_s}{l_1}\right)^2 \frac{r_s^2 v^2}{x^2}\left(w_1^2 + \frac{3}{2}\right) \tag{3.107}$$

Here r_s is the radius of the skimmer entrance, and n_s is the gas density at the skimmer; $w_1 = \mathcal{M}(\gamma/2)^{1/2}$ is the speed ratio after the distance l_1 from the nozzle. By contrast, for an oven beam effusing from the skimmer,

$$\mathcal{J} = \frac{n_s r_s^2}{x^2}\left(\frac{8kT_s}{\pi m}\right)^{1/2} \tag{3.108}$$

For nozzle beams, w_1 and l_1 simply become w_s and l_s if $l_s \ll l_1$. When $n_s \sim l_s^{-2}$, beam intensity is independent of l_s provided $l_s > l_1$ and no collisional losses occur. The ratio $\mathcal{R}$ of nozzle source to oven source intensities is

$$\mathcal{R} \simeq \left(\frac{\pi C_p}{k}\right)^{1/2} \gamma \left(\frac{l_s}{l_1}\right)^2 \mathcal{M}^2 \tag{3.109}$$

for specific heat C_p at constant pressure. At $\mathcal{M} = 10$ for nitrogen, the nozzle beam is almost 500 times as intense as the oven beam.

The Mach number depends primarily upon the nozzle design, in particular upon the ratio of nozzle exit area A_e to nozzle throat area A_t. Thus for $A_e/A_t \simeq 540$, $\mathcal{M} \simeq 10$, and for $A_e/A_t \simeq 15\,000$, $\mathcal{M} \simeq 20$. In the original proposals of Kantrowitz and Grey[128] it was believed necessary to design a converging-diverging nozzle according to standard gas-dynamical practice, but it now appears that the diverging section is not entirely necessary. Since most of the gas is removed before the skimmer, it is the region between nozzle and skimmer that the fastest pumping is required ($\geqslant 1000$ litre sec^{-1} for preference). It should be emphasized that intensive research into the theory and practice of nozzle beams is at present in progress, and the ideas presented here are far from complete. For example, the vibrational and rotational temperatures of molecules in a nozzle beam are not yet predictable, although it is known that $T_{source} > T_{vib} > T_{rot} > T_{trans}$.

Many nozzle beams are underexpanded in the sense that they do not achieve their maximum Mach number until well downstream from the nozzle. According to computations of the Owens–Thornhill type[129, 130], a Mach 1 beam at the nozzle can reach Mach 8 at a distance of 10 nozzle diameters downstream. But at slightly larger distances, the axial intensities are improved.

It is at present necessary for a collision physicist making use of a nozzle beam to analyse its velocity distribution, using methods discussed in Section 3.12. Anderson and Fenn[131], using the time-of-flight technique, found the maximum Mach number $\mathcal{M}_{max}$ achieved in an argon beam to be given by

$$\mathcal{M}_{max} = 1 \cdot 17 K^{-0 \cdot 4} \qquad (3.110)$$

where K is the Knudsen number, or ratio d/l_f. It is expected that gases other than argon will obey a similar relation, with exponent $(1 - \gamma)\gamma$.

Some experimental verification of theory has been obtained. The intensity of the beam at the collimator $n_c \bar{v}$, is related to the computed Mach number[129] as follows:

$$n_c \simeq \frac{n_0 \mathcal{M}^2 \gamma a^2 / 8 d_1^2}{[1 + \frac{1}{2}(\gamma - 1) \mathcal{M}^2]^{1/(\gamma-1)} + c a n_0} \qquad (3.111)$$

with skimmer diameter a, pre-nozzle gas density n_0, skimmer–collimator distance d_1, and

$$c = \frac{\pi \varsigma^2}{2^{1/2}}$$

The collision diameter ς at the appropriate low temperature is usually not available, but in the experiments with nitrogen a realistic value of $8 \cdot 8$ Å was obtained.

It may be necessary to test for molecular aggregation in the nozzle beam, with the aid of an ionizer and a mass-spectrometer. Condensation in supersonic flows is a well-known phenomenon in high Mach number wind tunnels and steam turbines; mass-spectrometric observations[132] have demonstrated dimerization of polar molecules such as CO.

There is one highly attractive aspect of nozzle beams which has been pointed out by Anderson, Andres and Fenn[111]; this is their ability to produce fast beams if gas mixtures are used. The oven beam is only a relatively small factor $(kT/2)$ hotter than the gas temperature; this is because of the selectivity of the orifice for molecules of higher velocity. The nozzle beam is only slightly different, in that the final beam energies which can be attained still depend essentially on source temperature. Because the expansion process converts total enthalpy into beam translational energy, the final beam energies are higher than pre-nozzle gas temperature only by a factor which for $\mathcal{M} > 4$ is approximately $2\gamma/3(\gamma - 1)$.

In nozzle beams it is possible to achieve high temperatures by pre-mixing the gas with high proportions (100 : 1) of a much lighter gas, such as hydrogen. In the expansion process, there are sufficient collisions to maintain energy and momentum equilibrium between the two species. The mean jet velocity will be almost that of pure hydrogen, provided that $\mathcal{M} > 4$. The dilution with unwanted hydrogen, however, is an experimental inconvenience.

There is a separation effect in gas mixtures[133], whereby the lighter molecules migrate radially from the beam axis, leaving a higher concentration of heavier molecules. Experiments[134] lead to the conclusion that there are shock-like skimmer-jet interactions in typical nozzle beam experiments.

In order to achieve beam energies greater than 1 eV, it would be necessary even with a nozzle beam to maintain source temperature substantially above 3000°K. Attempts have been made to achieve this by pre-heating the gas upstream of the nozzle by means of a reflected shock wave[135]. Such heating is transient, so that the nozzle itself does not become too hot. However the transience clearly poses additional problems in the analysis of the experiment. A different approach is to cool the nozzle and heat the source gas by means of an arc[136]; this produces surface problems, and can also introduce metastables.

A shock-heated nozzle source should be shuttered so that the beam only passes into the experiment for a short time (milliseconds); this has the very great advantage that the fast mechanical or booster pumps necessary in nozzle beam research are avoided. The development of shutter valves is of great importance.

Nozzle beams have been exploited[137, 138] for the production of high intensities of condensable species such as sodium. Skimmers of special design are necessary.

Fast Atomic Beam Production

There are three contrasted techniques of producing fast neutral beams which are in competition with the hydrogen-diluted nozzle beam. The first is the comparatively neglected, but recently revived, technique of accelerating oven beams by tangential collisions with a moving surface[139]. The second is the sputtering technique, which is discussed below. The third technique, which has had considerable success, is to neutralize an ion beam by charge transfer collisions[140, 141]. Since there is very little momentum transfer in such an inelastic collision, the neutral beam will maintain almost the same velocity distribution as that in the ion beam. The techniques of ion velocity analysis are of importance, but in many experiments intensity requirements preclude the use of ion velocity or mass analysis before the neutralization. It should be remembered that many ion sources produce velocity distributions several electron volts wide; in general, the electron impact ion sources operating with smallest electric field strengths and highest gas pressures produce the narrowest velocity distributions.

Resonant charge transfer collisions are normally used for neutralization, since it is likely that, in the lower velocity range, the large cross-section (Chapter 13) militates against the production of metastables and against the neutralization of impurity ions. Furthermore, the polar scattering angle θ for a charge transfer collision of energy defect ΔE at impact energy E is

$$\theta = \frac{1}{2}\left(\frac{\Delta E}{E} - \frac{m_e}{m_+}\right) \qquad (3.112)$$

In general, therefore, resonant collisions ($\Delta E = 0$) have smaller scattering angles than those with impurity ions, which are deflected from the beam during neutralization. Possibly the most important reason for using neutralization in the ion-forming gas is one of convenience. An apparatus can readily be constructed[90, 142] in which the source chamber is separated from the acceleration and neutralization chamber by only a hemispherical grid through which gas diffuses.

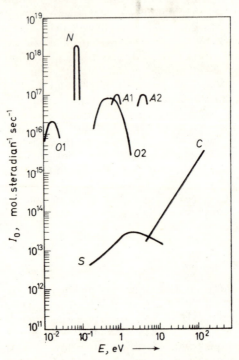

Fig. 3.14. *Molecular beam intensities obtainable with various sources*[145]*: C, ion neutralization by charge transfer; S, sputtering (potassium); O1, low-temperature oven; O2, oven T = 2500°K; N, nozzle beam T = 300°K; A1, argon with 99 per cent helium in nozzle beam; A2, argon with 99 per cent helium in nozzle beam T = 2000°K*

Unneutralized ions must be removed from the beam by means of a transverse electrostatic field; metastables may possibly be produced by ion or electron bombardment of the surfaces of the electrodes producing the field.

A contained plasma device especially suitable for producing hot neutral beams[143] is as follows. A low-pressure radiofrequency ion source produces plasma which diffuses, without acceleration, through a relatively wide orifice (~ 1 cm diameter). The plasma is confined in the form of a beam by a longitudinal magnetic field, which is provided by a long solenoid. Bending of this solenoid, with a large (20 cm) radius of curvature, allows the plasma 'beam' to follow round the axis of the solenoid, whilst the un-ionized part of the effusing gas continues straight into the throat of a pump. The confinement of the plasma beam raises its potential a few volts above ground, presumably because the electrons diffuse radially outwards faster than do the positive ions. The plasma beam is now crossed with a gas jet for neutralization by resonance charge transfer, and a transverse electric field applied to remove the electrons and ions. A flux of hot neutrals, $\sim 10^{14}$ atoms cm^{-2}, with an average energy of 1–2 eV, is produced; this is an order of magnitude more intense than the conventional neutralization source. It is also more intense than the sputtering source, in which a kilovolt ion beam is directed onto a metal surface, from which the sputtered atoms emerge with low energies.

The sputtering technique[144] has been used for the production of potassium beams which by velocity selection have been shown to have velocities between 0.22×10^5 cm sec^{-1} and 9.4×10^5 cm sec^{-1} (0.01–17.5 eV). A solid potassium target is bombarded with a 10 keV Ar^+ beam of 500 μA intensity. The useful energy range which it is now possible to produce is 0.5–35 eV. The angular distribution of the sputtered flux is $\cos^2 \theta$. It is necessary to focus the ion beams with great refinement and, while the target is maintained at earth potential, to rotate and watercool it. The possibility of using oxide, nitride or lithium hydride targets for atomic oxygen, nitrogen or hydrogen beams is attractive.

A graphical comparison of the intensities obtained with different systems at different energies[145] is given in *Figure 3.14*.

A recently proposed technique of fast atom production makes use of the photodetachment of negative ion beams[146]. With 50 W of white light from a 5000 W xenon–mercury discharge lamp, it is possible to neutralize one ion in 10^4 of an H^- beam. Interruption of the light beam produces an alternating current of H^- ions, detection of which enables the flux of H to be inferred absolutely. Use of high-power CO_2 laser beams to photodetach negative ions such as NO^- might provide molecules in the $v = 0$ vibrational state.

3.7 SOURCES OF ATOMIC GASES AND OF RADICLES

A source of atomic hydrogen or similar atomic gas or radicle consists of a gas-containing chamber into which energy is fed for dissociating the molecules. The neutral atoms are allowed to effuse from an orifice, and, if necessary, charged particles are removed; excited atoms or molecules cannot easily be eliminated where they occur. The different methods of applying energy are as follows.

Glow Discharge Source

Atomic hydrogen can be produced by a d.c. glow discharge, as used in the well-established 'Wood's tubes'[147]. Typically, these consist of glass tubes (~ 50 cm in length and ~ 3 cm in diameter) containing gas, through which a discharge (~ 0.05 A) is passed between aluminium electrodes maintained at a potential difference ~ 10 kV. The atoms are extracted at a point well removed from the electrodes, so as to avoid sputtering impurities. Wall reassociation may be minimized by using hydrogen with added water vapour, and, if necessary, by coating the tube walls[148] with a glaze of orthophosphoric acid or polytetrafluoroethylene. Almost complete dissociation of hydrogen can be obtained. A similar coating of boric acid is suitable for atomic chlorine, which can also be produced in a discharge.

Effusion of Gas from a Heated Furnace

Where furnace sources are practicable, they represent the most effective method of atomic gas production in that they produce neither ultra-violet photons nor atoms in long lifetime excited states of high energy; excited atoms in ground state fine structure, such as exist in the periodic table groups III and VII, cannot be eliminated from thermal source atomic beams and must be calculated from the Boltzmann factor. The sources do not demand the use of large radiofrequency powers, which might interfere with low-level circuits used in the experiment. However, they are valuable only where the molecular dissociation energy is sufficiently low for the required temperatures to be achieved without undue experimental difficulty; thus they are suitable for hydrogen but not for nitrogen. Nevertheless, concentrations of atomic nitrogen as high as 20 per cent can be obtained by the use of a furnace. Atomic chlorine has been produced in a graphite furnace. For atomic oxygen, a furnace can be made from refractory oxides.

The dissociation of a diatomic molecule AB is governed[149] by an equilibrium of the type:

$$\ln K = \ln \left(\frac{p_A p_B}{p_{AB}} \right) = -\frac{D}{kT} + \ln \left[\frac{\mu \Sigma}{10^{40} I} \left(\frac{T}{100} \right)^{3/2} \right]$$

$$-\ln \left[1 - \exp \left(-\frac{\Theta_v}{T} \right) \right] + 6 \cdot 94 + \ln \left(\frac{g_A g_B}{g_{AB}} \right) \qquad (3.113)$$

In this equation the pressures p_{AB}, p_A and p_B are measured in atmospheres, D is the dissociation energy at $0°K$ in kcal mole^{-1}, Σ is a symmetry number (2 for a homonuclear molecule and 1 for other molecules), I is the moment of inertia in g cm^2, whilst μ is the reduced mass measured in atomic units; the symbols g_{AB}, g_A and g_B represent statistical weights. The characteristic vibrational temperature Θ_v is related to the fundamental frequency of vibration by the relation

$$\Theta_v = \frac{h \omega c}{k} \qquad (3.114)$$

The quantity $6 \cdot 94$ in equation 3.113 is made up of pressure conversion factors, and also of the quantities h, k and N_0. In the equation, there are implicit a large number of assumptions but only slight errors.

It is not always easy to determine the pressure within a high-temperature oven, but it may be achieved by measuring the mass M of gas effusing through an orifice of area $\mathcal{A}$ in a time t. This is given by the expression:

$$M = \left(\frac{m_{AB}}{2\pi RT} \right)^{1/2} \mathcal{A} t \left(p_{AB} + \frac{p_A + p_B}{\sqrt{2}} \right) \qquad (3.115)$$

where the m_{AB} represents the molecular mass in atomic units. It can be seen that, for homonuclear molecules, dissociation is more complete at lower pressures; this feature limits the attainable intensity.

The first serious attempt at atomic hydrogen beam production from a furnace was made by Lamb and Retherford[150]. An improved type of source developed in the studies of Fite and Brackman[151] has become popular; it consists of a cylindrical tungsten tube ~ 8 cm long and 0.5 cm in diameter, supported on water-cooled electrodes and joule-heated to temperatures between $1500°K$ and $3000°K$. The tube is plugged at both ends with tungsten wool for thermal equilibration of the gas which passes through it. The atom beam issues from a small orifice in the tube wall. Although appropriate calculations might be made of the ratio of atoms to molecules from equation 3.113, it is essential to monitor this ratio mass-spectrometrically. Under the conditions described above, a dissociation ~ 90 per cent can be obtained.

Atomic hydrogen furnace sources do not appear to function in thermodynamic equilibrium, since they are capable of producing high proportions of atoms without these having undergone many collisions in the gas phase, or even with the furnace walls. The mechanism responsible for the dissociation is not well understood, for the dissociation of hydrogen in the tungsten furnace is much more efficient than the temperature warrants. This could be because molecular hydrogen is chemisorbed as the molecular species WH, whose dissociation energy might be considerably smaller than that of molecular hydrogen. Although the interior of tungsten furnaces is frequently used for scattering experiments demanding a proportion of atomic hydrogen, the actual H/H_2 ratio in the furnace interior is not very well known. It might be monitored[152] by passing kilovolt protons through the oven under single collision conditions; a post-collision electrostatic analyser measures the H^- intensity. Under single collision conditions, this ion can only be formed by two-electron capture from molecular hydrogen.

Radiofrequency Sources

Radiofrequency sources have been successfully used for dissociating N_2, O_2, Cl_2 and H_2. The application of the power, usually 10–50 W, and the general principles, follow the lines of the discussion of radiofrequency ion sources. However, the beam extraction is easier because of the absence of space charge. From the particles which effuse from an orifice, those with charges must be removed by transverse electric fields, so that only photons, excited atoms and neutrals remain. For nitrogen[153], a 5 per cent mixture of the gas in helium is effective.

By far the most easily available radiofrequency sources are those which make use of high-power microwave valves and circuitry. For the induction of the largest potentials in the gas, the applicator must possess the highest possible impedance; this is achieved by means of a resonant cavity with the highest selectivity. Since the cavities illustrated in the first edition of this book, experimentation has resulted in improved designs. In particular, the cavity illustrated in *Figure 3.15* is the most successful of five investigated by Fehsenfeld, Evenson and Broida[154]; it is commercially available.

The walls of the tube in which gas is passed through the cavity must be made of material of low dielectric loss and high heat resistance; fused silica and Vycor are the most successful. Continuous-wave magnetrons providing power outputs in the region of 200 W are readily available commercially, since there are wide applications in diathermy and dielectric heating. At this power, some water-cooling or air-cooling is required, but if necessary the power input to the cavity may be as low as 25 W or as high as 1500 W. It is usual to allow the gas temperature to reach $\sim 1000°$K.

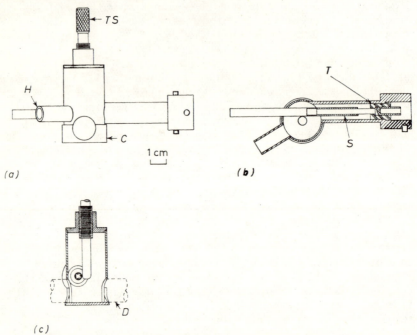

Fig. 3.15. *Design of cavity resonator for microwave discharge source of plasma:* (a) *master plan (C, removable cap; H, air-hose connexion; TS, tuning stub); (b) section (S, coupling slider; T, teflon); (c) section (D, discharge tube)*

When the flow line is unconstricted by orifices, there is evidence[155] that some of the emerging atoms (in this case nitrogen or oxygen) are in long-lifetime excited states. Reduction of these excited states by wall collisions still leaves a large fraction of the gas in the atomic state. Thus, glass wool plugs have been used to remove the excited states from atomic nitrogen beams. Hydrogen metastables have no laboratory stability, but wall collisions produce effective recombination of the atoms. Wall collisions can be avoided by striking the discharge in the skimmer region of a supersonic nozzle beam; provided that care is taken to avoid the disturbance of the flow by the discharge[156], this technique of atomic hydrogen production is competitive with the tungsten furnace.

'Plasma Dissociation' Sources

Caesium-seeded hydrogen discharges are particularly effective hydrogen atom producers on account of three-body collisions of the type

$$Cs^+ + H_2 + e \rightarrow Cs + H + H$$

Caesium vapour is added to the hydrogen, and the caesium ions preponderate over hydrogen ions, and carry the electric current. The recombination energy of the caesium ions is of suitable magnitude to dissociate the hydrogen.

Chemical Sources

Naturally there are large numbers of purely chemical techniques for producing streams of radicles. It is not proposed to discuss them in detail, since they are more appropriate to chemistry textbooks. They are powerful techniques in flowing systems, for such atomic gases as oxygen, which is conveniently produced by the reaction

$$N + NO \rightarrow N_2 + O$$

Atomic nitrogen[157] is produced by microwave discharge and 'titrated' by the addition of NO. The technique is greatly superior to the well-known thermal decomposition of ozone.

Charge Transfer Neutralization of an Ion Beam

There remains a technique for hot molecule beam production which can be used to produce atomic gas and radicle beams: namely, charge transfer neutralization of an ion beam. This is discussed in Section 3.6. The fact that symmetrical resonance charge transfer cannot be used for atomic gas and radicle beams is a great disadvantage as far as intensity is concerned. It is necessary to choose and investigate a suitable neutralizing gas capable of converting an atomic ion beam (for example N^+) into fast atoms in the ground state. Difficulty can arise from contamination of the ion beams not only by long lifetime states, but by multiply charged ions impossible to mass-separate (for example, N_2^{2+} in N^+ beams).

3.8 SOURCES OF ELECTRONS

The electron sources used in collision experiments include thermionic, photoemission, field emission and β-ray sources. Of these, the first is by far the most important, and will be discussed in some detail. Some excellent reviews of the subject exist, and that by Nottingham[158] remains valuable.

Under conditions such that space charge can be neglected, with emitted electrons in equilibrium with those in the metal, the energy distribution of emitted electrons is given by the equation:

$$dN_e(E_e) = \frac{4\pi m_e \mathscr{A} E_e}{h^3} \exp\left(-\frac{\phi + E_e}{kT}\right) dE_e \qquad (3.116)$$

where dN_e is the number of electrons emitted per second with energies between E_e and $E_e + dE_e$; T is the absolute temperature, $\mathscr{A}$ is the surface area of the emitter, and ϕ is its work function. Thus for a small plane collector at a potential $-V_c$ volts with respect to the cathode, the electron current density should vary as follows:

$$J_e = J_o \exp\left(-\frac{11\,600 V_c}{T}\right) \tag{3.117}$$

where J_0 is the current density observed when $V_c = 0$. The temperature variation of emission is given by the well-known Richardson equation, which is written in the form

$$J_0 = AT^2 \exp\left(-\frac{11\,600\phi}{T}\right) \tag{3.118}$$

with ϕ in electron volts and T in degrees Kelvin. The constant A should in solid state theory have the value 120 A cm^{-2} deg^{-2}. However, complications make it necessary for a semi-empirical approach to be adopted; tabulated values of 'thermionic constants' A and ϕ are found to fit the experimental data for different emitters. Among these complications is the fact that surfaces are sufficiently non-uniform to have different macroscopic and microscopic areas, and the fact that work functions are temperature variant. The work function is related to the Fermi level μ_s, and to the potential energy difference W_a between the bottom of the conduction band and the electron at infinite distance from the surface:

$$\phi = \frac{W_a - \mu_s}{e} \tag{3.119}$$

For metals, μ_s is positive and decreases with rising temperature, whilst for semiconductor emitters, principally the oxide cathode, μ_s is negative and increases in magnitude with increasing temperature. Unless electro-positive layers are absorbed in the metal, W_a decreases with increasing temperature. Empirical thermionic constants, within limited temperature ranges, are given[158] in Table 3.3.

From the Richardson equation (equation 3.118), the familiar exponential increase of emission with temperature is found; only for certain high melting point materials with reasonably small work functions will there exist a temperature range with emission adequate for experiments. The 2000°K temperatures necessary for emission from pure metals possess the disadvantages of producing wide electron energy spread and large heat dissipation. Emission at lower temperatures has been achieved by the following methods.

Oxide–coated Cathodes (BaSrO)

Wehnelt[159] discovered that barium oxide and other alkaline earth oxides, applied to a metal surface, will emit at temperatures far below the range of 'metallic' emitters. These semiconductor cathodes have enormous commercial value, but are less frequently used in laboratory experiments, because of the ease with which they can be 'poisoned', with loss of emission and discoloration, by organic and other vapours. Oxide-coated cathodes are found[160] to emit various negative ions.

The most detailed descriptions of oxide-coated cathode manufacture are those given by Herrmann and Wagener[161]. A solid solution of barium–

Table 3.3. THERMIONIC CONSTANTS

Emitter	T_{min} (°K)	T_{max} (°K)	A (cm^{-2} deg^{-2})	ϕ (eV)
C	1300	2200	30	4·34
	1300	2200	15	4·38
β Fe	1040	1180	26	4·48
γ Fe	1180	1680	1·5	4·21
Mo	1300	2100	115	4·37
	1350	2000	55	4·15
Ni	1300	1700	30	4·61
	1150	1700	50	5·24
Pt	1700	2100	32	5·32
Ta	1200	2000	52	4·19
W	1400	2400	72	4·52
	1200	2000	60	4·51
W–Th	1200	2000	3	2·63
W–Ba	1300	1700	1	1·8
	1300	1700	15	2·0
La–LaB	1100	1900	29	2·66
BaSrO	600	1200	0·5	1·0
Re				5·2
WO$_2$				5·9–6·2
Pt–W				5·5

strontium carbonate in approximately equal proportions by weight is suitable. The crystals are finely divided, mixed with an organic binder and sprayed on the base metal (for example, platinum or nickel). Under moderately good vacuum conditions (not ultra-high vacuum), the carbonate breaks down at a temperature in the range 1200°–1300°K, with the evolution of CO_2 and CO, after which it will emit at somewhat lower temperatures. The initial heating is known as 'activation' and may sometimes be repeated as an antidote to 'poisoning'.

An impression is still prevalent that the energy distribution of electrons from oxide-coated cathodes is in some way anomalous, but the analysis of experiments due to Hung[162] is in good agreement with calculations on the basis of an idealized Fermi distribution of emitted electrons, as computed for the appropriate tube geometry, with thermionic constants as in Table 3.3. Space-charge limited emission should result in a Maxwellian electron energy distribution, and the validity of the Richardson equation shows that this is approximately correct[163].

The principal obstacle in the path of the atomic collision physicist wishing to use oxide-coated cathodes is the poisoning which is caused not only by gases used in the experiment, but by atmospheric gases when the equipment is let up to atmospheric pressure; therefore some further remarks on poisoning may be apposite. Some reversible renovation of a poisoned cathode may be achieved by heating it in the presence of hydrogen, but this is not a lasting effect. Poisoning of a cathode can take place in the presence of most common oxygen-containing gases, such as CO, O_2, CO_2, H_2O.

Water vapour seems to be the most effective poisoner of all, since it plays the greatest part when a cathode is let up to atmospheric pressure. To a large extent, the poisoning effect of this operation can be avoided by maintaining the cathode at $\sim 120°C$ while it is carried out; the atmospheric water vapour cannot condense under these conditions.

Dispenser cathodes (W–Ba)

Dispenser cathodes[164] consist of porous sintered tungsten, molybdenum or nickel, into which barium is absorbed and through which it can diffuse fast enough to maintain the surface composite. They do not poison nearly as easily as oxide-coated cathodes, and are of particular value for the achievement of high current densities. They are often referred to as 'L' or 'D' cathodes.

Thoriation (W–Th)

Tungsten into which a small quantity of thorium had been inserted was found by Langmuir and Rogers[165] to have a work function much lower than that of pure tungsten. This is a result of the creation of a composite surface in which the absorbed polarizable atoms produce a large 'dipole moment per unit area', to which the work function can be related. Caesium and other alkali metals can also be used.

In general, the work function of a metal is supposed to increase when atoms of ionization potential higher than that of the metal are present. When atoms with ionization potential lower than that of the metal are in the surface, the work function is lowered.

For general purposes, the most popular electron emitter is still pure tungsten, or thoriated tungsten, which combines a high melting point with low cost. Although its evaporation rate is smaller than other emitters for the same emission rate, it has the property of evaporating a considerable quantity of pseudo-metallic substance when not thoroughly degreased and cleaned; it is also extremely brittle when once heated to high temperatures. Its vapour pressure at 2200°C is 4×10^{-8} torr; at 2500°C it is 4×10^{-6} torr. Its thermal conductivity at $\sim 1600°C$ is 0·36 cal cm^{-1} °C^{-1} sec^{-1}, and its resistivity at $\sim 1800°C$ is 60·06 $\mu\Omega$ cm^{-1}. It oxidizes and carburizes when heated in appropriate gases (oxygen, CO, CO_2, hydrocarbons, nitrogen oxides, SO_2). It is also attacked by halogens and by nitrogen above 2300°C.

Iodine vapour increases the lifetime of tungsten filaments, apparently by the formation of a space-charge of negative ions which inhibits the evaporation or sputtering of the metal.

The special filament problems encountered in ionic physics are problems of filament operation in atmospheres of reactive gases such as oxygen. Filament lifetimes may be increased by the use of materials such as thoriated iridium[166], which is superior to platinum and tungsten in resisting oxygen and halogens.

A more serious problem is set by the changes in work function resulting from surface chemical reactions. Not only do these influence the temperatures necessary for electron production, but they complicate the study of surface equilibria in the production of, for example, negative ions.

In mass-spectrometry, this filament sensitivity is widely known as 'gas-sensitivity', that is, sensitivity of instrument behaviour to previous infusion of gas into the source. Changes in filament work function are not the only factor; there is also the evaporation of WO_3 on the source electrodes. This oxide is an insulator and may give rise to surface charges, but it can be converted to the form W_2O_4, which is a conductor.

Gas-sensitivity can be largely avoided by the use of other emitters, such as carbon[167], and rhenium[168], which has an unstable carbide. Carburization not only affects the filament work function, but may be partly responsible for accidental burn-out of tungsten, whose carbides melt below the melting point of the metal (W_2C at 2857°C and WC at 2777°C). The rhenium filament has a higher work function and a lower melting point than tungsten, with higher evaporation rate for the same emission. However, it operates very cleanly under normal vacuum conditions.

The operation of indirectly heated emitters in residual pressures of reactive gases also presents problems. In vacuum systems which are likely to contain organic vapours, the conventional barium oxide cathodes are less successful than materials such as barium aluminate, boron lanthanide and, most recently, barium zirconate deposited electrolytically on platinum; these materials are resistive to poisoning and can be exposed to atmosphere without subsequent loss of emission. Barium zirconate can actually be heated for thermionic emission in atmospheric air.

The maintenance of a filament surface at a measurable uniform potential which can be taken as a reference potential in electron energy measurements is a problem whose solution could be important in electron impact studies, but which presents serious difficulties; simpler alternatives offer themselves, for example, the velocity analysis of the emitted electrons. The difficulties are largely those of avoiding surface effects and space charges.

The thermionic emission of cathodes is modified in very strong electrostatic fields by the 'Schottky effect'. This consists of a reduction of work function by the effects of field emission, which are superimposed upon those of thermionic emission. The modified current I' is given in practical units by the equation

$$\ln I' = \ln I + 0.4402 \left(\frac{X}{T}\right)^{1/2} \tag{3.120}$$

for extraction by fields of strength X V cm^{-1}.

There is also a departure from thermionic equilibrium in retarding fields; this in consistent with the 'reflexion' of a number of the slowest electrons at the surface potential boundary[169].

In pure polycrystalline tungsten filaments, the electron energy distributions are modified by patch effects[170].

Energy Distribution of Electrons in Beams

The full width at half maximum (FWHM) ΔV_M of the Maxwell–Boltzmann energy distribution is approximately 0·3 eV for thermionic emission from a filament at 1300°K. The value of ΔV_M depends strongly upon the absolute temperature. These FWHM values are not observed at current densities $\gtrsim 10^{-9}$ A mm^{-2}, since the space charge broadens the distribution. For a 100 μA cylindrical beam at energy $V = 20$ eV, the *total* spread due to space charge is in theory ~ 1 V. In theory it is proportional to current and $V^{-1/2}$.

In some fairly intense beams, a two-peaked ditribution is observed[171].

From a hairpin filament, the half-widths are found by experiment[172] to be as high as 1·0 eV at beam currents 25 μA, rising to 1·2 eV at 50 μA, 1·5 eV at 100 μA and 1·8 eV at 200 μA. The measurements are reported to be consistent with the expression

$$\frac{\Delta V_J - \Delta V_0}{\Delta V_0} = \frac{C J^{1/3}}{V^{1/6}} \tag{3.121}$$

where J is the current density at which the half-width is ΔV_J. The thermal half-width is ΔV_0, and C is a constant.

Energy spread can be reduced by cutting out from a beam the electrons which diverge to large angles θ. The half-width of the distribution of a pencil of electrons is proportional to their θ^2. The reduction is achieved by restricting the cathode area (as with hairpin cathodes, described below), and by inserting stops.

For low electron currents from directly heated filaments, the total energy distribution is, as predicted, Maxwellian, and possesses the correct width[173,174]. In addition to the total distribution, there is a 'normal distribution', describing the energy associated with momentum perpendicular to the cathode surface. This distribution function should have its maximum at $V = 0$, and have FWHM $\Delta V = 0·693\,kT$. In temperature-limited planar diodes the normal distribution is as predicted[175].

Measuiements with a hemispherical momentum analyser[176] of anomalous energy distributions in high-current guns show rather different behaviour from that indicated by equation 3.121. For current density J μA cm^{-2} and FWHM ΔV

$$\Delta V = \Delta V_0 + 1·4 \times 10^{-3} J V^{-3/2} \tag{3.122}$$

With increasing current, the distribution function departs from Maxwellian, becoming more symmetrical.

The theory of anomalous distributions is not yet well understood, but has been approached on the assumption that the beam is in thermodynamic equilibrium in a moving coordinate system[177]. The question of how this equilibrium is reached is still open, since Coulomb interactions are probably too small; it is possible that the equilibrium is only reached after a relaxation time, but this is not borne out by experiment.

Space-charge Limitation and Focusing of Electron Beams

Electron beams in collision experiments are sometimes of rectangular cross-section, sometimes in slit symmetry, and sometimes in orifice symmetry. Some general remarks about handling these systems may be of assistance.

It is convenient for discussions of electron optics, space-charge problems, etc. to be conducted in SI (rationalized MKS) units; this practice will be followed in this section *when stated*.

The current density in an electron beam is more usually limited by space charge than thermally. To calculate what might be expected from a planar diode carrying a potential difference V, it is sufficient to use the Child–Langmuir equation in the simple form in which thermal energy distribution is neglected

$$J = \frac{4\varepsilon_0 V^{3/2}}{9(m_e/2e)^{1/2}d^2} \quad \text{(SI units)} \tag{3.123}$$

or

$$J = \frac{2 \cdot 335 \times 10^{-6} V^{3/2}}{d^2} \text{ A cm}^{-2} \tag{3.124}$$

But it is inaccurate to use this model to calculate the potential distribution in the planar diode, since the effect of the distribution of electron velocities is significant[178]. In the presence of space-charge saturation, electrons starting from a planar cathode with zero velocity would never get started; but if all electrons left the cathode with the same normal velocity, then the potential gradient at the cathode would become negative and the potential function would possess a minimum of depth corresponding to the velocity of emission. This minimum would act as a virtual planar cathode at which electrons would come to rest. The actual Maxwellian distribution of thermionically emitted electrons complicates this picture; the analysis of Langmuir leads to a variation of potential in which there is a minimum of depth V_m below cathode potential, situated at a distance x_m cm from the cathode, where

$$x_m \simeq 0 \cdot 0156(1000J)^{-1/2} \left(\frac{T}{1000}\right)^{3/4} \tag{3.125}$$

and

$$V_m \simeq \frac{T}{5040} \log_{10}\left(\frac{1}{P}\right) \tag{3.126}$$

P is the fraction of emitted current transmitted to the anode, and J is the current density (A cm^{-2}). The transmitted current density is given by

$$PJ = \frac{2 \cdot 335 \times 10^{-6}(V_p - V_m)^{3/2}}{(x_p - x_m)^2}\left[1 - \frac{0 \cdot 0247T^{1/2}}{(V_p - V_m)^{1/2}}\right] \tag{3.127}$$

where the anode receiving the current is at potential V_p and distance x_p from the cathode. Close to the minimum, the potential variation along the direction of electron flow is as the four-thirds power of the distance from the cathode.

Space charge also limits the electron currents that can be obtained in slit symmetry and orifice symmetry. The general solutions of space-charge flow equations is complicated, and even for such simple geometry as the hollow cylinder a computer solution of the equations must be obtained for each set of required dimensions. However, relatively simple solutions for slit and orifice symmetry were published by Pierce[179]. The conditions of uniform space-charge flow can be achieved if a segment of such flow is utilized and the cathode and accelerating electrodes are shaped so as to maintain along the edge of the beam the same potential variation which would exist if there were a uniform extensive space-charge flow. Guns designed after such calculations, known as Pierce guns, are formed from a cathode and shaped paracathode, with a shaped accelerating electrode containing an orifice. The appropriate equipotential lines to which shape the electrodes should approximate are shown in *Figure 3.16* together with the appropriate data.

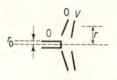

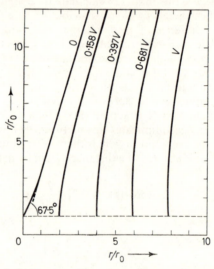

Fig. 3.16. Equipotentials for Pierce gun[179] with circular cathode of radius r_0 at potential zero and final aperture at potential V; emission is in the direction of the x-axis

The production of electron beams in orifice symmetry has been extensively studied at low energies; systems which compete in intensity and ease of control with the Pierce gun include the hairpin gun. Where only small currents are required, but with small beam diameters, a cathode is made from fine tungsten wire bent in the form of a sharp hairpin; the cathode is fronted by an electrode biased negatively and containing a circular aperture. For sufficiently high energy beams, focusing enables a beam diameter as small as 10^{-2} mm to be obtained. Small endfire oxide cathodes (3×10^{-2} mm effective diameter) can be constructed by filling a platinum capillary with oxide; endfire oxide cathodes are also available commercially. Single tungsten crystal cathodes, commercially available, enable the emitting area to be reduced still further. It is often sufficient to use only three electrodes in front of the cathode: the variable negative aperture G, the control electrode K and an aperture stop A at anode potential.

A version of the hairpin gun[180] is shown in *Figure 3.17*. The hairpin aperture G is held negative to the cathode, and the control electrode K is operated at less than one-tenth of the anode potential V_A. The potential V_K governs the distance L between the beam focus and the cathode. Typically, the current, $I = 5$ μA, spot diameter $= 0\cdot05$ mm, divergence $= 4$ mrad, $L = 23$ cm, $V_A = 30$ kV, $V_K = 0\cdot7$ kV and $V_G \simeq -50$ V.

Simpson and Kuyatt[181] have used this gun for lower energy beams, with $V_A < V_K$. For $V_A = 500$ V, $V_K = 4\cdot4$ kV, $I = 30$ μA, spot diameter $= 1\cdot5$ mm

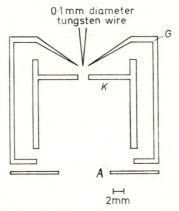

Fig. 3.17. Scale section of hairpin thermionic emitter: A, anode aperture stop; K, control electrode; G, hairpin aperture (loosely termed the 'grid' by analogy with the triode valve)

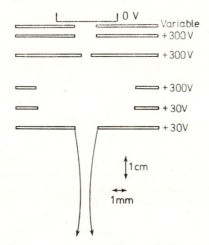

Fig. 3.18. Scale section of apertured planar oxide thermionic emitter

and divergence $= 3$ mrad. These authors also describe a planar oxide cathode with six apertures (*Figure 3.18*), which is capable of producing 8 μA current at 30 eV, in a 1 mm spot at 4 cm from the last aperture. The divergence is $3\cdot5$ mrad. For less stringent requirements, four electrodes would be sufficient. A similar gun[182], with potentials 0 V, $+2$ V, $+15$ V, $+100$ V, $+30$ V, $+10$ V and $+6$ V, produces an even lower energy beam.

A higher energy cathode system which is capable of producing a high-intensity beam in a small spot is the spherical system represented in *Figure*

3.19, which shows also the diameter and position of the minimum diameter as a function of cathode parameters, independent of potentials applied.

In any electron-optical system, the highest current intensities through collision chambers are obtained by minimizing the aberrations of the system. Collimation in a field-free region is used to remove electrons furthest from the optical axis. One often uses two collimating apertures, one of which is placed near a focus, to remove electrons from the outer (low-density) part of the beam. The other, a wide aperture, well-removed from the focus, ensures that only electrons within a certain range of solid angle pass into the image.

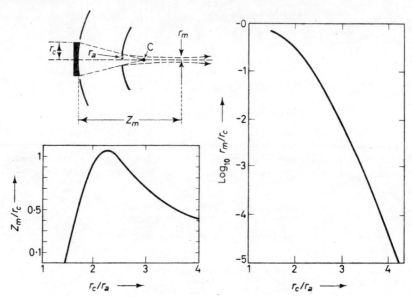

Fig. 3.19. *Data and dimensions of spherical cathode thermionic emitter of centre C*
(From Pierce[183], by courtesy of Van Nostrand)

The advances in low-energy gun design may well reduce the popularity of the technique of collimating electron beams by a magnetic induction B parallel to the axis. Such collimation causes electron paths to be increased due to their helical motion, but permits very high density beams to be obtained. The optimum confinement within a radius a at potential V is given by:

$$\frac{eB^2a^2}{8m_e} = \frac{2V}{3} \quad \text{(SI units)} \tag{3.128}$$

for which the maximum current is

$$I = \frac{16}{3\sqrt{6}} \pi\varepsilon_0 \left(\frac{e}{m_e}\right)^{1/2} V^{3/2} \quad \text{(SI units)} \tag{3.129}$$

or

$$I = 25\cdot4 \times 10^{-6} V^{3/2} \tag{3.130}$$

The parameter $(I/V^{3/2})$, which is a measure of the effectiveness of a system in confining the space charge of a beam, is termed the 'perveance'. Thus 25·4 micropervs is the maximum possible perveance for a cylindrical beam.

Magnetic confinement introduces a radial variation of potential:

$$V(r) = \frac{3B^2 a^2 e}{16 \, m_e} \quad \text{(SI units)} \tag{3.131}$$

which may sometimes be inconvenient. Non-uniformity of magnetic field introduces instability. In the absence of a confining magnetic field, the maximum current which can be passed through a tube of length l and diameter d is

$$I = 35 \cdot 6 \times 10^{-6} V^{3/2} \left(\frac{d}{l}\right)^2 \tag{3.132}$$

For a beam focused in orifice symmetry with semi-angle α, the maximum current is

$$I = 38 \cdot 5 \times 10^{-6} V^{3/2} \alpha^2 \tag{3.133}$$

Although it is impracticable to do any justice to a discussion of the principles of electron optics in this section, it may be of value to collision physicists to present a small selection of the computed data to be found in the textbooks, particularly the classic work of Pierce[183].

A thermionic electron beam accelerated through a potential V travels with a mean velocity

$$v = 5 \cdot 931 \times 10^7 \left(V + \frac{T}{11\,600}\right)^{1/2} \quad \text{cm sec}^{-1} \tag{3.134}$$

where T is the absolute temperature of the filament. The thermionic distribution of velocities is responsible for chromatic aberration in electron lens systems. Design data for two-element lens systems are to be found in the electron optics textbooks[184], but it will be remembered that these are thick lenses, having two principal planes and two focal lengths. They have not, therefore, the simplicity or advantages of three-element lens systems, the most popular of which is the three-aperture lens. This consists of a central aperture, midway between two other apertures separated by a distance $2L$; potentials V_1, V_2 and V_1, measured with respect to the cathode, are applied, with the result that the system functions as a thick convex lens of focal length f, with its two principal planes a distance D on either side of the system. The functions $L/f(R)$ and $D/L(R)$, where R can be either V_2/V_1 or V_1/V_2, are shown graphically in *Figure 3.20*. The three-aperture lens is known as an 'einzel' lens, because there is a single (einzel) value of the limiting potential. Slit-symmetry einzel lenses are also popular. Spherical aberration is severe when more than the central one-third of the lens aperture is filled by the beam.

A problem which frequently occurs in collision physics is the production of an image of variable energy but fixed position. For example, an electron momentum analyser operates with electrons of fixed energy, and it is then required to accelerate these into a collision region, through a potential

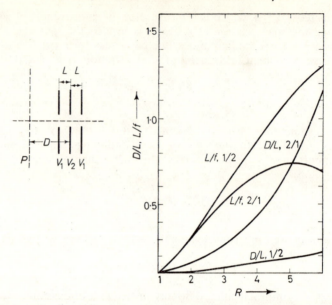

Fig. 3.20. *Data for the principal plane P and focal length f of a three-element electrostatic ens, for different potential ratios: $R = V_1/V_2$, and V_2/V_1 are written as 1/2 and 2/1 respectively*

(From Pierce[183], by courtesy of Van Nostrand)

variable over wide limits, but without changing the beam current passing through the collision chamber orifice. An empirical solution of this problem has been given[185], and suitable three-aperture lens data have been computed[186].

Data are presented for the solenoid magnetic lens in *Figure 3.21*, in terms of the variable

$$\Phi = \left(\frac{e/m_e}{8V} \right)^{1/2} B_{\text{max}} L \quad \text{(SI units)} \tag{3.135}$$

for electrons accelerated through a potential V; B_{max} is the maximum axial magnetic induction produced by the lens.

It sometimes happens that an electron optical system has been designed, but must be altered in physical dimension L to suit a redesigned experiment. Scaling can be carried out with the aid of the equation.

$$\frac{V}{B^2 L^2} = \text{const.} \tag{3.136}$$

The time t taken for the electron to pass through the system is scaled according to the relation

$$\frac{V t^2}{L^2} = \text{const.} \tag{3.137}$$

Current densities J will not remain constant during the scaling, since they are controlled by the relation:

$$\frac{V^{3/2}}{JL^2} = \text{const.} \tag{3.138}$$

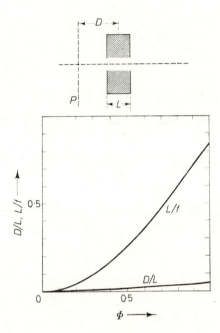

Fig. 3.21. Data for the principal plane P and focal length f of a solenoid magnetic lens as functions of

$$\Phi = \frac{0.186 \times \text{ampere turns}}{(\text{eV})^{1/2}}$$

(From Pierce[183], by courtesy of Van Nostrand)

Other Sources of Electrons

Although thermionic emission is by far the most frequently used technique for producing electron beams for collision experiments, there are situations when photoemission and β sources can be advantageous. The high temperatures associated with filaments accelerate unwanted chemical reactions in gases (for example, thermal dissociation). Certain β sources, such as tritium, have been used in high-pressure drift tube experiments.

Photoemission has been used for producing pulses of electrons for drift tube and attachment experiments. Radiation in the quartz ultra-violet impinges on a gold cathode, usually at oblique incidence, but more recently from the rear[187]; the gold is deposited as a 300 Å film on quartz, and there is less interaction between the radiation and the gas. The most satisfactory photon source for this purpose is a hot-cathode hydrogen lamp[188], which can be pulsed with rise and fall times as low as 200 nsec; these times are much shorter than those obtainable with a mercury discharge lamp.

The choice of gold as photocathode necessitates that the radiation be of much shorter wavelength than if alkali metals were used. However, the large atomic number of the gold ensures that the spectral response be broad. It is important that the radiation is not polarized with electric vector parallel to the surface, because in this case the photoemissive sensitivity is much reduced.

Sources of Spin-polarized Electrons

Spin-polarized electrons are produced by: scattering electrons at certain angles off atoms such as mercury; photoionizing polarized alkali atoms; scattering electrons from neon atoms at the energy of the appropriate compound state; and ionization of polarized H $2s$ atoms.

Scattering of electrons from mercury atoms[189] produces electron fluxes with polarization vector (spin axis) perpendicular to the plane containing incident and scattered electron. The polarization

$$P(E_e, \theta) = \frac{N_\uparrow - N_\downarrow}{N_\uparrow + N_\downarrow} \tag{3.139}$$

of the scattered electron beam depends upon the electron energy E_e and the polar scattering angle θ. Data[190] for energy dependence at $\theta = 90°$ are in good correspondence with theory[191]. Efficient source design[192] enables a current of 2×10^{-7} A, 20 per cent polarized, to be obtained. For this purpose, a 15 mA beam at energy 100 eV is scattered through 70° from mercury vapour at density 10^{14} cm^{-3}. Molecular targets may be used[193], and also crossed-beam techniques.

Polarized electrons are also produced[194] by the photoionization of a spin-polarized lithium beam. A pulsed high-voltage spark between tungsten electrodes in nitrogen is suitable as a light source. With a weak magnetic field (150 G) in the photoionization region, polarization $P \sim 90$ per cent can be achieved. Lithium has the highest threshold photoionization cross-section of any alkali metal, but others have been used.

Polarized electrons may also be produced[195] by 90° elastic scattering in the p-wave compound-state resonance region of neon. For scattering to the right at the maximum of the $P_{3/2}$ resonance, or for scattering to the left at the maximum of the $P_{1/2}$ resonance, the spin will point upwards. The technique requires that the electrons be well monochromated in momentum, but this may in some circumstances be advantageous.

Collisional ionization of spin-polarized H $^2S_{1/2}$ atoms has also been proposed[196] as a technique for the production of polarized electrons. The metastable atoms are produced by charge transfer of protons or deuterons with caesium. The collisional ionization of H $1s$, which is also present in the beam, will produce a background of non-polarized electrons.

3.9 SOURCES OF PHOTONS IN THE VISIBLE AND ULTRA-VIOLET

Photon sources in the visible and the ultra-violet are necessary in collision physics for the performance of photoabsorption, photoionization and photo-detachment experiments. This brief discussion will include pulsed, continuous and amplitude-modulated sources, sources for calibration of detectors, for the measurement of oscillator strengths, and for excitation of plasma; also lasers and bremsstrahlung sources.

Sources for the Measurement of Oscillator Strengths

These sources produce atomic or molecular spectral lines in emission; from the relative intensities of these lines, oscillator strengths may be calculated (Chapter 9) provided that the relative populations of the excited states are known. To achieve this condition, the vapour in the source must be in 'local thermodynamic equilibrium' over the emitting region, so that the Boltzmann equation may be applied. Since the energies of many atomic levels are considerable when expressed on a temperature scale, the problem is one of producing equilibrium at high temperatures, and at present only the following three methods have proved successful:

Furnaces

Furnaces containing metal vapour are used in the classic experiments[197] for comparatively low energy levels. Relative oscillator strengths have been measured, but absolute determinations are limited by uncertainties in the vapour pressure data. Glass windows in furnaces are attacked by alkali metal vapours. For measurements on alkali metals (and other reactive metals), single crystals of MgO or Al_2O_3 are suitable transparent and inactive windows.

High-current arcs

These have been used principally at Kiel[198] for atoms with rather higher energy levels (O, N, C, Si, Cl; both I and II). Relative oscillator strengths are usually obtained.

Shock tubes

Shock tubes show great promise for this application[199]. The conventional shock tube[200] consists of a stout metal tube (typically 2–3 m in length and 5 cm in diameter) partitioned at about one-third of its length, by a thin metal diaphragm, behind which the pressure of a gas such as hydrogen can be raised to a value of 25–600 atm. In front of the diaphragm the gas under study is maintained at a pressure of perhaps 10 torr. When the pressure of the 'driver' gas is raised sufficiently, the diaphragm breaks and the expansion of the high-pressure gas generates a shock wave which travels forward in the low-pressure gas. The latter is heated to a temperature which may be $\sim 15\,000°$K, and its pressure rises to the order of an atmosphere, in a period between 10 μsec and 1000 μsec. The shock-heated gas is separated from the driver gas by a contact surface moving with the wavefront velocity. It is possible either to avoid reflexion of the wave by allowing it to expand into a large chamber, or to allow the wave to reflect from the closed end of the tube, with a fraction of its initial velocity, the system gas being designed dynamically in such a way that the reflected wave introduces inappreciable unwanted heating ('tailored interface'). The gas behind the reflected wave is brought nearly to rest and is further compressed and heated to perhaps $10\,000°$K.

This type of shock tube is only efficient when the ratio of sound velocities in the driving gas and in the driver gas is large. For obtaining a shock wave in such gases as helium[201], it is possible to replace the driver gas and diaphragm with a solid or gaseous high explosive; temperatures $\sim 20\,000°$K may be produced, at much higher pressures than in the conventional tubes. A third method, the magnetic shock tube, produces shock-heated plasmas at temperatures $\sim 20\,000°$K utilizing the ohmic dissipation of high-pulsed currents and the action of the Lorentz $I \times H$ forces that are set up by plasma currents and magnetic fields. Although conical pinch[202, 203] and annular magnetic field[204] geometries have been investigated, the configuration most suitable for spectroscopic study is the T-tube[205]. This consists of a pulsed high-current discharge struck between two electrodes at the top of a T-shaped tube. The resulting expansion of the ohmically heated gas into the side arm is supplemented by the effect of the magnetic field of the current in a return conductor situated perpendicular to the side arm, parallel to the discharge current, on the top of the T. The discharge of a high-voltage condenser bank through a low inductance path will drive H_2 and He plasmas with velocities up to 15 cm μsec^{-1}, producing thermodynamically equilibrated plasmas with electron densities $n_e \sim 10^{17}$ cm^{-3}, at temperatures $20\,000°$–$40\,000°$K.

The temperatures behind all of these shock waves can be measured by spectral line reversal[206], by electrical conductivity[207] and by Stark broadening of lines[208].

Glow and Capillary Discharges as Line Sources

The glow discharge, both positive column and negative glow, is a valuable source of line radiation in the visible and ultra-violet. When a large amount of energy is applied to gas confined in a capillary tube, populations of higher energy levels are produced, and lines are observed in the vacuum ultra-violet and extreme ultra-violet (XUV). Continua are also emitted. Capillary discharges in flowing gases are valuable sources of higher order atomic spectra, which are being investigated and analysed at shorter and shorter wavelengths[209].

The higher the order of spectra required, the hotter must be the gas temperature (although there is probably not local thermodynamic equilibrium); the wall cooling problem becomes severe, and in the XUV it is necessary to pulse the capillary by discharging a condenser bank across it.

Continuous energy can be fed into a discharge by exciting it inside a microwave resonant cavity. A miniature quartz capsule under microwave excitation is most effective. The principal radiation emitted is of course resonance radiation (of all orders); reference to the Appendix will yield the first order resonance wave numbers of common atoms.

At wavelengths shorter than the limit of transparency of glass (3000 Å), of quartz (1800 Å), or of a clean lithium fluoride crystal window* ($\sim$ 1200 Å), the source must be in the same vacuum system as the rest of the experiment (monochromator and collision region). Radiation below 3000 Å is strongly absorbed by oxygen, and below 2000 Å other absorption functions become important. The source gas must be isolated from the experiment gas by means of a fast-pumped region between the capillary source and the experiment. Pressures as high as 1 bar have been used in capillary sources. An improvement is the 'flowing helium window', in which a stream of helium is directed into this pumping region, through holes in the inside wall of an annular gas source.

Thin aluminium windows ($\sim$ 1000 Å thickness) are also valuable for isolating discharge sources from the vacuum system. They will not stand pressures in excess of 1 torr, even though mounted on fine meshes. However, they will not pass metastable atoms, and their bandpass characteristics in the XUV are valuable. Their transmission is high at X-ray wavelengths ($<$ 1 Å), and continues high right up to 100 Å, where it falls off. At 140 Å is a broad window, which falls off slowly as 500 Å is reached. Aluminium windows are fabricated by evaporating the metal onto a solidified sugar solution that has been painted on glass. The metal film can be floated off very slowly onto a surface of warm water and then removed by raising the frame up through the water surface. Tin windows are also effective. Collodion films (transparent down to 500 Å) can be made by allowing a drop of solution of collodion in amyl acetate to fall onto a water surface.

It is important that a gas discharge source of line radiation should not contain thermal gradients, if it is optically thick. The cooler gas will absorb only the centre of the line, leaving the wings relatively unchanged. This 'line-reversal' can be avoided in discharge sources of resonance radiation[210] by using an inert gas to carry the discharge and excite the admixed gas atoms, which are present in small quantities. Absorption in layers of cool gas is minimized by ensuring temperature uniformity: quartz walls may be run hot, surrounded by a vacuum jacket. In one early tube, the discharge region was deflected toward the optical window with a weak magnetic field.

It is sometimes argued that a high-current discharge will so raise the gas temperature as to eliminate line reversal; but unfortunately the high temperatures caused Doppler line broadening. Nevertheless it may be necessary to use high gas temperatures to vaporize the element whose resonance radiation is required. In an early successful source produced by Cario and Lochte-Holtgreven[211], low temperatures were maintained except in the region of vaporization; heating of the specimen was achieved by ion bombardment, and condensation on the optical window was prevented by streaming of inert gas, with suitable geometry.

---

* The transparency of these windows deteriorates with continued exposure to ultra-violet radiation.

The stability of a discharge is aided by running it with a hot cathode, but on the other hand a cool cathode, especially liquid-air cooled, will sharpen the lines. Clouding of the optical windows by sputtering is a serious problem which may be minimized by the use of hollow cylindrical cathodes made of very pure iron or of aluminium.

Sources of Continua in the Vacuum Ultra-violet

The recent rapid development of vacuum ultra-violet continua sources considerably widens the scope of experiments that can be performed. When the first edition of this book was published, few sources were available other than the triggered vacuum spark (or Lyman) source, the Hopfield continuum in helium, and the hydrogen molecular spectrum in a positive column. But now these are supplemented by the Tanaka continua in rare gases, and, in addition to the pulsed capillary discharge, one may use a theta-pinch, a synchrotron, a shock tube or a sliding spark. The bands covered by the continua are as shown in Table 3.4.

Table **3.4.** BANDS COVERED BY CONTINUA OF
VARIOUS SOURCES

Source	*Band* (Å)
H_2	1650
He	580–900 (1100)
Ne	750–900
Ar	1200–1400
Kr	1300–1600
Xe	1500–1800

The Hopfield continuum[212] maximizes around 800 Å and possesses a subsidiary maximum around 700 Å. It is believed[213] to arise from transitions from upper bound states of He_2 ($A^1\Sigma_u^+$, $D^1\Sigma_u^+$, etc.) to the ground state $X^1\Sigma_g^+$, which is antibonding. It is at its most intense form a repetitive gas discharge[214] produced by discharging a capacitor bank by means of a hydrogen thyratron. Compact and non-inductive capacitors, which avoid ringing, are commercially available, constructed from double-spiral windings of foil interleaved with a form of barium titanate.

The Tanaka continua[215] can be produced in microwave discharges (800 W) and in pulsed capillary discharges[216], as well as in miniature linear pinches and shock tubes. The continua are believed to arise from downward transitions from the bonding excited states of the diatomic rare gas molecules to the ground antibonding state.

If a capacitor bank is discharged through a 1 mm diameter quartz capillary containing gas at pressure up to $\frac{1}{2}$ bar, then the severe erosion of the capillary will demand its continual replacement. The use of increased power from modern non-inductive capacitors enables the capillary diameter to be raised to 1 cm, thus improving its lifetime. A superior design makes use of a coaxial line[217] by means of which the discharge is 'pinched' by its own magnetic field, thus minimizing the corrosion. A pulse of gas is admitted via a magnetically controlled valve. The design (*Figure 3.22*) using a ceramic capillary (Garton source), is commercially available. At the exit slit of a 1 m monochromator with 0·2 Å resolution, the flux maximizes at 6×10^{11} photons sec^{-1}, falling to 10^{11} photons sec^{-1} at short (500 Å) and long (3000 Å) wavelengths. Fast pumping is necessary between the capillary discharge and the rest of the experiment[218, 219].

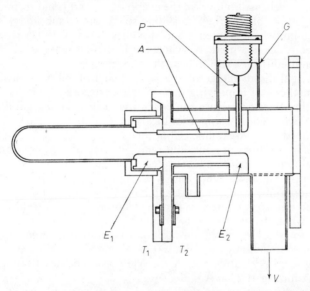

Fig. 3.22. Construction of Garton flash tube; G, glass extension piece; T_1 and T_2, plates of transmission line; E_1 and E_2, electrodes; P, tungsten pin; A, alumina ceramic tube (1 cm bore, 5 cm length); V, to pump

Pulsed discharge sources in the vacuum ultra-violet suffer from the disadvantage that they emit sputtered material which settles on reflexion gratings causing deterioration in reflectivity. This is to some extent true of the vacuum spark triggered by a subsidiary electrode (Lyman source), which uses similar hardware, at very low pressures, but produces the radiation continuum (down to ∼ 80 Å) by a different mechanism—possibly a bremsstrahlung spectrum caused by the stopping of electrons in the anode, enhanced by electron–ion interactions in the hot plasma. The intensity depends upon the electrode atomic number, being strongest for gold, platinum and, in particular, uranium. Repetition rates of 20 sparks sec^{-1} are feasible, with 20 kV high tension.

The conventional Lyman source consists of large metal electrodes (for heat dissipation), between which is a tube, often lined with Teflon, containing a vacuum side-arm down which the radiation passes[220, 221]. Thus the radiation does not come from the electrodes, but from the space between them.

Characteristic radiation in the soft X-ray region can be produced with low-continuum background by the bombardment of metal electrodes with high-intensity kilovolt ion beams. X-ray continua are produced in the normal way by electron bombardment. The so-called 'transition radiation' arises from the changing dipole field due to the electron and its image charge. For 24 keV electrons impacting silver or aluminium targets, the continuum is a maximum at about 900 Å, and falls to half intensity at 600 Å wavelength.

Another device is the 'sliding spark' in vacuum[222]. This is struck along a surface of quartz or glass, between two metal electrodes. It is started off in an atmosphere of gas, and is then evacuated; the spark moves around on the surface, and is believed to involve the evaporation and rapid recondensation of silica. It is possible to trigger a sliding spark with a subsidiary electrode. The continuum, which is intense and stretches down to 170 Å when heavy-element electrodes are used, or when the electrode material forms stable oxides, is emitted from a region near the anode[223].

The principal disadvantage of sparks is their instability of position, but there are methods of improving this. When a triggered spark is struck in argon at atmospheric pressure for use in the visible region, good stabilization can be achieved by using a jet of argon so that the spark is to some extent confined to a streamline. Continua in the visible are often marred by the diversity of lines present, and relatively uncluttered continua can be obtained by striking sparks under water, between uranium electrodes.

Pulsed hot-plasma machines and synchrotrons[224] are valuable sources of ultra-violet continua when they are available. The radiation from a circular synchrotron electron accelerator arises from the bremsstrahlung energy conversion when the electron forward momentum vector is reduced by the curvature of path. The maximum radiation is emitted around 0·5 Å wavelength for GeV machines, but for a 180 MeV machine, the maximum intensity is 8 erg sec^{-1} Å^{-1} at 500 Å. A 100 MeV machine has 10 times smaller output, maximizing at 2340 Å.

The power distribution per electron per unit wavelength λ is

$$P(\lambda) = \frac{3^{5/2}e^2c}{16\pi^2 r} \gamma_E^7 G(y) \tag{3.140}$$

with $y = \lambda_c/\lambda$, critical wavelength

$$\lambda_c = \frac{4\pi r}{3\gamma_E^3} = \frac{4\cdot 19 r}{\gamma_E^3} \times 10^{10} \text{ Å} \tag{3.141}$$

and

$$\gamma_E = \frac{E}{m_e c^2} \tag{3.142}$$

for electrons of energy E moving in circular orbits of radius r (in metres). The function $G(y)$ is illustrated in *Figure 3.23*. The power radiated at peak wavelength increases as E^7, and the peak wavelength ($0\cdot 42\ \lambda_c$) varies as E^{-3}. Since the photons are distributed in a very narrow angle cone tangential to the electron orbit, and since the electrons are approaching the observer with a velocity almost equal to c, the radiation is emitted in a pulse of width

$$\Delta t = \frac{r}{c}\left(\frac{m_e c^2}{E}\right)^3 \tag{3.143}$$

A linear accelerator beam of fast electrons will produce bremsstrahlung when the electrons are deflected magnetically through 90°, but it is necessary to use a superconducting magnet in order that the radius be sufficiently small for the peak power to be in the XUV.

Hot-plasma machines, such as the linear z-pinch and the thetatron, are valuable sources of radiation pulses in the XUV. The thetatron (named after the angle θ commonly used as a symbol in polar coordinates) is a continuous toroidal discharge suddenly pinched by a single-turn current pulse around it; this pulse is normally produced by the discharge of a capacitor bank through a strip line. Typically, the maximum electron density is around 10^{17} cm^{-3} and the emitted wavelength distribution (~ 25 Å) is

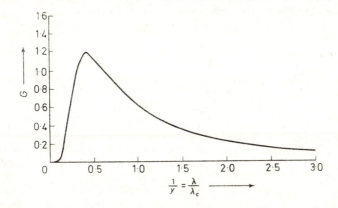

Fig. 3.23. G(y) function for continuum radiation from a synchroton

Maxwellian, with mean temperature around 300 eV. However, as one proceeds to shorter wavelengths along the emission function, there are one or more downward jumps, which correspond to the onset of free-bound ion continua. Exactly what elements they correspond to is difficult to say; but a high atomic number impurity (such as gold) extends the continuum to lower frequencies. The thetatron is among the most successful XUV sources, since unlike other hot plasma machines it is within the capabilities of a small laboratory to construct.

It has been reported[225] that corona discharges emit ultra-violet radiation continua peaked around 800 Å and 1000 Å in oxygen. They are of course absorbed by photoionization, but at pressures around 1 torr this is not too serious.

In some types of experiment, such as flash photolysis, a pulse of maximum flux of radiation has to be passed into a volume of gas. This may be achieved with the aid of a scattering box, in which both light source and gas chamber are placed. The walls of the box are coated with snowy magnesium oxide, which has a very low absorption coefficient and is capable of scattering radiation with high efficiency. The coating is made by igniting magnesium inside the box.

Sources for Calibration of Detectors

Black-body radiation sources for sensitivity calibration of detectors in the visible are well developed. Standard tungsten ribbon[226] lamps and quartz–iodine–tungsten coiled filament lamps are readily available. The most accurate standards make use of fixed temperatures such as that of melting copper. The calibration of photomultipliers presents problems because they are designed to detect very weak intensities. The best solution seems to lie in the availability of calibrated sources made of phosphorescent paint (^{14}C and phosphor) uniformly distributed on glass. This arrangement provides stability and accuracy of solid angle. There is need for a central calibration service to recalibrate every year or so, because of the slow decay of the phosphor. Electroluminescent sources are also available.

Modulation of Sources

Another problem of importance to radiation sources is their amplitude modulation. This is necessary in diagnostics by radiation absorption of excited states in time-varying plasmas. The plasmas emit radiation of the same wavelength as that used for absorption, but the probing radiation can be distinguished if it is modulated at a frequency which is large compared with the highest component frequency of the plasma time-variation. The modulation time must still be long compared with the decay time of the excited state.

Modulation can be carried out around 1 MHz by means of a fast Kerr cell fed with a radiofrequency signal. Lasers may be modulated by means of a hexamine crystal. A modulated plasma device[227] has been developed in the form of a high-frequency capillary discharge filled to a pressure of several torr with He–H_2 mixture. When the discharge is excited by a 300 W 1 MHz oscillator and cooled, the rapid changes in mean electron energy are such as to cause a substantial amplitude modulation in Balmer radiation intensity. The exact modulation waveform agrees with prediction from theory.

Pulse modulation is normal operation for many of the vacuum ultra-violet sources discussed above. Pulse modulation is also required for obtaining, by photoemission, electron pulses for drift velocity measurements. For such a purpose it would not be worthwhile to employ a powerful continuum source; a hot-cathode hydrogen lamp will suffice[228, 229].

Lasers

Monochromated light sources of great intensity, coherence and parallelism are valuable for the study of very weak transitions, such as two-quantum processes. Solid state lasers are commercially available in forms capable of developing an output of several joules, either continuous or in irregularly shaped pulses whose total length is around 500 μsec, although the spikes are of nanosecond length. Gas lasers are usually capable only of continuous operation. The Q-spoiling or switching of solid state lasers enables the pulse length to be reduced and the peak power to be increased considerably.

Table 3.5. AVAILABLE LASERS AT WAVELENGTHS BELOW 8500 Å

Species	Wavelength (Å)
Ne	3324
N_2	3371
Ar	4579
Ar	4765
Ar	4880
CdS	4950
Ar	4965
Ar	5017
Ar	5145
$CaF_2 : Ho^{3+}$	5512
Kr	5682
$LaF_3 : Pr^{3+}$	5985
$Y_2O_3 : Eu^{3+}$	6113
HgHe	6150
HeNe	6328
$Al_2O_3 : Cr^{3+}$	6929
$Al_2O_3 : Cr^{3+}$	6934
$SrF_2 : Sm^{2+}$	6969
$Al_2O_3 : Cr^{3+}$	7009
$Al_2O_3 : Cr^{3+}$	7041
$CaF_2 : Sm^{2+}$	7083
$Al_2O_3 : Cr^{3+}$	7670
GaAs	8400
$Ne-O_2 : Ar-O_2$	8446
Br_2-Ar	8446
Na–Ca–Si	9200
Li–Mg–Al–Si	10 150
$Y_3Al_5O_{12} : Yb^{3+}$	10 296
$CaF_2 : Nd^{3+}$	10 460
$CaWO_4 : Pr^{3+}$	10 468
K–Ba–Si	10 600
CO_2	10 600
$CaMoO_4 : Nd^{3+}$	10 610
$LaF_3 : Nd^{3+}$	10 633
$Y_3Al_5O_{12} : Nd^{3+}$	10 648
$CaF_2 : Tm^{2+}$	11 160

Visible wavelengths for which laser action has been obtained are summarized in Table 3.5[230-237]. More complete summaries of gas laser wavelengths are available[238, 239].

A high-intensity, widely tunable photon source deriving from the laser is a crystal of potassium niobate excited by a neodymium laser. There is also a system[240] capable of putting out 1 kW power with 14 cm^{-1} bandwidth, tunable in the range 4820–5790 Å. In this system, the output of a neodymium-doped glass laser is mixed with the continuum produced in CS_2 liquid by the same beam; the mixing is carried out in a crystal of potassium dihydrogen phosphate. Another system of tuning laser output is by means of a monochromator incorporated in a cavity containing a wide-band emitting dye. The cavity is pumped by means of a second laser operating in the high-energy side of the band. In this way coherent light is 'driven' through the absorbent material at a frequency tunable by operating the monochromator. Suitable dyes are available for the entire range 4000–13 000 Å.

3.10 SOURCES OF IONS

Since the first edition of this book, there has been some change of emphasis in the development of ion sources. As far as ionic collision physics is concerned, singly charged positive ion intensity has not been a problem for some years. But the achievement of the greatest intensity[241] is still of importance in ion propulsion motors, in thermonuclear machine injection, in tandem van der Graaf technology, in isotope separation, in ion implantation in semiconductors, in printing circuits by means of fast ion beams, and so on. The important problems for ionic collision experiments are now:

1. The design of ion sources which produce ions in the ground state (or in known proportions of a long-lifetime excited state);
2. The achievement of adequate intensities of ions of the smallest kinetic energy (1–10 eV) (this is more of a problem of ion optics than of source development, but the energy distribution of ions produced by the source must be minimal; this is linked with the problems of beam modulation and reduction of noise);
3. The production of high intensities of highly stripped ions.

These three problems are discussed individually after the following four principal types of ion source have been described: (*i*) electron bombardment sources, (*ii*) discharge of various kinds, (*iii*) surface ionization devices, and (*iv*) field ionization sources.

Electron Bombardment Sources

The Nier type mass-spectrometer source[242], sometimes called the Nier–Bleakney source (*Figure 3.24*) may be operated either in a low range of pressures (10^{-4}–10^{-6} torr) or in a higher pressure range (10^{-3}–10^{-2} torr). In the latter range, it operates as a low-voltage arc (see below), but in the

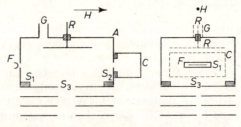

Fig. 3.24. Nier electron impact mass-spectrometer source (schematic views not to scale): cross-hatched areas represent insulation; potential differences can be held between chamber A and repeller R, electron collector C, extraction orifice S_3, and filament F; G represents gas inlet

former range it may correctly be regarded as a confined-beam electron impact device. Electrons from a filament F, confined by a magnetic field parallel to their optical axis, pass through collimation slits S_1 and S_2 into a collision chamber A, thence into a collecting cage C. In the gas-containing collision chamber, an electric field is applied transverse to the electron beam axis by means of the repeller electrode R; this aids the extraction of ions through slit S_3, through which the ion accelerating field also penetrates to assist further in extraction. In most ion-optical systems, a number of additional slits are interposed, as in the figure. They serve to focus the extracted beam and reduce chromatic aberration.

This type of source in low-pressure operation is not usually intended to produce ion currents larger than those which can be conveniently measured after mass analysis. It is intended, however, that the ions should be produced by electrons having a reasonably narrow and measurable energy distribution, and that the ions should be formed in a potential region sufficiently confined for their energy distribution to be dominated by the processes of their formation. Indeed, in the retarding potential difference technique (Section 3.13) which is sometimes employed in this ion source, the ions are swept out of the source by a pulsed repeller field which only operates after a pulsed electron beam has been collected.

At the lowest electron currents and collision chamber gas pressures, the ion energies, as measured in the first instance by Hagstrum[243] using retardation techniques, are dominated by source chamber electrode potentials; but in more intense electron beams, the electron space charge controls the potential configuration[244], and in particular can create a potential trough in the centre of the beam from which only energetic positive ions can escape. However, a relatively small positive-ion concentration is necessary to neutralize the faster-moving electrons. Above a certain gas pressure, the positive-ion space charge takes over and may distort the natural ion energy distribution by its own space-potential configuration.

The paths traced out by the electrons are cycloidal-helical. A purely helical path would result from parallel electric and magnetic fields, but the cycloidal contribution arises from the transverse electric field imposed by the repeller electrode. There is a drift of the electrons across the width of the extraction slit; the ions extracted therefore arise from regions of different potential.

The trapping of positive ions[245] in the space charge of the electron beam is a significant feature of the Nier–Bleakney source. Subsequent collisions of the electrons with trapped ions produce further ionization.

A problem for electron bombardment sources is set by the introduction of the gas molecules. The filling of the entire source chamber with gas may not only introduce unwanted inelastically produced particles into a region from which direct extraction does not take place, but may also attenuate the electron beam unnecessarily before it reaches the ion-extraction region. The remedy is to reduce the size of this operative region and, if possible, to apply the gas only where it is wanted. Gas flow from capillary channels has frequently been found effective; for example, the use of a 0·2 mm capillary leads to a five-fold increase in the efficiency[246], defined as the ratio of ion flux density in the beam to molecule flux density in the source feed pipe. However, in the region of molecular flow (0·02 mm capillary), the efficiency can be smaller, possibly owing to the difficulty of lining up the canal with the mass-spectrometer.

In electron bombardment ion sources operating as low-voltage arcs, the extracted ion beam increases with increasing source chamber gas pressure, until a maximum is reached; further increase of pressure above this point only attenuates the ion beam. This fact is in part due to attenuation of the electron beam before it reaches the collision region from which ions are extracted, and in part due to ion–molecule collisions in the course of extraction. Introduction of gas through a capillary raises the maximum pressure, so that more intense ion beams can be extracted. In a typical capillary injection source[247, 248], an improvement of a factor of 1000 over the conventional Nier source is achieved for low-energy extraction conditions (1–100 eV ions) by means of comparatively small magnetic fields to control the electrons. The electron beam is confined to a narrow channel across which the jet of gas is directed, and the alignment of the magnetic field and of the jet is fairly critical.

This type of source is valuable for ground-state ion production; for electron energies near to ionization threshold the ion intensities are still reasonably large. However, there is some trapping of the ions in the electron beam space charge.

An important stabilization technique applicable to electron impact sources has been developed in a number of laboratories[249]. The electron current produces a differential signal which is amplified and fed to the filament heating power supply; the filament is operated using thermal rather than space charge limitation, so that electron current stabilization is achieved. However, the thermal time constant of the filament must be minimized, so that it is smaller than typical fluctuations in emission; this is achieved by using thin tungsten ribbon rather than wire. This type of stabilization is valuable in low-voltage arc and sputtering sources; it appears that the ion beam brilliance and emittance are extremely sensitive to filament current.

Discharge Sources

Glow discharges, low-voltage arcs, radiofrequency discharges and even sparks have been utilized as ion sources. It has already been mentioned that the dividing line between the low-voltage arc and the Nier electron impact source is indistinct and consists only in a difference of gas pressure.

An ion source must not only produce ions with a high efficiency and in such a way that they can be extracted without unduly disturbing the source; it is also frequently required that the extracted ions possess a narrow energy distribution. This can only be achieved by maintaining the plasma of a discharge source at low X/p—that is, high pressures and low electric fields; in the low-voltage arc, it is arranged that the electrons are all accelerated through a sheath before entering the ionization region; thus an approximately equipotential region is maintained all over the volume from which ions are extracted. The low-voltage arc in the Nier form is fully treated in an important article by Bohm, Burhop and Massey[250]. In particular, it was shown by probe measurements that potential patterns were produced in the source chamber, indicating the presence of magneto-hydrodynamic flow.

The theory of probes in magnetic fields was developed, and the ion density available for extraction shown to be calculable by taking the orifice as a negative probe of area A equal to the aperture; thus

$$I_+ = 0{\cdot}40n_+ A\left(\frac{2kT_e}{m_+}\right)^{1/2} \qquad (3.144)$$

The arc source is very simple in its basic components—gas inlet, filament, anode and extraction electrode. A miniature version, formed of concentric tubes through the inner one of which the gas enters, has been described by Sidenius[251].

The capillary arc source[252] is a development of the low-voltage arc at even higher pressures. Powerful arcs, capable of producing 750 mA protons, albeit over a circle 8 cm in diameter, have been developed[253].

Low-voltage arc sources can be used for solid specimens which are made into a cathode for a sputtering process[254]. Initially a buffer gas must be introduced to start the discharge, but eventually it becomes possible, for some solids, to maintain a stable discharge with sputtering ions only. The sputtering source[255, 256] is now well developed, stable and versatile. The filament is in the form of a ring or flat helix; at one side is the anode with orifice and extraction electrode, at the other is the negative sputtering electrode. A magnetic field is applied parallel to the extraction.

The low-voltage arc is often constructed[257] with a heavy filament perpendicular to the extraction field but parallel to the lines of magnetic force. In this form, the source can produce 500 mA ion current, for use in an isotope separator. An oven may be mounted behind the filament, and it is reckoned that with a temperature of 500°–600°K available, any element or its chloride is sufficiently volatile to produce its positive ion. A source of similar construction is the 'magnetron discharge' (similar to the old split anode magnetron), from which equally large ion currents can be extracted[258]. The electrons are inhibited from attaining their full energy by the magnetic field, but the extent and mechanism of this 'inhibited plasma' is not entirely clear. There are two electron current cut-off values of the magnetic field, instead of the single cut-off which is well known in the high-vacuum magnetron. The ion current is a strongly peaked function of magnetic field, and, when end extraction is employed, there appears at higher magnetic fields a peak which consists almost entirely of the fundamental ion and contains ions with an unusually narrow energy distribution. The voltage drop along the filament is of great importance. For refractory elements such as iron, titanium, molybdenum, tungsten and iridium, it is possible either to coat the filament with the powdered element bound in collodion, or to run a strongly negative sputtering electrode within the source.

The *oscillating electron source* was developed by Heil[259] and by von Ardenne[260–265]. In order to increase the electron path in the ionization region, the electrons are made to oscillate, as in the Penning ionization gauge, along lines of magnetic force, within an electric potential well between cathode and anti-cathode. About one-third of the electrons emitted by the filament are formed by positive-ion bombardment rather than by the thermionic process. A balance must be observed in the operating conditions to ensure not only efficient ionization, but efficient extraction.

Two configurations are possible; extraction of ions by electric fields parallel, or perpendicular, to the magnetic field. In the parallel configuration, the magnetic field does not impede the extraction by confining the ions to transverse helical paths. However, ring magnets or coils are necessary, whereas for the perpendicular configuration a simple permanent magnet will suffice. The perpendicular extraction source has been widely used, sometimes in conjunction with 180° mass analysis, where the same magnetic field is employed for both source and analysis. Alternatively, a small source may be placed within the pole-pieces of a powerful magnet; such a source is illustrated schematically in *Figure 3.25*. The source, with negative cathode and anode but positive chamber, possesses a potential saddle at the centre.

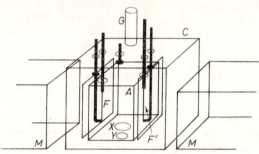

Fig. 3.25. Construction of oscillating electron sources, shown as though planar surfaces were transparent: M, magnetic pole-pieces; F and F', filament and spare filament (both are backed by screening plates biased negatively); A, open-ended cubical anode, containing hole X; C, source chamber containing gas inlet G and extraction orifice Y.

An experimental and theoretical investigation of the oscillating electron source has been given[261]. It appears that the source can operate under normal discharge conditions with weak electron currents, but when the currents are sufficiently strong a 'bipolar discharge' takes over; the condition has been termed a 'superstate'. The electron beam differs from normal beams in that the space-charge dispersion is counteracted by the magnetic confinement. At the lowest pressures and within a cylindrical boundary, the electron density is independent of distance from the central beam axis. As the pressure is raised, collisions interfere with these conditions, and the electron density remains uniform only up to a certain distance from the axis, above which it falls off as r^{-1}. The ion current is proportional not only to the square root of the pressure, but also to the square root of the emission current.

When the pressure is raised sufficiently for the superstate to be realized (the critical condition observed by other workers[262]), a brightly glowing plasma may be seen. The radial electron density function is now gaussian, and the radial potential function parabolic. The ion current becomes directly proportional to the emission current and is independent of pressure. Large proportions of multiply-charged ions may be produced.

An electron beam confined magnetically in the presence of gas will fill with ions at a rate proportional to pressure. At the moment of switching on it can, under suitable conditions, operate with a virtual cathode (axial potential more negative than filament), but as the ion density increases, it will pass into another mode of operation, which can readily be detected[263];

the onset time of this mode is directly proportional to pressure, and its measurement may be used as a manometer. This situation is similar to the superstate discussed above. The collision of electrons with the ions held in the space charge of the magnetically confined electron beam is responsible for the high proportions of multiply-charged ions produced in sources employing magnetic confinement.

An oscillating electron beam has been intersected with a gas jet from a supersonic nozzle[264].

One of the most popular modern sources is the 'duoplasmatron'[266–269]. It can yield 200 mA ion current, with only 60 mrad beam divergence. Retardation by an einzel lens system can achieve this intensity of positive ions

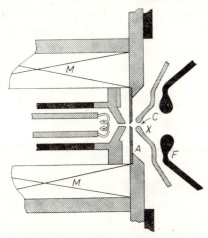

Fig. 3.26. Section through duoplasmatron ion source: M, magnet coils; A, anode, with refractory metal orifice; X, extraction electrode; C, channel; F, focusing electrodes; electrodes made of magnetic materials are shown black, those made of brass or copper shown shaded

at only 200 eV kinetic energy. A typical duoplasmatron has a 'brightness' of several hundred A cm^{-2} steradian^{-1}.

A diagram of duoplasmatron source electrode configurations is shown in *Figure 3.26*. The magnetic field requires critical adjustment to obtain the highest focused electron current through the positive channel C and into the high-pressure gas-containing region in front of the 0·1 mm diameter orifice A, which is maintained several kilovolts positive to the filament and to C. The extraction electrode X, spaced about 1 mm from A, has a somewhat larger orifice, 0·5 mm in diameter, and is maintained at −10 keV. The duoplasmatron is more of an arc than oscillating electron source.

The working volume of the source is unusually small and the extraction field intense. As with radiofrequency sources, the extraction conditions are critical and depend upon the geometry of the plasma boundary[270]. For a concave boundary, the extracted ions are brought by a potential V to a 'waist' situated at a distance Z_m from the boundary[271]:

$$\frac{1}{4} n_+ e v_+ = J_{\text{beam}} = J_{\text{plasma}} = \frac{1}{9\pi} \left(\frac{e}{m_+} \right)^{1/2} \frac{V^{3/2}}{Z_m^2} \qquad (3.145)$$

There is thus a limit to the positive ion density n_+ in the plasma for which the current density J is such that the waist will coincide with the position of the extraction electrode (*Figure 3.27a*). When the ion density n_+ in the plasma is large, Z_m becomes so small that the plasma is drawn out into the interelectrode space, forming a 'plasma cap' of the configuration shown in *Figure 3.27b*.

After the waist, the ion beam expands under its own space charge[272]; the angle of divergence (typically of the order of 6°) is governed by the ratio $R = r_m/r_0$ as a function of Z_m, with the quantities as defined in *Figure 3.27a*

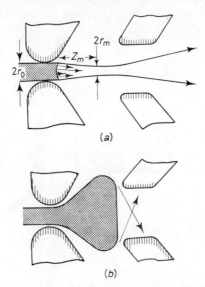

(a)

(b)

Fig. 3.27. (a) Normal mode of extraction from duoplasmatron source (plasma shown shaded), (b) formation of 'plasma cap' at extraction electrodes

Assuming now that the plasma boundary is planar, it can be shown that the expansion of the beam in a time t is governed by the equation:

$$\frac{Z}{r_0} = 2k^{1/2} \int_0^{\sqrt{\ln R}} \exp(t^2)\, \mathrm{d}t \qquad (3.146)$$

with

$$k^{1/2} = \frac{4\cdot942 V^{3/4}}{m_+^{1/4} I_+^{1/2}} \qquad (3.147)$$

for ions of mass m_+ in beam currents I_+ accelerated to an extraction potential V. Beam diameter measurements have shown that the plasma boundary is slightly concave, but that the initial angle at which the particles emerge from the orifice is of importance. It follows that there is an optimum value of orifice diameter r_0 for which r is minimum at a given position of the extraction orifice. This value of r_0 is so much larger than the customary duoplasmatron orifice that the gas effusion would be insuppo. tably large. Therefore the duoplasmatron is now often operated with a convex 'plasma cap'. This can be controlled by moving the extraction electrode several millimetres further back and filling the space with a short cylinder maintained at anode potential; this is known as the 'expansion cup' (not shown in the figure). A different gas may be introduced into the cup, and the ions from this gas, formed by collision with plasma electrons, will predominate in the beam. The ion beam will possess superior profile emittance (divergence $\times$ cross-sectional area) and brilliance when the expansion cup is correctly designed[273].

'Uniplasmatron' sources are also successful[267, 274] in producing beam intensities as high as 500 μA, but their brightness is smaller than that of a duoplasmatron. A uniplasmatron is, broadly speaking, a duoplasmatron without magnetic field.

Radiofrequency ion sources

Radiofrequency ion sources have achieved great popularity for the production of protons for use in accelerators. The extracted ion intensities can be very large, but the energy spread of the ions is so high (40–100 eV) that the sources are unsuitable for ionic collision experiments except at high energies. The sources are particularly valuable for the production of high proportions of protons ($\sim$ 70 per cent) in comparison with H_2^+ ions. The Nier low-pressure electron impact source produces only $\sim$ 1 per cent H^+; low-voltage arcs do rather better; but radiofrequency sources produce mostly protons and are also suitable for use with heavier ions. They will produce negative ions of hydrogen and other elements.

The first radiofrequency source was built by Getting[275]; after some years, advances of importance were made by Thoneman and his colleagues[276]. Although there are no fewer than 32 types of source reported in the literature, the physical differences between them are minor. Among the more successful of the modern sources are those of Morozov[277], of Moreau and Vienet[278] and of Goodwin[279]. A review of the radiofrequency sources is available[280] and includes a tabulation of the physical details of nearly all sources published at that time.

The radiofrequency source operates with gas at a pressure of between 10^{-3} torr and 10^{-1} torr; the application of radiofrequency power causes breakdown of the gas. A typical arrangement is shown in *Figure 3.28*. The vessel is made of silica or of glass, in order that wall processes will be reduced below the levels appropriate to metals. The wall recombination processes include not only neutralization of ions by electrons, but recombination of neutral atoms, and of neutrals and ions. The effect of metal surfaces in enhancing the recombination of hydrogen atoms is well known. On glass and silica surfaces, this reaction proceeds at a very much slower rate. Within the radiofrequency source there is a minimum of metal: only the extraction electrode, and an anode usually of wire. Between the electrode and anode, a large steady potential difference is maintained. A radiofrequency field, 20–100 MHz, is applied capacitatively or inductively to the vessel; the ion production of the source increases monotonically with the radiofrequency power applied. Up to 700 W power has been applied in some discharges, and as little as 30 W has been used in others.

In certain sources, a steady magnetic field H is applied, either along the steady electric vector[281] (longitudinal field), or transverse to it[282]. The greatest augmentation of ion densities is achieved for transverse field at the ion cyclotron frequency

$$v = \frac{He}{2\pi m_+ c} \tag{3.148}$$

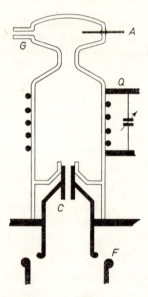

Fig. 3.28. *Section through radiofrequency ion source: G, gas inlet; A, anode; Q, inductive feed; C, extraction channel; F, focusing electrode*

Typically, the transversely applied magnetic field may be 10–60 G. But longitudinal fields have the effect of confining the ions to paths along the parallel electric and magnetic vectors. There is no resonance effect, and the ion beam augmentation increases with magnetic field strength, which may be of the order 300–1000 G. Longitudinal magnetic fields are most suitable for accompanying capacitatively coupled radiofrequency fields[283].

The frequency and the gas pressure are important variables in these sources. The choice of frequency is governed by the nature and efficiency of the coupling (as well as by the cyclotron frequency with steady transverse magnetic field), and by the dielectric heating of the plasma. For inductive coupling, a frequency of 100 MHz is suitable. The dissipation of heat is aided by the cooling of the vessel walls, which can be effected by the evaporation of organic liquids[284]. The source gas pressure is usually adjusted until the maximum ion beam can be extracted. The existence of an optimum pressure is probably associated with the best extraction conditions.

It is reported that increased beam intensity can be obtained when the silica container is constricted near to the extraction orifice.

Extraction of ions from discharge sources

Extraction of ions is achieved through tubes (as for the original canal rays), through orifices or through grids. Effusion has long been replaced by electric field draw-out, which is limited by the inability of the electric field to penetrate far into a plasma. There is thus a balance of effusing plasma and electric field. Potential conditions for a radiofrequency source are illustrated in

Fig. 3.29. *Configuration of equipotentials in the extraction system of the radiofrequency ion source*

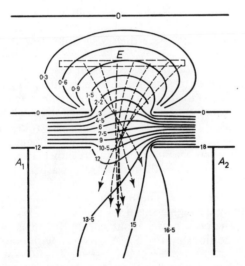

Fig. 3.30. *Equipotentials (full lines) and approximate ion trajectories (broken lines) in Nier source[243] without repeller electrode: electrodes A_1 and A_2 are maintained at different potentials for compensation of ion path curvature in magnetic field; E represents cross-section of electron beam*

(From Hagstrum[243], by courtesy of the American Physical Society)

Figure 3.29; potential configurations for a low-pressure Nier source, including a split extraction electrode for compensation of the effect of magnetic field, are shown in *Figure 3.30*. The duoplasmatron source extraction conditions have been discussed above. The field configuration for best extraction can usually be found for a particular source design by adjustment of the variables of geometry, pressure and potential.

14

The flux density J of ions of mass m_+ that can be extracted by an electrode at a potential V, a distance d from the plasma surface (ideal plasma geometry) is given[285] by the equation

$$J = \frac{K}{d^2} \left(\frac{V^3}{m_+} \right)^{1/2} \tag{3.149}$$

where K is a constant. Note that the lightest ions are the easiest to extract. Detailed studies of tube extraction systems[286-288] and of orifice extraction [272, 280, 289] have been given. The extraction potential is nearly always in the kilovolt region; potentials as large as 8 kV have been employed.

With good design of source and extraction electrodes, milliampere ion beam currents should be readily available at kilovolt energies. Some of the earlier sources produced currents of only fractions of a milliampere; most of the modern sources produce several milliamperes.

Cold-cathode discharge sources

Although the d.c. glow discharge as a source of ions is no longer of great importance on account of the wide ion energy distribution, the spark discharge has a future. It possesses the disadvantage of high noise level and comparative instability, but is nevertheless capable of producing intense ion beams.

A pulsed spark source of protons has been described[290]. It can only be used in pulses because of dissipation and instability problems. The source consists of a pair of electrode bars which terminate a twin parallel transmission line along which passes a voltage pulse, obtained by the discharge of an artificial line. A spark is struck between the electrodes; it is maintained by the hydrogen gas occluded in titanium washers which, interspersed with mica washers of smaller size, make up the region between the electrodes. In order to confine the spark to the inside of the washers where the gas density is greater, a subsidiary trigger discharge is struck in this region, between much smaller wire electrodes. The ions may be extracted from the spark region, either radially through a slit in the washers or axially through a hole in the electrodes. The peak power input during the pulse can be ~ 1 MW, and under these conditions up to 75 mA protons, with only 4 mA H_2^+, can be extracted.

A Penning discharge source has been described[241], working on the same principle as the Penning ionization gauge. The magnetic field is perpendicular to the ion extraction axis. A heavy current discharge is struck: 10 A at 1500 V. The source is particularly suitable for the production of multiply-charged ions. In a xenon discharge, ~ 5 per cent of the total ion beam is composed of Xe^{8+}.

Surface Ionization Ion Sources

The thermal equilibrium at a hot surface between gas atoms and their positive ions is not completely one-sided when the ionization potential is sufficiently small, the work function sufficiently large, and the surface temperature T_s sufficiently high. This is the case for alkali metals at tungsten filaments.

In writing the thermodynamic equilibrium equation, one must take account of the reflexion coefficients R_+ and R_0 for positive ions and neutrals at the metal surface:

$$\frac{n_+}{n_0} = \left(\frac{1-R_+}{1-R_0}\right)\frac{g_+}{g_0}\exp\left(\frac{\phi-E_i}{kT}\right) \tag{3.150}$$

The emission of positive ions from hot metal surfaces has been an established phenomenon for many years[291]. Many metallic ions can be produced by the heating of the metal oxide to a white heat. Metals will emit their own ions when the temperature is sufficiently high, and data exist for characteristic 'positive ion work functions' ϕ_+ for this phenomenon. Equation 3.150 assumes the complete dissociation of molecules into atoms, which is seldom achieved. Since dissociation is favoured by high temperatures, the optimum conditions for atomic positive ion production remain large work function, low ionization potential and high temperature. Surface ionization sources fall naturally into the two classes: those in which an attempt is made to achieve favourable thermodynamics, and those which produce ions by heating some special compound which has been found efficacious. The latter include:

1. The Kunsman source[292], in which a filament is coated with a mixture of ferric oxide and the hydroxide of the alkali metal whose ion is desired;
2. The production of alkali metal ions by heating their synthetic alkali aluminium silicates to 1000°K;
3. The production of lithium ions from the mineral β eucryptite, $Li_2O : Al_2O_3 : 2 SiO_2$.

The problems of obtaining favourable thermodynamics are not trivial. The necessary high temperatures will often vaporize rapidly the metal or compound which is being ionized. What is often needed is a system at two temperatures, one relatively low to vaporize the solid, which passes to an adjacent surface, and a second, high temperature suitable for ionization.

One device making use of these principles is the triple filament source[293] which is used in some commercial mass-spectrometers. Three tungsten ribbon filaments are mounted within one or two millimetres of each other, and separately heated. Two would suffice, but the third maintains potential symmetry. The device can readily be modified for electron impact usage. A simplification[294] uses a single filament, the temperature gradient along its length being such that both hotter and cooler regions can operate appropriately.

For the production of more intense ion beams, a superior arrangement is a stainless steel oven, fronted by a separately heated sintered tungsten plug. The maintenance of potential gradient is possible owing to the poor heat conductivity of stainless steel.

Modern surface ionization sources require that the work function of the ionizer metal should be as high as possible and stable. Rhenium (possessing an unstable carbide) is considered better than tungsten, and is satisfactory for the ionization of all the alkali metals except lithium; for the latter, a stream of oxygen is directed over the metal surface, since the work functions of the oxides are higher than those of the metals.

Fluxes of Ba^+ have been produced by evaporating barium from a stainless steel oven and allowing it to pass through a rhenium gauze joule-heated to 2500°K. Molybdenum thimble ovens at 650°K have also been used, with ionizer surface as cool as 1500°K.

A feature of surface ionization filaments is that they are sensitive not only to their surrounding gas but to gases previously in their surround. This memory is well demonstrated in the production of negative ions. When a platinum or similar filament is heated to 800°C in vacuo, the initial current of negative ions is almost as large as a typical electron thermionic emission current. As the gas is desorbed, the negative ion current falls to around 10^{-12} A. Exposure of the filament to the atmosphere rejuvenates the negative ion current; this effect might have application in mass-spectrometric sampling or in gas-analysis.

An interesting application of positive ion emission is in the 'halogen leak detector'[295]; this is a gas diode (occasionally a vacuum diode) with heated platinum cathode held strongly positive. The equilibrium positive ion current is enhanced by a minute trace of halogen, which, one may speculate, removes electrons from the anode and travels to the cathode, elevating the space-charge limited positive ion current.

An ion source which has some affinity to surface ionization sources, although its mechanism may be more similar to the low-voltage arc, consists of a high-power laser pulse directed at a metal surface. A high temperature is produced locally at the surface, vaporizing the metal and giving thermionic emission and positive ion emission. Ion currents are sufficient to perform mass-analysis on the metal or alloy, but too unstable for other purposes.

Bombardment of gases and of solids by α particles has been used as a source of ions in mass-spectrometers[296].

Field Ionization Sources

The ionization of atoms and molecules in sufficiently strong electric fields has been anticipated in theory for many years, but only recently put on a sound experimental basis, with the aid of techniques developed in field emission microscopy. The first review of the subject appeared in 1960[297]; a monograph[298] is available.

The ionization takes place by quantum-mechanical tunnelling through the potential barrier that is formed on the positive side of the atom by distortion of the potential function. For the hydrogen atom, a Jeffreys' approximation can be applied; a crude estimate of the time τ taken to ionize an atom in a field X is:

$$\tau = 10^{-16} \exp\left(\frac{0 \cdot 68 E_i^{3/2}}{X}\right) \tag{3.151}$$

with X in V Å^{-1}, and the ionization potential E_i in electron volts. Thus fields in excess of 10^6 V cm^{-1} are necessary for the observation of field ionization.

A tungsten point of radius r in the region of 500 Å, such as is used in field emission microscopy, is contained within a hollow electrode biased to a potential V of order $-20\,000$ V, and containing an exit orifice for the positive ions. Gas is introduced at a pressure which must be kept below the region of electrical breakdown; 10^{-4} torr is usually satisfactory. Positive ion currents of the order 10^{-8}–10^{-9} A are emitted, and may be related by means of theory to the experimental parameters. Wires of diameter ~ 1 μ and also razor blades have been successful.

When the field ionization of an atom takes place near a metal surface, the process is modified by the image force; there is also an effect whereby the polarized atom is sucked into the metal surface by the electric field. The total emitted ion current can be related to the potential V when this is sufficiently large for all atoms arriving at the tungsten point of radius r to be ionized:

$$I_+ = \frac{0{\cdot}1 e p \alpha V^2}{m_0^{1/2} k^{3/2} T^{3/2}} \text{ A} \tag{3.152}$$

where the gas pressure p is expressed in dyne cm^{-2}, the atomic polarizability α in cm^3, and mass m_0 in a.m.u.; V is in e.s.u. and e in coulombs. The measured ion currents are found to depend upon p and upon V^2. The 'appearance potential' is not well defined, but depends upon the sensitivity of the detector; for a given detector, the appearance potential depends upon E_i and upon r^{-2}. The dependence on gas temperature T is not known by experiment.

One important advantage of the field ionization source is that in the cracking of organic molecules for mass-spectrometric identification, the parent ion is found to predominate; indeed, for hydrocarbons the dissociated ions are less abundant by three or four orders of magnitude. A further advantage is that the ions originate from a very small and well-defined spatial region. The ion energy distribution is conveniently narrow; experiments with retarding field analysis of Ar^+ ions yield a FWHM of ~ 2 eV.

A serious disadvantage is that molecules with permanent dipoles exhibit anomalous behaviour, in that they are easily condensed on the point and are ionized by different mechanisms. Thus H_2O is found to yield H_3O^+ as its primary ion, and will condense on the point when its partial pressure exceeds 10^{-10} torr. If it is desired to avoid such ions, an ultra-high vacuum system is mandatory.

A technique is now available[299] for setting many points, each a few Ångstroms in diameter, on a flat surface, for use in field ionization sources.

Sources for the Production of Negative Ions

The processes of electron attachment to atoms (Chapter 8) differ from those of ionization in that they are exothermic, and the energy defects of the processes are smaller. This implies that thermal equilibria are likely to produce negative ions with efficiency. Furthermore, the electron impact processes which produce positive atomic ions from molecules will sometimes produce negative ions simultaneously.

Under suitable conditions, the sources designed to produce positive ions can also be used to produce the commoner types of negative ion, such as O^-.

However, there are certain negative ion states which arise from excited atoms or molecules[300]; these must be produced in single electron capture or in two-electron capture by positive ions. The conversion of positive to negative ions is of importance in electrostatic generators which double the available kinetic energy by sending the ions up and down the accelerator, changing the sign of their charge at the top. Among the most suitable gas targets for such conversion are streams of easily condensed mercury or water vapour. A commercially available H^- source utilizes all the common hydrogen ions (H^+, H_2^+ and H_3^+) without the need for mass-separation, since H^- is the only negative hydrogen ion stable in the laboratory.

A negative hydrogen ion source capable of producing milliamperes of the ions utilizes a jet of hydrogen gas crossed with an intense electron beam. The production of protons and H^- in a single collision is the dominating process.

Heated-filament surface sources are also effective for negative ions. The application of the Saha equation shows that, for atomic gases, negative ions would predominate over electrons at low temperatures; however, the situation can be reversed in molecular gases. A filament over which gas molecules are passed makes a suitable source for certain negative ions[301]; solids coated on filaments have also proved satisfactory. In fact, the study of negative ion–electron current ratios as a function of coated-filament temperature is still in use[302, 303] as a technique for electron affinity measurement.

Radiofrequency sources[304], oscillating electron sources and low-voltage arcs in magnetic fields are also successful in producing negative ions. In low-voltage arcs used as negative ion sources in photodetachment studies, the ions are extracted through a hole in the anode, parallel to the magnetic field and to the filament-anode axis.

Duoplasmatron sources designed specifically for the production of intense negative ion beams are now popular[305–308]. The final and middle electrodes are offset laterally by about 0·05 cm, in order to minimize the electron beam current. Beams of from 3×10^{-7} A to 10^{-6} A intensity at 2 keV have been produced at a 6 mm orifice 127 cm downstream. The plasma volume appears to possess a sheath of electrons which in its turn is surrounded by a sheath of negative ions.

High-intensity arc sources produce negative ions, but a common feature is that the optimum pressure is critical, since there is a catastrophic reduction of negative ions at pressures high enough for various two-body and three-body collisional detachment processes to become important. Since the temperature of an arc increases with pressure, thermodynamic arguments lead to the same conclusion.

The production of spin-polarized H^- ions is of importance in nuclear physics[309]. Protons are neutralized with caesium atoms, to produce H $2s$ atoms, which are selectively quenched in the $m = -\frac{1}{2}$ sub-state in a magnetic field of 575 G. These atoms capture electrons from an atom of suitable ionization potential (Ar):

$$\text{H } 2s + \text{Ar} \rightarrow \text{H}^- + \text{Ar}^+$$

At energies $\gtrsim 1$ keV, this collision will proceed by pseudo-crossing of potential energy curves, whereas the similar process for ground state H $1s$ is much less likely.

Sources of Positive Ion Beams in Pre-determined Energy States

It has already been mentioned that the study of atomic collisions demands beams with the ions all in the ground state. One also requires to produce beams of ions in known proportions of long-lifetime excited states. A list of metastable states of ions is given in Chapter 13.

Conventional ion sources produce proportions of ions in long-lifetime excited states, and it is impracticable to allow them to decay spontaneously before the collision experiment. Stimulated decay is possible, but easier alternatives exist; the most usual are ion production by energy-monochromated electrons and energy-monochromated photons.

A number of workers have produced ion beams by extraction from a region in which velocity-analysed electrons pass through a gas or a jet of atoms. The difficulties are those of obtaining sufficient ion beam intensity and sufficiently narrow impact energy distribution, in the face of the extraction field. Ionization functions of atoms possess onset functions approximately linear with energy; if the electron energy is maintained lower than the energy of the first long-lifetime excited state, the ionization rate will be small, but not prohibitively so; for multiply-charged ions with power law onset functions, ionization rates will be very small.

The minimization of extraction field is important, and is assisted by employing a nozzle or capillary molecular beam rather than a collision chamber filled with gas[310]. A pulsed extraction field will eliminate the fringing field contribution to impact energy distribution.

When the first long-lifetime excited sub-level is within a few hundred millivolts of the ground state (as with rare-gas ions), electron velocity selection must be refined, and the ion beam intensity will be very low. If the first state is several electron volts above ground, no special electron velocity analysis will be necessary, and more intense electron and ion currents will be obtained. If the electron beam is intense, trapping of ions in the space charge of the electron beam will occur. It can readily be detected from the form of the electron ionization functions. For multiply-charged ion production, the trapping can be an advantage, since all the onset functions will be linear with energy.

A different approach is to filter excited atomic ions by a suitable charge transfer process involving accidental resonance (Chapter 12), but there can be complications in the interpretation of charge transfer even for atomic ions.

Since the threshold functions for ionization by photons are superior in definition of structure and in reliability of interpretation to those for ionization by electrons, the use of monochromated ultra-violet photons for controlled ion production appears promising. The ion intensities are small, even with the modern photon source techniques described in Section 3.9; the gas pressures cannot be reduced to a point where the effects of gas collisions are negligible.

For the production of atomic ions from molecular gases, the interpretation of the threshold behaviour of the electron impact ionization function must be thoroughly understood. Molecular ion populations are more difficult to specify, because the processes leading to their production are complicated. The vibrational states are closely spaced, and, for non-polar molecules, have long lifetimes. The vibrational populations will probably follow the overlap of the wave functions of ion and molecule (the Franck–Condon factor), but near threshold there may be complications due to auto-ionization; if there are not, then at least the lowest vibrational state may be produced, even if the intensity is exceedingly small. Dependable results must await the reliable recording and interpretation of the threshold functions.

Diagnostics of the vibrational state populations of molecular ion beams are not yet in a developed state. It is possible that the infra-red techniques successful for neutral molecules[311] may be applicable. High vibrational quantum numbers of fast H_2^+ beams may be diagnosed by Lorentz disso-ciation in intense electrostatic fields, the protons being separately analysed after the Lorentz process. Thus it has been found[312] that the proportions of high vibrational quantum numbers in H_2^+ are higher when the mole-cular ions are produced not by electron impact but by dissociation of H_3^+ in gas impact. The equilibrium separation of the nuclei in this ion is larger than in H_2^+.

The two remaining problems of ion production in ionic collision physics have been specified earlier as those of producing low-energy beams and of producing highly stripped ions. Low-energy beam production is considered in Section 3.14. Highly stripped ions are important because of their pre-valence as impurities in hot hydrogen plasmas, and the diversity of their charge-changing collision processes which take place at low energies at potential energy curve pseudo-crossings.

It has been known for more than 10 years that arc and oscillating electron ion sources produce currents of highly stripped ions even when the energy necessary for their production direct from the atoms is higher than the potential applied to the source. This implies that the ions are produced in a sequence of ionizing collisions. To maximize the multiply-charged ion cur-rent, the lifetime of ions in the source must be maximized. The space-charge trap of a magnetically confined electron beam has already been mentioned; if the radial trapping potential is reinforced by a potential trap contrived along the beam axis[313], the ions spend sufficiently long in the beam to allow considerable sequential ionization to take place; the yields of multiply-charged ions are considerable.

3.11 SOURCES OF EXCITED ATOMS AND MOLECULES

Metastable Atoms

A list of common metastable atoms and molecules will be found in Chap-ter 13.

Metastable atom beams can be produced by techniques similar to those used for producing beams of atoms or radicals from molecular gases; energy may be transferred to the gas atoms thermally, in discharges, or by means of collisions with electrons or with photons. The problems of separating metastables from ground state atoms, or of estimating the proportions of the atomic beam that are excited, are more severe. Charged species in the beam are removed by electric fields.

Thermal source

The simplest source is thermal in operation. The population of an excited state separated by energy E from the ground state is, from thermodynamic arguments, a factor $\exp(-E/kT)$ below the population of the ground state. Where E is sufficiently small, the proportion of excited state atoms is adequate. This will be the case for the atomic ground state fine-structure doublets of boron, aluminium, gallium, indium, thallium, bromine and iodine. Oven sources may therefore be used for producing beams with populations in the higher state; for many of the atoms listed, the intensities will be small.

Hot-cathode low-voltage arc discharge

The hot-cathode low-voltage arc discharge has been used for rare-gas and nitrogen metastable production. Fluxes around 3×10^9 atoms cm^{-2} sec^{-1} effuse from an orifice in the anode, when a discharge current ~ 20 mA is maintained by a potential difference ~ 1000 V. High gas purity and ultra-high vacuum technique are necessary, for the avoidance of attenuation by Penning ionization and by quenching collisions. Quenching is most serious when the energy difference between the metastable state and the nearest superior non-metastable level is smallest; thus it is to be expected that krypton metastables are easier to produce than neon or argon; xenon metastable decay is enhanced by spin-orbit coupling. Neon metastables are destroyed effectively by hydrogen, less so by nitrogen, and slightly by helium[314]. The helium discharge is supposed to produce He 2^3S when it glows yellow, and He 2^1S when it glows red. Cold-cathode glow discharges have also been used.

In the helium low-voltage arc, the triplet metastable state is believed to predominate, since there are many higher triplet levels which can decay to it. Resonance photons are of course produced in comparable quantities; this remains the principal disadvantage of the arc source. The separation of the metastables by a rotating-sector velocity selector is possible only with considerable reduction in intensity. The only alternative is to separate the metastables from the photons by means of a very thin collodion film which transmits the photons in the ultra-violet. The effect of the metastables on the experiment is calculated either as a difference, or by utilizing the large values of certain photoabsorption coefficients.

The molecular nitrogen metastables are produced in a xenon discharge containing 1 per cent N_2. The oxygen metastable $O_2\,{}^1\Delta_g$ (and possibly O ^{1}D) is produced in high concentration in a radiofrequency discharge in pure oxygen[315]. Atomic oxygen can be removed by mercuric oxide[316] produced by distilling mercury through the discharge.

A considerable amount of work has been carried out with the Hg 6^3P_0 metastable state, which is formed in the mercury arc and glow discharges. Hydrogen impurities are particularly effective in destroying the mercury metastable population. Nitrogen is supposed to enhance the metastable population by reducing diffusive loss.

Electron bombardment of atomic beams

Metastable atoms are produced by the electron bombardment of a jet of atoms[317, 318]. When used under strict crossed-beam single collision conditions, this method is capable of producing, for example, pure triplet helium metastables, but with weak intensity. If the electron beam were to cross a fully collimated molecular beam, the atom collimation would be destroyed by recoil. Therefore the electron beam is directed magnetically along the molecular beam path. When the electrons are sufficiently energetic to produce ionization, plasma conditions may be approached.

Miscellaneous techniques

Atomic beams may be excited by intense resonance radiation[319] from vapour lamps. Alkali metals and thallium beams have been excited so that they pass into $^2P_{3/2}$ states, and Tl $6^2P_{3/2}$ may also be produced by the dissociation of thallium iodide by radiation of wavelength 2100 Å. The irradiation of mercury vapour with 2537 Å resonance radiation leads to the production of Hg 6^3P_0.

Since the suggestion made in the first edition of this book that metastable atoms might be produced from ion beams by suitably chosen electron capture processes, H $2s$ has been produced by the action of caesium on a proton beam[320]. The production of H $2s$ in proton collisions with rare gases, etc. has been investigated in experiment and theory and is discussed in Chapter 12.

The glancing incidence collision at kilovolt energies of ions with metal surfaces[321, 322] was reported to be capable of producing metastables with efficiency as high as 50 per cent. Negative results have since been reported but not published.

Electron bombardment of gases absorbed on metal surfaces

The electron bombardment of gases absorbed on metal surfaces is one of the most promising production techniques recently investigated. It has been shown to be successful for rare gases and carbon monoxide, being least efficient for helium, whose absorption energy is very small. It is supposed that, in addition to bound absorption states, there exist antibonding states, some of which will terminate in metastable levels and some in states of the positive ion, which is efficiently converted to a metastable atom at the surface. As high a proportion as 10^{-4} of atoms desorbed by electron bombardment can be in metastable states. A continuous flux of metastables is produced by bombardment with 50 mA of 100 eV electrons of a sintered block of silver, copper, or nickel through which the gas is allowed to leak. The sintered metal is made by the action of heat in vacuo on compressed powder (100 μ spheres), followed by further heat in an atmosphere of hydrogen. Possibilities exist of focusing the flux by shaping the surface to a concave form. An end-fire geometry utilizes a shaped sintered ring, with gas fed at the circumference, and an axial filament. The angular distribution of metastables with respect to the surface is not yet known, but it is found that glancing incidence electrons, such as can be achieved by applying a suitable axial magnetic field, are the most effective. The principal disadvantage of this metastable production technique lies in the high proportion of unexcited atoms streaming along with the metastables.

It has been reported that metastable atoms of metals can be produced by the impact of ultra-violet photons on a metal surface.

When more than one metastable species is produced in a source, separation can be effected magnetically. Thus, deflexion in an inhomogeneous magnetic field has been used[323] to determine the relative properties of He 2^1S and 2^3S in a beam. The magnetic resonance method has been used[324] to separate Hg $^3P_2$ from $^3P_3$, and inhomogeneous magnetic fields are also valuable.

State separation by beam deflexion in magnetic fields

An inhomogeneous magnetic field exerts a force F in the direction z on an atom with magnetic moment μ,

$$F = \mu_x \frac{\partial H_z}{\partial x} + \mu_y \frac{\partial H_z}{\partial y} + \mu_z \frac{\partial H_z}{\partial z} \qquad (3.153)$$

where μ_x, μ_y amd μ_z are the components of μ. The magnetic moment of an atom depends on its total angular momentum; for example, that of a triplet state is finite while that of a singlet state is zero (nearly all molecules are singlet states and are difficult to deflect).

The separation of metastable atoms from ground-state atoms of different total angular momentum may be achieved successfully by means of inhomogeneous magnetic fields. Initially these were produced[325] by magnets with wedge-shaped poles opposed by rectangular troughs (*Figure 3.31a*). The difficulty of measuring field gradients in these magnets led to the use of currents through parallel wires[327]. In this arrangement, the atomic beam passes near two wires carrying current I in opposite direction (*Figure 3.31b*). The field and gradient at a point defined by distances r_1 and r_2 from the wire centres are

$$H = \frac{4Ia}{r_1 r_2} \qquad (3.154)$$

and

$$\frac{\partial H}{\partial z} = \frac{-4Ia(r_1^2 + r_2^2)z}{r_1^3 r_2^3}$$

where a is the wire radius, and z is the coordinate perpendicular to the line joining the wire centres and also perpendicular to the wires. For $z = a$ and

$r_1 = r_2,$

$$\frac{1}{H}\frac{\partial H}{\partial z} = -\frac{1}{a} \tag{3.155}$$

But for $z = 1 \cdot 2a$, the gradient is most nearly constant over the height of the beam in the figure.

Because the two-wire field requires inconveniently large currents, with water-cooling of the wires, pole-pieces of equivalent shape may be employed[328] (*Figure 3.31c*). A typical magnet for deflecting molecules is made

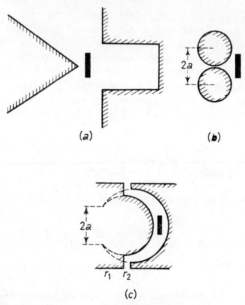

(a)   (b)

(c)

Fig. 3.31. Inhomogeneous magnetic fields for Stern–Gerlach separation: (a) wedge and slot poles of permanent magnet; (b) two wires carrying current in opposite directions; (c) permanent magnetic pole-pieces producing equivalent field to (b). In all cases the atom beam is defined by the black rectangle

(From King and Zacharias[326], by courtesy of Academic Press)

with Permendur pole-pieces, saturating at $2 \cdot 1 \times 10^4$ G, of radii $1 \cdot 25$ mm and $1 \cdot 47$ mm, having a 1 mm gap, and

$$\frac{1}{H}\frac{\partial H}{\partial z} = 8 \text{ cm}^{-1}$$

Focusing of atoms with magnetic moments is obtained by using a hexapole magnetic field (*Figure 3.32a*)[329, 330]. In such a field, formed by pole-pieces shaped in the form of hyperbolic cylinders of axes inclined at 60°, the field H is proportional to the square of the distance from the central z-axis. Since the potential of an atom with magnetic moment μ in a field H is $H\mu$, the restoring force is such as to make the radial motion of the atoms simple harmonic. This is termed 'first order Zeeman focusing'.

A suitable hexapole magnet[330] possesses magnetic field

$$H = H_0\left(\frac{r}{r_0}\right)^2 \tag{3.156}$$

where the orifice radius $r_0 = 1\cdot5$ mm, $H_0 = 9400$ G, and the magnet length is 76 mm.

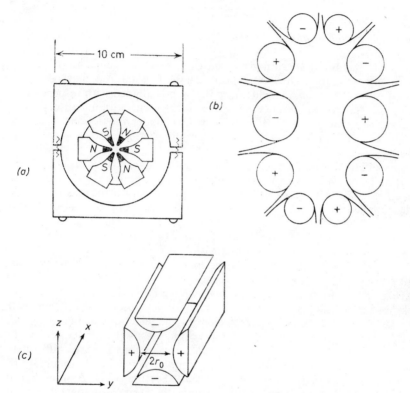

Fig. 3.32. (a) Hexapole magnet (pole-pieces N and S are made of magnetized Alnico, and black pole-tips are made of Armco soft iron); (b) dekapole electric field, constructed of ten cylindrical rods, approximating to the correct curved planes which are also shown; (c) quadrupole electric field for second order Stark focusing and Coulomb focusing

The focal length f of a magnet of length L, with field H_0 at the tip of each pole (radius r_0), is

$$f = \frac{1}{q \sin qL} \tag{3.157}$$

where

$$q = \left(\frac{\mu H_0}{kT}\right)^{1/2} \frac{\bar{\bar{v}}}{v} \frac{L}{r_0} \tag{3.158}$$

Here v is the velocity of the focused particle, which effuses from an oven at temperature T, with most probable molecule velocity $\bar{\bar{v}}$. The device can therefore be used as a velocity selector possessing the unique advantage of a duty cycle of unity.

Dipolar magnetic fields have been used for focusing of permanent magnetic dipoles[331]. Molecules with permanent electric dipoles have also been focused with dipole electric fields, corrections being applied by the addition of further poles to make a dekapole field[332] (*Figure 3.32b*).

Spin polarization of atomic particles

Spin polarization of an atomic particle (ion, atom or excited atom) signifies the alignment of the spin along a specified axis. This axis is in fact specified by a uniform weak magnetic field; once the spin is aligned with respect to this field, the spin vector is no longer free to re-orient without absorption of energy, and will only re-orient through 180° under the action of a radiofrequency field, tuned to the so-called 'spin-flip' frequency. Spin polarization of electrons is discussed in Section 3.13.

Spin polarization of atomic particles may be achieved by bringing about their collision with electrons, with atoms, or with photons, in a weak magnetic field. The last is termed 'optical pumping', but this term is also sometimes applied to the production of population inversion for lasers. It will be remembered that the term 'polarization of an atomic particle' refers to the distortion of the electronic wave functions by an electric field, resulting in an 'induced' electric dipole; in the present context, the term 'spin polarization' is therefore preferable to the more frequently used 'polarization'.

Alignment by optical pumping is based on the fact that polarized light will excite only those atoms with dipoles aligned in a certain direction. Spin polarization was first achieved by the irradiation of atoms of caesium with circularly polarized resonance radiation[333]. The technique has been applied to the spin polarization of metastable helium atoms[334] with the use of the transition of $2^3P \rightarrow 2^3S$. The spin polarization of ions is more difficult to achieve, since the necessary intensities can only be reached by maintaining the ions in a cage or quadrupole field trap (Section 3.17). A spin-aligned beam of caesium atoms is passed through the trap[335] so that alignment takes place by the fast process

$$Cs(\uparrow) + He^+(\downarrow) \rightarrow Cs(\downarrow) + He^+(\uparrow)$$

Vibrationally and Rotationally Excited Molecules

Vibrationally excited molecular beams might most simply be produced by flashing atom sources with sufficiently intense infra-red radiation, but owing to the low efficiency of the process, no reports of such a technique have been published. The usual method is to flash the source with intense visible or ultra-violet radiation, so that the molecule becomes electronically excited[336]. Either by fluorescence or by collision quenching, the molecule returns to its ground electronic state, but by application of the Franck–Condon principle, it may be that the population of the first few excited vibrational levels is increased by factors of 10^2–10^3. Thus for nitric oxide:

$$NO \; X^2\Pi + h\nu \; \rightarrow \; NO \; A^2\Sigma$$
$$\scriptstyle v=0 \qquad\qquad\qquad v=0,\,1,\,2\,\ldots$$

$$NO \; A^2\Sigma + M \; \rightarrow \; NO \; X^2\Pi$$
$$\scriptstyle v=0,\,1,\,2\,\ldots \qquad\qquad v=0,\,1,\,2\,\ldots$$

$$NO \; X^2\Pi + NO \; X^2\Pi \rightleftharpoons 2\, NO \; X^2\Pi$$
$$\scriptstyle v=2 \qquad\quad v=0 \qquad\qquad v=1$$

Vibrational exchange processes of the type 20/11 (the last of the three processes given above for nitric oxide) will proceed at a sufficiently high rate for the enhancement to be predominantly of the $v = 1$ level. The population distribution may be studied in optical emission.

Rotationally excited molecules proceed from thermal molecular beam sources without the necessity for applying additional excitation energy. Separation of the states is achieved[337] by means of an electrostatic quadrupole field (*Figure 3.32c*), formed by hyperbolic cylinder electrodes, with hyperbola axes inclined at 90° (approximately, four cylindrical rods). In such a system, the field E is proportional to the distance r from the central z-axis. For a molecule with negative polarizability α, the restoring force is proportional to αr^2, which makes the radial motion of the molecules simple harmonic. This is termed 'second order Stark focusing'.

A polar molecule of rotational and magnetic quantum numbers J and m possesses a positive induced dipole which increases monotonically with applied electric field when $J = m$, and is largest for $J = m = 0$. However, for $J = m$ the induced dipole is negative for weak fields, although positive when the field is sufficiently strong[338]. The largest negative values are achieved for $J = 1$, $m = 0$. Thus in a quadrupole field, the focal length (or wavelength of the harmonic motion) is shortest for $J = 1$, $m = 0$; longer for $J = 2$, $m = 0$, and so on. By a suitable placement of orifices and obstacles, molecules of different J may be separated.

First order Stark focusing[338] of polar molecules of permanent dipole moment μ_0 is possible in hexapole electrostatic fields. The first order Stark energy

$$W = -\mu E \tag{3.159}$$

with

$$\mu = \mu_0 \langle \cos \theta \rangle \tag{3.160}$$

where θ is the angle between field E and μ_0. Moreover

$$\langle \cos \theta \rangle = \frac{Km}{J(J+1)} = \frac{\mu_{\text{induced}}}{\mu_0} \tag{3.161}$$

where J, K and m are rotational and angular momentum projection quantum numbers. The force on the molecule

$$F = \mu \, \nabla E \tag{3.162}$$

The potential variation in an electrostatic hexapole is

$$V = V_0 \left(\frac{r}{r_0} \right)^3 \cos 3\phi \tag{3.163}$$

where ϕ is the polar angle of the hyperbolic cylinders, and

$$E = 3\,\frac{V_0 r^2}{r_0^3} \tag{3.164}$$

Therefore

$$F = \frac{6\mu V_0 r}{r_0^3} = m\ddot{r} \tag{3.165}$$

Molecules of axial velocity v are focused after a distance l, where

$$\frac{\pi v}{l} = \left(\frac{-6\mu V_0}{m r_0^3}\right)^{1/2} \tag{3.166}$$

or

$$-\langle\cos\theta\rangle = \frac{\pi^2 r_0^3 m v^2}{6 l^2 \mu_0 V_0} \tag{3.167}$$

Molecules with $J = K = m$ or $J = m$ are difficult to focus, because they are deflected away from the axis in a quadrupole or hexapole field. For these, an electrostatic dipole field between two horizontal cylinders, one above the other, achieves separation. In such a dipole field, the horizontal restoring force on molecules of positive polarizability is towards the centre. But there is vertical defocusing, since the field increases in intensity as either pole is approached. The dipole field is stigmatic in that molecules in one spin state are focused to a horizontal line image whilst those of opposite spin are focused to a vertical line image. Additional subsidiary poles[332] may be introduced, producing a dekapole field (*Figure 3.32b*), to correct the astigmatism in some measure; the resolution is poor.

Thus the separation of rotational and vibrational states of molecules is possible in principle, but the available beam intensities are low.

3.12 VELOCITY SELECTION OF ATOMIC AND MOLECULAR BEAMS

In the previous section, mention was made of the use of magnetic deflexion, particularly with hexapole configuration, to select atoms or molecules of finite magnetic moment and defined velocity. This technique of velocity selection, although not widely employed as yet, enjoys the advantage over mechanical selectors that its duty cycle is unity, but it is valueless for the large number of atoms and molecules in singlet states, whose magnetic moments are zero. A Stern–Gerlach magnetic deflector and orifice can readily produce a velocity resolution of ~ 10 per cent, and the hexapole configuration is probably better.

The most widely used method of velocity selection is by means of rotating shutters containing slots or holes. A single pair of such shutters will not only transmit particles within a certain velocity band, but will also pass certain velocity sidebands which can only be eliminated by means of more complex moving systems.

The earliest systems consisted of solid rotors in which slots or grooves were machined. The first such rotor was designed for the velocity analysis of slow neutrons[339] and consisted of a single helical groove, of pitch $dL/d\phi$ and width l, cut in a solid cylinder of length L and radius r_0 (*Figure 3.33a*) rotating with angular velocity ω. This rotor was used[340, 441] to velocity-select alkali metal beams, whose Maxwellian distribution was exactly verified. A particle of velocity $v_0 = \omega L/\phi_0$ will pass through the groove without changing its position relative to the edges. The limiting velocities transmitted are:

$$v_{\text{max, min}} = v_0(1 \mp \gamma) \qquad (3.168)$$

with

$$\gamma = \frac{l}{r_0\phi_0} \qquad (3.169)$$

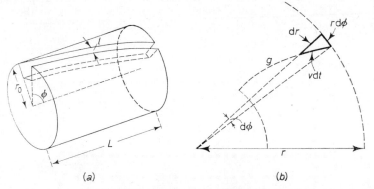

(a)   (b)

Fig. 3.33. Solid rotors for velocity selection of atomic beams: (a) with circumferential groove; (b) with radial groove g in end surface

The admittance A, or ratio of effective groove aperture for particles of velocity v to effective groove aperture for particles of velocity v_0, is given by:

$$A = \frac{1 - |(v_0/v) - 1|}{\gamma} \qquad (3.170)$$

The resolution R, defined for $A = 0.5$, is therefore

$$R = \frac{\Delta v}{v_0} = \gamma \qquad (3.171)$$

and if the velocity distribution is constant, the total transmitted intensity

$$I = -\frac{I_0 v_0}{\gamma}(1 - \gamma^2) \simeq I v_0 \gamma \qquad (3.172)$$

A different type of solid cylindrical rotor[342] possesses spiral grooves cut in the end surface in such a way that a particle of velocity v will travel radially inwards when the rotor has angular velocity ω. Examination of *Figure 3.33b* shows that the relevant polar coordinate equation of an infinitely thin groove is

$$\frac{d\phi}{dr} = \left(\frac{v^2}{\omega^2} - r^2\right)^{-1/2} \qquad (3.173)$$

The system possesses the peculiar advantage of suitability for simultaneous velocity analysis of two orthogonal colliding beams.

The solid velocity selectors have three main disadvantages:

1. Large mass of rotor, resulting in a limitation on maximum angular velocities;
2. Difficulties inherent in cutting the grooves;
3. Complications caused by collisions of atoms with the side walls of the slots (particularly collisions of condensable vapours).

Rotating thin disc sectors were first used in atomic physics by Hostettler and Bernstein[343] using several discs in order to avoid solid rotor problems and at the same time reduce sidebands. An 'unrolled' surface of the rotor is shown in *Figure 3.34* with the appropriate dimensions indicated.

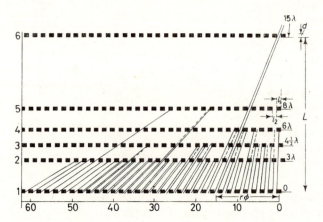

Fig. 3.34. *Unrolled circumferential surface of six coaxial discs comprising the beam velocity selector of Hostettler and Bernstein*

(From Hostettler and Berstein[343], by courtesy of the American Physical Society)

Molecules of the desired velocity are transmitted by passing through those slits in the end discs which are displaced by sector rotation through an angle ϕ. The slots in the intermediate discs are arranged, largely by trial and error, in the simplest manner consistent with the elimination of the sidebands. The geometry of the rotor may be characterized by parameters defined as follows:

$\beta = d/L$ is the ratio of disc thickness to length;
$\gamma = l_1/r\phi$ is the ratio of slit width to arc;
$\eta = l_1/(l_1+l_2)$ is the fraction of sector which is open to the beam.

This arrangement of the discs permits easy alignment.

The transmitted velocity spectrum is triangular (Fraunhofer distribution) with peak at $v_0 = L\omega/r\phi$ falling linearly to zero at v_{max} and v_{min}. The resolution R corresponding to half-intensity width is:

$$R = \frac{\Delta v}{v_0} \simeq \gamma - \beta \tag{3.174}$$

The total transmitted intensity is related to the pre-selection velocity distribution $I(v_0)$:

$$I = \eta \gamma \left(1 - \frac{\beta}{\gamma}\right)^2 v_0 I(v_0) \tag{3.175}$$

These results are valid only for an infinitely narrow beam. The treatment of a wide beam, applicable to this type of selector, has been given by Dash and Sommers[339]. When the beam is inclined to the rotor shaft at an angle α, in the tangential plane, the modified transmission becomes:

$$I' = I \left(1 - \frac{2\alpha L}{r\phi}\right) \tag{3.176}$$

The transmission of a beam divergent in the tangential plane differs only slightly from I, since the effects of positive and negative values of α cancel in part.

The rotating disc selector has proved the most successful of all techniques of atomic velocity analysis[344-346]. For an infinitely narrow beam, the product of optimum transmission and resolving power is equal to $rL\omega/v_0(l_1+l_2)$; therefore it is clear that both the selector length and the speed of rotation must be maximized. For beams effusing from orifices in ovens (Section 3.6), the beam intensity is maximum at the shortest distance from the orifice; therefore the achievement of maximum speed of rotation is of central importance. Motors capable of 50 000 revolutions per minute are commercially available[347]. A factor of two can be obtained by mounting the discs on two independent motors, rotating in opposite senses[348].

In experiments involving the collision of two velocity-selected (crossed) atomic beams, the problem of intensity is most severe. At present, it is necessary that the velocity selection on one of the beams be only minimal. A compact selector, ~ 2 cm in length, with only 26 per cent resolution but 45 per cent transmission, has been described[349].

For certain applications, other velocity selection techniques can be valuable. A pulsed source of radioactive atoms can be detected by deposition on a rotating drum[350]; subsequently an X-ray photographic film is wrapped around the drum and scanned densitometrically after development. Synchronization of the pulsed source is achieved by means of light reflected from a mirror mounted on the drum; the light pulse is converted to an electrical pulse, which operates an ion beam sputtering the radioactive atoms from a metal surface. Pulsed time of flight systems[351] have duty cycles which are inferior to those of rotating disc selectors. A combination of a single rotating disc with time-resolving beam detector has been used successfully[352].

By application of momentum and energy equations, it is possible to obtain information about beam particle velocities using the 'recoil technique' of measuring the scattering angle when collisions take place with a crossed beam[353, 354]. If the crossed beam consists not of atoms but electrons, then the velocity measurement problem can be solved by the methods described in the next section.

3.13 VELOCITY SELECTION OF ELECTRONS

In Section 3.8, it is shown that electrons in beams possess energy distributions which can frequently extend to ~ 1 eV. In collision experiments, it is often desirable that the electrons in the collision region possess an energy distribution much narrower than this. One additional source of distribution width is the penetration of electric field into the collision region, since the electrical screening can never be perfect; it can usually be made so good as to be a negligible source of extra width. Nevertheless, the screening electrodes can themselves introduce extra width because of the contact potential variation from point to point on the metal surface. Between two points, 2·5 mm apart, experiments[355] have shown that potential differences of the order of 0·1 V are typical. This value can be reduced by evaporating gold onto the metal surface. If narrow energy distribution at the collision region is required, then this region must be as far removed from metal surfaces as is necessary to ensure the reduction of potential inhomogeneity by smoothing. In general, it is important never to allow beams of charged particles to pass within, say, 1 cm of a metal surface, except where necessary at a collimating orifice.

Narrow collisional energy distributions can only be achieved by some method of velocity selection (monochromation) of the electron beam. The available methods[356] include:

1. Retarding methods of analysis: (*i*) retardation by negative collector, (*ii*) the einzel filter, (*iii*) the inverse retarding field analyser, and (*iv*) the retarding potential difference technique;
2. Analysis in uniform (and non-uniform) magnetic fields;
3. Analysis in electrostatic fields: (*i*) uniform, (*ii*) inverse first power, and (*iii*) inverse second power;
4. Analysis in crossed electric and magnetic fields;
5. Time of flight analysis.

The retarding methods have specialized applications, but otherwise methods 3(*ii*), 3(*iii*) and 4 are the most competitive. This is not so much because of their resolution constants (see Table 3.6, p. 222) but because of their good electron optics. An analyser not falling into any of these categories is described by Möllenstedt[357].

1. Retarding Methods of Analysis

(i) Retardation by negative collector

This is the oldest technique of electron energy distribution measurement. A current passing through an aperture stop is measured at a collector to which a negative retarding potential V is applied. Differentiation of the function $I_e(V)$ yields the electron energy distribution. Electronic differentiation techniques can be applied[358]. Errors can occur owing to space charge, owing to the efficient reflexion of slow electrons at metal surfaces, and owing to the fact that what is actually measured is the electron momentum normal to the equipotential lines. The aperture stop and the collector (a Faraday cage with its own entrance aperture) together form an immersion lens which diverges the beam. The divergence is greater, the greater the distance from the axis; at sufficiently large distances, the electrons do not reach the collector even when this is at the same potential as the aperture stop. Beam alignment at the aperture can be checked by means of a fluorescent screen. The optimum collector aperture is found to be 0·4 of its distance d from the aperture stop. The limiting apparent resolution is found[359] to be

$$\frac{\Delta E_e}{E_e} = \sin^2 \left(\frac{r_0}{4d} \right) \tag{3.177}$$

where r_0 is the radius of the aperture stop. Application of an axial magnetic field is sometimes used to increase the beam intensity, but spherical aberration is introduced by this unwise practice.

A more satisfactory retarding field analyser for wide-angle beams is formed of concentric spheres; the outer sphere, of radius b, is the retarding potential collector; the inner sphere, of radius a, is either made of grid mesh or contains an aperture of undefined radius through which the beam enters. The energy resolution is found to be dependent upon the position and dimensions of the electron optical object, which is taken as having a radius r_0 and separation p from the analyser aperture. The resolution

$$\frac{\Delta E_e}{E_e} = 2\left[1 - \frac{1}{4}\left(\frac{1}{1-(a/p)}\right)\right]\left(\frac{r_0^2}{r_0^2 + p^2}\right) \tag{3.178}$$

is optimum when the analyser is designed so that

$$\frac{p}{a} = \left[1 - \frac{1}{4}\left(\frac{1}{1-(a/p)}\right)\right]^{-1} \tag{3.179}$$

The concentric sphere retarding analyser has proved successful in photoelectron spectroscopy (Chapter 9) with the inner electrode in the form of a spherical grid. Such grids are commercially available and can be formed from tungsten or stainless steel mesh on a ball bearing with the aid of a pair of circular jigs between which the mesh is spot-welded.

The technique has the advantage of enabling electrons to be collected from a small collison region over a solid angle of very nearly 4π steradian.

(ii) Retardation by einzel filter

The theory of the einzel lens as a velocity filter is given by Hahn[360]. The lens consists of two apertures A and B maintained at the same potential $V_A = V_B$. Between them is an einzel aperture M held at the negative potential V_M, which plays the role of retarding electrode. The apertures A and B are smallest, while M is larger and is not knife-edged but thick. There is a saddle potential V_S at the centre of the einzel aperture; this potential is not equal to V_M, but can readily be calculated[361]; results are represented graphically in *Figure 3.35*. On the axis the saddle represents the actual retarding potential, but of course there is radial variation. It follows that the resolution is best for beams of smallest diameter: $\Delta V/V = 0·0001$ can be achieved[362] for a beam of 0·01 mm. A resolution of 30–40 mV has been reported. Flattening of the saddle field, and with it improved resolution, can be obtained[363] by using two einzel apertures. Simpson and Marton[364] use a five-electrode system, consisting of two short-focus immersion lenses, one each side of a central electrode. Thus the entrance aperture is imaged at the einzel electrode, and again at the exit aperture. This system is suitable for relatively wide beams; it has been confirmed[365] that a beam of 1 mm diameter is reduced in intensity to 10 per cent if $\Delta V/V = 5\times10^{-4}$.

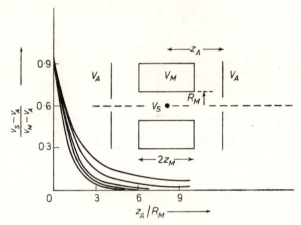

Fig. 3.35. *Data for saddle potential V_s at the centre of an einzel filter; for the five curves from top to bottom, $z_M/z_A = 0, 0\cdot1, 0\cdot2, 0\cdot4$ and $0\cdot6$*

(iii) The inverse retarding field analyser[366]

In this technique, the retarding electrode reflects a fraction of the beam, which is collected on an electrode at earth potential; the collecting electrode is apertured to admit the forward beam. The method is particularly suited to the investigation of the high-energy side of the distribution. For investigation of the low-energy side, retardation by negative collector is used, with the same electrode system.

(iv) Retarding potential difference technique (RPD)

This technique, devised by Fox and his colleagues[367–369] is included in this section mainly for historical reasons. Although still in use, it is not now a strongly competitive technique, but during the 1950s and early 1960s it contributed much to the subject. It was fully described in the first edition of this book. Unlike methods 1(*i*), 1(*ii*) and 1(*iii*) described above, which are techniques of analysis of energy distribution, the RPD technique can be used to provide cross-section functions appropriate to narrow energy distributions; however, techniques 2–5 can be used as real monochromators, whilst RPD is a difference technique.

A negative electrode control is provided whereby the low-energy end of the thermionic energy distribution can be chopped off; a cross-section measurement is made, and this control is varied by a small amount, so that the energy distribution is chopped off at a slightly different energy. Another measurement is made, and the difference between the two is taken to be the cross-section appropriate to the narrow energy band of the difference between the two settings of the control. Automated control is possible[370].

The criticism of RPD which has been made is that the chopping of the energy distribution is imperfect, so an unknown contribution arises from electrons outside the energy band. The axial magnetic field, which is applied for maximizing electron current, introduces complications. It is possible to surmount these criticisms and operate RPD successfully, but even so, techniques 3(*ii*), 3(*iii*) and 4 listed above, are superior. It is possible that in some of the published RPD experiments the energy distribution was inferior to that claimed and could have varied over the cross-section function. In other RPD experiments, the energy distribution was probably satisfactory.

Detailed analysis of the possible cross-section function distortion, including the appearance of negative signals, is available[371–373] and must be studied in detail if this technique is to be used in an experiment. It is shown that the apparent cross-section measured by RPD technique is actually $\sigma + k(\mathrm{d}\sigma/\mathrm{d}E_e)$, where k is a constant. Thus sharp resonances on cross-section functions have been seriously distorted in RPD measurements.

Comparison of analyser properties

Analysers 2, 3(*i*), 3(*ii*), 3(*iii*) and 4 all utilize spatial deflexion of the electron beam, followed by an aperture which selects those of appropriate energy. In methods 2, 3(*ii*) and 3(*iii*), deflexion is in circular paths of radius r which will in the discussion be regarded as beam axes. Suppose that, at the entrance aperture, the radial displacement of an electron from the axis is x_1, then it will pass through the analyser to a point displaced by a distance x_2, provided that its energy is $E + \Delta E$, where E is the energy of the electrons passing through the analyser along the beam axis. The electron is supposed to enter the analyser inclined at angles α and β to the beam axis (α in the plane of the trajectory, β in the tangential plane). Orbit analysis shows that these quantities are related by the equation

$$\frac{x_2}{r} = -\frac{A x_1}{r} + \frac{B \, \Delta E}{E} - C\alpha^2 - D\beta^2 \qquad (3.180)$$

where the constants A, B, C and D are integers tabulated in Table 3.6[374].

The negative sign on the first right-hand term of equation 3.180 indicates that the focusing action of the field is responsible for the displaced electron actually crossing the beam axis. The constant B is effectively a measure of the analytical power of an analyser relative to the others. Thus, on grounds of analytical power alone, there is no overwhelming difference between the different field configurations. Since equation 3.180 is linear in x and E, it follows that, for electrons entering on the axis (with negligible α^2), the exit aperture selects a slice of initial energy distribution. If the initial energy distribution is wide in comparison with the slice, then the slice width is

Table 3.6. MOMENTUM ANALYSER PARAMETERS IN EQUATION 3.180

Sub-section	Analyser	A	B	C	D	Remarks
2	180° magnetic	1	1	1	1	
3(*i*)	Uniform field, electrostatic	1	2	4	2	*r* taken as half distance between entrance and exit apertures
3(*ii*)	Inverse first power field, electrostatic	1	1	4/3	1	
3(*iii*)	Inverse second power field, electrostatic	1	2	2	0	
4	Crossed electric and magnetic fields	1	1	3	1	*r* is radius of trajectory in magnetic field alone

given by

$$\frac{B \, \Delta E}{E} = \frac{s}{r} \tag{3.181}$$

A slit of width s admits electrons of energies $E \pm \Delta E$. (Circular apertures introduce a complication.) To a first approximation, an equation similar to Equation 3.181 holds for electrons which are off-axis at the entrance slit, which is usually of identical width s. The overall energy distribution of the analyser is therefore a convolution of two identical slices, which is a triangular (Fraunhofer) distribution of FWHM

$$\frac{\Delta E_{1/2}}{E} = \frac{s}{Br} \tag{3.182}$$

The quantity $\Delta E/E$ is called resolution; its inverse is resolving power. Although the resolution is strictly speaking the quantity $\Delta E/E$, most workers loosely refer to the quantity ΔE as the resolution of an instrument. Perhaps this could be termed the 'absolute resolution', 'relative resolution' being reserved for $\Delta E/E$. It is important to make consistent use of the FWHM $\Delta E_{1/2}$ rather than other definitions of ΔE. To obtain the minimum $\Delta E_{1/2}$, the analysers are operated with E minimized, often as low as 1–2 V. For collisional energy loss studies with high impact energy, one must first 'monochromate' the electrons, accelerate them into the collision region, decelerate again, sometimes using the identical voltage source, and analyse again[375]. Deceleration lenses have been discussed[376].

The relations between the resolution obtained in a scattering experiment and the available current are important. Since the maximum current that can be passed through an optical system of given aperture is proportional to $V^{3/2}$, and since the quantity $\Delta E/E$ is a constant for a given momentum analyser, it follows that the absolute resolution ΔE is related to current by $I \propto \Delta E^{3/2}$. But since the monochromator selects a slice of width ΔE from a Maxwellian energy distribution, the current must be proportional to ΔE. Thus a proportionality $I \propto \Delta E^{5/2}$ is implied. In a scattering experiment with post-collision momentum analysis, a slice of width ΔE is taken from a continuum of post-collision energies, therefore an extra power of ΔE must be included in this type of experiment, and $I \propto E^{7/2}$.

2. Analysis in Uniform (and Non–uniform) Magnetic Fields

Momentum analysis in uniform magnetic fields (180° magnetic analyser) is a well established technique, but it has not been so frequently used as has electrostatic analysis in recent years. This is primarily because spherical aberration arising in the fringing field seriously affects the resolution. Shaped pole-pieces[377] and correcting coils[378] help towards minimizing this effect, and, by means of the latter, a resolution of 1/8000 has been achieved[379].

Magnetic sector analysis also suffers from the effects of fringing field, which can however be analysed quantitatively and used to eliminate aberrations[380]. A wedge-shaped pole-piece gap converts the 180° into a sector instrument[381]. Both 180° and sector magnetic analysis are more fully discussed in Section 3.14.

A 'telecentric' magnetic deflector has been reported[382] in which 90° deflexion by circular magnetic pole-pieces takes place with a single focus in the centre of the magnetic field. The beams converge on and diverge from two apertures similar to those appropriate to sector operation.

In addition to analysis in uniform magnetic fields, the magnetic lenses familiar in β-ray spectroscopy are occasionally used to momentum analyse electrons formed in collision processes[383]. The lenses possess comparatively poor resolution[384].

Analysis along lines of magnetic field

It is of value to be able to conduct energy monochromation or analysis when the beam is confined helically along lines of a magnetic field. Total excitation cross-sections might in this way be measured. Furthermore, the separation of negative ions from electrons becomes relatively easy under these conditions, so electron attachment cross-sections can be measured.

Experiments on attachment have been conducted for some years with the retarding potential difference technique. The resolution is not better than 100 mV, but ~ 20 mV has been obtained by Stamatovic and Schulz[385] using the following technique. The beam passes through two pinhole orifices and into a region of transverse electric field E. Under the action of this and the axial magnetic field B, the electrons pursue trochoidal paths, with guiding centre deflected in a direction perpendicular to E. In 1 cm pathlength, deflexions of a few millimetres can be obtained. The guiding centre travels in this region with velocity $v = E/B$. The deflected beam now passes through further pinhole orifices, arranged parallel to the original axis but displaced from it. The entire analyser is only about 2·5 cm in length.

3. Analysis in Electrostatic Fields

(i) Analysis in uniform electric field

The parallel-plate analyser[386] is simple to construct and is capable of a resolution ~0·001. A uniform electric field is set up between two parallel plates separated by d, across which a potential difference V is maintained. Electrons pass through a slit into the space between the plates along a path axis making an angle $\theta = \pi/4$ with the plates; they describe parabolic paths of height $h = x/4$ and arrive at an exit slit a distance x_0 along the plate which contains the entrance slit. For electrons of energy eE eV,

$$x = \frac{2Ed}{V} \sin 2\theta \qquad (3.183)$$

Differentiating,

$$\frac{dx}{d\theta} = \frac{4Ed}{V} \cos 2\theta \qquad (3.184)$$

Thus $dx/d\theta = 0$ for $\theta = \pi/4$, at which entrance angle first order focusing is obtained. Complete separation of two (triangular) energy distributions separated by ΔE is obtained when

$$\frac{\Delta E}{E} = \frac{s}{x_0} [(1 + \sec 2\ \Delta\theta) - (1 - \sec 2\ \Delta\theta) \qquad (3.185)$$

for half-divergence $\Delta\theta$ at entrance. For $\Delta\theta = 0$, this reduces to equation 3.182 with $r = 2x_0$ and $\Delta E = 2\ \Delta E_{1/2}$.

In order to keep the electron beam away from the metal surfaces, one must ensure that $d \geqslant 0·3x_0$. The necessity of collimating the beam parallel along the $\theta = \pi/4$ axis before entry is the principal disadvantage of this analyser. The principal advantage is that the beam enters the analyser through an equipotential surface, so no fringing field correction is necessary, contrary to most other analysers. There is no focusing action perpendicular to the plane of the parabola; it has been shown[387] that the introduction of a magnetic field parallel to the electric field produces such focusing, although it skews the orbits.

A double analyser (without magnetic field) has been reported[388].

When an electron beam passes through a gas or across another beam, electrons are often scattered with azimuthal symmetry but at a variety of polar scattering angles θ. An analyser will normally accept flux scattered through a particular small range of azimuthal scattering angles. However, greater intensity would be achieved if an analyser could be designed with cylindrical symmetry, able to accept all azimuthal scattering angles, although only one polar scattering angle. This is possible by means of either parallel discs with two annular orifices in the first, or by means of concentric cylinders with two annular orifices in the inner one. The parallel-disc device[389], known as the 'fountain spectrometer', will collect flux scattered at $\theta = 45°$. Modifications could presumably be made for use at other angles. The cylinder device is well known in β-ray spectroscopy, and has been used[390–393] in the study of electrons emitted in atomic collisions. The flux scattered at $\theta = \alpha_0 = 42\cdot30°$ can be analysed with second order focusing[393]. In this 'cylindrical-mirror' analyser[394], the flux enters through a circumferential slit in the inner of two concentric cylinders (radii $r_1 < r_2$) and is focused onto a second slit at a distance $L_0 = 6\cdot1r_1$ along the cylinder. It is not essential that the analyser be used with complete cylindrical symmetry through 2π. For electrons of energy E_e, the operating intercylinder potential V_c is given by

$$\frac{E_e}{eV_c} = 1\cdot3 \ln \left(\frac{r_2}{r_1}\right)$$

The maximum distance off-axis is $1\cdot8r_1$. The performance of the analyser is calculated from the equation

$$\Delta L = 5\cdot6r_1 \frac{\Delta E_e}{E_e} - 15\cdot4r_1(\Delta\alpha)^3 + 10\cdot3r_1 \frac{\Delta E_e}{E_e} \Delta\alpha \qquad (3.186)$$

A comparison of performance of this analyser with others is only possible numerically, but it is reported[394] that it compares favourably with the hemispherical analyser. Focusing onto the axis is also possible.

(ii) Analysis in inverse first power electric field

The electric field X between two cylinders, across which is maintained a potential difference V_{12}, is radial and varies inversely with the radius variable r:

$$X = -\frac{2V_{12}}{r} \left\{\frac{1 + \frac{1}{2} \ln [(a+b)/2a]}{\ln (a/b)}\right\} = -\frac{AV_{12}}{r} \qquad (3.187)$$

where a and b are the cylinder radii. Electrons entering at any radius with a velocity v travel in circular orbits provided they enter tangentially, that is, perpendicular to the lines of force, and provided

$$\frac{m_e v^2}{r} = -Xe = \frac{AeV_{12}}{r} \qquad (3.188)$$

equating electric and centrifugal forces. Thus electrons of energy E pass through the analyser when $E = \frac{1}{2}AeV_{12}$.

Consider electrons entering at a radius r_0, midway between the two cylinders, but with entrance paths making an angle α with the circular beam axis. Hughes and Rojansky[395, 396] have analysed the differential equations of orbit in polar coordinates r and ϕ, taking a reduced radius variable $y = r/r_0$:

$$\frac{d^2y}{d\phi^2} + y = \frac{c^2}{y} \qquad (3.189)$$

with

$$c^2 = \frac{Ae}{m_e v^2} = \frac{v_0}{v \cos \alpha} \qquad (3.190)$$

This equation is solved subject to the conditions $\phi = 0$ when $y = 1$, and $dy/d\phi = -\tan \alpha$. The orbit solutions are non-circular:

$$y = c + (1-c) \cos \sqrt{2}\phi - \frac{1}{\sqrt{2}} \tan \alpha \sin \sqrt{2}\phi \qquad (3.191)$$

However, two particles of the same velocity entering at angles $+\alpha$ and $-\alpha$ travel in orbits which re-cross at an angle

$$\Phi = \frac{\pi}{\sqrt{2}} = 127° \ 17' \qquad (3.192)$$

A focusing velocity selector may therefore be built with entrance and exit slits separated by the angle Φ. The resolving power is defined by the following equation:

$$r_\Phi - r_0 = (2\beta - 4\beta^2)r_0 \qquad (3.193)$$

with

$$\beta = \frac{v_0 - v}{v_0} \qquad (3.194)$$

The symbol r_Φ indicates the radial coordinates of the crossover. A slit of width $2(r-r_0)$ placed at this angle receives a velocity spectrum defined by these equations, which are a special case of equation 3.180.

Electrons of wide entrance angle α, but identical velocities v, are not all focused at the point defined by r_Φ, but at points separated by a distance

$$s = \frac{4\alpha^2 r_0}{3} \qquad (3.195)$$

The resolution is therefore

$$\frac{\Delta E_{1/2}}{E} = \frac{s}{r_0} + \frac{4}{3}\alpha^2 \qquad (3.196)$$

for an analyser with entrance and exit slits both equal to s.

When Hughes and his colleagues originally investigated the '127° analyser', using metal cylinders, they failed to monochromate to much better than 1 eV, at however low an energy the instrument was operated. Therefore the analyser was comparatively[397] neglected until the work of Marmet and Kerwin[398] who showed that this failure was due to the high efficiency of reflexion of slow electrons from metal surfaces. Multiple reflexions within the monochromator cause a build-up of space charge which distorts the electron orbits. When the metal cylinders were replaced by grids of high transparency backed by cylindrical surfaces held highly positive (~ 50 V), the reflexions were reduced to such an extent that resolutions $\sim 0·05$ eV became available. Those electrons which reach the cylindrical surfaces cannot return to cause space charge in the monochromator.

In order to minimize the effect of fringing field from the entrance and exit orifices on the electron orbits, the intergrid distance is kept as small as is consistent with sufficiently small perturbation due to field penetration from behind the grids (perhaps $0·1r_0$). However, this increases the number of electrons reflected from the grids themselves. Even though the electrons might be well aligned before entry into the analyser, their thermionic energy distribution will ensure that many pass in orbits which impinge on the grids. Those which impinge tangentially upon the outer grid will in fact see the grid much less transparent than it appears in normal incidence. Many will be reflected into such an orbit that they pass out of the exit slit, although their energy is perhaps 0·5 eV higher than that of the main analysed beam. Such an effect may be observed in experiments with two analysers in tandem scanned one against the other, and also in 'ghosts' in electron spectra[399]. The effect can be eliminated by narrowing the electron energy distribution before the analyser, for example by using two analysers in tandem.

Reflected electrons exert a powerful adverse influence upon attainable resolution, and their minimization by tandem analysers is necessary if theoretically predicted resolution is to be achieved in the laboratory[399]. The reflexion at the grid may actually take place at the periodic potential formed by perturbation of the outer field through the grid, but it is easily shown by measurement that a large fraction of the current passes out through the grids and is collected on the outer cylinders.

The effect of the entrance and exit electrode fringing fields upon the electron orbits cannot be entirely eliminated. The slit electrodes are normally held at the potential which corresponds to the electron beam axis—*not* $\frac{1}{2}(V_1 - V_2)$. Attempts can be made to neutralize the fringing field by means of biased wires very close to these electrodes, or by the use of slit pieces made of resistive material. Analysers with the grids subtending smaller angles (for example 120°) than the slit electrodes can be shown[400, 401] to produce smaller distortion of the orbits, and this 'Herzog correction' is often applied.

A further source of broadening lies in the penetration of an accelerating field which is usually applied outside the exit slit. Although this field does not seriously disturb the orbits, it gives rise to a potential distribution across the exit slit. Since electrons arriving at this slit fall within a certain energy distribution, they receive by the time they arrive at a collision chamber an additional energy which is not unique but possesses a distribution corresponding to the potential distribution across the slit. Experiments[402] have shown that the penetration broadening is proportional to the accelerating field and does not exceed 10 mV even for acceleration from 1 V to 20 V.

There is also an adverse effect on resolution arising from the finite length of slits; this is usually known as β-correction, following the notation of equation 3.180 (not the β defined in equation 3.194). Calculation of the β-correction[400] proceeds on the assumption that an axial electron passing through the entrance slit at one end, and through the exit slit at the other, possesses an orbit radius $r(l^2 + L^2)^{1/2}$ where l is the slit length and L the axial path length in the analyser. However, the full β-correction must only be applied when pre-analyser acceleration is sufficiently low for such a trajectory to be likely. In the limit of large pre-analyser acceleration, no β-correction will be applicable.

The Hughes analysis leading to equation 3.196 has been extended by Ballu[238] using computer calculation of trajectories from the equation:

$$y'' - \frac{2y'^2}{y} + y^2 \left[\frac{y}{(1+\beta)^2 (r_1/r_0)^2 \cos^2 \alpha} - \frac{1}{y} \right] = 0 \qquad (3.197)$$

where r_1 is the trajectory radius at entry. The single analyser treated is found to have a resolution approximately that of equation 3.196, but the value of α is rather large. However, the most significant result of this analysis is that the energy distribution is asymmetrical, being very approximately triangular, but 50 per cent wider on the high-energy side than on the low.

Analysers receiving only very weak fluxes of electrons need have no grids, the reflexion of electrons from the cylinders being unimportant. Such instruments have been used in photoelectron spectroscopy.

An analysis has also been made of the 127° analyser with crossed magnetic field. The stable orbit equation is

$$\frac{m_e v_0^2}{r_0} = Ee + v_0 B \tag{3.198}$$

and since this contains terms in both v_0^2 and v_0, there are two characteristic velocities which can pass through the analyser. This has been observed experimentally.

Operation of an analyser in which the electron energy is ~ 1 eV requires that the residual magnetic field be reduced to ~ 0.001 G. In addition to screening material, it is advisable to use Helmholtz coils at least 30 times the radius of the analyser. The inconvenience of building circular coils can be avoided by square or octagonal construction[403]. Collision events at energies as low as 0.05 eV can be studied under these conditions. Beam instability arising from fluctuating magnetic fields can be avoided by feeding back a signal from a sensitive gaussmeter into the Helmholtz coils.

For certain collision experiments, it is valuable to use inverse first power velocity analysis in sector configuration. The following relations[404, 405] connect the quantities:

s = distance from line source to sector entrance,
d = distance from line image to sector exit,
f = focal length,
Φ = lens angle,
$f^2 = (s-g)(d-g)$,
$f = r/(\sqrt{2} \sin \sqrt{2\Phi})$,
$g = (r/\sqrt{2}) \cot \sqrt{2\Phi}$.

Reflexion of electrons from the sector plates are a serious source of spurious signals. This may be minimized by Venetian blind construction[406], by construction from grid mesh, or from resistive material (orthogonal construction). The principal disadvantage of sector operation is that the widely separated electrodes, necessary to contain wide beams, possess large fringing fields[375, 376].

(iii) Analysis in inverse second power field

The electric field $X(r)$ between two spheres of radii R_1 and R_2, between which is a potential difference V_{12}, is

$$X(r) = \frac{V_{12} R_1 R_2}{(R_2 - R_1) r^2} \tag{3.199}$$

where r is the radius variable. This inverse second power field was first used for velocity analysis, under sector conditions, by Purcell[407], following a suggestion by Aston. For a particle of charge e travelling with velocity $v = (2eE_e/m_e)^{1/2}$ radius $a = \frac{1}{2}(R_1+R_2)$, there is a balance between centrifugal and electric forces

$$eX = \frac{amv^2}{r^2}.$$

(3.200)

Provided that

$$V_{12} = E_e\left(\frac{R_2}{R_1} - \frac{R_1}{R_2}\right)$$

(3.201)

the particle travels in a circular orbit of radius a. Graphical orbit analysis shows that focusing in the plane of the orbit is perfect at 360°, but only first order at 180°. The first use of the 180° device was by Simpson[408]. It will be seen from Table 3.6 that this is the only analyser for which there is focusing of particles diverging in a surface orthogonal to the plane of the beam axis. The analyser compares favourably with other deflexion analysers as regards intensity.

The most severe limitation on analyser beam intensities is set by the space charge. The maximum current which can pass through a field-free space is dependent upon $E_e^{3/2}$ but if this expression is combined with equation above, it is seen that the maximum current passing through an analyser of fixed energy spread is actually dependent on $E_e^{-1/2}$, that is, it is greatest for an analyser operating at the lowest energy.

The principal disadvantage of the hemispherical analyser is that sufficiently accurately constructed grids are not available, so the beam must be well aligned before entrance, and $R_2 - R_1$ must be made relatively large in order to avoid the electrons in the tail of the Maxwell thermionic energy distribution from striking the outer sphere and being reflected. But for large $R_2 - R_1$, the fringing field effects are correspondingly large; one method of avoiding these effects is by use of virtual entrance and exit apertures. Since the β-focusing is good, circular apertures may be used, so that cylindrical rather than slit symmetry is possible; this is much more satisfactory from the point of view of electron optics. An image of a real aperture is focused upon each end of the hemispherical analyser, in the 180° plane. A small transverse potential applied across a split cylinder lens assists the matching from cylindrical symmetry to inverse second power radial symmetry. The system has the additional advantage that electrons scattered from the real aperture, with loss of energy, cannot enter the analyser. However, such a system of virtual apertures requires careful adjustment so that the resolution is maximized.

In *Figure 3.36* is shown graphically the theoretical relationship connecting intensity and resolution given by Simpson. The inset shows the electrode configuration in a spectrometer of this type. For virtual slits, the intensity is more favourable than for real slits, but even so is as small as 10^{-14} A for $\Delta E_{1/2} = 0.003$ eV. It rises exponentially to 10^{-9} A for $\Delta E_{1/2} = 0.27$ eV. The observed performance of a cylindrical analyser[409] is also shown in this figure.

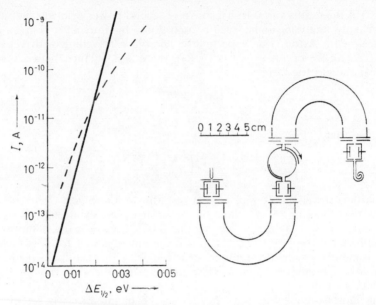

Fig. 3.36. Dependence of the electron current transmitted through a momentum analyser upon its resolution (FWHM) (full line, hemispherical analyser and virtual orifice; broken line, cylindrical analyser and real orifice). Inset: electrode system of double hemispherical spectrometer

Use of the concentric spherical analyser in sector geometry[191] is possible with high resolution and intensity. A spherical indirectly heated cathode is advantageous, with circular aperture in a spherical accelerating electrode. After passing between sector concentric spheres, the electrons of correct velocity are brought to focus at a circular aperture.

It is of interest that the electron-optical quality of the β-focusing of concentric spheres is seriously reduced at sufficiently high energies by the relativistic modification of trajectories[410].

4. Analysis in Crossed Electric and Magnetic Fields

In the 'Wien filter'[411–414], the electron beam axis is linear, and perpendicular to both the uniform crossed electric and magnetic fields X and H, the action of each of which balances that of the other:

$$eX = evH \tag{3.202}$$

and

$$E_e = \frac{mX^2}{2H^2} \tag{3.203}$$

Scanning of the electric field X gives a resolution in energy for constant H:

$$\frac{\Delta E_e}{E_e} = \frac{\Delta X}{2X} \tag{3.204}$$

The resolution in terms of apparatus dimensions is given by equation 3.182. These relations hold even when the fields are not directly superposed, so long as any deflexion is small compared with the path length l through either field, and provided the path lengths through the two fields are equal. If they are unequal, then a correction factor l_H^4/l_X^4 must be applied to v in equation 3.202.

The focusing properties of the filter are stigmatic. A point source is focused as a line. The crossed fields produce an electron-optical image with unit magnification provided

$$l = \frac{\pi X}{\omega_c H} \tag{3.205}$$

with

$$\omega_c = \frac{eH}{m_e c} \tag{3.206}$$

A dispersion of 4 mm for 1 eV has been reported[415] for operation at 300 eV, with deceleration from 50 keV. Thus a resolution of 50 meV can be achieved.

One of the principal problems of construction of the Wien filter is the maintenance of a uniform weak magnetic field over an extended rectangular region. This is achieved with the aid of two rectangular slabs of ferromagnetic material, each surrounded by a current-carrying coil. No return path is required.

5. Time–of–flight Analysis

This technique has been put into service in collision physics[189], and has been critically examined[416]. For 7 nsec time resolution of 1 eV electrons passing through 3 m drift space, 3 mV resolution is possible. Since 2 nsec resolution electronics is available, this technique possesses possibilities of very narrow monochromation, provided that the very long analysis region can be tolerated. The neutralization of stray magnetic field over such a path is almost prohibitive.

Usage of Velocity Analysers

Velocity analysers may of course be used either as monochromators or as spectrometers; that is, either for producing a beam with narrow energy resolution or for determining the energy distribution of a beam already in existence. Apart from the RPD technique, retarding techniques are normally only used spectrometrically, although it is possible to conceive of collision experiments conducted actually inside a retarding region[417]. For spectrometric studies, the analyser is inserted into the beam and may if necessary be traversed across it. A competitive method of studying the energy distribution in a beam is by means of the cross-section function it measures. This subject is treated in Section 1.5. However, it is difficult with this technique to obtain anything more refined than a space average of the distribution, whereas an analyser can actually be traversed through the beam. An analyser such as the parallel plate can also be used with a large number of exit orifices and detectors so that scansion in potential is unnecessary.

An important problem is to measure an absolute kinetic energy which corresponds to the mean of a distribution. Owing to the existence of contact potentials, there is only one way in which this can be done, namely by the study of a collision process whose cross-section function is known independently to possess structure (peak, step or onset) at a known impact energy. It is important that an excitation process should be used for energy scale setting, since the space charge produced in an ionization experiment will affect the energy scale. The excitation of a known energy level may be used to set the difference between two analysers. Since the chemisorption of gases (particularly oxygen) can change the contact potential at a surface, the setting of an energy scale by means of excitation of a gas A for the purposes of subsequent study of a gas B should actually be carried out in a mixture of A and B.

The phenomenon of contact potential can adversely affect the resolution of analysers by its variation from point to point on the surfaces which bound the analyser. A beam which passes close to a surface will traverse a path or orbit which is distorted by spatial variation of contact potential. The distortion is smaller the further the beam is from the surface. This affects the design of analysers, and in general is the basis of an argument for large sizes. Uniformity of magnetic field limits these sizes. The minimization of spatial variation of contact potential is achieved by suitable choice and treatment of the surfaces. Experimental investigations have been made[418] using a vibrating probe very close to the surface, and variations are as follows:

Unbaked gold $\sim$ 70 mV cm^{-1},
Baked gold $\sim$ 15 mV cm^{-1},
Unbaked colloidal graphite $\sim$ 40 mV cm^{-1},
Baked colloidal graphite $\sim$ 7 mV cm^{-1}.

Colloidal graphite coatings also have the advantage of absorbing pump oil vapour, thereby minimizing the effects of insulating films of oil.

It is important to remember that the setting of an energy scale to an absorption line can be carried out to an accuracy which is an order of magnitude better than the width of the analyser energy distribution, since one is concerned with measuring the energy at which the observed current maximizes.

Detection of Non-energetic Electrons

In electron collisions which are just sufficiently energetic to excite atoms or molecules to particular levels, the electrons lose nearly all of their kinetic energy. The detection of the resulting very slow electrons is an important method of study of such collisions. As the incident electron energy is raised, the appearance of very slow electrons marks the energy of the level excited, so this technique is one of the methods of electron spectroscopy by which energies of levels (particularly forbidden levels) can be determined.

The most refined method of detecting non-energetic electrons is by means of one of the momentum analysers described above. However, there exist more sensitive, although less refined, devices which are specific to non-energetic electrons.

The first of these is the trapped electron technique[419, 420], in which a potential trap of depths w is formed around the collision region; the incident electrons of energy eV_e are able to escape from the trap, but the non-energetic electrons remain between electrodes A and B (*Figure 3.37*), being able to spiral outwards through the grid G and reach electrode M, which is maintained at 25 V positive.

The trapped electron technique derives from the early high pressure experiments of Maier-Leibniz[421], which were relatively insensitive, although sufficiently refined for the technique to be used for detection (but not the discovery) of the helium 19·3 eV resonance[422]. A modern analysis of this type of experiment has been given[423].

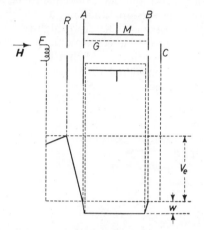

Fig. 3.37. *Schematic diagram of trapped electron technique: H, magnetic field; F, filament; R, retarding electrode; C, collector; AB, collision region; G, grid; M, trapped electron collector*

A second method of detecting non-energetic electrons is known as the 'scavenger technique'[424]. The scavenger molecule SF_6 is added to the collision region gas, and the negative ions SF_6^- are extracted and detected after mass-separation. The technique depends upon the fact that SF_6 two-body electron attachment is a resonance process, possible only with very low energy electrons[425]. Certain other scavenger molecules have been reported[426], including molecules able to resonance-attach electrons of non-zero energy.

Detection of Spin–polarized Electrons

The spin polarization of a beam of electrons is determined by 'Mott scattering'; that is, by anisotropy of its scattering from heavy nuclei[427, 428]. In order that the scattering be purely nuclear, unmodified by atomic electrons, the electron energy must be in the MeV region. Therefore, for determination of the spin polarization of electrons scattered in an atomic process, high voltage accelerating equipment must be available. The target is usually a thin metallic film. Asymmetry factors δ have been calculated for a variety of targets, energies and polar scattering angles θ. Values as high as $\delta \sim 0·1$

16*

are possible, typically with $\theta \sim 90°$ and $v_e = 0·6c$. In order to avoid the effects of multiple scattering, extremely thin foils are used, with incidence angle 45° and $\theta = 90°$[429]. Mott scattering detects transversely polarized electrons, whereas atomic scattering sources produce longitudinally polarized electrons; reorientation of the direction of motion is necessary.

An alternative proposal for detection of polarized electrons has been made[430].

3.14 VELOCITY AND MASS SELECTION OF IONS

Velocity and mass selection of ions may conveniently be considered together. Many ionic collision cross-section functions are relatively free from fine structure, except at low energies. Therefore high-resolution velocity analysis has in the past not been considered particularly important. It is possible that this situation will not continue, and that the problem of ion velocity selection will have to be considered more seriously: it is included in the following headings of this section:

1. Velocity selection of ions,
2. Production of very high energy ion beams,
3. Production of very low energy ion beams,
4. Magnetic mass-spectrometry,
5. Time of flight mass-spectrometry,
6. The quadrupole mass filter.

Velocity Selection of Ions

Since ion sources frequently produce ions with energy distributions of several electron volts (sometimes even several tens of electron volts) FWHM, simple acceleration of the ions is not always sufficient for a collision experiment. Finer energy resolution can be obtained by methods similar to those discussed in Section 3.13 above. The most frequently used analysers are an inverse first power electric 90° sector, and the Wien filter. The first of these is employed in large double-focusing mass-spectrometers, thereby increasing the mass resolution. The operation is as discussed in Section 3.13. The Wien filter, consisting of crossed electric and magnetic fields (X and H) in which the electric and magnetic deflecting forces are equal and opposite, will select velocities but pass ions of all masses. Alternatively it will select masses of ions of identical energies $E = \frac{1}{2}m_+v^2$, since for transmitted ions $eX = evH$.

Production of Very High Energy Ion Beams

When ions are accelerated through sufficiently large potential differences their original energy distribution might be regarded as insignificant, and velocity analysis as unnecessary.

For ion collision experiments above tens of kilovolts energy, standard high-voltage techniques must be applied. The most important consideration is the avoidance of breakdown in the vacuum system. At high vacuum, the breakdown conditions are usually below critical, that is to say, breakdown between two electrodes takes place at a lower potential difference when they are further apart (that is, when their separation is no longer smaller than the mean free path of ionization). It follows that an EHT potential drop (for example, 100 kV) must not be maintained by two well-separated electrodes, but by a series of closer spaced electrodes maintained by a resistance chain at smaller potential differences. At a pressure of 10^{-5} torr a 1 mm gap will support a 60 kV potential difference, but longer gaps will support smaller potential differences. The limit is set by field emission, so all electrodes must be rounded and polished.

Collisional production of ion–electron pairs in the beam is a serious cause of breakdown if these particles are allowed to travel long distances along the beam. Transverse fields are therefore applied, usually by slanting the apertured electrodes through which the beam passes.

Production of Very Low Energy Ion Beams

Simple ion collision experiments such as total symmetrical resonance charge transfer cross-section measurement can be conducted in a magnetic field at energies as low as 0·1 eV[431]; this requires the minimization of contact potentials by electrolytic deposition of amorphous copper on the entire aluminium electrode system; however, for reasons of intensity, more complicated beam measurements are not yet possible much below ~ 2 eV. Methods of beating the 'low energy barrier' by indirect means include:

1. The merged-beam technique discussed in Section 3.2;
2. Ion cyclotron resonance mass-spectrometry, discussed later in this section;
3. Drift tube techniques, discussed in Chapters 10 and 14;
4. Pulsed 'mass-spectrometer source' techniques discussed in Chapter 14.

Direct methods of handling ion beams at low energies are limited by the low intensities imposed by space charge[432]. For each geometry the quantity $(2m_+/e)^{1/2} (I/V^{3/2})$ is constant[433]. The $V^{3/2}$ dependence of available current densities from ion sources has been known for many years[434], and is analogous to Child's law. Since the energy spread of ions emerging from gaseous ion sources is at least several electron volts, energy analysis must be introduced; this is conveniently done at the same time as mass-analysis, since the uniform magnetic field has the property of momentum analysis.

For a 180° magnetic mass-spectrometer of radius of curvature ρ and potential V_a relative to the source, into which a homogeneous beam of ions enters with angular divergence α through an infinitely narrow slit,

$$\rho^2\alpha^2 \propto mV_a$$

For a source energy spread ΔV_0, two ion species of mass numbers m_1 and m_2 will be resolved if

$$m_1 V_a = m_2(V_a - \Delta V_0) \tag{3.207}$$

so that the resolution is:

$$\frac{m}{\Delta m} = \frac{V_a}{\Delta V_0} \tag{3.208}$$

For $\Delta V_0 = 10$ eV, a resolving power of 20 may only be obtained when $V_a \geqslant 200$ V. The ion current that can be extracted from a source can be taken as proportional to $V^{3/2}$, where V is the potential of the accelerating electrode immediately outside the source. Both of these factors make it difficult to perform ion mass-analysis at ion energies $\leqslant 200$ eV, unless the circumstances be exceptional. Therefore, in producing very low energy ion beams, the problem of retardation is frequently encountered. A strongly retarding electric field will act as a convex lens of such short focal length that serious ion-optical problems will arise; further difficulties stem from the space-charge repulsion of the slow ions and from instabilities in the ion source.

The 180° mass-spectrometer possesses the most suitable geometry for producing low-energy ion beams[435]. Not only is the resolving power greater, but also the stability. Suppose that fluctuations in the ion source result in a variation δV_a of energy of the beam entering the mass-spectrometer. In a 180° instrument, this results in a displacement $\delta\varrho$, where

$$\frac{\delta\varrho}{\varrho} = \frac{\delta V_a}{2V_a} \tag{3.209}$$

but in a 90° sector instrument, the displacement may be several times larger, being equal to $\delta\varrho + l\theta$, where l is the distance between exit slit and magnet, and θ is an angle which can be calculated from sector geometry.

There are two ways in which the 180° instrument can be used; the first is to velocity-analyse the beam at the low energy at which it is finally required, conducting the experiment in a collision chamber within the magnetic field, with a divergent beam traversing a curved path. The mass-analysis of beam impurities is largely sacrificed, and the method is undoubtedly limited, but it is capable of producing 1–3 eV ion beams with an energy spread of 0·25 eV[436].

The more general method is to allow the ion beam to travel out of the magnetic field and retard it by means of a specially designed lens, to about 3 per cent of its energy.

The focal length f of a single aperture at potential V with respect to the source, and field strengths X_1 and X_2 in the object and image spaces respectively is[437]

$$f = \frac{4V}{X_1 - X_2} \tag{3.210}$$

Consider the system comprising a field-free region followed by an aperture at 100 V, followed by an aperture at 1 V, followed by another field-free region. Since the potential of the second aperture is as low as 1 V, its convex focal length is short enough to dominate the lens and may be much less than 1 cm, reducing the beam intensity by perhaps 10^4 in a 10 cm path. The first aperture at potential $V \sim 100$ V has a much larger concave focal length, which is far from compensating for the second aperture. It is possible to conceive a decelerating field in which the focusing effects of the two apertures cancel, that is

$$f_1 = |f_2| \tag{3.211}$$

Such a field has a potential along the optical axis z of the form

$$V(z) = a \exp(-bz) \tag{3.212}$$

with *a* and *b* constants. In this field, most of the retardation is carried out when the ions have the largest kinetic energy, and so are least susceptible to focusing. Detailed calculation[434] of the ion paths confirms that the overfocusing effect is much reduced. Such a field cannot be reproduced with accuracy, but has been approximated with the aid of many-aperture lenses connected to a resistance chain, and is not unsuccessful. Gustafsson and Lindholm[438] claim that an exponential field is inferior to the empirically designed apertures shown in *Figure 3.38*. With the aid of these apertures, the mean energy of the ion beam can be reduced to 3 per cent of its former value.

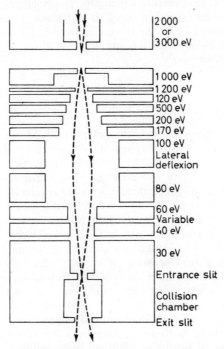

2 000
or
3 000 eV

1 000 eV
1 200 eV
120 eV
500 eV
200 eV
170 eV
100 eV
Lateral deflexion

80 eV

60 eV
Variable
40 eV

30 eV

Entrance slit

Collision chamber

Exit slit

Fig. 3.38. Retarding lens of Gustafsson and Lindholm
(From Hasted [435], by courtesy of Academic Press)

The problem of extracting an ion beam from a 180° magnetic field is not trivial. By using wedge-shaped pole-pieces[439] the image of the source aperture can be formed outside the field itself. For orthodox flat pole-pieces, the image is formed within the field at its edge. A tube of magnetic screening material is positioned so that the beam passes out of the field region in a comparatively straight path. It is unnecessary to place an aperture in the 180° plane provided that a real aperture and cylinder lens system is positioned in the tube in such a way that a virtual aperture, or image, is formed in the 180° plane. Since the magnetic field penetrates some distance into the tube, one component of the cylinder lens is divided and a transverse potential applied, so that a virtual aperture is not formed too far from the tube axis[440].

An alternative extraction technique, of greater value at higher energies, is to use a pair of strong focusing quadrupole lenses.

It shouid be pointed out yet again that slow ions are extremely sensitive to surface charges arising at insulating films on metal electrodes, particularly orifices. Colloidal graphite coating on electrodes may partly exert its well-known beneficial effect by absorbing films of diffusion-pump oil. Another technique which has been reported to be efficaceous is to spray orifices with electrons from a small filament suitably positioned.

Low-energy ion beams, and to a lesser extent high-energy ion beams, may gather a proportion of fast neutrals by charge exchange with background gas molecules during their mass and velocity selection. It is important to minimize the background gas pressure and analysis path length, but an additional precaution may be necessary: namely to introduce a neutral-component-remoevr (by small-angle deflexion in an electric field), after the analysis, just qefore the collision chamber or region.

Magnetic Mass–spectrometry[441]

Mass-spectrometers separate ions of different mass number m/e and determine this rati boy the measurement of two of the three quantities, momentum, energy and velocity ;the velocity is then eliminated. Momentum is measured by the deflexion of ions in a magnetic field, energy by the acceleration or deflexion of ions in an electric field, and velocity by the time taken for the ions to cover a curved or rectilinear path. In conventional single sector or $180°$ magnetic field mass-spectrometers, ions are accelerated electrically to a known energy, and deflected through a known radius of curvature r magnetically. In trochoidal[442] instruments the principles are essentially similar.

Magnetic mass-spectrometers are classified according to their 'order' and 'multiplicity' of focus. The 'order' is a quantity denoting the suppression of terms in the equation for line width w of the measured current of ions of single m/e value as a function of electromagnetic and geometric parameters. In a zero order instrument (very seldom used), w contains a term in α, the half-angle of divergence of the beam approaching the instrument. In a first order instrument, w contains no term in α, but only a term in α^2 and higher terms which are unimportant. Such an instrument is the $180°$ magnetic analyser, for which the paths of beams of identical mass number and velocity, but different angles of emergence from the same point, pass almost through the same point. At $360°$, the focusing is of a high order; this is the important feature achieved in the trochoidal instrument.

The 'multiplicity' of focus is ambiguously used to describe either the number of dimensions in which focusing in angle is achieved (for example, one dimension in the $180°$ magnetic analyser mentioned above) or to describe the number of different types of focusing that are achieved; thus a single focusing instrument focuses only in angle, whilst a double focusing instrument focuses in both angle and energy.

Magnetic mass-spectrometers make use of the fact that particles of mass m a.m.u. and charge e electron charges, having been accelerated through a potential V V will in normal uniform magnetic field H G, follow circular paths of radius r cm where

$$\frac{m}{e} = \frac{r^2 H^2}{20\,880\,V}$$

(3.213)

A graphical representation of this equation, which will be found convenient for design problems, is shown in *Figure 3.39*.

In magnetic instruments, the ions are made as monoenergetic as possible by the design of the ion source, and acceleration by suitably designed electrodes supplied with stabilized potentials; in the high-resolution commercial 'double' instruments, an electrostatic velocity selector, similar to those

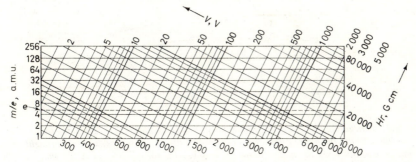

Fig. 3.39. Graph for obtaining the interdependence of m/e, V and Hr in problems of magnetic analysis; the broken line indicates the electron, for which the Hr scale should be divided by 100

(after Harnwell[443])

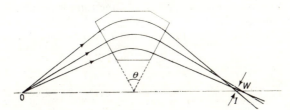

Fig. 3.40 Ray paths in sector field analysis: O, object; I, image; W, total effective beam width at exit

(From Kerwin[441], by courtesy of Academic Press)

described in Section 3.9, is incorporated. A popular type of first order magnetic analyser is the magnetic 'sector field' instrument, for which the beam paths are traced in *Figure 3.40*[444]. It will be seen that the source and focus lie on a straight line through the field apex. The fringing field produces an effect which makes it necessary to move the source–focus line away from the magnet by a distance which is experimentaly found for the best focusing to be 0·7 times the gap width *g*. The figure 1·0 has also been claimed[445]. The fringing field produces some defocusing action, but it can be reduced by bevelling the edges of pole-pieces. The resolution is impaired by the inclination of this line of focus to the beam, but a method has been found[446] of counteracting this effect by using staggered slit edges as in *Figure 3.41*. The ability to adjust the magnet relative to the ion path, with two translational and two rotational degrees of freedom, is a necessary design feature.

It is not actually necessary for the source, field apex and focus all to be in a straight line. In the sector instrument shown in *Figure 3.42* a first order

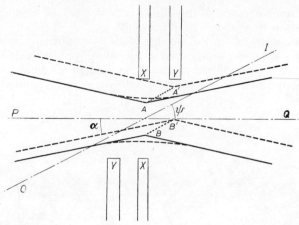

Fig. 3.41. *Staggering of mass-spectrometer slits for best resolution: OI represents the source-focus line and PQ the path of ions emerging normally from the source; ions are brought to a focus at AB, but when the magnetic field is changed the focus moves to A'B'; thus the slit XX normal to the beam is inadequate for resolution and should be replaced by staggered knife edges YY*

(From Kerwin[441], by courtesy of Academic Press)

image will be formed at a point which is a function of source distance s and parameters g_1 and g_2:

$$g_1 = \frac{r \cos \varepsilon_1 \cos (\theta - \varepsilon_2)}{\sin (\theta - \varepsilon_1 - \varepsilon_2)} \tag{3.214}$$

$$g_2 = \frac{r \cos \varepsilon_2 \cos (\theta - \varepsilon_1)}{\sin (\theta - \varepsilon_1 - \varepsilon_2)} \tag{3.215}$$

This unsymmetrical operation improves the resolution, but for 60° the gain is not large. The image distance d is related to a focal length f:

$$f^2 = (s - g_1)(d - g_2) \tag{3.216}$$

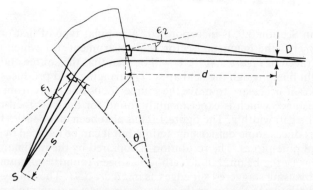

Fig. 3.42. *Sector field analysis without straight line between object, apex and image: S, source; D, dispersion*

(From King and Zacharias[447], by courtesy of Academic Press)

and f is given by the equation

$$f = \frac{r \cos \varepsilon_1 \cos \varepsilon_2}{\sin(\theta - \varepsilon_1 - \varepsilon_2)} \tag{3.217}$$

It is possible to conceive of a step boundary sector magnetic field which will produce second and even higher order focusing. The fringing field prevents the achievement of very good conditions, but there are available a number of valuable discussions of the achievement of second order focusing by shaping pole-pieces[448]. Perfect focusing could be achieved with a boundary (*Figure 3.43*) given by the equation:

$$y = \frac{x(a-x)}{(r^2 - x^2)^{1/2}} \tag{3.218}$$

but 'inflexion' pole-pieces of similar shape have often been used.

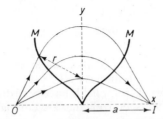

Fig. 3.43. *Magnetic sector field with curved boundaries for perfect focusing: O, object; I, image; MM, magnetic sector field*

(from Kerwin[441], by courtesy of Academic Press)

Focusing in the second dimension, across the pole-piece gap, is not easy to achieve. It is sometimes referred to as double focusing, or β-focusing. It appears that the curvature of the fringing field produces β-defocusing, which is worse for obtuse entry and for inflexion fields. Shaping of pole-pieces has been used to minimize these effects, and acute-angle entry will even produce β-focusing.

Extraction from a $180°$ magnetic system is greatly simplified by the use of inclined pole-pieces[449]. It is found that both focal points then fall outside the magnetic field, and in addition double focusing (α and β) is achieved for ions of unique momentum.

It has been shown that a high order of focusing may be achieved with $360°$ deflexion, but this arrangement is extremely inconvenient with a uniform magnetic field alone. The position of focus is independent of velocity, so the analyser is double focusing in the sense that both direction and velocity focusing are achieved. With trochoidal motion in crossed electric and magnetic fields, however, the $360°$ focusing becomes more convenient and mass analysis is achieved. This is the basis of high-resolution double focusing mass-spectrometers which have achieved great precision and success[450].

The trochoidal (cycloidal) system[451, 452] without an entrance slit is a valuable technique for collecting the total ion flux which is formed in a collision experiment.

Helicoidal paths[453] have also been employed in mass-spectrometry.

The important parameters of a mass-analysis device are its mass-resolution and resolving power, its mass-dispersion or discrimination, and the 'peak shapes' it produces.

The mass-resolution of a mass-analysis device is defined as the ratio $\Delta m/m$, where m and $m + \Delta m$ are the mass number of two ion beams which can just, and only just, be completely distinguished. Note that this is a different convention from that used in velocity analysis, where the FWHM and not the whole width of the peak is the important quantity. For a single focusing magnetic mass-analyser in which the radius of the beam axis is r,

$$\frac{\Delta m}{m} = \frac{s_1 + s_2}{r} + \alpha^2 + \frac{\Delta V}{V} \qquad (3.219)$$

where s_1 and s_2 are the widths of the entrance and exit slits; α (known as the spherical aberration) is the half-angle of divergence of the beam at entrance. The minimum width w to which such a beam is focused is

$$w = 2r(1 - \cos \alpha) \simeq r\alpha^2 \qquad (3.220)$$

for small α. The ion energy in this mass-analyser is eV, and the whole width of the ion source energy distribution is $e\,\Delta V$.

The focusing properties of sector fields have been discussed in detail[454]. The focusing properties of a 90° sector magnet are comparable to those of a 60° sector magnet; the former is, of course, more compact, and is equally popular in collision physics. The resolution of various instruments has been discussed in detail[455, 456].

The resolution of two instruments in tandem may be deduced using the methods of Chapter 1. Abundances $1/x$ of impurity ions proportional to the principal ion in the beam are reduced to $1/x^2$ by the addition of an identical mass-spectrometer in tandem.

The mass-dispersion D of a magnetic mass-analyser is defined as the change in beam displacement at the exit slit per unit change in mass number. The dispersion is a particularly important parameter when the instrument is used as a spectroscope rather than a spectrometer. If the total effective beam width at exit is W, then[456] the resolving power.

$$R = \frac{Dm}{W} \qquad (3.221)$$

When the total peak width is equal to the dispersion, the limit of resolution of the instrument is reached. The dispersion of a magnetic instrument depends upon the pole-piece design, whilst the peak width depends upon the source slit width, the α and β aberration, the energy variation, the space-charge defocusing, the fringing field effect, the slit width and inclination to the line of focus, and the time response of the detecting system, where automatic scanning is used.

Mass-discrimination, occasionally known as mass-sensitivity, is a term used to denote the property of certain instruments of displaying an efficiency of transmission which is not independent of mass. Such a property is a great drawback in isotope abundance studies, and a serious hazard in measuring the proportions of different ions formed in a collision process. In mass-spectrometers specially designed to minimize mass-discrimination, the ion source is built in a region free from magnetic field.

The term 'peak shape' refers to the functions $I_+(V)$ and $I_+(H)$ as the voltage or magnetic field are scanned through the region in which transmission of the beam through the instrument occurs. Equation 3.219 may be analysed in the same fashion that equation 3.180 was analysed to yield energy distributions for electron energy analyers. The triangular (Fraunhofer) peak shape is the simplest that can be produced. Detailed treatment has been given [457]. When the exit slit is widened, the peak shape becomes trapezoidal, or 'flat-topped'[458]. Within limits, variation of the electric or magnetic field produces no variation in the beam intensity. If a flat-topped condition can be achieved, then it can be regarded as proof of the absence of mass-discrimination. Such a test is often applied to the collection with mass-analysis of collision products from a beam–gas or beam–beam experiment.

Time of Flight Mass–spectrometry

Ion beams may also be mass-analysed by time of flight methods; these depend upon the fact that ions of identical energy but different mass number travel at different velocities. Over a path of 1 m, discrimination between 10 keV ions of $m/e = 100$ a.m.u. and 101 a.m.u can be achived by a time-discrimination of 150 nsec. Modern electronic pulse techniques must therefore be employed[459], but the low duty cycle can limit the usefulness of this method in ionic collision experiments. For monitoring of chemical and ionic reactions, however, the commercial instruments available have been most successful. Particle detection must be achieved at surfaces which are planar, otherwise space defocusing will decrease the resolution; trochoidal multipliers were developed for this purpose.

The advantages of time of flight mass-spectrometers over magnetic instruments are: firstly, speed, since for a 100 nsec ion pulse the whole mass spectrum can be scanned in a few microseconds; secondly, simplicity, since no accurately machined parts are required; the absence of an entrance slit is particularly suitable in collision experiments. The disadvantages include a low duty cycle (0·1–10 per cent), and a linear beam path, along which ultraviolet or XUV photons might pass and be registered on the detector. With space focusing, resolution of $\frac{1}{300}$ has been obtained. Ions are not formed at a single point, but over a collision volume. When the accelerating pulse is applied over this volume, the ions are formed at different potentials. Those which are accelerated the most have also the longest distance to travel; therefore there will be a bunching action, and, by positioning the detector correctly, advantage may be taken of this. It is usual to design the instrument with a source region s, an accelerating region d, and a drift region D, each separated by a grid. The accelerating field E_d is adjusted until the space focusing is optimum.

The mass-resolution is made up of two parts, one arising from the different locations of formation, and the other arising from the different potentials of formation. The first is

$$\left(\frac{m}{\Delta m}\right)_s = 16\left(\frac{s_0 E_s + dE_d}{s_0 E_s}\right)\left(\frac{s_0}{\Delta s}\right)^2 = 16 k_0\left(\frac{s_0}{\Delta s}\right)^2 \tag{3.222}$$

and the second is

$$\left(\frac{m}{\Delta m}\right)_E = \frac{1}{4}\left(\frac{E_t}{E_0}\right)^{1/2}\left[\frac{k_0+1}{k_0^{1/2}} - \frac{(k_0^{1/2}-1)d}{(k_0+k_0^{1/2})s_0}\right] \tag{3.223}$$

where

$$E_t = e s_0 E_s + e d E_d \tag{3.224}$$

The overall resolution is the sum of the two parts:

$$\left(\frac{\Delta m}{m}\right)_{total} = \left(\frac{\Delta m}{m}\right)_s + \left(\frac{\Delta m}{m}\right)_E \tag{3.225}$$

The term Δs is the half-width of the ion formation region, whose centre is at a distance s_0 from the source region exit; E_0 is the initial energy with which ions are formed (it is often in excess of the thermal energy).

The time of flight mass-spectrometer requires two pulses, the first of which accelerates the ions out of the source region, and the second is a gating pulse which keeps ions of incorrect velocity from entering the detector region. Alternatively, one may do without the gating pulse and amplify the detected signal for oscilloscopic display. For a linear scan in mass, the time base waveform must be parabolic.

A further time of flight device[460] makes use of a pulse of electrons to produce the ions; this pulse triggers a delay that is applied to a coincidence circuit, and this circuit counts ions in coincidence with the electrons which produce them.

A pulsed time of flight instrument can be operated in an axial magnetic field, which not only increases the ion path length by imposing a helical motion, but also introduces the direction focusing of a 360° magnetic analyser. This is the principle of the 'chronotron' instrument[461]. Another pulsed instrument of extremely high resolution and accuracy, and operating with flat 360° orbits in a magnetic field is the synchrometer[462].

Time of flight analysis is also the basic principle of the radiofrequency mass-spectrometer[463, 464] (sometimes known as the 'topotron'), which has proved valuable for monitoring ion populations in gas discharges and afterglows; the instruments can be made very small, with path lengths as low as ~ 2 cm and instrument diameters ~ 1 cm.

A series of grids or aperture electrodes (*Figure 3.44*) are fed with a radiofrequency signal of several megahertz. The ions are initially accelerated by a steady potential difference between SG_1, and can only be further accelerated if they pass the apertures at, or shortly after, the moments at which the aperture potentials are negative. The radiofrequency fields accelerate ions of mass number appropriate to the applied frequency, in the same manner as the fields in a linear accelerator. This acceleration only increases the ion velocity by a small amount compared with the original velocity; otherwise the ions would not pass all the apertures at approximately the correct phase. The extra acceleration imparted to ions of the correct mass number is sufficient to allow them to pass a final retarding potential difference between G_2G_3, which is maintained so large that no ions unaccelerated by the radiofrequency field can pass to the Faraday cage FC. The screen Sc serves to suppress secondary electrons.

In the Bennett instruments, only a small number of suitably spaced grids are employed. In the Boyd instrument a full complement of 12, 16 or 24 evenly spaced apertures C_1–C_{24} is used; grids are impracticable because even those of high transparency would still reduce the intensity seriously.

The generalized theory of the Boyd radiofrequency mass-spectrometer has been given by Redhead[465]. It is assumed that the energy gained in the radiofrequency fields is very small compared with the initial energy of acceleration V_a, and also that the electrodes act as though they were equipotentials in their plane of symmetry, there being no other equipotentials.

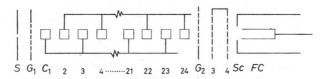

$$S \quad G_1 \quad C_1 \quad 2 \quad 3 \quad 4 \cdots\cdots 21 \quad 22 \quad 23 \quad 24 \quad G_2 \quad 3 \quad 4 \quad Sc \quad FC$$

Fig. 3.44. Schematic diagram of 24-stage radiofrequency mass-spectrometer

The radiofrequency potential is of the form $V_m \sin \phi$ at the moment the ion crosses the first equipotential: ϕ is the phase angle. For a distance s along the x-axis between the equipotential planes, the field between any two is

$$X = \frac{2V_m}{s} \sin (\omega t + \phi) \tag{3.226}$$

where ω is the angular frequency, and the time t is zero when the ion crosses the first equipotential. All potentials are taken with reference to the entrance orifice. The energy gain is

$$\Delta E = \tfrac{1}{2}X \int_0^s \sin (\omega t + \phi)\, \mathrm{d}x - X \int_s^{2s} \sin (\omega t + \phi)\, \mathrm{d}x$$
$$+ X \int_{2s}^{3s} \sin (\omega t + \phi)\, \mathrm{d}x \ldots + \int_{(n-1)s}^{ns} \sin (\omega t + \phi)\, \mathrm{d}x \tag{3.227}$$

for an instrument containing n accelerating electrodes. The ratio of time taken to travel a distance $2s$ to the radiofrequency period T is denoted by the symbol β:

$$\beta = \frac{2s}{vT} \tag{3.228}$$

for ion velocity v. For optimum phase angle, the maximum energy gain is

$$\Delta E = \frac{V_m}{\pi\beta}\left\{\cos\phi+4\sum_{i=1}^{(n/2)-1}\cos(2i\pi\beta+\phi)+3\cos(n\pi\beta+\phi)\right.$$

$$\left.-\cos(\pi\beta+\phi)-4\sum_{i=1}^{(n/2)-1}\cos[(2i+1)\pi\beta+\phi]-\cos[(n+1)\pi\beta+\phi]\right\}$$

$$(3.229)$$

The $\Delta E(\beta)$ function is found to have one large central maximum framed by smaller maxima, as shown in *Figure 3.45*, where it has been calculated[466]

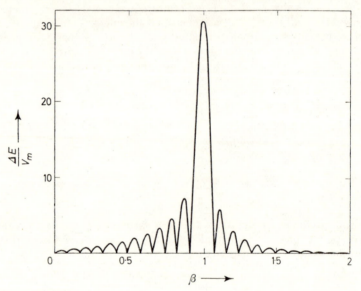

Fig. 3.45. Energy gain–frequency function computed for 24-stage radiofrequency mass-spectrometer

for the 24-stage Boyd instrument. The maximum value $\alpha = (E_m/V_m)_{max}$ occurs at $\beta = 1$ and is about 30·7. There is also a spectrum of smaller subsidiary maxima at $\beta = 3, 5, 7, \ldots$. If the retarder potential V_r is made sufficiently large, all ions whose mass numbers correspond to 'harmonics' in β may be rejected.

An expression for the mass-resolution, defined as:

$$\mathcal{R} = \frac{m_1-m_2}{2m_0} \tag{3.230}$$

is given as

$$\mathcal{R} = \frac{2}{\pi}\left[\frac{3\pi}{2n(n^2-1)}\frac{\alpha}{V_m}\right]^{1/2} \tag{3.231}$$

The mass of the accepted ion is m_0, and the masses of the two rejected ions framing the accepted ion are m_1 and m_2. Resolutions $\frac{1}{20}-\frac{1}{40}$ are typical.

In can also be shown that the efficiency, or ratio of emerging to entering particles of the accepted mass is

$$\mathcal{E} = \frac{T}{\pi} \arccos \left(\frac{V_r}{\alpha V_m} \right) \qquad (3.232)$$

which might, typically, be of the order of 0·1 for perfectly transparent electrodes. Another expression for the efficiency[463] is:

$$\mathcal{E} \simeq \frac{n\mathcal{R}}{4\sqrt{3}} \qquad (3.233)$$

Where grids are used instead of orifices, the departure from perfect transparency will greatly reduce these efficiencies.

For the above equations to hold, the ions entering the orifices are assumed to have zero energy. Ions possessing initial kinetic energies are more efficiently collected than those possessing smaller energies; a detailed calculation has been given by Boyd and Morris[463].

The principal advantage of the Boyd instrument is its short path length, typically 2–4 cm. Operation at pressures $\sim 10^{-4}$ torr is possible. A further reduction to path length 1 mm, facilitating ionosphere D region operation rather below 10^{-2} torr, is possible. For this purpose[467] one uses not the maximum current but the cusp-like change dI/df of current with frequency which occurs at minimum or zero current (see *Figure 3.45*). For a given grid spacing, the resolution is improved by an order of magnitude if the ions detected are those which gain no energy. Ions entering a uniform radiofrequency field gain no energy if they leave after a whole number of cycles, irrespective of phase of entry. The system has therefore a duty cycle of 0·5. The instrument consists of three electroformed grids and a collector plate, with total path length ~ 1 mm. Ions entering the first grid receive several hundred electron volts energy in their passage to the second. The third grid and the collector have radiofrequency potentials applied to them, but the third grid is at the same mean potential as the second grid, and the collector at a slightly more retarding potential than the first grid. The ion current usually arriving at the collector will be half the total current arriving at the first grid (less transparency losses), because half the ions gain energy in the radiofrequency field. If now the radiofrequency passes through the value for the resonance condition of one of the ion species present, the current will fall by the contribution due to that species. The resolution of the instrument depends on the quality and parallelism of the grids. Low-frequency modulation of the high-frequency signal is of value in improving the resolution.

The Bennett radiofrequency mass-spectrometer[468] makes use of only three stages of radiofrequency acceleration, each of which is achieved by means of a series of three equally-spaced grids, separated by distances s. Thus the transparency of the instrument is fairly good even though grids are substituted for the cylinder electrodes of the Boyd instrument. The radiofrequency potential $V_0 \cos \omega t$ ($\omega = 2\pi f$) is applied between the central and (coupled) outside grids of each stage, and the ions of unique mass number

$$m = \frac{0 \cdot 266 V_0}{s^2 f^2} \qquad (3.234)$$

17

will be accelerated when the separations D_i between the stages are

$$D_i = \frac{2s\pi\lambda}{2\cdot331}$$

(3.235)

where $\lambda = n_{i+1} - n_i$. The symbols n denote the number of radiofrequency periods which elapse while the ion travels between two stages. For a three stage instrument, i has the values 1 and 2, and typically $n_1 = 7$ and $n_2 = 5$. A number of analytical and experimental papers have appeared dealing with this mass-spectrometer[469].

It is reported[470] that a radiofrequency mass-spectrometer of high resolution can be made with two slits, the radiofrequency being applied across the two sides of each slit so that the ions can only pass through the slit at those instants when the potential difference across the slit is temporarily zero. The path between the two slits may be made as long as 1 m.

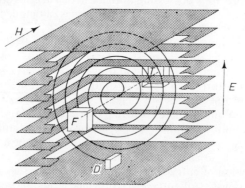

Fig. 3.46. Schematic diagram of omegatron: electrons from filament F are collected on target T; collisionally formed ions travel in a flat spiral path (solid line) to the collector D

(From Kerwin[441], by courtesy of Academic Press)

The combination of 360° magnetic field focusing with radiofrequency time of flight operation is used in the 'omegatron' instrument (*Figure 3.46*)[471]. The ions are formed in a region of crossed alternating electric and steady magnetic fields, similar to those employed in the cyclotron. The ion paths are flat spirals if the mass number is appropriate to the cyclotron frequency (~ 7 MHz in the first instrument); these ions are double focused on the detector D. Thus the omegatron is very attractive to ionic physicists on the grounds that the ions are collected with complete efficiency; however, the path length they travel is so long that pressures $\leqslant 10^{-7}$ torr are mandatory. The resolving power is proportional to the dispersion and may be made as large as 10 000 by increasing the number of path cycles and using high frequencies and intense magnetic fields. But on the whole, low-resolution instruments suit ionic collision experiments better.

The Quadrupole Mass-Filter

The superposition of steady and alternating electric fields upon four hyperbolic cylinders gives rise to interesting and valuable orbiting of charged particles; since only particles within a certain range of mass numbers will travel in stable quasi-helical orbits for a certain set of field parameters, the configuration may be used as a mass-spectrometer[472]. In this form it is sometimes known as the 'electric mass filter'.

The generalized potential for a quadrupole field is

$$\phi = U(t)(\alpha_x x^2 + \alpha_y y^2 + \alpha_z z^2) \qquad (3.236)$$

where $U(t)$ is the applied alternating potential difference. Since $\nabla^2 \phi = 0$, the constants $\alpha_{x,\,y,\,z}$ are related by the equation

$$\alpha_x + \alpha_y + \alpha_z = 0 \qquad (3.237)$$

and since for the hyperbolic cylinders $\alpha_z = 0$,

$$\alpha_y = -\alpha_x \qquad (3.238)$$

For the application of a composite potential difference $U + V \cos \omega t$, the potential configuration is $(U + V \cos \omega t)(y^2 - z^2)/r_0^2$; the equations of motion of a particle of charge e and mass m are:

$$m\frac{d^2 x}{dt^2} = 0 \qquad (3.239)$$

$$\frac{dx}{dt} = \text{const.} \qquad (3.240)$$

$$\frac{d^2 y}{dt^2} - \frac{2e}{mr_0^2}(U + V \cos \omega t)y = 0 \qquad (3.241)$$

$$\frac{d^2 z}{dt^2} + \frac{2e}{mr_0^2}(U + V \cos \omega t)z = 0 \qquad (3.242)$$

These are special cases of Mathieu's differential equations, for which two types of solution exist. One of these corresponds to an ion path with exponentially increasing amplitudes of oscillation of y and z but constant dx/dt. Ions proceeding along such paths will not pass through the instrument, but will be collected on one of the four cylinders. The other solution corresponds to stable orbiting paths, but still with constant dx/dt. Ions proceeding along these paths can pass through the electrode system and to a collecting electrode.

The stable orbit conditions are described in terms of parameters a and q, defined as

$$a = \pm \frac{8eU}{r_0^2 m \omega^2} \qquad (3.243)$$

and

$$q = \frac{4eV}{r_0^2 m \omega^2} \qquad (3.244)$$

For stable orbits these parameters must lie within limiting curves, given by the equations:

$$a = \frac{-q^2}{2} + \frac{7}{128}q^4 - \cdots \qquad (3.245)$$

17*

and

$$a = 1 - q - \frac{q^2}{8} + \frac{q^3}{64} - \frac{q^4}{1536} - \cdots \qquad (3.246)$$

In the representation of these curves in *Figure 3.47*, the region of stable a and q is ABC; both positive and negative values of a must lie within this region.

For ions of any mass,

$$\frac{a}{2q} = \text{const.} = \frac{U}{V} = u$$

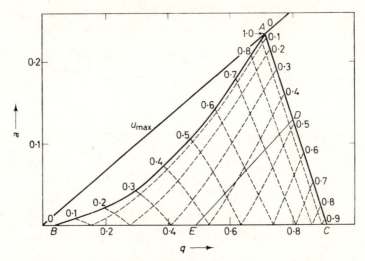

Fig. 3.47. *Quadrupole mass-spectrometer stable orbit characteristics: ABC, stable orbit region; ABDE stable orbits for three-dimensional ion trap. Lines passing through the origin represent loci of constant resolution, and the point A represents the best resolution obtainable; broken lines represent ion trap characteristics B_y and B_z, with B_y values written at left and B_z at right*

These ratios are represented by straight lines u passing through the origin. The band of stable q values has a width which varies with u, becoming infinitely narrow at $u_{max} = 0\cdot166$, where $q = 0\cdot706$. It is advantageous to work near this maximum, where the theoretical resolving power for mass number m is

$$\frac{\Delta m}{m} = \frac{4}{3}\left(1 - \frac{u}{u_{max}}\right) \qquad (3.247)$$

and the frequency bandwidth for given mass is

$$\Delta f = \frac{2}{3}f\left(1 - \frac{u}{u_{max}}\right) \qquad (3.248)$$

Scansion in mass number, represented by motion along a line u, is achieved by keeping the d.c./radiofrequency ratio constant, and varying both together.

A feeling for the frequencies f involved may be obtained from the equation:

$$\frac{V}{r_0^2 f^2} = 7 \cdot 15\left(\frac{m}{e}\right) \tag{3.249}$$

with m/e in atomic mass units, f in megahertz, and V in volts. For convenience, this equation is shown graphically in *Figure 3.48*.

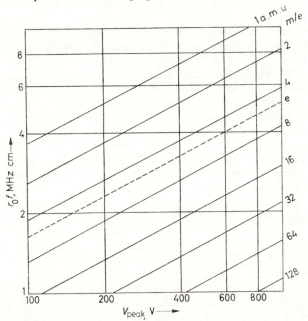

Fig. 3.48. *Values of* V_{peak} *and* $r_0 f$ *for mass-analysis of ions of specified* m/e *in quadrupole mass-spectrometer; the broken line represents the electron, for which values of* V_{peak} *must be divided by 100*

The mass-filter has the useful feature that resolution can be controlled electrically. Low-resolution flat-topped peaks can be used to prove the absence of mass-discrimination. A further advantage is that the selection of ions is independent of entrance conditions over a wide range of velocity, off-axis distance and direction of travel. There is no chromatic aberration. Unlike time of flight instruments, the mass-filter will accept ions at all phases.

It is important that the quadrupole rods are sufficiently long for the unstable orbiting ions to be laterally displaced far enough to strike them. Their length L is limited by the condition

$$L \gg r_0 \pi \left(\frac{2V_a q}{V}\right)^{1/2}$$

where V_a is the potential to which the ions have been accelerated before entering the mass-filter. The path lengths of ions in mass-filters are shorter than those in sector instruments of the same resolution.

Although theoretical prediction of the transmission efficiency is difficult experimentally, it is found that the efficiency can reach 10 per cent for a resolution of $\sim$ 1 per cent but only 0·01 per cent for a resolution of 0·2 per cent.

For ease of construction, the hyperbolic cylinders can, for all except the highest resolution instruments, be replaced by circular cylindrical rods of radius equal to 1·2 r_0.

The maximum attainable resolution is governed by the machining accuracy of the rods and by their mounting relative to the rod diameter. They must be made as large as possible (2 cm diameter) for resolution as high as 1/3000 a.m.u.

In recent years, the quadrupole mass-filter has proved important in two ways. First, there are a number of fine instruments available commercially. These possess electron impact sources, and some can handle mass numbers from 1 to 1000 with resolution as high as 1000 and scanning speed 0·01 sec decade^{-1} in mass number. Second, the mass-filter has proved to be a highly successful instrument for sampling ions from a plasma. The instrument operates best when the ions entering have only thermal energies (although acceleration up to 300 V is quite possible when the rods are sufficiently long). There are no magnetic fields or orifices, other than that which separates the plasma from the mass-filter. The path length in the mass-filter is shorter than in sector instruments, so for the same pumping speed the orifice size can be increased to make the sampled current greater. Mass-discrimination must be avoided in the ion sampling technique, and the lowering of resolution to achieve flat-topped peak conditions is easily achieved in the mass-filter, without mechanical adjustment of slits.

The efficiency of transmission of ions through the quadrupole mass-filter has been investigated in experiment and theory[473]. The measurements of transmission efficiency are made by comparison of the mass-analysed ion current with the ion current transmitted when d.c. but no radiofrequency is applied to the quadrupole. The ion beams used for this comparison must of course be of unique species. It is found that as a/q is raised, and with it the resolution, the transmission falls; for input energy 9·5 eV, the sharpest fall occurs around $a/q = 0·33$. High transmission efficiencies are obtained with high frequencies. Investigations were also made of the entrance conditions which the quadrupole mass-filter requires. The transmission is seriously affected by transverse energies possessed by the ions at entrance. The effect is far more serious in the monopole mass-filter, in which there is only one cylindrical rod and in which crossed metal surfaces are arranged so that the electric field is of identical configuration to that of the quadrupole. The monopole mass-filter is simple to construct, but because the instability at entrance is responsible for 'peak splitting', the quadrupole is to be preferred.

The possibility of using additional radiofrequency fields to discriminate against unwanted ions is under investigation.

Alternating quadrupole fields may be used not only for the mass-selection of beams of particles but for their containment or 'trapping' in a 'cage'. Such containment enables studies to be made of their decay or increase due to collisions with each other or with beams or gas. Trapping of charged particles in magnetic fields is well known in plasma physics ('pinch', 'sausage', 'picket fence' etc.), but in application to collision studies these are inferior to quadrupole fields because the latter can be designed so as to contain only particles with a small range of mass numbers.

The necessary quadrupole field configuration[474] is provided by a ring-shaped electrode (*Figure 3.49*), whose inside surface is formed by the rotation of one-half of a hyperbola; a hyperboloid is added so as to bound the remaining surfaces. Since in this case $\alpha_z = \alpha_y$ and $\alpha_x = -2\alpha_z$,

$$\phi = (U+V \cos \omega t)\, \alpha_z(z^2+y^2-2x^2) \qquad (3.250)$$

This field binds charged particles in the correct mass number range quasi-elastically to the central point. The conditions relating to the parameters a and q are included in the data of *Figure 3.47*. The presence of oscillating ions in the cage can be shown by their electromagnetic damping of oscillations at angular frequencies $\omega_d = \frac{1}{2}\beta_z\omega$ or $\frac{1}{2}\beta_y\omega$, where β_z and β_y values are given as dashed lines in *Figure 3.47*; the impedance of the electrode system may be shown to depend upon the number of particles contained.

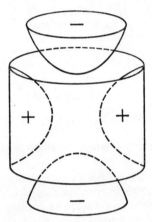

Fig. 3.49. Schematic diagram of three-dimensional quadrupole system for Coulomb trapping; the x-axis is vertical

Another method of determining this number is to measure the current to an electrode when a potential pulse is applied, sufficiently powerful to sweep all the charged particles to it. The efficiency of trapping is complete, but of course the charged particles can decay through collisions with background gas.

By the maintainance of background pressure as low as 5×10^{-10} torr, trapping times as long as 10^8 sec have been achieved[475]. The densities of ions in the active central region of the trap were probably of order 10^7 ions cm^{-3}; the achievement of higher densities is inhibited by space charge. When a particle multiplier is used to detect ions extracted from the trap, the assembly is an extremely sensitive partial-pressure manometer.

The injection of electrons into such cages may be achieved by means of an electron gun inserted into one of the electrodes, but these electrons will not be trapped since they do not satisfy the a, q conditions; ions in the cage are produced by electron impact ionization of slow gas atoms injected in a pulsed jet.

3.15 DETECTION AND WAVELENGTH MEASUREMENT OF PHOTONS

The requirement in atomic physics is for detectors with the aid of which optical signals in the visible (6500–3900 Å), ultra-violet (3900–1900 Å) and vacuum ultra-violet (1900–1000 Å) (*Figure 1.4*) can be converted into electrical signals, with a response time and sensitivity sufficient to permit the detection of radiation from the smallest possible number of atomic collisions. This requirement stretches the available detectors to their extreme limit. Infra-red radiation can only be detected with inferior sensitivity, and, despite the importance of the subject, relatively few collision experiments have yet been carried out. The present discussion is concentrated on vacuum ultra-violet detection technique.

In the visible and ultra-violet, radiation may be detected by:

1. Photographic plates,
2. Photoemission of electrons from surfaces,
3. Photoionization chambers and counters,
4. Thermocouples and thermistors,
5. Semiconductor devices,
6. The human eye.

Wavelength measurement may be achieved by:

1. Spectrometric techniques,
2. Transmission filters,
3. The variation of detector sensitivity with wavelength.

Photoemission detectors have been made available in forms many orders of magnitude more sensitive during the last decade. The opacity of glass photocell envelopes in the vacuum ultra-violet has led to the use of multipliers with quartz and lithium fluoride windows and of photoemitting surfaces built into the vacuum system of the experiment. Unfortunately, exposure to harmful vapours may lead to changes in the emissive properties of the surface. The surfaces are less exposure-sensitive when they are used for the detection of more energetic photons, so one is faced not with cumulative but with counter-balanced disadvantages. For ultra-violet photon detection, it is an advantage to use emitting surfaces whose thresholds are at short wavelengths, in order that intense radiation at longer wavelengths produces no signal. An important contribution has been the calibration of these materials in photoelectric efficiency throughout the entire ultra-violet spectrum[476]. In the infra-red region a number of semiconductor detectors have been developed[477].

Summaries of photoemission ultra-violet detector data are available[478, 479]. The quantum efficiency ε of a photoemitting surface does not exceed 0·3 electrons per photon even for the most modern materials; maximum values of order 0·1 are more usual. Although the wavelength band for maximum response is usually much wider than that for an ionization chamber, it is possible to utilize the 'bandpass' qualities of photoemitters, either with or without windows. Some characteristics of photocathodes are tabulated in Table 3.7.

Table 3.7. PHOTOCATHODE CHARACTERISTICS

Material	Peak quantum efficiency ε (electrons photon^{-1})	Long-wavelength threshold (μ)
SbCsO	0·15	0·67
SbCs	0·10	0·7
Sb(NaK)Cs	0·20	0·85
BiAgOCs	0·05	0·8
AgOCs	0·004	1·2
Mg	0·004	0·39
Au	10^{-4}–10^{-5}	0·27
CsI	0·3	0·15
RbTe	>0·06	0·27
CsTe	>0·02	0·26
KBr	0·1	0·16
CsBr	0·15	0·16
RbI	0·2	0·15
CuI	>0·07	0·18
Ni	0·001	$\varepsilon \propto \lambda^{-n}$
W	0·001	$\varepsilon \propto \lambda^{-n}$
BeO	0·001	$\varepsilon \propto \lambda^{-n}$
SrF$_2$	0·001	$\varepsilon \propto \lambda^{-n}$
Ta	10^{-5}	
Cd	10^{-5}	

Vacuum ultra-violet radiation is sometimes detected[480, 481] by conversion into visible radiation, which is much less easily absorbed in the glass windows of photomultipliers. The windows are coated with a fluorescent or scintillator material which emits in the visible when activated by ultra-violet. This technique has the disadvantage of leading to higher dark currents. Photographic plates may also be sensitized by fluorescent lacquer[482, 483] with emission band peak at 3150 Å; at this wavelength, the plates can be calibrated using monochromated light from the mercury arc. The lacquer, consisting of sodium salicylate in methanol, must be sprayed on as evenly as possible; it is available commercially. Salicylate detectors have efficiences in the region 1–5 per cent. It would be possible to combine this technique with fibre optics.

Between 2000 Å and 1000 Å, both sodium salicylate and Eastman ultra-violet fluorescing lacquer exhibit nearly constant efficiency, but the salicylate is superior to the lacquer in that it is much less sensitive to stray light. An important point in the calibration of the lacquer is that over the region 300–1300 Å the multiplier current is directly proportional to the logarithm of the product of light intensity and exposure time.

Photographic detection has its place in exploratory atomic collision experiments. The available slow plates include Ilford Q 1, 2 and 3, and much faster ones are Kodak WSR and SC7 emulsion. Calibration of photographic plates involves quantitative comparison of densities and also mechanical means of reducing radiation intensities while exposures remain constant; these means employ step or logarithmic sector discs specially designed for incorporation in vacuum spectrographs.

The calibration of detectors in the visible is important in the measurements of excitation functions (Chapter 5). The response to a standard tungsten strip lamp is measured. For lamp temperatures below 2500°K at wavelengths shorter than 1 μ, the emission can be represented by the Wien equation

$$\Phi_w(\lambda, T) = e_w(\lambda, T)c\lambda^{-4}\exp\left(-\frac{c_2}{\lambda T}\right) \text{ photons cm}^{-2}\text{ sec}^{-1}$$

$$\text{steradian}^{-1}\text{ (cm wavelength)}^{-1} \qquad (3.251)$$

where $e_w(\lambda, T)$ is the emissivity of tungsten[484], $c_2 = hc/k$ is the second radiation constant. When a photon flux ϕ_c photons sec^{-1}, produced by unit path length of impacting beam, produces a signal I_c, then $I_c = k_c\phi_c s_c$, where s_c is the monochromator entrance slit width, and k_c is the instrumental constant, which includes the observed solid angle, the magnification of the collector lens, the transmission of the optics, the quantum efficiency of the photocathode, and the gains of the photomultiplier and amplifiers. When the tungsten strip, of width s_w and instrumental constant k_w, replaces the collision chamber, it produces a signal

$$I_w = k_w\Phi_w(\lambda, T)hs_w\,\Delta\lambda$$

where h is the dimension of the tungsten strip in the direction parallel to the monochromator slit, and

$$\Delta\lambda = s'_w\frac{d\lambda}{ds} \qquad (3.252)$$

is the spectrum passband of the monochromator, whose exit slit is of width s'_w; the monochromator is assumed to have collimator and telescope of equal focal length. The absolute value of the collisional photon flux

$$\phi_c = \frac{I_c}{I_w}\frac{k_w}{k_c}\frac{s_w}{s_c}hs'_w\frac{d\lambda}{ds}\Phi_w(\lambda, T) \qquad (3.253)$$

The brightness temperature T_B of the tungsten strip is found by means of an optical pyrometer, and the true temperature

$$T = \left[T_B^{-1}+\frac{\lambda}{c_2}\ln e_w(\lambda, T)\right]^{-1} \qquad (3.254)$$

Photoemitters are used for detection of intense radiation with nothing more than a current amplifier at output. Simple arguments can be adduced[479] to show that the signal–noise ratio is greatly improved when electron multiplication is used. Photoelectrons may either be multiplied in sealed or windowless[485–487] photomultipliers by a system of high secondary emission dynodes, between which the electrons are accelerated and focused; or they may be accelerated in one stage to an energy high enough for them to emit photons from a scintillator. The noise associated with the initial dark current can be greatly reduced by pulse height discrimination techniques. The lowest dark currents ($\sim$ 1 count min^{-1}) have been achieved by the use of a high work-function tungsten cathode with strip and channel multi-pliers[488, 489].

The superior gain ($\sim 10^8$) of the parallel-strip multiplier rests on the large number of surface collisions made by the electrons. The cycloidal motion of the electrons in crossed magnetic and electric fields between the two strips is modified by the additional electric field produced by the potential drop along the strips, which are made of resistive material, so that a voltage can be applied to them. This ensures that the electrons strike the surface with non-zero energy.

The commercially available channel multiplier[490] consists of a capillary tube (~ 1 mm diameter, ~ 5 cm long) of resistive material, between the ends of which several thousand volts potential difference is applied. The electrons pass down the interior of the capillary, making frequent collisions with the walls. The ratio of length to diameter of the tube is important. Under suitable conditions, the electron current becomes space-charge limited, so the current pulse arising from a single electron is always of approximately the same magnitude. The adverse effect of ions, which arise in surface collisions and pass in the reverse direction along the channel multiplier, is minimized by introducing curvature into the capillary tube. When used as an ultra-violet photon detector[491], the photocathode is simply the resistive material with which the capillary is coated. The multiplier does not suffer from reduction of gain by contamination with gases in the way that dynode multipliers do. Arrays of hundreds of tiny capillaries are available for special applications. The channel multiplier is valuable not only in the vacuum ultra-violet but in the soft X-ray band.

All photomultiplier output currents or count rates drift slowly with time, although whether this effect is connected with a change in sensitivity or in dark current is not known.

Photomultiplier noise is made up of several components:

1. Glass envelope leakage current, which, as a steady current, may be eliminated in experiments by the use of low-frequency light modulation or chopping.
2. Thermionic emission from photocathode and dynodes, which is equal to $(2ei\,\Delta f)^{1/2}$ in bandwidth Δf for photocurrent i; this is multiplied by gain G and therefore produces a voltage $RG\,(2ei\,\Delta f)^{1/2}$ in multiplier load resistance R.
3. Thermal noise voltage $(4KTR\Delta f)^{1/2}$ in the load resistor, which will be dominated by shot noise if $GRI > 0.06$, where I is the total anode current; the shot noise is minimized if a phase-sensitive amplifier of narrow bandwidth is used to provide chopped light to the photo-multiplier.

For monitoring photon fluxes which vary with time, the electron multiplier dynodes may be pulse-gated. It is sufficient and convenient to gate the first dynode, but, if pulses shorter than 1 μsec are used, an advantage accrues when all dynodes are gated, in that the noise level is reduced. The rise time of the noise level is comparable with 1 μsec, possibly because the noise may arise from ion bombardment of the dynodes. Thus a gated multiplier can be operated with increased dynode potentials and increased gain.

Sealed-off photoionization chambers and counters are valuable ultra-violet detectors. They are sensitive only to photons sufficiently energetic to ionize a suitable molecular gas. The counters are constructed with a fine wire anode maintained at high positive potential; current pulses are counted at the cathode, as in a Geiger counter. The gas enclosed in the chambers must be at a comparatively high pressure (10 torr), and a reasonably robust window (for example, of Mylar) is necessary. This places limits upon the energy of the photons for which these detectors are suitable. With a lithium fluoride window and with the chambers filled with nitric oxide[492] or iodine[493], a response band of only 300 Å is made available. The important Lyman α radiation lies in this narrow band, and the gases have large absorption coefficients for this radiation. When used as self-quenching counters, the electro-negative nitric oxide and iodine must be diluted with inert gases. Features of the characteristics of other ionization chambers[478, 494, 495] are tabulated in Table 3.8.

Table 3.8. CHARACTERISTICS OF IONIZATION CHAMBERS

Gas	Window	ε_{max}	Approximate wavelength band (Å)
$(CH_2)_2O$	LiF	0·1–0·2	1050–1180
CS_2	LiF	0·5–0·6	1050–1240
$(CH_3)_2CO$	CaF_2	0·08–0·1	1230–1290
NO	CaF_2	0·1–0·3	1230–1350
NO	LiF	0·1–0·5	1050–1350
$(C_2H_5)_2S$	BaF_2	0·1–0·25	1350–1480

For the detection of X-rays, the counters are often used in the 'proportional' mode, in which the height of the pulse is proportional to the energy of the X-ray. The gas pressure is such that all the photon energy is converted into ionization and excitation, and there is a gas multiplication factor of $\sim 10^3$, producing a voltage pulse equal to the total charge per photon, divided by the capacitance. Careful control of the stability of gas pressure (for example, of Ar–CH_4 mixture) is necessary.

A popular non-absolute ultra-violet photon detector makes use of photo-emission of electrons from a metal surface. The electrons are then accelerated through 10 kV onto a scintillator; the visible radiation pulses from this are counted, with complete efficiency, on a photomultiplier. This technique avoids contamination of photodynodes in the vacuum system of the experiment.

With this and other detectors, the central problem is that of absolute calibration. The most successful method[496] makes use of photoionization of a gas whose ionization potential is sufficiently low; nitric oxide is popular. The radiation is passed through a differentially pumped gas chamber containing electrodes across which a potential is applied, so that all the ions or electrons formed can be collected and measured (collision chamber techniques are discussed in Chapters 6 and 11). Behind the ionization chamber is the ultra-violet detector. The importance of this technique lies in the fact that its employment does not rest on a knowledge of the photo-ionization cross-section. The principle, which could be of wider application, was first used in the detection of metastable atoms[497]. Between 10 per cent and 15 per cent efficiency is possible.

Suppose that the detector registers a photon flux I_0 when there is negligible gas pressure in the collision chamber, and I_1, I_2, ... when the collision conditions correspond to target parameters π_1, π_2, ... (equation 1.26). On the assumption that all the photon absorption is due to photoionization and that one photon produces one ion,

$$i_1 = I_0 - I_1$$
$$i_2 = I_0 - I_2$$

(3.255)

and so on, where i_1, i_2, ... are the collected ion currents. From equation 1.18,

$$\ln I_1 = \pi_1 \sigma \ln I_0$$
$$\ln I_2 = \pi_2 \sigma \ln I_0$$

(3.256)

and so on. It is valuable to check the internal consistency of the method by obtaining an over-determined set of equations. The term σ is readily eliminated, but may be measured if desired. Even the gas pressures need not be measured, since a variety of collector plates of known lengths l_1, l_2, ... will enable a single unknown gas pressure to be eliminated.

A further method (the 'branching ratio technique') of absolute ultra-violet detector calibration[498, 499] utilizes simultaneous measurement of two radiations from a collision region, one in the ultra-violet and one in the visible, where absolute calibration is standard practice. It is necessary that both radiations proceed from the same level, and that both the relevant oscillator strengths (or at least their ratio) are known.

A rather insensitive absolute ultra-violet detection technique is known as 'actinometry'[500]. The radiation is made responsible for a chemical reaction in the gas phase, and the gas product is titrated chemically; for example ozone is so produced from oxygen.

A grazing incidence photodetector has been reported consisting of a hollow polygon of photoemitting surfaces, at the centre of which is a positively charged electron-collecting wire. The photon flux is reflected at grazing incidence around the various surfaces. Whereas the normal photoemission detector has typically a maximum sensitivity around 800 Å, the grazing incidence detector maximizes around 400 Å. All ultra-violet photoelectric sensitivities of undegassed metals are identical within $\sim$30 per cent, so approximate calibration is automatic.

The use of thermocouples for absolute measurements of the intensities of spectral lines in the vacuum ultra-violet has been described by Packer and Lock[501]. Thermistors remain a potentially valuable but so far unexplored technique. The use of semiconductor detectors has hitherto been confined to the infra-red region of the spectrum, and therefore discussion of them will not be pursued here; reference may be made to a modern textbook[502]. Silicon barrier detectors, with copper, beryllium or iron barriers are now in use for ultra-violet radiation.

Detection of weak intensities of visible radiation by the human eye has been used in some recent experiments. The efficiency of the eye is the equal of some of the best photomultipliers; the chance of observing a single isolated photon is better than one in ten. The region of greatest sensitivity is the blue-green; perhaps this corresponds closest to moonlight, in which the clear sight so necessary to man's hunting may have been attained in the evolutionary process. Experiments have been conducted on the maximization of human sensitivity by the use of drugs.

The measurement of wavelength by spectrometric techniques, such as vacuum monochromators, is beyond the scope of this discussion; but it is relevant that reflexion grating surfaces, although prepared with carefully studied techniques[503], possess diminishing reflectivity in the vacuum ultra-violet. In the XUV (< 400 Å wavelength), grazing incidence reflexion gratings are necessary. But 400–1100 Å is still in the normal incidence region, and gratings ruled on plastic and coated with gold or platinum can be used; several reflexions are possible from coated mirrors, without prohibitive intensity loss. The grazing incidence operation makes use of blaze, or 'echelette' gratings

in which the reflexions from the different surfaces differ in phase, as opposed to conventional scratch gratings

in which they differ in amplitude. For the correct blaze, 30 per cent efficiency is possible. The incidence is about 88°, and spectral resolution down to 13 Å is possible. Since fluorescence of the background gas can interfere, ultra-high vacuum techniques are used in grazing incidence instruments. Monochromators have been reviewed by Samson[504].

It is possible to select wavelength by gas filters of a suitable narrow transmission. Unfortunately these must be bounded by windows, so the lithium fluoride low-wavelength limit (1250 Å) becomes a very important boundary. In the visible, it is possible to obtain graded transmission filters accurately calibrated in their region of variable absorption; these filters have been employed in photodetachment measurements[505] from data such as that published by Watanabe, Zelikoff and Inn[506]. For example, oxygen and carbon dioxide are both very suitable for the transmission of Lyman α radiation.

Interference filters[507–510] are readily available commercially. A bandwidth of 40 Å is typical, 20 Å fairly common, 10 Å rather more rare; filters of 1·5 Å bandwidth can still be purchased, and those of 0·1 Å bandwidth can be made in the laboratory, provided great care is taken with temperature control. The bandwidth depends upon the angle of incidence of the radiation. Some degree of tuning is possible by pressure variation when the filters consist of a layer of gas between two glass plates[511]. A recently developed window for gas filters is the lanthanum fluoride crystal, which can be bonded vacuum-tight to copper. It is transparent between 1240 Å and 3 μ.

Windows in the XUV can be made from very thin metal films ($\sim$ 1000 Å thickness). Aluminium is transparent in the region 170–835 Å, and magnesium will not transmit below 256 Å. Carbon has also been used as a window.

The decrease in reflectivity of mirrors as shorter wavelengths are reached may actually be utilized for defining one side of bandpass. Metal mirrors are reflective up to shorter wavelengths than silvered mirrors. As with metal gratings, the reflecting power can be improved by a coating of magnesium fluoride.

Organic liquids have been used for filters, for example at 2800 Å; bandpass widths of the order of 200 Å are possible[512].

Some narrow windows have been reported in the photoabsorption functions of gases and vapours, notably zinc, cadmium and helium; for example, the narrow window at 201 Å in helium, where the cross-section actually falls to zero. It is possible that these windows could be used for monochromation.

In problems of the detection of radiation from a collision region, the signal–noise ratio is all important. By multi-channel analysis technique[513], measurements can be made at far smaller ratios than normally (0·01 as opposed to 5). Essentially, this technique is an extension of the common practice of interruption or chopping of the radiation[514], but instead of on-off modulation, it is necessary to scan one of the variables of the experiment, usually the impact energy, and extract by storage and addition of the information in many channels the cross-section function from a large number of rapid scansions.

Another method of reducing background noise is by the use of a pulse-height analyser; it is only applicable to those photomultipliers in which the background pulses are a certain height, and detected radiation pulses a different height.

3.16 DETECTION AND COUNTING OF CHARGED PARTICLES AND FAST NEUTRALS

Ions, electrons and fast atoms can be detected by a wide variety of methods; the first to be considered here is the collection and measurement of charge by electrometer.

The measurement of very small, comparatively steady currents arising from beams of charged particles may be made by means of galvanometers, suspension electrometers, electrometer amplifiers, or vibrating reed and similar electrometers. Galvanometers are not normally used for currents below 10^{-9} A, and suspension electrometers, at one time very popular in nuclear physics for measuring rates of increase of charge, have been superseded because of their comparatively high sensitivity to mechanical vibration. Modern electrometer triodes and tetrodes are capable of measuring changes in grid potential of the order of 10^{-2} V. Thus currents of the order of 10^{-14} A can be measured when applied across a grid resistance $R = 10^{12}\ \Omega$, and it is possible to obtain resistances $R = 10^{13}\ \Omega$ and so gain in sensitivity. Not only must the leakage resistance of the electrometer tube and of the electrodes (and apparatus for applying potential to them) be large compared with the grid resistance, but also the grid current arising in the electrometer tube must be negligible compared with the measured current. Experiments at high temperatures are made more difficult by the fact that the resistivity of all insulators, particularly quartz and glass, falls something like exponentially with temperature. Photoelectrons are largely avoided by shielding the electrometer tube from light, and instability arising in the power supply is avoided by the use of bridge circuitry, often with two triodes contained in the same envelope, and by using the same cathode. Nevertheless the electrometer triode is rivalled in performance by devices in which the steady potential is converted by mechanical vibration of a capacitance electrode into an oscillating potential; low noise amplification may then be achieved by the use of tuned circuits. Such a vibrating reed or similar electrometer is usually capable of accurate measurement of potentials of 10^{-2} V; commercial availability of these instruments and of electrometer amplifiers has facilitated many collision experiments during the last decade. Solid state electronics can be used throughout the vibrating reed circuitry.

The two fundamental limits of all current measurement devices are those of time-constant and of noise. The high grid resistance R necessary in an electrometer device, together with the electrode and screened connexion capacitance C, inevitably several picofarads, produce time-constants CR of the order of seconds. These have the effect of integrating unstable signals and of lengthening the time taken for experiments. In comparing two unstable signals, it is necessary that the time-constants of measurement are approximately the same. Time-constants may be reduced by a factor $\sim 10^2$ by connecting a feedback line from the electrometer amplifier to the electrode connexion screen, which must therefore (inconveniently) be insulated from earth.

Noise arises from the background gas in the collision experiment itself, in the grid resistor and in the electrometer amplifier. The background gas noise is of particular importance in crossed-beam experiments, and must be reduced either by the use of fast getter-ion pumps or by mechanical chopping of one or both beams, followed by phase-sensitive detection. Grid resistor noise can be reduced by liquid nitrogen cooling and careful selection of the component. All white noise, including that arising in the electrometer amplifier from a variety of causes, is proportional to the amplifier bandwidth.

The ultimate limit to the measurement of a flux of particles is the statistical fluctuation caused by the random arrival of individual particles. Since the arrival of one electronic charge per second would produce an integrated current of 1.6×10^{-19} A, statistical fluctuation would only become serious below 10^{-16} A. Electrometers can be used to handle fluxes of this order by causing them to measure the accumulation of charge in a time much longer than one second. However, it is more usual to 'multiply' the fluxes by methods discussed in the following sections, and then to measure the integrated current, or to count the individual particles, measuring the number arriving in a sufficiently long period of time.

Reference should be made to discussions of the statistical techniques used in the counting of nuclear particles. A common difficulty which is encountered in collision physics is that of limiting the intensity of the flux to such an extent that the counting rate is sufficiently low to be handled by the equipment without accidental superposition of pulses. For pulses of the order of 10^{-8} sec, counting rates of 10^5 pulses sec^{-1} are statistically reliable.

Ion currents are normally collected in a 'Faraday cup'[515] or hollow metal container within which the electric field is sufficiently small for charged secondary particles not to be extracted. It is recommended that the entrance orifice subtend a solid angle $\leqslant 10^{-2}$ steradian at any point on the surface on which the beam impinges. Secondary electrons can be suppressed more efficiently by fronting the cup with a negatively biased aperture electrode (grids sometimes develop insulating films). Energetic ions ($\geqslant 10$ keV) produce not only energetic secondary electrons, but also a certain number of slow secondary ions; these may be effectively suppressed by transverse magnetic and transverse electric fields respectively; there remains, however, no known method of suppressing reflected metastable atoms, ultra-violet photons and soft X-rays, which will often produce electrons at metal surfaces. A test of Faraday cup efficiency for high-energy particles is to alter the electric bias upon it; the measured current should of course remain unchanged.

The Faraday collector (cup, cage) is normally surrounded by a metal screen which serves for both electromagnetic and particle screening. Designs have received some attention in recent years, especially where very high energy beams are concerned[516, 517]. The design of the rear surface of the collector is affected by back-scattering studies[374, 518]. The back-scattering of high-energy ions decreases as the atomic number of the target decreases, therefore light materials such as beryllium or carbon are suitable for the rear surface. This surface is often oblique to the beam. Back-scattered ions are much harder to suppress than secondary electrons, for which an applied magnetic field is adequate. However, this technique is unsuitable when the beam itself is composed of electrons; in this case electric fields applied in various configurations are effective. Invariance of apparent collected current with applied field is an indication of complete control of secondary electrons. The most popular Faraday cup electrode configurations[374] are as follows:

1. Parallel plates, with rear surface attached to one plate, and electric field transverse to the beams;
2. Charged wire stretched down centre of cylindrical collector;
3. Cylindrical collector with negative apertured electrode in front; to avoid secondary emission from this electrode, it is fronted by another electrode with smaller aperture, so that the beam does not impinge on the negatively charged electrode.

Fluxes of thermal electrons may be detected not only by measurement of their current when received in a Faraday collector, but by the following methods:

1. Measurement of their contribution to permittivity at high frequencies;
2. Measurement of absorption of radiofrequency power at the cyclotron resonance frequency, in a steady magnetic field;
3. Chemical titration of the nitrogen produced when the electrons are passed into nitrous oxide gas: $e + N_2O \rightarrow N_2 + O^-$.

A popular fast-particle detection technique (ions or atoms rather than electrons) is by means of calorimetry. The fast-particle flux is allowed to fall upon a metal foil attached to a thermocouple or a thermistor. The sensitivity of this method of detection depends directly upon the energy of the particle; the modern thermopile yields microvolt signals for milliwatt input. An absolute calibration of each detector is necessary (for example, by mounting them in a Faraday cage and using charged beams), but of course reliance can be placed upon the relative sensitivity to two beams of known energies.

In a calorimetric detector, the thermal capacity is maintained as small as possible, so that the detector reaches radiative equilibrium with as small a time-constant as is practicable. A typical detector[519, 520] might consist of a platinum disc, 1 cm in diameter and 0·0025 cm thick, held by quartz fibres within a 200 g copper housing. To the housing, and to the back of the metal disc, are attached either thermocouples (0·0075 cm copper–constantan wire) or thermistors (beads $\sim$ 0·04 cm diameter, of resistance 10–5000 Ω at laboratory temperatures, dropping $\sim$ 3 per cent for each 1°C rise). Mounting on quartz fibres has been used. For 10^{10} particles sec^{-1} at 20 keV, a temperature difference of $\sim$ 5°C is developed. Time constants cannot be made smaller than about a second. A combined Faraday cup, secondary emission detector, and thermocouple detector has been described[521].

Although thermocouples and thermopiles have been in use as detectors for many years[522], developments are still being made, particularly in the use of new inorganic materials for commercial thermopiles. Bolometers have been described[523] with sensitivity 10 μW and time-constant sufficiently short (10^{-7} sec) to permit the detection of beams modulated at low frequency. These consist of a film of aluminium over a film of aluminium oxide, over a film of germanium; the sensitivity is of the order of a few microwatts. Thermistors are of equal sensitivity, but possess much longer time-constants.

Normally in collision experiments a thermopile detector is fronted by an orifice, but it is possible to construct a thermopile detector of such fine wire (0·01 mm diameter)[524] that without any fronting orifice it has good resolution for differential scattering measurements.

Crystals of barium titanate sandwiched between gold films are relatively cheap detectors of fast particles. The change in polarization produced by the flux of fast particles produces a current when a potential difference is maintained across the crystal. Single crystals of this ferroelectric material can have sensitivities as low as 0·5 μW, but do not operate above 100°C. The time-constant is around 10^{-9} sec, and it is believed that very much higher sensitivities might be achieved. However, the linearity of these detectors has not yet been clearly established.

Fast neutral particles can be converted into ions either by electron-impact ionization or by stripping. Electron-impact ionizers are discussed in Section 3.13. The stripping method[525] depends for its absoluteness on a previous knowledge of the stripping cross-section (Chapter 11). This only becomes appreciable in the kilovolt energy region, but it increases right up to $\sim$ 10^5 eV. In the stripping technique of detection, the projectile particles are ionized by collision with gas atoms or molecules, and the resulting electrons are collected by the usual methods. Processes of the type $A + B \rightarrow A^+ + B^-$ are more probable than stripping at low energies and enable this technique to be extended downwards in energy to tens of electron volts.

Fast neutrals and ions can be detected by means of their secondary electron emission at metal surfaces; this process is discussed at the end of this section. Before the development of particle multipliers, secondary emission detectors, consisting of a metal surface with a positively biased electron collector nearby, were much used as relative detectors. But the calibration varies with particle species, charge state, impact energy, and the condition and nature of the surface. An ingenious method of overcoming this disadvantage, when comparing charged beam with parent neutral beam intensities, is to charge equilibrate both in 'gas converter cells' (collision chambers), making use of the charge-changing collision processes discussed in Chapters 11 and 12. Secondary emission detectors are now less commonly used than particle multipliers, which have much greater sensitivity.

The particle multiplier[374, 526] is usually a commercially available photon multiplier with the photocathode removed; it will respond to atoms, ions or electrons when the secondary emission coefficient at the first surface exceeds unity. It consists of a number of metal electrodes ('dynodes') whose secondary emission coefficients, usually enhanced by the Malter effect, may be between 3 and 5. The electrodes[527] can be shaped in various ways so that the emitted electrons will be focused on the next dynode; to each dynode a potential of 100–300 V positive to the previous one is applied. By the time the last dynode is reached, considerable multiplication has taken place, depending upon the magnitude of the positive potentials. The metals in normal use are 2 per cent beryllium–copper, barium oxide on nickel, beryllium on nickel, aluminium, silver–magnesium, and brass[528]; they can be sensitized by heating in vacuo, in hydrogen[529], in an inert atmosphere[530], or in oxygen. Beryllium–copper is the most stable, but silver–magnesium has the highest gain. Silver–magnesium dynodes lose gain on exposure to the atmosphere, and often, when they are mounted in a vacuum system which must be let up to atmospheric pressure, pure argon is admitted. The dynodes can be reactivated by heating for 30 min to 400°C in oxygen at 1 torr pressure. The steady running of multipliers at 150°C slows down their gradual deterioration. A multiplier should not deteriorate to a gain of lower than 10^3.

The most popular arrangements of dynodes are the linear array of curved dynodes, the squirrel cage or compact array of curved dynodes, the Venetian blind, and the array of micro-points (the 'box-and grid' is now rare). The array of micro-points is a new commercial multiplier with an extremely high gain (10^8) and an efficiency which is apparently constant over the entrance area of the multiplier (this is not the case for beams focused to 0·1 mm diameter). However, the pulse height analysis curves for this multiplier do not always indicate an efficiency of 100 per cent unless the grid in front of the entrance surface is sufficiently negatively biased to suppress all the secondary electrons. Amplification factors of multipliers have been systematically investigated[531].

The literature of secondary emission of electrons by ion bombardment of surfaces is of importance in understanding the operation of multipliers[532, 533]; the subject is discussed briefly at the end of this section.

When a particle multiplier is made in the laboratory from a commercial photomultiplier, the latter can be mounted in its correct position in the apparatus, and the vacuum system filled with dry inert gas. The rupture of the glass envelope and removal of the photocathode may then be carried out with a minimum of exposure of the surface to harmful moisture.

18*

The modern particle multiplier uses more stages than was considered necessary some years ago. It suffers a smaller reduction of gain when exposed to atmosphere. The electron secondary emission coefficient of a surface reaches a flat maximum between 3 and 5 for electron energy 300–500 eV. For positive ions, the maximum is at about 100 keV. After exposure to air this figure is nearer 2·5 keV. For a typical pulse amplifier, the noise level is ~ 500 μV, therefore the input pulse should be of order 10 mV. By means of a pre-amplifier, it is possible to reduce the input capacitance to about 15 pF. The input pulse is of charge $1·5 \times 10^{-13}$ C, which requires that the multiplier gain be of order 10^6. If the stage gain, even with completely efficient focusing, is only 2·5, then at least 14 stages are necessary. For nanosecond counting times, higher gains are required. At 50 Ω amplifier input impedance, a 5 mV pulse of duration 3×10^{-9} sec requires a multiplier gain 4×10^7. Multipliers of 20 stages have been made to such a specification[534]. The strip and channel multipliers (discussed below) also possess very high gains, owing to their operation with many surface collisions.

The number of electrons released from a dynode when one electron strikes it follows a Poisson distribution. For secondary emission coefficient 2·5, 11 per cent of the electrons produce no secondary at all. The particle multiplier is thus not completely efficient for electrons (usually less than 80 per cent efficient); for fast ions, the coefficient can be higher and the efficiency can be close to 100 per cent. The counting efficiency is stated[521] to be ~ 100 per cent for H^+, H_2^+, He^+, N^+, Ar^+ and N_2^+ between 10 keV and 200 keV. For N_2^+ the efficiency is 92 per cent at 4 keV, but for Li^+ at 2 keV the efficiency is 100 per cent. For 3 keV Ar^+ it is ~ 90 per cent. Calibration of the British EMI 9603 Venetian blind multiplier for many ions between 3 keV and 90 keV impact energy has been given[535]. Calibration is of special importance when the multiplier is used not with counting of the output pulses, but simply for recording the smoothed output current. In this mode of operation, the detector is of different sensitivity for each ion and at each impact velocity. But the experiments on multiply-charged rare gas ions showed sensitivity proportional to $m_+^{1/2}$ and to v_+ over quite a large range, but independence of charge (for identical m_+ and v_+). In the operation of multipliers with pulse counting, the sensitivity does not vary greatly with m_+ or v_+ provided that the pulse height sufficiently exceeds that of the minimum detectable pulse. Pulse-height discrimination reduces the background count.

For use as an absolute detector, the electron multiplier final dynode must be connected to a wideband (pulse) amplifier; the pulses due to the arrival of atomic particles are then counted with conventional techniques. By variation of the bias of the amplifier, the minimum registered pulse height is varied and a curve of counting rate against pulse height obtained. The first differential of this curve gives the distribution of pulse amplitudes. Only when this distribution function is well-separated from the noise pulses can the collection efficiency be estimated and ~ 100 per cent efficiency obtained. It is difficult to obtain this condition with atomic particles of < 5 keV energy unless the multiplier surfaces are carefully isolated from all moisture and handled only in pure, dry inert gas. Reactivation can only be achieved with complete reliability by heating to a dull red heat in oxygen.

In addition to the advantage of high efficiency, the individual counting technique makes it possible for the electronics to be isolated in d.c. potential from the dynodes by means of a blocking condenser through which the pulses pass.

For a variety of reasons, usually connected with the focusing in initial stages, the pulse-height distribution of a multiplier can be found to fall exponentially rather than show a characteristic peak. Such a multiplier must be rejected. For satisfactory operation, it is necessary that the count rate be independent of the setting of discriminator level over a reasonable range. In coincidence experiments, a correction for random coincidences must be applied[536]. An alternative to coincidence technique is the use of discrimination technique to count the double-height pulses which are produced when two particles arrive simultaneously at one multiplier.

The background count in particle multipliers is usually a few counts per minute. This can be reduced still further by cooling and by screening of the first dynode from stray particles within the vacuum system and from cosmic rays.

The main difference between focused dynode and Venetian blind multipliers is that the former have overall electron transit times $\leqslant 10^{-9}$ sec, whilst for the latter the time is $\sim 5 \times 10^{-9}$ sec. The Venetian blind multiplier will still operate in a magnetic field (< 30 G). In both systems (and particularly for Venetian blind dynodes), the sensitivity is dependent upon the exact point of the first dynode surface which the incident electron strikes. Deflecting or focusing electrodes are sometimes incorporated to counteract this effect. In the Venetian blind multiplier, the incident ion should possess more than 10 keV energy for completely satisfactory collection.

It should again be stressed that frequent recalibration of multipliers is necessary if they are to be relied on as absolute detectors. Recalibration is achieved by comparing the multiplier detection with the collection and charge measurement obtained with an efficient Faraday cup. Measurements have shown that the efficiency of a multiplier is independent of the charge state of the ion. Provided that the incident energy is sufficiently high (5–10 keV), efficiencies close to 100 per cent are reached. Any sensitivity to charge state may, with high-energy ions, be removed by charge equilibration in a gas or thin foil before the multiplier is reached; it is assumed that calibration of the multiplier for a charge-equilibrated beam has previously been carried out.

A variety of particle multipliers[537] depend upon continuous resistive surfaces of high secondary emission coefficient, rather than on an array of individual dynodes. More collisions with surfaces are possible, so the gain is very high despite a low secondary emission coefficient. The configurations are strip, parallel-plate and tube. The strip multipliers consist of two parallel plates separated in potential and parallel to lines of magnetic induction, so that electrons undergo trochoidal motion along one of the plates, making many surface collisions. The first collision surface may be oriented in several different ways. The parallel-plate multipliers also consist of two parallel plates, but they are at the same potential; a large potential difference, several thousand volts, is maintained along them, so that the electrons pursue a zigzag path between the plates.

The tube multiplier is similar; the highly successful channel multiplier ('channeltron')[538] consists of a hollow semiconducting capillary tube made of a compound containing mostly vanadium, but lesser amounts of phosphorus and and sulphur. Along this capillary tube, of internal diameter 1–2 mm, a potential gradient is developed by the application of 3–5 keV between the ends. Particles enter at one end and electrons multiply and travel in a zigzag path down the tube.

The channeltron is rugged from the point of view of loss of gain on exposure to atmosphere; no special precautions need be taken with these multipliers. When the applied potential is sufficiently high, the flux of secondary electrons becomes space-charge limited at some point in the tube, so the emerging pulse of electrons is the same height no matter how many electrons were produced in the first surface collision. This mode of operation (the 'Geiger mode') can actually be employed with output current measuring rather than output pulse counting. No change in sensitivity with impact velocity or ionic species is supposed to occur. The gain can be as high as 10^9 for 7 kV applied to the channeltron, 10^8 for 5 kV, and 10^7 at 2 kV. Unfortunately, the gain is dependent upon gas pressure. At 10^{-5} torr, a pressure rise of a factor of two leads to a gain rise of a factor of two. This effect is much reduced when the channeltron is constructed in the form of a segment of a circle or helix of a certain radius. This construction minimizes the effects of the reverse direction passage of positive ions formed in surface or gas collisions. It has been reported that at pressures lower than 10^{-8} torr the gain decreases.

The channeltron can be used either 'open-ended' or with a sealed output end, the current pulse being collected either at a specially mounted collector or at the channeltron connector itself. The sealed end is suitable for pulse counting. The current pulse at maximum is about one-fifth the magnitude of the ohmic current passing through the channeltron. At count rates in excess of 10^7 particles sec^{-1}, the channeltron ceases to operate as a linear device; the current which is passed becomes independent of the number of particles received.

Calibration of the efficiency of the channel multiplier has been described [539]. It is reported that for different ions the curve of relative efficiency versus $v_+Z^{0.4}$ is universal (Z being the atomic number). The efficiency reaches a value which is inferred to be unity when $v_+Z^{0.4}$ is 1.2×10^6 cm sec^{-1}, but falls to 90 per cent at 6×10^5 cm sec^{-1} and 60 per cent at 4×10^5 cm sec^{-1}. The pulse-height distribution curve is unusually good in this type of multiplier, and the gain is constant with particle entrance anywhere across the tube. Since the entrance aperture is inconveniently small, some multiplier entrances are flared out into a cone (the 'funneltron'); in this case, the sensitivity is dependent upon the point of first surface collision. In strip multipliers this also applies, and the strip multiplier pulse-height distribution curve is not well resolved. A uniform sensitivity almost to the edge of the funnel can be achieved by covering it with a highly transparent grid at front end potential; this has the effect of preventing penetration of funneltron field into the space in front of it, and so minimizes lens action.

Whilst the counting of individual atomic particles by the direct particle multiplier technique is not unsatisfactory, it has the disadvantage of lack of robustness of collecting surface. Experience with particle multipliers has led some physicists to prefer the enclosed photomultiplier. The electron gain inside a good photomultiplier can be as high as 10^8. For this type of detection, ions are directed[540, 541] onto a phosphor such as caesium iodide, and the emitted light pulses are piped through a guide and directed to a photomultiplier external to the vacuum system. Alternatively, the phosphor is fixed on the surface of the photomultiplier to avoid photon loss in the guide. The scintillator is enclosed in a truncated cone of reflecting material. However, phosphors are comparatively inefficient for incident ions, and their efficiency decays with time; high particle velocities are necessary for 100 per cent efficiency[542]. The lower energy limit for detection of protons by the caesium iodide detector is about 10 keV, although even at 25 keV some 10 per cent, of the protons are not detected. For heavier particles, higher energies are required; thus for H_2^+ ions 30 per cent are not detected at 45 keV.

A more satisfactory system[543, 544] is to convert the ions to secondary electrons which are accelerated to the phosphor. The ions entering the collector chamber need not have a particularly high energy; once inside, they are subjected to a transverse electric potential difference of ~ 20 kV, produced by the action of a single high-voltage electrode. They travel in a curved path to this electrode and emit perhaps five electrons per ion. The secondary emission coefficients are greater for acute angles of incidence. The secondary electrons are accelerated in the same potential difference, and fall upon a phosphor which emits photon pulses to a photomultiplier. The phosphor is coated on one side with a thin evaporated aluminium film, which prevents surface electrical charges when it is electrically connected; the film also reflects photons and minimizes degassing. The release of a photoelectron from a phosphor and photocathode requires 1–2 keV energy; the energy of the electron pulse per ion falling on the phosphor is $5 \times 20 = 100$ keV, so ~ 100 photoelectrons are released per ion. This is well within the capabilities of an electron multiplier and pulse amplifier; in fact, a recordable pulse can be produced with the photomultiplier operated at low gain, which implies a low dark current. The ion noise can be made as low as 4×10^{-20} A.

Other versions of these 'Daly–Afrosimov detectors' are successful. Their most important advantage is that of being able to reduce the ion and electron kinetic energy without losing 100 per cent efficiency or raising the noise level inconveniently. A second advantage is the great robustness of the collecting surfaces, whose secondary emission coefficients for ions are not affected by gas absorption, cleanliness, etc.; but it is uncertain whether fast molecular particles do not occasionally dissociate at the surface, yielding two electron pulses separated in time sufficiently for the counting efficiency to be greater than unity.

The threshold of secondary electron emission by protons is ~ 20 eV. This represents the lower theoretical energy limit for these detectors. The Daly–Afrosimov detector can be used as a coarse electrostatic analyser of resolution ~ 10 per cent, giving advantages in heavy-particle collision study. The disadvantage of this electrostatic analysis is that the potential applied to the high-voltage electrode is unique for a unique ion and kinetic energy, whilst the pulse height also depends upon this potential. An alternative arrangement which avoids this disadvantage allows the heavy-particle beam to fall directly onto the high-voltage electrode without electrostatic deflexion. A weak magnetic field is applied to deflect the electrons onto the scintillator. A Venetian blind photomultiplier must then be used because of its relative insensitivity to magnetic field. The distribution of electron energy distributions can upset the optics of this system.

As with the particle multiplier, the efficiency of the detector is close to 100 per cent only when the secondary emission coefficient γ is high. For a Poisson distribution of probabilities of emission, the efficiency is $1-\exp(-\gamma)$. In the kilovolt region, γ increases with increasing ion energy. Therefore the ion collector potential should be arranged to provide adequate impact energy. This detector is particularly useful, since a particle multiplier cannot always be conveniently floated at high potential; no such difficulty applies here.

A further modification[545] makes use of a 200 Å aluminium foil target, through which the electrons emerge in a forward direction. This device can be made to distinguish between such ions as H_2^+ and He^{2+}, since[546] the range

$$R = \frac{CE_+}{n^{1/2}} \tag{3.257}$$

where n is the ionic charge, C a constant, and E_+ the ion energy.

The silicon barrier detector, or semiconductor particle detector[547–549] consists of a p–n junction diode reverse biased so as to produce a space-charge region having an electric field sufficient to sweep out charge carriers produced in the solid by the ionizing radiation. The theory of this mechanism has been discussed[149]. The response of the system is linear to particle flux and to energy; it is sufficiently fast for single particle counting, and is unaffected by magnetic fields of order 100 G. The barrier detector can be used to detect electrons[550]. Lithium-drifted germanium has also been used.

Emphasis has been placed upon the energy resolution obtainable with these detectors when receiving γ-rays or X-rays; pulse-height analysis enables a resolution of a few thousand electron volts to be obtained, and use can be made of this feature in high-energy ion detection. A promising version of barrier detector for lower energy particles is the silicon surface-barrier detector developed by Blankenship and Borkowski[551]. A p-type inversion layer is produced on an etched surface of a lapped n-type single crystal of silicon sliced to ~ 1 mm thickness, by surface oxidation in laboratory air. The opposite surface has previously been nickel-plated and soldered to a pigtail wire, and the oxidized surface is now made conducting by a thin layer of gold ($< 100\ \mu g\ cm^{-2}$) deposited by vacuum evaporation. The crystal, which may be as small as 4 mm² or as large as 1 cm², is mounted in polystyrene cement and epoxy resin, and operated with the inverted surface, on which the particle flux falls, at earth potential. The pigtail wire is connected to a low-noise preamplifier, thence to an amplifier, pulse-height analyser and counting rate equipment; 20–100 V positive bias is provided by a battery, in series with a 1 MΩ resistance. Cooling improves the energy analysis, and 0·6 keV has been claimed for X-rays.

Whereas the energy loss necessary for the production of an ion pair in argon gas is 27 eV, the solid state detector is greatly superior; for diamond the figure is 9 eV, and for silicon 3·3 eV. The detector is, however, a low-impedance source of charge, which necessitates the use of special amplification techniques. At present, the best signal obtainable is a 400 μV pulse when a 1 MeV particle is detected. Since for a 5×5 mm counter with 20 V bias the reverse current or noise level is $0.05\ \mu V$, it is clear that a satisfactory counting efficiency could be achieved for particles with 1 keV energy. The low-energy limit of penetration of protons through the gold layers is ~ 1 keV for $1\ \mu g\ cm^{-2}$ coating; this sets the low energy limit on this type of detector at ~ 7 keV.

The depth D of the space-charge region is given by

$$D = \frac{\varepsilon V}{2\pi e N} \qquad (3.258)$$

for silicon dielectric constant $\varepsilon = 12$, applied bias potential V and carrier concentration

$$N = \frac{1}{\varrho \mu e} \qquad (3.259)$$

An appropriate resistivity $\varrho = 3500\ \Omega\ cm$ and mobility $\mu = 1200\ cm^2\ V^{-1}\ sec^{-1}$ lead to barrier depths of 200 μ, with capacitance 50 pF cm^{-2}.

Barrier detectors are commercially available, incorporating cooling equipment for reduction of background count. However, they are expensive, they have a relatively high energy threshold for heavy particle detection, and they exhibit reflectivity to electron fluxes.

Ionization of gas has been used for particle detection, in the form of the thin-window proportional counter[552].

Photographic detection of ions of energies as low as 4 keV has been developed[553]. Ions O^+, N^+ and Ar^+ are directed onto AgBr, and the resulting silver is fixed and developed.

There is a method of detecting slow ions without their ever striking a solid surface; this is the technique of ion cyclotron resonance[554]. It requires that the ions be travelling in cycloidal paths under the action of steady crossed electric and magnetic fields; the ions can be created by ionization by an electron beam passing down the lines of magnetic force. The ions drift in a direction perpendicular to both electric and magnetic fields (E and H), with velocity

$$v_d = \frac{Ec}{H} \qquad (3.260)$$

the cycloid radius being $r = v_d/2\pi f_c = v_d/\omega_c$, where c is the velocity of light. The ions are exposed to a radiofrequency field, parallel to the steady electric field, at the cyclotron frequency

$$f_c = \frac{eH}{m_\perp c} \qquad (3.262)$$

A frequency of 307 kHz requires a steady magnetic field of 100 G $(a.m.u.)^{-1}$. The radiofrequency field is scanned in frequency, and at f_c the ions absorb power from the field and travel in ever-increasing spiral paths. This absorption of power is measured, and constitutes the detection of the ions. The mass resolution of this system is given by $leH^2/2.7m_+c^2E$, where l is the length of the region over which the radiofrequency field is applied. More than one frequency can be applied to different electrodes, to enable several ions in a mixture to be detected simultaneously. Commercial instruments are available. The time average of the power absorption is

$$P(\omega_c) = \frac{P_0}{1+[(\omega-\omega_c)/\nu_0]^2} + \frac{P_0}{1+[(\omega+\omega_c)/\nu_0]^2} \tag{3.263}$$

where $P_0 = n_+e^2E_0/4m_+\nu_0$ is the peak power absorption at $\omega = \omega_c$, and $E = E_0 \sin \omega t$ is the applied intensity. Furthermore

$$\nu_0 = \frac{\nu_D m_0}{m_+ + m_0} \tag{3.264}$$

where

$$\nu_D = n_0\sigma_D\bar{v}_+\left(1+\frac{m_+}{m_0}\right)^{1/2} \tag{3.265}$$

is the collision frequency for momentum transfer of ions with gas molecules of mass m_0, at mean laboratory ionic velocity $\bar{v}_+$. Where $\nu_0 \ll \nu_c$, the ion cyclotron resonance line possesses Lorentzian shape, with FWHM ($\Delta\omega$) equal to $2\nu_0$. Thus not only the ion density n_+ but also the momentum transfer cross-section can be deduced from an ion cyclotron line measurement.

Secondary Emission of Surfaces Bombarded by Ions

It has been shown that when the secondary electron current I_e from a surface is measured, an absolute value of the impinging ion flux I_+ can be deduced when the secondary emission coefficient

$$\gamma = \frac{I_e}{I_+} \tag{3.266}$$

is known from separate experiments, such as these listed by Hagstrum[555]. It is found that γ is dependent upon the nature, charge and kinetic energy of the incident ion; the internal energy of the projectile also contributes and neutral atoms or molecules possess their appropriate value of γ; the nature of the surface also contributes. For the secondary electrons to be multiplied so that the pulse produced by a single incident ion is individually counted, γ must well exceed unity for complete efficiency of detection, but its exact magnitude need not be known for I_+ to be measured absolutely. Discrimination between particles of different γ may be achieved by the methods of pulse-height analysis.

Secondary emission under ion bombardment takes place by two processes, conveniently termed the 'potential' and the 'kinetic' processes. In the potential process, the internal energy of the ion is made available to the process of liberating one or more electrons from the conduction band. Neutralization[556] of the ion into a metastable state of the atom is followed by an Auger two-electron process of electron ejection. For singly charged ions, values of γ for the potential process are less than unity and decrease with increasing impact energy; above a few kilovolts, the kinetic process predominates. Since the energy of the metastable state E_{ex} must be sufficient to eject the electron when the atom falls to the ground state ($E_{ex} > \phi$, the metal work function), neither singly-charged alkali metal ions, any ground state neutral atoms nor low E_{ex} metastables will be able to emit by the potential process. Multiply-charged ions, possessing considerable internal energy, can by various potential processes produce values of γ much greater than unity.

In the kinetic process, the impact energy of the ion is made available to the process of liberating one or more electrons from the conduction band[390, 557]. For this process, the values of γ increase with increasing impact energy, at least into the 10^5 eV energy region. Owing to the complexities of electron capture and loss processes in the surface region, the charge of the ion does not play a dominant role. In fact the value of γ is largely independent of ionic charge.

Since the nature and condition of the metal are also important in determining the secondary emission coefficient, it is concluded that if the value of γ is required for a particular detector surface and impact, the only alternative is to measure it *in situ*.

3.17 DETECTION OF ATOMIC AND MOLECULAR BEAMS

The most important and sensitive detection techniques for atomic beams make use of ionization and detection of the electric charge. The older techniques include the measurement of gas pressure developed on storage. A number of less important techniques are also available. Absolute calibration of detectors can be made using an oven atomic beam source, whose operation can in turn be calibrated for a potassium beam, using the surface ionization detector (see below), which is absolute for potassium.

Pirani Gauge Detector

The Pirani gauge detector of molecular beams has been used for many years. Other pressure gauges might be used similarly, and the Bayard–Alpert gauge has been tested*[558].

Although the sensitivity of the Pirani gauge can be deduced from first principles with fair success, it remains essentially a relative detector. A Pirani gauge is exposed to the atomic beam, which enters it via a long narrow channel. The pressure in the gauge rises until the rates of entrance and of exit are equal.

For a rectangular channel of length l, width w and height h, with $l \gg h \gg w$, the ratio of the entrance and exit rates is:

$$\varkappa = \frac{l}{w[\frac{1}{2} + \ln{(2h/w)}]} \tag{3.267}$$

The equilibrium pressure is therefore

$$p_d = \frac{\varkappa n_b (M_0 T_d)^{1/2}}{3 \cdot 5 \times 10^{22} \pi r^2} \text{ torr} \tag{3.268}$$

for beams of molecular weight M_0, and

$$n_b \gg n_r$$

* The pumping action of ionization gauges is a disadvantage.

The subscript d refers to the detector, the subscripts b and r refer respectively to beam and residual gas. The beam radius is denoted by r.

In the Pirani gauge, the heat energy removed from the filament by conduction through the gas under steady state conditions[559] is

$$W_d = \frac{3 \cdot 5 \times 10^{22} p_d A_f k_a k f (T_f - T_d)}{2 (M_0 - T_d)^{1/2}} \qquad (3.269)$$

In addition to the previous symbols, A_f represents the filament surface area, k_a the gas–filament accommodation coefficient, f the number of degrees of freedom of the gas molecule, T_f and T_d the filament and gauge wall temperatures and k the Boltzmann constant.

For an electrical bridge made up of four filaments of resistance R equal to the galvanometer resistance, the unbalance current I_g is related to the total gauge current I_d by the approximate relation

$$I_g \simeq \frac{W_d}{I_d R} \qquad (3.270)$$

In addition to the thermal time-constant, which may be a few seconds, there is a gas time-constant τ given by:

$$\tau = \frac{\varkappa V}{hw} 2 \cdot 76 \times 10^{-4} \left(\frac{M_0}{T_d} \right)^{1/2} \qquad (3.271)$$

where V is the gauge volume. The value of this time-constant limits the channel dimensions and therefore the sensitivity of the gauge.

A sensitive gauge[560] with platinum filaments $7 \cdot 6 \times 0 \cdot 051 \times 0 \cdot 0001$ cm in a cavity $7 \cdot 9 \times 0 \cdot 1 \times 0 \cdot 48$ cm of brass, with channel $2 \cdot 5 \times 0 \cdot 79 \times 0 \cdot 0015$ cm, operates with $I_d = 0 \cdot 024$ A. The galvanometer (7 μV cm^{-1} sensitivity) comes to 90 per cent of its maximum response in 12 sec. A 1 cm deflexion is produced by $7 \cdot 7 \times 10^{10}$ hydrogen mol.sec^{-1}.

The Pirani detector can be made in the form of a metal or glass block B cut in the form shown in *Figure 3.50*. A flat metal plate P covers the channelled block, with soft metal ($0 \cdot 01$ mm aluminium) gasket F cut to extend the dimensions of channel C and provide a gas-tight seal. Neither detecting nor

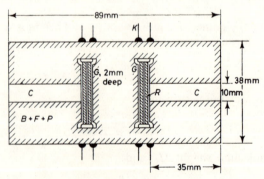

Fig. 3.50. Pirani detector for molecular beams
(From King and Zacharias[447], by courtesy of Academic Press)

compensating grooves G, containing ribbons R, should be exposed to electromagnetic radiation at radiofrequencies, since the filaments act as detectors. Even with these precautions, Pirani detectors are slow, and relatively insensitive.

Miscellaneous Detectors

Brief reference will now be made to a number of detectors which have been used from time to time in atomic beam researches, but have not the popularity of the Pirani gauge or of ionizing devices.

Deposition detectors[561] were used in many of the early experiments, and are reliable and semi-quantitative, but relative. Cooled surfaces condense a variety of beams, eventually giving visible deposits which may be examined densitometrically; alternatively, the time for obtaining a deposit at the threshold of visibility may be measured. Long times of collection, even hours, may be required. Chemical detection is more appropriate to radicles, and is discussed in the next section.

The sensitivity of a deposition detector is enormously increased when the beam can be produced in radioactive form, and the detector subsequently removed and placed near a radiation detector[562]. The targets may either be cooled to increase the accommodation coefficient, or ionization and electric field acceleration may be used[563] to force the atoms into the target surface. Data are available[564] for the expected accommodation coefficients of various beams and targets. An extremely sensitive radioactive deposition detector[565] employs as target a reel of continuous clean copper tape, which can be wound past a contact with a liquid nitrogen cooled surface, past the impinging beam, to an end window counter in the vacuum chamber.

It is also possible to utilize for detection purposes the impact energy of the atoms as they fall upon a sensitive balance.

Although semiconductor detectors are of value for electrons and fast particles, the only solid state detection of slow atoms so far achieved[566] is for iodine atom beams, which affect the conductivity of a thin cuprous oxide film maintained at 200°C. Work function changes have been used to detect oxygen[567], and might also be used for carbon-containing molecules. In these detectors, which must be distinguished from the Kingdon cage discussed below, the beam impinges upon a tungsten filament, thereby changing the work function by oxide formation (see Section 3.4). The electron emission change, measured by comparison with an unbombarded section of the filament, may be as much as 25 μA for $\sim 4 \times 10^{13}$ O_2 mol. sec^{-1}.

Another technique which may have application to particular collision experiments is the photography of atomic beams using Schlieren techniques,[568] with a crossed electron beam.

Ionizing Detectors

In these detectors, the atoms are ionized, or electron-attached, either by electrons, by photons, by electric fields or at metal surfaces; the positive or negative ions, or their effects, are measured.

The earliest ionizing detector was the 'Kingdon cage'[569]. It is so simple as to be of lasting value. The atomic beam is directed through an aperture in the

anode of a cylindrical space-charge limited diode with closed ends (the filament passes through a small hole in the end plates); in the course of its
passage through the space near to the filament, a proportion of the beam
is ionized; the positive ions partly neutralize the negative space charge, so
the diode current increases. For greatest sensitivity, the ionization potential
must be sufficiently small to allow the ionization to take place in a region of
slow-moving electrons, forming a deep potential trough. The major drawback of this detector is its lack of stability, which might be eliminated by
sufficient attention to the oxidation or carburation of the cathode surface
by impurity vapours. It has been largely replaced by the surface ionization
detector.

The ions in a radial electric field move in 'Kingdon orbits', which have
been analysed in detail. The Kingdon cage can also be used as an extremely
sensitive detector of positive ions.

In the surface ionization detector[570, 571], the atoms are directed onto a hot
filament, and the positive ions are collected and measured. The surface
ionization of atoms at hot filaments is discussed in Section 3.6, but some
further remarks may be of value here. The proportion of positive ions is
determined by temperature, ionization potential E_i and surface work function ϕ (see Table 3.3):

$$\frac{n_+}{n_0} = \frac{1 - R_0}{1 + A} \tag{3.272}$$

where

$$A = \frac{1 - R_+}{1 - R_0} \frac{g_+}{g_0} \exp\left(\frac{\phi - E_i}{kT}\right) \tag{3.273}$$

R_+ and R_0 are reflexion coefficients and g_+ and g_0 are statistical weights.
Thus the atoms with lowest ionization potentials (potassium, rubidium and
caesium halides and atoms) can be ionized on pure tungsten at 2900°K,
whilst lithium and sodium halides and atoms require the higher work function of oxidized tungsten. Atoms and salts of gallium, indium and thallium,
and possibly calcium and barium, may be ionized at oxidized tungsten which
is continually renewed by spraying gaseous oxygen over its surface. Barium
is difficult to ionize, possibly because it forms a compound which reduces the
work function and hinders ionization. Fluorine gas has been found to raise
the work function of tungsten to 5·6 eV, but is driven off above 2650°K.
Sulphur is reported to raise the work function of tungsten to the extent that
bismuth beams can be ionized.

The great limitation of the surface ionization detector, sometimes called
the Langmuir–Taylor detector[572], is also its great advantage. Because only
atoms of low ionization potential are ionized and detected, the background
gas molecules produce no noise. The efficiency of detection of thermal
potassium atoms at a tungsten filament is 100 per cent between filament
temperatures 800°K and 2500°K, since the reflexion coefficient of the neutral
potassium at tungsten is zero[573]. A surface ionization detector can in this
case have very small area, which is suitable for differential scattering experiments, but for fast potassium atoms (say 10 eV) a noticeable reduction
of efficiency might be expected.

Serious background noise arises from alkali impurities in the tungsten. Ultra-pure wires can be prepared[574] by methods such as the deposition of tungsten vapour. Spectrally pure platinum containing 8 per cent tungsten, and possessing a work function $\sim 5.5\,\text{eV}$, is commercially available. Background noise can be avoided by chopping the atomic beam; instead, or in addition, one may mass-analyse the positive ions formed. The background current consists of hydrocarbons in the mass range 50–225 a.m.u., but may also be reduced by surrounding the detector with surfaces cooled by liquid nitrogen.

The time-constant of the surface ionization filament is determined by the time taken to reach thermodynamic equilibrium. Times ranging from 10^{-2} sec to 3×10^{-5} sec have been observed in experiments with a rotating shutter[575], but times of the order of seconds have been observed[576] for praseodymium on molybdenum filaments; these may be associated with the extremely small ionization efficiency.

For negative ions of electron affinity E_a, the proportion of electron-attached particles is given by:

$$\frac{n_-}{n_0} \propto \exp\left(-\frac{\phi - E_a}{kT}\right)$$

Since work functions are usually larger than electron affinities, the efficiency of electron attachment is greatest for low work functions and high temperatures; the latter are a prerequisite for the prevention of surface absorption or compound formation, both of which raise the work function. Thoriated tungsten filaments are satisfactory except for chlorine, but in general all halogens and caesium halide beams may be detected by negative ion formation.

In construction, the surface ionizer consists of a tensioned filament contained within a cylindrical shield biased negative and cooled to collect positive ions; a small aperture admits the atomic beam. For electron attachment, the shield is biased strongly positive and the thermionic electrons are confined by a magnetic field parallel to the filament. It must be remembered that background positive and negative ion currents are observed from most tungsten filaments, except where the greatest pains are taken to ensure purity.

Where it is desired to mass-analyse the ions, the surface ionization chamber is made in the form of a pair of plates between which a uniform field is maintained, undisturbed by the presence of the filament therein; the particles enter and leave through orifices in the plates. Other geometries are possible. The formation of ions at thermal energies renders them particularly sensitive to the influence of surface charges on the electrodes. It must be remembered that caesium, and to a lesser extent rubidium and potassium, will ionize on almost all hot surfaces.

One important feature of the surface ionization detector is that its sensitivity depends upon the nature of the atomic beam. Thus, Taylor and Datz[577] found that although the tungsten surface detector is about equally sensitive to beams of potassium and potassium bromide, a platinum alloy detector is much more sensitive to potassium than to potassium bromide. This enables a separation to be made of an elastically scattered beam from a beam scattered by a collision process which involves the chemical

reaction

$$K + HBr \rightarrow H + KBr$$

This type of detector has proved of great value in studies of reactive scattering in the 1960s (Chapter 14). The alloy consists of 92 per cent platinum, 8 per cent tungsten; the selective behaviour apparently depends on a special state of surface contamination[578].

Electron impact ionizers for atomic beam detection have been designed in various ways. In their original form[579] they owed much to the oscillating electron ion source of Heil. The beam was passed through a region in which a transverse oscillating electron beam was confined by a magnetic field; a series of apertures collimated and focused the ion beam. The experimental efficiency of this ionizer was 1 in 3000 potassium atoms. A great improvement was effected[580] by arranging the electron-emitting filament parallel to the atom beam, so as to increase the collision volume. Another configuration has been reported[581].

High efficiencies can only be attained with careful attention to the ion optics. A signification advance in the development of an ionizer is reported by Weiss[582]. In the geometry shown in *Figure 3.51*, there is no electron oscillation, but the heavy electron current produces a space-charge potential gradient sufficient to push the ions forward from the ionizing region. An efficiency of 1 in 40 for argon was claimed; the elastic scattering is kept to the low level of 0·1 per cent by the maintenance of pressures of the order of 10^{-9} torr. The ionizer is operated with a positive potential of $V = 75$ V upon the grid, so that the emission is thermally limited; but the plate is operated at the same potential as the grid, so that a large space-charge potential minimum of about $V_{min} = 30$ V is built up, according to the expression:

$$\frac{V_{min}}{V} \simeq 1 - \frac{3d^2}{40} \left(\frac{J}{a^2 V^{3/2}} \right) \tag{3.274}$$

for

$$0{\cdot}4 \leqslant \frac{V_{min}}{V} \leqslant 1$$

where d is the plate-grid distance, $a^2 = 2{\cdot}335 \times 10^{-6}$ AV$^{-3/2}$ and J is the electron current density in A cm^{-2}. The difference in potential ΔV_{min}

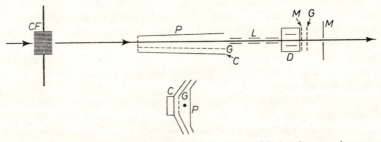

Fig. 3.51. *Electron impact ionizer for atomic beam: CF, crinkly foil directional entrance for beam; C, cathode; G, accelerating grid; P, electron collecting plate; L, ion-focusing lens system; D, deflecting plates; MGM, mass-spectrometer grid and apertures*
(From Weiss[582], by courtesy of the American Institute of Physics)

between the minima at the two ends of the ionizer, is

$$\frac{\Delta V_{\min}}{V_{\min}} = -\frac{2\theta}{3}\left(\frac{J}{a^2 V^{3/2}}\right)^{1/2}\frac{[1-(V_{\min}/V)^{1/2}]^{1/2}}{(2V_{\min}/V)-(V_{\min}/V)^{1/2}} \qquad (3.275)$$

where θ is the angle in radians between the plate and grid. This potential difference causes the positive ions to travel out of the ionizer, so that they can be directed into a mass-spectrometer. (There is an arithmetical error of a factor of 2π in the original paper, which lowers the efficiency of ionization to 0·3 per cent, a figure subsequently shown to be approximately correct.)

A new oscillating electron ionizer has been reported by Brink[583]. The atomic beam passes along the axis of a cylinder, which is lined with a cylindrical grid, with a filament between the two. The outer cylinder and the filament F are held at earth potential, and the cylindrical grid at $+250$ V. A quadrupole mass-filter is held at $+105$ V, so that ions formed within the grid are accelerated into the mass-spectrometer. The potential variation inside the grid is such as to force the ions out into the mass-filter. The electrons follow a radially oscillating path, giving an efficiency which, even after the mass-filter, is as high as 1 in 80 for krypton. For argon it is 1 in 600.

Ionizing detectors are all much improved by phase-sensitive detection to reduce the background noise. The latter is dependent on the frequency bandwidth $\Delta \nu$, and arises from the following causes[584]:

1. Random fluctuations in the cosmic ray intensity, giving rise to a noise signal equal to $(2e\,\Delta \nu \mathcal{I}_c)^{1/2}$ where $\mathcal{I}_c$ is the cosmic ray current;
2. Multiplier noise, arising from statistical distribution in the amplification factors for individal dynodes; the noise signal is equal to $[G/(G-1)]^{1/2}N_1$, where G is the average stage gain and N_1 the noise at the first dynode;
3. Fluctuations of background pressure in the vacuum system, which can be reduced to the order of 10^{-9} torr;
4. Statistical fluctuation of the signal $\mathcal{I}_s$ giving rise to a noise current $(2e\,\Delta \nu \mathcal{I}_s)^{1/2}$.

Use of a mass-spectrometer behind the ionizer reduces the first of these sources. Particle counting, as opposed to measurement of output current of the multiplier, reduces multiplier noise. Beam chopping and phase-sensitive detection, together with reduction of beam path length, reduces the third source of noise, and increase of beam intensity and measuring time reduce the statistical fluctuation of the signal.

An ionizer of the duoplasmatron type having 1 per cent efficiency has been reported. Another system makes use of a cylindrical cathode, down the centre of which the beam is fired. Electrons are accelerated inwards by means of a cylindrical grid within the cathode.

3.18 DETECTION AND IDENTIFICATION OF ATOMIC SPECIES, OR RADICLES

Detection of atomic species such as hydrogen, oxygen, nitrogen and other chemical radicles, can be achieved by calorimetric, chemical, diffusion and ionizing techniques. It is necessary to distinguish between detection of beams and the monitoring of flowing or stationary gases containing radicles.

19

Chemical Techniques

Chemical development of depositions produced by beams of radicles is a long-established technique. Hydrogen atoms can be detected by colour change of MoO_3; oxygen atoms on PbO; and nitrogen atoms an $AgNO_3$.

Titration of radicle species in flowing gases is also established practice; a known flow rate of gas is added which will react with the radicle. The product, formed in an excited state, is detected by optical emission (chemiluminescence). Atomic hydrogen is titrated by C_6H_{12} cyclohexane, yielding H_2. Atomic oxygen is titrated[585] by NO, yielding NO_2 and radiation. Atomic nitrogen is titrated by NO, yielding N_2 first positive bands[224]. In these titration experiments, the flow rate of the added gas is increased until the amount of light (length of the visible tail) ceases to increase. At this 'end-point', the flow rate of the added gas is measured.

Calorimetric Techniques

Calorimetric detection can be a reliable technique in bulk, but the problems that are posed in detecting localized beams (microcalorimetry) have not yet been solved.

Bulk-flow calorimetry[586–588] can be used to detect flowing atomic hydrogen gas, whose atoms recombine at a metal surface with the emission of heat. The accommodation coefficient is not unity, so it must be arranged that the atoms strike the metal surface of the calorimeter several times. Nevertheless it has been reported[589] that a 50 cm length of 34 s.w.g. platinum plus 13 per cent rhodium wire coiled in a spiral is an 80 per cent efficient detector. A further spiral mounted downstream checks this figure. Such a wire is conveniently mounted in a bridge circuit, so that its temperature may readily be inferred from change of resistance. Silver (plated on platinum wire) is used as a calorimeter surface for atomic oxygen, and cobalt for atomic nitrogen.

Detection by Diffusion

Detection by diffusion is known as the Wrede–Harteck technique[590]. Suppose that a gas chamber contains a mixture of atomic and molecular hydrogen. One wall of the chamber is made of a glass film 10–15 μ thick and contains 50 μ diameter holes, through which the gas flow can be expected to be purely effusive. The atoms effuse faster than molecules, but the atoms transport mass less effectively than the molecules. Therefore a pressure difference Δp builds up across the glass film, and the degree of dissociation

$$\alpha = \frac{n(\text{H})}{n(\text{H}) + n(\text{H}_2)} = \frac{3 \cdot 41 \, \Delta p}{p} \qquad (3.276)$$

Detection by Ionization

Detection by ionization follows much the same lines as described in Section 3.17 above. Electron impact ionization cross-sections for radicles are available in the literature.

Electron Paramagnetic Resonance

Atomic oxygen has been monitored by electron paramagnetic resonance (EPR). The atom concentration in the gas is obtained by calibrating the EPR detection cavity with molecular oxygen[591]. The EPR spectrum of the oxygen atom consists of six lines, generally unresolved. The four main lines result from transitions between the five M_J levels of the ground state $^3P_2$, and the two smaller lines on either side are due to transitions between the three M_J levels of the $^3P_1$ state. In the calibration, an integration is performed over the six-line composite of the oxygen atom spectrum and over the E line in the O_2 molecular spectrum[592]. The spectrum is usually recorded[593] as the derivative of the absorption signal, which may be related[594] to the absorption resonance, and thence[591] to the atomic oxygen density.

EPR spectrometry has also been used to detect radicle products of the bombardment of ammonia by electrons[595]. Condensation of the radicles proved necessary.

3.19 DETECTION OF EXCITED ATOMS, MOLECULES AND IONS

Excited species populations are monitored by the radiation they produce. This cannot easily be done for metastable species of long lifetime, but it is still possible for vibrationally excited species[596].

A new method of detecting vibrationally excited nitrogen molecules[597] is by means of Penning ionization by He 2^3S to $N_2^+ B^2\Sigma_u^+$, which passes radiatively to $N_2^+ X^2\Sigma_g^+$. Inference of the gas vibrational temperature can be made from the band intensities, assuming calculated Franck–Condon factors.

Metastable atoms and molecules are detected by electron ejection from metals, by radiation absorption, by Penning ionization, by ionization using electron impact, Lorentz or surface processes, and by calorimetric techniques.

Electron Ejection from Metals

Electrons are ejected from metal surfaces by the impact of metastable atoms which possess sufficient internal energy. The necessary energy condition for this process to be possible is

$$E_{\text{ex}} > \phi < D_e - \phi$$

where E_{ex} is the energy of excitation of the atom to the metastable state, ϕ is the work function of the metal, and D_e is the depth of the lowest electron level in the metal. This condition is met for rare gas and mercury metastables, but not, for example, for oxygen metastables, unless alkali metal surfaces, formed by continuous evaporation, are used.

The absolute yield ζ of electrons per incident metastable atom was at one time believed to be of the order of 0·5; that is, an electron is always associated with the collision, but it has an even chance of actually leaving the surface of the metal. Dorrestein[598] and Lamb and Retherford[599] produced data which lent support to this view, but the first experiment designed principally

19*

to measure this quantity was performed by Stebbings[600], who found a somewhat lower yield. In this experiment, a thermal energy beam of metastable He 2^3S atoms, emerging from a hot-cathode low-voltage arc, was passed through a collision chamber containing a low pressure of argon (single collision conditions) and to a metal target fronted by a positively biased grid to collect electrons. The argon positive ions, formed by Penning ionizing collisions which also deactivate the metastables, were collected at negatively biased wires in the collision chamber. If every deactivated metastable produces a positive ion, then ζ can be deduced from the equation

$$I_1 - I_2 = \zeta I_3 \qquad (3.277)$$

where I_1 is the target electron current in the absence of argon, I_2 is the target electron current in the presence of argon, and I_3 the positive ions simultaneously produced. Corrections were made for elastic scattering and for the presence of photons in the beam.

A table of values of ζ, measured relative to gold in subsequent experiments[601], is given in Table 3.9. However, no attempt was made to separate singlet from triplet metastables, although the source produced a predomination of triplets. Recent unpublished experiments[624] yield $\zeta = 0.49 \pm 0.09$ for He 2^1S and 0.67 ± 0.10 for He 2^3S on chemically cleaned but undegassed gold.

The electron ejection detector has usually been designed upon the assumption that a metastable atom can sometimes be reflected from a metal surface without undergoing deactivation. It may be arranged that the metastable flux falls into a cone or wedge, so that a number of collisions can be made. However, experiments show that reflexion without deactivation is a comparatively rare process, and no great error is involved in using a planar metal surface fronted by a grid of known transparency. With suitable precautions, either the emission current from the target or the electron collection current at the grid can be measured.

Table 3.9. VALUES OF ζ FOR HELIUM METASTABLES

Species	ζ
Gold, undegassed	0.29
Tungsten, undegassed	0.17
Tungsten, degassed	0.14
Molybdenum, undegassed	0.19
Molybdenum, degassed	0.11
Platinum, undegassed	0.26
Platinum, degassed	0.25

For surface detectors it is necessary that the surface condition is standard from one experiment to another; for example, an atomically clean surface may be used. The only way to keep a cold surface atomically clean under bombardment from gas is to evaporate metal onto it at a rate sufficient to cause it to grow faster than it is being contaminated.

Metastable atom beams issuing from a gas ionization source contain both positive ions and electrons. These must be removed very efficiently when detection by electron ejection is used. Assuming that the source and detector are at zero potential, then the simplest grid system which efficiently deals with all unwanted secondary species consists of three grids, with potentials negative, positive, negative.

In order that low-lying metastable states be detectable at a metal surface, a coating of low work function metal, such as caesium, may be continually evaporated onto it. To ensure efficient accommodation, the metal surface, for example the first dynode of a multiplier, may be cooled by conduction from liquid nitrogen via a sapphire finger[602]. Distinction may be made between the metastable states of mercury[603] by the inability of the lower pair to be detected at an uncoated surface.

Detection by Absorption and Emission of Radiation

Metastable atoms have large absorption coefficients to radiation of wavelength suitable to raise them to higher states. For example[604], the Ne $^3P_0$, $^3P_1$ and $^3P_2$ state populations can be monitored by their absorption of the 6402 Å and 6143 Å lines.

When metastable atoms are exposed to such radiation, some of them are raised to a level from which an allowed downward transition will occur; a known part of the emitted radiation is then measured. The optical emission technique is perhaps the most refined of all detection methods. It can be applied to low energy states, such as O ^1D and O ^1S, which will not eject electrons from metals. The quenching of a metastable state can sometimes be achieved by the Stark broadening action of an electric field. Thus H $2s$ and also He$^+$ $2s$ are quenched[605] to $2p$, and the Lyman α radiation is observed.

The radiation absorption technique is combined with phase-sensitive detection[606], which is necessary because the plasma or gas containing metastables frequently radiates the same line as that which is used for absorption monitoring. Radiation from a source can either be interrupted mechanically; or the source can be specially designed to possess amplitude modulation[607], to which the phase-sensitive detector is locked. Frequencies as high as 1 MHz can be used, enabling the time variation of metastable population in afterglows to be studied. An alternative method[608] is to provide the monochromator detector with two input slits; a part of the radiation from the source is focused onto one slit via a path which does not pass through the volume containing metastables. The absorption coefficient at the centre of the line is

$$k_0 = \text{const.} \times \frac{\lambda f n_m}{\bar{v}} \qquad (3.278)$$

where $\bar{v}$ is the root mean square gas atom velocity, f is the appropriate oscillator strength, and n_m is the metastable atom density. There is a correction factor depending upon the ratio of radiation source Doppler width to metastable-containing volume Doppler width. The fraction of radiation transmitted

$$S = \frac{1}{l(\pi k_0)^{1/2}} \int_{-\infty}^{\infty} \{1 - \exp\left[-k_0 l \exp\left(-\omega^2\right)\right]\} \, d\omega \qquad (3.279)$$

where l is the path length in the metastable-containing volume. The integral has been tabulated[609].

Penning Detection

Metastable atoms will ionize atoms of lower ionization potentials by the Penning process (Chapter 13)

$$X^m + Y \rightarrow X + Y^+ + e \qquad (0^m0/01e)$$

The ion or electron currents are measured at suitably biased electrodes.

Ionization Detection

Detection of metastables by ionizing them has been developed by various workers; the first use of electron impact was by Foner and Hudson[610]. Since the ionization potential of a metastable is often substantially lower than that of the ground state, a flux of metastables in the presence of their parent gas may be detected by an electron impact ionizer operating at a suitable low energy. The detection of molecular oxygen metastables can be achieved by their resonance conversion to O^- by very low energy electrons[611]. Surface ionization techniques at hot filaments can also be used.

Lorentz ionization of fast-moving high principal quantum number states of H and H_2 is a standard technique for their detection. The usual electrode configuration is two aperture electrodes separated by a high potential, but slits with electric field transverse to the beam have also been used. Since the ions are formed in this field, they are deflected, and the device may be used as a spectrometer; the deflexion is dependent upon the principal quantum number. For theoretical treatment of this process, it is necessary that the space variation of electric field be taken into account[612-615]. The beam may be passed between a scintillator and a high-voltage electrode, so that the electrons are accelerated onto the scintillator, as in the Daly–Afrosimov detector.

Calorimetric Detection

Calorimetric detection of excited molecules has been achieved using cobalt surfaces[588].

Detection of Fast Metastables

Fast metastable atom beams are produced from ion beams which have passed through collision chambers containing gas (Chapters 11 and 12). Such beams contain proportions of neutral atoms, and the detection of the metastable component poses a special problem. It can be solved[616] by utilizing a further gas collision chamber containing a transverse electric field for positive ion removal, followed by a fast-particle detector such as a particle multiplier. The gas collisional ionization processes for metastables ($0^m0/10e$) and for neutrals ($00/10e$) have different probabilities. As the gas pressure in the collision chamber is raised, the detector current attenuates exponentially. For two beam components, two distinct regions of different logarithmic slope are found in these graphs. Without a knowledge of the two relevant cross-sections, the fraction of excited atoms in the beam can be calculated, provided that the two cross-sections are appreciably different and provided it is known which is the larger.

Detection of Metastable Ions

There are two methods of detecting metastable ions: by means of their ejection of electrons from metals, and by means of 'Aston band techniques', including the new retardation technique.

The efficiency of ejection of electrons from metals is different for ground state and metastable ions[617]. Provided that a control experiment is used, this difference can be utilized for metastable ion detection. When both coefficients (γ_{+m} and γ_+) are known, the electron yield together with the total ion current provide information about the proportion of metastables in the beam. For an atomically clean tungsten surface, the values of the coefficients for rare gas ions are as shown in Table 3.10.

Table 3.10. EMISSION COEFFICIENTS FOR RARE GAS IONS ON TUNGSTEN

Rare gas	Helium	Argon	Krypton	Xenon
$\dfrac{\gamma_{+m}-\gamma_+}{\gamma_+}$	1·62	2·42	4·25	10·9

The Aston band technique can only be used to detect molecular metastable ions which dissociate spontaneously to fragment ions of smaller mass, with lifetimes of the order 10^{-5}–10^{-6} sec. These give rise to mass-spectrometer peaks which are anomalously broad and usually at non-integral mass number; they were noticed in Aston's original work.

For unambiguous identification of the mass numbers of both the metastable ion and the fragment, a mass-spectrometer with electrostatic and magnetic deflexion is necessary. If the metastable ion of mass number m_1, after being accelerated through an energy eV in the electrostatic section, dissociates before magnetic deflexion with kinetic energy T of fragment ion, of mass number m_2, then a peak will be observed[618] at mass number

$$m^* = \frac{m_2^2}{m_1}\left[1\pm\left(\frac{MT}{eV}\right)^{1/2}\right]^2 \qquad (3.280)$$

where $M = (m_1 - m_2)/m_2$. The peak width

$$W = \frac{4m_2^2}{m_1} \left(\frac{MT}{eV} \right)^{1/2}$$
(3.281)

The double mass-spectrometer avoids detection of peaks arising from dissociation in the electrostatic or magnetic deflexion regions, or in the accelerating region. Only ions are detected which dissociate after having been accelerated and deflected electrostatically but before being deflected magnetically. Dissociations without kinetic energy should yield peaks of normal instrumental width; but the observed peaks are nearly always broad Gaussians, indicating a distribution of kinetic energies of dissociation.

Advantages of unambiguity can be obtained by detecting dissociations which take place before electrostatic deflexion[619-621]. The ion of mass number m_1 is accelerated through eV_0, and those ions which have not dissociated are detected. The energy of the fragment ion is $eV_0 m_2/m_1$. Therefore when an additional accelerating potential is applied, raising the total energy to eV_1, a metastable dissociation spectrum will be observed when

$$V_1 = V_0 \frac{m_1}{m_2}$$
(3.282)

Thus m_1 and m_2 are identified unambiguously, and normal ions do not interfere.

'Metastable mass spectra' give information about the dissociation paths of complex molecules. Such mass spectra may contain dozens or hundreds of metastable dissociation processes. Kinetic energy release in the transitions can be derived from the peak shapes[622].

A retardation technique by which metastable dissociated molecular ions can be recorded whilst normal ions are not registered has been reported[623]. Ions are passed through an orifice electrode into a retarding field which is terminated by a metal-coated scintillator. Normal ions are able to impinge upon the scintillator, but the photons produced by them are negligible, because the ions have insufficient energy in these experiments. However, lighter fragment ions, formed by dissociation after acceleration but before entering the retardation region, possess after dissociation insufficient energy to reach the scintillator and turn back to impinge on the orifice electrode, emitting secondary electrons which pass onto the scintillator and produce photons. The detector is also capable of registering ions formed by dissociation in the retardation region.

3.20 CONTROL OF EXPERIMENTS

Advances in techniques of automation and data handling are at present being exploited in the service of more efficient and rapid atomic collision experimentation.

The simplest form of continuous automatic data recording is the pen-recorder; its time-constant is normally not much shorter than 1 sec for full-scale traverse of the pen (but, at expense, a time-constant two orders of magnitude shorter is available). Normally the paper chart is set to move at a constant rate, while, at a constant rate of change, potential, corresponding to impact energy, is applied to an electrode. Calibration marks are entered manually or automatically. On some instruments ($X_1 X_2$ recorders), two pens record simultaneously, so that, for example, primary beam and scattered beam can be recorded simultaneously. Another form of recorder (XY recorder) plots two variables simultaneously, as ordinate and abscissa; all types of spectra may readily be recorded on such a device.

For many purposes a pen-record, possessing, say, ± 2 per cent accuracy, is satisfactory, but it is tedious to reduce a pen-record to digital form; this is necessary if mathematical operations (such as deconvolution, or application of a correction) are necessary. For digital recording there exist 'dataloggers' which record analogue data (voltages) at regularly spaced intervals, using punched paper tape, or magnetic tape. If punched paper tape (which is cheapest) is used, then seven or eight holes are to be preferred, as this provides a parity check, protection against errors. The data-logger can select data from a number of different channels in turn. Typically, the potential corresponding to impact energy is scanned at a constant rate, and simultaneous recordings are made of this potential and of, say, two potentials corresponding to primary beam and scattered beam currents. Instructions are then punched onto the tape, enabling a computer to handle the data.

There is a problem concerning the constant rate scansion of the impact energy potential. Often it is necessary to repeat the experiment under different conditions, for example, measuring the beam currents when no gas (or crossed beam) is present. How are the two sets of data to be compared in the computer? Comparison of individual data points cannot be made, since although the impact energy potential is scanned at the same rate in both experiments, the energies at which the two data sets are taken do not necessarily correspond. It is possible to programme the computer to solve this problem by curve-fitting both data sets, but a simpler solution is to tackle the problem at the impact energy potential end by arranging for exactly reproducible scansion in steps. Even so, the contact potentials at the electrodes may change slightly between the two experiments. It is clear that the situation in which the collision chamber must be alternatively filled and emptied of gas, which can take minutes, is to be avoided. Crossed-beam experiments are much easier to handle because either beam can be interrupted by a movable chopper or 'flag' in a fraction of a second.

The ability of data-loggers to receive information successively from a large number of channels has led to the development of the 'multi-channel analyser', a piece of equipment designed for measuring a signal in the presence of a high noise level. Suppose that a fast stepwise scansion of impact energy potential is conducted, and at each step the signal (plus noise) is recorded in a channel corresponding to that step. The operation is repeated many times, and each time the signal should be approximately the same although the noise is different. The record in each channel is accumulated, so that, as time progresses, the signal stands out ever more clearly from the noise. Signals as small as one-hundredth of noise may be measured in this way.

These facilities and many more are afforded when the entire experiment is linked to a small computer. In particular, the multi-channel scan need not be made over the entire range, but only over those regions in which signal shows signs of appearing. Automated control of the parameters of the experiment can also be achieved. The computer is asked to vary certain parameters, such as focusing potential, until the beam intensity achieves a certain magnitude. There is a case for split-beam operation, with one beam used only as a monitor. As of 1970 there is not very wide experience in the automated control of atomic collision experiments. The important components are the interfaces, or digital-to-analogue conversion circuits, which control the parameters of the experiment.

REFERENCES

1. Cook, C. J. and Ablow, C. M. Stanford Research Institute, Palo Alto, Report ARC RC-TN 59-472, 1959.
2. Trujillo, S. M., Neynaber, R. H. and Rothe, E. W. *Rev. sci. Instrum.* 37 (1966) 1655.
3. Belyaev, V. A., Brezhnev, B. G. and Erastov, E. M. *Soviet Phys. J.E.T.P. Lett.* 3 (1966) 207.
4. Trujillo, S. M., Neynaber, R. H. and Rothe, E. W. *Rev. sci. Instrum.* 37 (1966) 1655.
5. Fite, W. L. and Brackmann, R. T. *Phys. Rev.* 112 (1958) 1141.
6. Harrison, M. F. A. (Chap. 1.4) and Bederson, B. (Chap. 1.3) *Methods of Experimental Physics, Vol. 7*, Ed. B. Bederson and W. L. Fite, 1968. New York; Academic Press.
7. Dolder, K. T., Harrison, M. F. A. and Thonemann, P. C. *Proc. Roy. Soc.* A264 (1961) 367; 274 (1963) 546.
8. Harrison, M. F. A. *Brit. J. appl. Phys.* 17 (1966) 371.
9. Smith, R. A., *J. Instn elect. Engrs* 98 (1951) 43.
10. Lipsom, H. G. and Littler, J. R. *Appl. Optics* 5 (1966) 472.
11. Dance, D. F., Harrison, M. F. A. and Smith, A. C. H. *Proc. Roy. Soc.* A290 (1966) 74.
12. Busch, von F., Strunck, H. J. and Schlier, C. *Z. Phys.* 199 (1967) 518.
13. Lorents, D. C. and Aberth, W. *Phys. Rev.* 139 (1965) A1017; Verba, J. and Hawlyrak, R. *Rev. sci. Instrum.* 32 (1961) 1037.
14. Lassettre, E. N., Berman, A. S., Silverman, S. M. and Krasnow, M. E. *J. chem. Phys.* 40 (1964) 1232.
15. Jordan, E. B. and Brode, R. B. *Phys. Rev.* 43 (1933) 112.
16. Fuls, E. N., Jones, P. R., Ziemba, F. P. and Everhart, E. *Phys. Rev.* 107 (1957) 704.
17. Carbone, R. J., Fuls, E. N. and Everhart, E. *Phys. Rev.* 102 (1956) 1524.
18. Lorents, D. C. and Aberth, W. *Phys. Rev.* 139 (1965) A1017.
19. Wittkower, A. B. Thesis, University of London, 1966.
20. Critchfield, C. L. and Dodder, D. C. *Phys. Rev.* 75 (1949) 419.
21. Briggs, O. R. Ph.D. Thesis, University of Illinois, 1953.
22. Herb, R. G., Kerst, D. W., Parkinson, D. B. and Plain, G. J. *Phys. Rev.* 55 (1939) 998; Breit, G., Thaxton, H. M., Eisenbud, L. *Phys. Rev.* 55 (1939) 1018; Zimmerman, E. J., Kerman, R. O., Singer, S., Kruger, P. G. and Jentschke, W. *Phys. Rev.* 96 (1954) 1322.
23. Gagge, A. P. *Phys. Rev.* 43 (1933) 776; 44 (1933) 808.
24. Westin, S. *K. norske vidensk. Selsk. Skr.* 2 (1946) 58.
25. Kieffer, L. J. and Dunn, G. H. *Rev. mod. Phys.* 38 (1966) 1.
26. Barber, M., Farren, J. and Linnett, J. W. *Proc. Roy. Soc.* A274 (1963) 293.
27. Rusch, M. and Bunge, O. *Z. tech. Phys.* 13 (1932) 77.
28. Politiek, J., Los, J. and Ikelaar, P. G. *Vakuum-Tech.* 6 (1968) 150.

29. COBIC, B., CARTER, G. and LECK, J. H. *Vacuum* 11 (1961) 247.
30. ISHII, H. and NAKAYAMA, K. *Vacuum Symp. Trans.* 8 (1961) 518; *J. Jap. vac. Soc.* 7 (1957) 113.
31. GAEDE, W. *Ann. Phys., Lpz.* 46 (1915) 357.
32. ERNSBERGER, F. M. and PITMAN, H. W. *Rev. sci. Instrum.* 26 (1955) 584.
33. MEINKE, CH. and REICH, G. *Vacuum* 13 (1963) 579.
34. DE VRIES, A. E. and ROL, P. K. *Vacuum* 15 (1965) 135.
35. ROTHE, E. W. *J. vacuum Sci. Tech.* 1 (1964) 66.
36. UTTERBACK, N. G. and GRIFFITH, T. G. *Rev. sci. Instrum.* 37 (1966) 866.
37. TAKAISHI, T. *Trans. Faraday Soc.* 61 (1965) 840.
38. BANNON, J. *Rev. sci. Instrum.* 14 (1943) 6.
39. COMPTON, K. C. and VAN VOORHIS, C. C. *Phys. Rev.* 26 (1925) 436.
40. SCHRAM, B. L. *Physica, Eindhoven* 31 (1965) 94; *J. chem. Phys.* 44 (1966) 49.
41. FRYBURG, G. C. and SIMONS, J. H. *Rev. sci. Instrum.* 20 (1949) 8, 541.
42. BENNEWITZ, H. G. and DOHMANN, H. D. *Vakuum-Tech.* 14 (1965) 8; BUREAU, A. J., LASLETT, L. J. and KELLER, J. M. *Rev. sci. Instrum.* 23 (1952) 683.
43. BANNERBERG, J. G. and TIP, A. *Proc. 4th Int. Congr. Vacuum Sci. Technol.* p. 609, 1968.
44. SCHULZ, G. J. *Rev. sci. Instrum.* 28 (1957) 1051.
45. PETERMANN, L. A. and BAKER, F. A. *Brit. J. appl. Phys.* 16 (1965) 487; *J. vacuum Sci. Tech.* 3 (1966) 285.
46. KLOPFER, A. *Trans. 8th Vacuum Symp.* p. 439, 1961. London; Pergamon.
47. SCHUMANN, W. C. *Rev. sci. Instrum.* 34 (1963) 700.
48. YARWOOD, J. *High Vacuum Technique*, 1956. London; Chapman and Hall.
49. DUSHMAN, S. *Scientific Foundations of Vacuum Technique*, revised J. M. Laferty, 1962. London; Chapman and Hall.
50. REDHEAD, P. A., HOBSON, J. P. and KORNELSON, E. V. *The Physical Basis of Ultrahigh Vacuum*, 1968. London; Chapman and Hall.
51. WEINRICH, G. and HUGHES, V. *Phys. Rev.* 95 (1954) 1541.
52. EDMONDS, P. H. and HASTED, J. B. *Proc. phys. Soc., Lond.* 84 (1964) 99.
53. HUDDLESTONE, R. H. and LEONARD, S. L. *Plasma Diagnostic Techniques*, 1965. New York; Academic Press; HEALD, M. A. and WHARTON, C. B. *Plasma Diagnostics with Microwaves*, 1965. New York; Wiley.
54. LANGMUIR, I. and MOTT-SMITH, H. *Gen. Elect. Rev.* 27 (1924) 7, 449, 538, 616, 762, 810.
55. DRUYVESTEYN, M. J. *Z. Phys.* 64 (1930) 793.
56. SLOANE, R. H. and MCGREGOR, E. I. R. *Phil. Mag.* 18 (1934) 193.
57. BOYD, R. L. F. *Proc. Roy. Soc.* A201 (1950) 329.
58. LOEB, L. B. *Fundamental Processes of Electrical Discharges in Gases*, 1939. New York; Wiley.
59. VON ENGEL, A. and STEENBECK, M. *Elektrische Gasentladungen*, 1932. Berlin; Springer-Verlag.
60. ALLEN, J. E., BOYD, R. L. F. and REYNOLDS, P. *Proc. phys. Soc., Lond.* B70 (1957) 297.
61. BOHM, D. *The Characteristics of Electrical Discharges in Magnetic Fields*, Chap. 3, Ed. A. Guthrie and R. K. Wakerling, 1949. New York; McGraw-Hill.
62. BILLS, D. G., HOLT, R. B. and MCCLURE, B. T. *J. appl. Phys.* 33 (1962) 29.
63. WAYMOUTH, J. F. *Gaseous Electronics Conf.* American Physical Society, 1964.
64. HEIL, H. and BLANKEN, R. A. *Gaseous Electronics Conf.* American Physical Society, 1964.
65. MOTT-SMITH, H. M. and LANGMUIR, I. *Phys. Rev.* 28 (1926) 727.
66. TAKAYAMA, K., IKEGAMI, H. and MIYAZAKI, S. *Phys. Rev. Lett.* 5 (1960) 238.
67. BOYD, R. L. F. and TWIDDY, N. D. *Proc. Roy. Soc.* A250 (1959) 53.
68. BOYD, R. L. F. and THOMPSON, J. B. *Proc. Roy. Soc.* A252 (1959) 102.
69. OSKAM, H. J. Thesis, University of Leiden, 1957; OSKAM, H. J. and MITTLESTADT, V. R. *J. appl. Phys.* 31 (1960) 940.

70. CHANTRY, P. J., WHARMBY, J. and HASTED, J. B. *Proc. 5th Int. Conf. Ioniz. Phenom. Gases.* p. 630, 1961. Amesterdam; North Holland Publishing Company.
71. LANGSTROTH, G. F. O. and HASTED, J. B. *Disc. Faraday Soc.* 33 (1962) 698; LANGSTROTH, G. F. O. Thesis, University of London, 1962.
72. THOMPSON, J. B. *Proc. phys. Soc., Lond.* 73 (1959) 818.
73. COHEN, I. M. *Phys. Fluids* 6 (1963) 1942.
74. WAYMOUTH, J. F. *Phys. Fluids* 7 (1964) 1843.
75. COURT, G. R. Ph.D. Thesis, University of Birmingham, 1953.
76. WAYMOUTH, J. F. *J. appl. Phys.* 30 (1959) 1404.
77. ECKER, F. and McCLURE, B. T. *Gaseous Electronics Conf.* American Physical Society, 1964.
78. SCHULZ, G. J. and BROWN, S. C. *Phys. Rev.* 98 (1955) 1642.
79. BETTINGER, R. T. and WALKER, E. H. *Phys. Fluids* 8 (1965) 4, 748.
80. JOHNSON, G. O. and MALTER, L. *Phys. Rev.* 80 (1950) 58.
81. BOND, R. H. *Gaseous Electronics Conf.* American Physical Society, 1965.
82. DERFLER, H. Stanford University Report, 1961.
83. MARGENAU, H. *Phys. Rev.* 69 (1946) 508.
84. HARP, R. S. and CRAWFORD, F. W. *J. appl. Phys.* 35 (1964) 12.
85. GOLDSTEIN, L. *Elect. Commun.* 29 (1952) 243; GOLDSTEIN, L. *Advanc. Electron.* 7 (1955) 474.
86. BRYANT, J. H. and FRANKLIN, R. N. *Proc. phys. Soc., Lond.* 81 (1963) 531, 790.
87. BRYANT, J. H. and FRANKLIN, R. N. *Proc. phys. Soc., Lond.* 83 (1964) 971.
88. IRISH, R. T. and BRYANT, G. H. *Proc. phys. Soc., Lond.* 84 (1964) 975.
89. HEROUX, L. *Proc. phys. Soc., Lond.* 83 (1964) 121.
90. GRIEM, H. R. *Plasma Spectroscopy*, 1964. New York; McGraw-Hill.
91. GRIEM, H. R., KOLB, A. C. and SHEN, K. Y. United States Naval Research Laboratory Report 5805, 1962; 5455, 1960; GRIEM, H. R. United States Naval Research Laboratory Report 6084, 1965.
92. DeSILVA, A. W., EVANS, D. E. and FORREST, M. J. *Nature, Lond.* 199 (1963) 1281.
93. THOMPSON, J. B. Thesis, University of London, 1957; LANGSTROTH, G. F. O. Thesis, University of London, 1962.
94. BÖHME, D. K. and GOODINGS, J. M. *Rev. sci. Instrum.* 37 (1966) 362.
95. NUTT, C. W. and RIFAI, M. S. *Abstr. Mass Spectroscopy Conf.* University College London, 1965.
96. ROBERTS, J. *Abstr. Mass Spectroscopy Conf.* University College London, 1965.
97. AFROSIMOV, V. V. *Soviet Phys. tech. Phys.* 5, No. 12 (1961).
98. GABOR, D., ASH, E. A. and DRACOTT, D. *Nature, Lond.* 176 (1955) 916.
99. LIN, S.-C., NEAL, R. A. and FYFE, W. I. *Phys. Fluids* 5 (1962) 1633.
100. GOLDSTEIN, L., ROUX, M. and DAYBON, J. A. *Proc. 6th Int. Conf. Ioniz. Phenom. Gases*, p. 115, 1963. Paris; SERMA.
101. BERLANDE, J., GOLDAN, P. D. and GOLDSTEIN, L. *Appl. Phys. Lett.* 5 (1964) 51.
102. BARTH, C. A. *Z. Phys.* 158 (1960) 85.
103. FERGUSON, E. E. *Advanc. atomic molec. Phys.* 4 (1970) 1.
104. SAUTER, G. F., GERBER, R. A. and OSKAM, H. J. *Rev. sci. Instrum.* 37 (1966) 572.
105. GRAY, E. P. and KERR, D. E. *Bull. Amer. phys. Soc.* 5 (1960) 372; *Ann. Phys., New York* 17 (1962) 276.
106. CHEN, C. L., LEIBY, C. C. and GOLDSTEIN, L. *Phys. Rev.* 121 (1961) 1391.
107. KERR, D. E. and HIRSCH, M. N. *Bull. Amer. phys. Soc.* 3 (1959) 258.
108. KERR, D. E. and LEFFEL, C. S. *Bull. Amer. phys. Soc.* 4 (1959) 113.
109. FROMMHOLD, L. and BIONDI, M. A. University of Pittsburgh, SRCC Report 71, 1968.
110. FOX, J. N. and HOBSON, R. M. *Phys. Rev. Lett.* 17 (1966) 161.
111. ANDERSON, J. B., ANDRES, R. P. and FENN, J. B. *Advanc. chem. Phys.* 10 (1966) 275.
112. CAMPANGUE, R. *Rarefied Gas Dynamics, Vol. 2*, Ed. J. H. de Leeuw, p. 279, 1966. New York; Academic Press.
113. RAMSEY, N. F. *Molecular Beams, Vol. 1: Experimental Nuclear Physics*, Ed. E. Segre, 1953. New York; Wiley.

114. SMITH, K. F. *Molecular Beams*, 1955. London; Methuen.
115. KING, J. G. and ZACHARIAS, J. R. *Advanc. Electron.* 8 (1956) 1.
116. HOWARD, W. M. *Phys. Fluids*, 4 (1961) 521.
117. GUSTAFSEN, W. A. and MIEL, R. E. *Phys. Fluids*, 7 (1964) 472.
118. ZAPATA, R. M., PARKER, H. M. and BODINE, J. H. *Proc. 2nd Int. Symp. Rarefied Gas Dynamics*, p. 67, 1960. Berkeley, California; Academic Press.
119. BECKER, E. W. and HENKES, W. *Z. Phys.* 146 (1956) 320.
120. DITCHBURN, R. W. *Proc. Roy. Soc.* 117 (1928) 486.
121. FOSTER, P. J., LECKENBY, R. E. and ROBBINS, E. J. *J. Phys.* B2 (1969) 478.
122. SMITH, J. M. *J. Amer. Inst. Aeronautics Astronautics* 3 (1965) 648.
123. TRISCHKA, J. W. *Methods of Experimental Physics, Vol. 3*, Ed. B. Bederson and W. L. Fite, p. 594, 1962. New York; Academic Press.
124. PAUN, J. and HASTED, J. B. Unpublished data, 1962.
125. GIORDMAINE, S. A. and WANG, T. C. *J. appl. Phys.* 31 (1960) 463.
126. HANES, G. R. *J. appl. Phys.* 31 (1960) 2171.
127. BECKER, G. *Z. angew. Phys.* 13 (1961) 59; *Z. Phys.* 162 (1961) 290.
128. KANTROWITZ, A. and GREY, J. *Rev. sci. Instrum.* 22 (1951) 328.
129. SHERMAN, F. S. *Rarefied Gas Dynamics, Vol. 2*, Ed. J. A. Laurmann, p. 228, 1963. New York; Academic Press. SHERMAN, F. S. Lockheed Missile and Space Company Technical Report, Fluid Mechanics 6-90-63-61, 1963.
130. OWENS, P. L. and THORNHILL, C. K. British Ministry of Supply, Aerodynamics Research Council Technical Report R+M No. 2616, 1952.
131. ANDERSON, J. B. and FENN, J. B. *Phys. Fluids*, 8 (1965) 780.
132. BECKER, E. W., BIER, K. and HENKES, W. *Z. Phys.* 146 (1956) 333; BECKER, E. W., KLINGELHOFER, R. and LOHSE, P. *Z. Naturf.* 17a (1962) 432; HENKES, W. *Z. Naturf.* 16a (1961) 842; 17a (1962) 786.
133. BECKER, E. W., BIER, K. and BURGHOFF, H. *Z. Naturf.* 10a (1955) 565; BECKER, E. W., and SCHUTTE, R. *Z. Naturf.* 11a (1956) 679; BECKER, E. W. *Z. Naturf.* 12a (1957) 609; *Proc. 2nd Int. Conf. peaceful Uses atom. Energy* 4 (1958) 455; BECKER, E. W. and SCHUTTE, R. *Z. Naturf.* 15a (1960) 336; BECKER, E. W., KLINGELHOFER, R. and LOHSE, P. *Z. Naturf.* 15a (1960) 644; BIER, K. *Z. Naturf.* 15a (1960) 714; *Fortschr. Phys.* 11 (1963) 325; BECKER, E. W., BIER, K. and BIER W. *Z. Naturf.* 16a (1961) 1393; 17a (1962) 778; 18a (1963) 246; ZIGAN, F. *Z. Naturf.* 17a (1962) 772; WATERMAN, P. C. and STERN, S. A. *J. chem. Phys.* 31 (1959) 405; STERN, S. A., WATERMAN, P. C. and SINCLAIR, T. F. *J. chem. Phys.* 33 (1960) 805; CHOW, R. R. University of California, IER Technical Report HE-150-175, 1959.
134. REIS, V. H. and FENN, J. B. *J. chem. Phys.* 39 (1963) 3240.
135. SKINNER, G. T. *Phys. Fluids*, 4 (1961) 1172; SKINNER, G. T. and FETZ, B. H. *Rarefied Gas Dynamics*, Ed. J. H. de Leeuw, 1963. New York; Academic Press; OMAN, R. A. Grumman Research Department Report RE 166, 1963.
136. KNUTH, E. L. University of California, Department of Engineering Report 63-30, 1963.
137. HUNDHAUSEN, E. and PAULY, H. *Z. Naturf.* 20a (1965) 625; HUNDHAUSEN, E. and HARRISON, H. *Rev. sci. Instrum.* 38 (1967) 131.
138. EMINYAN, M. E. Joint Institute of Laboratory Astrophysics Report 94, 1968.
139. BULL, T. H. and MOON, P. B. *Disc. Faraday Soc.* 17 (1954) 54.
140. UTTERBACK, N. G. *Phys. Rev.* 129 (1963) 219.
141. HOLLSTEIN, M. and PAULY, H. *Z. Phys.* 196 (1966) 353.
142. AMDUR, I. and JORDAN, J. E. *Advanc. chem. Phys.* 10 (1966) 29.
143. HUSHFAR, F., ROGERS, J. W. and WEBB, D. United States Air Force Cambridge Research Laboratories Report, 1966.
144. POLITIEK, J., ROL, P. K., LOS, J. and IKELAAR, P. G. *Rev. sci. Instrum.* 39 (1968) 1147.
145. PAULY, H. and TOENNIES, J. P. *Methods of Experimental Physics, Vol. 7*, Ed. B. Bederson and W. L. Fite, p. 254, 1968. New York; Academic Press.
146. VAN ZYL, B. and UTTERBACK, N. G. *Proc. 6th Int. Conf. Phys. electron. atom. Colli-*

sions, p. 393, 1969. Cambridge, Mass.; Massachussetts Institute of Technology Press.
147. WOOD, R. W. *Proc. Roy. Soc.* 97 (1920) 455.
148. MARSHALL, T. C. *Phys. Fluids*, 5 (1962) 743; WITTKE, J. P. and DICKE, R. H. *Phys. Rev.* 103 (1956) 620.
149. FOWLER, R. H. and GUGGENHEIM, E. A. *Statistical Thermodynamics*, 1939. London; Cambridge University Press.
150. LAMB, W. E. and RETHERFORD, R. C. *Phys. Rev.* 79 (1950) 549.
151. FITE, W. L. and BRACKMAN, R. T. *Phys. Rev.* 112 (1958) 1141.
152. LOCKWOOD, G. J., HELBIG, H. F. and EVERHART, E. *J. chem. Phys.* 41 (1964) 3820.
153. FAIRCHILD, C. E. and CLARK, K. C. *Bull. Amer. phys. Soc.* 7 (1962) 458.
154. FEHSENFELD, F. C., EVENSON, K. and BROIDA, H. P. United States National Bureau of Standards Report 8701, 1964.
155. NUTT, C. W. and BIDDLESTONE, A. T. *Nature, Lond.* 191 (1961) 798; *Trans. Faraday Soc.* 59 (1961) 1363, 1368, 1376.
156. FLEISCHMANN, R. *Nucl. Instrum. Meth.* 11 (1961) 112.
157. FERGUSON, E. E. *Advanc. atomic molec. Phys.* 4 (1970) 1.
158. NOTTINGHAM, W. B. *Handb. Physik* 21 (1956); HERRING, C. and NICHOLS, M. H. *Rev. mod. Phys.* 21 (1949) 185.
159. WEHNELT, A. *Ann. Phys., Paris* 14 (1904) 425.
160. SLOANE, R. H. and PRESS, R. *Proc. Roy. Soc.* A168 (1938) 284.
161. HERRMANN, G. and WAGENER, S. *The Oxide-Coated Cathode, Vol. 1*, 1951. London; Chapman and Hall.
162. HUNG, C. S. *J. appl. Phys.* 21 (1950) 37, 143.
163. IVY, H. F. *Advanc. Electron.* 6 (1954) 137.
164. HULL, W. A. *Phys. Rev.* 56 (1939) 86; LEMMENS, H. J., JANSEN, M. J. and LOOSJES, R. *Philips tech. Rev.* 11 (1950) 341.
165. LANGMUIR, I. and ROGERS, W. *Phys. Rev.* 4 (1914) 544.
166. MELTON, C. E. *Rev. sci. Instrum.* 29 (1958) 250.
167. DUVAL, X., LE GOFF, P. and VALENTIN, R. *J. Chim. phys.* 53 (1956) 757.
168. ROBINSON, C. F. and SHARKEY, A. G. *Rev. sci. Instrum.* 29 (1958) 250.
169. HUTSON, A. R. *Phys. Rev.* 98 (1955) 889.
170. NOTTINGHAM, W. B. *Phys. Rev.* 49 (1936) 78.
171. SPEIDEL, R., and GAUKLER, K. H. *Z. Phys.* 208 (1968) 419.
172. DIETRICH, W. *Z. Phys.* 151 (1958) 519; 152 (1958) 306.
173. YOUNG, R. D. *Phys. Rev.* 113 (1959) 110.
174. LINDSAY, P. A. *Advanc. Electron.* 13 (1960) 202.
175. SHELDON, H. *Phys. Rev.* 107 (1957) 1553.
176. SIMPSON, J. A. and KUYATT, C. E. *J. appl. Phys.* 37 (1966) 3805.
177. HARTWIG, D. and ULMAN, K. *Z. Phys.* 173 (1962) 294; ULMAN, K. and ZIMMERMAN, B. *Z. Phys.* 182 (1964) 194.
178. SPANGENBERG, K. R. *Vacuum Tubes*, 1948. New York; McGraw-Hill.
179. PIERCE, J. R. *J. appl. Phys.* 11 (1940) 548.
180. BRAUCKS, F. W. *Optik*, 16 (1959) 304.
181. SIMPSON, J. A. and KUYATT, C. *Rev. sci. Instrum.* 34 (1963) 265.
182. HART, P. A. and WEBER, C. *Philips Res. Rep.* 16 (1961) 376.
183. PIERCE, J. R. *Theory and Design of Electron Beams*, 1954. Princeton, New Jersey; Van Nostrand.
184. COSSLETT, V. E. *Electron Optics*, 1950. London; Oxford University Press.
185. WHITEHOUSE, J. and HASTED, J. B. to be published.
186. IMHOF, R. E. and READ, F. H. *J. sci. Instrum.* 2 (1968) 859.
187. MORUZZI, J. L. *Rev. sci. Instrum.* 38 (1967) 1284.
188. ALLEN, A. J. and FRANKLIN, R. G. *J. opt. Soc. Amer.* 29 (1939) 452; WEEKS, R. F. *J. opt. Soc. Amer.* 49 (1959) 429.
189. BALDWIN, G. C. General Electric, Schenectady, Report 66-C-246, 1966.
190. EITEL, W., JOST, K. and KESSLER, J. *Abstr. 5th Int. Conf. Phys. Electron. atom. Collisions*, p. 549, 1967. Leningrad; Akademii Nauk.

191. FRANZEN, W. *Abstr. 5th Int. Conf. Phys. electron. atom. Collisions*, p. 545, 1967. Leningrad; Akademii Nauk.
192. JOST, K. and ZEMAN, H. D. *Proc. 6th Int. Conf. Phys. electron. atom. Collisions*, 1969. Cambridge, Mass.; Massachusetts Institute of Technology Press.
193. HILGNER, W. and KESSLER, J. *Abstr. 5th Int. Conf. Phys. electron. atom. Collisions*, p. 546, 1967. Leningrad; Akademii Nauk.
194. HUGHES, V. W., LUBELL, M. S., POSNER, M. and RAITH, W. *Abstr. 5th Int. Conf. Phys. electron. atom. Collisions*, p. 544, 1967. Leningrad; Akademii Nauk; LONG, R. L., RAITH, W. and HUGHES, V. W. *Phys. Rev. Lett.* 15 (1965) 1.
195. FRANZEN, W. *Abstr. 5th Int. Conf. Phys. electron. atom. Collisions*, p. 545, 1967. Leningrad; Akademii Nauk; FRANZEN, W. and GUPTA, R. *Phys. Rev. Lett.* 15 (1965) 819.
196. DONALLY, B. *Abstr. 5th Int. Conf. Phys. electron. atom. Collisions*, p. 543, 1967. Leningrad; Akademii Nauk; DONALLY, B., CLAPP, T., SAWYER, W. and SCHULTZ, M. *Phys. Rev. Lett.* 12 (1964) 502.
197. KING, A. S. *Astrophys. J.* 40 (1914) 205; 56 (1922) 318.
198. JURGENS, G. *Z. Phys.* 138 (1954) 613; MOTSCHMANN, H. *Z. Phys.* 143 (1955) 77; MASTRUP, F. and WIESE, W. *Z. Astrophys.* 44 (1958) 259; RICHTER, J. *Z. Phys.* 151 (1958) 114; HEY, P. *Z. Phys.* 157 (1959) 79.
199. WILKERSON, T. D. Dissertation, University of Michigan.
200. KOLB, A. C. and GRIEM, H. R. *Atomic and Molecular Processes*, Ed. D. R. Bates, 1962. New York; Academic Press.
201. SEAY, G. L. Dissertation, University of Oklahoma.
202. JOSEPHSON, V. and HALES, R. W. *Phys. Fluids* 4 (1961) 373.
203. LASEK, A. and BRUN, E. A. *Trans. aerospace electron. Systems* 3 (1967) 460.
204. PATRICK, R. M. *J. appl. Phys.* 2 (1959) 589.
205. McLEAN, E. A., KOLB, A. C. and GRIEM, R. H. *Phys. Fluids* 4 (1961) 1055; WIESE, W., BERG, H. F. and GRIEM, R. H. *Phys. Rev.* 120 (1960) 1079; *Phys. Fluids* 4 (1961) 250; WIESE, W. *Rev. sci. Instrum.* 31 (1960) 943.
206. CLOUSTON, J. G., GAYDON, A. G. and GLASS, I. I. *Proc. Roy. Soc.* A248 (1958) 429; FAIZULLOV, F. S., SOBOLEV, N. N. and KUDRYAVTSEV, E. M. *Optika Spektrosk.* 8 (1960) 585, 761.
207. COHEN, R. S., SPITZER, L. and McROUTLY, P. *Phys. Rev.* 80 (1950) 230; LIN. S. C., RESLER, E. L. and KANTROWITZ, A. *J. appl. Phys.* 26 (1955) 95.
208. DOHERTY, L. R. *Bull. Amer. phys. Soc.* 5 (1960) 131; Dissertation, University of Michigan.
209. EDLEN, B. *Rep. Progr. Phys.* 26 (1963) 181.
210. MITCHELL, A. C. G. and ZEMANSKY, M. W. *Resonance Radiation and Excited Atoms*, 1961. London; Cambridge University Press.
211. CARIO, G. and LOCHTE-HOLTGREVEN, W. *Z. Phys.* 42 (1927) 22.
212. HOPFIELD, J. J. *Astrophys. J.* 133 (1930).
213. HUFFMAN, R. E., LARRABEE, J. C. and TANAKA, Y. *J. opt. Soc. Amer.* 52 (1962) 851; 55 (1965) 101.
214. HUFFMAN, R. E., LARRABEE, J. C. and CHAMBERS, D. *Appl. Optics* 4 (1965) 1145.
215. HUFFMAN, R. E., LARRABEE, J. C. and TANAKA, Y. *Appl. Optics* 4 (1965) 1581; *Jap. J. Appl. Phys.* 4, Suppl. 1 (1965) 494; TANAKA, Y. and JURSA, A. S. *J. opt. Soc. Amer.* 50 (1960) 1118.
216. WILKINSON, P. G. and BYRAM, E. T. *Appl. Optics* 4 (1965) 581.
217. GARTON, W. R. S. *J. sci. Instrum.* 36 (1959) 11; WHEATON, J. E. G. *Appl. Optics* 3 (1964) 1247; PARKINSON, W. H. and REEVES, E. M. *Proc. phys. Soc., Lond.* 80 (1962) 516, 860; SAMSON, J. A. R. *J. opt. Soc. Amer.* 56 (1966) 769.
218. HUFFMAN, R. E., HUNT, W. W. and TANAKA, Y. *J. opt. Soc. Amer.* 51 (1961) 487.
219. COOK, G. R. and METZGER, P. H. *J. chem. Phys.* 41 (1964) 321.
220. GARTON, W. G. *Proc. 5th Int. Conf. Ioniz. Phenom. Gases*, p. 206, 1961. Amsterdam; North Holland Publishing Company.
221. NEWBERG, R. G., HEROUX, L. and HINTEREGGER, H. E., *Appl. Optics* 1 (1962) 733.

222. EDLEN, B. *Rep. Progr. Phys.* 26 (1963) 181; VODAR, B., ROMAND, J. and BALLOFFET, G. *C. R. Acad. Sci., Paris*, 240 (1955) 412.
223. BALLOFFET, G. and ROMAND, J. *C. R. Acad. Sci., Paris*, 246 (1958) 733; TORESSON, Y. G. *Ark. Fys.* 17 (1960) 179; FREEMAN, G. H. C. *Proc. phys. Soc., Lond.* 86 (1965) 117.
224. YOUNG, R. A. and BLACK, G. *J. chem. Phys.* 44 (1966) 3741.
225. SROKA, W. *Phys. Lett.* 14 (1965) 301.
226. DE VOS, J. C. Thesis, University of Leiden, 1953; RUTGERS, G. A. W., and DE VOS, J. C. *Physica, Eindhoven* 20 (1954) 715.
227. HAMBERGER, S. M. *Plasma Phys.* 5 (1963) 73.
228. ALLEN, A. J. and FRANKLIN, R. G. *J. opt. Soc. Amer.* 29 (1939) 453.
229. WEEKS, R. F. *J. opt. Soc. Amer.* 49 (1959) 429.
230. SCHAWLOW, A. L. *Solid St. J.* 2 (1961) 27; MAIMAN, T. H. *Phys. Rev.* 123 (1961) 1145, 1151; NELSON, D. F. and BOYLE, W. S. *Appl. Optics* 1 (1962) 181.
231. PATEL, C. K. N., BENNETT, W. R., FAUST, W. L. and MCFARLANE, R. A. *Bull. Amer. phys. Soc.* 7 (1962) 444.
232. SOROKIN, P. P. and STEVENSON, M. J. *IBM J. Res. Dev.* 5 (1961) 56; KAISER, W., GARRETT, C. G. and WOOD, D. L. *Phys. Rev.* 123 (1961) 766.
233. WHITE, A. D. and RIGDEN, J. D. *Appl. Phys. Lett.* 2 (1963) 211.
234. BENNETT, W. R., FAUST, W. L., MCFARLANE, R. A. and PATEL, C. K. N. *Phys. Rev. Lett.* 8 (1962) 470.
235. PATEL, C. K. N., MCFARLANE, R. A. and FAUST, W. L. *J. opt. Soc. Amer.* 53 (1963) 522.
236. MATHIAS, L. E. S. and PARKER, J. T. *Appl. Phys. Lett.* 3 (1963) 16.
237. BENNETT, W. R. *Phys. Rev. Lett.* 17 (1966) 987.
238. BALLU, Y. *Rev. Phys. appl. (Fr.)* 3 (1968) 46; *C. R. Acad. Sci., Paris* 264 (1967) 599.
239. PATEL, C. K. N. *Lasers, Vol. 2*, Ed. A. K. Levine, p. 1, 1969. New York; Marcel Dekker.
240. CARMAN, R. L., HANUS, J. and WEINBURG, D. L. *Appl. Phys. Lett.* 11 (1967) 250.
241. *Abstr. Ion Sources Symp.* June, 1969. Saclay, France; Secretary, M. Guy Gautherin, Batiment 220, Faculte des Sciences, 91 Orsay, France.
242. NIER, A. O. *Phys. Rev.* 50 (1936) 1041.
243. HAGSTRUM, H. D. *Rev. mod. Phys.* 23 (1951) 185.
244. PIERCE, J. R. *Theory and Design of Electron Beams*, 1954. Princeton, New Jersey; Van Nostrand.
245. BAKER, F. A. and HASTED, J. B. *Phil. Trans.* A261 (1966) 33.
246. BARNARD, G. B. *Mass Spectrometer Researches, Report (DSIR)*, 1956. London; Her Majesty's Stationery Office.
247. EDMONDS, P. H. and HASTED, J. B. *Proc. 3rd Int. Conf. Phys. electron. atom. Collisions*, p. 88, 1963. Amsterdam; North Holland Publishing Company.
248. ONG, P. P. and HASTED, J. B. *Proc. phys. Soc.* 2 (1969) 91.
249. RUSSELL, R. D. and KOLLER, F. *Canad. J. Phys.* 38 (1960) 616; EDMONDS, P. H. Thesis, University of London, 1963.
250. BOHM, D., BURHOP, E. H. S. and MASSEY, H. S. W., *Gaseous Discharges in Magnetic Fields*, Chap. 2, Eds. A. Guthrie and R. K. Wakerling, 1949. New York; McGraw-Hill.
251. SIDENIUS, G. *Nucl. Instrum.* 38 (1965) 19.
252. TUVE, M. A., DAHL, O. and HAFSTAD, L. R. *Phys. Rev.* 48 (1935) 241.
253. LAMB, W. A. S. and LOFGREN, E. J. *Rev. sci. Instrum.* 27 (1956) 907.
254. BERNAS, R. and CASSIGNOL, P. *Onde elect.* 35 (1955) 1029; CAMERON, A. E. *Rev. sci. Instrum.* 25 (1954) 1154.
255. HILL, K. J. and NELSON, R. S. *Nucl. Instrum.* 38 (1965) 15, 97.
256. DRUAUX, J. and BERNAS, R. *Electromagnetically Enriched Isotopes and Mass Spectrometry*, Ed. M. L. Smith, p. 30, 1956. London; Butterworths.
257. FREEMAN, J. H. *Nucl. Instrum.* 22 (1963) 306.

258. COBIC, B., TOSIC, D. and PEROVIC, B. *Nucl. Instrum. Meth.* 24 (1963) 358.

259. HEIL, H. *Z. Phys.* 120 (1944) 212; FINKELSTEIN, A. T. and SMITH, L. P. *Rev. sci. Instrum.* 11 (1940) 94.

260. VON ARDENNE, M. *Phys. Z.* 43 (1942) 91.

261. HOYEAUX, M. M. *J. Phys. Radium* 15 (1954) 264.

262. HASTED, J. B. and CHONG, A. Y. *J. Proc. phys. Soc., Lond.* 80 (1962) 893.

263. LLOYD, O. Atomic Energy Research Establishment, Harwell, Private Communication, 1965.

264. VANCE, D. W. and BAILEY, T. L. *Rev. sci. Instrum.* 34 (1963) 925.

265. FINKELSTEIN, A. T. *Rev. sci. Instrum.* 11 (1940) 94.

266. VON ARDENNE, M. *Tabellen der Elektronenphysik, Ionenphysik and Übermikroskopie*, 1956. VEB Deutsch Verlag der Wissenschaften.

267. VON ARDENNE, M. *Z. angew. Phys.* (1963).

268. ILLGEN, J. *Sessions d'études sur la physique et la production des ions lourds*, La Plagne, 1969.

269. TAWARA, H. *Jap. J. appl. Phys.* 3 (1964) 342.

270. COLLINS, L. E. and BROOKER, R. J. *Nucl. Instrum. Meth.* 15 (1962) 193.

271. PIERCE, J. R. *J. appl. Phys.* 11 (1940) 548.

272. ROSE, P. H., WITTKOWER, A. B., BASTIDE, R. P. and GALEJS, A. *Nucl. Instrum. Meth.* 14 (1962) 79.

273. CHAVET, L. and BERNAS, R. *Nucl. Instrum. Meth.* 47 (1967) 51.

274. KISTEMAKER, J., ROL, P. K. and POLITIEK, J. *Nucl. Instrum. Meth.* 38 (1965) 1.

275. GETTING, Y. A. *Phys. Rev.* 59 (1941) 467.

276. THONEMAN, P. C., MOFFATT, J., ROAF, D. and SANDERS, J. H. *Proc. phys. Soc., Lond.* 61 (1948) 483; *Proc. Nucl. Phys.* 3 (1953) 219.

277. MOROZOV, V. M. *Dokl. Akad. Nauk. SSSR* 102 (1955) 61.

278. MOREAU, J. and VIENET, R. *Rapp. CEA* AC-4972 (1957).

279. GOODWIN, L. K. *Rev. sci. Instrum.* 24 (1953) 635.

280. BLANC, D. and DEGEILH, A. *J. Phys. Radium* 22 (1961) 230.

281. RUTHERGLEN, J. G. and COLE, J. F. I. *Nature, Lond.* 160 (1947) 545; ALLISON, J. K. and NORBECK, F. *Rev. sci. Instrum.* 2 (1956) 285.

282. KOCH, B. and NEUERT, H. *Z. Naturf.* 4a (1949) 456; KOWALEWSKI, V. J., MAYANS, C. A. and HAMMERSCHLAG, M. *Nucl. Instrum. Meth.* 5 (1959) 90.

283. GABOVICH, M. D. *J. tech. Phys., Moscow* 28 (1958) 872.

284. BUDDE, R. and HUBER, P. *Helv. phys. Acta* 25 (1952) 459.

285. LANGMUIR, I. and CROMPTON, K. T. *Rev. mod. Phys.* 3 (1931) 237.

286. GABOVICH, M. D. and PJANKOV, G. N. *Ukr. fiz. Zh.* 3 (1958) 419.

287. SERBINOV, A. N. *Pribory Tekh. Eksp.* 3 (1958) 391.

288. REIFENSCHWEILER, O. *Ann. Phys., Lpz.* 6 (1954) 34; *Z. Naturf.* 6a (1951) 331.

289. THONEMANN, P. C. and HARRISON, E. R. Atomic Energy Research Establishment, Harwell, Report CP/1190, 1953.

290. EHLERS, K. W. *Rev. sci. Instrum.* 29 (1958) 615.

291. SMITH, L. P. *Handbook of Physics*, Ed. E. U. Condon and H. Odishaw, 1967. New York; McGraw-Hill.

292. KUNSMAN, C. H. *Science* 62 (1925) 269; POWELL, C. F. and BRATA, L. *Proc. Roy. Soc.* A141 (1933) 463.

293. INGHRAM, M. G. and CHUPKA, W. G. *Rev. sci. Instrum.* 24 (1953) 518; IVANOV, R. N. and KUKAVADZE, G. M. *Pribory Tekh. Eksp.* 1 (1957) 106.

294. KENDALL, B. R. F. *Rev. sci. Instrum.* 29 (1958) 1089.

295. WHITE, W. C. *Proc. Inst. Radio Engrs.* 38 (1950) 852.

296. HOGG, A. M. and KEBARLE, P. *J. chem. Phys.* 43 (1965) 449.

297. MÜLLER, E. *Advanc. Electron. electron Phys.* 13 (1960) 83.

298. GOMER, R. *Field Emission and Field Ionization*, 1961. Cambridge, Mass.; Harvard University Press.

299. Stanford Research Institute, California.

300. FOGEL, YA. M., MITIN, R. V., KOZLOV, V. F. and ROMASHKO, N. D. *J. exp. theor.*

Phys. 35 (1958) 565; FOGEL, YA. M., KOZLOV, V. F., KALMYKOV, A. A. and MURATOV, V. I. *J. exp. theor. Phys.* 36 (1959) 1312; DUKELSKII, V. M., AFROSIMOV, V. V. and FEDORENKO, N. V. *J. exp. theor. Phys.* 30 (1956) 792.

301. HASTED, J. B. *Proc. Roy. Soc.* A222 (1954) 74; KHITNIJ, J. M. *Pribory Tekh. Eksp.* 2 (1958) 51; PHILLIPS, J. L. and TUCK, J. L. *Rev. sci. Instrum.* 27 (1956) 97.
302. ANSDELL, D. A. and PAGE, F. M. *Trans. Faraday Soc.* 48 (1962) 1084.
303. METLAY, M. and KIMBALL, G. E. *J. chem. Phys.* 16 (1948) 774.
304. SMITH, S. J. and BRANSCOMB, L. M. *Rev. sci. Instrum.* 31 (1960) 733.
305. ABERTH, W. and PETERSON, J. R. Stanford Research Institute Report, 1966.
306. LAWRENCE, G. P., BEAUCHAMP, R. K. and McKIBBEN, I. L. *Nucl. Instrum.* 32 (1965) 357.
307. COLLINS, L. E. and GOBBET, R. H. *Nucl. Instrum.* 39 (1965) 277.
308. BETHGE, K. and RAU, G. *Nucl. Instrum.* 39 (1966) 157.
309. DONNALLY, B. L. and SAWYER, W. *Phys. Rev. Lett.* 15 (1965) 439.
310. BOHME, D. K., NAKSHBANDI, M. M., ONG, P. P. and HASTED, J. B. *Proc. 8th Int. Conf. Ioniz. Phenom. Gases*, p. 16, 1967.
311. POLANYI, J. C. *Chem. in Brit.* 2 (1966) 151.
312. WIND, H. Ph.D. Thesis, University of Utrecht, 1965; Culham Laboratory Report CLM R46, 1965. London; Her Majesty's Stationery Office.
313. REDHEAD, P. A. *Abstr. Mass Spectrometry Conf.* American Society for Testing and Materials, 1966.
314. ECKSTEIN, L. *Ann. Phys., Lpz.* 87 (1928) 1003.
315. ELIAS, K., OGRYZLO, E. A. and SCHIFF, H. I. *Canad. J. Chem.* 37 (1959) 1680.
316. OGRYZLO, E. A. *Disc. Faraday Soc.* (1964) 218.
317. LAMB, W. E. and RETHERFORD, P. C. *Phys. Rev.* 79 (1950) 549; 81 (1951) 222.
318. MUSCHLITZ, E. E. and SHOLETTE, W. P. University of Florida Report G5967, 1962.
319. PERL, M. L., RABI, I. I. and SENITSKY, B. *Phys. Rev.* 98 (1955) 611.
320. DONNALLY, B. T., CLAPP, T., SAWYER, W. and SCHULTZ, M. *Phys. Rev. Lett.* 12 (1964) 502.
321. OLIPHANT, M. L. E. *Proc. Roy. Soc.* A124 (1929) 228.
322. GREENE, D. *Proc. phys. Soc., Lond.* B63 (1950) 876.
323. RICHARDS, H. L. and MUSCHLITZ, E. E. *J. chem. Phys.* 41 (1964) 559.
324. McDERMOTT, M. N. and LICHTEN, W. L. *Phys. Rev.* 119 (1960) 134.
325. STERN, O. *Z. Phys.* 7 (1921) 249.
326. KING, J. G. and ZACHARIAS, J. R. *Advanc. Electron. electron Phys.* 8 (1956) 1.
327. RABI, I. I., KELLOGG, J. M. B. and ZACHARIAS, J. R. *Phys. Rev.* 46 (1934) 157.
328. KUSCH, P. *Lecture Notes on Molecular Beams*, 1950. New York; Columbia University.
329. FOGEL, YA. M. and CLAUSNITZER, G. *Z. Phys.* 153 (1959) 609.
330. CHRISTENSEN, R. L. and HAMILTON, D. R. *Rev. sci. Instrum.* 30 (1959) 356.
331. FRIEDMAN, H. *Z. Phys.* 161 (1961) 74.
332. WAECH, T. G., KRAMER, K. H. and BERNSTEIN, R. B. *J. chem. Phys.* 48 (1968) 3978.
333. BROSSEL, J., KASTLER, A. and WINTER, J. *J. Phys. Radium* 13 (1952) 668.
334. COLEGROVE, F. D. and FRANKEN, P. A. *Phys. Rev.* 119 (1960) 680.
335. DEHMELT, H. G. and MAJOR, F. G. *Phys. Rev. Lett.* 8 (1962) 213.
336. BASCO, N., CALLEAR, A. B. and NORRISH, R. G. W. *Proc. Roy. Soc.* A260 (1961) 459.
337. BENNEWITZ, H. G., PAUL, W. and SCHLIER, C. *Z. Phys.* 141 (1955) 6.
338. SMITH, K. F. *Molecular Beams*, 1955. London; Methuen.
339. DASH, J. G. and SOMMERS, H. S. *Rev. sci. Instrum.* 24 (1953) 91.
340. MILLER, R. C. and KUSCH, P. *Phys. Rev.* 99 (1955) 1314.
341. COLGATE, S. O. University of Florida, private communication, 1961.
342. GROSSER, A. E., ICKZKOWISKI, R. P. and MARGRAVE, J. L. *Rev. sci. Instrum.* 34 (1963) 117.
343. HOSTETTLER, H. U. and BERNSTEIN, R. B. *Rev. sci. Instrum.* 31 (1960) 872.
344. TRUJILLO, S. M., ROL, P. K. and ROTHE, E. W. *Rev. sci. Instrum.* 33 (1962) 841.
345. KINSEY, J. L. *Rev. sci. Instrum.* 37 (1966) 61.

346. PLATKOV, M. A. and ILLARIONOV, S. V. *Instrums. exptl. Tech., Wash.* 2 (1962) 353.
347. Globe Industries, 2275 Stanley Avenue, Dayton, Ohio (45404), U.S.A.
348. MARCUS, P. M. and McFEE, J. H. *Recent Researches in Molecular Beams*, Ed. I. Esterman, p. 43, 1959, New York; Academic Press.
349. GROSSER, A. E. and BERNSTEIN, R. B. University of Wisconsin Report WIS TCI 117GX, 1965.
350. THOMPSON, M. W. United Kingdom Atomic Energy Authority, Harwell, Report HL63/3809 (C14), 1963.
351. STUART, R. V., BROWER, K. and MAYER, W. *Rev. sci. Instrum.* 34 (1963) 425.
352. DECKERS, J. and FENN, J. B. *Rev. sci. Instrum.* 34 (1963) 96.
353. RUBIN, K., PEREL, J. and BEDERSON, H. *Phys. Rev.* 117 (1960) 151.
354. SOMPER, R., M.Sc. Thesis, University College London, 1963.
355. PARKER, J. H. and WARREN, R. W. *Rev. sci. Instrum.* 33 (1962) 948.
356. KLEMPERER, O. *Rep. Progr. Phys.* 28 (1965) 77.
357. MÖLLENSTEDT, G. *Optik, Stuttgart,* 5 (1949) 499.
358. FORST, G. *Z. Phys.* 10 (1958) 546.
359. SIMPSON, J. A. *Rev. sci. Instrum.* 32 (1961) 1283.
360. HAHN, E. *Jenaer Jb.* (1961) 326.
361. REGENSTRIEF, E., *Ann. Radioelect.* 6 (1951) 114.
362. FROST, G. *Z. angew. Phys.* 10 (1958) 546.
363. BRACK, K. *Z. Naturf.* 17a (1962) 1066.
364. SIMPSON, J. A. and MARTON, L. *Rev. sci. Instrum.* 32 (1961) 802.
365. KESSLER, J. and LINDNER, H. *Z. angew. Phys.* 18 (1964) 7.
366. BOERSCH, H. and SCHWEDA, S. *Z. angew. Phys.* 16 (1962) 2.
367. FOX, R. E., HICKAM, W. M., KJELDAAS, T. and GROVE, D. J. *Phys. Rev.* 84 (1951) 859.
368. FOX, R. E., HICKAM, W. M., KJELDAAS, T. and GROVE, D. J. *Phys. Rev.* 98 (1953) 555.
369. FOX, R. E. *Rev. sci. Instrum.* 26 (1955) 1101.
370. CHANTRY, P. J. Westinghouse Scientific Paper 69-9EA-CO GAS-Pl, 1969.
371. ANDERSON, N. and EGGLETON, P. P. *Int. Electron.* 22 (1967) 496; ANDERSON, N., EGGLETON, P. P. and KESSING, R. G. W. *Rev. sci. Instrum.* 38 (1967) 924, 1624.
372. BRIGLIA, D. D. and RAPP, D. *Gaseous Electronics Conf.* American Physical Society, 1964.
373. GROB, V. *J. chem. Phys.* 39 (1963) 972.
374. KUYATT, C. E. *Methods of Experimental Physics, Vol. 7*, Ed. B. Bederson and W. L. Fite, 1968. New York; Academic Press.
375. BLACKSTOCK, A. W., BIRKHOFF, A. D. and SLATER, M. *Rev. sci. Instrum.* 26 (1955) 274; PLOCH, W. and WALCHER, W. *Z. Phys.* 127 (1950) 274.
376. HERZOG, R. *Z. Naturf.* 10a (1955) 887.
377. SIEGBAHN, K. and SWARTHOLM, V. *Ark. Mat. Astr. Fys.* A33 No. 21 (1946).
378. VOGES, H. and RUTHEMANN, G. *Z. Phys.* 114 (1939) 709.
379. RUTHEMANN, G. *Ann. Phys., Lpz.* 2 (1948) 113.
380. SIDAY, R. E. *Proc. phys. Soc., Lond.* 59 (1947) 509.
381. SIMPSON, J. A. and MARTON, L. *Rev. sci. Instrum.* 29 (1958) 567.
382. KLEMPERER, O. and SHEPHERD, J. P. G. *Brit. J. appl. Phys.* 14 (1963) 85.
383. BLAUTH, E. *Z. Phys.* 147 (1957) 228; KUYATT, C. Ph.D. Thesis, University of Nebraska, 1961.
384. KLEMPERER, O. *Phil. Mag.* 20 (1935) 45.
385. STAMATOVIC, A. and SCHULZ, G. J. Yale University, Mason Laboratory Report, 1968.
386. HARROWER, G. A. *Rev. sci. Instrum.* 26 (1955) 850.
387. KUREPA, M. and TOSIC, D. Boris Kidric Institute of Nuclear Sciences, Belgrade, Report IBK 104, 1964.
388. AFROSIMOV, V. V., GORDEEV, YU. S., LAVROV, V. M. and SHCHEMELININ, S. G. *Abstr. 5th Int. Conf. Phys. electron. atom. Collisions*, p. 127, 1967. Leningrad; Akademii Nauk.

389. EDELMAN, F. and ULMER, K. *Z. angew. Phys.* 18 (1965) 308.
390. STERNGLASS, E. J. *Phys. Rev.* 108 (1957) 1.
391. BLAUTH, E. *Z. Phys.* 147 (1957) 228; KUYATT, C. Ph. D. Thesis, University of Nebraska, 1961; MEHLHORN, W. *Phys. Lett.* 21 (1966) 155.
392. SAR-EL, H. *Z. Rev. sci. Instrum.* 38 (1967) 1210.
393. ZASHKVARA, V. V., KORSUNSKII, M. J. and KOSMACHEV, O. S. *Soviet Phys. tech. Phys.* 11 (1966) 96.
394. HAFNER, H., SIMPSON, J. A. and KUYATT, C. E. *Rev. sci. Instrum.* (1967)
395. HUGHES, A. L. and ROJANSKY, V. *Phys. Rev.* 34 (1929) 284.
396. HUGHES, A. L. and MCMILLEN, J. H. *Phys. Rev.* 39 (1937) 585.
397. CLARKE, E. M. Ph.D. Thesis, University Laval, Quebec, 1955.
398. MARMET, P. and KERWIN, L. *Canad. J. Phys.* 38 (1960) 787.
399. BONESS, M. J. W., HASTED, J. B. and LARKIN, I. W. *Proc. Roy. Soc.* A305 (1968) 493.
400. ANDRICK, D. and EHRHARDT, H. *Z. Phys.* 192 (1966) 99.
401. WILLMANN, K. and EHRHARDT, H. *Z. Phys.* 203 (1967) 1.
402. HASTED, J. B. and AWAN, A. W. *Proc. Phys. Soc.* 2 (1969) 367.
403. BLOOM, A. L. *J. appl. Phys.* 36 (1965) 2560.
404. HERZOG, R. *Z. Phys.* 89 (1934); 97 (1935) 596.
405. HENNEBERG, W. *Ann. Phys., Lpz.* 19 (1934) 345.
406. ALAM, G. D. Thesis, University of London, 1967.
407. PURCELL, E. M. *Phys. Rev.* 54 (1938) 818.
408. SIMPSON, J. A. *Rev. sci. Instrum.* 35 (1964) 1968.
409. SIMPSON, J. A. and KUYATT, C. *Rev. sci. Instrum.* 34 (1963) 265.
410. ROGERS, F. T. *Rev. sci. Instrum.* 22 (1951) 723.
411. WIEN, W. *Verh. dtsch. phys. Ges.* 16 (1892) 165.
412. HERZOG, R. *Z. Phys.* 89 (1934) 447.
413. HENNEBERG, W. *Ann. Phys., Lpz.* 19 (1934) 335.
414. RUNDELL, R. D. Thesis, University of Washington, 1965.
415. BOERSCH, H., GEIGER, J. and HELLWIG, H. *Phys. Lett.* 3 (1962) 64; BOERSCH, H., GEIGER, J. and STICKEL, W. *Z. Phys.* 180 (1964) 415.
416. LA BAR, C. Oak Ridge National Laboratory Report 3484, 1964.
417. SCHLIER, C. H. German Physical Society, Spring Meeting, 1967.
418. PARKER, J. H. and WARREN, R. W. *Rev. sci. Instrum.* 33 (1962) 948.
419. SCHULZ, G. J. *Phys. Rev.* 125 (1962) 229.
420. SCHULZ, G. J. *Phys. Rev.* 116 (1959) 1141.
421. MAIER-LEIBNIZ, H. *Z. Phys.* 95 (1935) 489.
422. FLEMING, R. J. and HIGGINSON, G. S. *Proc. phys. Soc., Lond.* 81 (1963) 974.
423. CHANTRY, P. J. and PHELPS, A. V. Westinghouse Scientific Laboratory Paper 66-9E3-113-P1, 1966.
424. COMPTON, R. N., HUEBNER, R. H., REINHARDT, P. W. and CHRISTOPHOROU, L. G. *J. chem. Phys.* 48 (1968) 901.
425. HICKAM, W. M. and FOX, R. E. *J. chem. Phys.* 25 (1956) 642.
426. NAFF, W. T., COMPTON, R. N., COOPER, C. D. and REINHARDT, P. W. Oak Ridge National Laboratory Report ORNL-TM-2260, 1968.
427. MOTT, N. F. *Proc. Roy. Soc.* A124 (1929) 425; A135 (1932) 429.
428. HILGNER, W. and KESSLER, J. *Proc. 5th Int. Conf. Phys. electron. atom. Collisions*, p. 546, 1967. Leningrad; Akademii Nauk.
429. SHULL, C. G., CHASE, C. T. and MYERS, F. E. *Phys. Rev.* 63 (1943) 29; SPIVAK, P. E. *ZETP* 39 (1961) 574.
430. FARAGO, P. S. and WYKES, J. S. *Proc. 6th Int. Conf. Phys. electron. atom. Collisions*, p. 778, 1969. Cambridge, Mass.; Massachusetts Institute of Technology Press.
431. BULLIS, R. H. *Abstr. 4th Int. Conf. Phys. electron. atom. Collisions*, 1965. Hastings-on-Hudson, New York; Science Bookcrafters.
432. IVERY, H. F. *Advanc. Electron electron Phys.* 6 (1954) 138.
433. PIERCE, J. R. *Theory and Design of electron Beams*, 1949. Princeton, New Jersey; Van Nostrand.

434. WILLMORE, A. P. Thesis, University of London, 1955.
435. HASTED, J. B. *Atomic and Molecular Processes*, Ed. D. R. Bates, 1962. New York; Academic Press.
436. EDMONDS, P. H. and HASTED, J. B. *Proc. 3rd Int. Conf. electron. atom. Collisions*, 1963. Amsterdam; North Holland Publishing Company.
437. COSSLETT, V. E. *Electron Optics*, 1950. London; Oxford University Press.
438. GUSTAFSSON, E. and LINDHOLM, E. *Ark. Fys.* 18 (1960) 219; VON KOCK, H. and LINDHOLM, E. *Ark. Fys.* 19 (1961) 123.
439. O'CONNELL, J. S. *Rev. sci. Instrum.* 32 (1961) 1314.
440. ALAM, G. D. Thesis, University of London, 1967.
441. BARNARD, G. B. *Modern Mass Spectrometry*, 1953. London; Institute of Physics; KERWIN, L. *Advanc. Electron. electron Phys.* 8 (1956) 187; INGHRAM, M. G. and HAYDEN, R. J. *Handbook of Mass Spectrometry*, National Academy of Sciences, National Research Council Publication 311, 1954; MAYNE, K. I. *Rep. Progr. Phys.* 15 (1952) 24; THODE, H. G. and SHIELDS, R. B. *Rep. Progr. Phys.* 12 (1949) 1; MATTAUCH, J. *Naturwissenschaften*, 39 (1952) 557; INGHRAM, M. G. *Advanc. Electron. electron Phys.* 1 (1948) 219; BEYNON, J. *Mass Spectrometry and its Application to Organic Chemistry*, 1963. Amsterdam; Elsevier.
442. MONK, C. W. and WERNER, G. K. *Rev. sci. Instrum.* 20 (1949) 93; BLEAKNEY, W. and HIPPLE, J. A. *Phys. Rev.* 53 (1938) 521; HIPPLE, J. A. and SOMMER, H. United States National Bureau of Standards Circular 522, 1953.
443. HARNWELL, G. P. *Principles of Electricity and Electromagnetism*, 1949. New York; McGraw-Hill.
444. STEPHENS, W. E. *Phys. Rev.* 45 (1934) 513.
445. NIER, A. O. *Rev. sci. Instrum.* 11 (1940) 212; COGGESHALL, N. D. *J. appl. Phys.* 18 (1947) 855.
446. DELEVA, J. and PETERLIN, A. *Rev. sci. Instrum.* 26 (1955) 399.
447. KING, J. G. and ZACHARIAS, J. R. *Advanc. Electron. electron Phys.* 8 (1956) 1.
448. HINTERBERGER, J. *Z. Naturf.* 3a (1948) 125; KERWIN, L. *Rev. sci. Instrum.* 20 (1949) 36; CARTAN, M. L. *J. Phys. Radium* 8 (1937) 453; PERSSON, R. *Ark. Fys.* 3 (1951) 455.
449. O'CONNELL, J. S. *Rev. sci. Instrum.* 32 (1961) 1314.
450. MONK, C. W. and WERNER, G. K. *Rev. sci. Instrum.* 20 (1949) 93; BLEAKNEY, W. and HIPPLE, J. A. *Phys. Rev.* 53 (1938) 521; HIPPLE, J. A. and SOMMER, H. United States National Bureau of Standards Circular 522, 1953.
451. SCHRAM, B. L., ADAMCZYK, B. and BOERBOOM, A. J. *J. sci. Instrum.* 43 (1966) 638.
452. BRIGLIA, D. D. and RAPP, D. *J. chem. Phys.* 42 (1965) 3201.
453. GOUDSMIT, S. *Phys. Rev.* 74 (1948) 622.
454. PRAHALLADA RAO, B. S. United Kingdom Atomic Energy Authority, Harwell, Report AERE-R5 186, 1966.
455. KERWIN, L. *Advanc. Electron. electron Phys.* 8 (1956) 187.
456. KERWIN, L. *Canad. J. Phys.* 30 (1952) 503.
457. PRAHALLADA RAO, B. S. United Kingdom Atomic Energy Authority, Harwell, Report AERE-R5 187, 1966.
458. FEDORENKO, N. V. and AFROSIMOV, V. V. *J. tech. Phys., Moscow* 26 (1957) 1872; *J. exp. theor. Phys., Moscow* 34 (1958) 1398.
459. KATZENSTEIN, H. S. and FRIEDLAND, S. S. *Rev. sci. Instrum.* 26 (1955) 324; WILEY, W. C. and McLAREN, I. H. *Rev. sci. Instrum.* 26 (1955) 1150.
460. ROSENSTOCK, H. S. United States Patent 2999157.
461. GOUDSMIT, S. A. *Phys. Rev.* 74 (1948) 622; RICHARDS, P. I., HAYS, E. E. and GOUDSMIT, S. A. *Phys. Rev.* 76 (1949) 180; 84 (1951) 24.
462. SMITH, L. G. United States National Bureau of Standards Circular 522, 1953.
463. BOYD, R. L. F. *Nature, Lond.* 165 (1950) 142; BOYD, R. L. F. and MORRIS, D. *Proc. phys. Soc., Lond.* A68 (1955) 1; PETERLIN, A. *Rev. sci. Instrum.* 26 (1955) 398; WHERRY, T. C. and KARASEK, F. W. *J. appl. Phys.* 26 (1955) 682; KERR, L. W. *J. Electron. Control* 2 (1953) 179.
464. BENNETT, W. H. *J. appl. Phys.* 21 (1950) 143.

465. REDHEAD, P. A. *Canad. J. Phys.* 30 (1952) 1; REDHEAD, P. A. and CROWELL, C. R. *J. appl. Phys.* 24 (1953) 331.
466. LANGSTROTH, G. F. O. Ph.D. Thesis, University College London, 1962.
467. ROGERS, A. J. and BOYD, R. L. F. *J. sci. Instrum.* 43 (1966) 791.
468. BENNETT, W. H. *J. appl. Phys.* 21 (1950) 143.
469. CANNON, W. W. and TESTERMAN, M. K. *J. appl. Phys.* 27 (1956) 1283; MARTISOVITS, V. *Mat.-fyz. Cas.* 13 (1963) 72; PETERLIN, A. *Rev. sci. Instrum.* 26 (1955) 398; SHCHERBAKOVA, M. YA. *Z. tech. Phys.* 28 (1957) 599; TOWNSEND, J. W. *Rev. sci. Instrum.* 23 (1952) 538; WHERRY, T. C. and KARASEK, F. W. *J. appl. Phys.* 26 (1955) 682; MARTISOVITS, V. and SZARKA, S. *J. sci. Instrum.* 1 (1968) 326; HOLMES, J. C. *Rev. sci. Instrum.* 28 (1957) 290; DEKLEVA, J. and RIBARIC, M. *Rev. sci. Instrum.* 28 (1957) 365.
470. VAN WIJNGAARDEN, K. University of Windsor, Ontario, 1966.
471. HIPPLE, J. A., SOMMER, H. and THOMAS, H. A. *Phys. Rev.* **76** (1949) 1877; SMITH, L. G. and DAMM, C. C. *Phys. Rev.* 90 (1953) 324; EDWARDS, A. G. *Brit. J. appl. Phys.* 6 (1955) 44; BRUBAKER, W. M. and PERKINS, G. D. *Rev. sci. Instrum.* 27 (1956) 720.
472. PAUL, W. and RAETHER, M. *Z. Phys.* 140 (1955) 2621; PAUL, W., REINHARDT, H. P. and VON ZAHN, U. *Z. Phys.* 152 (1958) 143; PAUL, W. and STEINWEDEL, H. *Z. Naturf.* 8a (1953) 448; WUERKER, R. F., SHELTON, H. and LANGMUIR, R. V. *J. appl. Phys.* 30 (1959) 342.
473. BRUBAKER, W. M. and TUL. J. *Rev. sci. Instrum.* 35 (1964) 1007; BRUBAKER, W. M. *Proc. 5th Int. Instrum. Meas. Conf.*, 1961. New York; Academic Press.
474. FISCHER, E. *Z. Phys.* 156 (1959) 1; WUERKER, R. F. *J. appl. Phys.* 30 (1959) 342.
475. DAWSON, P. H. and WHETTEN, N. R. General Electric Research and Development Center Report 67-C-282, 1967.
476. WALKER, W. C., WAINFAN, N. and WEISSLER, G. L. *J. appl. Phys.* 26 (1955) 1367; HINTEREGGER, H. E. and WATANABE, K. *J. opt. Soc. Amer.* 43 (1953) 604.
477. MILLER, G. L., BROWN, W. L., DONOVAN, P. V. and MACKINTOSH, I. M. *Trans. Inst. Radio Engrs.* 7 (1960) 185; MACKENZIE, J. M. and WAUGH, J. B. S. *Trans. Inst. Radio Engrs.* 7 (1960) 195; BLANKENSHIP, J. L. and BORKOWSKI, C. J. Oak Ridge National Laboratory Reports 1960, 1961; *Bull. Amer. phys. Soc.* 5 (1960) 38.
478. DUNKELMAN, L. *J. quantve. Spectros. Radiat. Transf.* 2 (1962) 533.
479. SHARPE, J. *Electron. Technol.* (1961).
480. JOHNSON, F. S., WATANABE, K. and TOUSEY, R. *J. opt. Soc. Amer.* 41 (1951) 702.
481. CORRIGAN, S. J. B. and VON ENGEL, A. *Proc. Roy. Soc.* A245 (1958) 335.
482. SCHNEIDER, E. G. *J. opt. Soc. Amer.* 30 (1940) 128; *J. chem. Phys.* 5 (1937) 106.
483. PO LEE, and WEISSLER, G. L. *J. opt. Soc. Amer.* 43 (1953) 512.
484. LARRABEE, R. D. *J. opt. Soc. Amer.* 49 (1959) 619.
485. TOUSEY, R. *Appl. Optics* 1 (1962) 679.
486. HINTEREGGER, H. E. *Space Astrophysics*, Ed. W. Liller, p. 108, 1961. New York; McGraw-Hill.
487. LUKIRSKII, L. P., RUMSH, M. A. and SMIRNOV, K. A. *Optika Spektrosk.* 9 (1960) 265.
488. GOODRICH, G. W. and WILEY, W. C. *Rev. sci. Instrum.* 32 (1961) 846.
489. GOODRICH, G. W. and WILEY, W. C. National Aeronautics and Space Administration, Washington, Report SP2, 1961.
490. BURROWS, C. N., LIEBER, A. J. and ZAVIANTSEFF, V. T. *Rev. sci. Instrum.* 38 (1967) 1477.
491. ANGEL, D. W., COOPER, H. W., HUNTER, W. R. and TOUSEY, R. *Image Intensifier Symposium*, Fort Belvoir, U.S.A., 1961.
492. CHUBB, T. A. and FRIEDMAN, H. *Rev. sci. Instrum.* 26 (1955) 493.
493. BRACKMAN, R. T., FITE, W. L. and HAGEN, K. E. *Rev. sci. Instrum.* 29 (1958) 125.
494. STOBER, A. K., SCOLNIK, R. and HENNES, J. P. *Appl. Optics* 2 (1963) 735.
495. EDERER, D. L. and TOMBOULIAN, D. H. *Appl. Optics* 3 (1964) 1073.
496. SAMSON, J. A. R. *J. opt. Soc. Amer.* 54 (1964) 6.
497. STEBBINGS, R. F. *Proc. Roy. Soc.* A241 (1957) 270.

498. GRIFFIN, W. G. and McWHIRTER, R. W. P. *Proc. Conf. opt. Instrum.*, p. 14, 1962. London; Chapman and Hall.
499. HINNOV, E. and HOFFMAN, F. W. *J. opt. Soc. Amer.* 53 (1963) 1259.
500. WARNECK, P. *Appl. Optics* 1 (1962) 721.
501. PACKER, D. M. and LOCK, C. *J. opt. Soc. Amer.* 41 (1951) 699.
502. KRUSE, P. W., GLAUCHLIN, L. D. and McQUISTAN, R. B. *Elements of Infrared Technology*, 1962. New York; Wiley.
503. HASS, G. and HUNTER, W. R. *J. quantve. Spectros. radiat. Transf.* 2 (1962) 637.
504. SAMSON, J. A. R. *Techniques of Vacuum Ultraviolet Spectroscopy*, 1967. New York; Wiley.
505. SMITH, S. J. and BRANSCOMB, L. M. *Rev. sci. Instrum.* 31 (1960) 733.
506. WATANABE, K., ZELIKOFF, M. and INN, E. C. Y. *J. chem. Phys.* 21 (1953) 1021.
507. SCHROEDER, D. J. *J. opt. Soc. Amer.* 52 (1962) 1380.
508. BRADLEY, D. J., BATES, B., JUULMAN, C. O. L. and MAJUMDAR, S. *Nature, Lond.* 202 (1964) 579.
509. HUNTER, W. R., ANGEL, D. W. and TOUSEY, R. *Appl. Optics* 4 (1965) 891.
510. FRIEDMAN, H. *Space Astrophysics*, Ed. W. Liller, p. 53, 1961. New York; McGraw-Hill.
511. WEISSLER, G. L. *Handb. Physik* 21 (1956) 320.
512. PELLICORI, S. J. *Appl. Optics* 3 (1964) 361.
513. BEDERSON, B. *Methods of Experimental Physics, Vol. 7A*, Ed. L. Marton, p. 77. New York; Academic Press.
514. LIPSON, H. G. and LITTLER, J. R. *Appl. Optics* 5 (1966) 472.
515. PERRIN, J. *C. R. Acad. Sci., Paris* 121 (1895) 1130.
516. PRUITT, J. S. *Nucl. Instrum. Meth.* 39 (1966) 329.
517. BROWN, K. L. and TAUTFEST, G. W. *Rev. sci. Instrum.* 27 (1956) 696.
518. BRONSHTEIN, I. M. and DENISOV, S. S. *Soviet Phys. solid St.* 7 (1966) 1819.
519. MILATZ, J. M. W., VREEDENBERG, H. A. and BREAK, J. W. *Physica, Eindhoven* 10 (1943) 433.
520. CHAMBERS, E. S. *Rev. sci. Instrum.* 35 (1964) 95.
521. BARNETT, C. F. and GILBODY, H. B. *Methods of Experimental Physics, Vol. 7A*, Ed. L. Marton, p. 396, 1968. New York; Academic Press.
522. HARRIS, L. *J. opt. Soc. Amer.* 36 (1946) 597.
523. GORELIK, L. L. and SINITZYN, V. V. *Soviet Phys. tech. Phys.* 9 (1964) 393.
524. AMDUR, I., GLICK, C. F. and PEARLMAN, H. *Proc. Amer. Acad. Sci.* 76 (1948) 101.
525. AFROSIMOV, V. V., GLADKOVSKII, I. P., GORDEEV, YU. S., KALINKEVICH, I. F. and FEDORENKO, N. V. *Soviet Phys. JETP* 5 (1961) 1378.
526. AKISHIN, A. I. *Soviet Phys. Usp.* 1 (1958) 113.
527. SMITH, L. G. United States National Bureau of Standards Circular 522, 1953.
528. MARPLE, D. T. F. *Rev. sci. Instrum.* 26 (1955) 1205.
529. DARE, J. A. and ROWEN, W. H. Massachusetts Institute of Technology Laboratory Nuclear Science and Engineering Technical Report 6, 1947.
530. OSBORNE, F. J. F. *Canad. J. Phys.* 30 (1952) 658.
531. SCHRAM, B. L., BOERBOOM, A. J. H., KLEINE, W. and KISTEMAKER, J. *Physica, Eindhoven* 32 (1966) 749.
532. INGHRAM, M. G. and HAYDEN, R. J. *Mass Spectrometry*, 1954. Washington; National Academy of Science; PLOCK, W. and WALCHER, W. *Rev. sci. Instrum.* 22 (1951) 1028; OSBORNE, F. J. F. *Canad. J. Phys.* 31 (1953) 1189; HAGSTRUM, H. D. *Phys. Rev.* 96 (1954) 336; HAGSTRUM, H. D. *Phys. Rev.* 89 (1953) 244; 96 (1954) 325; 104 (1956) 672; FLAKS, I. P. *J. tech. Phys., Moscow* 25 (1955) 2467, 2463; DUNEAV, YU. A. and FLAKS, I. P. *Dokl. Akad. Nauk. SSSR* 91 (1953) 43; IZMAILOV, S. V. *Soviet Phys. JETP.* 3 (1958) 2031; STERNGLASS, E. J. *Phys. Rev.* 108 (1957) 1; HAGSTRUM, H. D. *Phys. Rev.* 96 (1954) 336.
533. KAMINSKY, M. *Atomic and Ionic Impact Phenomena at Metal Surfaces*, 1967. Berlin; Springer-Verlag.
534. DIETZ, L. A. *Rev. sci. Instrum.* 36 (1965) 1763.

535. SCHRAM, B. L., BOERBOOM, A. J. H., KLEINE, W. and KISTEMAKER, J. *Physica, Eindhoven* 32 (1965) 749.
536. EVANS, R. D. *The Atomic Nucleus*, p. 791, 1955. New York; McGraw-Hill.
537. GOODRICH, G. W. and WILEY, W. C. *Rev. sci. Instrum.* 32 (1961) 846; WILEY, W. C. and HENDEE, C. F. *IRE Trans. Nucl. Sci.* 9, No. 3 (1962) 103; SPINDT, C. A. and SHOULDERS, K. R. *Rev. sci. Instrum.* 36 (1965) 775.
538. GOODRICH, G. W. and WILEY, W. C. National Aeronautics and Space Administration, Washington, Report SP2, 1961; SPINDT, C. A. and SHOULDERS, K. R. *Rev. sci. Instrum.* 36 (1965) 775; FARNWORTH, J. United States Patent 1,929,399, 1930; SCHEPKOV, O. *Pribory Tekh. Eksp.* 4 (1960) 89; GOODRICH, G. W. and WILEY, W. C. *Rev. sci. Instrum.* 33 (1962) 761; EVENS, D. S., ADAMS, J. and MANLEY, B. W. *Electron. Engng.* (1965) 180; EVENS, D. S. Goddard Space Flight Centre Report 64-154, 1964.
539. BURROUS, C. N., LIEBER, A. J. and ZAVIANTSEFF, V. T. *Rev. sci. Instrum.* 38 (1967) 1477.
540. RICHARDS, P. I. and HAYS, E. E. *Rev. sci. Instrum.* 21 (1950) 99.
541. RIVIERE, A. C. and SWEETMAN, D. R. *Rev. sci. Instrum.* 34 (1963) 1286.
542. LOCKWOOD, G. J. and CANO, G. L. *IEEE Trans. nucl. Sci.* 13 (1966) 716; KAZLOV, V. F., KOLDT, V. YA. and DOVBRYA, A. N. *Instrums exptl. Tech., Wash.* 6 (1965) 1385; PORTER, F. T., FREEDMAN, M. S., WAGNER, F. and SHERMAN, I. S. *Nucl. Instrum. Meth.* 39 (1966) 35.
543. DALY, N. R. *Rev. sci. Instrum.* 31 (1960) 264.
544. AFROSIMOV, V. V. *Soviet Phys. tech. Phys.* 5 (1961) 1378.
545. DALY, N. R. *Rev. sci. Instrum.* 31 (1960) 7, 720.
546. YOUNG, J. R. *J. appl. Phys.* 26 (1955) 1302.
547. BARNETT, C. F. and RAY, J. A. Oak Ridge National Laboratory Reports 3836 (1965) and 4063 (1966).
548. TAYLOR, J. M. *Semiconductor Particle Detectors*, 1968. London; Butterworth.
549. BARNETT, C. F. Oak Ridge National Laboratory Report 3836, p. 70, 1965.
550. DELANEY, C. F. G. and WALTON, P. W. *IEEE Trans. nucl. Sci.* 13 (1966) 742.
551. BLANKENSHIP, J. L. and BORKOWSKI, C. J. Oak Ridge National Laboratory Reports 1960, 1961.
552. MCCLURE, G. W. *Phys. Rev.* 130 (1963) 1852.
553. TRILLAT, C. R. *Acad. Sci., Paris* 242 (1956) 1294; *Sci. Inds. photogr.* 27 (1956) 5; *J. Lab. Bellevue*, 43 (1958) 156.
554. WOBSCHALL, D. *Rev. sci. Instrum.* 36 (1965) 466; WOBSCHALL, D., GRAHAM, J. R. and MALONE, D. P. *Phys. Rev.* 131 (1963) 1565; *J. chem. Phys.* 42 (1965) 3955.
555. HAGSTRUM, H. D. *Phys. Rev.* 89 (1953) 244; 96 (1954) 325; 104 (1956) 672; FLAKS, I. P. *J. tech. Phys., Moscow*, 25 (1955) 2463, 2467; DUNAEV, YU. A. and FLAKS, I. P. *Dokl. Akad. Nauk. SSSR* 91 (1953) 43.
556. HAGSTRUM, H. D. *Phys. Rev.* 96 (1954) 336.
557. IZMAILOV, S. V. *Soviet Phys. JETP* 3 (1958) 2031.
558. DECKERS, J. and FENN, J. B. *Rev. sci. Instrum.* 34 (1963) 96.
559. JULIAN, R. Ph.D. Thesis, Massachusetts Institute of Technology, 1947.
560. KOLSKY, H. G., PHIPPS, T. E., RAMSEY, N. F. and SILSBEE, H. B. *Phys. Rev.* 87 (1952) 395.
561. HUGHES, A. L. and DU BRIDGE, L. A. *Photoelectrical Phenomena*, 1932. New York; McGraw-Hill.
562. GOODMAN, L. S. and WEXLER, S. *Phys. Rev.* 99 (1955) 192.
563. BELLAMY, E. H. and SMITH, K. *Phil. Mag.* 44 (1953) 33.
564. WEXLER, S. *Conf. atom. Beams*, 1954. New York; Brookhaven National Laboratories.
565. HOLLOWAY, J. H. S.B. Thesis, Massachusetts Institute of Technology, 1950.
566. FRICKE, G. and FRIEDBURG, H. *Z. Phys.* 141 (1955) 171.
567. REMLER, E. A. S.B. Thesis, Massachusetts Institute of Technology, 1955.
568. MARTON, L., SCHUBERT, D. C. and MIELCZAREK, S. R. *J. appl. Phys* (1956-7).
569. KINGDON, K. H. *Phys. Rev.* 21 (1923) 408.

570. Kusch, P. and Hughes, V. W. *Handb. Physik* 37 (1959).
571. Datz, S. and Taylor, E. H. *Recent Research in Molecular Beams*, Ed. T. Estermann, 1959. New York; Academic Press.
572. Taylor, J. B. *Z. Phys.* 57 (1929) 242; *Phys. Rev.* 35 (1930) 375.
573. Schroen, W. *Z. Phys.* 176 (1963) 237.
574. Greene, E. F. *Rev. sci. Instrum.* 32 (1961) 860.
575. Knauer, F. *Z. Phys.* 125 (1949) 279.
576. Lew, H. *Phys. Rev.* 91 (1953) 619.
577. Taylor, E. H. and Datz, S. J. *J. chem. Phys.* 23 (1955) 1711; 25 (1956) 295, 389.
578. Touw, T. R. and Trischka, J. W. *J. appl. Phys.* 34 (1963) 3635.
579. Wessel, G. and Lew, H. *Phys. Rev.* 92 (1953) 641.
580. Fricke, G. *Z. Phys.* 141 (1955) 166.
581. Quinn, W. E. *Rev. sci. Instrum.* 29 (1958) 935.
582. Weiss, R. *Rev. sci. Instrum.* 32 (1961) 397.
583. Brink, G. P. *Rev. sci. Instrum.* 37 (1966) 857.
584. Pauly, H. and Toennies, J. P. *Advanc. atom. molec. Phys.* 1 (1965) 195.
585. Kaufman, F. *Proc. Roy. Soc.* A247 (1958) 123.
586. Linnett, J. W. and Marsden, D. G. H. *Proc. Roy. Soc.* A234 (1956) 489.
587. Dingle, J. R. and le Roy, A. *J. chem. Phys.* 18 (1950) 1632.
588. Elias, L., Ogryzlo, E. A. and Schiff, H. I. *Canad. J. Chem.* 37 (1959) 1680.
589. Larkin, F. S. and Thrush, B. A. *Disc. Faraday Soc.* (1964).
590. Greaves, J. C. and Linnett, J. W. *Trans. Faraday Soc.* 55 (1959).
591. Westenberg, A. A. and de Haas, N. N. *J. chem. Phys.* 40 (1964) 3087.
592. Krongelb, S. and Strandberg, M. W. P. *J. chem. Phys.* 31 (1959) 1196.
593. Francis, P. D. *Proc. Phys. Soc.* 2 (1969).
594. Halbach, K. *Phys. Rev.* 119 (1960) 1230.
595. Magat, M. Private communication, 1969.
596. Dressler, K. *J. chem. Phys.* 30 (1959) 1621.
597. Schmeltekopf, A. L., Fehsenfeld, F. C., Gilman, G. I. and Ferguson, E. E. *Planet. Space Sci.* 15 (1967) 401.
598. Dorrestein, R. *Proc. Acad. Sci., Amst.* 41 (1938) 725.
599. Lamb, W. E. and Retherford, R. C. *Phys. Rev.* 79 (1950) 549; 81 (1951) 222.
600. Stebbings, R. F. *Proc. Roy. Soc.* A241 (1957) 270.
601. Hasted, J. B. and Mahadevan, P. *Proc. Roy. Soc.* 249 (1958) 42.
602. Levine, J. Ph.D. Thesis, New York University, 1966.
603. Lichten, W. *Phys. Rev.* 109 (1958) 1191.
604. Meissner, K. W. and Graffunder, W. *Ann. Phys., Lpz.* 84 (1927) 1009.
605. Harrison, M. F. A., Dance, D. F., Dolder, K. T. and Smith, A. C. H. *Rev. sci. Instrum.* 36 (1965) 1443.
606. Hamberger, S. M. *Proc. 5th Int. Conf. Ioniz. Phenom. Gases*, p. 1919, 1961. Amsterdam; North Holland Publishing Company.
607. Nodwell, R. A. and Robinson, A. M. *Proc. 7th Int. Conf. Ioniz. Phenom. Gases*, Belgrade, 1965.
608. Colville, C. and Allen, J. E. *Proc. 5th Int. Conf. Ioniz. Phenom. Gases*, 1961. Amsterdam; North Holland Publishing Company.
609. Ladenburg, R. and Levy, S. *Z. Phys.* 65 (1930) 189.
610. Foner, S. S. and Hudson, R. D. *J. chem. Phys.* 21 (1953) 1374; Faust, W. L. and McDermott, M. N. *Phys. Rev.* 123 (1963) 198.
611. Fite, W. L. *Disc. Faraday Soc.* (1964) 215.
612. Riviere, A. C. *Methods of Experimental Physics, Vol. 7*, Ed. L. Marton, p. 208, 1968. New York; Academic Press.
613. Price, M. H. and Good, R. H. *J. opt. Soc. Amer.* 52 (1962) 239.
614. Bailey, D. S., Hiskes, J. R. and Riviere, A. C. *Nucl. Fusion* 5 (1965) 41
615. Lanczos, G. *Z. Phys.* 68 (1931) 204.
616. Wittkower, A. B. and Gilbody, H. B. *Proc. phys. Soc.* 90 (1967) 353.
617. Hagstrum, H. D. *J. appl. Phys.* 31 (1960) 897.

618. BEYNON, J. H., SAUNDERS, R. A. and WILLIAMS, A. E. *Nature, Lond.* 204 (1964) 67; *Z. Naturf.* 20a (1965) 180.

619. BARBER, M. W. and ELLIOT, R. M. *12th Amer. Soc. Testing. Mater. Conf.* Montreal, 1964.

620. JENNINGS, K. R. *J. chem. Phys.* 43 (1965) 4176.

621. FUTRELL, J. H., RYAN, K. and SIECK, L. W. *J. chem. Phys.* 43 (1965) 1832.

622. BARKER, M., JENNINGS, K. R. and RHODES, R. *Z. Naturf.* 22 (1967) 15.

623. DALY, N. R., McCORMICK, A. and POWELL, R. E. Atomic Weapons Research Establishment Report, 1967; *Rev. sci. Instrum.* 39 (1968) 1163.

624. DUNNING, F. B. and SMITH, A. C. H. University College London, Private Communication, 1970.

Chapter 4

THE ELASTIC SCATTERING OF ELECTRONS BY ATOMS AND MOLECULES

4.1 TOTAL AND DIFFERENTIAL CROSS-SECTIONS

The elastic scattering of electrons by atoms is of importance not only in plasmas, but in situations where only electrons and gas are present, so it is possible for the electrons to escape from the relatively sluggish ion motion that dominates ambipolar diffusion. In atomic gases, the electron velocity distribution is strongly influenced by the momentum loss experienced in elastic collisions; but inelastic processes can dominate in molecular gases.

The passage of electron beams through gases is strongly influenced by elastic collisions. Thus there are good reasons for studying these collisions even apart from the tests they provide of quantum theory; this has progressed to the stage where the exact solution of the three-body problem (e–H) is available for comparison with experiment. Although the correspondence is satisfactory, the four-body problem (e–He) is not yet completely understood.

Electrons of energy much greater than that of outer atomic electrons are scattered in collision with atoms through very small angles or through angles close to 180°. Slow electrons are scattered more isotropically (centre-of-mass and laboratory angles are almost identical). Intuitive classical arguments lead to the supposition that, as the electron impact velocity v decreases, the total scattering cross-section should increase as v^{-1}! An infinite cross-section at zero energy is not realistic, but, if the atom were to possess zero electron affinity, a very large cross-section might be observed. It is clear that the total scattering cross-section defined in terms of finite lateral deflexion Δx from pre-collision path would have no limit in classical mechanics except the limit of resolution of the apparatus used for its measurement. In quantum theory, however, the intrinsic uncertainty of the Heisenberg principle

$$\Delta x \, \Delta v_e \simeq \frac{h}{m_e} \tag{4.1}$$

limits the resolution which is significant in theory, placing a finite limit upon the cross-section.

One can calculate[1] the resolution that is necessary for the experimental measurement of total collision cross-sections. For 1 per cent accuracy it is as follows: 11° for 1 eV electrons, 2·3° for 100 eV electrons, and 0·2° for

10^4 eV electrons. These angles reduce proportionally to the mass of the projectile; thus total proton collision cross-sections can only be determined at energies of a few electron volts.

The Ramsauer–Townsend Effect

In the quantum theory of elastic scattering of electrons, the incident electron travelling in the z direction is taken to be a planar de Broglie wave of the form $C \exp (ikz)$. The constant C defines the amplitude of the wave, and is related to the flux density A (number of particles passing in unit time through unit area of the xy-plane) by the equation

$$C = \left(\frac{A}{v_e}\right)^{1/2} \tag{4.2}$$

where v_e is the electron velocity.

The flux density is related to the wave function ψ by the equation

$$A = \frac{h}{2m_e}(\psi \operatorname{grad} \psi^* - \psi^* \operatorname{grad} \psi) \tag{4.3}$$

where m_e is electron mass. The momentum

$$k = \frac{2\pi}{\lambda_e} = \frac{(2m_e E_e)^{1/2}}{\hbar} = \frac{2\pi m_e v_e}{h} \tag{4.4}$$

where λ_e represents wavelength and E_e electron energy.

The electron scattered by an atom is taken to be an outgoing spherical wave of the form $Cr^{-1}f(\theta) \exp (ikr)$, where r is the radius variable and θ is the polar scattering angle. The differential cross-section

$$\sigma(\theta) = |f(\theta)|^2 \tag{4.5}$$

For an atomic force field $V(r)$, the wave equation is

$$\nabla^2\psi + \left[k^2 - \frac{8\pi^2 m_e V(r)}{h^2}\right]\psi = 0 \tag{4.6}$$

The wave function ψ is expanded as Legendre polynomials

$$\psi = r^{-1}\sum_{l=0}^{\infty}\phi_l P_l(\cos\theta) \tag{4.7}$$

where the azimuthal quantum number $l = 0$ for s wave scattering, 1 for p wave, 2 for d wave, and so on. The solution of equation 4.6 must have the asymptotic form $\exp (ikz) + \exp (ikr) f(\theta)/r$.

The total scattering cross-section σ_l is obtained from a solution of the wave equation in terms of the phase shift η_l experienced by the wave of specified l:

$$\sigma_l = \frac{4\pi}{k^2}(2l+1)\sin^2\eta_l \tag{4.8}$$

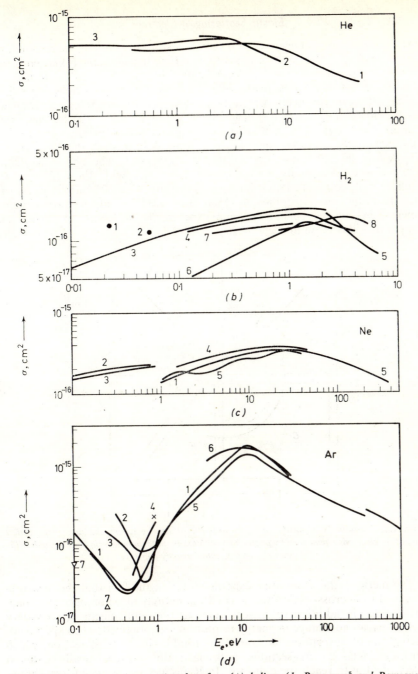

Fig. 4.1. Total electron cross-section data for: (a) helium (1, Ramsauer[5] and Ramsauer and Kollath[6,7]; 2, Normand[8]; 3, Gould and Brown[9]); (b) molecular hydrogen (1, Phelps, Fundingsland and Brown[10]; 2, Varnerin[11]; 3, Bekefi and Brown[12]; 4, Crompton and Sutton[13]; 5, Brode[14]; 6, Rusch[15]; 7, Ramsauer and Kollath[7]; 8, Bruche[16]); (c) neon (1, Bruche[17]; 2, Bowe[18]; 3, Barbiere[19]; 4, Ramsauer and Kollath[6]; 5, Normand[8]); (d) argon (1, Ramsauer[5] and Ramsauer and Kollath[6]; 2, Normand[8]; 3, Rusch[20]; 4, Drawin[21]; 5, Brode[14]; 6, Bullard and Massey[22]; 7, Phelps, Fundingsland and Brown[10])

At the lowest impact energies, it is only necessary to consider the zero order phase shift, with $l = 0$. As k tends to zero (low impact velocity) η_0 tends to $s\pi$, where s is an integer. For weak scattering fields such as are found in helium, $s = 0$ and $\sin \eta_0$ tends to zero at the same rate as does k, so the total collision cross-section remains finite at low impact velocities (*Figure 4.1*). However, for stronger fields such as are found in the heavier rare gases, $s \geqslant 1$ and the impact energy variation of η_l can lead to low velocity minima in the cross-section functions. This is known as the Ramsauer–Townsend effect[2-4] and is illustrated in *Figure 4.1*, where total electron scattering cross-sections for argon are contrasted with those for neon, helium and hydrogen. Its explanation constituted the first major success of the quantum theory of collisions. It was discovered not only in swarm experiments, but in the beam scattering experiments of Ramsauer, using an apparatus illustrated in *Figure 4.2*.

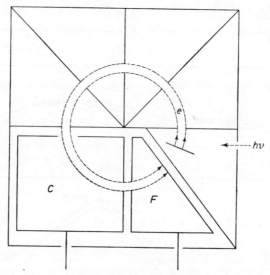

Fig. 4.2. Ramsauer's apparatus for measurement of total electron collision cross-section in gases: electrons e produced by photoemission, pass in uniform magnetic field through collision chamber C into Faraday cup F

The merit of the Ramsauer experiment lay in the fact that the measurement of total cross-section was carried out entirely in a homogeneous magnetic field; in addition to such features as the confinement of secondary electrons, an advantage in energy resolution was obtained. The experiment, being one of absorption, measured the total collision cross-section, which includes inelastic cross-sections; both elastically and inelastically scattered electrons fail to reach the collector. In Ramsauer's experiment, this failure was aided by the magnetic field, which reduces the path curvature of inelastically scattered electrons. The elastically scattered electrons, on the other hand, fail to reach the collector on account of their deflexion by scattering. The most satisfactory modern alternative to the Ramsauer experiment is the crossed-beam technique.

The differential cross-section for elastic scattering can be expressed as a sum of partial waves:

$$\frac{d\sigma(\theta)}{d\Omega} = \left| \frac{1}{2ik} \sum_{l=0}^{\infty} [\exp(2i\eta_l) - 1](2l+1)P_l \cos\theta \right|^2 \qquad (4.9)$$

This result was first obtained by Faxen and Holtsmark[23] and is essentially the same as that introduced by Rayleigh[24] to the problem of scattering of sound by a spherical obstacle. The differential scattering is characterized by the Legendre polynomials $P_l \cos\theta$ whose character is discussed in Section 4.2. The relative importance of the $s, p, d, \ldots$ scattered waves may be estimated by analysis of the differential cross-section.

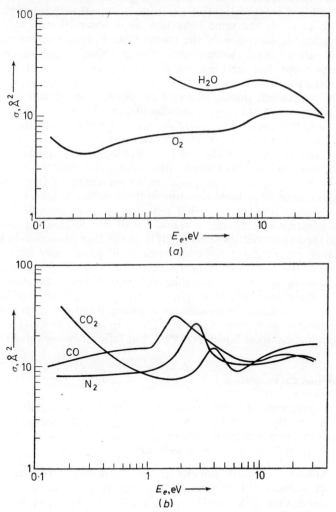

Fig. 4.3. Total electron cross-section data for: (a) O_2 (Ramsauer and Kollath[25]) and H_2O (Bruche[26]); (b) N_2 (Bruche[27]), CO (Ramsauer and Kollath[25]) and CO_2 (Ramsauer and Kollath[25])

A feature of this scattering formula is that it leads to non-isotropic scattering of electrons by impenetrable spheres of radius a. For scattering angles $\theta_c > \pi/ka$, the scattering is isotropic and is close to the classical limit

$$\sigma(\theta_c) = \tfrac{1}{2}a^2$$

but for $\theta_c \ll \pi/ka$, $\sigma(\theta_c)$ rises linearly with decreasing θ_c, to a value $\tfrac{1}{4}k^2a^4$. The total cross-section in quantum theory (in the limit $k \to 0$) is twice the classical total cross-section $2\pi a^2$ (in the limit $k \to \infty$). Since interatomic forces, with their strongly repulsive short-range and weakly attractive long-range components, may very crudely be approximated by an impenetrable sphere, this cross-section is not completely unrealistic.

Total collision cross-sections of electrons in impact with other atoms do not shown precisely the same behaviour as is observed with rare gases; however with the exception of the atoms from hydrogen to neon, whose force fields are too weak, some complexities of collision cross-section function always occur. For methane, the small cross-section at ~ 1 eV impact energy, rising to a maximum around 7 eV, has been attributed to Ramsauer effect; the molecule is similar to a rare gas atom. The characteristic cross-section functions persist in higher paraffin hydrocarbons.

Some molecular cross-sections are shown in *Figure 4.3;* a distinction may be made between those (H_2O, H_2 and O_2) in which there is no marked structure, and those (N_2, CO_2 and CO) in which there appears a bump. This structure is usually associated with a compound negative ion state, or 'resonance' of the molecule, and high-resolution scattering functions show oscillations due to vibrational structure in these resonant states (see Section 4.2 and *Figure 4.10*). However, the potential or non-resonant scattering may also contribute structure[130, 131], and it is of interest that the resonances in H_2 and O_2 do not contribute markedly to the total cross-section functions.

Ramsauer experiments are being repeated in present-day laboratories; for example, the experiments of Golden and Bandel[28] on helium yield cross-sections in agreement with effective range theory calculations[29].

Time of flight momentum-analysed electrons[30] have been used to study rare-gas total scattering cross-sections showing Ramsauer characteristics. The minimum cross-section in argon is found to be at 0·25 eV. The helium cross-section is constant down to 0·4 eV, but falls off below this.[129]

Total Collision Cross-sections

Beam measurements of total cross-section have been successful down to a lower energy limit of 0·3 eV, and electron spectrometers investigating resonances (Section 4.2) have been successful down to 0·1 eV. At thermal energies there are four alternative techniques:

1. Measurements of drift velocity and diffusion of electrons in gases at higher pressure (0·1–10 torr); the experiments yield momentum transfer rather than total cross-section, and are discussed in Section 4.4.
2. Electron collision frequency measurements by means of plasma conductivity also yield momentum transfer cross-sections at thermal energy, and are discussed in more detail in Section 4.5.

3. A further collision frequency technique, as yet unexploited[31], rests on the collisional de-phasing of electrons which have been set in oscillation in a plasma contained in a static magnetic field by a pulse of electromagnetic radiation at the cyclotron frequency.

4. Pressure-broadening of emission lines of very high principal quantum number states is a technique capable of yielding total collision cross-sections well below thermal energy (Chapter 13).

Total collision cross-sections for electrons in atomic hydrogen have been measured by Brackman, Fite and Neynaber[32, 33] using the crossed-beam technique. An atomic hydrogen beam emerges from an orifice in a tungsten furnace, and is mechanically chopped at a low frequency before being crossed with an electron beam. The absorption of the electron beam is observed as a low-frequency electron current whose phase is determined by the beam chopper; the H_2/H fraction is monitored by electron impact ionization and sector mass-spectrometer. The crossed-beam technique greatly reduces the noise from background gas molecules in the collision region (a factor of 10^4 in signal–noise ratio).

The importance of e–H total scattering lies in the test that it provides of more exact quantum theory of collisions than has hitherto been possible. A review of the quantum theory of the elastic scattering of electrons has been given by Moiseiwitsch[34], and the general problem for many-electron atoms has been formulated[35]. For atomic hydrogen, there is a good agreement between experiments and calculations[36–40] (*Figure 4.4a*), and the agreement with theory has been reinforced by the angular scattering data[41] (*Figure 4.4c*). Contributions to the scattering from the higher order waves $l = 1$, $2, \ldots$, although small, must be taken into account, as in the calculations of Temkin and Lamkin[37], using the method of polarized orbitals. Since the higher order scattering takes place through much smaller angles, it does not contribute strongly to the observed total collision cross-sections. Brackmann, Fite and Neynaber[32] consider that, for their experimental cone of observation, a good approximation is

$$\sum_l \sigma_l = 6{\cdot}85\sigma_{exp} + 0{\cdot}60\sigma_1 \qquad (4.10)$$

Treatment of e–H scattering has also been given by the optical model familiar in nuclear collision theory[47].

Calculations have been reported[48] for the total cross-section for lithium, but experimental comparison has not yet been made. Total collision cross-section for electrons scattered by atomic oxygen have been measured by crossed-beam technique[33], as well as in shock tubes[43]. Complete dissociation of the oxygen emerging from a radiofrequency source is not achieved; the fraction O_2/O is measured by electron impact ionization and mass-analysis. The electron beam, crossed with the chopped molecular-plus-atomic beam, is attenuated by scattering out of the collision region. The cross-sections are obtained relative to those for molecular oxygen, being converted into absolute values by means of earlier O_2 data[49]. The atomic oxygen cross-section function is shown in *Figure 4.4b*. Calculations due to Smith, Henry and Burke[44] are in good agreement except near threshold; this partly results from neglect of polarization[50], and partly from the low energy cross-section

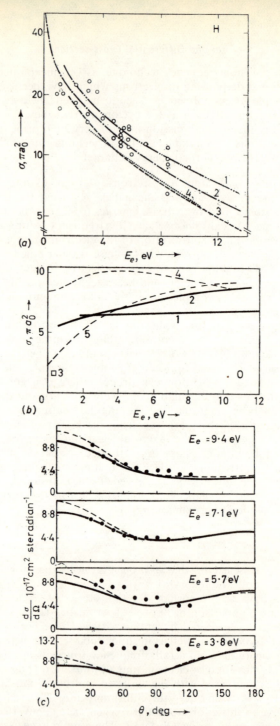

Fig. 4.4. Total electron cross-section data for: (a) H (1, Bransden et al.[36], calculations; 2, Temkin and Lamkin[37], calculations; 3, McEachran and Fraser[38], Smith[39] and Geltman[40], calculations; 4, Neynaber et al.[33], experiments; open circles represent the experiments of Brackmann, Fite and Neynaber[32] (From Neynaber et al.[33], by courtesy of W. A. Benjamin); (b) O (1, Neynaber et al.[33], experiments; 2, Sunshine, Aubrey and Bederson[42], experiments; 3, Lin and Kivel[43], experiments; 4, Smith, Henry and Burke[44], calculations; 5, Bauer and Browne[45], calculations); (c) differential scattering data points for e–H[41] compared with calculations[46] (solid lines include coupling 1S–2S; broken lines include coupling 1S–2S–2P)

being rather higher near threshold, as was reported by Lin and Kivel[43]. Atomic beam recoil technique has also been used for atomic oxygen total cross-section measurement[51].

Scattering by Polar Molecules

Altschuler[52] has applied the Born approximation to the scattering of electrons by a fixed point dipole, and obtained the relation for momentum transfer cross-section

$$\sigma_d = \frac{8\pi}{3k^2}\left(\frac{Dem_e}{h^2}\right)^2 = \frac{4\cdot928D^2}{E_e} \qquad (4.11)$$

with σ_d in Å², and D the dipolemoment in Debyes (10^{-18} e.s.u. cm). A graphical comparison of data[53] with this relation is given in *Figure 4.5*. Subsequently, an exact solution of the scattering problem was given[54] and was

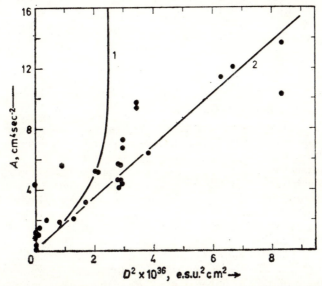

Fig. 4.5. Momentum transfer cross-section coefficient $A = \sigma_d v^2$ related to dipole moment (D) squared for various gases, determined by electron swarm experiments: 1, calculations of Mittleman and van Holdt[54]; 2, calculations of Altshuler[52]. For key to individual points, see Crawford, Dalgarno and Hays[53]

criticized[53]; when account is taken of the long-range interaction between electron and quadrupole moment q, a corrected expression can be written:

$$\sigma_d = \frac{8\pi}{3k^2}\left(\frac{Dem_e}{h^2}\right)^2\left(1+\frac{4kq}{5D}\right)^2 \qquad (4.12)$$

To this must be added the contribution from the spherically symmetric part of the potential, which in the low-velocity limit is

$$\sigma_d = 4\pi\left[a^2+\left(\frac{4\pi}{5a_0}\right)\alpha ka+\left(\frac{8}{3a_0}\right)\alpha k^2a^2\ln(ka_0)+\ \cdots\right] \qquad (4.13)$$

where a is the scattering length, and α the polarizability. These corrections may explain the apparent absence of resonances in such molecules as H_2O and H_2S, whose dipole moments lie above the critical value for which resonant capture of an electron by a polar molecule is supposed always to be possible[54].

Differential Cross-sections

The angular distribution of electrons elastically scattered by atoms, expected classically to be isotropic, was first shown to deviate from such behaviour in experiments of Ramsauer and Kollath[2], and also of Bullard and Massey[55]. The distributions exhibit minima which occur when a finite number of electron wavelengths fit into the spherical potential well which is taken for the atomic field. For a potential well of radius a and depth D, the minima occur at two scattering angles θ, given in radians by:

$$\theta = 4 \cdot 5 \left(\frac{8\pi^2 m_e D}{h^2} \right)^{1/2} a \qquad (4.14)$$

and

$$\theta = 7 \cdot 6 \left(\frac{8\pi^2 m_e D}{h^2} \right)^{1/2} a \qquad (4.15)$$

More refined types of calculation have been applied to this problem. It is found that $d\sigma/d\Omega$ does not increase indefinitely with decreasing θ, but passes through a maximum. The resulting finite total collision cross-section is an important result of quantum theory.

The analysis of angular scattering data in terms of equation 4.9 is a problem which was solved as long ago as 1946 by analogue computer[56], but digital regression analysis programmes are now available[57]. Fitted data are shown in *Figures 4.6* and *4.7*.

The existence of special problems in the experimental study of forward and backward scattering has long been recognized; 180° scattering experiments were first carried out by Gagge[58, 59].

In measurements of forward scattering, it is difficult to define the scattering angle when a collision chamber is used, since the exit slit must be made larger than the entrance orifice, and one is forced to integrate over a collision path length which is not accurately defined. This disadvantage is avoided by the use of crossed-beam technique[41].

Differential measurements should be carried out by means of post-collision energy analysis of the electrons; in this way, elastic and inelastic scattering are separated. Experiments of this type are described in Chapter 5, and mention must be made here of the 'fountain spectrometer' suitable for such measurements[60].

It is now understood that the Born approximation is inapplicable to forward scattering through small angles; departures from experiment have been found, and a 'dispersion relation analysis' has been applied[61].

Since the angular distribution is not isotropic at energies of a few electron volts, momentum transfer cross-sections σ_d differ from total collision cross-sections σ_t. However, electrons of the very lowest energies are scattered

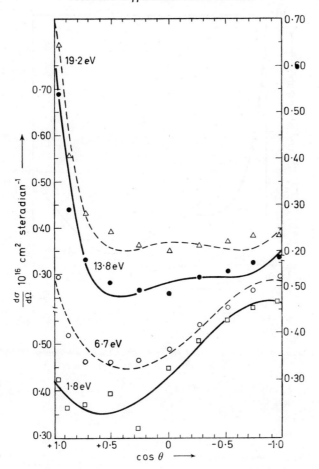

Fig. 4.6. Comparison of Ramsauer and Kollath differential scattering data for helium with phase shifts (degrees) fitted by least squares[57]

E_e (eV)	δ_0	δ_1	δ_2	δ_3
1·8	*332·5*	*181·6*	*−0·32*	*−0·30*
6·7	*303·4*	*182·8*	*−2·01*	*−1·16*
13·8	*294·3*	*197·0*	*+5·60*	*+1·80*
19·2	*289·1*	*197·3*	*+6·62*	*+2·48*

Left-hand scale refers to full lines; right-hand scale refers to broken lines

isotropically, and in this case $\sigma_t = \sigma_d$. At near-thermal energies there is a large volume of experimental data for electron swarms drifting through gases under the action of a uniform electric field; from this data is obtained the momentum transfer cross-section σ_d. There is also an increasing amount of microwave data from which collision frequencies v_c are deduced from the conductivity variation of the afterglow of a gas discharge (Section 4.4).

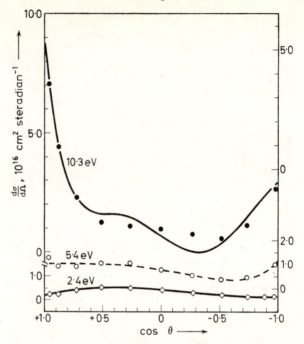

Fig. 4.7. *Comparison of Ramsauer and Hollath differential scattering data for argon with phase shifts* (degrees) *fitted by least squares*[57]

E_e(eV)	δ_0	δ_1	δ_2	δ_3
2·4	333·3	177·6	1·75	0·80
5·4	294·3	165·7	7·08	−0·83
10·3	276·1	141·4	37·51	11·26

Left-hand scale refers to full lines; right-hand scale refers to broken lines

4.2 RESONANCES IN ELASTIC SCATTERING

It has been known since the calculations of Burke and Schey[62] and the experiments of Schulz[63] that there exist compound states of the electron and atom or electron and molecule. At the energy of such states, the elastic scattering function shows structure, and there is also structure in inelastic scattering functions, which are discussed in Chapter 5. Atomic resonances have been reviewed[64, 65] and also resonances in molecules[66].

At energies below that of an excited atomic level, one or more bound states of the electron may be found. These are variously known as 'closed channel', 'core-excited', 'potential' or 'Feshbach' resonances, similar to those found in nuclear physics. The states can decay into the continua formed by lower states, including the ground state, and the free electron; lifetimes are between 10^{-12} sec and 10^{-14} sec. It is as though the excited state of the atom possessed a real electron affinity. The channel of decay to the excited state is 'closed', being energetically impossible.

Another sort of resonance is found, possessing an energy actually higher than that of the state of the atom. Owing to the centrifugal repulsive term $h^2 l(l+1)/2m_e r^2$, the dynamic interaction energy between an electron and an atom in a ground or excited state can be repulsive at large separations; it will often pass through a maximum and become attractive at closer distances. The barrier can support 'open-channel' resonances, so-called since they can decay to the original state plus electron; this property affects the width of the resonance, since the lifetime is decreased by decay tunnelling through the maximum in the interaction energy. The wider the tunnel, the narrower the resonance and longer the lifetime. These resonances are also known as 'coupled-channel' resonances, since it is the coupling to higher excited states that produces the repulsive interaction energy which sustains the barrier, and therefore the resonance; they are often called 'shape resonances', since the shape of the dynamic interaction energy gives rise to them. Excited states, as well as the ground state, of the atom can support shape resonances, and several are known in helium.

It is possible for the energy of the resonance to correspond exactly with the energy of the state of the atom from which it is formed. This rare condition, intermediate between the closed-channel and the shape resonance, is known as a 'virtual state'.

As the impact energy of the electron–atom collision passes through the resonance energy, the phase of the scattered wave undergoes a sudden discontinuity which can often be close to π. The cross-section function at resonance exhibits the Beutler–Fano[67] shape which is common to all states which are in a continuum. This can be understood in terms of configuration interaction. It is only rarely possible for two different wave functions to exist at the same energy; therefore the continuum states are repelled or perturbed in energy, and at the resonance energy there is a 'hole in the continuum', which gives rise to the 'anomalous line profile' which can be expressed by the equation

$$\sigma = \sigma_a + \sigma_b \frac{(q+\varepsilon)^2}{1+\varepsilon^2} \qquad (4.16)$$

where

$$\varepsilon = \frac{E_e - E_r}{\frac{1}{2}\Gamma} \qquad (4.17)$$

for a resonance at energy E_r of line width Γ, and line profile parameter q (see below). The total electron scattering cross-section σ at energy E_e is made up of background cross-section σ_a and contribution σ_b arising from excitation of the resonance state. The form of this cross-section function is shown in *Figures 4.8a* and *4.8b*: the computed values in *Figure 4.8a*, and in *Figure 4.8b* the historic helium scattering measurement in the region of the 19·3 eV helium plus electron compound state $1s^2\,2s\,^2S_{1/2}$.

The function in *Figure 4.8b* is measured 'in transmission', that is by measuring the electron current transmitted through the gas confined in a chamber with narrow entrance and exit orifices. Under single collision conditions, the transmitted current is $I_0(1-\pi\sigma)$, where π is the target parameter and I_0 is the electron current entering the chamber. It is here assumed that no scattered electrons can pass through the exit orifice. At higher

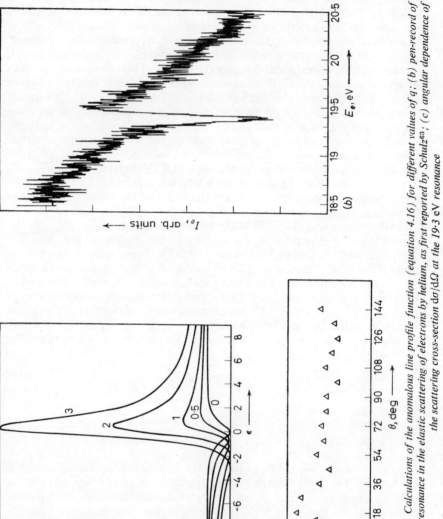

Fig. 4.8. (a) Calculations of the anomalous line profile function (equation 4.16) for different values of q; (b) pen-record of the 19.3 eV resonance in the elastic scattering of electrons by helium, as first reported by Schulz[63]; (c) angular dependence of the scattering cross-section $d\sigma/d\Omega$ at the 19.3 eV resonance

pressures, the current is $I_0 \exp(-\pi\sigma)$, and the structure is enhanced in the transmission function. A higher pressure experiment is therefore particularly suitable in searching for very weak (that is, narrow) resonances.

Figure 4.8c shows angular scattering data for the 19·3 eV $1s\,^2 2s^2\,S_{1/2}$ helium resonance. The angular dependence enables identification of the azimuthal quantum number L' of the resonance state to be made. Suppose that the state decays into the elastic channel, yielding an electron scattered wave of wavenumber k, angular momentum quantum number l, spin s:

$$A^{-*}(n', L', s') \rightarrow A(n, L, s) + e\left(k, l, s = \tfrac{1}{2}\right)$$

The azimuthal quantum number L of the atom is known, so $L' = L \pm l$ can be deduced from the angular momentum of the scattered electron wave. The angular dependence of the scattered electron is as follows:

$l = 0$, s wave, isotropic in θ;
$l = 1$, p wave, $P_1(\cos\theta)$ maximum at $0°$;
$l = 2$, d wave, $P_2(\cos\theta)$ minimum at $54°$ and maxima at $0°$ and $90°$.

The intensities of the scattered electron wave are as shown in Table 4.1.

Table 4.1. INTENSITIES
OF SCATTERED ELECTRONS

Scattering angle	P_0	P_1	P_2
$0°$	1	1	1
$54°$	1	$< \tfrac{1}{2}$	0
$90°$	1	0	$\tfrac{1}{2}$

For the helium resonance (*Figure 4.8c*), the angular dependence of the area under the resonance curve is almost isotropic, consistent with the identification of the resonance as $^2S_{1/2}$.

Partial wave analysis leads to the following expression (equation 4.9) for the differential elastic scattering cross-section:

$$\frac{d\sigma}{d\Omega}(\theta) = \frac{1}{4k^2}\left| \sum_l (2l+1)\left[\exp(2i\delta_l) - 1\right] P_l(\cos\theta) \right|^2 \qquad (4.18)$$

where P_l is the Legendre polynomial of order l. The cross-section is proportional to $\sin^2 \delta_l$, with

$$\delta_l = \delta_{lp} + \delta_{lr}$$

$$\delta_{lr} = -\operatorname{arc\,cot} \frac{E_e - E_r}{\tfrac{1}{2}\Gamma}$$

and where δ_{lp} represents the scattering phase shift close to the resonance, due not to the resonance but to the interaction potential. The line profile parameter

$$q = -\cot \delta_{lp}$$

As has already been mentioned, the angular dependence of the magnitude of the resonance peak is governed by the proportions of $l = 0, 1, 2, \ldots$ scattered waves. Only the lowest few values of l need be considered, since for collision length R it can be shown that $l_{max} \sim kR$. Thus $l_{max} \sim 3$ for the helium 19·3 eV resonance.

The asymmetry of the resonance is dependent on angle, since δ_{lp} can possess different values for different l. The angular dependence of the helium 19·3 eV resonance is shown in *Figure 4.8c*. The relative intensity of resonant and non-resonant scattering is invariant over angle, but the 'form' of the resonance (line profile) changes with angle. The resonance is very largely in the *s*-wave. Values $\delta_0 = 109°$, $\delta_1 = 22°$ and $\delta_2 = 2°$ are consistent with the angular dependence.

The line profile q for the helium 19·3 eV resonance is actually much less than unity. Such resonances can be described by a more approximate expression

$$\sigma \propto (E_e - E_r + \tfrac{1}{2}i\Gamma)^{-2}$$

due to Breit and Wigner; they are sometimes known as 'window resonances'. Another example is at 11·48 eV in N_2.

An instructive example of the interpretation of differential scattering data is to be found in the case of atomic hydrogen[68]. Following an earlier low-resolution experiment[69], both transmission measurements[70] and angular-dependence measurements[71] have been made. Calculations are made of the scattering at different angles predicted for ^1S ($q = 1$) and ^3P (Breit–Wigner), and the instrumental resolution is convoluted. A number of different line widths are assumed, and after convolution the calculations are compared with experiment (*Figure 4.9a*).

A tabulation of helium resonances observed experimentally in different channels of decay is given in *Figure 4.9b*. Not all of these resonances have been observed in the elastic channel; an open-channel resonance can decay into an excited state of the atom (plus a free electron), so the structure often appears strongly in the excitation function (Chapter 5); in some cases not at all in the elastic scattering function. Auto-ionizing states (Chapter 6) are similar in lifetime and in line shape to compound states. Although not included in *Figure 4.9b*, He^- compound states are reported[72] as high as 57·1 eV and 58·2 eV; all three electrons are in excited orbitals.

Molecules are able to support compound states, but transitions to them are complicated by variation of line width with internuclear distance, and by the Franck–Condon principle.

Suppose that the resonance state of a diatomic molecule possesses energy ε and vibrational frequency ω_- and quantum number λ. There is interference between the waves scattered from each vibrational level of the resonance state. The elastic cross-section function is rather complicated, being

$$\sigma = \frac{4\pi}{k^2} \sum_l \sin^2 \delta_{lnr} \left| 1 - \cot (\delta_r + i) \sum_\lambda \frac{\langle \phi(\lambda) | \phi(v) \rangle^2 \tfrac{1}{2}\Gamma}{E_e - \varepsilon - \lambda h \omega_- c + \tfrac{1}{2}i\Gamma} \right|^2 \quad (4.19)$$

where the bracketed term is the Franck–Condon overlap. The non-resonant phase shifts δ_{lnr} may vary considerably over the energy range to which the vibrational levels of the resonance contribute; this makes the Fano line

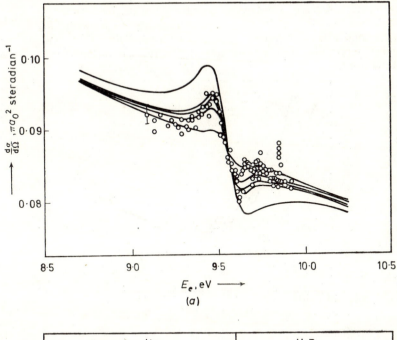

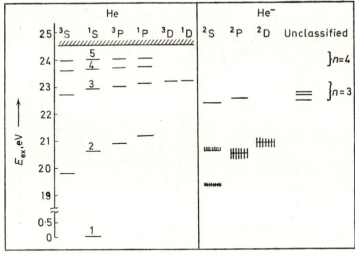

Fig. 4.9. (a) Comparison[71] of experimental e–H elastic scattering data points ($\theta = 90°$, $\Delta E_{1/2} = 0.06$ eV (with calculation for E_r (^{1}S) = 9.56 eV, Γ(^{1}S) = 0.01 eV, 0.03 eV, 0.04 eV, 0.08 eV, 0.11 eV (top to bottom), and for E_r(^{3}P) = 9.73 eV, Γ = 0.01 eV; (b) term diagram of He and He$^-$ levels

profile parameter interpretation unrealistic for molecular resonances. The above equation is probably a fairly satisfactory representation of molecular resonance scattering. The typical appearance of the resonance is a series of regular oscillations on the cross-section function, as shown in *Figure 4.10*. According to the compound state theory the variation of line width Γ with nuclear separation is proportional to the energy separation of the adiabatic interactions.

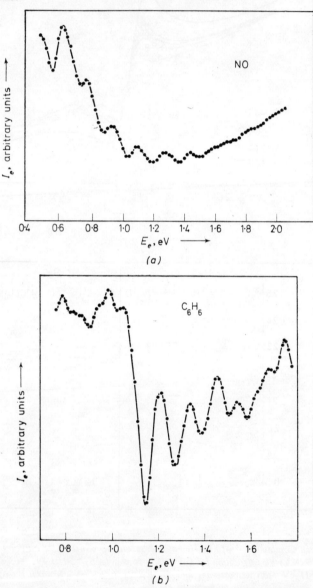

Fig. 4.10. *Resonance structure in the total elastic scattering of electrons from: (a)* NO *(Hasted and Awan*[74]*); (b)* C_6H_6 *(Boness and Hasted*[75]*)*

There exist H_2^- $^2\Sigma_g^+$ resonances around 11·5 eV and a further hydrogen resonance[73] around 14 eV. The first molecular resonance to be discovered, N_2^- $^2\Pi_g$, lies at 1·90 eV, and there are more than a dozen others known.

4.3 ELECTRON SWARM EXPERIMENTS[4, 76–84]

The drift of electrons in a gas under the action of a uniform electric field can be analysed to yield the momentum transfer cross-section function, together with information about the mean energy loss per collision. In molecular gases, a more detailed analysis in terms of the electron excitation functions is possible.

Two experimental quantities are measured: the drift velocity v_d of electrons in the direction of the field X, in a gas at pressure p or number density n; and the electron diffusion coefficient D_e. A bibliography of measurements is available[85].

Measurement of Drift Velocity[83, 86–89]

The drift velocity is usually measured with a pair of electrical 'shutter-grids'. The application of alternating electric potential differences to alternate wires of a specially constructed shutter-grid has the effect of sweeping the electrons to these wires, so that they can only pass through the grid during those periods of time when the potential differences are substantially zero. The variation of the frequency of the alternating potential, together with a study of the dependence of electron swarm current transmitted by the pair of shutter grids upon this frequency, enables the drift velocity to be deduced.

An innovation introduced[84] to minimize the disturbance of the drift field in experiments at low X/p consists of operating the shutters in a transmitting condition, with zero bias across the grid wires. The diminution of transmitted current during short periods of bias application serves to measure the drift velocity.

There also exists a method for measuring v_d based upon the study of the lateral deflexion of a drifting electron swarm in a steady magnetic field. In its modern form[88], the method makes use of a symmetrically split collecting electrode at which unbalanced currents are measured on the application of the magnetic field.

In most swarm experiments, the electrons are produced by photoelectric emission from a continuously illuminated metal plate.

Measurement of Electron Diffusion Coefficient

The ratio v_d/D_e is retermined by a lateral diffusion experiment, in a drift tube of design similar to that shown in *Figure 4.11*. The electron swarm enters through an orifice O the region of uniform field X, defined by the guard rings G, and diffuses to collecting electrodes C_1 and C_2 in cylindrical symmetry. The currents I_1 and I_2 to these electrodes are given by the

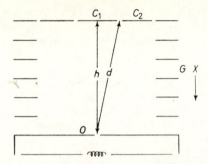

Fig. 4.11. Schematic diagram of cylindrical drift tube, with vertical axis, for the determination of divergence of electron swarm from orifice O, in electric field X

equation

$$\frac{I_1}{I_1+I_2} = 1 - \frac{h}{d} \exp\left[-\frac{v_d h}{2D_e}\left(\frac{d}{h}-1\right)\right] \qquad (4.20)$$

with h and d defined in *Figure 4.11*.

Modern drift tubes are designed in such a way that the drift velocities can be studied in the smallest possible electric fields. In this way, the closest approach to laboratory temperature electron–atom collisions can be made. Two important problems must be faced: the fields arising from surface charges on insulators, and those arising from contact potential variation over a metal surface. The former are avoided by constructing the guard rings in the form of cylinders, with minimal spacing between them. The distortion of the axial electric field can be shown to be negligible provided that the ring diameters are sufficiently large. The contact potentials on metal surfaces can be investigated by a surface probe attached to an electrometer. By evaporating metal onto a surface, the contact potential variation can be kept as low as 8 mV over a disc 8 cm in diameter. It is the electron-collecting surface over which the variations of potential are most harmful.

Simplified Electron Drift Equations

When the electron mean free path l_f is independent of its energy ('constant mean free path') it may be represented as

$$l_f = \frac{1}{n\sigma_d} \qquad (4.21)$$

In a drift of distance x, the actual path traversed is $x\bar{v}_e/v_d$, where $\bar{v}_e$ is the root mean square electron random velocity, and v_d is the drift velocity. If the mean fractional energy loss per collision is $\lambda\bar{E}_e$, where $\bar{E}_e = \frac{1}{2}m\bar{v}_e^2$, then the energy gained Xex is equal to the energy lost in traversing the actual path:

$$Xex = \frac{\lambda\bar{E}_e}{l_f}\frac{\bar{v}_e x}{v_d} \qquad (4.22)$$

so that

$$\frac{\lambda \bar{v}_e^3}{v_d} = 2 \frac{Xel_f}{m_e} \qquad (4.23)$$

In a rigid treatment, the right-hand side would be multiplied by a factor 0·88. In atomic gases of mass m_g, the fractional energy loss per collision is $(m_e/m_g)(1-\cos\theta)$ or, averaged isotropically over angle, m_e/m_g. In molecules, the energy loss is dominated by rotational and vibrational excitation; λ is often measured in units of m_e/m_g for convenience.

When the interval between collisions can be written as

$$\delta t = \frac{l_f}{\bar{v}_e} \qquad (4.24)$$

independently of electron energy ('constant mean free time'), then the mean distance traversed in the direction of the field in this time is $v_d \, \delta t$. The momentum equation is then

$$Xe\,\delta t = m_e v_d \qquad (4.25)$$

so that

$$v_d \bar{v}_e = \frac{Xel_f}{m_e} \qquad (4.26)$$

This is the 'Townsend equation'. In a rigid treatment, the right-hand side would be multiplied by a factor 0·815.

Combining equations 4·23 and 4·26,

$$\lambda = 2\cdot356 \frac{v_d^2}{\bar{v}_e^2} \qquad (4.27)$$

Thus the ratio of drift to random velocities is determined by the energy loss parameter λ. The electron energy is conveniently described by a parameter k_e:

$$k_e = \frac{T_e}{T_g} = \frac{m_e \bar{v}_e^2}{m_g \bar{v}_g^2} \qquad (4.28)$$

so

$$\bar{v}_e = 1\cdot15 \times 10^7 k_e^{1/2} \qquad (4.29)$$

$$\bar{E}_e = \frac{k_e}{27} \qquad (4.30)$$

and

$$l_f = 8 \times 10^{-9} \frac{v_d k_e^{1/2} p}{X} \qquad (4.31)$$

with p in torr at 300°K.

The electron diffusion can be considered in terms of a balance of electric pressure and diffusion pressure:

$$n_e \bar{v}_e = -D_e \, \text{grad} \, n_e + n_e v_d \qquad (4.32)$$

where $\bar{v}_e$ is the electron flow velocity vector, for which

$$\text{div} \, n_e \bar{v}_e = 0 \qquad (4.33)$$

The momentum lost by electrons per unit volume per unit time in collision with gas atoms is $p_e\bar{v}_e/D_e$, where $p_e = n_e kT_e$ is the electron pressure. The momentum transferred per unit volume per unit time by the flow of electrons across the unit surface, is $-\mathrm{grad}\, p_e$. The momentum gained per unit volume per unit time by the electrons from the field is therefore

$$\frac{p_e v_d}{D_e} = \frac{p_e \bar{v}_e}{D_e} + \mathrm{grad}\, p_e \tag{4.34}$$

This is equal to $n_e eX$, which yields the Einstein equation

$$\frac{v_d}{D_e} = \frac{Xe}{kT_e} = \frac{40 \cdot 3X}{k_e} \tag{4.35}$$

An extensive collection of early electron drift data has been given[72] in the form of tables of v_d and D_e as functions of X/p, together with tabulated l_f, λ, $\bar{v}_e$ and k_e. It has been pointed out[90] that the parameter X/p should really be replaced by X/n_0, since experiments conducted at the same X/p but different temperatures will be appropriate to different number densities. The parameter X/n_0 is measured in units of 10^{-17} V cm^2 mol.$^{-1}$, and it is suggested[91] that this unit be known as a Townsend:

$$1 \text{ Townsend} = 1 \cdot 0354 \times 10^{-2} \text{ T V cm}^{-1} \text{ torr}^{-1} \tag{4.36}$$

or 1 Townsend $\simeq 3$ V cm^{-1} torr^{-1} at 300°K.

Results of More Rigorous Analysis of Electron Drift

The undoubted success of the simplified analysis (that is, a realistic value of l_f) may be attributed[92] to the fact that D_e is relatively insensitive to the form of the electron energy distribution, provided that $\sigma_d(E_e)$ is not a rapidly varying function.

When the diffusion cross-section is not assumed to be energy invariant, a more complex analysis is necessary[93]. For a power law dependence of σ_d upon v_e, a corresponding form of the electron energy distribution can be deduced. For $\sigma_d \propto v_e^{-r}$, the energy distribution $f(v_e)$ is

$$4f(v_e)v_e^2 \, dv_e = \frac{4v_e^2}{a^3 \Gamma(3/n)} \exp\left(-\frac{v_e^n}{\alpha^n}\right) dv_e \tag{4.37}$$

where $n = 4 - 2r$ and

$$\alpha^2 = \frac{2\bar{v}_e^2}{3} \tag{4.38}$$

In particular cases, standard distributions are found. For $\sigma_d \propto v_e^{-1}$, a Maxwellian distribution follows, with

$$\frac{v_d}{D_e} = \frac{26 \cdot 9 \times 1 \cdot 5X}{k_e} \tag{4.39}$$

and

$$\lambda = \frac{7 \cdot 65 \times 10^{-15} \times 2 \cdot 37 v_d^2}{k_e} \qquad (4.40)$$

When σ_d is independent of velocity, a Druyvesteyn distribution is found, with

$$\frac{v_d}{D_e} = \frac{26 \cdot 9 \times 1 \cdot 213 X}{k_e} \qquad (4.41)$$

and

$$\lambda = \frac{7 \cdot 65 \times 10^{-15} \times 2 \cdot 54 v_d^2}{k_e} \qquad (4.42)$$

A trial $\sigma_d(v_e)$ function can be taken, and iteratively applied to the $v_d(X/n_0)$ and $D_e(X/n_0)$ data, until the best fitting $\sigma_d(v_e)$ is found.

The equations used for this purpose are written in the form[94]:

$$v_d = \frac{1}{3} \frac{e}{m_e} \frac{X}{p} \left\langle v_e^{-2} \frac{\mathrm{d}}{\mathrm{d}v_e} \left(\frac{v_e^2}{\sigma_e} \right) \right\rangle \qquad (4.43)$$

$$\lambda = \frac{6 v_d^2}{v_e^2} \langle v_e \sigma_d \rangle^{-1} \left\langle v_e^{-2} \frac{\mathrm{d}}{\mathrm{d}v_e} \left(\frac{v_e^2}{\sigma_d} \right) \right\rangle^{-1} \qquad (4.44)$$

$$\frac{v_d}{D} = \frac{Xe}{m_e} \left\langle v_e^{-2} \frac{\mathrm{d}}{\mathrm{d}v_e} \left(\frac{v_e^2}{\sigma_d} \right) \right\rangle \left\langle \frac{v_e}{\sigma_d} \right\rangle^{-1} \qquad (4.45)$$

In these equations

$$\frac{m_e}{e} = \frac{9 \cdot 1 \times 10^{-28} \times 300}{4 \cdot 8 \times 10^{-10}} \qquad (4.46)$$

When σ_d is not strongly energy-variant, and when a Maxwellian energy distribution is assumed, as is valid for low X/p, the equations simplify to:

$$\bar{\sigma}_d = 2 \cdot 464 \times 10^{-10} \frac{(X/p) T^{1/2}}{v_d k_e^{1/2}} \ \mathrm{cm}^2 \qquad (4.47)$$

$$\lambda = 5 \cdot 19 \times 10^{-12} \frac{v_d^2}{k_e T} \qquad (4.48)$$

$$\frac{v_d}{D} = 11 \cdot 59 \times 10^3 \frac{X}{k_e T} \ \mathrm{cm}^{-1} \ \mathrm{torr}^{-1} \qquad (4.49)$$

$$\bar{v}_e^2 = 45 \cdot 5 \times 10^{10} k_e T \qquad (4.50)$$

For non-Maxwellian energy distributions, the numerical constants in these equations assume different values.

The standard forms of the electron energy distribution given above may not be realistic, because the inelastic collisions are not taken into account in detail, but only insofar as they contribute to the energy loss parameter λ. For molecules, this parameter is strongly energy dependent, and some

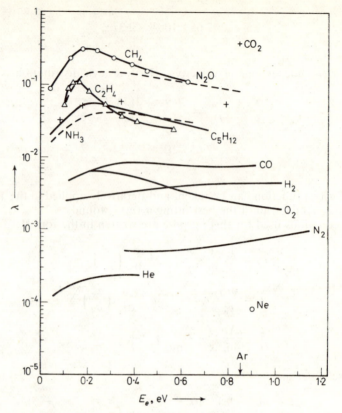

Fig. 4.12. *Energy loss parameters λ for various gases*[95]

actual data are shown in *Figure 4.12*[95]. Since λ is a measure of the effective-ness of a molecule in removing the energy from an electron, it is likely to be related to the efficiency of the molecule as third body in three-body electron attachment and recombination processes.

In order to improve further on this analysis, the real electron energy distribution must be used in place of the simplified form. To determine this, a solution of the Boltzmann transport equation is required, and this in turn requires a detailed knowledge of the elastic and inelastic cross-section functions. The drift velocity[82, 84] can be written

$$v_d = \mu_e n_0 \frac{X}{n_0} = \frac{4\pi e}{3m_e} \frac{X}{n_0} \int_0^\infty f_0 \frac{d}{dv_e}\left(\frac{v_e^3 n_0}{v}\right) dv_e \qquad (4.51)$$

where $\mu_e = v_d/X$ is the electron mobility, $v = n_0\langle\sigma v_e\rangle$ is the collision fre-quency, and f_0 is the spherically symmetrical term in the expansion of the electron velocity distribution, expanded in spherical harmonics:

$$f_0 = \left(\frac{m_e}{2\pi kT}\right)^{3/2} \exp\left(-\frac{m_e v_e^2}{2kT}\right) \qquad (4.52)$$

An assumed expansion of the cross-section

$$\sigma^{-1} = \frac{m_e}{e} \sum_j \frac{\Gamma(5/2)}{\Gamma(5/2-j/2)} B_j v_e^{1-j}$$

(4.53)

substituted into equation 4.51 leads to the solution

$$\mu_e = n_0^{-1} \sum_j B_j \left(\frac{2kT}{m_e}\right)^{-j/2}$$

(4.54)

A cross-section constant with energy, such as the momentum transfer cross-section for electrons in helium, would show only $j = +1$; in this case $\mu_e n_0$ varies as $(2kT/m_e)^{-1/2}$. However, the data for argon require the assumption of series with $j = +1, -2$ or $j = +1, -3$ or $j = -1, -2$. These lead to the three cross-section functions shown in *Figure 4.13*, which are used to interpret the drift velocities of *Figure 4.14*.

For molecular gases, the analysis is more complicated[94, 100], since it is necessary to supplement the elastic momentum transfer cross-section $\sigma_d(E_e)$ by further cross-sections $\sigma_{+j}(E_e)$ in which electrons take up $(+j)$ and lose

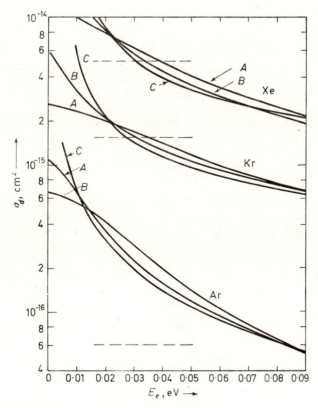

Fig. 4.13. Momentum loss cross-section functions[84, 96] for e–Ar, e–Kr and e–Xe, computed from swarm data of Fig. 4.14: A, j = +1, −2; B, j = +1, −3; C, j = −1, −2; broken lines represent microwave data[97]

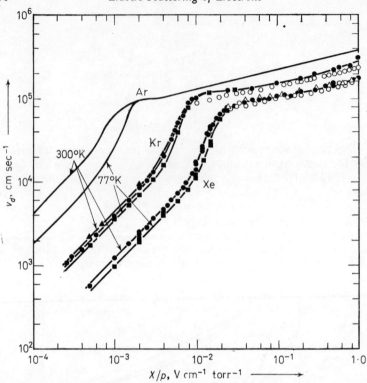

Fig. 4.14. Drift velocities[84, 96] of electrons in argon, krypton and xenon: squares, 195°K; filled circles, 300°K; filled triangles, 368°K; open circles, Bowe[98]; open triangles, English and Hanna[99]

$(-j)$ energy E_j. The steady state distribution $f(E_e)$ of electron energies is given by the Boltzmann transport equation:

$$\frac{X^2}{3}\frac{d}{dE_e}\left(\frac{E}{n_0\sigma_d}\frac{df}{dE_e}\right)+\frac{2m_e}{m_g}\frac{d}{dE_e}(E_e^2n_0\sigma_d f)+\frac{2m_e kT}{m_g e}\frac{d}{dE_e}\left(E_e^2n_0\sigma_d\frac{df}{dE_e}\right)$$
$$+\sum_j(E_e+E_j)f(E_e+E_j)\,n_0\sigma_{+j}(E_e+E_j)-E_e f(E_e)n_j\sum_j\sigma_{+j}(E_e)$$
$$+\sum_j(E_e-E_j)f(E_e-E_j)\,n_0\sigma_{-j}(E_e-E_j)-E_e f(E_e)n_j\sum_j\sigma_{-j}(E_e)=0 \qquad (4.55)$$

Normalized variables assist in the solution:

$$z=\frac{eE_e}{kT_g} \qquad (4.56)$$

$$\theta=\frac{\sigma_d(E_e)}{\sigma_{d0}} \qquad (4.57)$$

$$\alpha=\frac{.m_g}{6m_e}\left(\frac{eX}{n_0\sigma_{d0}kT_g}\right)^2 \qquad (4.58)$$

$$\eta_{\pm j}=\frac{m_g\sigma_{\pm j}}{2m_e\sigma_{d0}} \qquad (4.59)$$

where σ_{d0} is the value of σ_d at some reference energy, and T_g is the gas temperature. With the aid of the principle of detailed balancing and the Boltzmann relation for relative populations of excited states, equation 4.55 can be written:

$$\left(\frac{\alpha}{\theta}+z\theta\right)z\frac{\mathrm{d}f}{\mathrm{d}z}+z^2\theta f+\sum_j\int_z^{z+z_j}z\eta_j(z)\left[f(z)-\exp\left(-z_j\right)f(z-z_j)\right]\mathrm{d}z=0 \quad (4.60)$$

The experimentally available quantities are diffusion coefficients D_e, mobilities μ_e, and 'perpendicular mobilities' $\mu_\perp$; the latter are measured at right angles to a mutually perpendicular electric and weak magnetic field of strength B[82, 88]. These three quantities are related to the variables of equation 4.60:

$$D_e=\frac{(2kT/m_e)^{1/2}}{3n_0\sigma_{d0}}\int_0^\infty\frac{z}{\theta}f(z)\,\mathrm{d}z \quad (4.61)$$

$$\mu_e=\frac{v_d}{X}=-\frac{e(2kT/m_e)^{1/2}}{3n_0\sigma_{d0}}\int_0^\infty\frac{z}{\theta}\frac{\mathrm{d}f}{\mathrm{d}z}\,\mathrm{d}z \quad (4.62)$$

$$\mu_\perp=-\frac{e^2B}{3m_e\sigma_{d0}^2n_0^2kT}\int_0^\infty\frac{z^{1/2}}{\theta^2}\frac{\mathrm{d}f}{\mathrm{d}z}\,\mathrm{d}z \quad (4.63)$$

The quantity D_e/μ_e is determined from lateral diffusion experiments in the absence of magnetic field; it is a measure of the average electron energy. The quantity $\mu_\perp/\mu_e$ is equal to the tangent of the angle of deflexion of a swarm of electrons drifting along an electric field when a weak magnetic field is applied perpendicular to it.

Frost and Phelps[84] have used various mathematical techniques to solve equation 4.60 for the experimental data. For sufficiently weak electric fields, giving near-thermal electron energies, an exact solution is possible[101]. For stronger electric fields, a solution can be found neglecting collisions of the second kind; but it is more satisfactory to obtain a solution by approximating the effect of a large number of rotational excitations as a continuous function. For this purpose, one must use selection rules, and theoretical expressions for rotational excitation cross-section functions[102]. These are discussed in Chapter 5.

The Boltzmann fraction of molecules in the Jth rotational level is:

$$f_J=\frac{(2t+1)(t+a)(2J+1)\exp\left(-eE_J/kT_g\right)}{\sum_J(2t+1)(t+a)(2J+1)\exp\left(-eE_J/kT_g\right)} \quad (4.64)$$

where t is the nuclear spin, $a=0$ for even J and $a=1$ for odd J. The functions $f(E+E_J)$ and $f(E-E_J)$ can be expanded in Taylor series expansions when the rotational energy losses are sufficiently small, and under these

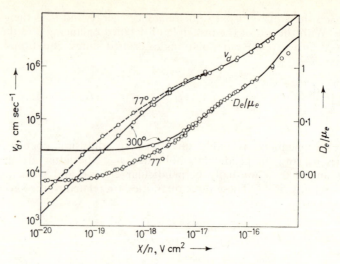

Fig. 4.15. Drift velocity and diffusion data points for electrons in hydrogen compared with computations from cross-sections of Fig. 4.16

(From Frost and Phelps[84], by courtesy of Westinghouse Electric Corporation)

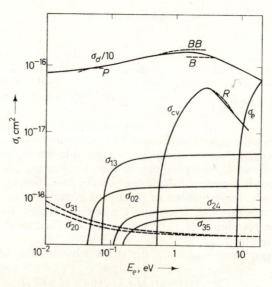

Fig. 4.16. Momentum loss cross-section function and relevant inelastic cross-section functions for interpretation of hydrogen drift velocities: σ_{cv}, vibrational excitation (R, Ramien[103]) σ_{ij}, rotational excitation quantum numbers $i \to j$; σ_e, electronic excitation; σ_d momentum loss cross-section function, compared with other data (P, Pack and Phelps[84]; BB, Bekefi and Brown[104]; B, Brode[3])

(From Frost and Phelps[84], by courtesy of Westinghouse Electric Corporation)

conditions the rotational excitation functions become energy-independent over the important energy range. The electron energy distribution then becomes:

$$f(E_e) = A \exp \left[-\int \frac{\theta(E_e\theta + B_0\eta_r)}{\varepsilon} \, dE_e \right] \quad (4.65)$$

with

$$\eta_r = \frac{2m_g\sigma_{r0}}{m_e\sigma_{d0}} \quad (4.66)$$

and

$$\varepsilon = \frac{m_g X^2}{6m_e n^2 \sigma_{d0}^2} \quad (4.67)$$

A is a normalization constant, B_0 the rotational constant, and σ_{r0} is defined similarly to σ_{d0}.

The analysis procedure is one of successive approximations. Trial sets of cross-sections are proposed, and the values of μ_e, $\mu_\perp$ and D_e are computed for comparison with experiment; corrections are then made to obtain better agreement. The final agreement for v_d and D_e/μ_e is shown for H_2 in *Figure 4.15*, and the cross-sections in *Figure 4.16*. The calculated electron energy distributions are found to lie between Maxwellian and Druyvesteyn, that is, $f(E_e)$ lies between $A \exp[-E_e(D_e/\mu_e)^{-1}]$ and $B \exp[-E_e^2(\pi D_e/\mu_e)^{-2}]$. The values of $\mu_\perp$ calculated are in satisfactory agreement with those measured by Townsend and Bailey[4, 77], and it will be seen that the correspondence with diffusion and mobility data is encouraging. For hydrogen, the best fit is obtained by retaining the threshold energy and shape of the theoretical rotational excitation cross-sections, multiplying them by a factor of 2·5. It has been suggested[105] that there should be a correction to the value of the quadrupole moment, which reduces the discrepancy to 1·7. Similar analyses have been made by Phelps and his colleagues for the simple molecular gases.

A simpler method of arriving at trial $\sigma_j(E_e)$ functions has also been suggested[106]. A Maxwellian distribution is taken and the assumption is made that $\lambda(E_e)$ is dominated by a single mode of vibrational excitation; rotational excitation is neglected.

4.4 COLLISION FREQUENCIES DETERMINED BY MICROWAVE TECHNIQUES

Diffusion cross-sections at thermal energies can be determined not only in drift tubes but also by conductivity measurements. These may be conducted in arcs[107], but are more usually made at microwave frequencies in the afterglow of a pulsed discharge.

In a weakly ionized gas, the electron collision frequency v_m is proportional to the electron collision cross-section and to the pressure. Absorption of microwave radiation is used to measure the imaginary part of the permittivity of electrons in the gas; since this is related to the rate of loss of momentum due to collisions, the collision cross-section derived from the measurements is that for momentum transfer, σ_d:

$$\sigma_d = \frac{v_m}{n_0 \bar{v}_e} \tag{4.68}$$

with $\bar{v}_e$ and n_0 in appropriate units. In practice, the cross-section is taken to be $\sigma_d = v_m/p_0\bar{v}_e$, with σ_d in cm^2 cm^{-3} torr^{-1}, and p_0 the reduced pressure equal to $273p/T_g$ with p in torr. Collision frequency determination can be used for the monitoring of electron temperatures in plasmas once the form of the $\sigma_d(\bar{v}_e)$ function is known.

The collisions act as a damping force, which can be expressed in terms of a damping constant k; under the action of an alternating electric field $X_0 \exp(j\omega t)$, the equation of motion of the electrons is[108]

$$m_e \frac{dv_e}{dt} + kv_e = -eX_0 \exp(j\omega t) \tag{4.69}$$

with

$$k = m_e v_m = n_0 \bar{v}_e \sigma_d m_e \tag{4.70}$$

When k is assumed independent of electron velocity, the solution is

$$v_e = -\frac{eX_0 \exp(j\omega t)}{j\omega m_e + k} \tag{4.71}$$

and the current density

$$J = -n_e e v_e = \frac{n_e e^2 X_0 \exp(j\omega t)}{j\omega m_e + k} \tag{4.72}$$

The complex conductivity $\hat{\sigma}$ can be written:

$$\hat{\sigma} = \sigma' + j\sigma'' = \frac{J}{X_0 \exp(j\omega t)} \tag{4.73}$$

so that

$$\left| \frac{\varepsilon''}{\varepsilon'} \right| = \frac{k}{m_e \omega} = \left| \frac{\sigma'}{\sigma''} \right| \tag{4.74}$$

where the symbols ε represent parts of the complex permittivity. A determination of both parts of the permittivity leads to v_m.

A more general expression is:

$$|\sigma'/\sigma''| = -\int_0^\infty \frac{v_m/\omega}{1+(v_m/\omega)^2} v_e^3 \, df_0 \Bigg/ \int_0^\infty \frac{1}{1+(v_m/\omega)^2} v_e^3 \, df_0 \tag{4.75}$$

where v_e is the random velocity, and f_0 the spherically symmetrical part of the velocity distribution function.

For a thermally ionized plasma in LTE, the conductivity for a steady applied potential difference is given approximately by the expression

$$\sigma' \simeq \frac{n_e e^2 n_0 \sigma_d}{(2m_e kT)^{1/2}} \tag{4.76}$$

Another simplifying assumption which is successful is:

$$v_m = a p_0 v_e^h \tag{4.77}$$

with a and h constants. A relation can be derived in terms of a parameter γ which measures the average number of collisions that an electron makes per radian of electric field:

$$\gamma = \frac{\frac{1}{2} a p_0}{\omega} \left(\frac{2kT_g}{m_e} \right)^{(h+1)/2} \tag{4.78}$$

A solution of this unspecified relation can be obtained in the form of a function:

$$\left| \frac{\sigma'}{\sigma''} \right| = f(\gamma, h) \tag{4.79}$$

Single values of a and h can only be obtained by making measurements at different pressures and gas temperatures.

The microwave techniques that have been used in collision frequency determinations are more varied than those used in studies of recombination, ambipolar diffusion and attachment in afterglows. They may be summarized as follows:

Cavity Studies of Afterglows[104,109]

These are similar in many respects to the recombination studies more fully described in Chapter 7. A radiofrequency discharge of microseconds duration is struck in a quartz bottle contained within a cylindrical metal cavity whose resonant frequency may be determined by means of a microwave monitoring signal. This signal is sawtooth frequency-modulated repetitively, and reflected from the cavity in such a way that a series of cavity resonance curves are displayed on an oscilloscope timebase through the afterglow. The shift of the resonant frequency f_r is proportional to the space average of the real part of the plasma permittivity, hence to the space average of electron density:

$$\Delta f_r \propto \overline{\varepsilon'} \propto \overline{n_e} \tag{4.80}$$

The selectivity Q is inversely proportional to the unreal part of the permittivity:

$$\Delta(Q^{-1}) \propto \overline{\varepsilon''}$$

For a microwave angular frequency ω:

$$\left| \frac{\varepsilon''}{\varepsilon'} \right| = \frac{v_m}{\omega} \tag{4.81}$$

so that

$$v_m \propto \frac{\omega \, \Delta(Q^{-1})}{\Delta f_r}$$

at any moment during the afterglow, when the electron temperature is known. A microwave circuit suitable for the application of breakdown pulse and monitoring signal is shown in *Figure 3.9*. The frequency-variation of the cavity impedance yields the selectivity:

$$\left| \frac{Z_f}{Z_r} \right| = \frac{1 \cdot 414}{Q} \qquad (4.82)$$

at $f = 0 \cdot 707 \, f_r$, and

$$\left| \frac{Z_f}{Z_r} \right| = \frac{90}{Q} \qquad (4.83)$$

at $f = 0 \cdot 996 \, f_r$, where Z represents the cavity impedance, and the subscripts f and r refer respectively to the frequency f and to the resonance condition. Filled-cavity techniques were first developed in microwave dielectric researches[110], and at that time the cavity method was employed for collision frequency studies by Brown and his colleagues[109, 111–113]. The time-dependence of v_m in the afterglow can be used as a measure of the thermalization rate of the electrons.

Cavity measurements were extended by Gould and Brown[111] to higher electron temperatures, produced by continuous heating of the plasma by radiation injected into the cavity, which was designed to resonate simultaneously in different modes at different frequencies and selectivities. A low Q is mandatory for the heating signal.

Microwave Interferometry

The transmission of an unguided microwave signal through a discharge provides a method of studying the ratio of imaginary to real parts of the permittivity[114]. This method is particularly suited to the application of interferometer technique[115]. The signal (*Figure 3.8*) is split by means of a directive feed into two parts, one of which is passed with the aid of horn antennae through the plasma. The other part passes through a phase-shifter and a calibrated attenuator, after which it joins the first part of the signal through a magic T or directive feed. The combined signal shows a minimum when the two parts are out of phase, so the wavelength in the plasma, hence the real part of the permittivity, can be measured. The attenuation experienced between successive minima, where the intensities are I_1 and I_2, is related to the unreal part of the permittivity:

$$\ln I_1 - \ln I_2 = \frac{\pi \varepsilon''}{n} \qquad (4.84)$$

for plasma refractive index n. The ratio of real to imaginary parts of the permittivity is unaffected by the configuration of the wave in the region of the plasma. The method is suitable for investigation of high-density plasmas, which may be studied in decay by means of oscilloscope display of the combined signals. During the afterglow, the interferometer pattern passes through maxima and minima, but the momentum loss causes the ratio to increase, at a rate from which v_m can be calculated.

Cross-modulation

The cross-modulation technique developed by Anderson and Goldstein[116, 117] makes use of the same cross-modulation which has been observed in the ionosphere, under the name 'Luxemburg effect'. The absorption of a radio wave depends upon the collision frequency of the electrons, and this in turn depends upon their mean energy. The mean energy can be changed by the application of a powerful radio wave of different frequency. If this wave is amplitude-modulated at a frequency sufficiently small for the electron energy to follow it faithfully, the absorption of the second wave is proportional to some function of the applied modulation energy. Thus the cross-modulation can be used quantitatively for the measurement of electron collision probability functions; the technique has been applied to the atmospheric gases.

An important feature of Anderson and Goldstein's measurements was the separation of the electron–ion from the electron–atom components of the collision frequency. It was found that the total collision frequency depended almost linearly upon the measured electron density, which can be assumed equal to the positive ion density if negative ions are neglected.

Classical scattering in an attractive coulomb field, cut off at the Debye shielding radius, leads to an expression for the electron–ion collision frequency:

$$\nu_{m+} = An_+ T_e^{3/2} \ln \left(\frac{BT_e^{3/2}}{n_+} \right) \tag{4.85}$$

for temperatures $T_e = T_+$. The constants A and B are given respectively as: 3·59 and 3320 by Ginsburg and Vilenski[118]; and 2·25 and 8400 by Burkhardt, Elwert and Unsold[119]. The experiments lead to the values 3·6 and 3700. Smaller electron–ion collision frequencies have been predicted[120], and have been found in shock-tube experiments[121].

Electron Cyclotron Resonance Broadening

A collision frequency measurement technique devised by Kelly, Margenau and Brown[122, 123] depends upon the collision broadening of the electron cyclotron resonance curve. The simultaneous application of a steady magnetic field H to a plasma, and of a high-frequency electric field perpendicular to it, causes the electrons to move in 'cyclotron' orbits, which are circular or slightly helical. The cyclotron frequency f_C at which this occurs is given by the well-known equation:

$$f_C = \frac{\omega_c}{2\pi} = \frac{eH}{2\pi m_e c} \tag{4.86}$$

The electron collisions modify the efficiency of transfer of energy from the radiofrequency field to the plasma; from the frequency dependence of this efficiency the collision frequency can be calculated.

Purely Resistive Plasma

Another method utilizing a steady magnetic field is due to Hirschfield and Brown[124]. In crossed alternating electric and steady magnetic fields applied to a plasma, the phase of the electron current with respect to the electric field is determined by the strength of the magnetic field. Only at the magnetic field strength corresponding to the cyclotron resonance frequency will the phase difference be zero and the plasma appear purely resistive to transmitted radiation. Electron collisions have the effect of shifting the condition of pure resistivity in the same way that the addition of damping to an oscillator shifts its resonant frequency. Under this condition, the real part of the plasma permittivity is unity, and for a microwave cavity containing only plasma the frequency shift Δf will be zero. From the value of the steady magnetic field for which this condition is realized the collision frequency can be calculated.

Table 4.2. ELECTRON COLLISION PROBABILITIES FOR MOMENTUM
TRANSFER, DETERMINED BY MICROWAVE MEASUREMENTS

(Mean electron temperatures 300°K)

Gas	P_m (cm^{-1} torr^{-1})	Reference
H_2	28·5	104
	42	125
	46	109
He	18·8 ± 0·9	126
	19	109
	∼18	111–113
	25	116
Ne	3·3	109
	1·8	111–113
Ar	2·1	109
Kr	66·2 ± 2·2	126
	54	109
Xe	156 ± 5	126
	180	109
N_2	15	109
	∼60	116

In Table 4.2 are summarized the available results of those microwave experiments in which the electron temperature may be taken to be that of the laboratory. The data are usually expressed in terms of a collision probability P_m, or number of momentum loss collisions made by an electron travelling 1 cm in a gas at 1 torr and 273°K:

$$P_m = l_f^{-1} = n_0 \sigma_d = \frac{n_0 \nu_m}{\bar{v}_e} \qquad (4.87)$$

Radiation Emitted by a Plasma

An experiment designed specifically to study the parameter $\lambda(E_e)$ was performed by Formato and Gilardini[127], who measured the incoherent radiation emitted by a plasma at microwave frequencies. The experiment is performed by sending through the plasma the microwave radiation of a black body at a known temperature and finding the temperature at which the absorbed radiation is equal to the emitted radiation. The 'plasma radiation temperature' T_r is related[128] to the plasma parameters, and may be deduced from the microwave black-body radiation under specific conditions. It can be shown that when an additional heating field of strength E_h and radial frequency ω_h is applied, the radiation temperature rise is related to $\lambda(E_e)$ as follows:

$$\lambda(E_e) = \frac{2e^2 E_h}{3 m_e \omega_h^2 k \, \Delta T} \qquad (4.88)$$

$$\Delta T = T_{rh} - T_{\text{gas}} - \frac{\varepsilon_0'' / \varepsilon_0'}{\varepsilon_h'' / \varepsilon_h'} (T_{r0} - T_{\text{gas}}) \qquad (4.89)$$

Here the subscript h refers to the conditions when the heating field is applied and the subscript 0 refers to the conditions when it is not. Simultaneous measurements of radiation, heating field strength, and ε' and ε'' are required, so the microwave circuitry is complicated. It is found that in nitrogen $(m_e/m_{\text{gas}}) \lambda(E_e) = 3 \cdot 95 \times 10^{-4}$ between 3000°K and 5000°K, and in oxygen $(m_e/m_{\text{gas}}) \lambda(E_e) = (4 \cdot 8 \pm 1) \times 10^{-3}$ between 850°K and 1900°K, decreasing slightly with temperature. The oxygen values are much smaller than those of drift tube measurements[76], but the nitrogen values are in approximate agreement[86].

In the graphical representation of collision probabilities, no systematic divergence between beam, swarm and microwave methods is immediately apparent, but the discrepancies between different experiments are much greater in molecular than in atomic plasmas. This would suggest that the effect of the parameter $\lambda(E_e)$ on the different situations encountered in experiments is insufficiently understood.

It will be seen from *Figure 4.13* that the microwave diffusion cross-sections in rare gases differ from those measured in swarm experiments. It has been suggested[127] that the electron temperatures are raised in these afterglow experiments by excitation processes involving metastable atoms.

REFERENCES

1. MASSEY, H. S. W. and BURHOP, E. H. S. *Electronic and Ionic Impact Phenomena*, 1952. London; Oxford University Press.
2. RAMSAUER, C. *Ann. Phys., Lpz.* 64 (1921) 513; 66 (1921) 546; 72 (1923) 345; RAMSAUER, C. and KOLLATH, R. *Ann. Phys., Lpz.* 3 (1929) 536; KOLLATH, R. *Phys. Z.* 31 (1930) 985.
3. BRODE, R. B. *Phys. Rev.* 25 (1925) 636; RUSCH, M. *Ann. Phys., Lpz.* 80 (1926) 707; BRUCHE, E. *Ann. Phys., Lpz.* 84 (1927) 279; NORMAND, C. E. *Phys. Rev.* 35 (1930) 1217.
4. TOWNSEND, J. S. and BAILEY, V. I. *Phil. Mag.* 43 (1922) 593.
5. RAMSAUER, C. *Jb. Radioakt.* 19 (1923) 345.
6. RAMSAUER, C. and KOLLATH, R. *Ann. Phys., Lpz.* 3 (1929) 536.
7. RAMSAUER, C. and KOLLATH, R. *Ann. Phys., Lpz.* 12 (1932) 529.
8. NORMAND, G. E. *Phys. Rev.* 35 (1930) 1217.
9. GOULD, L. and BROWN, S. C. *Phys. Rev.* 95 (1954) 896.
10. PHELPS, A. V., FUNDINGSLAND, O. T. and BROWN, S. C. *Phys. Rev.* 84 (1951) 559.
11. VARNERIN, L. J. *Phys. Rev.* 84 (1951) 563.
12. BEKEFI, G. and BROWN, S. C. *Phys. Rev.* 112 (1958) 159.
13. CROMPTON, R. W. and SUTTON, D. J. *Proc. Roy. Soc.* A215 (1952) 467.
14. BRODE, R. B. *Phys. Rev.* 25 (1925) 636; 39 (1942) 547.
15. RUSCH, M. *Phys. Z.* 26 (1925) 748.
16. BRUCHE, E. *Ann. Phys., Lpz.* 81 (1926) 537.
17. BRUCHE, E. *Ann. Phys., Lpz.* 84 (1927) 279.
18. BOWE, J. C. *Phys. Rev.* 117 (1960) 1416.
19. BARBIERE, D. *Phys. Rev.* 84 (1951) 653.
20. RUSCH, M. *Ann. Phys., Lpz.* 80 (1926) 707.
21. DRAWIN, H. W. *Z. Phys.* 146 (1958) 295.
22. BULLARD, E. C. and MASSEY, H. S. W. *Proc. Roy. Soc.* A130 (1931) 579.
23. FAXEN, H. and HOLTSMARK, J. *Z. Phys.* 45 (1927) 307.
24. RAYLEIGH, LORD *Theory of Sound, Vol. 2*, p. 323, 2nd edn, 1896. London; Macmillan.
25. RAMSAUER, C. and KOLLATH, R. *Ann. Phys., Lpz.* 4 (1930) 91.
26. BRUCHE, E. *Ann. Phys., Lpz.* 1 (1929) 93.
27. BRUCHE, E. *Ann. Phys., Lpz.* 82 (1927) 912.
28. GOLDEN, B. E. and BANDEL, H. W. *Phys. Rev.* 138 (1965) A14.
29. O'MALLEY, T. F. *Phys. Rev.* 130 (1963) 1020.
30. BALDWIN, G. C. General Electric Company Report GE 66-C-246, 1966.
31. BRUCE, L., CRAWFORD, F. W. and HARP, R. S. Stanford University Institute of Plasma Research Report 186, 1967.
32. BRACKMANN, R. T., FITE, W. L. and NEYNABER, R. H. *Phys. Rev.* 112 (1958) 1157.
33. NEYNABER, R. H., MARINO, L. O., ROTHE, E. W. and TRUJILLO, S. M. *Proc. 2nd Int. Conf. Phys. electron. atom. Collisions*, University of Colorado, 1961.
34. MOISEIWITSCH, B. L. *Atomic and Molecular Processes*, Ed. D. R. Bates, 1962. New York; Academic Press.
35. SMITH, K. and MORGAN, L. A. *Phys. Rev.* 165 (1968) 110.
36. BRANSDEN, B., DALGARNO, A., JOHN, T. and SEATON, M. J. *Proc. phys. Soc., Lond.* 71 (1958) 877.
37. TEMKIN, A. and LAMKIN, J. C. *Phys. Rev.* 121 (1961) 788.
38. McEACHRAN, R. and FRASER, P. *Canad. J. Phys.* 38 (1960) 317.
39. SMITH, K. *Phys. Rev.* 120 (1960) 845.
40. GELTMAN, S. *Phys. Rev.* 119 (1960) 1283.
41. GILBODY, H. B., STEBBINGS, R. F. and FITE, W. L. *Phys. Rev.* 121 (1961) 794.
42. SUNSHINE, G., AUBREY, B. B. and BEDERSON, B. *Abstr. 4th Int. Conf. Phys. electron. atom. Collisions*, p. 130, 1965. Hastings-on-Hudson, New York; Science Bookcrafters.

43. LIN, S. C. and KIVEL, B. *Phys. Rev.* 114 (1969) 1026.
44. SMITH, K., HENRY, R. J. W. and BURKE, P. G. *Phys. Rev.* 157 (1967) 51.
45. BAUER, E. G. and BROWNE, H. N. quoted in reference 42.
46. BURKE, P. G. and SCHEY, H. M. *Phys. Rev.* 126 (1962) 147.
47. LIPPMANN, B. A., MITTLEMAN, M. H. and WATSON, K. M. *Phys. Rev.* 116 (1959) 920.
48. STONE, P. M. *Abstr. 4th Int. Conf. Phys. electron. atom. Collisions*, p. 151, 1965. Hastings-on-Hudson, New York; Science Bookcrafters.
49. BRUCHE, E. *Ann. Phys., Lpz.* 83 (1927) 1065.
50. TEMKIN, A. *Phys. Rev.* 107 (1957) 1004.
51. SUNSHINE, G., AUBREY, B. B. and BEDERSON, B. *Phys. Rev.* 154 (1967) 1.
52. ALTSHULER, S. *Phys. Rev.* 107 (1957) 114.
53. CRAWFORD, O. H., DALGARNO, A. and HAYS, P. B. *Molec. Phys.* 13 (1967) 181.
54. MITTLEMAN, M. H. and VAN HOLDT, R. E. *Phys. Rev.* 140 (1965) A726.
55. BULLARD, E. C. and MASSEY, H. S. W. *Proc. Roy. Soc.* A130 (1931) 579; CHILDS, E. C. and MASSEY, H. S. W. *Proc. Roy. Soc.* A141 (1933) 473; A142 (1933) 509.
56. WESTIN, S. *K. norski vidensk. Selsk.* 2 (1946) 58.
57. HOEPER, P. S., FRANZEN, W. and GUPTA, R. Boston University Report, 1967.
58. GAGGE, A. P. *Phys. Rev.* 43 (1933) 776; 44 (1933) 808.
59. WESTIN, S. *K. norski vidensk. Selsk.* 2 (1946) 48.
60. EDELMANN, F. and ULMER, K. *Z. angew. Phys.* 18 (1965) 308.
61. MOTT, N. F. and MASSEY, H. S. W. *The Theory of Atomic Collisions*, 3rd edn, 1965. London; Oxford University Press.
62. BURKE, P. G. and SCHEY, H. M. *Phys. Rev.* 126 (1962) 147.
63. SCHULZ, G. J. *Phys. Rev. Lett.* 10 (1963) 104.
64. SMITH, K. *Rep. Progr. Phys.* 29 (1966) 373.
65. BURKE, P. G. *Advanc. Phys.* 14 (1965) 521.
66. BARDSLEY, J. N. and MANDL, F. *Rep. Progr. Phys.* 31 (1968) 471.
67. FANO, U. *Phys. Rev.* 124 (1961) 1866.
68. McGOWAN, J. W. *Phys. Rev.* 156 (1967) 165; *Phys. Rev. Lett.* 17 (1966) 1207.
69. SCHULZ, G. J. *Phys. Rev. Lett.* 13 (1964) 583.
70. KLEINPOPPEN, H. and RAIBLE, V. *Phys. Lett.* 18 (1965) 24.
71. McGOWAN, J. W., CLARKE, E. M. and CURLEY, E. K. *Phys. Rev. Lett.* 15 (1965) 917; 17 (1966) 66E.
72. SIMPSON, J. A., MENENDEZ, M. G. and MIELCZAREK, S. R. *Phys. Rev.* 150 (1966) 76.
73. WEINGARTSHOFER, A., EHRHARDT, H. and LINDER, F. *Proc. 6th Int. Conf. Phys. electron. atom. Collisions*, p. 91, 1969. Cambridge, Mass.; Massachusetts Institute of Technology Press.
74. HASTED, J. B. and AWAN, A. M. *J. Phys.* B2 (1969) 367.
75. BONESS, M. J. W., HASTED, J. B. and LARKIN, I. W. *Proc. Roy. Soc.* A305 (1968) 493; BONESS, M. J. W. and HASTED, J. B. *Phys. Lett.* 21 (1966) 526.
76. HEALEY, R. H. and REED, J. W. *The Behaviour of Slow Electrons in Gases*, 1941. Sydney; Amalgamated Wireless of Australia.
77. TOWNSEND, J. S. *Motion of Electrons in Gases*, 1925. London; Oxford University Press; *J. Franklin Inst.* 200 (1952) 563; *Phil. Trans.* 193 (1899) 129; 195 (1900) 259; *Proc. Roy. Soc.* A80 (1908) 207; A81 (1908) 464; A120 (1928) 511; TOWNSEND, J. S. and TIZARD, H. T. *Proc. Roy. Soc.* A87 (1912) 357.
78. BRADBURY, N. E. and NIELSEN, R. A. *Phys. Rev.* 49 (1936) 388; 50 (1936) 950; LOEB, L. B. and CRAVATH, A. M. *Phys. Rev.* 33 (1929) 605; LOEB, L. B. *Phys. Rev.* 48 (1935) 684.
79. MORSE, P. M., ALLIS, W. P. and LAMAR, E. S. *Phys. Rev.* 82 (1935) 412; MORSE, P. M. and ALLIS, W. P. *Phys. Rev.* 44 (1933) 269; TONKS, L. and ALLIS, W. P. *Phys. Rev.* 52 (1937) 710; ALLIS, W. P. and BROWN, S. C. *Phys. Rev.* 87 (1952) 419.
80. DRUYVESTEYN, M. J. *Physica, Eindhoven* 10 (1930) 69; DRUYVESTEYN, M. J. and PENNING, F. M. *Rev. mod. Phys.* 12 (1940) 87.
81. DAVYDOV, B. *Phys. Z. Sowjet.* 8 (1935) 59; SMIT, J. A. *Physica, Eindhoven* 3 (1937)

543; HOLSTEIN, T. *Phys. Rev.* 70 (1946) 367; MARGENAU, H. *Phys. Rev.* 73 (1948) 297, 309.

82. LOEB, L. B. *Basic Processes in Gaseous Electronics*, 1955. Berkeley, California; University of California Press; ALLIS, W. P. *Handb. Phys.* 21 (1956) 383.

83. HUXLEY, L. G. H. and ZAAZOU, A. A. *Proc. Roy. Soc.* A196 (1949) 402; HUXLEY, L. G. H. *J. atmos. terr. Phys.* 16 (1959) 46; *Proc. phys. Soc., Lond.* B64 (1951) 844; *Nuovo Cim. Suppl.* 9 (1952); *Aust. J. Phys.* 9 (1956) 44; 10 (1957) 118; 13 (1960) 578, 718; HUXLEY, L. G. H. and CROMPTON, R. W. *Proc. phys. Soc., Lond.* B68 (1955) 381.

84. PHELPS, A. V., PACK, J. L. and FROST, L. S. *Phys. Rev.* 117·(1960) 470; PACK, J. L. and PHELPS, A. V. *Phys. Rev.* 121 (1961) 798; PACK, J. L., VOSHALL, R. E. and PHELPS, A. V. *Phys. Rev.* 127 (1962) 2084; FROST, L. S. and PHELPS, A. V. Westinghouse Research Laboratory Paper 62-908-113-P1, 1962.

85. DUTTON, J. Joint Institute of Laboratory Astrophysics, Information Centre Report 4, 1967.

86. CROMPTON, R. W. and SUTTON, D. J. *Proc. Roy. Soc.* A218 (1953) 507; CROMPTON, R. W., HALL, B. I. H. and MACHLIN, W. C. *Aust. J. Phys.* 10 (1957) 366; HUXLEY, L. G. H., CROMPTON, R. W. and BAGOT, C. H. *Aust. J. Phys.* 12 (1959) 303.

87. HALL, B. I. H. *Proc. phys. Soc., Lond.* B68 (1955) 334.

88. HALL, B. I. H. *Aust. J. Phys.* 8 (1955) 468.

89. CROMPTON, R. W., ELFORD, M. T. and GASCOIGNE, J. *Aust. J. Phys.* 18 (1965) 409.

90. HUXLEY, L. G. H., CROMPTON, R. W. and ELFORD, M. T. *Bull. Inst. Phys., Lond.* 17 (1966) 251.

91. HUXLEY, L. G. H., CROMPTON, R. W. and ELFORD, M. T. *Brit. J. appl. Phys.* 17 (1966) 1237.

92. HEYLEN, A. E. D. *Proc. phys. Soc., Lond.* 76 (1960) 779; 79 (1062) 508.

93. HUXLEY, L. G. H. *Aust. J. Phys.* 10 (1957) 118; 13 (1960) 578, 718.

94. PHELPS, A. V. *Rev. mod. Phys.* 40 (1968) 399.

95. WALKER, I. C. Thesis, University of Edinburgh, 1965.

96. PACK, J. L., VOSHALL, R. E. and PHELPS, A. V. Westinghouse Scientific Laboratory Paper 62-908-113-P4, 1962.

97. PHELPS, A. V., FUNDINGSLAND, P. T. and BROWN, S. C. *Phys. Rev.* 84 (1951) 559.

98. BOWE, J. C. *Phys. Rev.* 117 (1960) 1411, 1416.

99. ENGLISH, W. N. and HANNA, G. C. *Canad. J. Phys.* 31 (1953) 768.

100. FROST, L. S. and PHELPS, A. V. *Phys. Rev.* 127 (1962) 1621; ENGELHARDT, A. G., PHELPS, A. V. and RISK, C. G. *Phys. Rev.* 135 (1964) A1566; HAKEM, R. D. and PHELPS, A. V. *Phys. Rev.* 158 (1967) 70.

101. SHERMAN, B. *J. math. Analys. Applic.* 1 (1961) 342.

102. GERJUOY, E. and STEIN, S. *Phys. Rev.* 97 (1955) 1671.

103. RAMIEN, H. *Z. Phys.* 70 (1931) 353.

104. BEKEFI, G. and BROWN, S. C. *Phys. Rev.* 112 (1958) 159.

105. DALGARNO, A. and MOFFETT, R. J. *Proc. Symp. Collision Processes*, Indian National Academy of Sciences, 1962.

106. POLLOCK, W. J. Thesis, University of Edinburgh, 1967.

107. FINKELNBURG, W. and MAECKER, H. *Handb. Phys.* 22 (1956) 378.

108. MARGENAU, H. *Phys. Rev.* 69 (1946) 508.

109. PHELPS, A. V., FUNDINGSLAND, O. T. and BROWN, S. C. *Phys. Rev.* 84 (1951) 559.

110. JACKSON, W. *Trans. Faraday Soc.* 42A (1946) 91; HORNER, F., TAYLOR, T. A., DUNSMUIR, R., LAMB, J. and JACKSON, W. *J. Instn elect. Engrs* 93 (1946) 53; COLLIE, C. H., HASTED, J. B. and RITSON, D. M. *Proc. phys. Soc., Lond* 60 (1948) 71.

111. GOULD, L. and BROWN, S. C. *Phys. Rev.* 95 (1954) 897.

112. GILARDINI, A. L. and BROWN, S. C. *Phys. Rev.* 105 (1957) 25, 31.

113. BEKEFI, G. and BROWN, S. C. *Phys. Rev.* 112 (1958) 159.

114. WHITMER, R. F. *Phys. Rev.* 104 (1958) 572.

115. BUCHANAN, T. J. *Proc. Instn elect. Engrs* 99 (1952) 61.

116. ANDERSON, J. M. and GOLDSTEIN, L. *Phys. Rev.* 100 (1955) 1037; 102 (1956) 388, 933.

117. RAO, K. V. N., VERDEYEN, J. T. and GOLDSTEIN, L. *Proc. Inst. Radio Engrs.* 49 (1961) 1877.
118. GINSBERG, V. L. *J. Phys., Moscow* 8 (1944) 253; VILENSKI, I. M. *J. exp. theor. Phys.* 22 (1952) 544.
119. BURKHARDT, G., ELWERT, H. and UNSOLD, A. *Z. Astrophys.* 25 (1948) 310; BURKHARDT, G. *Ann. Phys., Lpz.* 5 (1950) 373.
120. SPITZER, L. and HARM, R. *Phys. Rev.* 89 (1953) 977.
121. LIN, S. C., RESLER, E. L. and KANTROWITZ, A. *J. appl. Phys.* 26 (1955) 95.
122. KELLY, D. C., MARGENAU, H. and BROWN, S. C. *Phys. Rev.* 108 (1957) 1367.
123. VEATCH, G. E., VERDEYEN, J. T. and CAHN, J. H. *Abstr. 18th Gaseous Electronics Conf.*, p. 13, 1965; INGRAHAM, J. C. *Abstr. 18th Gaseous Electronics Conf.* p. 18, 1965.
124. HIRSCHFIELD, J. L. and BROWN, S. C. Massachusetts Institute of Technology, Research Laboratory of Electronics Technical Report, July 15, 1957.
125. VARNERIN, L. *Phys. Rev.* 84 (1951) 563.
126. CHEN, C. L. *Bull. Amer. phys. Soc.* 7 (1962) 487.
127. FORMATO, D. and GILARDINI, A. *Proc. 5th Int. Conf. Ioniz. Phenom. Gases*, 1961. Munich; North Holland Publishing Company.
128. BEKEFI, G., HIRSCHFIELD, J. L. and BROWN, S. C. *Phys. Fluids* 4 (1961) 173.
129. SCHUBERT, E. K. *Abstr. 5th Int. Conf. Phys. electron. atom. Collisions*, p. 263, 1967. Leningrad; Akademii Nauk.
130. FISK, J. B. *Phys. Rev.* 49 (1935) 167.
131. FISK, J. B. *Phys. Rev.* 51 (1937) 25.

Chapter 5

EXCITATION OF ATOMS AND MOLECULES BY ELECTRONS

5.1 THEORY OF ELECTRON EXCITATION

This chapter describes the excitation of atoms, of ions and of molecules by electrons. These inelastic processes ($e0/e0'$, ^{m}, en/en', ^{m}, $e00/e00'$, $^{m, v, r}$) require kinetic energy of impact and are of great importance in weakly ionized gases and in hot plasmas, whose electron temperatures can often be inferred by monitoring radiation and computing with the aid of known excitation functions. Review articles on excitation are available[1-4].

At the threshold electron energy, equal to the energy difference E_{ex} between the initial and final levels, the cross-section will in general be zero, but should rise proportionately to the velocity v possessed by the impacting electron after the collision. The energy range over which this threshold law is expected to hold is not known. Neighbouring resonances, energy levels and also degeneracy complicate the threshold function, sometimes producing finite cross-section at $E_e = E_{ex}$; finite cross-sections at threshold are also found for the electron excitation of positive ions.

As the impact energy increases, the electron–atom cross-section rises, until for $E_e \sim 1\cdot5$–$2\ E_{ex}$ for optically allowed (electric dipole) transitions a maximum is reached. The maximum cross-section can be as large as 10^{-16} cm². For optically forbidden transitions, the maximum is much sharper and may be reached within one electron volt of threshold. For allowed transitions, the high-energy cross-section falls off as $E_e^{-1} \ln E_e$; for forbidden transitions, as E_e^{-1}.

Many quantum theory calculations of excitation cross-sections have been made. For high energies of impact, weak coupling approximations are suitable; in these, the coupling from the initial to the final states, and to all other states, is neglected. The first Born approximation is widely used, and an even simpler form, the Bethe approximation, is also effective at high and moderate impact energies. A more complicated weak coupling approximation is that of the distorted wave, which represents the free electron by functions describing its motion in the static fields of the atom before and after excitation. For excitation processes in which there is an important contribution from electron exchange, the Born–Oppenheimer approximation is unsatisfactory, since certain first order terms in the interaction energy are neglected; a different expression, simple to apply and yielding satisfactory results, is the Ochkur approximation.

Strong coupling approximations are applicable at lower electron energies than weak coupling approximations. They include the modified Bethe approximation, the impact parameter method and the unitarized Born approximation. When the level to be excited is very close in energy to the state before excitation, the exact resonance approximation can be applied. When the coupling between the initial and final states is weak, but that to a third state is strong, one can use the close coupling approximation. The semi-classical perturbation–impact parameter treatment, sometimes called the 'dipole approximation', has also proved valuable[5, 6].

The Born and Bethe Approximations

The Born approximation for electron excitation of atoms[7] from state m to state n is conveniently expressed in terms of the momentum exchange variable

$$K^2 = k_1^2 + k_2^2 - 2k_1k_2 \cos \theta \tag{5.1}$$

where $k_{1, 2} = mv_{1, 2}/\hbar$ are the wave vectors of the incident and the scattered electron respectively.

On the Born approximation the expression for the differential cross-section is

$$\frac{d\sigma_{mn}}{dK} = \frac{128\pi^5 m_e^2 e^4}{k_1^2 h^4 K^3} |\varepsilon_{mn}(K)|^2 \tag{5.2}$$

where the matrix element of the transition dipole moment is

$$\varepsilon_{mn}(K) = \int \sum_{s=1}^{s=Z} \exp(iKx_s)\psi_m\psi_n^* \, d\mathbf{r} \tag{5.3}$$

To obtain this expression, integration of the wavefunctions ψ_m and ψ_n of the two levels is performed over the coordinates x_s of all atomic electrons along the path of impact.

Now the optical oscillator strength f_0 is defined as

$$f_0 = \frac{m_e^2 e^4 |x_{mn}|^2 v_{mn}}{R\hbar^4} \tag{5.4}$$

where the matrix elements

$$x_{mn} = \int \psi_m x \psi_n^* \, d\tau$$

Suppose that by analogy the generalized oscillator strength[8] is defined as

$$f_{m, n}(K) = (E_n - E_m)\frac{2}{K^2} \varepsilon_{mn}\varepsilon_{mn}^* \tag{5.5}$$

the differential cross-section becomes

$$\frac{d\sigma_{mn}}{dK} = \frac{16\pi^3 m_e e^4 f_{mn}(K)}{k_1^2 h^2 (E_n - E_m)K} = \frac{2k_2 f_{mn}(K)}{(E_n - E_m)k_1} \tag{5.6}$$

23*

The form of the function $f(K)$ is an important description of the transition; it can be deduced from a measurement of scattered electron current as a function of angle θ; for an allowed electric dipole transition f is a rapidly decreasing function of K; for a forbidden transition it is small for $K = 0$ but increases and passes through a maximum; it can also pass through a minimum, at a value of K which is related to a nodal surface in the wave-function; this might be characteristic of a Rydberg state in a molecule.

Total excitation cross-sections are given by the expressions

$$\sigma_{mn} \simeq \frac{64\pi^5 m_e^2 e^4}{k_1^2 h^4} |x_{mn}|^2 \ln \left(\frac{2m_e v^2}{E_n - E_0} \right) \tag{5.7}$$

for allowed transitions, and

$$\sigma_{mn} \simeq \frac{128\pi^7 m_e^3 e^4}{k_1^2 h^6} |x_{mn}|^2 E_e \tag{5.8}$$

for forbidden (electric quadrupole) transitions. Thus at electron energies sufficiently high for the Born approximation to be valid, the rate of fall-off of total excitation cross-section is $E_e^{-1} \ln E_e$ for allowed transitions and E_e^{-1} for forbidden, or multipole transitions. There are cases where contributions from both can be expected to yield a composite rate of fall-off. Processes involving two electrons are expected[9] to fall-off with E_e^{-1}.

According to equation 5.6, the generalized oscillator strength is proportional to the differential cross-section for scattering in angle:

$$f(K) = \frac{K^2 h^2}{8\pi^2 m_e e^4} (E_n - E_m) \frac{k_1}{k_2} \frac{d\sigma}{d\Omega} (\theta) \tag{5.9}$$

The change of momentum Δp of the impacting electron can be represented in terms of the ante-collision and post-collision momenta $p_a = \hbar k_1$ and $p_p = \hbar k_2$:

$$(\Delta p)^2 = p_a^2 + p_p^2 - 2p_a p_p \tag{5.10}$$

for zero-angle scattering. Equation 5.9 is often written

$$f(K) = \frac{E_{ex}}{2} \frac{p_a}{p_p} (\Delta p)^2 \frac{d\sigma}{d\Omega} (0) \tag{5.11}$$

for excitation of a level of energy $E_{ex} = E_e - \frac{1}{2} p_p^2 / m_e$ by an electron of energy $E_e = \frac{1}{2} p_a^2 / m_e$. Since

$$(\Delta p)^2 = \frac{E_e}{2} \left(\frac{E_{ex}}{E_e} \right)^2 \left[1 + \frac{1}{2} \left(\frac{E_{ex}}{E_e} \right) + \frac{5}{16} \left(\frac{E_{ex}}{E_e} \right)^2 + \ldots \right] \tag{5.12}$$

and

$$\frac{p_a}{p_p} = 1 + \frac{1}{2} \left(\frac{E_{ex}}{E_e} \right) + \frac{3}{8} \left(\frac{E_{ex}}{E_e} \right)^2 + \ldots \tag{5.13}$$

one can write

$$f(K) = \frac{E_{ex}^3}{4E_e^2} (E_e + E_{ex}) \frac{d\sigma}{d\Omega} (0) \tag{5.14}$$

neglecting terms in $(E_{ex}/E_e)^2$ throughout. Thus a high-energy measurement of the zero scattering angle cross-section for loss of energy E_{ex} by the impacting electron yields the generalized oscillator strength; in the limit of zero momentum exchange, this is equal to the optical oscillator strength. Note that the factor E_{ex}^3 in this equation results in a very rapid fall-off of this cross-section with increasing E_{ex}, which presents difficulties in the study of continuum and inner shell excitation.

Equation 5.2 can be written

$$\frac{d\sigma}{d\Omega}(0) = \frac{4p_p}{p_a}\frac{1}{(\Delta p)^4}|\varepsilon_{mn}|^2 \tag{5.15}$$

which when combined with equation 5.11 yields

$$f(K) = 2E_{ex}[\varepsilon_1^2 + (\varepsilon_2^2 - 2\varepsilon_1\varepsilon_3)(\Delta p)^2] \tag{5.16}$$

with the dipole moment matrix element

$$\varepsilon_l = \frac{1}{l!}|\langle m|\sum_s x_s^l|n\rangle| \tag{5.17}$$

In the limit $\Delta p \to 0$ (that is, $E_e \to \infty$) and $f(K) \to 2E_{ex}\varepsilon_1^2$, which is exactly the optical oscillator strength; this possesses a finite value for optically allowed transitions, but is almost zero for electric dipole forbidden transitions. Electric quadrupole transitions have been defined in different ways for electron and optical spectra[10]. As has been shown, the behaviour of the generalized oscillator strength $f(K)$ as a function of K is related to the wave functions of the two energy levels[11-13].

Since measurements of differential cross-sections cannot be conducted strictly at zero angle but must apply to all electrons scattered through angles smaller than those subtended by a certain orifice, it is important to estimate the contribution to equation 5.15 from scattering to an angle θ. Whilst the contribution to p_a/p_p is small, that to $(\Delta p)^2$ is large:

$$(\Delta p)^2 \simeq 8\left(E_e - \frac{1}{2}E_{ex}\right)\left\{\sin^2\frac{\theta}{2} + \left[\frac{E_{ex}}{4(E_e - \frac{1}{2}E_{ex})}\right]^2\right\} \tag{5.18}$$

Other Approximations

In addition to the Born–Bethe approximation, a number of theoretical methods have been used for calculations of excitation functions. Among the first to be used were distorted-wave calculations of Seaton[14] and of Vainshtein[15].

The Ochkur approximation[16] shows that the cross-section for exchange scattering falls off as the cube of electron energy. The exchange process accounts for most of the excitation of triplet levels from the singlet ground state of a light atom, but it is only a correction term in excitations described by the Bethe–Born approximation.

There are two approximate methods of calculating cross-sections, which are of value for astrophysics and plasma physics applications. The first is the method of eigenfunction expansions[17, 18]. The wave function for the electron-plus-atom system is expanded in terms of the atom eigenfunctions and functions F_n for the impacting electron:

$$\Psi = \Omega \sum_n \psi_n F_n \qquad (5.19)$$

where Ω is an operator which makes Ψ antisymmetric. The sum is made over discrete atomic states of principal quantum number n, and can include an integration over the continuum. The procedure is difficult, so that some calculations are made only with $n = 0$ (the static field approximation); or with $n = 0$ but exchange taken into account; or with only a few low-lying terms taken into account, using coupled equations; or using variational methods, in which it is possible to make allowance for the continuum. The technique is in general rather poor for calculating excitation cross-sections for levels which are widely separated in energy from their neighbours, but is good for groups of strongly coupled levels which are only weakly coupled to other levels.

The second approximate technique is really a combination of two methods, each of which is used only in the region where it is most suitable. It is known as the 'dipole approximation'[5], and it is most satisfactory for excitation processes of large oscillator strengths and small energy differences. It has also been used for ionization calculations. The two methods are: (*i*) the purely classical approximation, but including exchange; and (*ii*) the semi-classical impact parameter technique, using time-dependent perturbation theory. The former is most satisfactory for close electron–electron encounters, the latter for distant encounters. A suitable cut-off for each method is chosen and the two cross-sections summed. The semi-classical technique assumes classical scattering of the impact electron; time-dependent perturbation theory is used to calculate the probability of transition due to the field produced by the virtual photons emitted during the trajectory[19]. The method requires a knowledge of the oscillator strength and yields the correct high-energy cross-section function fall-off, unlike the purely classical method.

The incorrect (logarithmic) term in the high-energy fall-off predicted classically arises from the incorrect distribution of transferred energy; in quantum theory the transferred energy is precisely the energy of the level excited, but classically a continuum of energy is possible. It has been shown by Percival, using a method originally due to Heisenberg, that the correct fall-off is obtained when the correct transferred energy is taken. For exchange excitation, the E^{-3} fall-off predicted in the Ochkur approximation is also obtained classically[16].

A number of semi-empirical expressions for excitation functions have been given by various authors[19-24]. One such expression[18] is

$$\sigma = \frac{4\pi f}{v_e^2 E_{\text{ex}}} \ln \left(\frac{v_e + v_{ep}}{|v_e - v_{ep}|} \right) \qquad (5.20)$$

where v_{ep} represents the post-collision velocity of the impact electron. The excitation of electric quadrupole transitions can be treated empirically provided the cross-section is known at two high impact energies[25], but the formula yields results in excess of the Born approximation.

An empirical expression due to Drawin[26] for forbidden transitions without exchange is

$$\sigma = \sigma_{\max} \frac{4E_{ex}}{E_e}\left(1 - \frac{E_{ex}}{E_e}\right) \tag{5.21}$$

Experimental $\sigma_{\max}$ are in general to be preferred to theoretical $\sigma_{\max}$ for use in this formula. Similar expressions are given for allowed transitions and for ionization.

5.2 EXPERIMENTAL METHODS OF MEASUREMENT OF EXCITATION CROSS-SECTIONS

The most common type of experiment for excitation cross-section measurement involves the bombardment of a gas or atomic beam with a velocity-selected electron beam, and the measurement of a known part of the optical radiation. Only recently has this technique been understood in sufficient detail to allow accurate and reliable data to be obtained. It must be admitted that only data reported since about 1965 can be treated with confidence.

In the case of excitation to a level from which downward transitions are all forbidden, the metastable atom usually has time to travel out of the collision region to the surrounding surfaces before radiation takes place. The detection and measurement of a known part of the metastable atom flux is therefore necessary. Such detection is described in Section 3.14. An alternative experiment suitable both for allowed and forbidden transitions is the measurement of the fraction of the electron beam which suffers inelastic energy loss.

Consider the excitation by an electron beam of a level j from the ground state 0; radiation emitted during the downward transition $j \to k$ is measured and is distinguishable spectroscopically from all other radiation emitted. Its density is therefore determined by the population of the state j, but this state may also be produced by downward transitions from a higher state i; therefore

$$\frac{dn_j}{dt} = -A_j n_j + n_0 \sigma_j n_e v_e + \sum_i A_{ij} n_i \tag{5.22}$$

where A_{ij} is the probability of the atom making a radiation $i \to j$ transition in 1 sec, and A_j represents $\sum_k A_{jk}$. The symbol σ_j represents the electron excitation cross-section for the transition $0 \to j$. In equilibrium,

$$n_j = \frac{n_0 \sigma_j v_e n_e + \sum_i A_{ij} n_i}{A_j} \tag{5.23}$$

The total quantum emission $j \to k$ per unit path length is

$$J_{jk} = A_{jk} n_j S \tag{5.24}$$

where S is the cross-sectional area of the electron beam, which is assumed to be homogeneous. The electron current

$$I_e = v_e S e n_e \tag{5.25}$$

One may calculate σ_j from the observed J_{jk} provided n_0, all n_i, I_e, S and all A coefficients are known. The state densities n_i are found by employing the relation

$$J_{il} = A_{il}n_i S \tag{5.26}$$

that is, the quantum emission must also be measured for one line function arising from every level i that can combine with j. The states l that are chosen for this measurement are not restricted; one may choose j for this purpose. It is usually sufficient to apply only first order cascade corrections. A more complete discussion of transition probabilities (lifetimes, A coefficients, oscillator strengths, line strengths) appears in Chapter 9.

A study of the literature shows that only in a minority of transitions is all the information available for the deduction of true electron excitation functions; often one must be content with studying only the variation of line emission with electron energy and normalizing this at high energies to a Born approximation calculation of the excitation cross-section.

In the analysis outlined in equations 5.22–5.26, the following processes have been neglected: electron excitation from all states except ground; collisions of the second kind; collisions between excited and ground state atoms; and absorption of resonance radiation. Sufficiently low gas pressures and electron density conditions must be imposed for such neglect to be permissible; an experimental check of the linear relations between line intensity and gas pressure, and between line intensity and electron beam current, is necessary. The sensitivity of radiation detection must be adequate for these conditions to be met, and in some of the early experiments this was not so; effects due to both radiation transfer and to excited atom diffusion were observed. Long-exposure photography with stable beam currents and gas pressures has been superseded by modern photomultiplier technique.

The electron beam is directed through a gas collision chamber which must be substantially free of electric fields in its interior. The mean energy and energy distribution of the electrons must be known, and it is necessary for the distribution to be narrow compared with the structure of the excitation function. Any secondary electrons produced by impinging of the beam on the collector must be excluded from the collision region by a potential barrier.

It is important that the electron and gas densities should be uniform over the entire path length from which radiation is collected, but it is not possible to realize electron flux density constancy across the width of the beam, owing to space charge; therefore an optical scan of the beam width is desirable, so that the integrated emission may be measured.

Both spectrometers and filters have been employed for wavelength selection in electron excitation experiments. In the calibration of the radiation detector, the conventional tungsten strip standard lamps are used, and the solid angle subtended by the detector at the source must be known. Rotating shutters are sometimes employed to diminish intensities by known ratios. The very large factors which are necessary impose difficult electronic problems. Neutral filters of known attenuation are unsuitable. It is advisable where possible to avoid geometrical errors in the photomultiplier calibration by covering the detector with light-diffusing glass[27].

Pressure dependence of the form of excitation functions is often observed.

It can arise from imprisonment of resonance radiation[28], from collisional excitation transfer with impurity gases[29-31], or from threshold effects arising from ionization[32, 33].

Imprisonment of resonance radiation is unimportant in the study of excitation functions using emission lines between two excited states, but when the line terminates on the ground state, it is important that the collision chamber gas be optically thin. Collisional excitation transfer has been demonstrated in helium excitation studies, producing, for example, increased populations of ^1D and ^3D states via collisional transfer to higher n^1P states, followed by cascading. The effect of resonance radiation imprisonment is to produce photon emission from areas outside the beam; the emission is unpolarized, and must be shown to be negligible in accurate measurements.

Multiple scattering of electrons by gas can destroy the spatial definition of the beam and cause photometric complications.

For an absolute cross-section determination, the total radiation emitted must be deduced from a measurement over a small solid angle. The angular distribution of radiation depends upon the type of transition that the radiation accompanies. Dipole transition radiation, which includes almost all radiation from atomic collisions, has an angular distribution given by

$$I(\theta) = 3\bar{I}\left(\frac{1 - P\cos^2\theta}{3 - P}\right) \tag{5.27}$$

where $\bar{I}$ is the intensity per unit solid angle averaged over all solid angle, and the polarization factor

$$P = \frac{I_{||} - I_{\perp}}{I_{||} + I_{\perp}} \tag{5.28}$$

where $I_{||}$ and $I_{\perp}$ are the intensities observed at 90° to the electron beam axis with electric vectors respectively parallel and perpendicular to the electron beam axis. To eliminate P, the measurement of radiation at two angles is sufficient. In measurements of Lyman α radiation from the excitation of atomic hydrogen to the $2p$ level, Fite, Stebbings and Brackman[34, 35] used a single measurement at

$$\theta = \text{arc cos} \frac{1}{\sqrt{3}} \tag{5.29}$$

and checked the validity of equation 5.27 by making measurements at three angles. Theoretical values of P may also be used[36]. In some measurements, only 90° data are taken and only an 'apparent' cross-section σ_1 is given. Since the polarization factor is only large near the threshold of excitation, the cross-sections deduced on the assumption of isotropic radiation are only seriously in error near threshold.

There is an unfortunate error neglected in much published work which arises from the modification of the state of polarization of radiation by diffraction gratings or prism surfaces in monochromators. This should be measured and a correction applied[37]. Discussions of depolarization in monochromators and the errors arising therefrom are available[1, 2, 38, 39]. One solution suggested by Barrows and Dunn is to observe radiation at 90° to the electron beam axis, but by the use of polaroid sheet to allow only the component of polarization at 35° 14' to the electron beam axis to pass into the monochromator. It can be shown that this radiation is equal to $\frac{1}{2}\bar{I}$, and furthermore the output is independent of instrumental polarization effects.

There is also a possible error for long-lived states due to Larmor precession of electrons about residual magnetic fields. These fields are capable of curving the paths of low-energy electron beams, so it is often advisable to neutralize the residual by Helmholtz coils. For a magnetic field H perpendicular to the electron beam and along the direction of observation, the axis of polarization is rotated through angle Φ, where

$$\tan 2\Phi = 4\pi L\tau \qquad (5.30)$$

where the Larmor precession frequency

$$L = \frac{g\mu_0 H}{2\pi\hbar} = 1{\cdot}4 \times 10^6 gH \text{ Hz} \qquad (5.31)$$

with Bohr magneton μ_0, Landé g factor and lifetime τ. The polarization is reduced from a value P_0 to

$$P = P_0[1 + (4\pi L\tau)^2]^{-1/2} \qquad (5.32)$$

A number of factors have so complicated the optical technique of excitation function measurement that only the experimental work of the last few years is to be relied on in detail. Increased sensitivity of photomultipliers, coupled with techniques of background noise reduction, has enabled the collision chamber gas to be maintained at much lower pressures than in earlier work. The sensitivity of excitation function structure to pressure variation can now be greatly reduced, if not effectively eliminated. There is considerable real fine structure in excitation functions, but it is possible that unreal structure can arise owing to cascading from levels lying at slightly higher energy. Thus the emission function He 4^3S → 2^3P has structure influenced by 3^3P → 2^3S. It has also been suggested that impurities can give rise to unreal cross-sections below threshold, but this interpretation has been criticized.

Effects can also arise from slow electrons formed collisionally, particularly in the high-energy excitation of forbidden levels whose cross-sections decrease very fast with increasing energy.

In molecular excitation processes, the large cross-sections for transfer of excitation between vibrationally excited states and the gas make necessary the application of a pressure-dependent correction[40].

The Deduction of Cross-sections from Measurements of Optical Emission by Plasma

Considerable data are available for the intensities of lines emitted from discharges whose operating conditions, electron density and energy distribution, gas pressure and temperature are known. The complications in the analysis of the data are such that, although excitation functions have been deduced[41] (for example, for Hg 6^3P), the method is greatly inferior to electron beam technique where this is available. However, in discharge experiments it is sometimes possible to deduce excitation cross-sections for collisions of electrons with excited atoms; this must remain the principal justification for the method today. The excitation cross-sections for Hg 7^3S from 6^3P and 6^1S levels by 3 eV electrons were found in this way[42] to be 13 times greater than those for their excitation from the ground state.

Although many electrical discharge experiments have been carried out under d.c. conditions, afterglow techniques are probably more effective. Studies of the time-dependence of radiation emitted from and absorbed by the afterglows of discharges were made by Phelps and Molnar[43] (see also Chapter 13). The rate of conversion of He 2^1S $\rightarrow$ 2^3S by collisions with near-thermal energy electrons was deduced from measurements of the time-dependence of absorption of radiation of wavelength appropriate to convert one or the other state to 3^1P. The measured rate corresponds to the extremely large cross-section, $\sim 3 \times 10^{-14}$ cm^2.

The excitation functions for states of highly ionized species have in recent years been inferred from optical emission studies using hot-plasma machines. A knowledge of these cross-sections can be used in the diagnostics of electron temperature. In order that the cross-section for a single impact velocity may be inferred, the vacuum ultra-violet resonance radiation intensity from a volume of optically thin plasma must be measured absolutely at a time at which the electron temperature and ion number density are known. The electron temperature ($\sim 10^5$°K) is diagnosed by measuring the Doppler profile of scattered laser light. The ion density is inferred from separate experimental study of the machine. The measurement of N V $2s$–$2p$ excitation agrees with quantum theory calculations.

5.3 TECHNIQUES OF EXCITATION STUDY WITHOUT DETECTION OF PHOTONS

Non-optical techniques are of two types: those in which the measured energy loss of a fraction of the impacting electrons serves as an indication of the fraction of electrons that have suffered collision; and those in which the excited atoms formed in the collision are detected by means other than optical emission. The latter will only be considered briefly.

Although detection of metastables by optical absorption is a valuable technique[44, 45], the more popular method of detecting metastable atoms[46, 47] is by means of the electrons which they can eject from metal surfaces. Accounts of some electron excitation experiments in which metastable atoms were so detected were given in the first edition of this book, and some more recent studies can now be instanced[48–52]. Particle multipliers have been used for metastable atom detection, and separation of He 2^3S and 2^1S in inhomogeneous magnetic fields has been achieved; He 2^1S can be quenched in a sufficiently strong electric field (225 kV cm^{-1}). The effects of elastic scattering and excitation exchange of the metastable atoms must be avoided by minimization of the gas pressure. The metastable atom H $2s$ can be quenched in a relatively weak electric field, and the emitted Lyman α radiation detected with a photon counter.

Inner shell excitation processes in which the excitation process is followed by Auger processes yielding ionization and characteristic energy electrons can be studied by measuring the momentum-analysed electrons and the charge-analysed ions in coincidence.

Experiments directed solely at detecting electrons inelastically scattered in collisions have become increasingly important, both for total and differential cross-sections, particularly at zero angle (transmission). This is the field of research known as 'electron spectroscopy'. It is directed partly at total cross-section measurement and partly at:

1. The discovery of characteristic energy losses, or energy levels, particularly forbidden levels of molecules (triplet states) which do not show up in optical spectroscopy[53, 54]. The latter are searched for near the threshold of the excitation function, which is very sharp; this technique has been called 'threshold spectroscopy'.
2. The deduction of generalized oscillator strengths, using the Bethe–Born approximation, from differential measurements[76] of inelastically scattered electrons at high-impact energy[55-68].
3. The discovery of inelastic channel resonances appearing as structure in the zero-angle or fixed-angle inelastically scattered electron cross-section functions.

Either crossed-beam or beam–gas techniques can be used in these studies. Post-collision momentum analysis of scattered electrons is achieved by one of the methods outlined in Section 3.13. Scattering intensities at large angles are sufficiently small for single electron counting to be necessary. It is necessary to beware of multiple scattering effects arising in the following way: since the decrease of elastic scattering differential cross-section with increasing angle is much less rapid than that for inelastic scattering, the flux received at a large angle may include a background of double events: a large-angle elastic event followed by a very small-angle inelastic event. The removal of this background requires a careful analysis of the data with varying gas pressure or atomic beam intensity.

Measurement of total excitation cross-sections using post-collision electron momentum analysis requires collection of electrons scattered over all angles. But at impact energies sufficiently high for the Born–Bethe approximation to be valid, collection of zero-angle electrons is sufficient; when the collision path length, the atomic number density, and the maximum scattering angle for the inelastically scattered electrons in the apparatus are known, then a calculation can be made of the cross-section, correcting the oscillator strength according to equations 5.18 and 5.11 and then applying equations 5.4 and 5.8.

The discovery of forbidden energy levels by threshold electron spectroscopy is principally of value in the physics of polyatomic molecules, since forbidden atomic energy levels are usually accessible optically by intercombinations. Triplet states of molecules have been investigated[69-72] by means of momentum analysers, retarding potential difference technique, and scavenger technique for the detection of very slow electrons. Since the excitation functions for forbidden levels are sharply peaked, it is the very slow electrons which are of greatest importance; their detection implies that the impacting electrons are close to excitation threshold. Identification of these levels is assisted by measurements of angular distribution of scattered electrons.

At high impact energies, electrons suffering energy loss in excitation are mostly scattered through sufficiently small angles to enable them to be collected at a forward-placed momentum analyser. This enables the high-energy fall-off of total cross-section to be studied, using the Bethe–Born approximation in the form:

$$\sigma = \frac{A}{E_e} \ln cE_e + \frac{B}{E_e} \qquad (5.33)$$

where the first term is appropriate to allowed electric dipole transition, the

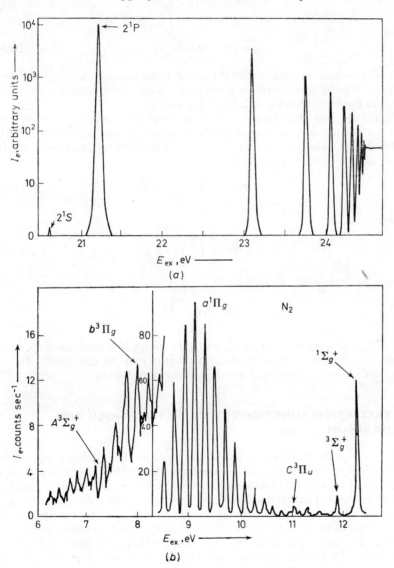

Fig. 5.1. Electron impact spectra: (a) helium at zero scattering angle and 25 keV impact energy[74]; (b) nitrogen at scattering angles 3° (left-hand) and 12° (right-hand) and 48 eV impact energy[75]

second to forbidden transition, and

$$A = \frac{4\pi a_0^2 R^2 f}{E_{ex}} \qquad (5.34)$$

In the case where both terms contribute, it is possible to write

$$\sigma = \frac{A}{E_e} \ln c'E_e \qquad (5.35)$$

with

$$c' = c \exp (B/A) \qquad (5.36)$$

An approximate expression for the constant c, namely $c = 4/E_{ex}$ has been found[21, 73] to be unsuccessful in many cases. A quantity of data has been analysed in the form of values of cE_{ex}, $c'E_{ex}$ and A.

When the post-collision momentum analyser is of sufficiently small aperture for the measurement to be regarded as one of only zero-angle inelastically scattered electrons, then the analysis of the data follows equation 5.14. This may be written in the form

$$\Delta\sigma(0) = \frac{4(E_e - \frac{1}{2}E_{ex}) f\beta}{E_{ex}^3} \qquad (5.37)$$

where

$$\beta = \left\langle \left(\frac{\theta}{\alpha}\right)^2 + 1 \right\rangle^{-1} \qquad (5.38)$$

and

$$\alpha = \frac{E_{ex}}{2(E_e - \frac{1}{2}E_{ex})} \qquad (5.39)$$

Since β varies only slowly with E_e, a scan of energy loss produces a spectrum from whose peak heights oscillator strengths may be calculated. Typical electron spectra so produced are shown in *Figure 5.1*.

5.4 EXCITATION FUNCTIONS CLOSE TO THRESHOLD AND MAXIMUM

It is easily seen from binary encounter classical theory that a zero cross-section is expected at threshold. Consider the classical bound electron possessing orbital velocity and phase angle ϕ. In a collision, the energy transferred from the incident electron will be some function $E_T(\phi) \leqslant E_e$, and for a transition to take place $E_T \geqslant E_{ex}$. For $E_e = E_{ex}$, E_T will only become equal to E_{ex} for a single value of ϕ, that is, an infinitely small range of ϕ values; since these values are equally probable, the cross-section should be zero at threshold. This is not the case for the excitation of positive ions, for which it is no longer possible to neglect the force field experienced by the incident electron. The considerable interaction potential so distorts the above energy relations that even at threshold the bound electron may receive energy E_{ex} over a finite range of ϕ values. A finite threshold cross-section function is shown in *Figure 5.12* (see page 370).

It is customary to denote the *direct* scattering amplitude* of an electron of velocity v_e accompanied by a transition from state a to state a' by the notation $f(a'v_{ep}, av_e)$, where the post-collision electron velocity is denoted by v_{ep}. The *exchange* scattering amplitude, appropriate to scattering with electron exchange, is denoted by $g(a'v_{ep}, av_e)$, and mixed scattering is also possible.

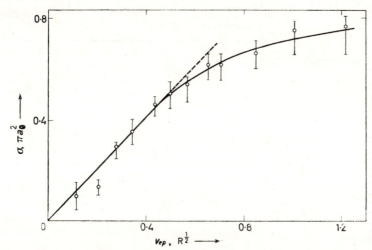

Fig. 5.2. Absolute electron excitation function[34] *for* H $1s \rightarrow 2p$; *the post-collision velocity* v_{ep} *of the projectile electron is in units of* $(Rydberg)^{1/2}$; *threshold structure (see Fig. 5.3) does not appear in these measurements*

(From Seaton[4], by courtesy of Academic Press)

For the hydrogen atom, with quantum numbers n, l and m, the first Born approximation expression for the cross-section for direct excitation to a state n, l, m is

$$_{nl}\sigma_{n'l'} = \frac{v_{ep}}{v_e} \frac{1}{2l+1} \sum_{m, m'} \int |f(a'v_{ep}, av_e)|^2 \, dv_{ep} \qquad (5.40)$$

For exchange excitation, the f is replaced by g.

The scattering amplitude is finite at the excitation threshold. For $v_{ep} = 0$, the cross-section is zero, but it can be shown that f is constant; thus the cross-section increases linearly with electron post-collision velocity:

$$\sigma \propto \frac{v_{ep}}{v_e}$$

where $v_{ep} \ll v_e$ and $\sigma \propto v_{ep}$. Moreover, if the electron velocities are measured in atomic units, such that for unit velocity v_e, $\frac{1}{2}m_e v_e^2 = R$ (R denoting one Rydberg), the cross-section for $1s$–$2p$ excitation of the hydrogen atom

---

*The reader should distinguish between f as used to denote direct scattering amplitude, and as used earlier to denote generalized oscillator strength.

is simply $v_{ep}\pi a_0^2$ at sufficiently low energies. Apart from the perturbation of the threshold discussed below, the measurements of Fite and Brackman[34] are in reasonable conformity with this expression (*Figure 5.2*). For more complex atomic systems, the higher partial wave summations converge rapidly, and sufficiently close to threshold the linear velocity dependence is applicable for *s* wave scattering. However, this may have little practical importance since higher power components arise extremely close to threshold; for example, the calculated cross-section for the excitation process O $2p^4$ ^{3}P → $2p^4$ ^{1}D is found to be virtually v_{ep}^3 dependent[77]. Many experi-

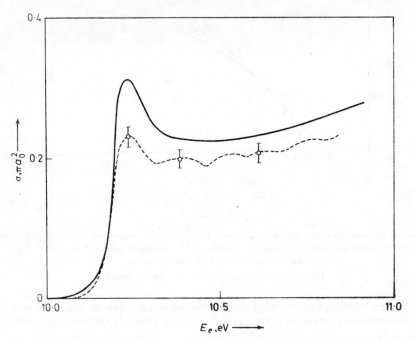

Fig. 5.3. H 2p excitation function[78] perturbed by neighbouring H 2s energy level; broken line data are of higher energy resolution than full line

ments show effectively a linear increase with energy for one or two electron volts above threshold. The range over which the $E_{ep}^{1/2}$ threshold law is predicted to hold is not known, and may sometimes be very small. On the other hand, the ^{3}D levels of helium show v_{ep} dependence for more than 3 eV, despite complications arising from resonances.

The v_{ep} threshold dependence of the excitation function is seriously perturbed by the presence of near-degenerate levels. The excitation of H 2p shows[78] in effect a finite threshold, connected with the presence of the level H 2s (*Figure 5.3*). Resonances are also present.[79] Calculations[80] predict an infinite number of resonances just before onset which could also perturb the threshold. Measurement of He 4^3S and 4^1S functions show that these reach a significant value within 30 meV of threshold[81] (4^1D and 4^3D lie very close to these levels).

The general form of observed excitation functions above threshold is that of a rise up to a maximum value which is reached, for allowed transitions, at an impact energy several times the excitation energy. This form is perturbed by the presence of structure which occurs at the energies of resonances which decay into the channel (level) which is excited in the experiment. It is true that much structure has appeared in experimental functions which is instrumental in origin; this is inevitable for weak transitions, since there are competing routes by means of which the level can be reached. Nevertheless, techniques have advanced sufficiently during the last few years for

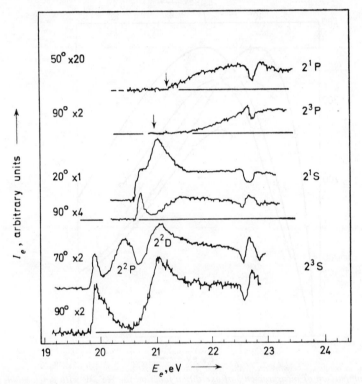

Fig. 5.4. *Resonances appearing in the excitation functions for* He 2^1P, 2^3P, 2^1S and 2^3S, *with post-collision electrons scattered through different angles*[91]

'inelastic channel resonances' to be observed by both optical and energy loss techniques[82–90]. Examples are to be found in *Figure 5.4*. In general, resonances appear stronger in inelastic than in elastic channels.

Experimentally observed emission functions are occasionally structured at energies corresponding to other levels; these effects might arise from cascading, but sometimes are not straightforward. For example, the function for radiation He $4^3S \rightarrow 2^3P$ is structured with a peak at the same energy as one occurring in the radiation function He $3^3P \rightarrow 2^3S$. Triplet helium structures can appear on singlet functions, presumably proceeding via the F configuration, which has characteristics of both singlet and triplet series.

24

One would not expect the Born approximation to yield accurate cross-sections except in the high-energy limit. The early data can be analysed to show that for allowed transitions the Born approximation holds with reasonable accuracy at impact energies above seven times the excitation energy[92]. Below this impact energy, the calculated cross-sections may be as much as twice those derived from experiment. For comparison of the excitation of p levels from s levels with theory[93], there is data for the excitation of He ^{1}P terms (from He ^{1}S ground) as well as data for allowed transitions

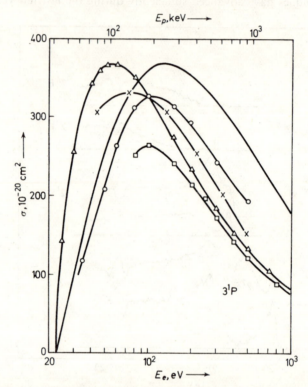

Fig. 5.5. Comparison of measurements of excitation function for He 3^1P *with calculations: solid line, Woolsey; open circles, St John et al.; squares, Moustafa; triangles, calculations of Bell et al.; crosses, proton experiments of Thomas and Bent (energy scale* E_p)

in sodium and mercury. The He 3^1P excitation function is shown for comparison with theory in *Figure 5.5*.

The $1s$–$2p$ transition in the hydrogen atom, resulting in the emission of Lyman α radiation, $\lambda = 1216$ Å, has been studied in a crossed-beam experiment[34]. Since the measurements are relative, because the absolute intensity of the hydrogen atom beam is indeterminate, the cross-section function is normalized with confidence at $E_e = 250$ eV to the Born approximation calculations. Great reliance can be placed on high-energy Born approximation calculations of inelastic electron–atom cross-sections for the simplest atomic systems. There is a complication in the experiment, namely the possibility of downward cascading and of quenching of the $2s$ states by electric fields in the apparatus. The experimental cross-section function is compared with the Born approximation and exchange distorted wave calculations in *Figure 5.6a*. The corrections made for anisotropy of the radiation seem to have been successful in this experiment.

Excitation transitions which involve a change of multiplicity are dominated by electron exchange, except in heavy atoms where spin–orbit coupling is not negligible. For the excitation of triplet levels of helium, the Born–

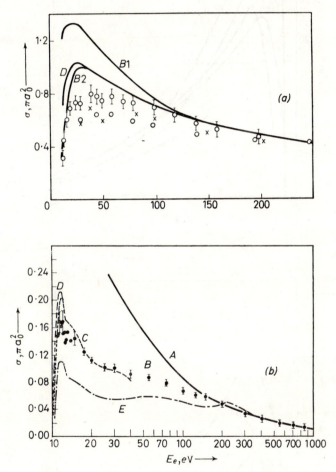

Fig. 5.6. (a) Excitation function[34] for H $1s \rightarrow 2p$ (experimental points: bracketed open circles are those taken at 90° to the electron beam, and unbracketed open circles have been corrected for polarization; crosses refer to data taken at 54·5°; B1, first Born approximation calculations[3]; B2, second Born approximation calculations[94]; D, exchange distorted wave calculations[95]); (b) excitation function H $1s \rightarrow 2s$ (A, Born approximation $\sigma(1s \rightarrow 2s) + 0.23 \ \sigma(1s \rightarrow 3p)$[98]; B, experiments of Kauppila, Ott and Fite[99]; C, experiments of Lichten and Schultz[96] normalized to B at 25 eV; D, close coupling approximation due to Burke; E, Experiments of Hils, Kleinpoppen and Koschmieder, normalized to Born approximation at 500 eV)

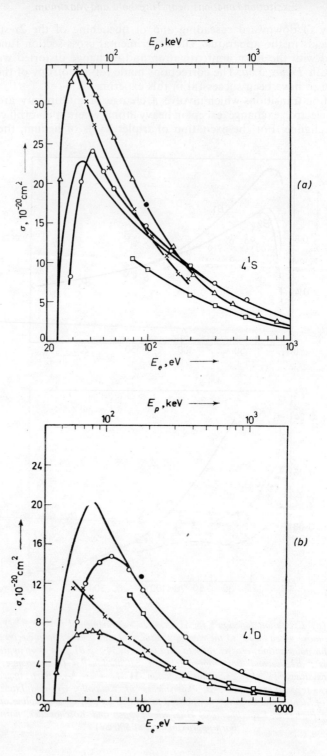

Oppenheimer approximation yields high results[100, 101] for the higher levels 4^3S and 5^3S. Comparisons of excitation of singlet and triplet levels of helium with calculations[4, 100, 102, 103] are given in *Figures 5.5, 5.7* and *5.8*. In measurements[104] on 3^3D or 4^3D, the effects are evident of competing excitation exchange processes producing the same radiation:

$$1^1S + 4^1P \rightarrow 4^3D + 1^1S$$

The exchange distorted wave theory yields cross-sections for He 2^3S excitation which are of the same order as experiment[47], but a detailed comparison is difficult close to threshold, because of the problem of separating 2^3S from 2^1S and also because of the contributions from resonances. There exist measurements over an energy range of several hundred electron volts. The He $1^1S \rightarrow 2^3P$ transition has also been calculated using the exchange distorted wave method[105]. Although Born–Oppenheimer calculations and even distorted wave calculations[106] do not seem to explain the excitation cross-sections where changes of multiplicity are involved, there is one theoretical result which is well borne out by experiment; namely that exchange-dominated transitions involving no change of azimuthal quantum number are stronger than those involving such a change. Thus *s–s* transitions, such as the excitation of He 2^1S and of H $2s$ are the strongest, and show the sharpest cross-section function maxima. Measurements[96, 97] for H $2s$ are illustrated in *Figure 5.6b*, being contrasted with H $2p$ measurements in *Figure 5.6a*. It is now believed that the original measurements[97] were rendered inaccurate at very high energies by the presence of slow electrons trapped in the magnetic field.

Since exchange-dominated excitation functions have much sharper maxima than those for the optically allowed transitions, the sharpness may be used as a method of studying the 'quality' of a transition, for example, in nitrogen[107]. Levels such as Hg 6^3P do not behave as purely triplet states[108], probably owing to the importance of spin–orbit coupling.

Forbidden transitions which do not involve a change of multiplicity have sharp cross-section function maxima, but are more satisfactorily interpreted by Born–Oppenheimer approximation than are those involving multiplicity change. Examples include n^1S and n^1D states of helium, particularly 4^1D and 5^1D (*Figure 5.7*). However, the cross-sections for these processes are much smaller than those for the corresponding optically-allowed transitions. In general, the Born approximation holds satisfactorily down to an impact energy which is smaller the more strongly the transition is forbidden.

The most accurate cross-sections available for helium excitation are probably those of St John, Miller and Lin[109], and there is good agreement between these and the earlier work of Gabriel and Heddle[103].

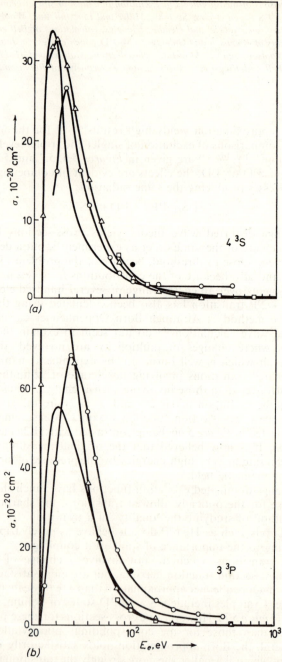

Fig. 5.8. Comparisons of measurements of excitation functions of triplet helium levels with calculations: (a) 4^3S, (b) 3^3P, (c) 4^3D; open circles, St John, Miller and Lin; full line, Woolsey; squares, Moustafa; closed circle, Gabriel and Heddle; triangles, calculations of Ochkur and Brattsev

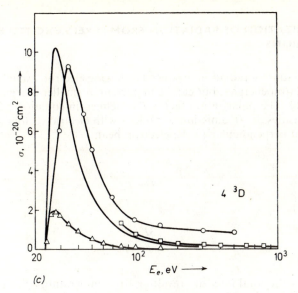

(c)

Fig. 5.8. continued

Distorted wave calculations for a transition involving strong coupling, the excitation of the Na D lines, were carried out by Seaton[14]. Calculations for this transition were also made[110] using the Bethe approximation, but including a relation between the scattering and reactance matrices which is appropriate to weak coupling conditions. Comparison with experiment is made in *Figure 5.9*[106, 111]. Calculations and experiments have been carried out for mercury, in which spin–orbit coupling is important.

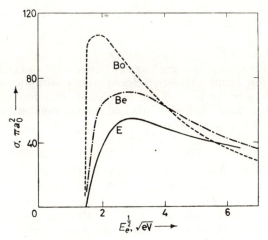

Fig. 5.9. Excitation functions for Na $3s \rightarrow 3p$: *Bo, Born approximation;* **Be,** *Bethe approximation; fullline, experimental data*

(From Seaton[4], by courtesy of Academic Press)

5.5 POLARIZATION OF RADIATION FROM LEVELS EXCITED BY ELECTRONS

The polarization of radiation produced following electron excitation is small at high electron energies, but can be important at smaller energies and close to threshold. The polarization factor P is defined in equation 5.28 in terms of the intensities of 90° emission $I_{\parallel}$ and $I_{\perp}$ with electric vectors respectively parallel and perpendicular to the electron beam:

$$P = \frac{I_{\parallel} - I_{\perp}}{I_{\parallel} + I_{\perp}} \tag{5.41}$$

Alternatively,

$$P = \frac{3I - \bar{I}}{I_{\parallel} + \bar{I}} \tag{5.42}$$

where $4\pi\bar{I}$ is the total intensity resulting from integrating $I(\theta)$ over all solid angles. Thus $\bar{I} = I_{\parallel} + 2I_{\perp}$.

A measurement of P is made by placing in the light path a polarization analyser which can be oriented in either of the two principal directions. The orientation must be changed rapidly to overcome the effect of non-constancy of excitation conditions and photomultiplier gain. It is possible[112,113] to rotate the analyser, thus modulating the output signal, the mean value of which is proportional to $I_{\parallel} + I_{\perp}$, at twice the frequency of rotation, with an amplitude proportional to $|I_{\parallel} - I_{\perp}|$.

The theoretical expression[36] for P involves the cross-section σ_m for exciting the sub-states of the level with quantum number m_l. For a $p \to s$ transition

$$P = \frac{\sigma_0 - \sigma_1}{\sigma_0 + \sigma_1} \tag{5.43}$$

For an s state the system has zero angular momentum and at threshold the scattered electron beam has zero velocity and contributes zero angular momentum, so only the $m_l = 0$ can be excited; therefore $\sigma_1/\sigma_0 = 0$ and $P = 1$. In the excitation collision, spin and orbital momenta are conserved separately, and it is possible to write

$$\begin{array}{ccccc} m_l & + & M_l & = & m_l' & + & M_l' \\ \text{Atom} & \text{Ante-collision} & \text{Excited} & \text{Post-collision} \\ & \text{electron} & \text{atom} & \text{electron} \end{array} \tag{5.44}$$

At threshold M_l' must be zero, and, because of the selection rule $\Delta m_l = 0$, it follows that $M_l = 0$, so $P = 1$. The taking into account of hyperfine structure, using Racah formalism, yields a smaller value (for example, $P = 0.43$ for He $n^1P \to n^1S$). But despite the prediction of large polarization at threshold, certain measured $P(E_e)$ functions for helium and hydrogen[114], although not for alkali metals, have been believed for many years to pass through a broad maxima and fall to small values with energy decreasing to threshold.

However, recent measurements on helium[115] with more finely momentum-analysed electron beams have shown that still closer to threshold the predicted polarization values can be realized; there is a very sudden fall, almost to $P = 0$, with increasing energy, followed by a rather slower rise to a maximum value of P, and then a gradual fall to negligible values at several

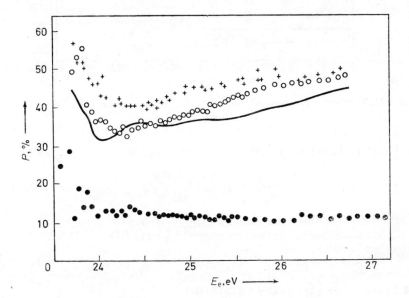

Fig. 5.10. Measured polarization functions $P(E_e)$ for He $4^1D \rightarrow 2^1P$ (crosses, McFarland; open circles, Heddle and Keesing[115]; full line, Federov and Golovanevskaya) and for He $3^3P \rightarrow 2^3S$ (closed circles, Heddle and Keesing[115])

hundred electron volts (*Figure 5.10*). This structure probably arises from severe differences between the excitation functions of magnetic substates of the level which is excited. Calculations made for Lyman α radiation exhibit the downward structure[116, 117]. There are instances, such as the polarization of resonance radiation from alkali metals, where the sudden fall and rise fails to appear[118]. In these cases, the polarization function (*Figure 5.11*) falls linearly with electron energy from its low energy value; this is characteristic of the situation when the excited state is well separated in energy from its neighbours. The downward structure is associated with coupling to the neighbouring states.

Resonance structure is occasionally observed in a polarization function[120].

At high electron energies the polarization function should be energy-invariant, but there exist data for which P goes negative and then rises again at higher energy; this behaviour could possibly be due to cascading.

For electric dipole radiation, quantum theory predicts that the variation

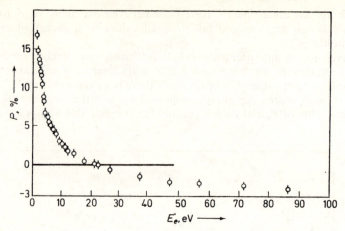

Fig. 5.11. Measured polarization functions $P(E_e)$ for Li $2p \rightarrow 2s$, due to Hafner and Klein-poppen[119]

of P with radiation polar angle θ can be expressed as

$$P(\theta) = \frac{P(\pi/2) \sin^2 \theta}{1 - P(\pi/2) \cos^2 \theta} \tag{5.45}$$

The harmony of this equation with early experiments[81] is probably fortuitous, but some more recent measurements[121] for He $4^1D \rightarrow 2^1D$ have shown consistency with it.

5.6 STUDIES OF ELECTRON EXCHANGE

In recent years it has become possible to study the electron exchange process by making use of the spin of the atomic electron. An atomic beam directed through an inhomogeneous magnetic field suffers deflexion proportional to the spin, so different spin states may be separated.

It has been seen that the scattering cross-section for electrons by atoms is made up of direct, exchange and mixed components:

$$\sigma = \frac{v_{e2}}{v_{e1}} \left\{ \frac{1}{2} |f|^2 + \frac{1}{2} |g|^2 + \frac{1}{2} |f-g|^2 \right\} \tag{5.46}$$

where the subscripts 1 and 2 refer to the incident and scattered electron. The term $\frac{1}{2} |f|^2$ signifies the direct scattering amplitude, $\frac{1}{2} |g|^2$ the exchange scattering amplitude[122], and $\frac{1}{2} |f-g|^2$ the mixed scattering amplitude. The spin states corresponding to these amplitudes are given in Table 5.1[123].

It is possible to study the spin state populations either in an elastic collision in which there is no excitation process but only a change of spin quantum number, or in a collision which includes excitation, such as the $1s \rightarrow 2s$ transition in the hydrogen atom. In both cases it is found that the exchange process occurs in a significant fraction of the collisions.

Table 5.1. SPIN STATES CORRESPONDING TO COLLISIONS WITH DIFFERENT SCATTERING COMPONENTS

Scattering component	Before collision		After collision	
	Incident electron	Atomic electron	Incident electron	Atomic electron
Mixed	$\frac{1}{2}$	$\frac{1}{2}$	$\frac{1}{2}$	$\frac{1}{2}$
Mixed	$-\frac{1}{2}$	$-\frac{1}{2}$	$-\frac{1}{2}$	$-\frac{1}{2}$
Direct	$-\frac{1}{2}$	$\frac{1}{2}$	$-\frac{1}{2}$	$\frac{1}{2}$
Direct	$\frac{1}{2}$	$-\frac{1}{2}$	$\frac{1}{2}$	$-\frac{1}{2}$
Exchange	$-\frac{1}{2}$	$\frac{1}{2}$	$\frac{1}{2}$	$-\frac{1}{2}$
Exchange	$\frac{1}{2}$	$-\frac{1}{2}$	$-\frac{1}{2}$	$\frac{1}{2}$

For alkali metals, the lowest energy level possesses two spin states, which may be separated either by the use of inhomogeneous magnetic fields or by the absorption of resonance radiation. The optical technique was used in the experiments of Dehmelt[124] and later of Franken, Sands and Hobart[125]; the primary concern of the experiment was to produce a swarm of polarized electrons and measure their spin g-factor. However, the rate of exchange of polarization between sodium atoms and thermal electrons could also be deduced. Sodium atoms in a steady magnetic field were illuminated with circularly polarized resonance radiation, which depopulated their ground Zeeman substates unequally ($m = -\frac{1}{2}$ being eliminated), without their being replenished by an unbalance of excited substates; buffer gas collisions ensured equality of excited substate populations. The optically pumped $m = +\frac{1}{2}$ sodium atoms were subjected to pulses of radiofrequency discharge, and the substate populations were monitored as a function of time in the afterglow, by the absorption of the resonance radiation. Application of a radiofrequency magnetic field at the gyro-magnetic frequency served to compare the g-factor of the free electrons and sodium atoms; additional information from the experiment was the rate of exchange between electrons and $m = \frac{1}{2}$ ground state sodium atoms. The measured rate indicated a cross-section $> 2\cdot3\times10^{-14}$ cm^2 for sodium and $> 3\times10^{-14}$ cm^2 for potassium. The corresponding total electron collision cross-sections are less than one order of magnitude greater than these.

A different type of experiment[126] made use of two inhomogeneous magnetic fields, one to polarize a potassium atom beam and one to analyse the scattered atoms after collision with a crossed electron beam. The polarizing field served also to velocity-select the beam. The unscattered atoms were separated from the scattered by collimation, transverse momentum being acquired during the collision process. Detection of the two analysed components of the scattered beam was achieved by a surface ionization detector. In the range 0·5–4 eV, the ratio of exchange to total collisions was found to be about 0·3, so the exchange cross-section may be taken to be between 2×10^{-14} cm^2 and 1×10^{-14} cm^2.

Actually, experiments of this type measure the spin-flip and not the purely exchange cross-section. Spin-flip can occur not only by exchange, but at high impact energies by spin–orbit coupling. Use might be made of the recoil technique to obtain the differential exchange cross-section. For the total pure exchange cross-section it would be necessary to spin-polarize an electron beam, scatter it off an unpolarized atom beam, and spin-analyse it using Mott scattering. Alternatively a spin-polarized beam could be scattered off a spin-polarized atom beam, and the atom beam spin-analysed after the collision.

For the hydrogen atom, it is possible with the aid of Zeeman effect to study the exchange contribution to the excitation to the $2s\,{}^2S_{1/2}$ metastable state[96]. The Zeeman effect on the levels $n = 1, j = \frac{1}{2}$ is such that the separation between the ${}^2S$ substates increases with increasing magnetic field, whilst that between the ${}^2P$ states also increases, but less strongly. In the absence of magnetic field the ${}^2S$–${}^2P$ separation corresponds to a frequency of 1058 MHz, but when a field of 575 G is applied the ${}^2S$, $m_s = -\frac{1}{2}$ and ${}^2P$, $m_s = +\frac{1}{2}$ levels coincide. The admixing of the P state causes the atom to decay with emission of Lyman α radiation.

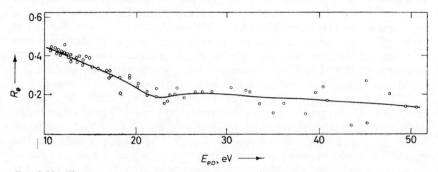

Fig. 5.12. *Electron energy dependence of ratio R_e of exchange to total electron collision cross-section (after Lichten and Schultz[96])*

Lichten and Schultz[96] selected hydrogen atoms from a thermal dissociation source in either $m_s = +\frac{1}{2}$ or $m_s = -\frac{1}{2}$ states, with the aid of an inhomogenous magnetic field. In order to avoid moving the post-field collimating slit, the oven source was made movable. The polarized beam was crossed with an electron beam in a region of steady 575 G magnetic field, and the metastable atoms so produced were detected by electron ejection from a metal surface. Those with $m_s = -\frac{1}{2}$ were magnetically quenched before reaching the detector. The ratio of exchange to total cross-section was equal to the ratio of ejected electron current for the $m_s = +\frac{1}{2}$ beam to the sum of currents obtained with both beams (see Table 5.1). This ratio is shown as a function of electron energy in *Figure 5.12*, and is not very different from the other exchange ratios discussed in this section.

Zeeman level crossings in hyperfine structure have now been observed for many atoms[127]; use may be made of isotopic shift to place the crossings in a workable experimental region, and crossing at zero field ('Hanle effect') is also known. The scope for carrying out sophisticated collision experiments is thereby widened.

It may also be mentioned that the He $2^1S \to 2^3S$ transition, which is only possible with electron exchange but involves a small energy transfer, has been found[43] to be $\sim 10^{-14}$ cm², which is a significant fraction of the total collision cross-section.

5.7 EXCITATION OF POSITIVE IONS BY ELECTRONS

Calculations for the excitation of hydrogenic positive ions[128, 129] and other one-electron systems[130] such as Ca^+ have been made with the aid of a Coulomb–Born approximation, and for exchange processes with a Coulomb–Born–Oppenheimer approximation. The scattering of an electron by a

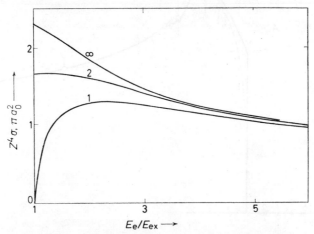

Fig. 5.13. Coulomb–Born calculations, due to Burgess[128], of excitation functions of the 2p level of stripped nuclei of charge Z

(From Seaton[4], by courtesy of Academic Press)

Coulomb potential[131] gives rise to electron waves which are not planar, but may be described as 'Coulomb-waves'. In these approximations, the incident electrons are treated not as plane waves but as undistorted Coulomb waves, and cross-section functions of the types shown in *Figure 5.13* are obtained. The cross-sections are shown multiplied by the fourth power of the nuclear charge Z.

The outstanding feature of the $e1/e1'$ ionizing collision cross-section is that it is non-zero at threshold (*Figure 5.14*), and may either rise to a maximum or decrease monotonically with increasing electron energy. In simple terms this feature can be regarded as being associated with the additional energy acquired by the electron during the collision, by virtue of the Coulomb attraction. In the limit of large electron velocity, the Coulomb wave approximates to a plane wave and the cross-sections are proportional to Z^{-4}.

Some of the importance of the excitation functions of positive ions arises from their use in electron temperature diagnostics in hot plasmas[133]. Absolute values of the excitation cross-section are required, therefore both experiment and concurring theoretical calculation must be sought.

Crossed-beam experiments have been carried out[132] on the excitation process He^+ $1s \rightarrow 2s$. The 40·8 eV radiated photons pass through a 10^3 Å aluminium film, which stops electrons but permits the photons to fall onto a photocathode. The $2s$ $^2S_{1/2}$ state is quenched to $2p$ $^2P_{1/2}$ under the action of an inhomogeneous electric field. A finite cross-section at threshold is inferred from the data by convolution of a trial function, as shown in *Figure 5.14*.

A study of this cross-section function close to threshold[134] has been made using the trapped-ion technique[135]. In this type of experiment, an electron

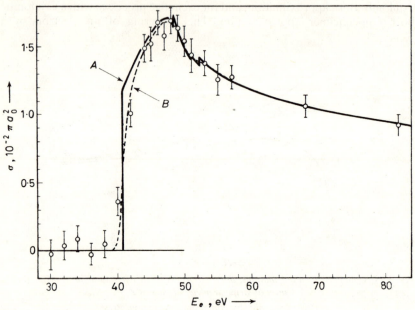

Fig. 5.14. *Excitation function for* He^+ $1s \rightarrow 2s$, *due to Dance, Harrison and Smith*[132]: *curve B is fitted to experimental data, and is a convolution of curve A with electron energy resolution*

beam is set up in such a way as to maximize the space-charge trapping of positive ions. The threshold production of He^{2+} formed in the sequence of processes

$$e + He \rightarrow He^+ \; 1s + 2e$$
$$e + He^+ \; 1s \rightarrow e + He^+ \; 2s$$
$$e + He^+ \; 2s \rightarrow He^{2+} + 2e$$

is dominated by the process having the highest threshold, in this case the He^+ $2s$ excitation. Three resonances appear, in agreement with calculations[136, 137], and are shown in *Figure 6.24* (p. 417).

An inference of the excitation cross-section, at a single energy, of resonance radiation in N^{5+} has been made from studies of the Zeta toroidal pinch hot-plasma machine. The electron density at an instant in time is diagnosed by laser beam scattering, the ion density is deduced from separate ionization balance experiments, and the absolute radiation density is measured. The cross-section is in reasonable agreement with calculations.

5.8 EXCITATION AND DISSOCIATION OF MOLECULES

Electronic Excitation

The electronic excitation of molecules by electrons[100, 138-147] is fundamentally similar to the excitation of atoms; typical data are shown in *Figure 5.15*. Much of these data refer to the excitation of spectra of order II, that is, to simultaneous ionization and excitation; there is no distinction between the

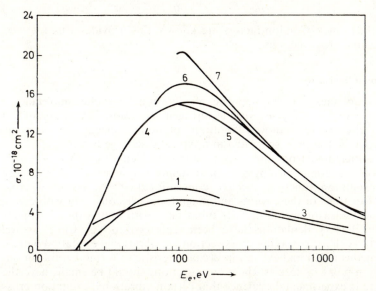

Fig. 5.15. *Measured excitation functions for 0,0 band of first negative system of* N_2^+ *(3914 Å) from nitrogen: 1, Stewart[139]; 2, Sheridan, Oldenberg and Carlton[138]; 3, Hayakawa and Nishimura[148]; 4, McConkey, Woolsey and Burns[147]; 5, Holland; 6, Srivastava and Mirza; 7, Aarts and de Heer*

experiments, although of course the cross-section functions for the two processes are expected to differ. When an electron beam is passed through molecular gases, sometimes the first order and sometimes the second order spectra predominate.

The excitation of electronic levels of molecules is complicated by the existence of vibrational levels, whose relative proportions are governed by the application of the Franck–Condon principle. The intensity of the radiation $J_{v'v''}$ photons sec^{-1} from the v' vibrational level of the upper electronic state to the v'' vibrational level of the lower state is

$$J_{v'v''} = N_{v'}A_{v'v''} \tag{5.47}$$

where $N_{v'}$ is the number of molecules in the upper state, $A_{v'v''}$ the molecular A coefficient is

$$A_{v'v''} = E_{v'v''}^3 r_e^2 q_{v'v''}^2 \tag{5.48}$$

and $E_{v'v''}$ is the energy of the transition between the two vibrational levels; the term r_e^2 is independent of nuclear separation r, so the Franck–Condon factor

$$q_{v'v''} = |\int \psi_{v'}^* \psi_{v''} \, dr|^2 \qquad (5.49)$$

dominates the vibrational populations. Since the apparent cross-section $\sigma_{v'v''}$ for the excitation of the $v'v''$ band is proportional to $J_{v'v''}$ two such cross-sections may be compared by the relation

$$\frac{\sigma_{v'v''}}{\sigma_{v'v'''}} = \frac{\lambda_{v'v'''}^3 q_{v'v''}}{\lambda_{v'v''}^3 q_{v'v'''}} \qquad (5.50)$$

Where the Franck–Condon factors are known, this provides a useful check on measured cross-sections.

Vibrational Excitation[149]

There is an unexpected abundance of vibrationally excited molecules in certain weakly ionized gases, such as nitrogen. Vibrational excitation of CO_2 can give rise to laser action of vibration–rotation wavelengths in the region 10 mμ; 1 mW power is generated in a typical discharge tube.

It is anticipated from Born approximation calculations that, when the transition is accompanied by rotational quantum number change, the vibrational excitation functions behave conventionally[150-153]. Assumed cross-section functions of these forms and magnitudes are consistent with certain energy loss measurements in drift tubes both with and without magnetic deflexion[154, 155]. Calculations have been reported for NO, CO[156] as well as for H_2. A sharply peaking cross-section for vibrational excitation without simultaneous rotational excitation is characteristic of forbidden transitions.

For non-polar molecules, these cross-sections should be small; nevertheless there is experimental evidence that certain vibrational excitation cross-sections are very much larger. This is to be expected for permanent dipole molecules and molecules in which transition moments can be generated, since in such cases the transitions are no longer forbidden. But in non-polar diatomic molecules, the principal cause of large vibrational excitation cross-sections is likely to be the formation of compound negative ion states, decaying into vibrational channels.

The low-energy cross-sections for vibrational excitation of nitrogen, and of certain other molecules, are known to be large and to be structured in a way that can be interpreted as arising from compound state formation[157-164]. Data for nitrogen are shown in *Figure 5.16*. Experiments have been conducted using both retardation technique and electron momentum analysers, both in transmission and at angles. There is no doubt that the cross-sections are orders of magnitude greater than those at energies well-removed from the vibrational levels of the resonance.

The exact energy structure of the excitation functions is complicated. As the vibrational quantum number v increases, the maxima, which can be regarded as arising from interference between waves scattered from different vibrational states λ of the compound state, shift to higher energies (see *Figure 5.16*). This has been interpreted[160] as resulting from changes in magnitudes of nuclear matrix elements in the transition matrix. The N_2^-

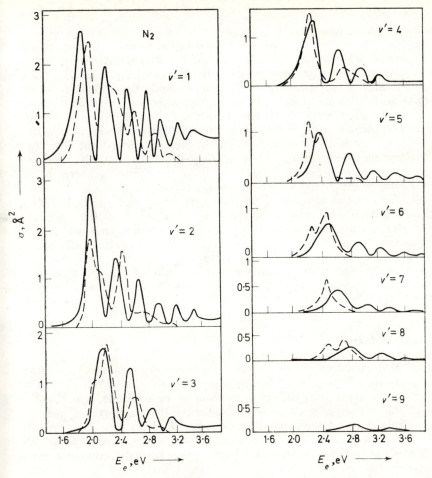

Fig. 5.16. Vibrational excitation functions for nitrogen from vibrational quantum number zero to specified values: full lines represent measurements (Schulz, Ehrhardt, Hasted); broken lines represent compound state calculations following the method of Chen, Herzenberg and Mandl[161], but with fixed compound state line width Γ

compound state $^2\Pi_g \sim 2\,\text{eV}$ is identified on theoretical grounds[165], on differential scattering grounds[164], and using isoelectronic series arguments[166]. For simple interpretation of the vibrational excitation data in nitrogen, it is possible[161] to assume a compound state mean line width $\Gamma = 0 \cdot 15\,\text{eV}$; the width varies with nuclear separation, being proportional to the energy difference between the compound state and ground state adiabatic interaction energies.

On the compound state model of vibrational excitation via resonance state, the cross-section is approximately proportional to

$$\sigma \propto \left| \sum_{\lambda} \frac{\langle \phi(v') \mid \phi(\lambda) \rangle \langle \phi(\lambda) \mid \phi(0) \rangle}{E_e - \varepsilon - \lambda h \omega_{-} c + \frac{1}{2} i \Gamma} \right|^2$$

25

The expressions in the numerator represent the Franck–Condon overlap integrals for excitation from ground vibrational level $v = 0$ to vibrational level λ of the compound state and thence to the v' level of the molecule. The compound state has energy ε, line width Γ and vibrational frequency ω_-. The interference between waves scattered from different levels λ of the compound state is responsible for the complex and shifting structure of the vibrational excitation function. However, the latter is too regular to be explained by interference effects with constant Γ, and a 'potential barrier' model has been invoked[167] with some success; the resonance state is represented only by the standing wave formed by the interaction of the expanding wave and the first reflexion at the potential barrier. Alternatively, Γ may be taken to vary with nuclear separation, in proportion to the energy difference between the adiabatic potential energy curves.

Similar structure appears in vibrational excitation data of H_2, CO, NO, CO_2, O_2, NO_2 and other molecules. In H_2 the width of the resonance (~ 5 eV) is such that only a broad maximum appears, and the angular dependence of electron scattering shows[168, 169] the important influence of $H_2^-\,{}^2\Sigma_u$. For oxygen[166, 170, 171], the vibrational excitation cross-section is much smaller that that in nitrogen, but structure in the $v = 1$ function arises from resonance scattering from the vibrationally excited levels of the $O_2^-\,{}^2\Pi_g$ ground state.

Rotational Excitation[172]

Rotational excitation cross-sections have been calculated using the Born approximation[173], and also other approximations and interactions[174]. The electron couples to the quadrupole moment of the molecule, which exerts an interaction of range sufficiently long for the Born approximation to be applicable. For rotational quantum number J, there is a selection rule $\Delta J = \pm 2$, so the excitation cross-sections ${}_J\sigma_{J+\Delta J}$ may be neglected for all values of ΔJ except ± 2. The energy levels of the rotating molecule are given approximately by

$$E_J = J(J+1)B_v \qquad (5.51)$$

where B_v is the rotational constant[175]. The non-zero cross-sections are given in the Born approximation[173] by:

$$_J\sigma_{J+2}E_e = \frac{(J+2)(J+1)}{(2J+3)(2J+1)}\,\sigma_0\left[1 - \frac{(4J+6)B_v}{E_e}\right]^{1/2} \qquad (5.52)$$

$$_J\sigma_{J-2}E_e = \frac{J(J-1)}{(2J-1)(2J+1)}\,\sigma_0\left[1 - \frac{(4J-2)B_v}{E_e}\right]^{1/2} \qquad (5.53)$$

$$\sigma_0 = \tfrac{8}{15}\pi q^2 a_0^2 \qquad (5.54)$$

where q is the electric quadrupole moment (see tables in the Appendix). Thus the cross-sections are zero at the appropriate onset energy, and they rise to 'saturation' at sufficiently large electron impact energy E_e. These equations are valid for ${}^1\Sigma$ molecules with $v = 0$ and $E_e = 0\cdot6$ eV. The analysis must be corrected for polarization effects.

Hitherto, the only experimental information on rotational excitation cross-sections has come from drift-tube measurements (see Chapter 4). Born approximation cross-sections are fed into the transport analysis from which the drift velocity functions are calculated[176]. For N_2, CO, NO and H_2O the agreement is satisfactory. The drift-tube data[154, 177] for H_2 (see *Figure 4.16*) require an adjustment of the established value of the quadrupole moment by a factor of $\sqrt{2\cdot5}$. Additional polarization terms[178] and short-range terms[179] must also be included. The importance of rotational excitation by electrons is shown by the fact that the behaviour of electron swarms in ortho-hydrogen and para-hydrogen is qualitatively different. For N_2 there is no established value of the quadrupole moment with which to compare the value consistent with the drift-tube data.

Differential and total beam measurements of electrons scattered by H_2 with rotational energy losses have recently been reported by Ehrhardt and Linder[180], and from these it is possible to deduce the differential and total cross-sections. The angular dependence shows that s, p and d wave scattering are involved, and the total cross-section for excitation $J = 1 \to J = 3$ ($v = 0$) maximizes at $\sim 8\times10^{-17}$ cm² around 4 eV. By 10 eV it has fallen to $\sim 5\times10^{-17}$ cm². Superelastic processes are also observed, and it is possible to separate pure vibrational from vibrational–rotational excitation. The former is four times as probable as the latter at 0°, but they are equal at 90°.

Dissociation

The dissociation of molecules by electron impact can appropriately be discussed in this chapter, since the process may generally be understood as a transition to an anti-bonding state. Investigations have been made of the dissociation of molecules by electrons in electric discharges, but there are virtually no beam studies. The detection of atoms formed in the dissociation process requires techniques outlined in Section 3.17. Alternatively, atoms formed in excited states can be detected by the line radiation they emit. Thus, Balmer α radiation has been detected from electron collisions with methane. The high-energy fall-off ($E_e^{-1} \ln E_e$ or $\ln E_e$) is an indication of the nature of the transition involved.

An electron impact experiment on hydrogen has been reported[181], using detection by atomic hydrogen absorption and pressure measurement. A closed glass vacuum system was used, containing a filament closely surrounded by an accelerating grid; the bulk of the tube was in a region of equipotential. The wall of the tube was lined with molybdenum trioxide, which is unaffected by molecular hydrogen but is completely efficient at absorbing atomic hydrogen. The resulting fall of pressure enabled the dissociation cross-section to be calculated. But the competing process of ionization produced H_2^+ which is either trapped directly on the molybdenum trioxide, or is trapped via the process:

$$H_2^+ + H_2 \to H_3^+ + H$$

The cross-section could not therefore be accurately estimated above 20 eV, but nevertheless the data are displayed in *Figure 5.17*.

25*

The dissociation of molecular ions by electrons has been studied both in crossed-beam experiments[182] and using trapped-ion technique[135]. The crossed-beam experiments on H_2^+ yield a cross-section falling from a value $\sim 10\,\pi a_0^2$ with increasing energy above 10 eV. The $E_e^{-1}\ln E_e$ high-energy fall-off holds down to much lower energies than is expected. The cross-section is larger than would be anticipated if the H_2^+ beam were entirely in the lowest vibrational level $v = 0$. But there is no detailed knowledge

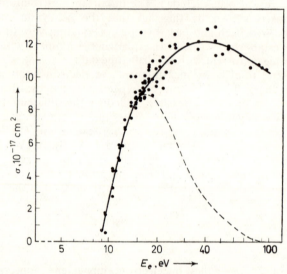

Fig. 5.17. *Dissociation cross-section[181] for the process $e + H_2 \rightarrow H + H + e$: points are experimental data, through which a smooth full line has been drawn; the broken line represents the cross-section function, derived by subtraction of the known cross-section for ionization*

about the vibrational state populations of molecular ion beams emerging from ion sources.

A discussion has been given by Dunn[183] of the angular distribution of fragments of molecular dissociation. The symmetry properties of the system (diatomic molecule plus electron travelling in direction k) are examined in terms of group theory. For a scalar operator, the matrix elements are non-zero only for transitions between states belonging to the same irreducible representation of the group of symmetry operations.

For an internuclear axis parallel to k, the probability of dissociation along this axis will only be finite when the selection rules for the transition to the antibonding states are obeyed: $\Delta\Lambda = 0,\ + \nleftrightarrow -$. For an internuclear axis perpendicular to k, finite probabilities of dissociation along k are indicated by the symbols X in Tables 5.2 and 5.3, whilst vanishing probabilities are indicated by the symbol O. Alignment of the internuclear and molecular symmetry axes parallel to each other is indicated by the symbols to the left of the sloping lines, whilst perpendicular axes are denoted by symbols to the right of the lines. Measurements and calculations for dissociative ionization processes are in harmony with these ideas[184].

Table 5.2. HOMONUCLEAR DIATOMIC MOLECULE TRANSITIONS

	Σ_g^+	Σ_g^-	Σ_u^+	Σ_u^-	Π_g	Π_u	Δ_g	Δ_u
Σ_g^+	X/X	O/O	O/X	O/O	O/O	X/O	X/O	O/O
Σ_g^-		X/X	O/O	O/X	O/O	X/O	X/O	O/O
Σ_u^+			X/O	O/O	X/O	O/O	O/O	X/O
Σ_u^-				X/X	X/O	O/O	O/O	X/O
Π_g					X/X	O/X	O/O	X/O
Π_u						X/X	X/O	O/O
Δ_g							X/X	O/X
Δ_u								X/X

Table 5.3. HETERONUCLEAR DIATOMIC MOLECULE TRANSITIONS

	Σ^+	Σ^-	Π	Δ
Σ^+	X/X	O/O	X/O	X/O
Σ^-		X/X	X/O	X/O
Π			X/X	X/O
Δ				X/X

5.9 INNER SHELL EXCITATION OF ATOMS

The process of exciting an inner shell electron into a vacant outer shell orbital is necessarily followed by the replenishment of the inner shell orbital. This is most likely by means of an Auger process—one outer shell electron falls into the inner shell orbital but distributes its energy amongst one or more outer shell electrons, which are emitted, giving rise to a positive ion. The process will therefore be considered in more detail under the heading of ionization in Chapter 6. What is necessary for an experimental study is a coincidence measurement in which the count of the impacting electron which has lost a measured amount of energy coincides with the count of a singly or multiply-charged ion.

Nevertheless, single particle detection experiments are adequate to detect the energy losses of electrons when they excite energy levels which can take part in Auger or Coster–Kronig transitions. An axial cylindrical electron spectrometer has been used by Mehlhorn[185] for identifying the levels of krypton, and a 360° magnetic spectrometer[186] has been used for studies of argon.

The Born approximation is expected to break down for inner shell excitation. The cross-sections will be small, being reduced from outer shell excitation cross-sections by a geometrical factor. The departure of inner shell excitation cross-sections from hydrogenic excitation may be due to three factors: (*i*) the existence of hybrid orbital states in unfilled inner shells; (*ii*) the thickness of shells, resulting in different screening effects on different electrons; and (*iii*) the Auger mechanisms of electron escape.

REFERENCES

1. MOISEIWITSCH, B. L. and SMITH, S. J. *Rev. mod. Phys.* 40 (1968) 238.
2. HEDDLE, D. W. O. and KEESING, R. G. W. *Advanc. atom. molec. Phys.* 4 (1968) 267.
3. MASSEY, H. S. W. *Handb. Physik* 26 (1956).
4. SEATON, M. J. *Atomic and Molecular Processes*, Ed. D. R. Bates, 1962. New York; Academic Press.
5. SEATON, M. J. *Proc. phys. Soc., Lond.* 79 (1962) 1105.
6. SARAPH, H. E. *Proc. phys. Soc., Lond.* 83 (1964) 763.
7. MILLER, W. F. and PLATZMANN, R. L. *Proc. phys. Soc., Lond.* A70 (1957) 299.
8. BETHE, H. A. *Ann. Phys., Lpz.* 5 (1930) 325.
9. SCHRAM, B. L. Thesis, University of Amsterdam, 1966.
10. LASSETTRE, E. N., SKERBELE, A. and MEYER, V. D. *J. chem. Phys.* 45 (1966) 3214; 49 (1968) 482.
11. LASSETTRE, E. N. *J. chem. Phys.* 43 (1965) 4479.
12. VRIENS, L. *Phys. Rev.* 160 (1967) 100; HEIDEMAN, H. G. M. and VRIENS, L. *J. chem. Phys.* 46 (1967) 2911.
13. KRAUSS, M. and MIELCZAREK, S. R. *J. chem. Phys.* 51 (1969) 5241.
14. SEATON, M. J. *Proc. phys. Soc., Lond.* A68 (1955) 457.
15. VAINSHTEIN, L. Moscow, Lebedev Physics Institute, Report A33, 1960.
16. OCHKUR, V. I. *Soviet Phys. JETP* 18 (1964) 503.
17. BURKE, P. G. and SMITH, K. *Rev. mod. Phys.* 34 (1962) 458; BURKE, P. G., MCVICAR, D. D. and SMITH, K. *Proc. phys. Soc., Lond.* 83 (1964) 397; MCCARROLL, R. *Proc. phys. Soc., Lond.* 83 (1964) 509.
18. MCCARROLL, R. *Proc. phys. Soc., Lond.* 83 (1964) 509.
19. ALDER, K., BOHR, A., HUUS, T., MOTTELSON, B. and WINTHER, A. *Rev. mod. Phys.* 28 (1956) 432.
20. NAGATA, T. and TOHMATSU, T. *Rep. Ionosph. Space Res. Japan* 14 (1960) 301.
21. VRIENS, L. *Physica, Eindhoven* 31 (1965) 1333.
22. ALLEN, C. W. *Astrophysical Quantities*, 1955. London; University of London Press.
23. HEISENBERG, W. *Phys. Z.* 32 (1931) 737.
24. MARTON, L. and SCHIFF, L. I. *J. appl. Phys.* 12 (1949) 749.
25. SCANLON, J. H. and MILFORD, S. N. *Astrophys. J.* 134 (1961) 724.
26. DRAWIN, H. W. Euratom Report EUR-CEA-FC-383, 1967.
27. READ, T. B. *Proc. phys. Soc.* 1 (1968) 795.
28. PHELPS, A. V. *Phys. Rev.* 110 (1958) 1362.
29. GABRIEL, A. H. and HEDDLE, D. W. O. *Proc. Roy. Soc.* A258 (1960) 124.
30. ST JOHN, R. M. and FOWLER, R. G. *Phys. Rev.* 122 (1961) 1813.
31. BOGDANOVA, I. P. and GEITSI, I. I. *Optika Spektrosk.* 17 (1964) 80.
32. BOGDANOVA, I. P. and MARUSIN, V. P. *Optika Spektrosk.* 20 (1966) 113.
33. HEDDLE, D. W. O. *Proc. phys. Soc., Lond.* 90 (1967) 81.
34. FITE, W. L. and BRACKMAN, R. T. *Phys. Rev.* 113 (1958) 1151;
35. FITE, W. L., STEBBINGS, R. F. and BRACKMAN, R. T. *Phys. Rev.* 116 (1959) 356.
36. PERCIVAL, I. C. and SEATON, M. J. *Phil. Trans.* A251 (1958) 113.
37. MCFARLAND, R. H. and SOLTYSIK, E. A. *Phys. Rev.* 127 (1962) 2090.
38. WOOLSEY, J. M. and MCCONKEY, J. W. *J. opt. Soc. Amer.* 58 (1968) 1309.
39. MCFARLAND, R. H. and SOLTYSIK, E. A. *Phys. Rev.* 127 (1962) 2090.
40. BURNS, D. J., SIMPSON, F. R. and MCCONKEY, J. W. *Proc. phys. Soc., Lond.* 2 (1969) 52.
41. MOHLER, F. *J. Res. nat. Bur. Stand.* 9 (1932) 493.
42. FABRIKANT, W. A. and CIRG, I. *Dokl. Akad. Nauk SSSR* 25 (1939) 663.
43. PHELPS, A. V. and MOLNAR, J. P. *Phys. Rev.* 89 (1953) 1203.
44. WOUDENBERG, J. P. and ORNSTEIN, L. S. *Physica, Eindhoven* 2 (1935) 355.
45. WOUDENBERG, J. P. and MILATZ, J. M. *Physica, Eindhoven* 8 (1941) 871.
46. DORRESTEIN, R. *Physica, Eindhoven* 9 (1942) 433, 447.

47. SCHULZ, G. J. and FOX, R. E. *Phys. Rev.* 106 (1957) 1179.
48. OLMSTED, J., NEWTON, A. S. and STREET, K. *J. chem. Phys.* 42 (1965) 2321.
49. DOWELL, J. T. *Gaseous Electronics Conf.*, American Physical Society, 1965.
50. DUGAN, J. L. G., RICHARDS, H. L. and MUSCHLITZ, E. E. *J. chem. Phys.* 46 (1967) 346.
51. HOLT, H. K. and KROTKOV, R. *Phys. Rev.* 144 (1966) 82.
52. CHAMBERLAIN, G. E. and HEIDEMAN, H. G. M. *Phys. Rev. Lett.* 15 (1965) 337.
53. SIMPSON, J. A. and MIELCZAREK, S. R. *J. chem. Phys.* 39 (1963) 1606.
54. KUPPERMANN, A. and RAFF, L. M. *J. chem. Phys.* 39 (1963) 1607.
55. LASSETTRE, E. N. and JONES, E. A. *J. chem. Phys.* 40 (1964) 1222; 40 (1964) 1218.
56. LASSETTRE, E. N. and KRASNOW, M. E. *J. chem. Phys.* 40 (1964) 1248.
57. LASSETTRE, E. N., SILVERMAN, S. M. and KRASNOW, M. E. *J. chem. Phys.* 40 (1964) 1261
58. SILVERMAN, S. M. and LASSETTRE, E. N. *J. chem. Phys.* 40 (1964) 1265
59. SKERBELE, A. M. and LASSETTRE, E. N. *J. chem. Phys.* 40 (1964) 1271
60. LASSETTRE, E. N., BERMAN, A. S., SILVERMAN, S. M. and KRASNOW, M. E. *J. chem. Phys.* 40 (1964) 1232
61. LASSETTRE, E. N., KRASNOW, M. E. and SILVERMAN, S. M. *J. chem. Phys.* 40 (1964) 1242.
62. LASSETRE, E. N. and SILVERMAN, S. M. *J. chem. Phys.* 40 (1964) 1256
63. LASSETRE, E. N. and FRANCIS, S. A. *J. chem. Phys.* 40 (1964) 1208
64. WHIDDINGTON, R. and PRIESTLEY, H. *Proc. Roy. Soc.* A145 (1934) 462
65. MILLER, W. F. and PLATZMAN, R. L. *Proc. phys. Soc., Lond.* A70 (1957) 299.
66. SIMPSON, J. A., MIELCZAREK, S. R. and COOPER, J. *J. opt. Soc. Amer.* 54 (1964) 269.
67. SIMPSON, J. A., CHAMBERLAIN, G. E. and MIELCZAREK, S. R. *Phys. Rev.* 139 (1965) A1039.
68. VRIENS, L. G., SIMPSON, J. A. and MIELCZAREK, S. R. *Phys. Rev.* 165 (1968) 7.
69. SCHULZ, G. J. *Phys. Rev.* 116 (1959) 1141.
70. KUPPERMANN, A. and RAFF, L. M. *Disc. Faraday Soc.* 35 (1963) 30.
71. BOWMAN, C. R. and MILLER, W. D. *J. chem. Phys.* 42 (1965) 681.
72. BRONGERSMA, H. H. and OOSTERHOFF, L. J. *Chem. phys. Lett.* 1 (1967) 169.
73. VRIENS, L. *Physica, Eindhoven* 31 (1965) 385; SCHRAM, B. L. and VRIENS, L. *Physica, Eindhoven* 31 (1965) 1431.
74. BOERSCH, H., GEIGER, J. and SCHRÖDER, B. *Gustav Hertz in der Entwicklung der Moderne Physik*, p. 15, 1948. Berlin; Akademik Verlag.
75. SKERBELE, A., DILLON, M. A. and LASSETTRE, E. C. *J. chem. Phys.* 46 (1967) 4161.
76. MOHR, C. B. O. and NICOLL, F. H. *Proc. Roy. Soc.* A138 (1932) 229, 469; A142 (1933) 320, 647.
77. SEATON, M. J. *Phil. Trans.* A245 (1955) 469.
78. CHAMBERLAIN, G. E., SMITH, S. J. and HEDDLE, D. W. O. Joint Institute of Laboratory Astrophysics Report 9, 1964.
79. WILLIAMS, J. F. and McGOWAN, J. W. *Phys. Rev. Lett.* 21 (1968) 719.
80. GAILITIS, M. K. and DAMBURG, R. *Proc. phys. Soc., Lond.* 82 (1963) 192.
81. HEDDLE, D. W. O. and KEESING, R. G. W. *Proc. phys. Soc., Lond.* 91 (1967) 510.
82. SMIT, C. Thesis, University of Utrecht, 1961; SMIT, C., VREDENBERG, W. and SMIT, J. A. *Physica, Eindhoven* 24 (1958) 380.
83. CHAMBERLAIN, G. E. *Phys. Rev. Lett.* 14 (1965) 581.
84. SCHULZ, G. J. and PHILBRICK, J. W. *Phys. Rev. Lett.* 13 (1964) 477.
85. HEIDEMAN, H. G. M., KUYATT, C. E. and CHAMBERLAIN, G. E. United States National Bureau of Standards Report, 1965; *Phys. Rev. Lett.* 12 (1964) 293; *J. chem. Phys.* 44 (1966) 440.
86. ANDERSON, R. J., LEE, E. T. P., LIN, C. C. and ST JOHN, R. M. *Gaseous Electronics Conf.* 1964, 1965.
87. SMIT, C. and FIJNAUT, H. M. *Phys. Lett.* 19 (1965) 121.
88. SIMPSON, J. A., MENENDEZ, M. G. and MIELCZAREK, S. R. United States National Bureau of Standards Report, 1966.

89. CHAMBERLAIN, G. E. *Phys. Rev.* 155 (1967) 46.
90. EHRHARDT, H. and WILLMANN, K. *Z. Phys.* 203 (1967) 1.
91. EHRHARDT, H., LANGHANS, L. and LINDER, F. *Z. Phys.* 214 (1968) 179.
92. MASSEY, H. S. W. and BURHOP, E. H. S. *Electronic and Ionic Impact Phenomena*, 1952. London; Oxford University Press.
93. SEATON, M. J. *Proc. phys. Soc., Lond.* 82 (1962) 1105.
94. BURKE, V. M. and SEATON, M. J. *Proc. phys. Soc., Lond.* 77 (1961) 199.
95. KHASHABA, S. and MASSEY, H. S. W. *Proc. phys. Soc., Lond.* 71 (1958) 574.
96. LICHTEN, W. and SCHULTZ, S. *Phys. Rev.* 116 (1959) 1132; LICHTEN, W. *Phys. Rev. Lett.* 6 (1961) 12.
97. STEBBINGS, R. F., FITE, W. L., HUMMER, D. G. and BRACKMAN, R. T. *Phys. Rev.* 119 (1960) 1939.
98. KINGSTON, A. E., MOISEIWITSCH, B. L. and SKINNER, B. G. *Proc. Roy. Soc.* A258 (1960) 245.
99. KAUPPILA, W. A., OTT, W. R. and FITE, W. L. SRCC Report 102, University of Pittsburgh, 1969.
100. THIEME, O. *Z. Phys.* 86 (1933) 646; 78 (1932) 413.
101. LEES, J. H. *Proc. Roy. Soc.* A137 (1932) 173.
102. WIGNER, E. P. *Phys. Rev.* 73 (1948) 1002; GERJUOY, E. *Rev. mod. Phys.* 33 (1961) 544.
103. GABRIEL, A. H. and HEDDLE, D. W. O. *Proc. Roy. Soc.* A258 (1960) 124.
104. HEDDLE, D. W. O. and LUCAS, C. *Proc. Roy. Soc.* A271 (1963) 129.
105. MASSEY, H. S. W. and MOISEIWITSCH, B. L. *Proc. Roy. Soc.* A258 (1960) 147.
106. MASSEY, H. S. W. and MOISEIWITSCH, B. L. *Proc. Roy. Soc.* A227 (1954) 38.
107. CERMAK, V. *Czech. Chem. Commun.* 27 (1962) 406.
108. PENNEY, H. C. *Phys. Rev.* 39 (1962) 467.
109. ST JOHN, R. M., MILLER, F. L. and LIN, C. C. *Phys. Rev.* A134 (1964) 888.
110. SEATON, M. J. *Proc. phys. Soc., Lond.* 77 (1961) 174.
111. HAFT, G. *Z. Phys.* 82, (1033) 73; CHRISTOPH, W. *Ann. Phys., Lpz.* 23 (1935) 51.
112. HEDDLE, D. W. O. and LUCAS, C. B. *Proc. Roy. Soc.* A271 (1963) 129.
113. DALGOV, G. G. *Optika Spektrosk.* 6 (1959) 469.
114. KLEINPOPPEN, H. and KRAISS, E. *Phys. Rev. Lett.* 20 (1968) 361.
115. HEDDLE, D. W. O. and KEESING, R. G. W. *Proc. Roy. Soc.* A299 (1967) 212.
116. BURKE, P. G., SCHEY, H. M. and SMITH, K. University of California, Lawrence Radiation Laboratory Report 10382, 1962.
117. PETERKOP, R. *Proc. phys. Soc., Lond.* A77 (1961) 1229.
118. KRUGER, H. and KLEINPOPPEN, H. *Conf. atom. Spectra & radiat. Processes*, The Institute of Physics and The Physical Society, 1965.
119. HAFNER, H. and KLEINPOPPEN, H. *Z. Phys.* 198 (1967) 315; *Phys. Lett.* 18 (1965) 270.
120. WHITTEKER, J. H. and DALBY, F. W. *Canad. J. Phys.* 46 (1968) 193.
121. McFARLAND, R. H. and SOLTYSIK, E. A. *Phys. Rev.* 127 (1962) 2090; *Phys. Rev.* 128 (1962) 1758; *Proc. Gaseous Electronics Conf.* Boulder, Colorado, 1960.
122. OPPENHEIMER, J. R. *Phys. Rev.* 32 (1928) 361.
123. SARAPH, H. E. *Proc. phys. Soc., Lond.* 82 (1964) 763.
124. DEHMELT, H. G. *Phys. Rev.* 105 (1957) 1487; 109 (1958) 381.
125. FRANKEN, P., SANDS, R. and HOBART, J. *Phys. Rev. Lett.* 1 (1958) 52.
126. RUBIN, K., PEREL, J. and BEDERSON, B. *Bull. Amer. phys. Soc.* 4 (1959) 234.
127. McDERMOTT, M. N. and THADDEUS, P. *Bull. Amer. phys. Soc.* 7 (1962) 433 (papers B4, B5, B6).
128. BURGESS, A. *Mém. Soc. r. Sci. Liège* 4 (1961) 299.
129. TULLEY, J. M.Sc. Dissertation, University College London, 1960.
130. VAN REGEMORTER, H. *Mon. Not. R. astr. Soc.* 121 (1960) 213.
131. MOTT, N. F. and MASSEY, H. S. W. *Theory of Atomic Collisions*, 1963. London; Oxford University Press.
132. DANCE, D. F., HARRISON, M. F. A. and SMITH, A. C. H. *Proc. Roy. Soc.* A290 (1966) 74.

133. HEROUX, L. *Proc. phys. Soc., Lond.* 83 (1964) 121.
134. DALY, N. *Phys. Rev. Lett.* 19 (1967) 1165.
135. BAKER, F. A. and HASTED, J. B. *Phil. Trans.* A261 (1966) 33.
136. ORMONDE, S., WHITTAKER, W. and LIPSKY, L. *Phys. Rev. Lett.* 19 (1967) 1164.
137. BURKE, P. *The Physics of Electronic and Atomic Collisions*, p. 128, 1968. JILA.
138. SHERIDAN, W. F., OLDENBERG, O. and CARLETON, N. P. *Proc. 2nd Int. Conf. Phys. electron. atom. Collisions*, 1961. University of Colorado.
139. STEWART, A. L. *Proc. phys. Soc., Lond.* A69 (1956) 437.
140. STEWART, D. T. *Proc. phys. Soc., Lond.* A68 (1935) 404.
141. LANGSTROTH, G. O. *Proc. Roy. Soc.* A146 (1934) 166.
142. BUNDY, F. P. *Phys. Rev.* 52 (1937) 698.
143. BERNARD, R. *C. R. Acad. Sci., Paris* 205 (1951) 193.
144. HAYAKAWA, S., NISHIMURA, H. and OTSUKA, M. *J. Geomagn. Geoelect., Kyoto* 16 (1964) 1; *Ann. Rev. Inst. Plasma Phys. Nagoya* (1965) 53.
145. McCONKEY, J. W. and LATIMER, I. D. *Proc. phys. Soc., Lond.* 86 (1965) 463.
146. NISHIMURA, H. *J. Phys. Soc. Japan* 21 (1966) 564; 24 (1968) 130.
147. McCONKEY, J. W., WOOLSEY, J. M. and BURNS, D. G. *Planet. Space Sci.* 15 (1967) 1332.
148. HAYAKAWA, S. and NISHIMURA, N. *J. Geomagn. Geoelect., Kyoto* 16 (1964) 72.
149. PHELPS, A. V. *Rev. mod. Phys.* 40 (1968) 399.
150. MASSEY, H. S. W. *Trans. Faraday Soc.* 31 (1935) 556.
151. CARSON, T. R. *Proc. phys. Soc., Lond.* A67 (1954) 908.
152. TA YOU WU, *Phys. Rev.* 71 (1947) 111.
153. MORSE, P. M. *Phys. Rev.* 90 (1953) 51.
154. FROST, L. W. and PHELPS, A. V. Westinghouse Research Report 62-908-113-P1, 1962.
155. RAMIEN, H. *Z. Phys.* 70 (1931) 353.
156. BREIG, E. L. and LIN, C. C. *Gaseous Electronics Conf.*, 1964.
157. HAAS, R. *Z. Phys.* 148 (1957) 177.
158. SCHULZ, G. J. *Phys. Rev.* 125 (1962) 229.
159. SCHULZ, G. J. *Phys. Rev.* 116 (1959) 1141.
160. CHEN, J. C. Y. *J. chem. Phys.* 45 (1966) 2710.
161. HERZENBERG, A. and MANDL, F. *Proc. Roy. Soc.* A270 (1962) 48.
162. HEIDEMAN, H. G. H., KUYATT, C. E. and CHAMBERLAIN, G. E. United States National Bureau of Standards Report, 1965.
163. SCHULZ, G. J. *Phys. Rev.* 135 (1964) A988.
164. EHRHARDT, H. and WILLMAN, K. *Z. Phys.* 204 (1967) 462.
165. GILMORE, F. R. *J. quantve Spectros. radiat. Transf.* 5 (1965) 369.
166. BONESS, M. J. W., HASTED, J. B. and LARKIN, I. W. *Proc. Roy. Soc.* A305 (1968) 493.
167. HERZENBERG, A. *Proc. phys. Soc.* B1 (1968) 548.
168. CHEN, J. C. Y. and MAGEE, J. L. *Proc. 2nd Int. Conf. Phys. electron. atom. Collisions*, 1961. Hastings on Hudson, New York; Science Bookcrafters.
169. TAKAYANAGI, K. *J. Phys. Soc. Japan* 20 (1964) 562.
170. HASTED, J. B. and AWAN, A. M. *Proc. phys. Soc.* B2 (1969) 367
171. SCHULZ, G. J. and DOWELL, J. T. Westinghouse Research Report 62-908-113-P3, 1962.
172. SMIT, J. *Physica, Eindhoven* 2 (1935) 104
173. GERJUOY, E. and STEIN, S. *Phys. Rev.* 97 (1955) 1671
174. LANE, N. F. and GELTMAN, S. *Phys. Rev.* 160 (1967) 53
175. HERZBERG, G. *Molecular Spectra and Molecular Structure, I: Spectra of Diatomic Molecules*, 2nd edn, 1951. New York; van Nostrand
176. PHELPS, A. V. Westinghouse Research Report 67-1E2-Gases-P2, 1967
177. DALGARNO, A. Communication mentioned in reference 154
178. SAMPSON, D. H. and MJOLSNESS, R. C. *Phys. Rev.* 140 (1965) A1466
179. GELTMAN, S. and TAKAYANAGI, K. *Phys. Rev.* 143 (1966) 25
180. EHRHARDT, H. and LINDER, F. *Phys. Rev. Lett.* 21 (1968) 419

181. CORRIGAN, S. J. B. *J. chem. Phys.* 43 (1965) 4381
182. DUNN, G. H. and VAN ZYL, B. *Phys. Rev.* 154 (1967) 40
183. DUNN, G. H. *Phys. Rev. Lett.* 8 (1962) 62
184. SASAKI, V. N. and NAKAO, T. *Proc. imp. Acad. Japan* 11 (1935) 138, 413; 17 (1941) 75.
185. MEHLHORN, W. *Z. Phys.* 187 (1965) 21.
186. HALE, G. C. and HASTED, J. B. *Nature, Lond.* 217 (1968) 945.

Chapter 6

IONIZATION BY ELECTRONS

6.1 INTRODUCTION: CLASSICAL AND QUANTUM CALCULATIONS

The ionization of atoms and molecules by electron impact is a process of great importance in ionized gas physics, particularly that of weakly ionized gases. Many of the earliest experimental studies were in fact made in terms of the ionization rate in a plasma, which can be written as

$$\frac{dn_+}{dt} = n_0 n_e \sqrt{2m_e} \int_{E_i}^{\infty} \sigma_i(E_e) f(E_e) E_e^{1/2} \, dE_e \qquad (6.1)$$

where σ_i is the total ionization cross-section, and $f(E_e)$ is the normalized electron energy distribution. However, modern measurements of ionization rate[1] in a plasma are usually directed at obtaining the electron energy distribution rather than the cross-section function. Beam studies of ionization processes are superior to ionization rate measurements in plasma.

The single ionization process

$$e + X \rightarrow e + X^+ + e \qquad\qquad e0/1e^2$$

dominates over multiple ionization processes. The total ionization cross-section is defined in such a way as to refer to the total electron emission:

$$\sigma_i = \sum_{n=1}^{n} n \,_{e0}\sigma_{ne^{n+1}} \qquad (6.2)$$

The impacting electrons generally lose in the collision an energy only slightly greater than the appropriate ionization potential, and at high energies are scattered mostly in the forward direction[2, 3]. The emitted electrons, mainly possessing small energies, are emitted at larger angles.

Accurate beam measurements of total and individual charge state cross-section functions have been available for some years. In particular, the original total ionization studies of Tate and Smith[4-8] have stood the test of time very well.

Theoretical studies of ionization have not been entirely successful in predicting total cross-sections until recent years. The Born approximation overestimates the cross-section except at the highest energies; this is because it takes no account of the exclusion principle. An ionization cross-section is more difficult to calculate than an excitation cross-section, because the emitted electron disturbs the scattering. Classical calculations are effective in the region of maximum cross-section, which for single ionization occurs typically at an electron energy of the order of five times the ionization potential; this energy is well below that at which the Born approximation is accurate.

Gryzinski's classical expression[9] for the differential cross-section for an electron impact process in which an energy exchange ΔE occurs is

$$\sigma(\Delta E) = \frac{2\pi(\varrho_1\varrho_2)^2}{m_1 v_2^2\,\Delta E} \left(\frac{v_2^2}{v_1^2+v_2^2}\right)^{3/2} \left(1-\frac{E_1}{E_2}+\frac{4E_1}{3\,\Delta E}\right) \tag{6.3}$$

for $E_2 \geqslant E_1 + \Delta E$, and

$$\sigma(\Delta E) = \frac{2\pi(\varrho_1\varrho_2)^2}{m_1 v_2^2\,\Delta E} \left(\frac{v_2^2}{v_1^2+v_2^2}\right)^{3/2} \frac{1}{3} \left(1+\frac{4E_1}{\Delta E}+\frac{2\,\Delta E-E_1}{E_2}\right)$$
$$\times \left(1+\frac{\Delta E}{E_1}\right)^{1/2}\left(1-\frac{\Delta E}{E_2}\right)^{1/2} \tag{6.4}$$

for $E_2 \leqslant E_i + \Delta E$. The terms v_1, E_1 and m_1 refer to the incident electron, and E_2 and v_2 to the orbital electron. (In this classical calculation of escape orbits, the Coulomb interactions between orbital and impacting electrons are considered but the interaction between the impacting electrons and the nucleus is neglected.) The total cross-section for a process in which ΔE exceeds E_i is therefore

$$\sigma(E_i) = \int_{E_i}^{\Delta E_{\max}} \sigma(\Delta E)\,\mathrm{d}E \tag{6.5}$$

and $\Delta E_{\max}$ is given by the inequality

$$4a^2X^2 - [b^2 + \Delta E(1+X)^2]^2 \geqslant 0$$

where

$$X = \frac{\varrho\mu V^2}{\varrho_1\varrho_2}$$

$$a = \mu v_1 v_2 \sin\theta$$

$$b = K_{12}[E_1 + E_2 + \tfrac{1}{2}(m_1-m_2)v_1 v_2 \cos\theta]$$

$$K_{12} = \frac{4m_1 m_2}{(m_1+m_2)^2}$$

This leads to

$$\sigma(E_i) = \frac{m_2}{m_1}(\varrho_1\varrho_2)^2 \frac{6\cdot6\times10^{-14}}{E_i^2}\,\mathcal{G}_j\left(\frac{E_2}{E_i}\ ;\ \frac{E_1}{E_i}\right) \tag{6.6}$$

with

$$\mathcal{G}_j = \left(\frac{v_2^2}{v_1^2+v_2^2}\right)^{3/2}\frac{2E_1}{3E_2}+\frac{E_i}{E_2}\left(1-\frac{E_1}{E_2}\right)-\left(\frac{E_i}{E_2}\right)^2 \tag{6.7}$$

for $E_i + E_1 \le E_2$, and

$$\mathcal{G}_j = \left(\frac{v_2^2}{v_1^2 + v_2^2}\right)^{3/2} \frac{2}{3} \left[\frac{E_1}{E_2} + \frac{E_i}{E_2}\left(1 - \frac{E_1}{E_2}\right) - \left(\frac{E_i}{E_2}\right)^2\right] \left(1 + \frac{E_i}{E_1}\right)^{1/2} \left(1 - \frac{E_i}{E_2}\right)^{1/2}$$

(6.8)

for $E_i + E_1 \ge E_2$. The function $\mathcal{G}_j(E_2/E_i; E_1/E_i)$ is illustrated in *Figure 6.1*.

The classical calculation leads to an incorrect high-energy fall-off of single ionization cross-section; the correct form, as in the case of excitation to allowed levels, is $E_e^{-1} \ln E_e$, both in the Born approximation and in experiment. The classical form arises from incorrect treatment of distant electron–electron interactions. However, as for excitation, it is profitable to combine

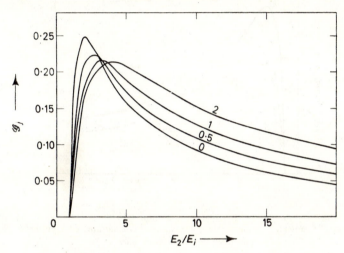

Fig. 6.1. *Function $G_j(E_2/E_i; E_1/E_i)$ for different values of E_1/E_i*
(From Gryzinski[9], by courtesy of the American Physical Society)

the classical calculation (over small and medium impact parameters) with time-dependent perturbation theory[10] (over large impact parameters) to give composite but realistic cross-section functions. Time-dependent perturbation theory is used to calculate the probability of transition due to the field of the electron, which is assumed to move in a classical orbit. The probability is determined by the behaviour of the virtual photons emitted during the trajectory. It is necessary for the calculation that the photoionization cross-section (oscillator strength for transition into the continuum) be known. The combined calculations have been carried out for various systems[11], and the method has been discussed in Section 5.1 under the title of the 'dipole approximation'.

Classical theory also leads to a simple expression which is capable of reproducing the experimental data within about a factor of two:

$$\sigma = \left(\frac{E_i}{R}\right)^2 \xi \sigma_0(v) \pi a_0^2$$

(6.9)

where $\sigma_0(v)$ is a universal function of impact velocity, and ξ is the number of electrons in the shell from which the ionized electron comes. This is not in conformity with the apparent findings of ionization-gauge calibrations[12], which are that at 50 eV impact energy the apparent cross-section is proportional to Z, except in the lightest atomic systems. An entirely empirical formula has also been reported[13].

The earliest classical theory of J. J. Thomson (which neglected orbital motion) allows the scaling of ionization cross-sections of isoelectronic species, so the ionization of positive ions can be compared with those of

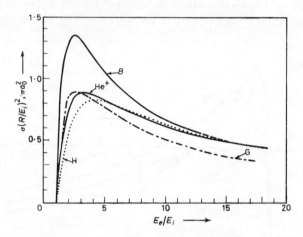

Fig. 6.2. *Comparison of classically scaled experimental e–H and e–He$^+$ ionization functions (lines labelled* H *and* He$^+$ *respectively); G, classical calculations of Gryzinski; B, Coulomb–Born calculations of Burgess*

atoms. For two isoelectronic species of ionization potentials E_{i1} and E_{i2}, the cross-sections are related by

$$\frac{\sigma_1}{\sigma_2} = \left(\frac{E_{i2}}{E_{i1}}\right)^2 \tag{6.10}$$

at the equivalent impact energies, which must be measured in units of ionization potential. The success of this scaling relation is illustrated in *Figure 6.2*. The additional Coulomb interaction experienced in electron–ion collisions but not in electron–atom collisions exerts a significant effect only at energies lower than about five times the ionization energy. The success of the scaling relation may be because it can also be proved by quantum theory in the limit of high Z (excess nuclear charge). It is significant that Na$^+$–Mg^{2+} scales better than the unusual case of Ne–Na$^+$, which scales poorly.

More detailed quantum theory calculations of ionization have been described[14–16]. Distorted-wave calculations have been made of the velocity distribution of emitted electrons, which maximizes around one-half of the velocity of the orbiting electron before ejection.

6.2 MEASUREMENT OF TOTAL AND INDIVIDUAL CROSS-SECTIONS

The measurement of total ionization cross-section was first achieved in a number of experiments[17-21] which are not in very good agreement. But the experiments of Tate and Smith[4-8] have stood the test of time and may be compared with modern beam–gas and crossed-beam measurements, which will also be described. In the Tate and Smith experiment, a magnetically confined (helical paths) current I_0 of electrons is directed through a gas cell containing a 'condenser plate' electrode system which supplies a transverse electric field and collects the positive ion current I_+ formed under single collision conditions over a path length l. Under single collision conditions, $I_+ \simeq I_0 n_0 l \sigma$. The electrons are not collected at the ion collector, being confined by the magnetic field. A number of problems are encountered in modern measurements of I_+ and these are now described.

Only the ions formed from a specified path of beam are collected. The beam passes centrally between parallel plates, at each end of which are guard plates maintained at the same potentials, so that over the region of the collecting plates the electric field is as uniform as possible. When the kinetic energy of ions formed is negligible in comparison with the energy they possess on striking the negative collecting plate, the specified beam path for collection is equal to the length of the parallel plates. It is necessary to increase the transverse field until constant current is collected (saturation conditions), when this equality is nearly always justified. The usual limits of accuracy set by all geometrical factors are applicable. Since atomic ions are formed with approximately thermal energies, very small transverse fields may be used; but fragment ions formed from antibonding states of molecules may possess several electron volts kinetic energy, so saturation conditions are more difficult to achieve. The expected angles of fragment ion emission from molecule ionization are of importance.

The electron beam path along the lines of magnetic force is not straight but helical, so the true distance travelled L is longer than the specified collection path l, being given by

$$\frac{L}{l} = \left(1 - \frac{v_y^2}{v_{\text{tot}}^2}\right)^{1/2} \tag{6.11}$$

The radius of the beam helix is not equal to the radius of the collision chamber orifice. The electron-optical equation is

$$\frac{v_y}{v_x} = \tan\theta = \frac{r}{4}\left(\frac{E-E'}{V_s}\right) \tag{6.12}$$

where E and E' are the electric fields in the ante-collision and collision chambers; V_s is the potential of the orifice with respect to the cathode. The beam axis is taken to be the x-axis. The 'helical correction' to the Tate and Smith experiment, calculated from these equations, is usually only about 2 per cent; it is independent of magnetic field and can be minimized by electron-optical alignment of the electron beam, with ante-collision and collision regions maintained at equal potentials. Collimating grids and honeycomb orifices have been used to this end[22]. Gas scattering might increase the helical correction.

In the presence of a transverse electric field[23] the electron motion is trochoidal-helical, and the correction is given by

$$L = l\left[1 + \frac{2v_d^2}{v^2} - \frac{2v_d^4}{v^4} + \alpha_{\max}\left(\frac{v_d}{v} - \frac{2v_d^3}{v^3}\right) + \alpha_{\max}^2\left(\frac{1}{6} - \frac{v_d^2}{3v^2} - \frac{4v_d^4}{3v^4}\right) + \cdots\right] \quad (6.13)$$

where $v_d = E/H$ is the transverse drift velocity, and $\alpha_{\max}$ is the maximum angle subtended between electron path and axis; this angle must be deduced from a measurement of the beam diameter near an orifice. The treatment assumes an isotropic distribution in angle at the filament.

The measurement of gas density in the collision region presents problems which are discussed in Section 3.3. Crossed-beam techniques have been employed in total ionization experiments[24, 25].

There is an appreciable back-emission of secondary electrons from the main beam-collecting electrode into the collision region; they may possess sufficient energy to cause unwanted ionization, particularly in experiments at high energies. It is necessary to impose a potential barrier between the collecting electrode and the collision region, sufficient to prevent the return of these secondary electrons.

Measurement of Individual Cross-sections

By the term individual cross-section is meant the cross-section for the production of an individual charge state, that is, a singly charged or multiply charged ion. The total ion current measured in the Tate and Smith experiment corresponds to a sum of cross-sections

$$\sum_{n=1}^{n} n \; e0\sigma_{nen+1}$$

The higher charge state cross-sections are orders of magnitude lower than that for the singly charged ion and become progressively smaller with increasing charge. In order to measure the individual cross-sections it is necessary to separate the ionized products in mass-number. A group of ions emerging from a defined collision path length is extracted by transverse electric field through an orifice in the condenser electrode. It is necessary to focus an image of the ion production region onto the entrance orifice of a mass-spectrometer. If a sector magnetic or a time of flight mass-spectrometer is used, the ions must be accelerated and focused, usually by einzel lens, onto an input orifice. It is possible to collect all the ions formed in a defined collision path at the detector orifice following the mass-spectrometer, but, in order to test this efficiency, the input and detector orifices must be widened to the point where the collected current is independent of focusing potentials and of the mass-scan variable (flat-topped peak conditions). Saturation conditions must be maintained, and it is necessary to calibrate the individual cross-sections, which are only relative, against a known cross-section or sum of cross-sections. The condenser-plate experiment is carried out together with the mass-analysis experiment, and all cross-sections may be obtained absolutely, provided that no individual ion species is missed[26].

Since complete collection of energetic ions is not easy to achieve, some more sophisticated techniques have been proposed. One is to cross the electron beam with a fast molecular beam. Another is a cycloidal system, in which the collision takes place in crossed electric and magnetic fields, and no orifices are necessary, only a small collector.

6.3 MASS-SPECTROMETRY AND FRAGMENT ION KINETIC ENERGY MEASUREMENT

The importance of measuring the kinetic energy of fragment ions derives from the fact that a molecular ion formed in an antibonding state dissociates spontaneously, distributing its excess internal energy amongst the fragments in the inverse ratio of their masses. The energetics of the process are described by equations for the appearance potential E_e of the fragment of kinetic energy T from a molecule AB of dissociation energy D, the atomic ionization potentials being E_i:

$$E_e(B^+) = D_{AB} + E_{iB} + \frac{m_A + m_{B+}}{m_A} T_{B+} \qquad (6.14)$$

$$E_e(A^+) = D_{AB} + E_{iA} + \frac{m_{A+} + m_B}{m_B} T_{A+} \qquad (6.15)$$

It is possible to measure kinetic energy, without mass-analysing the ions, by means of a technique first used by Lozier[27] and illustrated in *Figure 6.3*. It is a retarding analyser in cylindrical symmetry. Between the two coaxial sets A and B of annular discs is maintained a 5–10 V potential difference: ions formed in collision travel outwards from the magnetically collimated electron beam C and are guided by the electric field through the inter-disc spaces, whilst all electrons are confined to the region C by the magnetic field. Adjustment of the retarding potentials on the collector D enables the initial kinetic energies of ions to be measured. The operating pressure is $\sim 10^{-4}$ torr. The anisotropies of collision products which are expected in dissociative ionization may affect this type of experiment. The Lozier apparatus has been used for total cross-section measurement, but probably gives only 10 per cent accuracy.

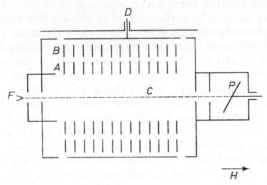

Fig. 6.3. Lozier's apparatus: F, filament; C, electron beam; P, electron-collecting plate; A, B, ion-guiding annular discs; D, ion-collecting cylinder

Kinetic energy analysis of mass-analysed ions can be achieved by:

1. Mathematical analysis of mass-spectrometer peak shapes of ions as the mass-number scansion is moved through the peak[28-33]; this method is mainly of historical interest.

2. Retarding analysis with the aid of potential applied to a retarding grid[34]; careful attention to the ion optics in the retardation region is important, and spherical grid and collector are best. The ion energy distribution is the first differential of the retardation characteristic.

3. Momentum analysers, described in Section 3.13, can be used for kinetic energy analysis. In double mass-spectrometers, sector cylinder momentum analysers (usually 90°) are incorporated in the instrument, and may be used for this purpose.

Recent crossed-beam investigations of fragment ion kinetic energies are in disagreement with early work[35]. The most recent crossed-beam investigations of molecular fragment ions have included studies of angular distribution as well as kinetic energy[36].

It is valuable to use coincidence-counting technique to study simultaneously two different collision products. For example:

1. The angular distribution of two post-collision electrons;

2. The kinetic energy of the emitted electron and the charge state of the ion produced;

3. The delayed fragmentation of molecular ions.

Experiments to measure simultaneously the angular distributions of the two post-collision electrons have been described by Ehrhardt[3]; in the region of maximum total cross-section, the two electrons are emitted in very approximately opposite directions, as is expected on the basis of the Lippmann–Schwinger equation. Close to threshold, however, the experiments should provide a sensitive test of the applicability of classical theory.

From mass-spectrometers, much important information about the ionization cross-sections of molecules can be obtained. It is important to remember that a commercial mass-spectrometer does not yield an accurate cross-section (even relative) for an ionization process. The ionizer usually consists of an electron impact ion source (Section 3.10), containing a magnetically confined electron beam and a 'repelling' electrode which assists in the extraction of the ions by application of an electric field transverse to the electron beam. Superficially, the conditions are not dissimilar from those of the Tate and Smith experiment, but numerous complicating factors arise. The 'repeller characteristic', that is, ion current as a function of repeller potential, does not always exhibit saturation conditions and, in particular, sharp maxima sometimes appear in this characteristic. Baker and Hasted[101] have shown how these maxima can arise from collisional effects during trapping of the ions in the space charge of the electron beam. These collision sequences can contribute significantly to fragment ion functions observed in mass-spectrometers, particularly in the region close to threshold. An analysis was given of the conditions under which space-charge trapping occurs in cylindrical beams; for ions produced with only thermal energy the trapping is likely to be serious in most commercial instruments at near-threshold electron energies.

Mass-spectrometers are now highly-developed commercial instruments, and it is valuable for physicists and chemists interested in the process of ionization of atoms and molecules by electrons to use them in their commercially available form to gain whatever knowledge is possible. There is need for detailed consideration of the ionization efficiency of the sources and also for the analysis, by means of the Born–Bethe approximation, of high-energy cross-section functions observed with mass-spectrometers. Mass-analysis technique is discussed in Section 3.14.

6.4 TOTAL AND INDIVIDUAL CROSS-SECTION DATA

A collection of ionization cross-section data has been compiled by Kieffer and Dunn[37]. Some typical atomic and molecular cross-section functions are shown in *Figures 6.4–6.11*. There are small discrepancies ($\sim$ 5 per cent) between different workers, but it is remarkable how well-established the absolute magnitudes of these cross-section functions appear to be, and how close is the agreement between Tate and Smith and the modern experiments [46, 49].

In the region of maximum cross-section the Born approximation is invariably high.

A feature of interest in the Ar, Kr and Xe cross-sections (*Figures 6.9–6.11*) is the minor structure that appears in the region of maximum cross-section (not usually claimed as structure by the experimenter). Effects due to more than one electron may be involved.

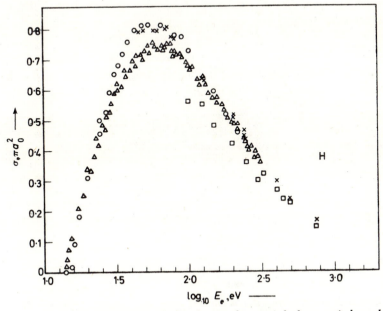

Fig. 6.4. Measured ionization cross-section function for atomic hydrogen: circles and crosses, Fite and Brackmann[24]; squares, Neynaber et al.[25]; triangles, Boksenberg[38]

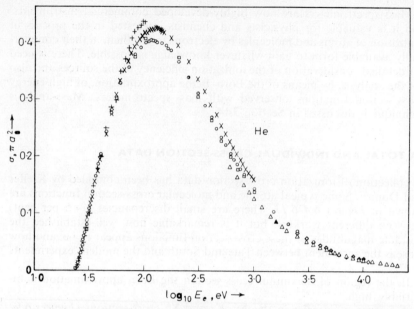

Fig. 6.5. Measured ionization cross-section function for helium: open circles, Smith, Tate and Bleakney[4-8]; diamonds, Liska[39]; squares, Harrison[40]; crosses, Rapp and Golden[22]; plus signs, Asundi and Kurepa[22]; triangles, Schram et al.[41-44]

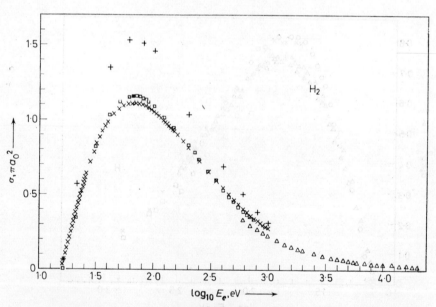

Fig. 6.6. Measured ionization cross-section function of molecular hydrogen: squares, Smith, Tate and Bleakney[4-8]; crosses, Rapp and Golden[22]; plus signs, Harrison[40]; triangles, Schram et al.[41-44]

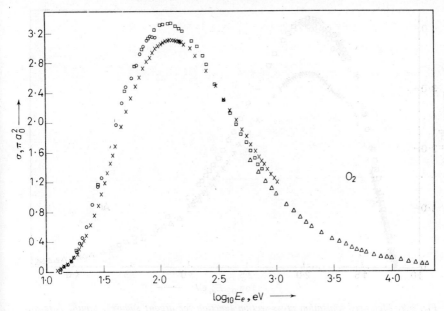

Fig. 6.7. *Measured ionization cross-section functions of molecular oxygen; squares, Smith, Tate and Bleakney[4-8]; crosses, Rapp and Golden[22]; plus signs, Schulz[45]; circles, Asundi, Craggs and Kurepa[46]; triangles, Schram et al.[41-44]*

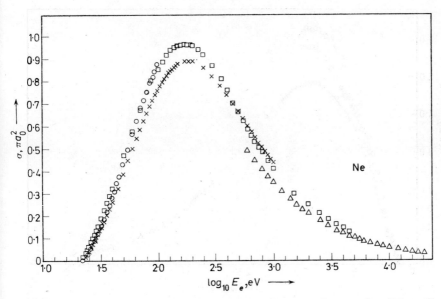

Fig. 6.8. *Measured ionization cross-section function of neon: circles, Asundi and Kurepa[22]; squares, Smith, Tate and Bleakney[4-8]; crosses, Rapp and Golden[22]; triangles, Schram et al.[41-44]*

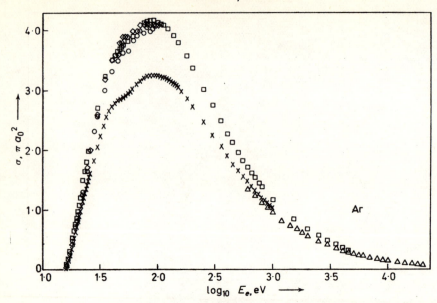

Fig. 6.9. Measured ionization cross-section function for argon: squares, Smith, Tate and Bleakney[4-8]; diamonds, Tozer and Craggs[47]; circles, Asundi and Kurepa[22], crosses, Rapp and Golden[22]; triangles, Schram et al.[41-44]

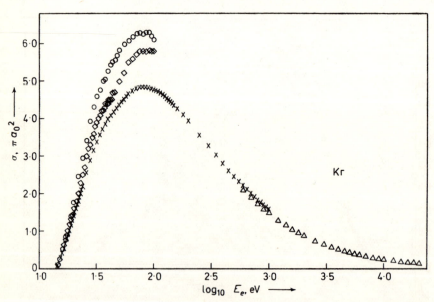

Fig. 6.10. Measured ionization cross-section function for krypton: circles, Asundi and Kurepa[22]; diamonds, Tozer and Craggs[47]; crosses, Rapp and Golden[22]; triangles, Schram et al.[41-44]

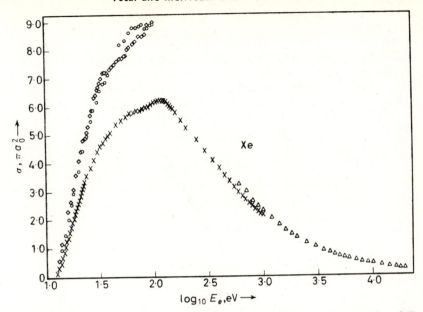

Fig. 6.11. *Measured ionization cross-section functions for xenon: circles, Asundi and Ku-repa[22]; diamonds, Frost and McDowell[48]; crosses, Rapp and Golden[22]; triangles, Schram et al.[41-44]*

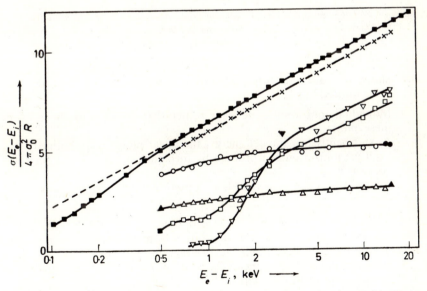

Fig. 6.12. *Plot of $\sigma(E_e - E_i)/4\pi a_0^2 R$ versus $\ln(E_e - E_i)$ for the total and partial ionization cross-sections of neon: closed squares, total cross-section; crosses, Ne^+; circles, $Ne^{2+} \times 20$; triangles $Ne^{3+} \times 200$; open squares, $Ne^{4+} \times 5000$; inverted triangles $Ne^{5+} \times 10^5$*

According to the Bethe–Born approximation, the high-impact energy fall-off of optically allowed transitions is as $E_e^{-1} \ln E_e$, and of disallowed transitions as E_e^{-1}. In classical theory, the distinction is not made and in all cases the behaviour is predicted to be E_e^{-1}. A graphical representation of data is often made[41-44] in the form of plots of σE_e against $\log_{10} E_e$; a linear dependence should then be found for electric dipole transitions (which include ionization), but for forbidden transitions there is independence (a horizontal as opposed to a rising 'Bethe plot' or 'Platzman plot'). *Figure*

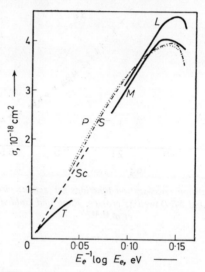

Fig. 6.13. Plot of σ versus $E_e^{-1} \log_{10} E_e$, for ionization cross-sections of two-electron systems: L, Li^+ (Lineberger et al.); M, Li^+ (Peart and Dolder); S, scaled He (Smith); Sc, scaled He (Schram et al.); T, scaled H^- (Tisone and Branscomb)

6.12 shows data for neon due to Schram, in which the following features should be noticed:

1. Individual ionization state data rather than totals must be used in these plots;
2. At sufficiently low impact energy, the linear plot is not obtained;
3. Multiple ionization can proceed by forbidden processes, but need not necessarily do so;
4. As the impact energy is raised, an 'onset' of a different mechanism can sometimes be observed.

Examples of forbidden processes are the double ionization of He, of N_2 and of alkali metals, the production of protons from H_2 (presumably via the $H^+–H^+$ antibonding state), and the production of excited states of He^+.

The slope of the Bethe plot yields the continuum oscillator strength, which is of value for sum rule calculations[50].

An alternative graphical representation of ionization data is as σ plotted against $E_e^{-1} \log_{10} E_e$. Two-electron system data is so represented in *Figure 6.13*.

Complexities in the form of multiple ionization functions of many-electron atoms have been reported[51], and are discussed under the heading of threshold processes. *Figure 6.14* illustrates orthodox multiple ionization processes in argon.

The ionization functions of atomic hydrogen[24] and also of atomic oxygen[25] have been measured in crossed-beam experiments; the total positive ion product is drawn out by a strongly negative electric field which penetrates into the collision region and extracts all ions formed with kinetic energies

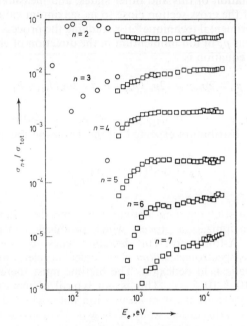

Fig. 6.14. *Fractional multiple ionization cross-sections* $\sigma_{n+}/\sigma_{\text{total}}$ *for argon: circles, Bleakney*[8]; *squares, Schram et al.*[41-44]

smaller than several electron volts. Invariance of the ion current with the strength of the electric field is proof that only a negligible number of ions are formed with energies in excess of these values. The ratio of H_2^+ to H^+ ions is obtained using a sector mass-spectrometer, and the relative cross-sections are normalized to the molecular hydrogen data of Tate and Smith[5].

There exist unexplained discrepancies as large as 80 per cent in the total ionization cross-sections for caesium[52, 53], and for lithium[53], for both of which calculations are available[54].

6.5 SIMULTANEOUS IONIZATION AND EXCITATION

The electron–atom collision in which the positive ion is formed in an excited state ($e0/1'e^2$) can make a significant contribution to the total ionization. The cross-section functions are of similar shape to those for simple ionization.

As has been mentioned in Section 5.8, they can be measured by observing the optical emission from the excited state of the positive ion[55]. Certain other information is also necessary in order to obtain absolute measurements.

The process is of importance in providing the pumping mechanism for positive ion lasers, which in general radiate at shorter wavelengths than neutral species lasers. For example, the Ar^+ laser operates on 4880 Å, $4p\ ^2D^o_{5/2} \rightarrow 4s\ ^2P_{3/2}$. The impulse approximation has been applied to calculating the excitation of this and other states, and measurements confirm the magnitude of the cross-section close to its maximum value.

A further experimental technique[56] for studying the process is by monitoring the component $\overline{p_x}$ of ion momentum in the direction of electron impact. The momentum equation is

$$2p^2_{ep} + p^2_+ + 2p_{ep}p_+ - 2p_{ea}(p_{ep} + p_+) + 2m_e(E_{ex} + E_i) = 0 \qquad (6.16)$$

Assuming the average $\overline{p_e} = 0$, neglecting $\overline{p^2_e}$ and $\overline{p^2_+}$, and taking an average over the angular distribution expected on the Thomson theory, one obtains

$$\overline{p_x} = 0.85(E_{ex} + E_i)\left(\frac{m_e}{E_e}\right)^{1/2} \qquad (6.17)$$

A technique for studying long-lifetime states embedded in the continuum is as follows. An ionization process which produces a metastable ion is detectable as an 'Aston band'. This detection relies on a change of mass-number in a mass-spectrometer *after* electric field acceleration or deflexion, but *before* magnetic field deflexion. The lifetime must therefore lie in the general region 10^{-5}–10^{-6} sec. This is not typical of metastable states of atomic ions, but there are many molecular ion states which dissociate spontaneously in times of this order; study of these paths of dissociation by Aston band technique is discussed further in Section 6.11 below.

Claims have been made[57] for the existence of states of singly charged atomic ions of such lifetimes, at energies just above the appearance potential of the doubly charged ion. These are presumably autoionization states whose transition into the continuum of the doubly charged ions is forbidden, so the lifetime is not $\sim 10^{-14}$ sec but $\sim 10^{-6}$ sec (2.5 μsec for xenon and 0.8 μsec for argon are reported). In these experiments, great care was taken that the spontaneous second ionization could not be taking place by collision with gas or with metal surfaces. Such collisional second ionization processes of states lying just below the doubly charged ions appearance potential have in fact been reported[58], and it has been pointed out[59] that high principal quantum number states of ions might have lifetimes sufficiently long for such effects to be observed. The effects of long-lifetime excited states of ions in various experimental situations must be carefully studied. Ionization of the high principal quantum number states of the one electron system He^+ has been achieved[60], not only collisionally but by the Lorentz process, which involves the action of a strong electric field. The detailed analysis of this ionization allows the effects of individual quantum number states to be separated.

6.6 AUTOIONIZING STATES

The existence of autoionizing states of atoms has appreciable influence upon the cross-sections for ionization by electron impact. When the impact energy is equal to the energy of the state, ionization becomes possible via the auto-ionizing state and a bump appears on the ionization function. At high impact energies, the slope of the Bethe plot can demonstrate the contribution to the total ionization cross-section made by such processes. However, it is in a different type of collision measurement that the autoionizing transitions appear most strongly: the measurement of the characteristic energies of electrons emitted in the ionization process shows up the energies of autoionizing levels; the ionization is not usually carried out by electrons but by ultraviolet or X-rays, fast protons, or metastable atoms.

The possible types of autoionizing state include:

1. Double (or multiple) excitation; two (or more) electrons are simultaneously excited, as in He $2s\,2p$.
2. Inner shell or subshell excitation; an inner shell or subshell electron is excited into a higher energy orbit, as in Ne $1s^2\,2s\,2p^6\,3s$.
3. Excitation with core rearrangement; an outer shell is excited whilst the inner core is rearranged, as in O $2p^3\,(^2D)\,4s\,(^3D)$ [the ground state is O $2p^3\,(^4S)\,2p\,(^3P)$].
4. Inner shell vacancy; the initial excited state is a state of the ion: Ne$^+$ $1s\,2s^2\,2p^6 \rightarrow$ Ne^{2+} $1s^2\,2s^2\,2p^4$, and the autoionizing transition leaves the atom doubly ionized. This is an Auger transition, but if one of the two final state vacancies is in the same shell as the primary vacancy, it is a Coster–Kronig transition.

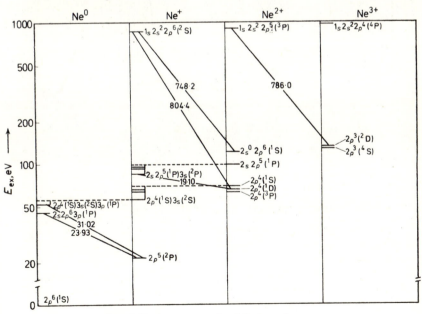

Fig. 6.15. Auger energy level diagram for neon

5. Vibration–rotation with electronic excitation; a molecule in a state of electronic excitation near the ionization potential receives vibrational and/or rotational energy sufficient for autoionization to be possible.

A type of energy level diagram suitable for representing autoionizing transitions has been devised by Rudd. *Figure 6.15* shows some levels of neon. An autoionizing event is represented by a transition from a state in one column into a lower state in the next right-hand column. The relation between a given Rydberg series and its parent and grandparent configuration, for example Ne $2p^4$ (^1S) $3s$ (^2S) nl, Ne$^+$ $2p^4$ ($1s$) ns (^2S), Ne^{2+} $2p^4$ (^1S), is clearly seen. Experiments by Mehlhorn[61] measured the energy spectra of the electrons emitted in ionization, and yielded the Auger spectra shown for krypton in *Figure 6.16*. Further discussion of autoionizing states may be found in Chapter 9.

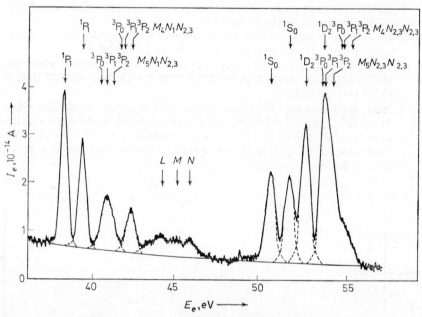

Fig. 6.16. Part of the Auger spectrum for krypton: impact electron energy, 2 keV; current, 220 μA; energy scale approximate (after Mehlhorn[61])

6.7 THRESHOLD IONIZATION AND STRUCTURE

The form of the ionization cross-section function within a few electron volts of threshold is of importance for three reasons:

1. The 'threshold law' expected in theory is of intrinsic importance.
2. The onset of simultaneous ionization and excitation processes can be detected by electron impact.
3. Autoionization processes can be detected.

The threshold law for single ionization of atoms is usually taken to be linear:

$$\sigma = \frac{\mathrm{d}\sigma}{\mathrm{d}E_e}(E_e - E_i) \qquad (6.18)$$

with $\mathrm{d}\sigma/\mathrm{d}E_e$ constant. Although this threshold law has been widely accepted for many years, as a result both of experiment and of theory[62], it was not until comparatively recently[63] that the problem was solved rigidly by quantum theory. But this solution contains a singularity[64] and is at variance with the classical treatment[65]; the latter leads to a threshold law that is not precisely linear but is a 1·127 power law:

$$\sigma \propto (E_e - E_i)^{1·127}$$

This law arises from a consideration of the mutual interaction of the two electrons. It has been argued that there must be a region or spherical shell in which the classical approximation is valid; since the electrons must at some stage pass through this region, the 1·127 power law should operate.

It cannot be said that the exact threshold law is a fully resolved problem, but it can be said that it is fairly close to linear. It is difficult to predict theoretically the exact range of energy over which the threshold law should hold: it could be only a few meV.

During the 1950s, low-resolution experiments conducted with atomic hydrogen[66] showed linearity up to about 5 eV above threshold; the value of $\mathrm{d}\sigma/\mathrm{d}E_e$ was higher than that calculated[67]. Retarding potential difference measurements[68–71] with helium showed linearity over a similar 5 eV range (*Figure 6.17a*). The function for production of the doubly charged ion was found in these experiments to be square law (the mass-number being separated from H_2^+ impurity by the use of He^3 isotope).

Numerous experiments have been conducted with other atomic species, but are complicated by simultaneous ionization and excitation. In modern work, it is necessary to take account of gas atom thermal energies.

However, the most recent and refined studies of atomic hydrogen using momentum-analysed electrons (50 mV FWHM)[72] show a power law clearly greater than unity over the lowest 0·3 eV. For 3 eV above this, the function is linear. A 1·13 power law gives a good fit to the data in the lowest 0·3 eV, but it is only possible to place an accuracy ±0·03 on the number 1·13. The experiment consists of crossed electron and atom beams, the ions being collected by a lens system and quadrupole mass-spectrometer, movable in angle. The differential cross-section (in angle) is measured, and an integration made over the entire range of angle to yield a total cross-section. In this way, the serious problem of recoil momentum, which can easily cause incorrect cross-sections to be recorded close to threshold, is solved. A careful setting of the energy scale and convolution of the 50 mV energy distribution reveals that ions are actually formed at an energy 0·02 eV lower than would be the case if the power law were exactly linear. *Figure 6.17b* illustrates the data. An interesting criticism of this experiment is that Rydberg states of the hydrogen atom will contribute to the measured ion current, being ionized by low-energy background photons; but the ion current was shown to be independent of this photon background.

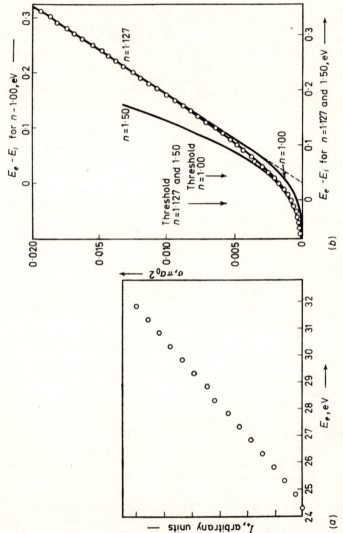

Fig. 6.17. *Threshold ionization functions: (a) for helium[71]; (b) for atomic hydrogen, showing experimental data points and calculated convolutions[72] for power laws 1·00, 1·127 and 1·50 ($\Delta E_{1/2} = 0·05$ eV)*

Recent experiments on helium[73] have also yielded a threshold power law greater than unity—greater even than 1·127. A more sensitive test of the rival proposals of a linear and supra-linear power law would lie in the relative angular distribution of the two electrons emitted; coincidence counts of these two electrons (see Section 6.3) are in course of measurement.

The single ionization function threshold law can often be taken as approximately linear over a range of several electron volts. (Note that the Born approximation yields an incorrect 3/2 power law.) There is no reason why the simultaneous ionization and excitation process should not show a similar approximately linear threshold law. Therefore, if the ion possesses low-lying excited states, the overall cross-section function will be piecewise linear (as in *Figure 6.18a*). These 'breaks' in ionization thresholds have

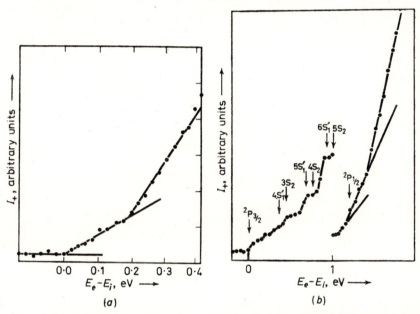

Fig. 6.18. Threshold ionization functions: (a) for argon[74]; (b) for xenon[75, 76] compared with optically identified autoionization levels

frequently been used in molecular studies for the investigation of ionic excited states or 'higher ionization potentials' of the molecule. The importance of these higher potentials arises from the fact that they correspond to energies of removal of various orbital electrons from the molecule; thus molecular orbital calculations can be checked. However these 'break-manship' studies do not represent the easiest method of investigating inner ionization potentials, since instrumental breaks can arise at metal surfaces and orifices—a disadvantage not present in crossed-beam experiments—and in any case the piecewise linear form is not easy to observe. Momentum analysis of the impact electrons (or alternatively high-resolution photo-ionization) is a valuable aid. According to the Bethe approximation (which does not of course apply close to threshold) the first differential of the photo-

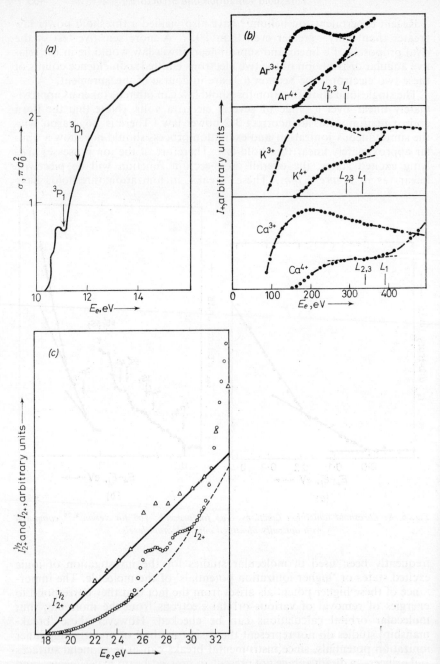

Fig. 6.19. Ionization functions: (a) absolute ionization function of mercury, showing structure attributed to autoionizing states; (b) anomalous ionization functions for $Ar \to Ar^{3+}$, $Ar \to Ar^{4+}$, $K \to K^{3+}$, $K \to K^{4+}$ *and* $Ca \to Ca^{4+}$ *(inner shell ionization energies are shown); (c) threshold ionization function of* $Ca \to Ca^{2+}$ *showing both* I_{2+} *and* $I_{2+}^{1/2}$ *(structure attributed to autoionization[87] appears around 25 eV)*

ionization function is proportional to the electron ionization function. The former is (very crudely) a staircase function; but since both functions are complicated by autoionization and other features, the most suitable investigating technique for inner ionization potentials is that of photoelectron spectroscopy; discussion of this is postponed until Chapter 9.

Autoionization structure is observed in krypton and xenon threshold functions[75-77], but the piecewise linear structure is also present (*Figure 6.18b*). The neon function is linear, but the argon function has been reported to possess a long tail[78] as well as autoionization structure[73, 79, 80]. The long tail is reported[81, 82] to be made up of piecewise linear sections appearing at 0·18 eV, 0·56 eV, 0·78 eV, 0·96 eV, 1·34 eV, 1·84 eV, 2·2 eV, 2·7 eV and 3·3 eV, but some of these energies are in dispute. Only the lowest corresponds to a level of Ar^+. Data are displayed in *Figure 6.18*. More recently, Marmet, using crossed-beam technique, has reported no structure whatever, apart from the $J = \frac{1}{2}$ level. The extent to which the heavy rare-gas threshold functions can be interpreted as autoionization processes is still uncertain. There is a legendary break in the helium function around 0·7 eV.

Since many autoionizing states arise from the simultaneous excitation of two electrons, it is to be expected that the phenomenon will be particularly marked in Group II elements. For many years, the anomalous threshold functions of zinc and mercury have been known and interpreted as 'ultra-ionization potentials'. More recently, alkali metal, alkaline earth metal and rare gas multiple ionization functions have been investigated in both single and double ionization[83-86], and very large effects have been found, which are illustrated in *Figure 6.19*.

The threshold functions for ionization of molecules are complicated by the production of the molecular ion in vibrationally excited states. Apart from autoionization, the relative proportions of different vibrational states are determined by the Franck–Condon factor. Where the equilibrium separation of molecule and molecular ion are identical, the $v = 0$ ion will dominate, but in the more usual case of unequal separations, there will be a skewed bell-shaped distribution of vibrational quantum numbers[88]. The threshold function should appear piecewise linear, the slopes being determined by this distribution. In addition to vibrational structure, there might arise breaks at the onset of electronically excited ions, presumably vibrationally structured as well. However, no undisputed example of such behaviour has been reported.

Piecewise linear H_2^+ data[89] have been reported, but subsequent retarding potential difference measurements showed a linear threshold without structure[90] (*Figure 6.20*). The most recent measurements, and some of the most reliable[91, 92], show a curved function between 15·38 eV and 15·50 eV, then a linear section lasting until 15·70 eV. This is interpreted as being to the $v = 0$ state of H_2^+, but is complicated by much autoionization. The first differential of the function corresponds to the photoionization function (*Figure 6.20*), but there is some additional structure which is supposed to arise from states which require electron exchange to form them.

In hydrogen the autoionization is believed to arise from vibrationally excited Rydberg states of the neutral molecule[93-95]. Other diatomic molecules have been studied with high-energy resolution[75, 76, 92, 96]. Diatomic molecule threshold autoionization processes have been reviewed[97-99].

27

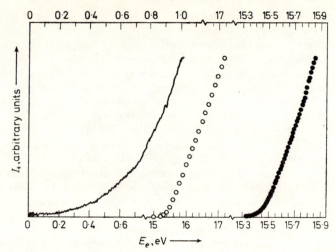

Fig. 6.20. Threshold ionization function of H_2: *pen-record, data interpreted as piecewise linear function*[89]; *open circles, data interpreted as linear onset*[90]; *closed circles, data interpreted as structured threshold function*[91]

It cannot be stressed too strongly that there are many instrumental contributions to threshold functions:

1. Electron energy distributions which are not invariant with energy; these can arise in incorrectly set up retarding potential difference experiments.

2. Surface effects, such as electron scattering at slits, and the possibility of ionization potentials becoming reduced when the molecule is absorbed on, or close to, a metal surface; such effects are only avoided by conducting crossed-beam experiments.

3. Breaks in molecular (parent) ion threshold functions can arise from ion–molecule reactions. Thus it has been proposed[100] that the 19·4 eV break in the CH_4^+ function from methane arises from the onset of the process

$$CH_2^+ + CH_4 \rightarrow CH_4^+ + CH_2$$

4. Sequences of processes involving ions trapped in the space charge of the electron beam can contribute to fragment ion threshold functions[101]. These need not always commence with an ionization; contributions of the type

$$e + RX \rightarrow R + X + e$$
$$e + R \rightarrow R^+ + 2e$$

may also be involved.

5. Spontaneous rearrangements of molecular ions may contribute to fragment ion functions; these can hardly be described as instrumental contributions or indeed be avoided.

Since the experimental measurement of appearance potentials is of importance in molecular structure studies, the technical difficulties of energy scale calibration are worth consideration. The experiments are often performed in commercial mass-spectrometers whose electron energy distributions are Maxwellian and cover a range of 0·3–0·5 eV; other experiments are conducted with momentum-analysed electrons.

In either case, the ion current I_+ observed at electron energy scale-setting E_0 is proportional to the integral

$$I_+(E_0) = \frac{\int \sigma_i(E_e) f(E_e + E_0) \, dE_e}{\int f(E_e + E_0) \, dE_e} \qquad (6.19)$$

where $\sigma_i(E_e)$ is the threshold ionization function, which is usually assumed linear above onset. The electron energy distribution at scale-setting E_0 is $f(E_e + E_0)$. The function $\sigma_i(E_e)$ can be deconvoluted or 'unfolded' from this equation, provided that $I_+(E_0)$ and $f(E_e + E_0)$ are known. The electron energy distribution may be obtained by observing the onset function $I_+(E_0)$ for He^+ from helium; since $\sigma_i(E_e)$ has been reported to be approximately linear for several electron volts above threshold,

$$\frac{d^2 I_+}{dE_0}(E_0) \propto f(E_e + E_0)$$

The computation of second differential yields both electron energy distribution and energy scale setting, using the known ionization potential of helium. The mean energy

$$\bar{E} = \frac{\int (E_e + E_0) f(E_e + E_0) \, dE_e}{\int f(E_e + E_0) \, dE_e} \qquad (6.20)$$

of the energy resolution is equal to the ionization potential.

Since the mathematical deconvolution techniques of Chapter 1 are not trivial to operate, simplified techniques have still a role to play. The simplest is to produce the line drawn through the supra-threshold linear section of $I_+(E_0)$ (provided it is sufficiently long) until it crosses the abscissa. This crossing is the ionization energy; the energy scale has previously been set by a similar procedure in helium. (Since contact potentials can change with change of gas in the ionization chamber, use of a mixture of gases is to be preferred.)

Another technique is to plot the onset functions in a semi-logarithmic fashion, that is, $\log_{10} I$ versus E_0. When the electron energy distribution is Maxwellian, and has an exponential high-energy tail, then the lowest energy part of the semi-logarithmic function will be linear (and vice-versa). The section which is linear on a conventional plot is curved on a semi-logarithmic plot[102]. When it is required to compare an unknown ionization potential with a known one, the two functions are plotted and one is fitted to the other by multiplying by an empirically determined constant factor and shifting the energy scale. The amount by which the scale must be shifted is equal to the difference between the ionization potentials. An improved application of this technique has been reported[103].

The technique of fitting a piecewise linear function to experimental data sometimes involves a rather subjective use of the straight edge, and is thus open to criticism. A mathematical technique[104] is to compute least squares fits of data points taken in groups of (say) five points, starting at consecutive data points. A plot is made of the variance to these fits against the energy of the starting points. This procedure is repeated with the number five varied in either direction. A search is made for the smallest variances, there being several orders of magnitude variation possible. In this way, the best fitting straight lines are obtained.

A threshold function which contains both breaks and step-functions arising from autoionization can be investigated by computing second differentials of the data. The breaks appear as peaks and the step-functions as 'anomalous dispersion' functions passing through both positive and negative values. Deconvolution of the energy distribution from the second differential is a mathematically sound procedure and can assist in separating overlapping levels.

The power law thresholds of multiple ionization of atoms are not universally substantiated. The He^{2+} and Ne^{2+} thresholds are square law[68], but certain linear thresholds (possibly owing to autoionization) have been claimed for argon, krypton and xenon up to six times ionized[51, 68]. However, power law functions have also been claimed for these cases[105–111], various techniques of data analysis being used. Naturally, the inferred ionization potentials for highly charged ions are widely different according to whether linear functions or high-power functions are assumed. Unfortunately, optical ionization potentials are not available in all cases, so to some extent the issue rests upon the preferred values of the optically unknown ionization potentials. Recent experiments[112] with the trapped-ion technique have yielded energies for the processes $e(n-1)/e^2n$, which have linear onsets and

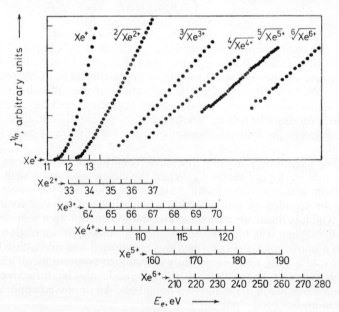

Fig. 6.21. Threshold functions for multiple ionization processes in xenon[111], interpreted in terms of nth rooth law

so give independent values of the ionization potentials. These support the power law measurements, adding some weight to what is already probably the preferred interpretation. Thresholds for xenon are illustrated in *Figure 6.21*.

6.8 IONIZATION PROCESSES IN MOLECULES

Some features of molecular ionization by electrons (namely, Franck–Condon factors, kinetic energies of fragments, and threshold functions) have already been discussed.

The mechanism of dissociative ionization can be inferred from measurements both of angular distribution and of kinetic energy distribution of fragments[113, 114]. For example, the kinetic energy distribution of protons from hydrogen gives a Franck–Condon overlap in disagreement with calculations. The fragmentation of molecules during impact ionization is basic to mass-spectrographic analysis and therefore plays an important part in structural chemistry.

Fragmentation ionization of diatomic molecules can take place either:

1. By transition to an antibonding state, yielding energetic fragment ions (from the energy distribution function, it is possible to deduce the relevant part of the antibonding interaction energy);
2. By vertical transition to a bonding state at, or near, the dissociation limit;
3. By vertical transition to highly vibrationally excited states which can, by thermal energy collisions, subsequently dissociate.

Isotope effects are found in dissociative ionization processes affected by curve-crossing or predissociation (for example, HOD, H_2O, H_2S and HSD).

The fragmentation processes of molecular ionization can be studied in detail with the aid of newly available techniques; an example is the simultaneous detection of the fragment ion and neutral in coincidence, which requires that the molecule be introduced into the electron impact region in the form of a fast molecular beam.

Another time of flight coincidence experiment[115] is designed in such a way as to detect two ions produced simultaneously from an electron–molecule collision, for example

$$e + CO_2 \rightarrow O^+ + CO^+ + 3e$$

It appears that such processes are responsible for the production of up to 25 per cent of some atomic ions.

A statistical approach to fragmentation has been attempted by Eyring and his colleagues[116, 117], and appears to be applicable at least to aliphatic hydrocarbons. The system is considered as a cascade of competing and consecutive unimolecular reactions from one state to another. The excess energy is assumed to be converted to vibrational energy by many radiationless transitions; the methods of statistical mechanics are applied to give an expression for the rate constant k for the decomposition of a particular state:

$$k = z\left(E_i - \frac{\Delta E_{\text{act}}}{E}\right)^p (E - \Delta E_{\text{act}})^q \qquad (6.21)$$

with

$$p = N - \tfrac{1}{2}L - 1 \qquad (6.22)$$

$$q = \tfrac{1}{2}(L - L^{\ddagger}) \qquad (6.23)$$

$$z = \varsigma(2\pi)^{3q} \frac{\Gamma(N - \tfrac{1}{2}L) \prod\limits_{i=1}^{L^{\ddagger}} (I_i^{1/2}/n_i) \prod\limits_{k=L+1}^{N} v_k}{\Gamma(N - \tfrac{1}{2}L^{\ddagger}) \prod\limits_{j=1}^{L} (I_j^{1/2}/n_j) \prod\limits_{l=L^{\ddagger}+1}^{N-1} v_L^{\ddagger}} \qquad (6.24)$$

The molecule is considered as L internal rotors and $N-L$ harmonic oscillators, where

$$N = 3N' - 6 \qquad (6.25)$$

for a non-linear molecule of N' atoms.

The ith rotor has a reduced moment of inertia I_i, and n_i equivalent positions. The total energy of the molecule is E, and the energy of activation of each reaction is ΔE_{act}. The frequency of vibration of the ith normal mode is v_i, and the symbol ς represents the number of equivalent ways of choosing the reaction coordinate, considering each atom labelled and no free rotation; the rotational energy levels are doubly degenerate and are affected by considerations of symmetry. The superscript $\ddagger$ indicates the activated complex.

There exists a large range of activation energies[99] for which the theory can show consistency with observed appearance potentials; but the frequency of breaking of a CH bond must be assumed to be impossibly large if agreement with the propane and deuterated propane experimental data is to be obtained. Other examinations of the theory[118] show that the statistical theory of mass-spectra is qualitatively but not quantitatively satisfactory. There is much to be said for the view[119] that in electron impact a particular chemical bond is excited in such a way that the excitation (or extra local charge) is not free to travel through the molecule; in this case, the relative abundances of fragments are governed by the relative probabilities of different excitations rather than by statistical distribution of the excitation energy.

Detailed investigation of the routes of fragmentation is possible using the Aston band technique (Section 6.11). Commercial double mass-spectrometers are best suited to these studies, since the ambiguities of interpretation are removed by the use of both electrostatic and magnetic deflexion. The observed Aston peak corresponds to a fragment ion of known mass-number produced by spontaneous decay of a fragment ion, also of known mass-number.

From a molecule such as benzene more than a hundred such Aston peaks are observed. It should be possible to plot a complete route map of production of fragment ions, provided that account is taken of the spontaneous rearrangements which are known to take place, especially where the hydrogen atom is concerned. However, the range of decay times that can be studied with this technique is comparatively small (10^{-5}–10^{-6} sec).

Fragmentation ionization is of importance as the technique of structure-analytical mass-spectrometry. Commercial mass-spectrometers have been developed for the study of (mostly organic) molecules, which are introduced usually as vapours into an electron bombardment source; for analytical purposes, the electron energy is held at a value in the region of cross-section function maxima, perhaps 80 eV. The ion currents detected by the mass-spectrometer are not accurately proportional to the relevant cross-sections. Nevertheless, they are qualitatively governed by the magnitude of the cross-sections, by the kinetic energy of the fragment ion and, perhaps, the time taken for the molecule to fragment. A particular mass-spectrometer should yield a certain intensity pattern of different fragment mass-numbers, specific to that particular instrument. These intensity patterns are measured for pure vapours, and the information may afterwards be fed into sets of linear equations for the purpose of analysing mixtures of vapours, such as petroleum hydrocarbons. For analytical purposes, it is often possible to combine mass-spectrometry with gas chromatography. Since the ratio of mass numbers 41–43 is particularly sensitive to the type of organic compound, a crude mass spectrometer capable of measuring only this ratio is sometimes fitted to a gas chromatograph.

Fragment intensities are often represented as a matrix; for example, for N_2O_4, the intensities corresponding to

$$N_2O_4^+ \quad N_2O_3^+ \quad N_2O_2^+ \quad N_2O^+ \quad N_2^+$$
$$NO_4^+ \quad NO_3^+ \quad NO_2^+ \quad NO^+ \quad N^+$$
$$O_4^+ \quad O_3^+ \quad O_2^+ \quad O^+$$

One of the important questions to be solved about molecular ions may be stated as follows: in the ionization of a molecule, a bonding or antibonding electron pair is split, leaving a single electron; so the ion is really an 'ion radicle', represented sometimes as AB^+; but does the emitted electron come from a localized orbital or from a non-localized region of electrons?

6.9 MULTIPLY-CHARGED MOLECULAR IONS

A number of doubly charged diatomic molecular ions have been observed in electron impact studies; they are found in both ground states and unstable excited states, similar to the unstable electronic states of the isoelectronic diatomic molecule. Some potential energy curves have been calculated[120, 121] using an integral form of the quantum-mechanical virial theorem. The functions for N_2^{2+} are shown in *Figure 6.22*. The energy of the atomic ions $N^+ + N^+$ is taken as zero in the figure, and the two sets of curves shown are appropriate to two assumed values of the dissociation energy D_e of the isoelectronic molecule C_2.

With the aid of these and similar calculations, several vertical ionization potentials have been calculated and are tabulated for comparison with experimentally determined potentials in Table 6.1. In the calculation, it is necessary to assume a value for the dissociation energy D_e of the isoelectronic molecule X_2, which is not always known accurately. The assumed value of D_e

and the molecule X_2, are also tabulated. In some cases, a strong vibrational excitation is expected in the Franck–Condon transition to the molecular ion.

The nitrogen ion N_2^{2+} is of particular interest, since the molecule C_2 has a low-lying excited state. However the $^3\Pi_u$ state of N_2^{2+} lies below the stable $^1\Sigma_g^+$ state (*Figure 6.22*). Electron impact studies[107] have provided evidence for the existence of two states, but the appearance potentials are not yet satisfactorily interpreted. The doubly charged ion NH_3^{2+} is observed only in small proportions in ionization of ammonia; presumably the Franck–Condon overlap is smaller, because highly vibrationally excited states are involved. Certain other polyatomic multiply-charged ions are known.

Part of the importance of doubly charged molecular ions is that pairs of singly charged ions each possessing kinetic energy arise from their dissociation[115].

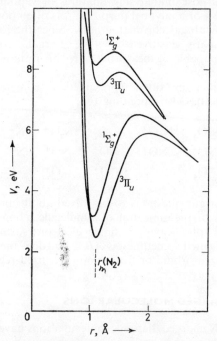

Fig. 6.22. *Approximate potential energy curves of* N_2^{2+} *states*

(From Hurley and Maslen[120], by courtesy of the American Institute of Physics)

6.10 IONIZATION OF POSITIVE IONS

The differences between ionization of neutral atoms and of positive ions by electron impact arise from the Coulomb attractive field. It is possible to make use of the Born approximation using Coulomb distorted waves instead of plane waves for the incident electron; a number of such 'Coulomb–Born' approximation calculations have been carried out[123–128], with results as in *Figure 6.23*. Allowance is made for the exchange contributions and, in

Table 6.1. DOUBLY CHARGED MOLECULAR ION VERTICAL IONIZATION ENERGIES

Ion	Isoelectronic neutral molecule	Dissociation energy of neutral molecule D_e (eV)	Calculated E_{iv} (eV)	Experimental E_{iv} (eV)	Reference
F_2^{2+}	O_2	5·114	43·26	–	120
Cl_2^{2+}	S_2	3·6	32·57	32·6	122
I_2^{2+}	Te_2	2·3	25·74	–	120
$N_2^{2+3}\Pi_u$	C_2	3·6	46·31	–	120
$N_2^{2+1}\Sigma_g$	C_2	6·50	41·32	42·7	107
N_2^{2+}	C_2	3·6	47·0	–	120
N_2^{2+}	C_2	6·50	42·1	43·5	107
O_2^{2+}	N_2	9·756	35·88	–	120
NO^{2+}	CN	7·6	38·17	39·8	107
CO^{2+}	BN	4	42·80	–	120
CO^{2+}	BN	5	41·17	41·8	107
CO^{2+} (excited)	BN	4	45·7	45·9	107
ClF^{2+}	SO	5·357	37·20	–	120
HCl^{2+}				35·5	107
DBr^{2+}				33·2	107
HI^{2+}				30·0	107
NH_3^{2+}				33·7, 36·8	107
$C_6H_5D^{2+}$				26·0	107
$C_6H_5CH_3^{2+}$				24·5	107
$C_{10}H_8^{2+}$				22·8	107
$C_6H_5D^{3+}$				44, 61 ?	107
$C_6H_5CH_3^{3+}$				42	107
$C_{10}H_8^{3+}$				40	107

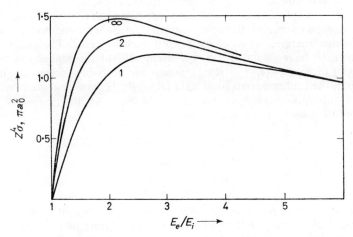

Fig. 6.23. Scaled ionization functions $Z^4\sigma$ for stripped nuclei of charge Z, calculated by Coulomb–Born approximation (after Burges[125])

strongly allowed transitions, for the strong coupling effects between different states of the system, which can reduce the cross-sections by as much as one-half. For ionization, the approximate radial wave functions for the ejected electron may be determined by the extrapolated quantum defect method.

The ionization of positive ions by electrons can be studied by crossed-beam or trapped-ion techniques. Crossed-beam techniques are discussed in Section 3.2. An increasing amount of cross-section data is becoming available from them, following the pioneer[129-135] experiments of Dolder and Harrison[136, 138]. Comparison with Coulomb–Born calculations is made in *Figure 6.23*. The crossed-beam technique is superior for absolute cross-section measurement, but for threshold studies the trapped-ion technique[101] enjoys the advantage of greater ion density in the collision region.

Certain technical problems[138] are peculiar to crossed electron–ion beams but are not general to the crossed-beam discussion of Section 3.2. These arise from space-charge effects:

1. The electron space charge can deflect ions, which are thereby lost to the mass-spectrometer which would normally detect them.
2. The electron space charge can decrease the scattering that accompanies the stripping collisions of ions with background gas (10/20e and 10/21); these stripping collisions are a normal background to the detected doubly charged ions, and are discriminated against by methods outlined in Section 3.2.
3. The ion beam space charge can deflect the electron beam, and thereby affect the background arising from it; the most important such background is radiation, which affects only ion excitation experiments.

Trapped-ion technique can be used for study of ionization of positive ions in either of two ways: either a single electron beam, usually of circular cross-section, is employed to form the ions, to hold them in their space charge, and to ionize them; or, in the more recent hollow-beam technique[139], a beam of annular cross-section is used to form and hold the ions, whilst a second beam is directed down the centre to ionize them. In both cases, the ion density can be calculated only very approximately, so absolute cross-sections cannot be measured; normalization is necessary, either to Coulomb–Born at high energies, or to crossed-beam measurements. The recent trap experiments[140] have included the measurement[112] of the higher ionization energies of rare gases ($R^{n+} \rightarrow R^{n+1}$), which are probably of greater importance than the multiple ionization data ($R \rightarrow R^{n+1}$), in calculating proportions of multiply-charged ions in hot plasmas. Another experiment[141] measures the onset of the sequences:

$$e + \text{He} \qquad \rightarrow \text{He}^+ \; 1s + 2e$$
$$e + \text{He}^+ \; 1s \rightarrow \text{He}^+ \; 2s + e$$
$$e + \text{He}^+ \; 2s \rightarrow \text{He}^{2+} + 2e$$

Of these, the second possesses the highest appearance potential, 41 eV. The measured He^{2+} current is proportional to the product of the three cross-sections, and, just above 41 eV, the first and third cross-sections vary only slowly with energy; therefore structure can be detected in the second,

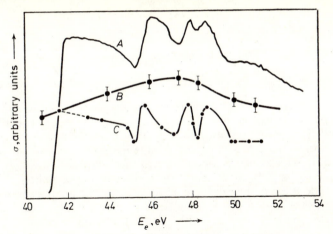

Fig. 6.24. Structure in the trapped-ion production of He²⁺ from helium, corresponding to resonances in the excitation He⁺ 1s–2s: A, deconvoluted data; B, crossed-beam measurements of He⁺ 1s–2s; C, calculated excitation function

threshold function. The $(sp, 3l)$ ¹S, ¹P, ¹D resonances exhibited are compared with calculations[142] in *Figure 6.24*.

An important feature of electron–ion collision experiments is the role played by long-lifetime metastable ions, which can be further ionized by electron collisions[131, 132]. Not only are metastable ions important, but high principal quantum number states can also be sufficiently long-lived to show similar effects. The ion sources used in crossed-beam experiments must be capable of providing ion beams with only ground state populations. In the trapped-ion technique, it is customary to operate the electron beam at the lowest possible energy.

6.11 FRAGMENTATION OF METASTABLE MOLECULAR IONS. ASTON BANDS

Molecules ionized by electron impact can be formed in states which are metastable to spontaneous fragmentation[143, 144]. Provided that the lifetimes of these are of the order $10^{-5} - 10^{-6}$ sec the decay process can be detected within a magnetic mass-spectrometer. When the metastable ion is accelerated and then dissociates before magnetic deflexion, the observed 'peak' appears at a non-integral mass-number. Such peaks were first observed and discussed in the pioneer mass-spectrometer studies of Aston. They arise because an ion dissociating in collision with a gas molecule, after acceleration, will retain a velocity appropriate to the energy of the undissociated ion. In conventional mass-spectrometers, the peaks are usually broadened because the region in which collisions occur is not equipotential. By separate pumping of the source and analysis chambers of a sector instrument, the two being connected only by a very fine slit and the source pressure being kept low, it may be arranged that collisions occur only in the analysis chamber, free of electric field, but before the beam undergoes magnetic deflexion. Under these conditions, the apparent ratio m^*/e^* is related to the primary molecular mass m_p and charge e_p and the fragment ion mass m_f and charge e_f, by the equation:

$$\frac{m^*}{e^*} = \frac{m_f^2 e_p}{m_p e_f^2} \tag{6.26}$$

On a mass-spectrometer with a logarithmic mass scan, m^*, m_i and m_f are equally spaced. It will be noticed that this equation does not identify the initial and final mass numbers uniquely; but more refined techniques are possible with a modern commercial double mass-spectrometer, that is, an instrument in which there is sector electrostatic analysis followed by sector magnetic analysis[145]. In these instruments, there is no possibility of a dissociation inside the regions of magnetic or electrostatic analysis being registered as a peak. An ion dissociating between the two analysers, with fragment kinetic energy T, appears with apparent mass-number

$$m^* = \frac{m_f^2}{m_i}\left[1\pm\left(\frac{\mu T}{eV}\right)^{1/2}\right]^2 \tag{6.27}$$

where $\mu = (m_i - m_f)/m_f$. The peak is square, of width

$$\Delta m^* = \frac{4m_f^2}{m_i}\left(\frac{\mu T}{eV}\right)^{1/2} \tag{6.28}$$

In practice, Gaussian peaks are observed, indicating distributions of kinetic energy of fragmentation.

Metastable decay processes occurring before the electrostatic analysis have also been studied[146-148]. The kinetic energy of the fragment ion (neglecting that which arises from the dissociation) is $eV_0 m_i/m_f$, where V_0 is the initial mass-spectrometer accelerating potential. When the electrostatic analyser is set to receive ions of energy $eV_1 = eV_0 m_i/m_f$, then the ordinary mass-spectrum of fragmentation is not seen but only the 'metastable fragmentation spectrum', with m_i and m_f uniquely determined.

The metastable fragmentation spectrum of a molecule may actually contain more peaks than the ordinary fragmentation spectrum, and its study throws light upon the routes of fragmentation.

REFERENCES

1. Irish, R. T. and Bryant, G. H. *Proc. phys. Soc., Lond.* 84 (1964) 975.
2. Tate, J. T. and Palmer, R. T. *Phys. Rev.* 40 (1932) 731.
3. Ehrhardt, H., Hesselbacher, K. H. and Willmann, K. *Proc. 6th Int. Conf. Phys. electron. atom. Collisions*, p. 217, 1969. Cambridge, Mass.; Massachusetts Institute of Technology Press.
4. Smith, P. T. *Phys. Rev.* 36 (1930) 1293.
5. Tate, J. T. and Smith, P. T. *Phys. Rev.* 39 (1932) 270.
6. Bleakney, W. *Phys. Rev.* 34 (1929) 157; *Phys. Rev.* 35 (1930) 1180; *Phys. Rev.* 36 (1930) 1303.
7. Bleakney, W. and Smith, P. T. *Phys. Rev.* 49 (1936) 402.

8. BLEAKNEY, W. *Phys. Rev.* 33 (1930) 1180.
9. GRYZINSKI, M. *Phys. Rev.* 115 (1959) 374.
10. ALDER, K. *et al. Rev. mod. Phys.* 28 (1956) 432.
11. BURGESS, A. *Proc. 3rd Int. Conf. Phys. electron. atom. Collisions*, p. 237, 1963. Amsterdam; North Holland Publishing Company; Harwell Atomic Energy Research Establishment Report 4818, p. 63, 1964.
12. DUSHMAN, S. and LAFFERTY, J. M. *Scientific Foundations of Vacuum Technique*, 1962. New York; Wiley.
13. DRAWIN, E. E. *Z. Phys.* 164 (1961) 513.
14. PETERKOP, R. K. *J. exp. theor. Phys.* 41 (1962) 1938; 43 (1962) 616.
15. RUDGE, M. R. H. and SEATON, M. J. *Proc. phys. Soc., Lond.* 83 (1964) 680.
16. GELTMAN, S., RUDGE, M. R. H. and SEATON, M. J. *Proc. phys. Soc., Lond.* 81 (1963) 375.
17. COMPTON, K. T. and VAN VOORHIS, C. C. *Phys. Rev.* 26 (1925) 436.
18. HUGHES, A. L. and VAN ATTA, C. M. *Phys. Rev.* 361 (1930) 214.
19. LAWRENCE, E. O. *Phys. Rev.* 28 (1926) 947.
20. NOTTINGHAM, W. B. *Phys. Rev.* 55 (1939) 203.
21. FUNK, I. *Ann. Phys., Paris* 4 (1930) 149.
22. ASUNDI, R. K. and KUREPA, M. V. *J. Electron. Control* 15 (1963) 41; KEIFFER, L. J. and DUNN, G. H. Joint Institute of Laboratory Astrophysics Report 51, 1965; RAPP, D. and GOLDEN, P. *J. chem. Phys.* 42 (1965) 4081; 43 (1965) 1464.
23. SCHRAM, B. L. Thesis, University of Amsterdam, 1966.
24. FITE, W. L. and BRACKMANN, R. T. *Phys. Rev.* 112 (1958) 1141; 113 (1959) 815.
25. NEYNABER, R. H., MARINO, L. L., ROTHE, E. W. and TRUJILLO, S. M. *Proc. 2nd Int. Conf. Phys. electron. atom. Collisions*, University of Colorado, 1961; *Phys. Rev.* 125 (1962) 582.
26. RAPP, D., GOLDEN, P. E. and BRIGLIA, D. D. *J. chem. Phys.* 42 (1965) 4081.
27. LOZIER, W. W. *Physica, Eindhoven* 36 (1930) 1285, 1417.
28. HAGSTRUM, H. D. and TATE, J. T. *Phys. Rev.* 59 (1961) 354.
29. MOHLER, F. L., DIBELER, V. H. and REES, R. M. *J. chem. Phys.* 22 (1954) 394.
30. COGGESHALL, N. D. *J. chem. Phys.* 12 (1944) 19.
31. BERRY, C. E. *Phys. Rev.* 78 (1950) 597.
32. CARERI, G. and NENCINI, G. *J. chem. Phys.* 18 (1950) 897.
33. McDOWELL, C. A. and WARREN, J. W. *Trans. Faraday Soc.* 48 (1952) 1084.
34. HAGSTRUM, H. D. *Rev. mod. Phys.* 23 (1951) 185.
35. DUNN, G. H. and KIEFFER, L. J. *Phys. Rev.* 132 (1963) 2109; *Gaseous Electronics Conf.* American Physical Society, 1963.
36. KIEFFER, L. J. and VAN BRUNT, R. J. *J. chem. Phys.* 46 (1967) 2728.
37. KIEFFER, L. J. and DUNN, G. H. Joint Institute of Laboratory Astrophysics Report 51, 1965.
38. BOKSENBERG, A. Thesis, University of London, 1961.
39. LISKA, J. A. *Phys. Rev.* 46 (1934) 169.
40. HARRISON, H. Thesis, Catholic University of America Press, Washington, D.C. 1956.
41. SCHRAM, B. L. Thesis, University of Amsterdam, 1966.
42. SCHRAM, B. L., MUSTAFA, H. R., SCHUTTEN, J. and DE HEER, F. J. *Physica, Eindhoven* 32 (1966) 734.
43. SCHRAM, B. L., DE HEER, F. J., VAN DER WIEL, M. J. and KISTEMAKER, J. *Physica, Eindhoven* 31 (1965) 94.
44. SCHRAM, B. L., VAN DER WIEL, M. J., MOUSTAFA, H. R. and DE HEER, F. J. *J. chem. Phys.* 44 (1966) 49.
45. SCHULZ, G. J. *Phys. Rev.* 128 (1962) 178.
46. ASUNDI, R. K., CRAGGS, J. D. and KUREPA, M. V. *Proc. phys. Soc., Lond.* 82 (1963) 967.
47. TOZER, B. A. and CRAGGS, J. D. *J. Electron. Control.* 8 (1960) 103.
48. FROST, D. C. and McDOWELL, C. A. *Proc. Roy. Soc.* 232 (1955) 227; 236 (1956) 278; *Advances in Mass Spectrometry*, p. 413, 1959. London; Pergamon.

49. RAPP, D., GOLDEN, P. E. and BRIGLIA, D. D. *Gaseous Electronics Conf.* American Physical Society, 1964.
50. DALGARNO, A. and STEWART, A. L. *Proc. phys. Soc., Lond.* A76 (1960) 49.
51. FOX, R. E. *Advances in Mass Spectrometry*, p. 397, 1959. London; Pergamon.
52. SCOTT, B. W. and HEIL, H. *Gaseous Electronics Conf.*, American Physical Society, 1965.
53. MCFARLAND, R. H. and KINNEY, J. D. *Phys. Rev.* 137 (1965) A1058.
54. MCDOWELL, M. R. C., MYERSCOUGH, V. P. and PEACH, G. *Proc. phys. Soc., Lond.* 85 (1965) 703.
55. MCCONKEY, J. W., BURNS, D. J. and WOOLSEY, J. M. *Proc. phys. Soc., Lond.* 86 (1965) 745; *Proc. phys. Soc., Lond.* B1 (1968) 71.
56. SUMIN, L. V., GUREV, M. V. and TUNITSKII, N. N. *J. exp. theor. Phys.* 47 (1964) 452; *Soviet Phys. JETP* 20 (1965) 299.
57. DALY, N. R. *Proc. phys. Soc., Lond.* 85 (1965) 897.
58. MCGOWAN, W. and KERWIN, K. *Canad. J. Phys.* 41 (1963) 1535.
59. LATYPOV, Z. Z., KUPRIYANOV, S. E. and TUNITSKII, N. N. *Soviet Phys. JETP* 19 (1964) 570.
60. DALY, N. R. *Proc. phys. Soc., Lond.* 85 (1965) 897.
61. MEHLHORN, W. *Z. Phys.* 187 (1965) 21.
62. GELTMAN, S. *Phys. Rev.* 103 (1956) 171.
63. RUDGE, M. R. H. and SEATON, M. J. *Proc. Roy. Soc.* A283 (1965) 262.
64. PETERKOP, R. and LIEPINSCH, A. *Bull. Acad. Sci. Latv. SSR* 1 (1969) 17.
65. WANNIER, G. H. *Phys. Rev.* 90 (1953) 873.
66. FITE, W. L. and BRACKMANN, R. T. *Phys. Rev.* 113 (1958) 1141.
67. GELTMAN, S. *Phys. Rev.* 102 (1956) 171.
68. FOX, R. E. *J. chem. Phys.* 33 (1960) 200.
69. FOX, R. E., HICKAM, W. M., KJELDAAS, T. and GROVE, D. *J. Phys. Rev.* 84 (1951) 859; 89 (1953) 555.
70. FOX, R. E. *Rev. sci. Instrum.* 26 (1955) 1101.
71. HICKAM, W. M., FOX, R. E. and KJELDAAS, I. *Phys. Rev.* 96 (1954) 63.
72. MCGOWAN, J. W., FINEMAN, M. A., CLARKE, E. M. and HANSON, H. P. *Phys. Rev.* 167 (1968) 43.
73. BRION, C. E., FROST, D. C. and MCDOWELL, C. A. *J. chem. Phys.* 44 (1966) 1034.
74. MARMET, P. and MORRISON, J. D. *J. chem. Phys.* 36 (1962) 1238; 35 (1961) 746.
75. PERESSE, J. and TUFFIN, F. *C. R. Acad. Sci., Paris* 265 (1967) 1314.
76. PERESSE, J., TUFFIN, F. and SINOU, G. *C. R. Acad. Sci., Paris* 165 (1967) 1234.
77. MORRISON, J. D. *J. chem. Phys.* 40 (1964) 2488.
78. KANEKO, Y. *J. phys. Soc., Japan* 16 (1961) 1587.
79. STUBER, F. A. *J. chem. Phys.* 42 (1965) 2639.
80. SEMAN, M. L. *Gaseous Electronics Conf.*, American Physical Society, 1965.
81. PERESSE, J., TUFFIN, F. and SINOU, G. *Phys. Lett.* 25A (1967) 773.
82. KERWIN, L., MARMET, P. and CLARKE, E. M. *Advances in Mass Spectrometry, Vol. 2*, p. 522, 1963. London; Pergamon.
83. FIQUET-FAYARD, F. and LAHMANI, M. *J. chim. Phys.* (1962) 1050.
84. KANEKO, Y. *J. phys. Soc., Japan* 16 (1961) 2288.
85. KANEKO, Y. *J. phys. Soc., Japan,* 16 (1961) 2288.
86. KANEKO, Y. and KANOMATA, I. *J. phys. Soc., Japan* 18 (1948) 1822
87. FIQUET-FAYARD, F. and ZIESEL, J. *Proc. 6th Conf. Ionized Gases*, p. 37, Paris, 1963.
88. KRAUSS, M. and KROPF, A. *J. chem. Phys.* 26 (1957) 1776.
89. MARMET, P. and KERWIN, L. *Canad. J. Phys.* 38 (1962) 972.
90. BRIGLIA, D. D. and RAPP, D. *J. chem. Phys.* 42 (1965) 3201.
91. MCGOWAN, J. W. and FINEMAN, M. A. *Phys. Rev. Lett.* 15 (1965) 179.
92. MCGOWAN, J. W., CLARKE, E. M., HANSON, H. P. and STEBBINGS, R. F. *Phys. Rev. Lett.* 13 (1964) 620.
93. DIBELER, V. H., REESE, R. M. and KRAUSS, M. *J. chem. Phys.* 42 (1965) 2045.
94. BARDSLEY, J. N. *Chem. Phys. Lett.* 1 (1967) 229.

95. BERRY, R. S. *J. chem. Phys.* 45 (1966) 1228.
96. BRION, C. E. *J. chem. Phys.* 40 (1964) 2995.
97. CRAGGS, J. D. and WARREN, J. W. *Hb. Physik* 37 (1959) 399.
98. CRAGGS, J. D. and MASSEY, H. S. W. *Hb. Physik* 37 (1959)
99. FIELD, F. H. and FRANKLIN, J. L. *Electron Impact Phenomena and the Properties of Gaseous Ions*, 1957. New York; Academic Press.
100. SJOGREN, H. and LINDHOLM, E. *Abstr. 4th Int. Conf. Phys. electron. atom. Collisions*, 1965. Hastings-on-Hudson; New York; Science Bookcrafters.
101. BAKER, F. A. and HASTED, J. B. *Phil. Trans.* 261 (1966) 33.
102. HONIG, R. E. *J. chem. Phys.* 16 (1948) 105.
103. CUTHBERT, J., FARREN, J., PRAHALLADA RAO, B. S. and PREECE, E. R. *Proc. phys. Soc., Lond.* 88 (1966) 91.
104. SCOTT, J. T. and HASTED, J. B. *Mass Spectrometry Symp.* British Institute of Petroleum/American Society for Testing and Materials, 1964.
105. MORRISON, J. D. *J. chem. Phys.* 22 (1954) 1219.
106. MORRISON, J. D. and NICHOLSON, A. J. C. *J. chem. Phys.* 31 (1959) 1320.
107. DORMAN, F. H. and MORRISON, J. D. *Canad. J. Phys.* 35 (1961) 575.
108. DIBELER, V. H. and REESE, R. M. *J. chem. Phys.* 31 (1959) 282.
109. KRAUSS, M., REESE, R. M. and DIBELER, V. H. *J. Res. nat. Bur. Stand.* 63A (1959) 201.
110. KISER, R. W. *J. chem. Phys.* 36 (1962) 2964.
111. DORMAN, F. H., MORRISON, J. D. and NICHOLSON, A. J. C. *J. chem. Phys.* 31 (1959) 1335.
112. REDHEAD, P. A. *Canad. J. Phys.* 45 (1967) 1791.
113. JENNINGS, K. R. *J. chem. Phys.* 43 (1965) 4176; *Chem. Commun.* 2 (1966) 283.
114. BEYNON, J. H., SAUNDERS, R. A. and WILLIAMS, A. E. *Nature, Lond.* 204 (1964) 67; *Z. Naturf.* 20a (1965) 180.
115. McCULLOH, K. E., SHARP, T. E. and ROSENSTOCK, H. M. *J. chem. Phys.* 40 (1964) 2114, 3501.
116. ROSENSTOCK, H. M., WAHRHAFTIG, A. L. and EYRING, E. H. *J. chem. Phys.* 23 (1955) 2200.
117. KRAUSS, M., WAHRHAFTIG, A. L. and EYRING, H. *Ann. Rev. nucl. Sci.* 5 (1955) 241.
118. FOX, R. E. and LANGER, A. *J. chem. Phys.* 18 (1950) 460.
119. REED, R. L. *Ion Production by Electron Impact*, 1962. New York; Academic Press.
120. HURLEY, A. C. and MASLEN, V. W. *J. chem. Phys.* 34 (1961) 1919.
121. HURLEY, A. C. *J. molec. Spectrosc.* 9 (1962) 18.
122. HERRON, J. T. and DIBELER, V. H. *J. chem. Phys.* 32 (1960) 1884.
123. BURGESS, A. *Mém. Soc. Sci. Liège* 4 (1961) 299.
124. VAN REGEMORTER, H. *Mon. Not. R. astr. Soc.* 121 (1960) 213.
125. BURGESS, A. *Astrophys. J.* 132 (1960) 503.
126. HILL, E. R. *Aust. J. sci. Res.* 4A (1961) 437.
127. SCHWARTZ, S. and ZIRIN, H. *Astrophys. J.* 130 (1959).
128. MALIK, F. B. and TREFFTZ, E. *Z. Naturf.* 563 (1961).
129. HARRISON, M. F. A., DOLDER, K. T. and THONEMANN, P. C. *Proc. phys. Soc., Lond.* 82 (1963) 368.
130. WAREING, J. B. and DOLDER, K. T. *Proc. phys. Soc., Lond.* 91 (1967) 887.
131. LATYPOV, Z. Z., KUPRIYANOV, S. E. and TUNITSKII, N. N. *J. exp. theor. Phys.* 46 (1964) 833.
132. KUPRIYANOV, S. E. and LATYPOV, Z. Z. *Soviet Phys. JEPT* 18 (1964) 558.
133. LINEBERGER, W. C., HOOPER, J. W. and McDANIEL, E. W. *Phys. Rev.* 141 (1966) 151.
134. HOOPER, J. W., LINEBERGER, W. C. and BACON, F. M. *Phys. Rev.* 141 (1966) 165.
135. PEART, B. and DOLDER, K. T. *Proc. phys. Soc., Lond.* 1 (1968) 872.
136. DOLDER, K. T. and HARRISON, M. F. Harwell Atomic Energy Research Establishment, Private communication.
137. DOLDER, K. T., HARRISON, M. F. A. and THONEMANN, P. C. *Proc. Roy. Soc.* A264 (1961) 367; A274 (1963) 546.

138. HARRISON, M. F. A. *Methods of Experimental Physics, Vol. 7A*, Ed. L. Marton, p. 95. New York; Academic Press.

139. HARTNAGEL, H. *Electronics* 18 (1965) 431.

140. CUTHBERT, J., FARREN, J., PRAHALLADA RAO, B. S. and PREECE, E. R. *Proc. phys. Soc., Lond.* 88 (1966) 91.

141. DALY, N. *Phys Rev. Lett.* 19 (1967) 1165.

142. ORMONDE, S., WHITTAKER, W. and LIPSKY, L. *Phys. Rev. Lett.* 19 (1967) 1164.

143. HIPPLE, J. A. and CONDON, E. U. *Phys. Rev.* 68 (1945) 54.

144. HIPPLE, J. A., FOX, R. E. and CONDON, E. U. *Phys. Rev.* 69 (1946) 347.

145. BEYNON, J. H., SAUNDERS, R. A. and WILLIAMS, A. E. *Nature, Lond.* 204 (1964) 67; *Z. Naturf.* 20a (1965) 180.

146. BARBER, M. W. and ELLIOTT, R. M. *12th Mass Spectrometry Conf.*, American Society for Testing and Materials, 1964.

147. JENNINGS, K. R. *J. chem. Phys.* 43 (1965) 4176.

148. FUTRELL, J. H., RYAN, K. and SIECK, L. W. *J. chem. Phys.* 43 (1965) 1832.

Chapter 7

POSITIVE ION RECOMBINATION

7.1 INTRODUCTION

The neutralization of the charge on a positive ion may be accomplished either by an electron or by a negative ion, in a variety of collision processes which form the subject of this chapter. In both cases the attractive Coulomb forces between the impacting particles increase the probability of interaction, which may be large provided that the excess of energy can be properly distributed.

In ionized gases, it is unusual for the negative ions greatly to outnumber the electrons. Since the electrons, even under thermal equilibrium with the ions and atoms, move much faster than either, it will be usual for electron–ion collisions to contribute far more to the ion charge neutralization than ion–ion recombination. The only conditions under which the ion–ion collisions come into their own are in gases ionized at high pressures by α-particles or X-rays, in certain high-temperature ionized gases and flames, and in charge-balanced plasmas where electrons are much less common than negative ions (such as the atmosphere of the sun). Electron–ion collisions are considered first.

The electron–ion recombination process is described by a rate coefficient α_e, defined so that:

$$\frac{\mathrm{d}n_e}{\mathrm{d}t} = -\alpha_e n_e n_+ \tag{7.1}$$

and where $n_e \simeq n_+$, as in plasmas where $n_e \gg n_-$,

$$\frac{\mathrm{d}n_e}{\mathrm{d}t} \simeq -\alpha_e n_e^2 \tag{7.2}$$

Thus the recombination process is likely at large electron and ion densities to dominate over electron-destroying collision processes, such as attachment, which involve only electrons and neutral molecules whose rates are therefore proportional to n_e. Recombination processes have the effect of smoothing spatial non-uniformities, or high electron density regions in plasmas.

Recombination rate coefficients are described in units of $cm^3\ sec^{-1} ion^{-1}$, often written as $cm^3\ sec^{-1}$.

7.2 CLASSIFICATION OF ELECTRON–ION RECOMBINATION PROCESSES

The state of knowledge of the various possible processes of recombination is described in a number of reviews[1-4]. Unfortunately, the measurement of a rate of loss of electrons, yielding an experimental recombination coefficient α_e appropriate to a certain electron temperature and gas, is insufficient information to enable one to decide which electron recombination process is dominating. It is often difficult to make comparisons between theory and experiment for this reason. In this section, the possible recombination processes are enumerated and the quantum and classical approaches to them are summarized.

Radiative Recombination

Radiative recombination processes are of the type

$$e + X^+ \rightarrow X' + h\nu \qquad\qquad e1/\phi0'$$

$$e + X^+ \rightarrow X + h\nu \qquad\qquad e1/\phi0$$

In this process, the capture of an electron from the positive energy continuum into a ground or excited level of the atom is accompanied by radiation; since the electron energy distribution is comparatively wide, the spectrum takes the form of a continuum. Where the process terminates in an excited state which can combine with lower levels, an atomic line spectrum is also produced. Where the ground state is formed, the process is the reverse of photoionization (Chapter 9) and the principle of detailed balancing may be applied in the approximate form

$$\alpha_e(T) \simeq \frac{g_0}{g_+} \left[\frac{\sqrt{2} E_i^2}{(\pi kT)^{1/2} m_e^{3/2} c^2} \right] \sigma_\phi(\nu) \qquad\qquad (7.3)$$

$$\alpha_e(T) \simeq 1 \cdot 3 \times 10^6 \, E_i^2 T^{-1/2} \frac{g_0}{g_+} \, \sigma_\phi(\nu) \qquad\qquad (7.4)$$

with E_i in Rydbergs, the photoionization cross-section σ_ϕ in cm^2, and T in °K. Sufficient data for photoionization processes are available to enable radiative recombination coefficients to be deduced not only for the ground state, but, employing the quantum defect method, for the lowest excited states. Higher excited states may be regarded as hydrogenic, so that the original calculations[1, 5] applicable to hydrogen ions of nuclear charge Z may be invoked. These calculations have been much improved[6] by the use of the first term of an asymptotic expression for the Kramers–Gaunt g factor:

$$g(n, \varepsilon) = 1 + 0 \cdot 1728 n^{-2/3} (u+1)^{-2/3} (u-1) + \; \ldots \qquad (7.5)$$

with $u = n^2 \varepsilon$ and $\varepsilon = E_e / Z^2$ Rydberg units; n is the principal quantum number.

The coefficient is given by the equation

$$\alpha_e(Z, T, n) = 5{\cdot}197 \times 10^{-14} Z \left(\frac{\Lambda}{n^2}\right)^{3/2} \exp\left(\frac{\Lambda}{n^2}\right) \mathscr{E}i\left(\frac{\Lambda}{n^2}\right) \qquad (7.6)$$

where $\Lambda = 157\,890 Z^2/T$, $\mathscr{E}i$ denotes the exponential integral, and T is in $°K$. Calculations have been made by Bates and Dalgarno[3] for H^+ ions at temperatures from $250°K$ to $64\,000°K$ and for $n = 1$ to 12; it is found that, at $250°K$:

$$\alpha_e = 1{\cdot}02 \times 10^{-12} \text{ cm}^3 \text{ sec}^{-1} \quad \text{for} \quad n = 1,$$
$$\alpha_e = 5{\cdot}66 \times 10^{-13} \text{ cm}^3 \text{ sec}^{-1} \quad \text{for} \quad n = 2,$$
$$\alpha_e = 3{\cdot}90 \times 10^{-13} \text{ cm}^3 \text{ sec}^{-1} \quad \text{for} \quad n = 4,$$
$$\alpha_e = 1{\cdot}11 \times 10^{-13} \text{ cm}^3 \text{ sec}^{-1} \quad \text{for} \quad n = 10,$$
$$\alpha_e = 4{\cdot}84 \times 10^{-12} \text{ cm}^3 \text{ sec}^{-1} \quad \text{summed over all } n.$$

When the mean thermal energy is small compared with the ionization potential of the level n,

$$\alpha_e(Z = 1, T, n) \propto n^{-1} T^{-1/2}$$

It is seen that levels with $n \gtrsim 10$ can be neglected to a first approximation. Scaling for higher Z nuclei can be carried out by multiplying temperatures by Z^2 and α_e by Z.

These recombination coefficients are several orders of magnitude smaller than those for dissociative recombination and various other recombination processes; nevertheless, radiative recombination of H^+ is an important process in gaseous nebulae. In the high ionization densities and temperatures obtained in shock tubes, a radiative coefficient $\alpha \sim 10^{-12}$ cm³ sec⁻¹ has been observed in hydrogen, in reasonable agreement with theory[7-10]. Discrepancies between calculations and experimental data have been reported[11] for H $6s$ and $6p$. In caesium plasmas, the coefficient is sufficiently large ($\sim 5 \times 10^{-14}$ cm³ sec⁻¹ for recombination into the $6p$ level) for the continuum to be observed in emission[12].

Recombination Involving Levels in the Positive Continuum

Many atoms and molecules possess energy level series which converge to continua situated above the ionization potentials. Thus there may be terms (denoted *) in such series whose energies exceed E_i. The photoexcitation of such terms is discussed in Chapter 9; the subsequent radiationless transition into the continuum, accompanied by the release of an electron, is the process of autoionization (see Chapter 6). The reverse process, radiationless recombination, must be considered possible. However, the radiationless transition is reversible in a plasma, so that equilibrium is maintained and the actual recombination rate is small. One must consider competing processes capable of removing the bound electron excited states before the electron is released; such processes are: (*i*) a downward transition to a level x below E_i (this series of processes is known as 'dielectronic recombination');

and (*ii*) a superelastic stabilizing collision with a free electron, resulting in a level of the atom lower in energy than E_i.

The dielectronic recombination coefficient α_{ed} depends upon the relative magnitudes of the lifetimes of level * to the radiationless and the downward transitions. It can be shown to be

$$\alpha_{ed} = CT^{-3/2} \frac{g^*}{g_+} \exp\left(-\frac{E_* - E_i}{kT}\right) f_{*x} \bar{\nu}_{*x} \tag{7.7}$$

with

$$C = \frac{(2\pi)^{1/2} e^2 h^3}{c k^{3/2} m_e^{5/2}} = 1 \cdot 4 \times 10^{-16} \text{ cm}^5 \text{ deg}^{3/2} \text{ sec}^{-1} \tag{7.8}$$

The terms $\bar{\nu}$ and f refer respectively to the wave number and oscillator strength for the emitted radiation. Bates and Dalgarno[3] have shown that the radiative recombination coefficient for radiation of wave number $\bar{\nu}_{\infty x}$ from level x with released electron energy E_e may be expressed in a similar form

$$\alpha_{er} = CT^{-3/2} \frac{g_x}{g_+} \int \exp\left(-\frac{E_e}{kT}\right) \bar{\nu}_{\infty x} \, \mathrm{d}f_{\infty x} \tag{7.9}$$

For a particularly favourable case, that is, where $E_* - E_i$ is very small and the exponential in equation 7.9 is taken as unity, the dielectronic recombination coefficient can be as large as $\sim 10^{-10}$ cm^3 sec^{-1} at 250°K; no such case has been found, and it is likely that most dielectronic recombination coefficients will be about two orders of magnitude smaller than this. Detailed balance calculations can be made from photoexcitation and autoionization data[13].

The collisional-stabilization radiationless recombination coefficient depends upon the relative magnitudes of the lifetimes of level * to the radiationless transition and to the collision process. It is only as effective as dielectronic recombination when

$$\lambda_{*x}^4 n_e \sim 10^{29}$$

with n_e in cm^{-3} and λ in Ångstrom units. But in such dense plasmas, neither form of recombination is likely to be effective.

It is apparent that the importance of dielectronic recombination lies in the fact that it is applicable in situations where radiative recombination is particularly inapplicable; that is, when the mean energy of the collisions is large. The discovery, by means of XUV photoabsorption and electron spectroscopy, of the important series of two-electron excited states in helium, has enabled the situation in helium to be clarified[14]. For the process

$$\text{He}^+(1s) + e \rightarrow \text{He}(2pnl) \rightarrow \text{He}(1snl) + h\nu$$

the dielectronic recombination coefficient $\alpha_{ed} \sim 10^{-12}$ cm^3 sec^{-1} at $T \sim 10^6$ °K. This is two orders of magnitude larger than the radiative recombination rate at the same temperature, and this fact has important consequences in astrophysics.

Dissociative Recombination

Dissociative recombination has for some years[15, 16] been recognized as capable of proceeding at a much faster rate than the previously considered recombination processes. It is likely to be the main contribution to the measured recombination rates in weakly ionized gases and in the earth's upper atmosphere. The process is

$$e + XY^+ \to X' + Y' \qquad\qquad e1/0'0'$$

In direct dissociative recombination, a radiationless transition occurs, in collision with an electron, to an antibonding state of the molecule in which the constituent atoms move apart, gaining energy by virtue of their mutual repulsion, so that the transition is prevented from proceeding in the reverse direction by the action of the Franck–Condon principle. The lifetime τ_a of this transition will usually determine the rate process, and, on the assumption that only the $v = 0$ vibrational level of XY^+ is concerned, the rate coefficient can be shown[3] to be

$$\alpha_e(XY^+) = \frac{2 \cdot 1 \times 10^{-16}\, T^{-3/2} g_{XY}}{\tau_a g_{XY^+}} \int |\psi_0(R)|^2 \frac{dR}{dE_e} \exp\left(\frac{-E_e}{kT}\right) dE_e \ \text{cm}^3\ \text{sec}^{-1}$$

$$(7.10)$$

where $\psi_0(R)$ is the normalized $v = 0$ vibrational wave function of XY^+ written as a function of nuclear separation; E_e represents the electron kinetic energy necessary to make the transition possible. The symbols g represent statistical weights. Presumably, the largest recombination coefficients are found when as many $X'Y'$ antibonding potential energy curves as possible pass through or near to the XY^+ potential energy minimum. For complex atom systems such as Xe_2^+ the recombination coefficient can be as large as 10^{-6} cm^3 sec^{-1}. Quantum theory calculations have been made for H_2^+[17, 18], and for NO^+[19].

Usually the expression inside the integral in equation 7.10 does not vary much with E_e over the relevant range; therefore the integral is fairly closely proportional to E_e and the recombination coefficient varies as $T^{-1/2}$. However, as is discussed in Section 7.5 below, the experimentally determined variation is sometimes substantially faster, of order $T^{-3/2}$.

It has been proposed[20, 21] that the higher rate of temperature variation arises from an indirect process in which the incident electron is captured by giving up its kinetic energy to the vibrational or rotational motion of the molecular nuclei. The incident electron moves in a hydrogen-like orbital with high principal quantum number (Rydberg state). This state then predissociates by pseudo-crossing with a non-Rydberg antibonding state $A'B'$, which then passes to $A' + B'$.

The indirect process may be important at 300°K if there exist Rydberg states differing from the initial ionic energy by $\sim kT$, which decay by predissociation faster than by autoionization. The recombination coefficient is

$$\alpha_e = \frac{2\pi^2 \hbar^2 g}{(2\pi m_e^3 kT)^{1/2}} \Gamma_c f(0) \qquad\qquad (7.11)$$

where Γ_c is the capture width, and $f(0)$ is the probability of finding the nuclei with appropriate separation; g is the ratio of multiplicities of the resonant molecular state and the initial state of the molecular ion.

Three-body Electron–Ion Recombination

Three-body electron–ion recombination can be written as

$$e + X^+ + Y \rightarrow X + Y \qquad\qquad e10/00$$

or

$$e + X^+ + Y \rightarrow X' + Y \qquad\qquad e10/0'0$$

In this process, the excess internal energy of the recombination collision is removed by a neutral atom or molecule. The collision is amenable to classical treatment; in the original analysis by Thomson[22], recombination was assumed to take place if a collision with a third body (neutral atom) was suffered by either ion or electron when its distance from the third body was

$$r \leqslant \frac{2e^2}{3kT}$$

This assumption leads to the following expression for the recombination coefficient in terms of the electron momentum transfer cross-section σ_d:

$$\alpha_{e3} = \alpha_T = \frac{64(3\pi^2)^{1/2}}{81} \frac{e^6 m_e^{1/2} \sigma_d n_0}{m_0(kT)^{5/2}} \qquad (7.12)$$

where the subscript zero refers to the neutral atoms. Representative values are $\sim 10^{-11}p$ cm^3 sec^{-1} for helium and $\sim 2\times10^{-10}p$ cm^3 sec^{-1} for air, with pressure p measured in torr. The theory is equally applicable to ion–ion recombination (Section 7.7 below), and it may be seen that

$$\alpha_{i3} = 2\left(\frac{m_-}{m_e}\right)^{1/2} \alpha_{e3} \qquad (7.13)$$

A treatment in terms of classical diffusion in energy space was given by Pitaevskii[23]; application of the Fokker–Planck equation leads to

$$\alpha_{e3} = \frac{1}{4\mathcal{J}}\left(\frac{\pi e^4}{kT}\right)^{3/2} \qquad (7.14)$$

where

$$\mathcal{J} = \int_0^\infty \left(\frac{\partial \Delta^2}{\partial T}\right)_E |E|^{5/2} \exp\left(\frac{E}{kT}\right) dE \qquad (7.15)$$

and

$$\Delta = \tfrac{1}{2}m_e(v_{ep}^2 - v_{ea}^2) \qquad (7.16)$$

or

$$\Delta = m_e G(v_{ea} - v_{ep}) \qquad (7.17)$$

where

$$G = \frac{m_0 v_{0a} + m_e v_{ea}}{m_e + m_0} \qquad (7.18)$$

or

$$G = \frac{m_0 v_{0p} + m_e v_{ep}}{m_e + m_0} \qquad (7.19)$$

The subscripts a and p refer to ante-collision and post-collision velocity vectors.

The Pitaevskii recombination coefficient is exactly

$$\alpha_p = \frac{9}{2} \left(\frac{6}{\pi} \right)^{1/2} \alpha_T \qquad (7.20)$$

It is also[24] the low-temperature limit of the collisional radiative coefficient[25] discussed in the next subsection; at higher temperatures, this coefficient is smaller. The three-body electron–ion process plays a significant part in the process of collisional radiative decay.

Collisional Radiative Recombination

Collisional radiative recombination is of the type

$$X^+ + e + e \rightarrow X' + e \qquad \qquad 1ee/0'e$$

This process can be regarded as the reverse of the ionization of an atom, excited or ground state, by the impact of an electron. It has been investigated theoretically by several workers[26–28], and is dominant at high temperatures in spark-channel processes, in the very early helium afterglow, and in highly ionized gas systems such as the stellarator[28, 29].

Recombination processes lead to the production of relatively large excited state populations of atoms. These can then be changed by upward or downward radiative or electron-collisional processes, so that an electron may go through a complicated history before eventually reaching the ground state. It is suggested[26] that the overall electron loss mechanism be termed 'collisional radiative decay'.

Following the treatment of Bates, Kingston and McWhirter for a one-electron system of high principal quantum number n, and nuclear charge Z, one assumes $n_+ \gg n_n \ll n_1$, so that n_n swiftly reaches the collisional equilibrium value in its relatively long spontaneous decay time; the Saha equation is written

$$n_n = n^2 n_e n_+ \left(\frac{h^2}{2\pi m_e kT} \right)^{3/2} \exp \left(\frac{Z^2 R}{n^2 kT} \right) \qquad (7.21)$$

The state populations n_n so computed enable rates for the formation and destruction of the ground state to be calculated, using spontaneous transition probabilities, radiative recombination coefficients and collisional excitation and ionization cross-sections. The overall rate dn_1/dt is assumed to be equal to the rate of depopulation of positive ions $-dn_+/dt$, with dn_n/dt ($n \neq 1$)

negligibly small. This rate of depopulation, less the radiative recombination rate, is taken to be the rate of collisional radiative radiation, α_{ee}; some representative values are shown in *Figure 7.1*. It is found that α_{ee} is a slowly decreasing function of T_e when n_e is small, but a very rapidly decreasing function when n_e is large. It is a rapidly increasing function of n_e when T_e is low, but a slowly increasing function when T_e is high. It is also found that the rate is not particularly sensitive to species of singly charged ion,

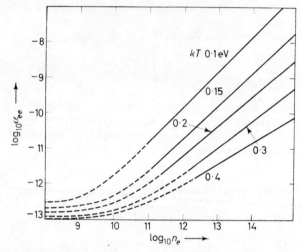

Fig. 7.1. *Collisional radiative recombination rates as functions of electron density, calculated for proton and electron systems by Bates, Kingston and McWhirter*

but that for multiply-charged bare nuclei a scaling factor must be used:

$$Z^{-1}\alpha_{ee}(Z; Z^7 n_e; Z^2 T) = \alpha_{ee}(1; n_e; T) \qquad (7.22)$$

The actual collisional radiative recombination coefficient C is defined[25] as

$$C(T_e) = \lim_{n_e \to \infty} n_e^{-1}\alpha_{ee}(n_e, T_e) \qquad (7.23)$$

Using classical cross-sections, it is found to be equal to

$$C = 3\times 10^{-19}\left(\frac{250}{T_e}\right)^{4\cdot83} \quad cm^6 \ sec^{-1} \qquad (7.24)$$

where T_e is in °K. With newer cross-sections[30], the coefficient is found[31] to be:

$$C = 7\times 10^{-21}\left(\frac{250}{T_e}\right)^{3\cdot625} \quad cm^6 \ sec^{-1} \qquad (7.25)$$

Calculations continue to be reported[32, 33]. In the high electron density plasmas ($n_e \sim 10^{15}$ cm^{-3}) where collisional radiative recombination dominates, the gas and electron temperatures are usually of the same order.

The characteristic relaxation time for temperature is much smaller than that for electron density; however, the case where electron temperature exceeds the gas temperature has also been considered[34].

Collisional radiative recombination has been investigated in caesium-seeded plasmas[35] and is proposed as the dominant process in the typical helium afterglow, time-dependent or flowing[36]; this proposal is not without difficulties[37], since relative insensitivity to temperature has been found. In the 'recombiner' experiments mentioned in Section 7.3, the collisional radiative process contributes to the Cs^+ recombination.

7.3 THE MEASUREMENT OF ELECTRON-ION RECOMBINATION COEFFICIENTS

The recombination process is of greatest importance at thermal energies; therefore it is conveniently studied in plasmas. Stability in time implies a balance of ionizing and recombining processes in a plasma, but if the ionizing processes are checked it is possible to study the electron and ion loss processes in isolation. This is achieved by measurements of the electron density $n_e(t)$ as a function of time in the afterglow of a discharge that has been switched off. It is also possible to study the time variation of the positive ion density, but the rate of volume decay of a given species of positive ion can be due to processes other than recombination (Chapter 14). Electron density decay is also governed by other processes than electron–ion recombination in the volume of the gas; nevertheless, these are more easily separated than are the positive ion processes. The data may be supplemented by measurements of emitted line radiation, from which excited neutral species densities are inferred.

The electron density in an afterglow decays according to three important processes: ambipolar diffusion (Chapter 10) with subsequent recombination at the walls, 'volume' electron–ion recombination, and attachment of electrons to atoms or molecules (Chapter 8). The decay rates are combined in the equation

$$\frac{dn_e}{dt} = D_a \nabla^2 n_e - \alpha_e n_e n_+ - h\nu_c n_e \qquad (7.26)$$

where D_a is the ambipolar diffusion coefficient, α_e the recombination coefficient, h the probability of attachment per collision, and ν_c the electron collision frequency. Wall recombination is assumed completely efficient.

Under conditions where the diffusion and attachment can be neglected, the recombination equation is:

$$\frac{dn_e}{dt} = -\alpha_e n_e n_+ \qquad (7.27)$$

and when there is a balance of charge

$$\frac{dn_e}{dt} = -\alpha_e n_e^2 \qquad (7.28)$$

of which the solution is

$$\frac{1}{n_e} - \frac{1}{n_{e0}} = \alpha_e t \qquad (7.29)$$

The electron density after time t has decayed from its original value n_{e0} to a value which is inversely proportional to t. This is in contrast to the diffusion-governed decay, $n_e = n_{e0} \exp(-t/\tau)$, and to the attachment-governed decay, both of which are exponential. Ambipolar diffusion is discussed in Chapter 1, and is considered here only in so far as it is relevant to the measurement of recombination.

Probe Measurements in Afterglows

The oldest established method of electron density measurement is the Langmuir probe (Section 3.16). In the hands of Kenty[38] and Mohler[12], this technique once provided valuable data about radiative recombination in mercury and other metal vapours. More recently, it has been employed in neon[39], argon[40], nitrogen, and oxygen[41], and in shock-tube experiments. In the absence of negative ions, a positively charged cylindrical probe will, under suitable conditions, collect a current directly proportional to the electron density. For this to hold, it is necessary that the mean free path greatly exceeds the probe dimension. It is therefore essential that the gas pressure be below about 1 torr. By contrast, the other well-established electron density monitoring technique, microwave permittivity measurement, can be used in afterglows at pressures more than an order of magnitude greater; under these conditions, it is much easier to separate the effects of diffusion from those of recombination since pD_a is constant. This may have a bearing on the fact that recombination coefficients as measured by probe techniques[12, 38–42] are sometimes as much as two to three orders of magnitude smaller than those measured by microwaves. But probe measurements have now been made in flowing afterglows, and there is no disagreement with microwave recombination measurement[127].

Probe measurements have been subject to various criticisms[42]. In many studies, a reference electrode of large area is employed to collect positive ions at the same rate as the probe collects electrons; in this way, an attempt is made to keep the plasma space potential constant during the plotting of the probe curve. This method can lead to a doubtful measured value of the electron density; a more satisfactory alternative, which has gained acceptance in rocket and satellite measurements, is to contact the plasma with a heated filament which replenishes the electrons taken by the probe. This procedure greatly sharpens the break in the probe characteristic. However, the thermionic electrons may produce molecular dissociation—a fact which underlines the difficulty experienced by different workers in determining identical recombination coefficients when the gas is ionized by different methods. It would also appear that contamination of the probe limits the time of valid operation in a discharge, so 'single-shot' techniques are necessary. The problem of replenishment is greatly reduced when the afterglow is flowing as opposed to time-dependent.

The condition which must be fulfilled in order that electron loss to the probe will have negligible influence on the plasma potential and hence on the measurement of recombination is

$$I_e \ll e\bar{n}_e \mathcal{V} \tau^{-1}$$

where $\bar{n}_e$ is space averaged over volume $\mathcal{V}$. The cylindrical probe is made as small as is possible, and must be very thoroughly cleaned; for example, 0·002 cm diameter tungsten wire is nickel-plated to a diameter of 0·015 cm and, except for a 0·75 mm length, is sheathed with glass.

The danger of extension of the effective probe area by sputtering is minimized by the use of nickel for the probe wire. It is also necessary that the glass support for the probe does not disturb the diffusion modes; the surface charge on the glass must not disturb the electric field in the neighbourhood of the probe.

Microwave Studies of Recombination in Time-dependent Afterglows

A plasma containing a space average of $\bar{n}_e$ electrons cm^{-3} possesses a space-averaged complex permittivity $\bar{\varepsilon}$ related to $\bar{n}_e$, to the electron collision frequency v_e, and to the electromagnetic radiation frequency f. When the plasma is contained within a microwave cavity resonator whose resonance frequency empty is f_r, a resonance frequence shift Δf_r is observed:

$$\frac{\Delta f_r}{f_r} = \frac{\bar{n}_e}{2(1 + v_e^2 m_e/e^2)} \tag{7.30}$$

and

$$\bar{n}_e = \frac{\int\limits_{vol} n_e X^2 \, dV}{\int\limits_{vol} X^2 \, dV} \tag{7.31}$$

for cavity electric field component X integrated over the volume of the cavity. The most convenient wavelength λ for the measurement of cavity frequency shifts is ~ 10 cm. At shorter wavelength, the volume over which the electric field is sufficiently uniform is inconveniently small. At much longer wavelength, the cavity is cumbersome and expensive.

Because of macroscopic polarization of the plasma, equation 7.30 is invalid[43] at high electron densities ($\sim 10^{11}$ cm^{-3}).

The cavity technique was developed initially by Biondi and Brown[44] and has been reviewed[45]. Referring to *Figure 3.9*, a schematic diagram of one of the possible microwave circuits for this experiment, the plasma may conveniently be excited by a high-power radiofrequency pulse fed into the cavity C from a magnetron M. In some researches, other methods of breakdown have been employed, such as ionization of nitric oxide by Lyman α radiation. Waveguide circuitry such as that of *Figure 3.9* is used to separate the breakdown pulse from the low-power monitoring signal from klystron K. This signal is frequency-modulated with a sawtooth waveform synchronized to the breakdown pulse; use is made of the fact that the impedance of the cavity changes sharply in the resonance region. The reflected monitoring signal, balanced and rectified, is displayed as a function of time on a calibrated timebase. The resonance curve of the cavity appears as a peak, which can be synchronized with accurate time signals, producing a series of measurements of Δf_r at times throughout the afterglow. Average densities $\bar{n}_e$ as small as 2×10^6 electrons cm^{-3} can be recorded with good technique.

In modern experiments, a microwave cavity is designed to operate in more than one electromagnetic mode[46], so that a continuous heating signal can be applied to raise the electron temperature by a calculated[47] or measured[48, 49] amount; an arrangement of the type of *Figure 7.2* has been used by Biondi and his colleagues.

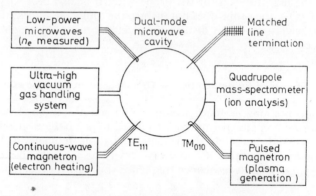

Fig. 7.2. *Schematic diagram of modern cavity apparatus resonant in two modes, for recombination measurement with electron heating*

At the pressures typical of cavity experiments (several torr), the electrons equilibrate thermally with the gas rather quicker than is the case in probe experiments, unless a heating signal is applied. To be certain that the diagnostic signal does not heat the electrons unintentionally, its power must be kept as low as a few microwatts. In a 300°K afterglow experiment, one must be sure that substantial thermalization of electrons has taken place by the time the measurements are conducted ($T_e = T_{gas}$). The rate of electron energy loss by inelastic collisions, where these are energetically possible, greatly exceeds that due to elastic collisions; therefore the electrons fall quickly to a temperature equivalent to the lowest excitation level of the gas. The slower thermalization below this temperature has been considered quantitatively by Oskam[50]. The difference between the lowest excitation energy E_{ex} and thermal energy E_T is related to the mean energy loss per collision, which is taken as $(2m_e/m_g)(E_{ex}-E_T)$. Two cases are considered; in the first it is assumed that the collision frequency $v_e = v_e/l_f$ is energy independent. The time of thermalization to within 10 per cent of the gas temperature is then given by

$$t_{10\%} = 1 \cdot 4 \times 10^{-3} \frac{l_f A}{pT_g^{1/2}} \ln \left[\frac{10(E_{ex}-E_T)}{E_T} \right] \quad \text{sec} \quad (7.32)$$

where l_f is the electron mean free path, A is the atomic weight, and the pressure p is measured in torr. In the second case, the mean free path is taken to be directly proportional to the impact velocity, which results in an expression:

$$t_{10\%} = 4 \cdot 5 \times 10^{-3} \frac{l_f A}{p T_g^{1/2}} \quad \text{sec} \tag{7.33}$$

It is necessary to apply such analysis to all afterglow measurements. Different thermal models of the afterglow have been considered in addition[51-54].

Other microwave techniques have been used for plasma permittivity measurement, for example, electrostatic column resonances occurring in plasma tubes across waveguides[55]. It is also possible to use a microwave interferometer (Chapter 3) in which the phase change and attenuation produced by a plasma are measured by balancing the signal against a reference wave transmitted from the same source. This can be applied to the study of plasma, using either a free wave[56, 57] or a guided wave[58]. Further examples of the microwave technique are to be found in Chapters 4 and 8.

Time-dependent afterglows suffer from the presence of excited species introduced by the ionizing pulse; unwanted secondary effects may interfere with the recombination processes. These effects can be avoided by conducting the experiment in a fast-flowing gas. The diagnostics are performed at different distances down the flow, where speed is measured. But if the experiment is still conducted with a pulsed source and time-dependent electronics are retained, then great flexibility is achieved in avoiding secondary processes. Flowing system recombination experiments have been performed on nitric oxide[59] and other gases.[127]

In Table 7.1 is presented a brief summary of results of microwave and optical measurements. A bibliography of afterglow measurements is available[60].

Table 7.1. RECOMBINATION COEFFICIENTS MEASURED BY MICROWAVE AND OPTICAL TECHNIQUES

Gas or ion	α (cm^3 sec^{-1})	Reference
He_2^+	$1 \cdot 5 \times 10^{-9}$, 4×10^{-9}, 2×10^{-11}	61, 62, 63
Ne_2^+	$1 \cdot 7 \times 10^{-7}$, $1 \cdot 8 \times 10^{-7}$	64, 65, 66
(Ar_2^+)	5×10^{-7}, $(7 \pm 2) \times 10^{-7}$, $(8 \cdot 5 \pm 0 \cdot 8) \times 10^{-7}$	65, 67, 68
(Kr_2^+)	$\sim 1 \cdot 2 \times 10^{-6}$	69
(Xe_2^+)	$\sim 1 \cdot 4 \times 10^{-6}$	69
N_2^+	$2 \cdot 8 \times 10^{-7}$, $(2 \cdot 7 \pm 0 \cdot 3) \times 10^{-7}$	66, 70
N_4^+	$\sim 1 \times 10^{-6}$	70
O_2^+	$2 \cdot 1 \times 10^{-7}$, $(2 \cdot 0 \pm 0 \cdot 5) \times 10^{-7}$, $(2 \cdot 2 \pm 0 \cdot 4) \times 10^{-7}$	70, 71
CO_2^+	$(3 \cdot 8 \pm 0 \cdot 5) \times 10^{-7}$	72
NO_2^+	$\sim 4 \times 10^{-7}, \left(4 \cdot 6 \left\{ {+0 \cdot 5 \atop -1 \cdot 3} \right\} \right) \times 10^{-7}, \left(4 \cdot 1 \left\{ {+0 \cdot 3 \atop -0 \cdot 2} \right\} \right) \times 10^{-7}$	60, 72, 73
Br_2^+	$\sim 10^{-7}$	
I_2^+	$\sim 10^{-7}$	
H_3O^+	2×10^{-7}	74
(Cs_2^+)	$\sim 10^{-6}$	75
(Hg_2^+)	$\sim 10^{-6}$	76
(H_2^+)	$< 3 \times 10^{-8}$	42, 77, 78

NOTE: Ions are identified mass-spectrometrically except where they are bracketed.

One of the principal sources of error in microwave recombination studies arises from the complexities of solution of equation 7.26; neglecting attachment,

$$\frac{dn_e}{dt} = D_a \nabla^2 n_e - \alpha_e n_e^2 \tag{7.34}$$

which is appropriate to the decay of electrons only by simultaneous volume recombination and ambipolar diffusion followed by wall recombination. This equation has been written[44] in the form

$$\frac{dn_e}{dt} = -\frac{D_a n_e}{\Lambda^2} - \alpha_e n_e^2 \tag{7.35}$$

with solution in the form of equation 3.90:

$$\frac{n_e}{1 + \alpha_e \Lambda^2 n_e / D_a} = \frac{n_{e0} \exp\left(-D_a t / \Lambda^2\right)}{1 + \alpha_e \Lambda^2 n_{e0} / D_a} \tag{7.36}$$

Here Λ is the diffusion length, and n_{e0} is the initial electron concentration. In cavity afterglow experiments, the frequency shift is proportional to a space-averaged electron density, and the symbol n_e in these equations refers to this average.

Sufficiently late in the afterglow (that is, for sufficiently small n_e) $\alpha_e \Lambda^2 n_e / D_a \ll 1$, so the electron density decay is purely exponential. A semi-logarithmic $n_e(t)$ is extrapolated into the earlier afterglow where the electron density is higher than the extrapolated value. The inverse of the difference between the two is plotted against time, the slope giving α_e. However, recombination coefficients so deduced by analysing data over one order of magnitude of n_e may be artificially large. The solution of equation 7.34 for parallel plane geometry was given by Oskam[50]. The infinite cylinder was treated numerically by Gray and Kerr[79]. Frommhold and Biondi[64, 80] have now developed a computer programme for the finite cylinder. Inverse time-dependence of n_e does not necessarily imply an accurate α_e, unless also the condition of equation 3.91 is fulfilled:

$$\beta = \frac{\alpha_e \Lambda^2 n_{e0}}{D_a} \geqslant 5 \tag{7.37}$$

The larger the afterglow parameter β (not to be confused with $\beta = n_- / n_e$) the better are the limits of accuracy that can be placed upon α_e. In some of the earlier helium experiments, β can only have been of the order of unity, but in recent measurements[47, 81, 82] careful attention has been paid to the achievement of high β by the use of intense discharge pulse and large cavity. Closer correspondence between the optical and the electronic behaviour of the afterglow is thereby obtained. In heavier rare gases, a high β value has usually been achieved.

The difficulty in extracting recombination coefficients from afterglow measurements in which diffusion is significant arises from the faster electron density decay in the higher diffusion modes; such modes are assisted by asymmetry in the initial electron density distribution. Probe studies of recombination are much less susceptible to inaccuracy arising from diffusion, since no space average is involved.

One can distinguish an afterglow which is recombination-controlled by 'afterglow quenching'; this phenomenon is the sudden decrease in electron density decay rate when a pulse of microwave heating is applied to the system. Recombination rates decrease when the temperature is raised, but diffusion rates increase.

Other Techniques for Measurement of Recombination

The availability of multichannel analysis technique has greatly facilitated the revival of optical afterglow studies[63]. A pulsed electrode discharge is struck in a quartz container, and the monochromated radiation from a window is detected by photomultiplier; it accumulates in successive phase increments, and is stored in a multichannel analyser. Atomic lines and molecular bands can be studied independently; the method is of particular value where a number of different processes contribute to the collisional radiative decay.

The merged-beam method, discussed in Section 3.2, is under development[83] for recombination study. A simpler beam technique for single temperatures was developed by Hammer and Aubrey[84]. An ion beam is directed through the centre of a cylindrical L cathode, which emits electrons inwards. These electrons are space-charge neutralized by the Ba^+ ions emitted by the cathode. Thus a high density of electrons, travelling with velocities appropriate to the cathode temperature, is formed. The system is known as a 'recombiner'; the heavy particle beam emerging from the recombiner contains both ions and fast neutrals which are separated electrostatically and counted. There is always a background current of neutrals emerging from the recombiner, arising from charge transfer with residual gas. This can be avoided by pulsing the recombiner electrons by means of a pulsed axial magnetic field which prevents them leaving the cathode.

7.4 INTERPRETATION OF AFTERGLOW RECOMBINATION DATA

Laboratory temperature recombination coefficients have been reliably measured in many gases; some of the coefficients measured in the early 1950s are now known to have been high, and only the more modern data, analysed with correct solution of the diffusion equations, is considered here; a tabulation appears in Table 7.1 (page 435).

Rare Gases other than Helium

The recombination coefficients for rare gases other than helium are sufficiently large to be confidently interpreted in terms of a dissociative rather than a radiative process. The rare gas diatomic positive ions R_2^+ (and certain mixed rare gas ions) are known, from mass-spectrometer studies, to dominate in weakly ionized rare gases and also in their afterglows; however, their formation will not be discussed at this stage. In modern experiments, such as those of Oskam and Mittelstadt[69], a mass-spectrometric identification of ions in the afterglow confirms the dominance of the diatomic ion under the conditions chosen for measurement.

It will be seen from *Figure 7.3* that there is a consistent dependence of α upon atomic number. Anderson[85] has pointed out that this may be due to the fact that atoms with more electrons possess, in general, more excited states suitable for the dissociative recombination process. This rather surprising state of affairs might indicate that states involving the excitation of more than one electron were involved.

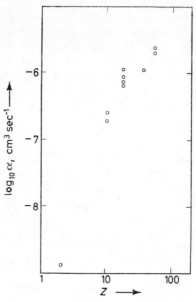

Fig. 7.3. Dependence of rare-gas dissociative recombination coefficients upon Z

The vibrational populations of the molecular ions are not usually specified in time-dependent afterglow experiments. Deactivation by atomic collisions is likely, but is not proved, since there is evidence[86] from the measured Doppler shifts in emission of radiation from excited neon atoms that their kinetic energy distribution is in the range 0–2·5 eV. This implies energetically that much of the Ne_2^+ must be vibrationally excited.

Helium

The situation in helium is more complicated, although at the time of the first edition of this book it was generally believed that dissociative recombination dominated. The weight of evidence now favours dominance of collisional radiative recombination even in microwave cavity afterglows, although there are incompletely understood features of the experiments, and there are situations where dissociative recombination contributes to some extent. The measured recombination coefficients are included in Table 7.1. In particular it may be mentioned that the value $\alpha = 9 \times 10^{-9}$ cm^3 sec^{-1} deduced by Chen, Leiby and Goldstein[87] from both optical and microwave measurements was taken under conditions when a collisional radiative coefficient equal to $1\cdot8 \times 10^{-8}$ would be expected.

Support for the collisional radiative proposal comes from two findings. First, Mulliken[88] has revised the calculations of the He$_2^+$ potential energy curves. He reports that no suitable states for dissociative recombination are present except when the He$_2^+$ is in a very high vibrational quantum number. The vibrational deactivation times are short ($\sim 10^{-2} v^{-1}$ msec^{-1} for vibrational quantum number v)[89]. Second, the Doppler broadening of the radiation from helium atomic states in the afterglow provides evidence against the kinetic energy that must arise from dissociative recombination to antibonding states[90].

However, electron temperature dependence of the recombination coefficient is reported to be $T^{-0.72}$. Collisional radiative temperature dependence is expected to be very fast ($T_e^{-9/2}$), and indeed the atomic line emission goes as $T_e^{-7/2}$.

The flowing afterglow measurements have been interpreted in collisional radiative terms[61, 91, 92], and there is a suggestion that a two-electron collisional dissociative recombination might take place. Molecular helium ion radiation is very strong, and the rotational population departs from the Boltzman distribution[62, 93]. Further support for the collisional radiative proposal comes from the observation[94] that the time dependence of 4650 Å helium radiation follows closely to that of the product of $n(\text{He}^+)$ and n_e^2. The molecular bands observed in this and other afterglows give rise to molecular levels which could take part in a collisional dissociative process. Nevertheless, inverse time dependence of the electron density over $1\frac{1}{2}$ orders of magnitude is observed in millimetre-wave diagnostic experiments[95]; this is not consistent with predominance of collisional radiative recombination.

As far as the method of formation of He$_2^+$ is concerned, mass-spectrometer studies of He$^+$ and He$_2^+$ decay rates[96] yield correct diffusion coefficients and a rate $78p^2$ sec^{-1} (pressure p measured in torr) for the process

$$\text{He}^+ + 2\text{He} \rightarrow \text{He}_2^+ + \text{He}$$

The original calculated rate[97] was $\sim 200p^2$ sec^{-1}, faster than experiments at that period[98]. It is also possible[56, 99] that molecular ions are formed by the Hornbeck–Molnar process

$$\text{He} + \text{He}^m\ 2^3\text{S} \rightarrow \text{He}_2^+ + e$$

The addition of neon to the helium afterglow does not alter the atomic radiation, although the dominant ion becomes HeNe$^+$; the atomic radiation does not arise from dissociative recombination.

Nitrogen, Oxygen and Other Molecular Gases

The most important feature in the interpretation of molecular gas recombination data is the mass-spectrometric identification of the dominant positive ion. It is now possible and necessary to achieve this identification in afterglow experiments, and the earlier studies in which this was not done are of little value where there is any doubt whatever about the identity of the dominant ion. Modern measurements with identified ions are listed in Table 7.1.

It is sometimes possible that under different conditions different molecular ions dominate in the same gas; different recombination coefficients may be observed. Thus in mixtures of neon and nitrogen, N_2^+ will normally dominate at low pressures, and N_4^+ at higher pressures[70, 100]. When simultaneous measurements of electron density time-dependence and positive ion density time-dependence are made, the recombination rates should agree; this is found to be the case for N_2^+, but not for O_2^+; detailed analysis[101] suggests that agreement should not always be expected.

The hydrogen afterglow provides a dramatic contradiction. All the microwave measurements[44, 102, 103] yielded high recombination coefficients ($\sim 2 \times 10^{-6}$ cm³ sec⁻¹) until extreme attention was paid by Persson and Brown[77] to gas–surface reactions and spatial inhomogeneities. Since the apparent recombination coefficient was found to be dependent upon the breakdown pulse length, it was proposed that the hydrogen discharge could perhaps attack a de-gassed silica cavity, producing water. 'Single-shot' microsecond pulse conditions were therefore used. Furthermore, the influence of higher diffusion modes, in fortuitous combination with the fundamental mode, was shown to be able to yield apparently linear $n_e^{-1}(t)$ plots and large apparent recombination coefficients[50]. In the published measurements[77, 78], no appreciable recombination was found in hydrogen ($\alpha < 3 \times 10^{-8}$ cm³ sec⁻¹).

This is not inconsistent with the quantum theory calculation[104] of a dissociative recombination cross-section $\sim 10^{-15}$ cm², although it is possible that H_3^+ may also be present in the hydrogen afterglow. Investigations[57] at extremely high electron densities ($\geqslant 5 \times 10^{13}$ electrons cm⁻³), using free space propagation of microwaves and inferometric technique, lead to the value $\alpha_e = 6 \times 10^{-11}$ cm³ sec⁻¹ in the very early afterglow; this is perhaps appropriate to a collisional radiative process.

7.5 TEMPERATURE DEPENDENCE OF RECOMBINATION COEFFICIENTS

Equation 7.10 yields the electron temperature dependence of dissociative recombination coefficients. It is expected that the cross-section be proportional to T_e^{-1}, and hence the rate coefficient to $T_e^{-1/2}$. However, the influence of Rydberg states of the molecule, which may pseudo-cross the antibonding level or levels, will produce a rather higher temperature variation[105].

The collisional radiative recombination temperature variation is expected to be very fast, $T_3^{-9/2}$. Although it is the electron temperature which dominates the collisional impact velocity, ion vibrational and rotational temperature can be important.

Experimental studies have been made with electron heating applied to the microwave cavity in the form of continuous power. It is usual for a different frequency to be used from that of the monitoring signal; a low-selectivity resonant mode is necessary[46]. The rise in electron temperature is calculated[47] or monitored by the measurement of radiated noise[48, 49]. It has been

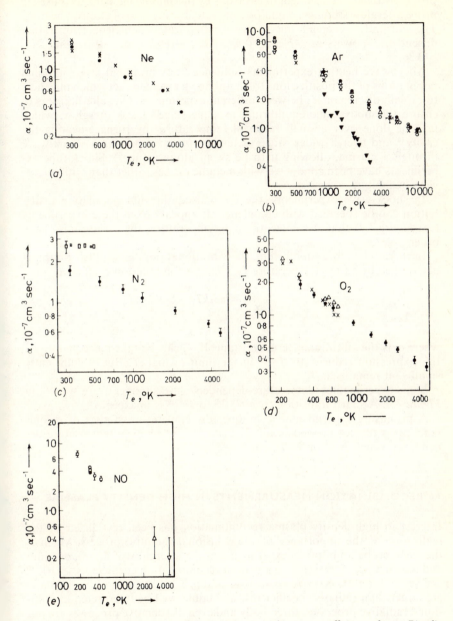

Fig. 7.4. Temperature dependence of dissociative recombination coefficients, due to Biondi, Hobson, Oskam, Lin, Gunton and Shaw, and others: (a) Ne_2^+ ($T_{gas} = 300°K$); (b) Ar_2^+ (inverted triangles, $T_e = T_{gas}$; all other data, $T_{gas} = 300°K$); (c) nitrogen (squares, $T_e = T_{gas}$; circles, $T_{gas} = 300°K$); (d) oxygen mixed with neon, and krypton ($T_{gas} = 300°K$); (e) nitric oxide ($T_{gas} = 300°K$)

proposed that replenishment of electrons by thermionic filament, as is used in probe afterglow studies[42], might prove a valuable method of electron heating. The currently accepted microwave heating rate coefficient temperature dependence powers are: Ne_2^+, -0.43; Ar_2^+, -0.67; O_2^+, -0.70; and N_2^+, -0.36.

Microwave heating experiments influence the vibrational and rotational populations of the positive ions to an unknown extent, a fact which may help to explain discrepancies between different experiments. One alternative is to ensure thermodynamic equilibrium at some stage of the afterglow, which may be done over a small range of temperatures by furnace-heating the cavity[44] and over a wider range of temperatures by passing a shock-wave through a plasma, allowing it to be swept along a tube[68]. Shock-tube experiments have been conducted with metallic probes rather than with microwave technique.

In shock-wave experiments, the thermal equilibrium population distribution can be specified with confidence. It appears from these experiments (see *Figure 7.4*) that at sufficiently high temperatures the variation is of the order $T^{-3/2}$.

In thermodynamic equilibrium, the vibrational population distribution is determined by an ion temperature T_i, so the recombination coefficient

$$\alpha \propto \frac{1 - h\omega/kT_i}{T_e^{1/2}}$$

where ω is the vibrational term constant. If T_i can be taken as equal to T_e, this expression reduces to $T^{-3/2}$ in the limit of high T. But at sufficiently small T, it remains as $T_e^{-1/2}$.

Some experimental temperature-dependence measurements are shown in *Figure 7.4*. A sharp transition from $T^{-1/2}$ to $T^{-3/2}$ is unexpected.

A promising experimental technique for temperature studies of recombination is that of merged beams[83]. Data for N_2^+ show a temperature dependence which is of order T^{-3} at 3 eV.

7.6 RECOMBINATION MEASUREMENTS IN HIGH-DENSITY PLASMAS

Interest in high-density plasma recombination has been revitalized by the realization of the importance of the collisional radiative process. Much of the early probe and optical experiments in the afterglow of argon, caesium and mercury arcs[12, 38, 75] led to apparent recombination coefficients $\gtrsim 10^{-10}$ cm^3 sec^{-1} for $1000°K < T_e < 4000°K$ and $n_e \sim 10^{12}$ cm^{-3}; although these are larger than radiative coefficients, and could well be interpreted as collisional radiative processes, three-body and even dissociative processes are not to be ruled out; the same applies in other experiments[57, 75].

The clearest evidence for collisional radiative recombination derives from observation of the correct variation of α_e with electron density n_e. One expects, from the 'collisional radiative recombination' processes considered in Section 7.2, that an apparent two-body coefficient (equation 7.2) should be observed, but that it should be proportional to $n_e^{1/2}$. This type of variation has been observed by Kuckes *et al.*[28] in the afterglow of discharges in a stellarator, using helium, in the range $10^{11} < n_e < 5 \times 10^{13}$ cm^{-3}.

The study of recombination in high-current spark channels[106] and in shock waves[107] leads one to suppose that here the recombination is complex and pressure-dependent; $\alpha \sim 10^{-11}$–10^{-12} cm^3 sec^{-1}. An interesting measurement[108] of the time dependence of the recombination continuum associated with the Balmer lines has been made in a shock-expanding field-free hydrogen plasma; the ion densities ($\sim 6 \times 10^{16}$ cm^{-3}) were determined by studying the Stark broadening of Hβ; at $T_+ = T_e \sim 4500°$K it was found that $\alpha \sim 10^{-12}$ cm^3 sec^{-1}. This is appropriate to purely radiative recombination.

7.7 RECOMBINATION OF POSITIVE AND NEGATIVE IONS

The Two-body Process

This process can be written as

$$X^+ + Y^- \rightarrow X + Y \qquad\qquad 1\bar{1}/00$$
$$X^+ + Y^- \rightarrow X' + Y \qquad\qquad 1\bar{1}/0'0$$

Two-body processes can take place with either atomic or molecular ions, and with remarkably high probability under certain circumstances; the excess energy is distributed as kinetic energy among the collision products. The Coulomb attraction between the reactants distorts the potential energy curves, so that pseudo-crossing is possible. The situation is very similar to that discussed in Section 12.7, where single electron capture processes of the type 20/11 are considered; curve-crossing theory will only be briefly recalled here.

The many possible product excited states complicate the situation. The energy defect of the reaction dominates the cross-section function, so at different impact energies different excited states may be formed; the thermal energy reaction rate is the most important, but even here the situation must be analysed taking into account all the possible product states.

Application of the Landau–Zener formula[109] leads to a cross-section for a process involving a single pseudo-crossover at nuclear separation R_x which may be summarized, for an exothermic collision of energy defect ΔE eV, by the equations

$$\sigma = 4\pi R_x^2 I(\eta) \qquad\qquad (7.38)$$

where

$$R_x \simeq \frac{27 \cdot 2}{\Delta E} \qquad\qquad (7.39)$$

$$I(\eta) = \int_1^\infty \exp(-\eta x)[1 - \exp(-\eta x)]x^{-3}\, dx \qquad\qquad (7.40)$$

and

$$\eta = \frac{247\mu^{1/2}\,\Delta U^2}{\Delta E^2 E_p^{1/2}} \qquad\qquad (7.41)$$

with the energy separation at crossover denoted by ΔU eV, the reduced mass denoted by μ, and where R_x is in a.u. The energy separation ΔU is a monotonically increasing function of R_x^{-1}, though it is also affected by other factors, whilst the function $I(\eta)$ increases to a maximum of ~ 0.1 at $\eta \sim 0.4$, subsequently falling off with increasing η. The energy of impact is denoted by E_p.

In mutual neutralization processes $1\bar{1}/0'0'$, the energy defect can take values over a wide range of energy. The cross-section is not negligible at thermal energies, and where ΔE is small, as in such cases as $Cs^+ + F^-$, it could be very large. The matrix element for the transition can be calculated for simple atomic systems; Bates and Lewis[110] have carried out calculations for the processes:

$$H^-(1s)^2 + H^+ \rightarrow H(1s) + H(2s \text{ and } p)$$

$$H^-(1s)^2 + H^+ \rightarrow H(1s) + H(3s, p \text{ or } d)$$

The second reaction, producing mainly the Balmer line, is shown to possess a cross-section rising from about 3×10^{-16} cm^2 at 10^3 eV to $> 7 \times 10^{-14}$ cm^2 at 0.1 eV. The room-temperature total recombination rate coefficient is calculated to be 1.3×10^{-7} cm^3 sec^{-1}. To date there has been no experimental data to compare with this calculation, but similar calculations yield rates of the same order, which correspond to very large cross-sections.

The assumption that transitions between the two potential energy curves can only take place at one nuclear separation (R_x) has been challenged[111]; a discussion of the resulting complications is given in Section 12.7. Experience with multiply-charged ion charge transfer collisions leads one to believe that the simplified picture is not too inaccurate. The H^+–H^- collisions, studied optically in merged beams, will prove the best experimental test of the situation. It is not expected that modifications to the Landau–Zener approximation will greatly affect the calculated rates of recombination.

The two-body ion–ion recombination coefficient α_{i2} is defined as a rate constant governing the rate of decay of positive ions:

$$\frac{dn_+}{dt} = \frac{dn_-}{dt} = -\alpha_{i2}n_-n_+ \simeq -\alpha_{i2}n_\pm^2 \tag{7.42}$$

$$n_\pm^{-1} \simeq {}_0n_\pm^{-1} + \alpha_{i2}t \tag{7.43}$$

where $n_\pm$ represents the ion density at a time t, and ${}_0n_\pm$ the initial density; there is usually charge balance. For a strictly two-body process, α_{i2} must be shown to be independent of gas pressure.

A two-body ion–ion recombination process can also take place associatively:

$$A^- + B^+ \rightarrow AB'$$

with subsequent radiation of a band spectrum. An attempt is being made to study this process using merged-beam technique.

Rate coefficients for ion–ion recombination have been measured (without ion identification) in the laboratory, for iodine and bromide ions (presumably $I_2^+I^-$ and $Br_2^+Br^-$). Sayers and his collaborators[112, 113], using a radiofrequency absorption technique for ion-density determination (Section 3.11), have deduced an iodine room-temperature coefficient of

$(1 \cdot 22 \pm 0 \cdot 03) \times 10^{-7}$ cm³ sec⁻¹. In a pure halogen vapour afterglow, the free electrons are captured by attachment in ~ 100 µsec, and probe measurements show that the atomic iodine ions disappear in a further 100 µsec. In the late afterglow, the ion density decays inversely with time and the recombination coefficient is independent of pressure between 0·01 torr and 1 torr. At higher pressures, the apparent coefficient is pressure dependent, as three-body ion–ion recombination processes begin to dominate. Between 0°C and 100°C the two-body recombination rate is found to decrease with increasing temperature.

A two-body rate for $O_2^- O_2^+$ recombination at 300°K has been reported[114] in an afterglow excited by a linac electron pulse.

Three-body Processes

In three-body recombination between positive and negative ions,

$$X^- + Y^+ + Z \rightarrow X + Y + Z \qquad 1\bar{1}0/000$$

a third body, atom or molecule, removes the excess internal energy.

This collision was investigated in the first experiments on ion neutralization ever carried out, namely the measurements made by Rutherford and Thomson[115] of the conductivity of atmospheric air under the action of an electric field and irradiation by X-rays. A rate of recombination α_{i3}, defined as in equations 7.42 and 7.43, but pressure dependent, was deduced. Soon afterwards Langevin[116] measured the quantity by another method, and devised a classical theory to explain the process in the high-pressure region. Thomson[117] arrived at a classical theory capable of explaining the process in the low-pressure region. For many years, independent measurements showed no quantitative agreement, since the gases were impure and the ions formed clusters. Not until the work of Sayers[118] and of Gardner[119], using probe techniques, was the subject put upon a sound quantitative basis.

At high pressures, the ion–ion recombination process comes into its own for two reasons: first, electrons can diffuse out of the ionized region faster than the negative ions; and second, the electron is a factor of m_e/m_i less efficient than the negative ion in removing excess internal energy. These two reasons are only applicable if there is an abundance of negative ions, which there can be in all cases except for very pure rare gases, and for nitrogen, and if the three-body ion recombination process is at all probable, which it will be because of the Coulomb attraction.

The apparent recombination coefficients α_{i3} are large, and markedly pressure dependent, as may be seen from *Figure 7.5* after Sayers[120] and Machler[121].

In the barest outline, the Thomson and the Langevin recombination theories[122] are as follows: For Coulomb interaction, there is a critical impact parameter at or below which two colliding particles orbit around each other. During orbiting, the ions are presumed only to recombine if the excess internal energy of collision, which can be large, is absorbed by a third body, a gas molecule. The critical impact parameter ϱ_c is such that, for ions in thermal equilibrium, the potential energy is equal to the mean energy of

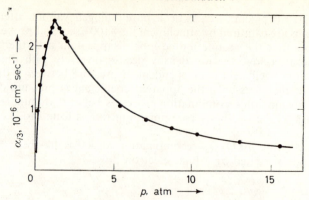

Fig. 7.5. *Comparison of pressure dependence of experimental three-body ion–ion recombination coefficients (points) with theory (full line)*

(From Sayers[120], by courtesy of Academic Press)

thermal agitation:

$$\frac{e^2}{\varrho_c} = \beta kT \qquad (7.44)$$

where $\beta = \frac{3}{2}$ in the original theory, but more recent values[123, 124] are $\beta = \frac{12}{5}$ and $\beta = 4$. Bates and Moffett[125] have treated the problem statistically, as was done in the case of collisional radiative recombination, and concluded that $\beta = 1{\cdot}55 \pm 10$ per cent. The Thomson theory is substantially confirmed.

When the probability of a stabilizing collision is very large, the cross-section can reach the maximum $\pi\varrho_c^2$ ($4{\cdot}3\times10^{-11}$ cm^2 at $300°$K), and the rate constant the maximum value:

$$\alpha_{i3\,max} = \frac{4\pi e^4 \sqrt{2}\bar{v}_+}{9k^2T^2} \qquad (7.45)$$

At sufficiently low pressures, where the mean free path l_f is much larger than ϱ_c, the probability of stabilization can be shown to be $8\varrho_c/3l_f$; therefore α_{i3} is directly proportional to pressure. At intermediate pressures an average must be made, over all paths through a sphere, of the collision equation $-dn/n = dx/l$, to give the probability of stabilization. The agreement between the experimental data, both for the pressure variation and for absolute values of α_{i3}, is remarkably good (*Figure 7.5*).

In this treatment, the mean free path l_f is assumed greatly to exceed ϱ_c; but at higher pressures the opposite condition will be found (Langevin theory). When the distance ϱ_c greatly exceeds the mean free path, the thermal motion can be neglected and the ion velocities are controlled by the Coulomb field:

$$\bar{v}_\pm = \mu X = (K_- + K_+)X = \frac{e}{r^2}(K_- + K_+) \qquad (7.46)$$

for mobilities K and inter-nuclear distance r. The number of negative ions approaching a single positive ion in one second is $4\pi n_- \bar{v} r^2$. Provided that every collision leads to a recombination, the loss of ions can be written:

$$\frac{dn_\pm}{dt} = -4\pi e(K_- + K_+)n_+ n_- \qquad (7.47)$$

$$\alpha_{i3} = 4\pi e(K_- + K_+) \qquad (7.48)$$

Since the mobility is inversely proportional to collision frequency (Chapter 10), and hence to pressure, the recombination coefficient will behave similarly (see *Figure 7.5*). This was found to be the case by Sayers. The temperature dependence of α_{i3} is governed by the temperature variation of mobility, discussed in Chapter 10; but there are no experimental data for the temperature variation of Langevin recombination.

The temperature variation of Thomson recombination has been studied by Gardner[119]. There is a T^{-2} variation in equation 7.45; also $\bar{v}$ varies as $T^{1/2}$, and the probability of stabilization varies with r_c/l_f, with r_c varying as T^{-1}; l_f may perhaps be taken to vary as T. Combining these effects $\alpha_{i3} \propto T^{7/2}$, which Gardner found to be in very approximate agreement with experiment. However Sayers[120] gives, for either air or oxygen,

$$\alpha_{i3} = CT^{-5/2}p \text{ cm}^3 \text{ sec}^{-1} \qquad (7.49)$$

with C an empirical constant equal to $1 \cdot 5 \times 10^{-2}$, with T in °K and p in torr.

A unified theory of three-body ion–ion recombination has been given by Natanson[123]. The theory yields an expression which is similar to the Thomson expression at low pressures. The agreement of this theory with experiment at laboratory temperature is satisfactory; the line drawn through the experimental points in *Figure 7.5* is, in fact, derived from it.

The Thomson theory is successful in the case where the third body is of identical mass with the two ions. But the case of unequal masses is much more complicated, especially since symmetrical resonance charge transfer can no longer help to equilibrate the mean ion energy to that of the gas temperature. Calculations have been made by Bates and Flannery[126].

REFERENCES

1. MASSEY, H. S. W. *Advanc. Phys.* 1 (1952) 395.
2. GOLDSTEIN, L. *Advanc. Electronics Electron. Phys.* 3 (1955) 399.
3. BATES, D. R. and DALGARNO, A. *Atomic and Molecular Processes*, Ed. D. R. Bates, 1962. New York; Academic Press.
4. DANILOV, A. D. and IVANOV-KOLODNYI, G. S. *Soviet Phys. Usp.* 8 (1965) 92.
5. MENZEL, D. H. and PEKERIS, C. L. *Mon. Not. R. astr. Soc.* 96 (1935) 77.
6. SEATON, M. J. *Mon. Not. R. astr. Soc.* 119 (1959) 81, 90.
7. FOWLER, R. G. and ATKINSON, W. R. *Phys. Rev.* 113 (1959) 1268.
8. OLSEN, H. and HUXFORD, W. *Phys. Rev.* 87 (1952) 927.
9. BOARDMAN, W. J. *Astrophys. J. Suppl.* 9 (1964) 185.
10. GLASCO, H. P. and ZIRIN, H. *Astrophys. J. Suppl.* 9 (1964) 193.
11. NORCROSS, D. W. and STONE, P. M. *Gaseous Electronics Conf.*, American Physical Society, 1965.
12. MOHLER, F. L. *Phys. Rev.* 31 (1928) 187; *J. Res. Nat. Bur. Stand.* 10 (1933) 771; 19 (1937) 447, 559.

448   Positive Ion Recombination

13. PERY-THORNE, A. and GARTON, W. R. S. *Proc. phys. Soc., Lond.* 76 (1960) 833.
14. BURGESS, A. *Astrophys. J. Suppl.* 139 (1964) 776.
15. BATES, D. R. and MASSEY, H. S. W. *Proc. Roy. Soc.* A192 (1947) 1.
16. BATES, D. R. *Phys. Rev.* 77 (1950) 718; 78 (1950) 492.
17. BAUER, E. and WU, T. *Canad. J. Phys.* 34 (1956) 1436.
18. DUBROVSKY, G. V. and OBEDKOV, V. D. *Abstr. 5th Int. Conf. electron. atom. Collisions*, p. 342, 1967. Leningrad; Akademii Nauk.
19. GIBBONS, J. J. *Abstr. DASA Reaction Rate Conf.* 1961. Boulder, Colorado; United States National Bureau of Standards.
20. BARDSLEY, J. N. *Abstr. 5th Int. Conf. Phys. electron. atom. Collisions*, p. 338, 1967. Leningrad; Akademii Nauk.
21. CHEN, J. C. and MITTLEMAN, M. *Abstr. 5th Int. Conf. Phys. electron. atom. Collisions*, p. 329, 1967. Leningrad; Akademii Nauk.
22. THOMSON, J. J. *Phil. Mag.* 47 (1924) 337.
23. PITAEVSKII, L. P. *Soviet Phys. JETP* 15 (1962) 919.
24. BATES, D. R. and KHARE, S. P. *Proc. phys. Soc., Lond.* 85 (1965) 231.
25. BATES, D. R. and KINGSTON, A. E. *Proc. phys. Soc., Lond.* 83 (1964) 43.
26. BATES, D. R. and KINGSTON, A. E. *Nature, Lond.* 189 (1961) 652; McWHIRTER, R. W. P. *Nature, Lond.* 190 (1961) 902; BATES, D. R., KINGSTON, A. E. and McWHIRTER, R. W. P. *Proc. Roy. Soc.* A267 (1962) 297.
27. D'ANGELO, N. *Phys. Rev.* 121 (1961) 505.
28. KUCKES, A. F., MOTLEY, R. W., HINNOV, E. and HIRSCHBERG, J. G. *Phys. Rev. Lett.* 6 (1961) 337; *Bull. Amer. phys. Soc.* 6 (1961) 199; HINNOV, E. and HIRSCHBERG, J. G. *Phys. Rev.* 125 (1962) 795.
29. ANDERSON, J. M., HINNOV, E., HIRSCHBERG, J. G., MOTLEY, R. W. and KUCKES, A. F. *5th Int. Conf. Ioniz. Phenom. Gases*, 1961. Amsterdam; North Holland Publishing Company.
30. SARAPH, H. E. *Proc. phys. Soc., Lond.* 83 (1964) 763.
31. STABLER, *Gaseous Electronics Conf.* American Physical Society, 1965.
32. HINNOV, E. and HIRSCHBERG, J. G. *Phys. Rev.* 125 (1962) 795.
33. DELOCHE, R. *C.R. Acad. Sci., Paris* 266 (1968) 664.
34. COLLINS, C. B. *Phys. Rev.* 177 (1969) 254.
35. HARRIS, L. P. *Gaseous Electronics Conf.* American Physical Society, 1964.
36. NILES, F. E. and ROBERTSON, W. W. *J. chem. Phys.* 40 (1964) 2909, 3568.
37. GERBER, R. A., SAUTER, G. F. and OSKAM, H. J. *Phys. Rev. Lett.* 19 (1966) 656.
38. KENTY, C. *Phys. Rev.* 32 (1928) 624.
39. EGOROV, V. S. and SHUKHTIN, A. M. *Optika Spektrosk.* 9 (1960) 419.
40. SAYERS, J. *Nature, Lond.* 159 (1947) 117; *Conf. Phys. Ionized Gases*, 1953. London; University College.
41. SAYERS, J. *J. atmos. terr. Phys. Suppl.* 6 (1956) 212.
42. COURT, G. R. Ph.D. Thesis, University of Birmingham, England, 1959.
43. PERSSON, K. B. *Phys. Rev.* 100 (1955) 729.
44. BIONDI, M. A. and BROWN, S. C. *Phys. Rev.* 75 (1949) 1700; BIONDI, M. A. *Rev. sci. Instrum.* 22 (1951) 500.
45. GOLANT, V. E. *Soviet Phys. JETP* 5 (1961) 1197.
46. GOULD, L. and BROWN, S. C. *Phys. Rev.* 95 (1954) 897.
47. CHEN, C. L., LEIBY, C. C. and GOLDSTEIN, L. *Phys. Rev.* 121 (1961) 1391.
48. HARRIS, D. B. *Microwave J.* 3 (1960) 41, 47.
49. FORMATO, D. and GILARDINI, A. *4th Int. Conf. Ioniz. Phenom. Gases*, 1959. Amsterdam; North Holland Publishing Company.
50. OSKAM, H. J. *Philips Res. Rep.* 13 (1958) 335.
51. SPENKI, E. and STENNBECK, M. *Wiss. Veröff. Siemens* 15 (1936) 18.
52. KNIEPKAMP, H. *Wiss. Veröff. Siemens*, 15 (1936) 2.
53. WASSERRAB, T. *Wiss. Veröff Siemens*, 19 (1940) 1.
54. GRANOVSKY, V. L. *J. Phys., Moscow* 8 (1944) 76.
55. DATTNER, H. *Ericsson Tech.* 2 (1957) 310.

56. YEUNG, T. H. Y. and SAYERS, J. *Proc. phys. Soc., Lond.* 70B (1957) 663.
57. WHITMER, R. F. *Phys. Rev.* 104 (1958) 572.
58. BIALECKE, E. P. and DOUGAL, A. A. *J. geophys. Res.* 63 (1958) 539.
59. YOUNG, R. A. and ST JOHN, G. *Phys. Rev.* 152 (1966) 25.
60. GUNTON, R. C. and SHAW, T. M. *Phys. Rev.* 140 (1965) A756.
61. FERGUSON, E. E., FEHSENFELD, F. C. and SCHMELTEKOPF, A. L. *Phys. Rev.* 138 (1965) A381.
62. SCHMELTEKOPF, A. L. and BROIDA, H. P. *J. chem. Phys.* 39 (1963) 1261.
63. COLLINS, C. B. and HURT, W. B. *Phys. Rev.* 167 (1968) 166; 179 (1969) 203; 177 (1969) 257.
64. FROMMHOLD, L., BIONDI, M. A. and MEHR, F. J. *Phys. Rev.* 165 (1968) 44.
65. BIONDI, M. A. *Phys. Rev.* 129 (1963) 1181.
66. KASNER, W. H. Westinghouse Research Laboratory Report 67-1E2-Gases-P3, 1967.
67. MEHR, F. J. and BIONDI, M. A. University of Pittsburgh SRCC Report 76, 1968.
68. FOX, J. N. and HOBSON, R. M. *Proc. 8th Int. Conf. Ioniz. Phenom. Gases,* 1967. Amsterdam; North Holland Publishing Company; *Phys. Rev. Lett.* 17 (1965) 451.
69. OSKAM, H. and MITTELSTADT, V. R. *Phys. Rev.* 132 (1963) 1435, 1445.
70. KASNER, W. H. and BIONDI, M. A. *Phys. Rev.* 137 (1965) A317.
71. MENTZONI, M. H. *J. appl. Phys.* 36 (1965) 57.
72. WELLER, C. S. and BIONDI, M. A. *Phys. Rev. Lett.* 19 (1967) 59; *Phys. Rev.* 172 (1968) 198.
73. DOERING, J. P. and MAHAN, B. H. *J. chem. Phys.* 36 (1962) 669.
74. CALCOTE, OSKAM, H. J. and MITTELSTADT, V. R. *Phys. Rev.* 132 (1963) 1435, 1445.
75. DANDURAND, P. and HOLT, R. B. *Phys. Rev.* 82 (1951) 278.
76. BIONDI, M. A. *Phys. Rev.* 90 (1953) 730.
77. PERSSON, K. B. and BROWN, S. C. *Phys. Rev.* 100 (1955) 729.
78. POPOV, N. A. and AFANASEVA, E. A. *Soviet Phys. JETP* 4 (1960) 764.
79. GRAY, E. P. and KERR, D. E. *Bull. Amer. phys. Soc.* 5 (1960) 372; *Ann. Phys.,* 17 (1962) 276.
80. FROMMHOLD, L. and BIONDI, M. A. *Ann. Phys.* 48 (1968) 407.
81. KERR, D. E. and HIRSCH, M. N. *Bull. Amer. phys. Soc.* 3 (1958) 258.
82. KERR, D. E. and LEFFEL, C. S. *Bull. Amer. phys. Soc.* 4 (1959) 113.
83. THEARD, L. P. *Proc. 6th Int. Conf. Phys. electron. atom. Collisions* p. 1042, 1969. Cambridge, Mass.; Massachusetts Institute of Technology Press.
84. HAMMER, J. M. and AUBREY, B. B. *Phys. Rev.* 141 (1966) 146.
85. ANDERSON, J. M. General Electric Report 61-RL2817 G, Schenectady, 1961.
86. CONNER, T. R. and BIONDI, M. A. *Phys. Rev.* A140 (1965) 778.
87. CHEN, C. L., LEIBY, C. C. and GOLDSTEIN, L. *Phys. Rev.* 121 (1961) 1391.
88. MULLIKEN, R. S. *J. Amer. chem. Soc.* 86 (1964) 3183; *Phys. Rev. Lett.* 13 (1964) A3; *Rev. mod. Phys.* 4 (1932) 1.
89. ROGERS, W. A. and BIONDI, M. A. *Phys. Rev.* 134 (1964) A1215.
90. TAKAYANAGI, K. Joint Institute of Laboratory Astrophysics Report 17, 1962.
91. NILES, F. E. and ROBERTSON, W. W. *J. chem. Phys.* 40 (1964) 2909.
92. COLLINS, C. B. and ROBERTSON, W. C. *J. chem. Phys.* 40 (1964) 701, 2202, 2208.
93. CALLEAR, A. B. and HEDGES, R. E. M. *Nature, Lond.* 215 (1968) 1267.
94. GERBER, R. A., SAUTER, G. F. and OSKAM, H. J. *Gaseous Electronics Conf.* American Physical Society, 1965.
95. Private communication. Department of Electrical Engineering, University of Liverpool.
96. SAUTER, G. F., GERBER, R. A. and OSKAM, H. J. *Gaseous Electronics Conf.* American Physical Society, 1965.
97. BATES, D. R. *Phys. Rev.* 77 (1950) 718; 78 (1950) 492.
98. PHELPS, A. V. and BROWN, S. C. *Phys. Rev.* 86 (1952) 102.
99. VON PAHL, M. and WEIMER, U. *Z. Naturf.* 14 (1959) 239.
100. KASNER, W. H. and BIONDI, M. A. Westinghouse Scientific Paper 64-928-113-P7, 1964; BIONDI, M. A. *Ann. Géophys.* 20 (1964) 5.

101. GUNTON, R. C. and SHAW, T. M. *Phys. Rev.* 140 (1965) A756.
102. VARNERIN, L. J. *Phys. Rev.* 84 (1951) 563.
103. RICHARDSON, L. M. and HOLT, R. B. *Phys. Rev.* 81 (1951) 153.
104. BAUER, E. and WU, T. Y. *Canad. J. Phys.* 34 (1956) 1436.
105. BARDSLEY, J. N. *Proc. 5th Int. Conf. Phys. electron. atom. Collisions* p. 338, 1967. Leningrad; Akademii Nauk.
106. CRAGGS, J. D. and MEEK, J. M. *Proc. Roy. Soc.* A186 (1946) 241; CRAGGS, J. D. and HOPWOOD, W. *Proc. phys. Soc., Lond.* 59 (1947) 771; FANG, TSUI- and CRAGGS, J. D. *J. Electron. Control*, 4 (1958) 493; McCHESNEY, M. and CRAGGS, J. D. *J. Electron. Control*, 4 (1958) 481; MITCHELL, E. E. L. Ph.D. Thesis, University of Liverpool, 1960.
107. GRIEM, H. R., KOLB, A. C. and SHEN, K. Y. *Phys. Rev.* 122 (1961) 1490; 116 (1959) 4; *Astrophys. J.* 132 (1960) 833.
108. FOWLER, R. G. and ATKINSON, W. R. *Phys. Rev.* 113 (1959) 1268.
109. BATES, D. R. and MASSEY, H. S. W. *Phil. Trans.* A 239 (1943) 269.
110. BATES, D. R. and LEWIS, J. T. *Proc. phys. Soc., Lond.* A68 (1955) 173.
111. BATES, D. R. *Proc. Roy. Soc.* A257 (1960) 22.
112. YEUNG, T. H. and SAYERS, J. *Proc. phys. Soc., Lond.* 71 (1958) 341; YEUNG, T. H. *J. Electron. Control* 5 (1958) 307.
113. GREAVES, C. Ph.D. Thesis, University of Birmingham, 1959; *J. Electron. Control* 17 (1964) 171.
114. HIRSCH, M., EISNER, P., SLEVIN, J. and HALPERN, G. *Gaseous Electronics Conf.* American Physical Society, 1965.
115. RUTHERFORD, J. J. and THOMSON, J. J. *Phil. Mag.* 42 (1896) 392; 44 (1897) 422.
116. LANGEVIN, P. *Ann. Chim. (Phys.)* 28 (1903) 433.
117. THOMSON, J. J. *Phil. Mag.* 47 (1924) 337.
118. SAYERS, J. *Proc. Roy. Soc.* A169 (1938) 83.
119. GARDNER, M. E. *Phys. Rev.* 53 (1938) 75.
120. SAYERS, J. *Atomic and Molecular Processes*, Ed. D. R. Bates, 1962; New York; Academic Press.
121. MACHLER, W. *Z. Phys.* 164 (1936) 1.
122. LOEB, L. B. *Fundamental Processes of Electrical Discharges in Gases*, p. 116, 1939. Wiley; New York.
123. NATANSON, G. L. *J. tech. Phys., Moscow* 29 (1959) 1373 *Soviet Phys. tech. Phys.* 4 (1960) 1263.
124. BRUECKNER, K. A. *J. chem. Phys.* 40 (1964) 439.
125. BATES, D. R. and MOFFETT, R. J. *Proc. Roy. Soc.* A291 (1966) 1.
126. BATES, D. R. and FLANNERY, M. R. *Proc. Roy. Soc.* A302 (1968) 367.
127. MAHDAHVI, C. Thesis, University of London, 1970.

Chapter 8

ELECTRON ATTACHMENT AND DETACHMENT

8.1 NEGATIVE IONS AND THEIR ELECTRON AFFINITIES

The collisions in which electrons attach to atoms to form stable negative ions[1, 2] must involve a release of energy, since a stable atomic negative ion has an energy somewhat below that of its parent atom and free electron. The binding energy of the extra electron is known as the electron affinity E_a; real (or positive) electron affinities of atoms forming stable negative ions are tabulated in Table 8.1. These energies are appropriate to the ground configurations of the negative ions, which are also tabulated. In addition to those cited in Table 8.1, stable ground configurations are known for the following ions: B⁻, Al⁻, As⁻, Sb⁻, Bi⁻, Se⁻, Te⁻, Cr⁻ and Tl⁻.

It is likely that any excited states below the detachment continuum will be rare, metastable, finite in number and close to the continuum; the forces

Table 8.1. ATOMIC ELECTRON AFFINITIES

Atom	Negative ion state	E_a (eV)	Reference
H	$1s^2$	$0\cdot754 \pm 0\cdot001$	3
Li	^1S	$0\cdot616$ (calc.)	4
Na	^1S	$0\cdot41$	5
K	^1S	$0\cdot22$	5
Rb	^1S	$0\cdot16$	5
Cs	^1S	$0\cdot13$	5
F	^1S	$3\cdot448 \pm 0\cdot005$	6
Cl	^1S	$3\cdot613 \pm 0\cdot003$	6
Br	^1S	$3\cdot363 \pm 0\cdot003$	6
I	^1S	$3\cdot063 \pm 0\cdot003$	6
C	^4S	$1\cdot25$	7
O	^2P	$1\cdot465$	8
S	^2P	$2\cdot07$	9
P	^3P	$1\cdot12$ (calc.)	10
Si	^4S	$1\cdot46$	3
Cu	^1S	$1\cdot6$	5
Ag	^1S	$2\cdot0$	5
Au	^1S	$2\cdot8$	5
He	$^4P_{5/2}$	$0\cdot08$	

binding the extra electron are short range, so excitation would result in the nuclear Coulomb force being so strongly screened that the extra electron would no longer be stably bound. The only ions for which there is experimental evidence of such an excited state are C^- and C_2^-, but extrapolation techniques[11] for determining energy levels lead one to suppose that such states will be found in Si^-, and possibly also in B^-, Al^- and P^-. Absorption spectra of negative atomic halogen ions have been observed[12], but they only display a continuum whose edge is given by the electron affinity.

Molecular negative ions are also known in many species, both those deriving directly from molecules and radicles, and those formed by more devious routes, including what are probably dipolar clusters (for example, CO_4^-, CO_3^-, HCO_3^-, $NO_3H_2^-$). Some molecular electron affinities are listed in Table 8.2; these 'adiabatic electron affinities' are defined as the difference between the electronic energies of the molecule and the negative ion state, and are thus different from vertical attachment or detachment energies.

Table 8.2. ADIABATIC MOLECULAR ELECTRON AFFINITIES

Molecule (or radicle)	E_a (eV)	Reference
H_2S	1·11	3
OH	1·83	13
SH	2·32	3
C_2	$3·1 \pm 1·0$	14
CN	2·8, $3·2 \pm 0·3$	3, 15
O_2	$0·43 \pm 0·02$	
CO_4	$1·22 \pm 0·07$	16
NO	0·09	
Br_2	2·6	
NO_2	2·75, 3·8, $4·0 \pm 0·2$	6, 17, 18, 19
O_3	$\sim 3·0$	20
	$1·9 \pm 0·4$	21
CCl_4	2·12	22
C_2H_6	1·47	22
HCl	2·64	23
HBr	3·03	23
NH_2	1·20	23
N_2O	< 1·465	24
SF_6	$\sim 1·5$	

Excited states of negative ions can be considered under three headings:

1. Regular levels, already mentioned, which are rare and lie below the continuum for electron emission.
2. Compound states of electrons and atoms or molecules, which normally decay into a continuum formed by the atom or molecule in ground or excited states, plus the electron; these are discussed in Chapters 4 and 5, and are not be considered in detail here; their lifetimes are in the range 10^{-12}–10^{-14} sec.
3. Negative ion metastable levels, which lie above the continuum but

which are forbidden to pass into it; their lifetimes are of the order of several microseconds, and they will therefore live sufficiently long to pass as beams at kilovolt energies through vacuum apparatus and be registered as ion counts or current.

Negative ion metastable levels can be formed by one-electron capture from fast metastable atoms; possibly two-electron capture from positive ions can contribute. The helium negative ion is found[25-29] in a two-electron excited $^4P_{5/2}$ state; its autoionization lifetime[30] is 1.8×10^{-5} sec. It can be formed with ~ 1 per cent efficiency by passing He^+ beams through alkali metal vapour. Two consecutive processes occur, involving the formation of He 2^3S metastables and possibly other states. The nitrogen negative ion N^- is not stable in the ground state ($E_a = -0.27 \pm 0.11$)[31], but is found[32] in a ^1D metastable state. The $(2p^2)$ ^3P level of H^- is believed to be stable, but has not been demonstrated experimentally.

Although the only regular negative ion excited levels reported are of C^- in the photodetachment data, and of C_2^- for which a band spectrum has been claimed, there is reason to suppose that eventually some further regular levels of other ions will emerge. This judgement is based on the advances made in extrapolation techniques for the estimation of magnitudes of excitation and ionization energies E_i[6, 33, 34]. These techniques are based on the premise that the energies of different species of identical electronic configuration but different nuclear charge Z are expressible in terms of Z and a number of parameters specific to the configuration. The spectroscopic ionization potentials of these members of an isoelectronic series are used to determine the parameters. The following expressions have been used:

$$E_i(Z) = a_2 Z^2 + a_1 Z + a_0 \tag{8.1}$$

and

$$E_i(Z) = \frac{(Z-\sigma)^2}{n^2} + a_0 + \frac{a_{\bar{1}}}{Z-\sigma} \tag{8.2}$$

where n is the principal quantum number and σ the screening parameter; also

$$E_i(Z) = a_2 Z^2 + a_1 Z + a_0 + \frac{a_{\bar{1}}}{Z-s} \tag{8.3}$$

and

$$E_i(Z) = a_2 Z^2 + a_1 Z + \Delta(Z) \tag{8.4}$$

where

$$\Delta(Z) = \Delta(Z_0) + \delta(Z) \tag{8.5}$$

$\Delta(Z_0)$ is the value of $\Delta(Z)$ for the neutral atom, and $\delta(Z)$ vanishes for $Z = Z_0$.

Equation 8.4 is the most successful. Equation 8.2 has been used to find the lowest levels of Al^- which are: ^1D, 0.39 eV; ^1S, 0.94 eV above the ground state ^3P. But $E_a = 0.52$ eV, therefore a ^1D metastable is probable. Similarly, a ^2D metastable of Si^- is probable, and C^- ^2D, Si^- ^2P and P^- ^1D are possible.

The application of such methods to molecular negative ions is not successful. It has been shown[35] that the problem of molecular negative ion stability can only be approached by using the nuclear kinetic operator. For stability of a negative ion, the wave function must contain at least as many molecular spin orbitals of given spin symmetry as does that of the neutral molecule.

There is an additional factor affecting the existence of stable negative ions, operative only in the case of polyatomic (particularly triatomic) molecules. The lowest stable state of a given system of nuclei and electrons must always possess a certain geometrical configuration. Simply formulated rules governing this geometry are known as the Walsh rules[36]. Should the addition of a single electron to a system imply a change in the geometrical configuration, then it is difficult, if not impossible, to form the negative ion by collisions of the molecule with electrons[37]. Triatomic molecules not containing hydrogen and possessing 16 or fewer valence electrons are linear in their ground electronic states (for example, CO_2, N_2O, C_3, N_3). Molecules with 17 electrons, (for example, NO_2) are bent, and those with 18 electrons are more bent (O_3, SO_2, S_2O, NF_2, ClO_2). When the formation of a negative ion requires deformation, a significant 'activation energy' of attachment may arise, even though the adiabatic electron affinity is positive (stable negative ion).

8.2 DETERMINATION OF ELECTRON AFFINITIES

There are four direct techniques for electron affinity determination: (*i*) photodetachment thresholds; (*ii*) thermodynamic techniques; (*iii*) electron impact thresholds; and (*iv*) lattice energies of ionic crystals.

Photodetachment Thresholds

Photodetachment is discussed in Section 8.6. A measurement is made of the electrons produced by detachment from a mass-separated negative ion beam exposed to monochromated photons. The threshold wavelength for electron production is thus an accurate measure of electron affinity or, in the case of molecules, the vertical detachment energy. Vibrational excitation of the negative ion beam may cause the measured vertical detachment energy to be smaller than either the $v = 0$ vertical detachment energy or the adiabatic electron affinity.

Thermodynamic Techniques

Thermodynamic techniques[38] are more suited to investigations of molecular electron affinities, since they allow the difference in energy of the $v = 0$ states of the molecular negative ion and the molecule plus electron to be inferred. When a sufficient density of molecules is in contact with a high-temperature thermionic emitter, a local thermodynamic equilibrium is established between molecular negative ions, molecules and electrons. From a measurement of the temperature dependence of the relative abundances of electrons and negative ions, the electron affinity is calculated using Boltzmann statistics. For this measurement, the space-charge technique[39, 40] and the magnetron technique[41] were at one time popular. Some years ago, it was pointed out that strongly electro-negative impurities in the filament could cause anomalous large electron affinities to be measured[41]; nevertheless, the methods have found a new lease of life[19].

In the magnetron device, a very pure metal filament is surrounded by two grids and a cylindrical anode. A magnetic field is applied parallel to the filament. The first grid is held several hundred volts positive to the filament, to minimize space charge current limitation; the second grid suppresses secondary emission from the anode. The total electron-plus-negative-ion current is measured in the absence of magnetic field, and the negative ion current is measured on its own when a magnetic field is applied; the field is sufficiently strong to confine electrons to helical or cycloidal-helical paths along the filament surface.

Electron Impact Thresholds

Measurement of an electron impact appearance threshold together with the kinetic energy of an atomic negative ion yields the atomic electron affinity, provided that the dissociation energy of the molecule is known. This technique has been applied to oxygen-containing diatomic molecules[42-47], for which the negative ion kinetic energy is close to zero; in many experiments, an incorrect electron affinity was deduced, owing to failure to take into account the recoil momentum imparted to the thermal molecule by the electron. Chantry and Schulz[48] analysed this effect, which could give rise to similar discrepancies. Its elimination strengthens the reliability of this technique of electron affinity determination. The kinetic energy distribution of the O^- ions is not exactly zero, but is a bell-shaped distribution function of FWHM $\Delta E = (11\beta kTE_0)^{1/2}$ where

$$\beta = \frac{m(O^-)}{m(O_2)} \qquad (8.6)$$

and E_0 is the most probable ion energy. This equation was checked experimentally at various gas temperatures T. The apparent onset of dissociative attachment is such as to lead to $E_a(O) = 1 \cdot 5 \pm 0 \cdot 1$ eV, in good agreement with the photodetachment value.

Lattice Energies of Ionic Crystals

The fourth method of electron affinity determination is from lattice energies[49]. In an ionic crystal M^+X^-, an energy equation may be written

$$E_a(X) = E_i(M) + D(MX) + S - U \qquad (8.7)$$

where S is the heat of sublimation of the crystal, and U is the lattice energy. When $E_i(M)$ and the dissociation energy $D(MX)$ are known, determination of S and U leads to E_a. It is not yet possible to make a critical experimental comparison of this technique with the other three.

Indirect inferences of electron affinities can be made from various collision studies. The most reliable are the theories of symmetrical resonance charge transfer[50] (Chapter 12), and collisional detachment[51, 52] (Section 8.7). Charge transfer spectra have also been considered, but probably do not give much accuracy.

A bracketing technique of some value is that of determining whether a charge transfer process between the negative ions of two species is endothermic or exothermic; this is done by determining whether or not it will proceed at near-thermal energies. A particularly important example is Cl^-NO_2, which is now reported not to proceed, in contradiction to earlier experiments.

8.3 ELECTRON ATTACHMENT PROCESSES. INTRODUCTION

In the atomic attachment collision the excess energy may be emitted as a quantum of radiation (radiative attachment):

$$A + e \rightarrow A^- + h\nu \qquad\qquad e0/\overline{1}0$$

Alternatively, a third body (atom, molecule or electron) can stabilize the attachment process, with transfer of the internal energy.

Since the elements listed in Table 8.1 do not normally exist as free atoms in the vapour phase, the measurement of atomic attachment collision cross-sections is difficult and has not been achieved except in shock-wave and high-current arc experiments.

For radiative attachment to atoms, there are quantum theory calculations, and detailed balance calculations from the reverse process, that of photo-detachment ($\phi\overline{1}/0e$). Some radiative attachment cross-sections so deduced can be found in Section 8.6.

Electron attachment to molecules proceeds by a vertical transition from the equilibrium position of the atoms. Some possible types of adiabatic potential energy curves for diatomic molecules and ions are given in *Figure 2.2* (page 72). Considerations similar to those governing the ionization of diatomic molecules determine whether dissociative or molecular attachment takes place. Corresponding to the 'vertical ionization energy', there is a 'vertical detachment energy', and this may have the opposite sign to the electron affinity of the atoms that compose the molecule. The true adiabatic electron affinity of a molecule is given by:

$$E_a(XY) = -D(XY) + E_a(X) + D(XY^-) \qquad (8.8)$$

A list of important molecule and radicle adiabatic electron affinities is given in Table 8.2, following earlier review articles in this field[49, 53]. Potential energy curves for molecular negative ions are depicted in the Appendix.

The molecular attachment process, yielding the negative ion of the parent molecule, can only proceed via an excited state, since it is a resonance process and the molecule has a real electron affinity. The excited state, which can be purely vibrational, may either be deactivated collisionally (three-body attachment) or pass radiatively to the ground state. A third possibility, for polyatomic molecules, is that the negative ion exists for periods of the order of microseconds, detaching spontaneously after this time. The lifetimes of such temporary polyatomic negative ions can be measured by time of flight mass-spectrometry, and can be related by means of statistical mechanics to the cross-section for formation, to the number of vibrational degrees of freedom, and to the electron affinity of the molecule. Reasonable estimates of the latter have been obtained. In very approximate terms, the logarithm of the lifetime is proportional to the number of vibrational degrees of freedom.

8.4 MEASUREMENT OF ATTACHMENT

Swarm Experiments

The classical measurements of attachment were made by studying the attenuation of an electron swarm drifting in a gas at pressures of several torr under the action of a uniform electric field X. The attenuation of a current I_e in a distance dx is

$$dI_e = -I_e \eta \, dx \tag{8.9}$$

where the 'attachment coefficient' η is the probability that an electron in passing unit distance in the direction of the field will attach to a gas molecule. Thus the ratio of electron current after two distances x_1 and x_2, will be

$$\frac{I_{2e}}{I_{1e}} = \exp\left[-\eta(x_2 - x_1)\right] \tag{8.10}$$

The attachment coefficient η is related to the attachment cross-section σ_a as follows:

$$\eta = \frac{n_0 \bar{\sigma}_a \bar{v}_e}{v_d} \tag{8.11}$$

where $\bar{v}_e$ and v_d are electron random and drift velocities respectively. Note that this attachment coefficient is not a conventional rate coefficient, but could be converted to a monoenergetic rate coefficient $k_{2a} = \bar{\sigma}_a \bar{v}_e$ provided n_0 and v_d were known. The coefficient η is defined in units of molecule cm^{-1}. Frequently the results of measurements are expressed in terms of the probability h of attachment per collision, given by

$$h = l_f \sigma_a n_0 \tag{8.12}$$

where l_f is the electron mean free path, which is assumed to be known from swarm or scattering experiments, and is related to the total electron collision cross-section $\sigma_e = (n_0 l_f)^{-1}$.

A swarm attachment experiment, therefore, consists in measuring electron current attenuations together with drift or random velocities. The latter are connected by the Townsend relation

$$v_d = \frac{0.815 X e l_f}{m_e \bar{v}_e} \tag{8.13}$$

In the absence of positive ions, electron densities n_e must be held $< 10^7 \, cm^{-3}$ in electron drift tubes, in order that space-charge effects be negligible.

30*

The measurement of electron current implies the separation of negative ions and electrons in the negative swarm. One method by which this has been achieved is the high-frequency filter grid developed by Loeb and Cravath[54], and used by Bradbury[55]. Alternate grid wires are fed from the output terminals of a high-frequency oscillator, and the mobile electrons are swept to the grid wires, whilst the less mobile negative ions may pass through the grid.

Modern filter grids are made photographically, and then cut and transferred onto two rings mutually insulated with epoxy-resin (which unfortunately absorbs water and could with advantage be replaced by a resin which does not).

For the application of equation 8.10 it is necessary to vary the distance between the electron source S and the filter grid G. In the apparatus sketched in *Figure 8.1*, either of two filter grids G_1 and G_2 can be moved into the path of the swarm. The total negative current, without the electrons that survive the distances SG_1 and SG_2, is collected on the plate P, whilst a uniform field is maintained by the guard rings A–E.

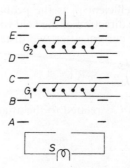

Fig. 8.1. Schematic diagram of electron swarm experiment, with filter grids, for measurement of attachment

This method largely replaces the less direct diffusion method, first devised by Bailey[56], subsequently used in the researches of Healey and co-workers[57], and simplified and improved by Huxley and his colleagues[58-59]. In Bailey's method, swarm currents are measured to a series of equally spaced annular plate electrodes, using two sets of homologous values of plate separation, pressure and field, both with and without a magnetic field parallel to the electric field. This magnetic field affects the lateral diffusion of the electrons, but not that of the negative ions. In the deduction of the equations relating the attachment probability to the current measurements, the assumption must be made that the probability h of attachment is independent of pressure; this may be invalidated by three-body processes. Nevertheless, the agreement between results so obtained and the results of filter grid measurements is not unsatisfactory, despite possible attachment to impurities arising from neglect of degassing prior to experiment.

In Huxley's method, the radial diffusion of a swarm of electrons is measured in the absence of a magnetic field; the swarm sets out from a small orifice and diffuses between field-forming guard rings to a central disc surrounded by two annular electrodes. Radial diffusion of negative ions is sufficiently small for all of them to be collected at the central disc. From the annular-to-disc current ratios at different diffusion distances, Huxley calculated not only the attachment rate η but also the ratio of drift velocity to electron diffusion coefficient, v_d/D_e, which is related[60] to the mean electron temperature T_e by the Einstein equation

$$\frac{v_d}{D_e} = \frac{40 \cdot 3 X T_g}{A T_e} \qquad (8.14)$$

where A is a numerical constant which depends upon the form of the electron energy distribution, and T_g is the gas temperature. The simultaneous determination of the two quantities η and T_e is the principal advantage of this method.

The shutter methods of attachment measurement have been further developed by Doehring[61] and by Chanin, Phelps and Biondi[62]. Doehring used a four-grid double-shutter method similar to that used in ionic mobility tubes (Chapter 10). Negatively charged particles can only pass through the shutter during the period when a short pulse is applied between the two grids that compose it. The pulse of electrons that emerges from the first shutter is partly converted to negative ions during passage through the gas. Under the action of a uniform applied electric field, the negative ions drift towards the collector with velocity v_{id}. A second pulse, delayed by a time t, is applied to the second shutter, so that the number of ions reaching the collector is:

$$n_- = A \exp{(\eta v_{id} t)} \qquad (8.15)$$

where

$$A = \eta n_e v_{id} e^{-d} \qquad (8.16)$$

for $0 < t < d/v_{id}$, and

$$A = 0 \qquad (8.17)$$

for $t \geqslant d/v_{id}$, the shutters being separated by a distance d. The electron drift velocity is so high that the electron transit time can be neglected in comparison with that of the negative ions. A plot of $\ln I_-$ versus t should provide a straight line of slope ηv_{id}, whilst the time at which I_- drops to zero provides the ion drift velocity. The electron drift velocity, necessary for converting η to σ_a, must be measured in a separate experiment, using the same shutters.

Chanin, Phelps and Biondi used a technique essentially similar to that of Doehring, although the apparatus was that of Bradbury. The first shutter was replaced by a pulsed source of photoelectrons, thus avoiding the distortion produced by the few negative ions which enter the inter-shutter space. The shutter was inserted for electron drift velocity measurements.

A technique employed by Herreng[63] avoids the use of shutters, relying on a collimated pulse of X-rays to form the electrons and an oscilloscope display of the collected current, both for the measurement of η and v_{id}; this was actually the first experiment in which the drift velocity was accurately determined in the same apparatus as that used for the measurement of the attachment rate. A similar technique was used by Bortner and Hurst[64], whose work evolved from the discovery that attachment appreciably modified the pulses obtained from ionization counters.

It should be noted that electron swarm studies of ionization are rendered inaccurate with a high X/p attachment. The situation has been analysed quantitatively by Geballe and Harrison[65]; the idea can be applied to the measurement of the attachment coefficient, provided ionization and attachment proceed at comparable rates, which is likely only over a limited range of X/p.

Single-collision Electron Beam Experiments

Similar principles apply to electron beam measurements of attachment as apply to single-collision experiments on the ionization of molecules by electrons. In view of the comparative simplicity of interpretation, the emphasis has been on total ion collection and on the kinetic energy measurement of the negative ions; but in modern experiments, mass-analysis and ion energy analysis have also been undertaken[66, 67].

Many experiments without mass-analysis have been performed with the apparatus originally designed by Lozier[68] (Chapter 6). The electron beam passes through a gas along the lines of force of a uniform magnetic field; the ions pass radially through the spaces between the discs which surround the beam axis and are collected on the outer cylinder, whilst the electrons are confined by the magnetic field and cannot be so collected. Retardation technique is used to measure the kinetic energy with which the ions are produced. The Lozier apparatus is ineffective when 90° scattering is minimum. The retarding potential difference technique was employed by some workers[20, 42, 46, 69], but not by all[43].

Experiments involving mass-analysis were pioneered by Hagstrum[45], and were performed with a Nier–Bleakney electron impact source, a 90° mass-spectrometer, and a retardation chamber. This type of experiment is most suited to the study of the onset potentials for dissociative processes, from which vertical detachment energies may be deduced.

In modern experiments, electrostatically momentum-analysed electrons are crossed with a molecular beam. It is necessary, following the work of Chantry and Schulz[48], to use a method of ion detection which takes account of the recoil momentum imparted by the electron.

Afterglow Measurements

The decay of electron density in the late afterglow of discharges in electronegative gases can be dominated by attachment processes; the time-dependence of the electron density is given by

$$\frac{\mathrm{d}n_e}{\mathrm{d}t} = -v_a n_e \qquad (8.18)$$

and

$$n_e = n_{e0} \exp\left(-v_a t\right) \qquad (8.19)$$

with attachment frequency

$$v_a = \sigma_a n_0 \bar{v}_e \qquad (8.20)$$

where n_0 is the density of the attaching species; σ_a is the attachment cross-section and $\bar{v}_e$ is the mean electron velocity.

This behaviour is complicated by the decay due to electron–ion recombination and to ambipolar diffusion. The three processes may be distinguished by their different loss rate dependences on the afterglow parameters; these are summarized in Table 3.2 (page 150).

The recombination process, requiring the collision of an electron and a positive ion, can only dominate at large electron densities, in the early afterglow. As the afterglow proceeds, the recombination t^{-1} electron density decay gives way to the exponential decay, characteristic both of diffusion and of attachment. It is only necessary in this chapter to consider the simultaneous decay due to attachment and diffusion:

$$\frac{dn_e}{dt} = D_a \nabla^2 n_e - v_a n_e \tag{8.21}$$

$$n_e = n_{e0} \exp\left(-\frac{t}{\tau}\right) \tag{8.22}$$

$$\frac{p}{\tau} = p\frac{D_a}{\Lambda^2} + \frac{v_a p^2}{p} \tag{8.23}$$

where D_a is the ambipolar diffusion coefficient, and Λ is the diffusion length. For two-body attachment, v_a/p is pressure independent, as is pD_a. A plot of p/τ versus p^2 should be linear, the gradient yielding v_a and the intercept D_a.

At high pressures, when loss by diffusion to the walls becomes small in comparison with the loss by the volume process of attachment, the simple equation 8.19 may be applied.

In afterglow studies of attachment, electron density has been measured by means of the microwave cavity technique outlined in Chapter 7 and developed by Biondi and Brown[70]. The first attachment data obtained in this way were those for iodine[71]. Since the attachment coefficient is extremely large, the iodine must be diluted with helium; strongly attaching molecules can be successfully studied in the presence of a buffer gas.

In weakly attaching gases, the process may be masked by diffusion; the ambipolar diffusion coefficient is enhanced by the presence of negative ions to a value

$$D_a' \simeq D_a(1+\beta) \tag{8.24}$$

where β is the ratio n_-/n_e.

In oxygen, an extremely small apparent attachment coefficient[72-75] is measured, and the special state of affairs that prevails in the late oxygen afterglow is discussed in Section 8.5 under the heading 'thermal energy processes'.

The principal difficulty in afterglow measurements is that the negative ions formed may take part in inelastic processes; where these involve detachment, the kinetics become complicated. Undoubtedly the best approach would be to eliminate the species giving rise to detachment. Such a species usually possesses internal excitation, formed in the pulse of energy which originally produces the electrons. The only methods of avoiding this would be:

1. To produce the original ionization by ultra-violet photons or by relativistic electrons from a linear accelerator.
2. To operate the afterglow not in the time-dependent mode, but as a flowing system, such as is discussed in Chapters 7 and 14. The gas could be introduced sufficiently far down the flow for it to avoid internal excitation. For example, in the study of attachment to oxygen, ionization would be applied to a buffer gas such as helium, and the oxygen introduced downstream.

A bibliography of swarm and afterglow measurements of attachment is available[76].

Optical Study of Attachment. The Radiative Process

Apart from optical studies of photodetachment, from which radiative attachment cross-sections can be inferred using the principle of detailed balance, detection of photons can serve as a measurement of radiative attachment. The emitted radiation is in the form of a continuum, reflecting the electron energy distribution.

It did not prove possible in early attempts to observe the 'affinity spectrum' or radiative attachment continuum in weakly ionized gases. However, high-current arcs emit excellent H^- continua[77, 78], and plasmas produced in shock tubes can also be studied[79]. The continuum from oxygen has been observed[80] in a well-stabilized arc at atmospheric pressure and known temperature; the radiative attachment cross-section $_{e0}\sigma_{\bar{1}\phi}$ deduced from this observation differs by only 30 per cent from that calculated from the accurately determined photodetachment cross-section $_{\phi\bar{1}}\sigma_{e0}$. The detailed balance equation is written in the form

$$_{\phi\bar{1}}\sigma_{e0} = \left(\frac{m_e v_e c}{h\nu}\right)^2 \left(\frac{g_0}{g_-}\right)\,_{e0}\sigma_{\phi\bar{1}} \tag{8.25}$$

where g_0 and g_- are the atom and negative ion statistical weights. Application of this equation has been made[2] to infer radiative attachment cross-sections from known photodetachment functions. For molecular processes, it is necessary to know the vibrational populations, as well as the vertical detachment energies.

The threshold laws for radiative attachment follow from equation 8.25. For atomic hydrogen the threshold law is $E^{1/2}$, and for oxygen it is $E^{-1/2}$.

Nitrogen arc continua[79, 81] have been attributed to N^- ^1D. Atomic halogen continua have been observed by Popp[82] in arcs. The correct low-energy threshold of the continuum, equal to the electron affinity, is in each case observed. In addition there is in each halogen function a structure of Beutler–Fano shape, indicating the presence of a 'resonance' or compound state of atom and electron. These are the only known instances of resonances observed in emission spectra.

8.5 INTERPRETATION OF ATTACHMENT DATA

Dissociative Attachment

Dissociative attachment proceeds by resonance capture of an electron by the molecule AB, into an antibonding quasi-stationary state AB^-. There is then competition between dissociation and the autodetachment process $AB^- \rightarrow AB + e$. The cross-section for resonance capture should be of the order of the square of the electron wavelength, that is, several $Å^2$ at 10 eV. However, the typical experimental cross-section is 10^{-19}–10^{-20} cm², the small size being attributed to two factors. First, there is uncertainty in the nuclear separation of the transition, and therefore an uncertainty ΔE in the energy of the transition. If this uncertainty in energy is much greater than the line width Γ of the level (decaying to $AB + e$), then the cross-section is reduced by the process of averaging over ΔE. Second, there is competition between the reverse process and the dissociation, which results in a probability of dissociation equal to

$$\exp \left[\frac{- \int_0^\infty \Gamma(r) \, dr}{v(r)} \right] = \exp \left(\frac{-2\bar{\Gamma}\tau}{h} \right) \qquad (8.26)$$

where τ is the time of dissociation. The average value $\bar{\Gamma}$ is an integration over nuclear separation r between the equilibrium r_{AB} and the crossing r_x of AB and AB^-. Typically τ is of order 10^{-13} sec, $\bar{\Gamma} \simeq 0.1$ eV and $2\bar{\Gamma}\tau/h \simeq 10$.

With certain restrictions[83], the cross-section for dissociative attachment may be written

$$\sigma \propto \frac{\Gamma}{E_e V'} |\chi(E)|^2 \exp \left(\frac{-2\bar{\Gamma}\tau}{h} \right)$$

where $\chi(E)$ is the vibrational wave function written in terms of the electron energy, and V' is the slope of the molecular negative ion potential energy curve.

The validity of this analysis is confirmed by the 'isotope effect'. For example, D_2^- possesses approximately the same interaction as H_2^-, but the increased mass inertially increases the time of dissociation by a factor $\sqrt{2}$. The corresponding decrease in dissociative attachment cross-section

$$e + D_2 \rightarrow D^- + D$$

was predicted independently by Bardsley, Herzenberg and Mandl[84], and by Demkov[85], and was observed at much the same time by Rapp, Sharp and Briglia[86]. It has been claimed[87] that the first observation of isotope effect was in the experiments of Hurst on H_2O and D_2O. The maximum cross-section is equal to $12\pi^{3/2} \Gamma(r_{AB})/\Delta k^2$, where the value of line width Γ, which is a function of nuclear separation r, is taken at the equilibrium separation of AB; Δ is the width of the Franck–Condon reflexion in the AB^- curve, and k is the electron wave number.

The dissociative attachment in hydrogen is an instructive illustration of several of the important features of this type of collision. The potential energy curves are illustrated in the Appendix, and the dissociative attachment function in *Figure 8.2*. The bonding H_2^- $^2\Sigma_u$ state[6, 88–90] has only a very short lifetime (10^{-15} sec) for spontaneous detachment to H_2 $^1\Sigma_g$. Its width is $\gtrsim 0.5$ eV, so the level cannot be observed as resonance scattering. However, Schulz and Asundi[91] report a dissociative attachment peak 1.6×10^{-21} cm² at 3·75 eV, which must arise from a vertical transition to the dissociation limit of this level.

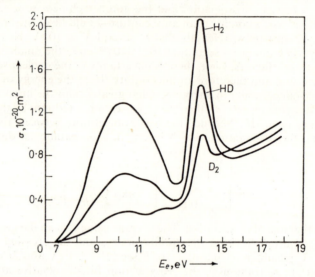

Fig. 8.2. *Dissociative attachment cross-section functions*[86] *for* H_2, *HD and* D_2, *showing isotope effect*

The major part of the dissociative attachment takes place via the antibonding level H_2^- $^2\Sigma_g$. The 'reflexion' of the H_2 $^1\Sigma_g$ ($v = 0$) wave function in this potential energy curve gives the form of the cross-section function[92]:

$$\sigma \propto \left| \int \psi_i(r)\, \psi_f(r)\, dr \right|^2$$

It is often sufficient to employ a Morse function for ψ_i and a delta function for ψ_f. In the simplest case, this procedure gives rise to a bell-shaped cross-section function, and, in the interesting case of a shallow bonding level, to a bell truncated at the lower-energy end. The cross-section function for H^- is bell-shaped but contains a marked dip at the crossing of the curves H_2^- $^2\Sigma_g$ and H_2 $^3\Sigma_u$. There are also some interference oscillations due to coupling between H_2^- $^2\Sigma_g$ and the resonance H_2^- $^2\Sigma_g^+$, possibly also the resonance just above it. This effect is an example of predissociation. Both the resonances are known in elastic scattering. The high-energy feature in the dissociative capture function arises from the resonance level or levels, and is an example of the truncated bell function, showing clearly the onset of H $^2S + H^-$ $1s^2$, which interact attractively to form the compound state H_2^- $^2\Sigma_g^+$ (σ_g^o $2s$) $(\pi_u^{+1}$ $2p)(\pi^{-1}$ $2p)$. Detailed discussions should be consulted[93–96].

The dissociative attachment process in oxygen has been widely investigated, particularly in respect of the deduction of the electron affinity of oxygen

discussed above[42-48]. The maximum cross-section[43] is $\simeq 2 \times 10^{-18}$ cm^2. It is
of interest that the dissociative attachment peak is markedly shifted to lower
energies when the oxygen is at temperature $\sim 2100°$K, or when it emerges
from a discharge; this would be expected on the basis of vibrational excitation of the molecule[97].

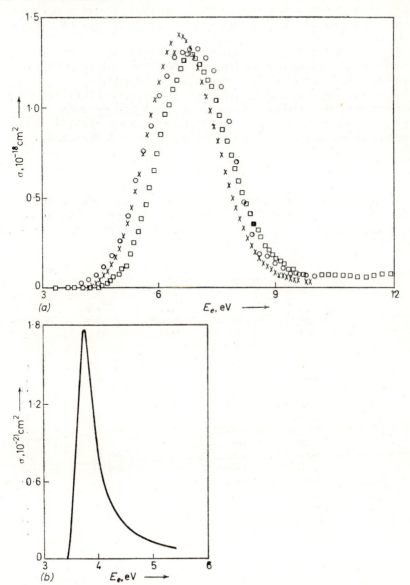

*Fig. 8.3. Dissociative attachment cross-sections typical of the 'reflexion' of the ground state
wave function in the molecular negative ion potential energy curve: (a) O$^-$ from O$_2$
(squares, Schulz[47]; circles, Asundi, Craggs and Kurepa[99]; crosses, Rapp and Briglia[100]);
(b) H$^-$ from H$_2$ (Schulz and Asundi[101]) showing sharp onset at threshold*

Dissociative attachment in CO produces[98] two O^- ion peaks, of which the larger, at smaller energy, arises from a state formed from $O^- + C\,^3P$; the smaller arises from $O^- + C\,^1D$. Both these states are bonding and, by vertical transitions, ions of zero kinetic energy can be produced at the theoretical threshold. The energy distribution of the ions extends upwards from zero. *Figure 8.3* shows dissociative attachment peaks typical of 'reflexion' of a molecular ground state wave function in the molecular negative ion potential energy function.

In nitric oxide a single structured O^- peak is observed, which may arise from one, or two, states derived from $O^- + N\,^2D^o$. It is possible that crossings with levels such as NO $G^2\,\Sigma^-$ contribute to the structure.

The most important dissociative attachment process in H_2O results in the formation of $H^- + OH$, which appear at 5·5 eV and maximize at 6·5 eV; there is also a smaller H^- peak at higher energy; above 7·5 eV, O^- appears. At high pressures, OH^- can be formed by the ion–molecule reaction

$$H^- + H_2O \rightarrow OH^- + H_2$$

Table 8.3 shows attachment cross-sections and the energies at which they occur. With the exception of SF_6, all these data refer to dissociative processes, although the ions have not always been identified. Further data are available[20, 99, 104-106].

Table 8.3. ATTACHMENT CROSS-SECTIONS MEASURED WITH ELECTRON BEAMS

Molecule	σ_{max} (1st peak) (Å²)	E_e (max) (1st peak) (eV)	σ_{max} (2nd peak) (Å²)	E_e (max) (2nd peak) (eV)	Reference
O_2	0·022	6·8			43
	0·013	6·2			69
	0·013	6·8			102
CO	0·027	9·85		11·02	42, 43
CO_2	0·0051	7·8			43
SF_6	5·7	0·00		0·1 (SF_5^-)	69
	≲10	0·00			103
CCl_4	1·3	0·02	1·0	0·6	69
CF_3I	0·78	0·05	0·32	0·9	69
CCl_2F_2	0·54	0·15			69
BCl_3	0·28	0·4			69
HBr	0·58	0·5			69
HCl	0·039	0·6			69
H_2O	0·048	6·4	0·013	8·6	69
NO		8–9			

In addition to the resonance dissociative attachment process, there occurs in molecules, such as O_2, CO[43, 66] and halogens, an ion pair formation process

$$e + AB \rightarrow A^+ + B^- + e$$

e0/1̄e

The cross-section rises monotonically with increasing electron energy for some tens of electron volts. The threshold is lower than that for positive ionization processes.

Thermal Energy Processes

For small molecules, the direct two-body attachment of a thermal energy electron hardly ever occurs. It is possible that the high-energy tail of electrons will take part in a dissociative attachment process. It is also possible for the three-body electron attachment process to occur, yielding the molecular negative ion

$$e + AB + X \rightarrow AB^- + X$$

The species X may be atomic or molecular.

The swarm data for oxygen can be interpreted in terms of the dissociative and three-body processes. Above about 1·5 eV, the two-body attachment rates deduced from impact and from swarm measurements are in agreement, as can be seen from *Figure 8.4*. At high X/p, detachment competes[107].

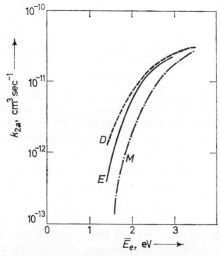

Fig. 8.4. Two-body oxygen attachment coefficient k_{2a} as a function of mean electron energy: E, deduced from the swarm experiments of Chanin, Phelps and Biondi[62]; D, calculated from the electron impact data of Craggs and Massey[66] assuming Druyvesteyn energy distribution; M, calculated from the electron impact data of Craggs and Massey[66] assuming Maxwellian electron energy distribution

(From Chanin, Phelps and Biondi[62], by courtesy of the American Physical Society)

In electron impact experiments[42], only a small number of O_2^- ions are observed, and these are attributed to secondary processes, in particular the ion–atom interchange process

$$O^- + O_2 \rightarrow O_2^- + O^-$$

for which a cross-section as high as 10^{-15} cm² has been suggested. An upper limit of 10^{-20} cm² is placed upon the two-body O_2 radiative attachment cross-section.

When swarm measurements are extended[62] to low X/p, the two-body attachment coefficients are found to be pressure dependent (*Figure 8.5*).

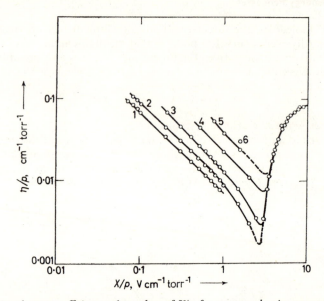

Fig. 8.5. Attachment coefficients at low values of X/p for oxygen, showing pressure dependence arising from three-body process: 1, pressure 7.6 torr; 2, pressure 10·5 torr; 3, pressure 15·0 torr; 4, pressure 25·0 torr; 5, pressure 44·0 torr; 6, pressure 54·0 torr

(From Chanin, Phelps and Biondi[62], by courtesy of the American Physical Society)

This had been overlooked in previous studies at rather larger X/p. It was therefore proposed that the three-body process

$$O_2 + e + O_2 \rightarrow O_2^- + O_2$$

predominates at low energies. As was first suggested by Bloch and Bradbury[108], the third body, in this case oxygen, stabilizes the excited molecular negative ion O_2^- before it has time to detach spontaneously. The three-body attachment coefficient k_{3a} for a particular third body X is defined by the equation:

$$\frac{dn_e}{dt} = -k_{3a}n_x n_e n_0 \tag{8.27}$$

with k_{3a} in units of cm⁶ sec⁻¹.

For thermal electrons at 300°C the three-body attachment coefficient (with O_2 the third body) measured by Chanin[62] is $\sim 2·8 \times 10^{-30}$ cm⁶ sec⁻¹. The interpretation of meteor trial decay in the atmosphere is only possible with a somewhat higher coefficient below 90 km height, namely $2·5 \times 10^{-29}$ cm⁶ sec⁻¹. Other drift-tube measurements[64, 109] lead to three-body coefficients in substantial agreement with those of Chanin. The $v = 3$ vibrational level of O_2^- probably lies within thermal energy of the $v = 0$ level of O_2, if some rotation is assumed.

However, the microwave experiments in pure oxygen[72–75] at thermal energies lead to a value of $v_a \sim 4 \times 10^2 \, \text{sec}^{-1}$, whereas at 10 torr pressure the swarm data[62] yield $v_a \sim 3 \times 10^5 \, \text{sec}^{-1}$. This large but well-established discrepancy was interpreted[110] as being due to a competing detachment process which is not operative in drift-tube measurements. The species involved in the competing detachment process could be vibrationally excited oxygen or, more probably, the $^1\Delta_g$ state:

$$O_2^v + O_2^- \rightleftharpoons O_2 + O_2 + e$$

or

$$O_2\,{}^1\Delta_g + O_2^- \rightleftharpoons O_2 + O_2 + e$$

The latter process could be of great significance in the lower ionized regions of the atmosphere[111], and its experimental study has now been reported. Afterglow experiments in which ionization is produced from linac electrons, thus avoiding secondary excitation processes, yield three-body attachment rates in agreement with swarm experiments[112, 113]. Dilution with buffer gas produces a similar effect[74].

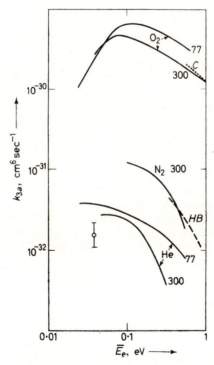

Fig. 8.6. Three-body oxygen attachment coefficients: full lines, measurements of Chanin, Phelps and Biondi[62] for third bodies O_2, N_2 and He, at temperatures denoted in °K; HB, drift-tube measurements of Bortner and Hurst[64] for third body N_2 at 300°K; C, drift-tube measurements of Chatterton[109] for third body O_2 at 300°K; single point, microwave data[74] for third body N_2

Afterglow attachment measurements have been reported in NO^{114}, NO_2, $SF_6^{65, \, 115, \, 116}$, $CO_2^{109, \, 117}$, CO^{118}, $F_2{}^{97}$, H_2O^{119}, and humid air[120]. The absence of thermal energy attachment in water is significant in view of the behaviour of the 'free' electron in liquid water subjected to linac irradiation. The afterglow studies of attachment to iodine vapour[71] yield an extremely large attachment rate, which is dissociative, since the I_2 molecule is weakly bound and the iodine electron affinity large. Swarm data[121–123] were discussed in the first edition of this book.

An important feature of three-body attachment data is the relative efficiencies of different third bodies. Data for oxygen[62, 64, 72–75, 109] are] shown in *Figure 8.6*, and more recent studies with other third bodies[124–127] are available. The effectiveness of a third body in removing excess energy from the electron could possibly be related to its energy loss parameter λ (Chapter 4), which is itself a function of energy. Detailed comparison has not yet proved possible. On the other hand, it might be that stabilization of the vibrationally excited negative ion by excitation exchange is the operative process.

8.6 PHOTODETACHMENT AND RADIATIVE ATTACHMENT

Photodetachment from an atomic negative ion is the reverse process of radiative attachment, in which an electron attaches to an atom causing emission of radiation. From a plasma this will be a continuum, owing to the distribution of electron energies. Observations of radiative attachment continua have been made in arcs[128] and more recently in shock-heated gases[129]; the thresholds enable electron affinities to be deduced. However, the reverse process is easier to study with precision, since photodetachment can readily be investigated by irradiating a negative ion beam with photons and collecting the electrons.

The direct measurement of the detachment of electrons from negative ions under the action of radiation was first achieved in a series of experiments carried out by Branscomb and his collaborators[2, 7–9]. A mass-analysed negative ion beam ($\sim 10^{-7}$ A) is crossed with a mechanically chopped filtered light beam from a carbon arc. The slow electrons produced ($\sim 10^{-14}$ A) are collected by means of weak transverse magnetic (50 G) and electric (15 V cm^{-1}) fields; the signal is detected by phase-sensitive amplification.

The ion beam moving in the x direction encounters transversely a flux of photons $\phi(x, \lambda) \, d\lambda$ erg cm^{-2} sec^{-1}. The probability of attachment in passing through a distance dx is:

$$P_x \, dx = \frac{dx}{v_-hc} \int \sigma(\lambda) \, \phi(x, \lambda)\lambda \, d\lambda \qquad (8.28)$$

A distance s is defined normal to the ion beam and greater than its width, but still sufficiently small for the light intensity to be regarded as constant along it. In practice, s is one dimension of the calibrated radiometer used to measure the light, the other dimension being l. The normalized radiant flux $\phi'(\lambda)$ is such that

$$\phi(\lambda) = \frac{\phi'(\lambda) \, \omega(x)}{s} \qquad (8.29)$$

where $\omega(x)/s$ is the power density of the photon beam in the region of interaction. The ratio of electron to ion currents is

$$\frac{I_e}{I_-} = P = \frac{W}{hcv_-s} \int \sigma(\lambda)\,\phi'(\lambda)\lambda\,d\lambda \qquad (8.30)$$

where W is the total power incident on the area sl.

With the aid of calibrated transmission filters, photons of different wavelength bands are directed at the ion beam; the electron current allows the deduction of

$$\int \sigma(\lambda)\,\phi'(\lambda)\,T(\lambda)\,d\lambda$$

for each filter; $T(\lambda)$ is its normalized transmission, determined in a separate experiment. A typical graphical representation of transmission $T(\lambda)$ of filters calibrated by the National Bureau of Standards in Washington is given in *Figure 8.7*; these transmissions give a good idea of the sharpness of photon wavelength analysis which is possible. However, the evaluation of the

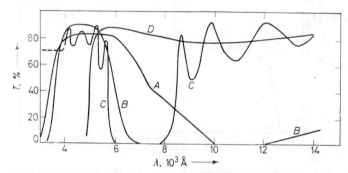

Fig. 8.7. *Transmission functions of four filters A, B, C and D*
(From Branscomb[2] by courtesy of Academic Press)

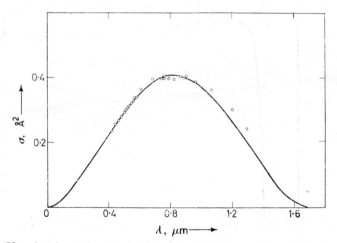

Fig. 8.8. *Photodetachment function for* H$^-$; *points indicate experimental data normalized at* $\lambda = 5280$ Å *to variational calculations*[2, 131] *(full line)*

31

cross-section function $\sigma(\lambda)$ from the experimental data must be carried out with the aid of a computer. In recent years, monochromated radiation having a bandwidth as low as 20 Å has been used in photodetachment experiments. A high-aperture monochromator is necessary[130]. Increased sensitivity is obtained with the use of single-particle counting techniques for detecting the electrons.

Photodetachment cross-section functions are illustrated in *Figures 8.8–8.10*. A part of their importance arises from the determination of detachment on-set, which is the electron affinity for atomic ions and the vertical detachment energy for molecular ions. In view of the difficulty of absolute calibration of electron energy, an optically determined energy is particularly valuable, although the bandwidth of the filters is not as narrow as that obtainable with

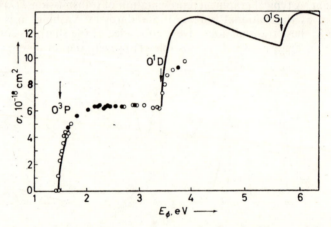

Fig. 8.9. Photodetachment cross-section function for O⁻ *showing experimental data[9], calculations[2], and onsets of* O ³P, O ¹D *and* O ¹S

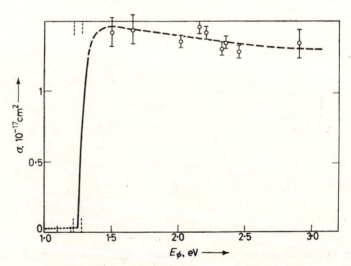

Fig. 8.10. Photodetachment cross-section function[7] for C⁻

diffraction monochromation. The electron affinities so determined are listed in Tables 8.1 and 8.2 (pages 451 and 452).

Another important feature of photodetachment data is the opportunity of calculating the probability of the reverse process, that of radiative attachment (Section 8.1); a comparison with quantum theory calculations is also possible[132-135].

For a negative ion of dipole moment μ interacting with an oscillating electric field of frequency ν, time-dependent perturbation theory gives for the probability P per unit time of a transition:

$$P \propto \nu |\langle \psi_a | \mu | \psi_b \rangle|^2 \varrho(E_e)$$

where $\varrho(E_e)$ is the density of states in the continuum per unit energy range corresponding to the energy $E_e = h\nu - E_a$ of the electron. Since $\varrho(E_e) \propto E_e^{1/2}$,

$$\sigma \propto \nu E_e^{1/2} |\langle \psi_a | \mu | \psi_b \rangle|^2$$

Three formally equivalent expressions for the dipole moment matrix element which can be written are designated as the dipole length, velocity and acceleration formulae. It is found that for H^- the velocity formula[136, 137] using the central-field continuum function with correction for exchange[138], or a variational procedure[132], gives best agreement with the experimental data normalized at 5280 Å (*Figure 8.8*). The normalization procedure allows of a smaller error than is appropriate to the absolute cross-section. Furthermore, the normalized data are consistent with the oscillator sum rules[139], which when applied to this process yield[140]

$$\frac{m_e c}{\pi e^2} \int_{\nu_0}^{\infty} \sigma_\nu \, d\nu \leqslant 2$$

and

$$\frac{m_e c}{\pi h a_0} \int_{\nu_0}^{\infty} \nu^{-1} \sigma_\nu \, d\nu \leqslant 14\cdot22$$

Use of the dipole length formula leads to threshold laws[1, 8, 141]

$$\sigma \propto \nu E_e^{3/2}(a_0 + a_1 E_e + a_2 E_e^2 + \ldots)$$

for H^-, and

$$\sigma \propto \nu E_e^{1/2}(a_0 + a_1 E_e + a_2 E_e^2 + \ldots)$$

for p electron shells such as C^- and O^-. Although it is not known over what energy range these laws hold, since the parameters a_n are unspecified, it is likely that the distinction between the gradual rise of $\sigma(H^-)$ and the sharp rise of $\sigma(C^-)$ and $\sigma(O^-)$ is a real one. For diatomic molecules, the power law depends upon the angular momentum projected along the molecular axis associated with the orbital from which the electron is ejected[142]. For C_2^- it is $E^{3/2}$, and for other molecular ions $E^{1/2}$.

The best photodetachment cross-section for O^- is reproduced in *Figure 8.9*. Particular attention was paid in the experiments to the region of onset,

31*

since it was at one time believed that the $2p^4 3s^4P$ level of O^- might be stable below the continuum. Early electron impact data supported this view, but collisional detachment data[143] have for a long time supported the view that no metastable O^- ion existed. The onset 'bump' in the photodetachment function may be interpreted in terms not of a metastable O^-, but of the multiplicity of O and O^- levels (O^- $^2P_{1/2}$, O^- $^2P_{3/2}$, O $^3P_2$, O $^3P_1$, O $^3P_0$). Whilst the O levels are very closely spaced, the O^- levels may be as widely separated as 160 Å (argument from isoelectronic extrapolation). Using the threshold law given above, and data from seven filters differing greatly in transmission in the threshold region, computational procedures led to the function of *Figure 8.9*. The pure one-half power threshold law holds for about 0·3 eV above onset. Born approximation calculations have been carried out at higher energies[144]; these calculations, the oscillator strength sum rules, and experimental data, have been used to predict a form of the total photodetachment function[2], including processes yielding O^1D and ^1S states. This is also included in *Figure 8.9*. Analogous behaviour is expected of S^-, but the experimental data have yielded only the threshold.

The photodetachment function[7] for C^- (*Figure 8.10*) shows a weak absorption in the wavelength region 1–2·6 μm, which may be attributed to the stability of the metastable C^- $1s^2 2s^3 2p^3$ ^2D.

In recent photodetachment measurements, the electron affinity of iodine has been found to be smaller by 0·1 eV than the previously accepted value, but this is in partial disagreement with the value deduced from the observation of the attachment continuum emission from shock tubes. The accepted halogen electron affinities were found to be slightly higher than the true values, and this was confirmed by the observations of Berry, Reimann and Spokas[79] (see Table 8.1).

The interpretation of the photodetachment function of OH^- is of interest for the light thrown on the OH electron affinity. Although the previously accepted value[49] was $E_a = 2·2$ eV, the threshold energy was found[2] to correspond to $E_a = 1·78$ eV, and it was suggested that a transition between vibrationally excited states was involved. The sharp onset of the photodetachment function, the comparatively large vertical detachment energy and the almost identical nuclear separation of the potential minima of OH and OH^- make this a particularly simple transition; the potential energy curves run almost parallel. However, the identical onset was determined for OH^- and OD^-, and no vibrational fine structure was found in either photodetachment function ($E^{1/2}$ threshold dependence is expected). The only hypothesis capable of explaining these facts is that the OH^- and OD^- ions are all formed in the vibrational ground state; it must be remembered that the vibrational level separations in OH and OH^- are almost identical, whilst those in OH and OD, and in OH^- and OD^-, are not. It is therefore likely that the onset energy observed is appropriate to ground vibrational states of ion and molecule. The unusual form of the photodetachment function, which shows a strong maximum about 300 Å above onset, is unexplained.

By contrast, the O_2^- photodetachment function shows no sharp onset, but rises gradually from almost zero energy. Experimental points were fitted to the threshold law with an onset of $0·15\pm0·05$ eV; however, it appears that in these experiments the molecular negative ions were in vibrational excited states. Electron affinities of 0·43 eV[145] and 0·58 eV[20] have been proposed, and the earlier value of 0·9 eV is now supposed to be in error[146]. It may be possible to conduct the photodetachment experiments using O_2^- ions which have drifted through oxygen, thereby ensuring deactivation to the ground vibrational state. Not only the threshold of the function is seriously affected by vibrational excitation; the entire function could be, perhaps, 50 per cent different.

Detachment of an electron from I^- by two 6934 Å photons from a Q-spoiled red ruby laser has been reported[135], and the process has been treated theoretically[147]. At the unique wavelength of the experiment, the agreement is reasonable. The probability of detachment is proportional to the square of the intensity of the radiation, and it is of such small magnitude that only with megajoule intensities can observations be made. The importance of this particular two-photon experiment is that no possibility exists for participation of any real intermediate state, since such states lie above the two-photon energy. Only virtual intermediate states can contribute. In two-photon ionization and excitation processes, the participation of a real intermediate state is nearly always possible.

Laser photodetachment is also of importance, even when only a single photon participates; the laser can produce a known, rotatable, pure linear polarization in the interaction region. Under these circumstances, it is possible to study the angular distribution of the photodetached electrons[148]. Where several angular momentum waves combine, departures from the classical dipolar $\cos^2 \theta$ limit are found. It may also be possible to use laser photodetachment to produce an atomic beam in a single rotational and vibrational level.

8.7 COLLISIONAL AND ASSOCIATIVE DETACHMENT

The detachment of electrons from H^- by electron impact is a collision process important in the understanding of stellar photospheres, since upon its probability can depend the question of whether the system is in local thermodynamic equilibrium. According to some calculations, the cross-section function maximizes as high as $1400\pi a_0^2$, but Born–Bethe calculations of McDowell and Williamson[149] maximize around 10 eV at $75\pi a_0^2$. This is not an unrealistic value, according to crossed-beam experiments[150, 151] which are in approximate accord; the fast neutral hydrogen atoms produced in the detachment process were detected. However, the cross-section function does not scale with the ionization functions of helium and Li^+. Above 20 eV, the following expression is a good approximation for the cross-section function (E_e eV):

$$\sigma \simeq \left[1 - \frac{1·6}{(E_e \log_{10} E_e)^{1/2}} \right] \frac{950}{E_e} \log_{10} \left(\frac{E_e}{0·92} \right) \pi a_0^2 \qquad (8.31)$$

The collisional detachment process

$$X^- + Y \rightarrow X + Y + e \qquad \qquad \overline{1}0/00e$$

differs from both capture and ionization collisions in that the electron is 'edged off' as the nuclei approach along their potential energy curves.

There is usually a pseudo-crossover at nuclear separations smaller than the potential minimum. When this region is reached, a transition can occur; the electron passes out of the collision region in a very short time on account of its smaller mass. The pseudo-crossover is found at larger separation than that for the electron loss process $00/10e$, because, in general, electron affinities are much smaller than ionization energies. Its separation is also larger than that for ionization $10/11e$, since in this case the collision product interaction curve is distorted by Coulomb repulsion. However, the pseudo-crossover at small separations can play an important part in all three electron loss processes, and may have the effect of preserving a sizable cross-section down to comparatively low impact energies. In the collisional detachment of negative ions, this feature is most noticeable.

It is seen that an atomic negative ion cannot in all cases approach within a close distance of a neutral atom without an electron being 'edged off' at a distance R_0[152]; the composite nucleus so formed may not be capable of binding the total of $Z_1 + Z_2 + 1$ electrons, that is, may not possess its own stable negative ion. The cross-section can then approach πR_0^2.

The collisional detachment process was first observed independently by Dukelskii and Zandberg[153] and by Hasted[143], using negative ion beam gas experiments in the single collision region. The slow negative charge, collected by an electrode system maintaining an electric field transverse to the ion beam, could not in general be produced by negative ion charge transfer, since the target atoms used in these studies were incapable of stable negative ion formation. Total electron production measurements of this sort have since been more widely made[154–156], and typical data are shown in *Figures 8.11–8.14*. In cases where a negative ion can be formed by the target, the negative ions may be separated from the electrons by conducting experiments both with and without a magnetic field parallel to the beam.

The collision cross-section functions of negative halogen ions with a light target ($m_1 \gg m_2$), are found substantially to superpose (*Figure 8.12*).

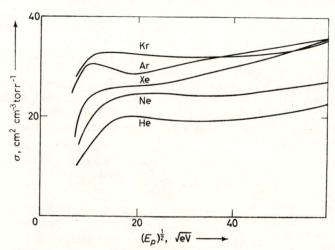

Fig. 8.11. Collisional detachment cross-section functions[143] measured for O^- ions in rare gases

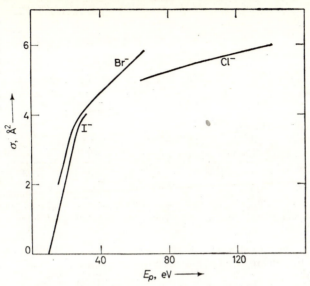

Fig. 8.12. Superposition of collisional detachment cross-section functions for halogen ions in helium[153]

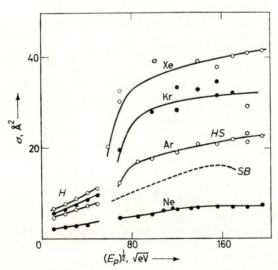

Fig. 8.13. Collisional detachment cross-section functions for H^- *in rare gases: HS, Hasted and Stedeford[155]; H, Hasted[143]; SB, Stier and Barnett's measurements[157] for* H^-–Ar

The threshold for I^- in He occurs at 8·8 eV, which indicates the lowest impact energy

$$V = \frac{8 \cdot 8 m_2}{m_1 + m_2} \text{ eV} \tag{8.32}$$

at which the pseudo-crossover is presumed to occur. Below this impact

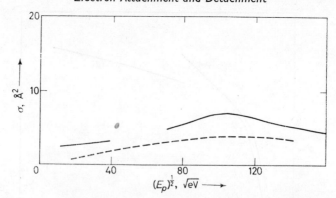

Fig. 8.14. H⁻–He *collisional detachment cross-section functions: full lines, experiments of Hasted and Stedeford*[155]*; broken line, calculations of Sida*[158]

energy, negative ion beams will produce no collisional detachment background in the electron currents detected in photodetachment and $e\bar{1}/0e^2$ experiments.

Recent collisional detachment studies[159, 160] include O_2^- and O^- in Ar and N_2, and O^- in O_2. Collisional detachment cross-sections have also sometimes been measured in photodetachment apparatus[8].

Theoretical treatment has been given by Smirnov and Firsov[161] and by Demkov[162]. The Smirnov and Firsov theory yields an expression for the collisional detachment cross-section

$$\sigma = \frac{\pi}{\varkappa_1 \varkappa_2} \qquad (8.33)$$

where $\varkappa_1 = \sqrt{2E_a}$ and $\varkappa_2 = 2\sqrt{\pi/\sigma_{el}}$, the elastic scattering cross-section σ_{el} being that for very low energy electrons. Thus this expression for σ is independent of impact energy. In the kilovolt region this is not unrealistic (for example, *Figure 8.11*), and one may write

$$\sigma \simeq \pi \left(\frac{13\cdot 6}{E_a}\right)^{1/2} a_0^2 \qquad (8.34)$$

This expression enables inferences to be made of unknown electron affinities, such as those of the alkali metals[163] (Table 8.1).

The conclusions of Demkov's theory are rather different; the average momentum of the emitted electrons is proportional to the one-third power of the impact velocity. The magnitude of the maximum cross-section is determined by the geometrical radii of the colliding systems[101].

The total collisional detachment functions for H⁻ ions in rare gases are illustrated in *Figure 8.13*. The apparent sudden rise around 5 keV was at one time[2] attributed to the onset of the two electron detachment process $\bar{1}0/10e^2$ in this region; but these cross-sections are reported[163] to be much too small to produce such an effect. However, measurements on H⁻ in Ar[157] using a technique which does not include the two-electron process, show no sudden rise.

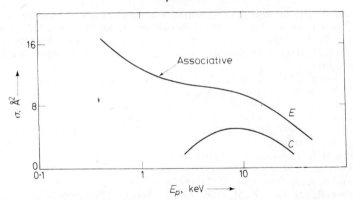

Fig. 8.15. H⁻–H *collisional detachment cross-section functions: C, collisional detachment calculations of McDowell and Peach*[165]*; E, experimental data of Hummer, Fite and Stebbings*[164]*, which include the associative detachment process*

The H⁻ in H total collisional detachment function is shown in *Figure 8.15*, and it appears that the experimental data[164] are in serious disagreement with the Born approximation calculations[165] at low energies. Although no identification of the neutral product or products of the collision was made in this experiment, it is presumed that there arises a strong contribution from the associative detachment process[166–168].

$$\mathrm{H^- + H \rightarrow H_2^-(^2\Sigma_u) \rightarrow H_2 + e} \qquad \overline{10}/(00)e$$

This process is the reverse of dissociative attachment, and a number of calculations[169, 170], flowing afterglow measurements[171] and drift-tube studies[172, 173] have been made in recent years. Since it is usually an exothermic process, it can take place at thermal energies, unlike collisional detachment. Owing to inward spiralling orbits, the thermal energy cross-sections can be very large, that for H⁻ in H being[174] $\sim 10^{-14}$ cm² at 400°K; the calculated cross-sections agree[164, 169].

Negative ion charge transfer processes are discussed in Chapter 12; in the passage of negative ions through gases, there are clustering and also interchange processes[175] considered in Chapters 10 and 14.

REFERENCES

1. MASSEY, H. S. W. *Negative Ions*, 2nd edn, 1950. London; Cambridge University Press.
2. BRANSCOMB, L. M. *Advanc. Electronics Electron Phys.* 9 (1957) 43; *Atomic and Molecular Processes*, Ed. D. R. Bates, Chap. 4, 1962. New York; Academic Press.
3. VARDYA, M. S. Joint Institute of Laboratory Astrophysics Report, 1966.
4. WEISS, A. W. quoted by L. M. Branscomb in *Atomic and Molecular Processes*, Ed. D. R. Bates, 1962. New York; Academic Press.
5. BYDIN, YU. F. *J. exp. theor. Phys.* 50 (1966) 35.
6. MOISEIWITSCH, B. L. *Advanc. atomic molec. Phys.* 1 (1965) 61.
7. BRANSCOMB, L. M. and SEMAN, M. United States National Bureau of Standards Report 7080, 1961; SEMAN, M., FINEMAN, M. A. and BRANSCOMB, L. M. *Bull. Amer. phys. Soc.* 6 (1961) 29.

8. BRANSCOMB, L. M. BURCH, D. S., SMITH., S. J. and GELTMAN, S. *Phys. Rev.* 111 (1958) 504; BURCH, D. S., SMITH, S. J. and BRANSCOMB, L. M. *Phys. Rev.* 112 (1958) 171; 114 (1959) 1652; SMITH, S. J. and BURCH, D. S. *Phys. Rev. Lett.* 2 (1959) 165; 116 (1959) 1125; HALL, J. H., ROBINSON, E. J. and BRANSCOMB, L. M. *Phys. Rev. Lett.* 14 (1965) 1013.

9. BRANSCOMB, L. M. and SMITH, S. J. *J. chem. Phys.* 25 (1956) 598; *Phys. Rev.* 98 (1955) 1028; SMITH, S. J. and BRANSCOMB, L. M. *J. Res. nat. Bur. Stand.* 55 (1955) 165; *Rev. sci. Instrum.* 31 (1960) 733.

10. GLOMBAS, P. and LADANYI, K. *Z. Phys.* 158 (1960) 261.

11. EDLEN, B. *J. chem. Phys.* 33 (1960) 98; JOHNSON, H. R. and ROHRLICH, F. *J. chem. Phys.* 30 (1959) 1608.

12. BERRY, H. S. *J. chem. Phys.* 42 (1965) 1541.

13. BRANSCOMB, L. M. *Phys. Rev.* 148 (1966) 11.

14. HONIG, R. E. *J. chem. Phys.* 22 (1954) 126.

15. NAPPER, R. and PAGE, F. M. *Trans. Faraday Soc.* 59 (1963) 1086.

16. PACK, J. L. and PHELPS, A. V. *J. chem. Phys.* 44 (1966) 1780.

17. BRANSCOMB, L. M. *Proc. 5th Int. Conf. Ioniz. Phenom. Gases,* 1962. Amsterdam; North Holland Publishing Company.

18. CURRAN, R. K. *Phys. Rev.* 125 (1962) 910.

19. FARRAGHER, A. L., PAGE, F. M. and WHEELER, R. C. *Disc. Faraday Soc.* 37 (1964) 203.

20. CURRAN, R. K. *J. chem. Phys.* 35 (1961) 1849.

21. WOOD, R. H. and D'ORAZIO, L. A. *J. phys. Chem.* 69 (1965) 2562.

22. GAINES, A. F., KAY, J. and PAGE, F. M. *Trans. Faraday Soc.* 62 (1966) 874.

23. PAGE, F. M. *Trans. Faraday Soc.* 56 (1960) 1742; 57 (1961) 359, 1254; ANSDELL, D. A. and PAGE, F. M. *Trans. Faraday Soc.* 58 (1962) 1084.

24. PAULSON, J. L. *Advanc. chem. Soc.* 58 (1966) 28.

25. HIBY, J. W. *Ann. Phys., New York* 34 (1939) 473.

26. DUKELSKII, V. M., AFROSIMOV, V. V. and FEDORENKO. N. V. *J. exp. theor. Phys.* 30 (1956) 792.

27. HOLØIEN, E. and MITDAL, J. *Proc. phys. Soc., Lond.* 68 (1955) 815.

28. WINDHAM, P. M., JOSEPH, P. J. and WEINMAN, J. A. *Phys. Rev.* 109 (1958) 1193.

29. SWEETMAN, D. R. *Proc. phys. Soc., Lond.* 76 (1960) 998.

30. SWEETMAN, D. R. *Proc. phys. Soc., Lond.* 76 (1960) 998.

31. CLEMENTI, E. and McLEAN, A. D. *Phys. Rev.* 133 (1964) A419.

32. FOGEL, YA. M., KOZLOV, V. F. and KALMYKOV, A. A. *J. exp. theor. Phys.* 36 (1959) 963; *Soviet Phys. JETP* 9 (1959) 963.

33. LAYZER, D. and BAHCALL, *Ann. Phys., New York* 17 (1962) 177.

34. LAYZER, D. *Ann. Phys., New York* 8 (1959) 271.

35. STANTON, H. E. *J. chem. Phys.* 32 (1960) 1348.

36. WALSH, A. D. *J. chem. Soc.* 3 (1953) 2260.

37. FERGUSON, E. E., FEHSENFELD, F. C. and SCHMELTEKOPF, A. L. *J. chem. Phys.* 47 (1967) 3085.

38. VIER, D. T. and MAYER, J. E. *J. chem. Phys.* 12 (1944) 28; SUTTON, P. P. and MAYER, J. E. *J. chem. Phys.* 3 (1935) 20; MAYER, J. E. *Z. Phys.* 61 (1930) 798; MAYER, J. E. and HELMHOLZ, L. *Z. Phys.* 75 (1932) 19; *J. chem. Phys.* 2 (1934) 245; DOTY, P. M. and MAYER, J. E. *J. chem. Phys.* 12 (1944) 323.

39. GLOCKLER, G. and CALVIN, M. *J. chem. Phys.* 3 (1935) 771; 4 (1936) 492.

40. METLAY, M. and KIMBALL, G. E. *J. chem. Phys.* 16 (1948) 774.

41. BRANSCOMB, L. M. *Advanc. Electronics Electron Phys.* 9 (1957) 43.

42. RANDOLPH, P. L. and GEBALLE, R. University of Washington, Department of Physics Technical Report 6, 1953.

43. CRAGGS, J. D., THORBURN, R. and TOZER, B. A. *Proc. Roy. Soc.* A240 (1957) 473; CRAGGS, J. D. and TOZER, B. A. *Proc. Roy. Soc.* A247 (1958) 337; A254 (1960) 229; TOZER, B. A. *J. Electron. Control* 4 (1958) 149.

44. BLACKMAN, V. *J. Fluid Mech.* 1 (1956) 61.

45. HAGSTRUM, H. D. *Phys. Rev.* 71 (1947) 376; HAGSTRUM, H. D. and TATE, J. T. *Phys. Rev.* 59 (1941) 354; HAGSTRUM, H. D. *Rev. mod. Phys.* 23 (1951) 185.
46. FROST, D. C. and McDOWELL, C. A. *Proc. Roy. Soc.* A237 (1955) 227.
47. SCHULZ, G. J. *Bull. Amer. phys. Soc.* 11, 7 (1962) 487; *Phys. Rev.* 128 (1962) 174, 178.
48. CHANTRY, P. J. and SCHULZ, G. J. *Phys. Rev. Lett.* 12 (1964) 449; Westinghouse Report 66-9E3-113-P3, 1966.
49. PRITCHARD, H. O. *Chem. Rev.* 52 (1953) 529.
50. BYDIN, YU. F. *J. exp. theor. Phys.* 46 (1964) 1612; 49 (1965) 1094.
51. BYDIN, YU. F. *J. exp. theor. Phys.* 50 (1966) 35.
52. SMIRNOV, B. M. and FIRSOV, O. B. *J. exp. theor. Phys.* 47 (1964) 232.
53. BUCHELNIKOVA, N. S. *Advanc. phys. Sci., Moscow* 65 (1958) 351.
54. LOEB, L. B. and CRAVATH, A. M. *Phys. Rev.* 33 (1929) 605.
55. BRADBURY, N. E. *Phys. Rev.* 44 (1933) 883; *J. chem. Phys.* 2 (1934) 827.
56. BAILEY, V. A. *Phil. Mag.* 50 (1925) 825.
57. HEALEY, R. H. and REED, J. W. *The Behaviour of Slow Electrons in Gases*, 1961. Sydney; Amalgamated Wireless Company of Australia.
58. HUXLEY, L. G. H., CROMPTON, R. W. and BAGOT, C. H. *Aust. J. Phys.* 12 (1959) 303; HUXLEY, L. G. H. *Aust. J. Phys.* 12 (1959) 171.
59. HURST, C. A. and HUXLEY, L. G. H. *Aust. J. Phys.* 13 (1960) 21.
60. HUXLEY, L. G. H. and ZAAZOU, A. A. *Proc. Roy. Soc.* A196 (1949) 402.
61. DOEHRING, A. *Z. Naturf.* 7a (1950) 253.
62. CHANIN, L. M., PHELPS, A. V. and BIONDI, M. A. *Phys. Rev. Lett.* 2 (1959) 344.
63. HERRENG, P. *Cah. Phys.* 38 (1952) 7.
64. BORTNER, T. E. and HURST, G. S. *Health Phys.* 1 (1958) 39.
65. GEBALLE, R. and HARRISON, M. A. *Phys. Rev.* 85 (1952) 372; HARRISON, M. A. and GEBALLE, R. *Phys. Rev.* 91 (1953) 1.
66. CRAGGS, J. D. and MASSEY, H. S. W. *Handb. Physik* 37 (1959) 314.
67. CHANTRY, P. J. Westinghouse Report 69-9E4-113-P1, 1968.
68. LOZIER, P. *Phys. Rev.* 46 (1933) 268.
69. BUCHELNIKOVA, N. S. *J. exp. theor. Phys.* 35 (1958) 1119.
70. BIONDI, M. A. and BROWN, S. C. *Phys. Rev.* 76 (1949) 1697.
71. BIONDI, M. A. *Phys. Rev.* 109 (1958) 2005.
72. BIONDI, M. A. *Phys. Rev.* 84 (1951) 1072.
73. SEXTON, M. C., MULCAHY, M. J. and LENNON, J. L. *Proc. 4th Int. Conf. Ioniz. Phenom. Gases*, 1959. Amsterdam; North Holland Publishing Company.
74. CHANTRY, P. J., WHARMBY, J. S. and HASTED, J. B. *Proc. 5th Int. Conf. Ioniz. Phenom. Gases*, 1961. Amsterdam; North Holland Publishing Company.
75. HOLT, E. H. *Bull. Amer. phys. Soc.* 4 (1959) 112.
76. DUTTON, J. Joint Institute of Laboratory Astrophysics, Information Centre Report 4, 1967.
77. WEBER, O. *Z. Phys.* 152 (1958) 281.
78. LOCHTE-HOLTGREVEN, W. *Naturwissenschaften* 38 (1951) 258.
79. BERRY, R. S., REIMANN, C. W. and SPOKAS, G. N. *Bull. Amer. phys. Soc.* 7 (1962) 69.
80. BOLDT, G. *Z. Phys.* 154 (1959) 319, 330.
81. BATES, D. R. and MOISEIWITSCH, B. L. *Proc. phys. Soc., Lond.* A68 (1955) 540.
82. POPP, H-P, *Proc. 8th Int. Conf. Ioniz. Phenom. Gases*, p. 448, 1967. Amsterdam; North Holland Publishing Company.
83. CHEN, J. C. Y. and PEACHER, J. L. *Phys. Rev.* 163 (1967) 103.
84. BARDSLEY, J. N., HERZENBERG, A. and MANDL, F. *Proc. phys. Soc., Lond.* 89 (1966) 305, 321.
85. DEMKOV, YU. N. *Phys. Lett.* 15 (1965) 235.
86. RAPP, D., SHARP, T. E. and BRIGLIA, D. D. Lockheed Missile and Space Company Report 6-76-64-45, 1964.
87. COMPTON, R. N. and CHRISTOPHOROU, L. G. *Phys. Rev.* 154 (1967) 110.
88. TAYLOR, H. S. and HARRIS, F. E. *J. chem. Phys.* 39 (1963) 1012.

89. DEMKOV, YU. N. *Soviet Phys. JETP* (1964) 762.
90. IONESCU, TH. V. *Rev. Physique* 1 (1956) 38.
91. SCHULZ, G. J. and ASUNDI, R. K. Westinghouse Report 65-9E3-113-P3, 1965.
92. RAPP, D. and SHARP, T. E. Lockheed Missile and Space Company Report 6-74-64-45, 1964.
93. SHARP, T. E. and DOWELL, J. T. *J. chem. Phys.* 46 (1967) 1530; *Phys. Rev.* 167 (1968) 124.
94. O'MALLEY, T. F. *Phys. Rev.* 150 (1966) 14; *J. chem. Phys.* 47 (1967) 5457.
95. DUBROVSKY, G. V., OBEDKOV, V. D. and JANEV, R. K. *Abstr. 5th Int. Conf. Phys. electron. atom. Collisions*, p. 342. 1967. Leningrad; Akademii Nauk.
96. KHVOSTENKO, V. E. and DUKELSKII, V. M. *Soviet Phys. JETP.* 6 (1958) 657.
97. FITE, W. L. and BRACKMANN, R. T. *Proc. 6th Int. Conf. Ioniz. Phenom. Gases*, 1963. Paris; SERMA.
98. CHANTRY, P. J. *Phys. Rev.* 172 (1968) 125.
99. ASUNDI, R. K., CRAGGS, J. D. and KUREPA, M. V. *Proc. phys. Soc., Lond.* 82, (1963) 967.
100. RAPP, D. and BRIGLIA, D. D. *J. chem. Phys.* 43 (1965) 1480.
101. SCHULZ, G. J. and ASUNDI, R. K. *Phys. Rev.* 158 (1967) 25.
102. SCHULZ, G. J. *Phys. Rev.* 128 (1962) 174.
103. HICKAM, W. M. and FOX, R. E. *J. chem. Phys.* 25 (1956) 642; SCHULZ, G. J. *J. appl. Phys.* 31 (1960) 1134.
104. REED, R. I. *Ion Production by Electron Impact*, 1962. New York; Academic Press.
105. RAPP, D. and BRIGLIA, D. D. Lockheed Missile and Space Company Report 6-74-64-40, 1964.
106. FROST, L. S. and McDOWELL, C. A. *J. chem. Phys.* 29 (1958) 503, 964; *J. Amer. chem. Soc.* 80 (1958) 6183; *Canad. J. Chem.* 36 (1958) 39.
107. HESSENAUER, H. *Z. Phys.* 204 (1967) 142.
108. BLOCH, F. and BRADBURY, N. E. *Phys. Rev.* 48 (1935) 689.
109. CHATTERTON, P. Ph.D. Thesis, University of Liverpool, 1961.
110. CRAGGS, J. D. *Proc. 3rd Int. Conf. Ioniz. Phenom. Gases*, 1957. Milan; Italian Physical Society.
111. MEGILL, L. R. and HASTED, J. B. *Planet. Space Sci.* 13 (1965) 339.
112. EISNER, P. N., HIRSH, M. N. and SLEVIN, J. A. *Bull. Amer. phys. Soc.* 10 (1965) 189.
113. VAN LINT, V. A. J., WILNER, E. G. and TRUEBLOOD, D. L. *Bull. Amer. phys. Soc.* 5 (1960) 125.
114. WELLER, C. S. and BIONDI, M. A. University of Pittsburgh, SRCC Report 72, 1968.
115. BHALLA, M. A. and CRAGGS, J. D. *Proc. phys. Soc., Lond.* 80 (1962) 151.
116. McAFEE, K. B. *J. chem. Phys.* 23 (1955) 1435.
117. BHALLA, M. S. and CRAGGS, J. D. *Proc. phys. Soc., Lond.* 76 (1960) 369.
118. BHALLA, M. S. and CRAGGS, J. D. *Proc. phys. Soc., Lond.* 78 (1961) 438.
119. CHANTRY, P. J. *Proc. 3rd Int. Conf. Phys. electron. atom. Collisions*, p. 565, 1963. Amsterdam; North Holland Publishing Company.
120. PRASAD, A. N. and CRAGGS, J. D. *Proc. phys. Soc., Lond.* 76 (1960) 223; KUFFEL, E. *Proc. phys. Soc., Lond.* 74 (1959) 297.
121. HEALEY, R. H. *Phil. Mag.* 26 (1938) 940.
122. FOX, R. E. *Phys. Rev.* 109 (1958) 2008; BIONDI, M. A. and FOX, R. E. *Phys. Rev.* 109 (1958) 2012.
123. BUCHDAHL, R. *J. chem. Phys.* 9 (1941) 146.
124. REES, J. A. Australian National University, Report 64-3, 1964.
125. BOUBY, L. and ABGRALL, H. *Abstr. 5th Int. Conf. Phys. electron. atom. Collisions*, p. 584, 1967. Leningrad; Akademii Nauk.
126. FIQUET-FAYARD, F. Private communication.
127. PACK, J. L. and PHELPS, A. V. Westinghouse Reports 66-6E2-GASES-P1, 1962 and 65-6E2 113 P1, 1965.
128. BOLDT, G. *Z. Phys.* 154 (1959) 319, 330.

129. BERRY, R. S., DAVID, C. W. and MACKIE, J. C. *J. chem. Phys.* 42 (1965) 1541; BERRY, R. S., MACKIE, J. C., TAYLOR, R. L. and LYNCH, R. *J. chem. Phys.* 43 (1965) 3067.
130. MANN, A. E. quoted by S. J. Smith, *Methods in Experimental Physics, Vol. 7*, ed. B. Bederson and W. L. Fite, 1968. New York; Academic Press.
131. GELTMAN, S. and KRAUSS, M. *Bull. Amer. phys. Soc.* (2) 5 (1960) 339.
132. KLEIN, M. M. and BRUECKNER, K. A. *Phys. Rev.* 111 (1958) 1115.
133. COOPER, J. W. and MARTIN, J. B. *Phys. Rev.* 123 (1962) 1402.
134. GARRETT, W. R. and JACKSON, H. T. *Gaseous Electronics Conf.* American Physical Society, 1965.
135. HALL, J. L., ROBINSON, E. J. and BRANSCOMB, L. M. *Phys. Rev. Lett.* 14 (1965) 1013; HALL, J. L. *J. quantve Electronics* 2 (1966) 361.
136. CHANDRASHEKHAR, S. *Astrophys. J.* 102 (1945) 223, 395; 128 (1958) 114.
137. BATES, D. R. and MASSEY, H. S. W. *Astrophys. J.* 91 (1940) 202; *Phil. Trans.* A239 (1943) 269; DOUGHTY, N. A. and FRAZER, P. A. *Proc. 3rd Int. Conf. Phys. electron. atom. Collisions*, p. 527, 1963. Amsterdam; North Holland Publishing Company.
138. JOHN, T. L. *Astrophys. J.* 131 (1960) 743; *Mon. Not. R. astr. Soc.* 121 (1960) 41.
139. CHANDRASHEKHAR, S. and KROGDAHL, M. K. *Astrophys. J.* 98 (1943) 205.
140. DALGARNO, A. and KINGSTON, A. E. *Proc. phys. Soc., Lond.* 73 (1959) 455.
141. WIGNER, E. P. *Phys. Rev.* 73 (1948) 1002.
142. GELTMAN, S. *Phys. Rev.* 112 (1958) 176.
143. HASTED, J. B. *Proc. Roy. Soc.* A212 (1952) 235; A222 (1952) 74.
144. SMITH, S. J. *Proc. 4th Int. Conf. Ioniz. Phenom. Gases*, 1960. Amsterdam; North Holland Publishing Company.
145. PHELPS, A. V. and PACK, J. L. *Phys. Rev. Lett.* 6 (1961) 111.
146. MULLIKEN, R. S. *Phys. Rev.* 115 (1959) 1225.
147. ROBINSON, E. J. and GELTMAN, S. *Phys. Rev.* 153 (1967) 4.
148. HALL, J. L. and SIEGEL, M. W. *J. chem. Phys.* 48 (1968) 943.
149. McDOWELL, M. R. C. and WILLIAMSON, J. H. *Phys. Lett.* 4 (1963) 159.
150. TISONE, G. C. and BRANSCOMB, L. M. *Phys. Rev. Lett.* 17 (1966) 236.
151. DANCE, D. F., HARRISON, M. F. A. and RUNDEL, R. D. *Proc. Roy. Soc.* 229A (1967) 525.
152. BATES, D. R. and MASSEY, H. S. W. *Phil. Mag.* (7) 45 (1954) 173.
153. DUKELSKII, V. M. and ZANDBERG, Y. *Dokl. Akad. Nauk. SSSR* 99 (1954) 947; 82 (1952) 33; *J. exp. theor. Phys.* 21 (1951) 1270.
154. BYDIN, Y. F. and DUKELSKII, V. M. *J. exp. theor. Phys.* 31 (1957) 474.
155. STEDEFORD, J. B. H. and HASTED, J. B. *Proc. Roy. Soc.* A227 (1955) 466; HASTED, J. B. and SMITH, R. A. *Proc. Roy. Soc.* A235 (1956) 349.
156. BAILEY, T. L., MAY, C. J. and MUSCHLITZ, E. E. *J. chem. Phys.* 26 (1957) 1446; MUSCHLITZ, E. E. *Proc. 4th Int. Conf. Ioniz. Phenom. Gases*, 1960. Amsterdam; North Holland Publishing Company.
157. STIER, P. M. and BARNETT, C. F. *Phys. Rev.* 103 (1956) 896.
158. SIDA, D. W. University of Otago, New Zealand, Private Communication, 1954.
159. DOEHRING, J. P. *J. chem. Phys.* 41 (1964) 1164.
160. FROMMHOLD, L. *Fortschr. Phys.* 12 (1964) 597.
161. SMIRNOV, B. M. and FIRSOV, O. B. *Soviet phys. JETP* 20 (1965) 156.
162. DEMKOV, YU. N. *Soviet phys. JETP* 19 (1964) 762.
163. BYDIN, YU. F. *J. exp. theor. Phys.* 50 (1966) 35.
164. HUMMER, D., FITE, W. L. and STEBBINGS, R. *Phys. Rev.* 119 (1960) 668.
165. McDOWELL, M. R. C. and PEACH, G. *Proc. phys. Soc., Lond.* 74 (1959) 463.
166. McDOWELL, M. R. C. *Observatory* 81 (1961) 240.
167. BATES, D. R. *Atomic and Molecular Processes*, 1962. New York; Academic Press.
168. DALGARNO, A. *Ann. Geophys.* 17 (1961) 16.
169. BARDSLEY, J. N. *Abstr. 5th Int. Conf. Phys. electron. atom. Collisions*, p. 340, 1967. Leningrad, Akademii Nauk.
170. CHEN, J. C. Y. *Phys. Rev.* 156 (1967) 12.

171. FEHSENFELD, F. C., FERGUSON, E. E. and SCHMELTEKOPF, A. L. *J. chem. Phys.* 45 (1966) 1844.
172. MORUZZI, J. L. and PHELPS, A. V. *J. chem. Phys.* 45 (1966) 4617; Westinghouse Report 66-6E2-GASES-P3, 1966.
173. MORUZZI, J. L., EKIN, J. W. and PHELPS, A. V. Westinghouse Report 67-1E2-GASES-P4, 1967.
174. HERZENBERG, A. Westinghouse Report 66/1EO/CNSUL/P1, 1966.
175. FROMMHOLD, L. *Fortschr. Phys.* 12 (1964) 597.

Chapter 9

PHOTON EMISSION AND ABSORPTION

9.1 THE EMISSION AND ABSORPTION OF RADIATION

The interaction of an electromagnetic wave with an atomic system is to a first approximation the interaction with the electric dipole moment. For detailed discussion the reader is referred to textbooks of atomic and molecular spectra[1-4]. Although the permanent electric dipole moment of an atom is zero in the absence of an electromagnetic field, the interaction energy introduced into the wave equation produces non-zero matrix elements R_{nm} of a (temporary) 'transition moment'. These matrix elements are calculated from the wave functions

$$R_{nm} = \left\langle n \left| \sum_{i=1}^{N} r_i \right| m \right\rangle = \int \psi_n^* \sum_{i=1}^{N} r_i \psi_m \, d\tau \qquad (9.1)$$

summing over all the atomic electrons $i = 1, 2, \ldots, N$. They can be expressed in units of cm $eV^{-1/2}$. If the matrix element differs from zero for two states n and m, the two states combine with each other with a certain probability; this is accompanied by the emission or absorption of radiation. If the matrix element is zero, the transition is forbidden as a dipole transition, but may still take place if the matrix elements of the magnetic dipole or electric quadrupole moment are non-zero. However, magnetic dipole transition probabilities are $\sim 10^5$ times less likely, and electric quadrupole transitions $\sim 10^8$ times less likely, than electric dipole transitions. The permissibility of an electric dipole transition is governed by the selection rules outlined in the Appendix; the present chapter is concerned almost entirely with allowed electric dipole transitions involving a single quantum of radiation.

The energy intensity $\mathcal{I}_\nu$, erg sec^{-1} of a spectral line spontaneously emitted in all directions from unit volume of gas at a frequency $\nu_{nm} = E_{nm}/h$ corresponding to the downward transition from level n to level m is given by

$$\mathcal{I}_\nu = n_n h \nu_{nm} A_{nm} \qquad (9.2)$$

where the emitted quantum energy $E_{nm} = E_n - E_m$, and n_n and n_m are the densities of atoms in states n and m. Note that ν is here expressed as a frequency and not a wave number; A_{nm} is the Einstein A coefficient, or transition probability of spontaneous transition by one atom in one second. The A coefficient is related to the transition matrix element:

$$A_{nm} = \frac{64\pi^4 \nu_{nm}^3}{3h} |R_{nm}|^2 \qquad (9.3)$$

A typical value of A_{nm} for an electric dipole transition is 10^8 sec^{-1}. A number N_{0n} of atoms in state n decays according to the equation

$$N_n(t) = N_{0n} \exp\left(-\frac{t}{\tau}\right) \qquad (9.4)$$

with

$$\tau = \sum_m \tau_{nm} = \sum_m A_{nm}^{-1}$$

The width of the spectral line is denoted by Γ, and the lifetime

$$\tau = \frac{h}{\Gamma} \qquad (9.5)$$

Thus broad lines correspond to short lifetimes.

A typical lifetime τ_{nm} for an excited state decaying by a single quantum electric dipole transition is $\sim 10^{-8}$ sec, but metastable states have lifetimes $\sim 10^{-3}$ sec (magnetic dipole) and ~ 1 sec (quadrupole). Lifetimes are shortened by the application of electric fields[5].

It often happens that there are g_m states i and g_n states j belonging to degenerate levels m and n, and in this case

$$A_{nm} = \sum_{ij} \frac{A_{ji}}{g_n} \qquad (9.6)$$

and g_m and g_n are known as the statistical weights.

For a parallel beam of light of constant intensity between frequencies ν and $\nu+d\nu$, the absorption in a path dx through gas containing densities of excited atoms n_n and n_m is

$$d(\mathcal{I}_\nu \, d\nu) = n_m \, dx \, h\nu_{nm} B_{mn} \frac{\mathcal{I}_\nu}{4\pi} - n_n \, dx \, h\nu_{nm} B_{nm} \frac{\mathcal{I}_\nu}{4\pi} \qquad (9.7)$$

or

$$d(\mathcal{I}_\nu \, d\nu) = \frac{\mathcal{I}_\nu}{4\pi} h\nu_{nm} \, dx \, (n_m B_{mn} - n_n B_{nm}) \qquad (9.8)$$

where $\mathcal{I}_\nu/4\pi$ is the intensity of the equivalent isotropic radiation for which the Einstein B coefficients are defined. The product $B_{mn}\mathcal{I}_\nu$ is the probability of an upward transition $m \rightarrow n$ per second per atom exposed to radiation of intensity $\mathcal{I}_\nu$. This is the transition probability of absorption. The product $B_{nm}\mathcal{I}_\nu$ is the probability of a stimulated downward transition $n \rightarrow m$ per second per atom exposed to radiation of intensity $\mathcal{I}_\nu$. The transition probability of stimulated emission is of importance in the understanding of lasers.

The original Einstein B coefficients were defined in terms of radiation density rather than intensity, and were therefore $c/4\pi$ times the modern B coefficients. The latter are related to the transition matrix elements:

$$B_{mn} = \frac{8\pi^3}{3h^2c} \, | \, R_{nm} \, |^2 \tag{9.9}$$

so

$$B_{mn} = \frac{A_{nm}}{8\pi hcv_{nm}^3} \tag{9.10}$$

The two B coefficients are in the ratio of the statistical weights:

$$\frac{B_{nm}}{B_{mn}} = \frac{g_m}{g_n} \tag{9.11}$$

From equation 9.7 it can be seen that the absorption coefficient μ_v, integrated over the whole line, at whose centre the frequency is v_0 and wavelength λ_0, is:

$$\int \mu_v \, dv = \frac{hv_0}{4\pi} (B_{mn}n_m - B_{nm}n_n) = \frac{\lambda_0 g_n}{8\pi g_m} \frac{n_m}{\tau} \left(1 - \frac{g_m n_n}{g_n n_m} \right) \tag{9.12}$$

or

$$\int \mu_v \, dv \simeq \frac{\lambda_0^2 g_n n_m}{8\pi g_m \tau} \tag{9.13}$$

where nearly all the atoms are in the state m (as in experiments where ground state atoms are irradiated).

Absorption coefficients μ_v are usually defined relative to gas at s.t.p. rather than at pressure 1 torr (see Chapter 1).

At this stage, there will be no discussion of the relation between this ideal absorption coefficient and the absorption coefficients that are appropriate to broadened lines. Lines are broadened by the finite lifetime of the excited state, by Doppler effects brought about by the motions of the atoms, by gas collisions, and by Stark effect arising from collisions with electrons and with ions.

In the classical electron theory of dispersion, the optical behaviour of n_0 atoms cm^{-3} was represented by the behaviour of n_d classically bound electrons, which were called 'dispersion electrons'. The ratio n_d/n_0 was found to be constant for a particular spectral line, and was denoted by the dimensionless symbol f_{nm}. The classical oscillator strength f_{nm} of a line is a measure of the degree to which the ability of the atom to emit this line resembles the ability of an oscillating electron to do so classically. It is defined by the equation

$$\int \mu_v \, dv = \frac{\pi e^2 n_d}{m_e c} = \frac{\pi e^2 n_0 f_{nm}}{m_e c} \tag{9.14}$$

so the oscillator strength is inversely proportional to the lifetime:

$$f_{nm}\tau = 1 \cdot 4993 \times 10^{-16} \frac{g_n}{g_m} \lambda_0^2 \tag{9.15}$$

3^2

It is related to the transition matrix element by the equation

$$f_{nm} = \tfrac{1}{3}(E_n - E_m)\,|R_{nm}|^2 \tag{9.16}$$

The refractive index n and electromagnetic theory absorption coefficient $\varkappa$ are related to μ_ν by the equation

$$\frac{4\pi n \varkappa}{\lambda_0} = \mu_\nu \tag{9.17}$$

Strictly speaking, the departure of refractive index from unity in the neighbourhood of a line should be taken into account in the calculation of oscillator strengths.

Some authors prefer to use 'line strengths' S_{ji}, which are related to the transition dipole moment by the equation

$$S_{ji} = |i| - \sum_s e r'_s\,|j|^2 \tag{9.18}$$

where the position of the electron s is specified by the vector r'_s. Thus

$$S_{nm} = \sum_{ij} S_{ji} \tag{9.19}$$

and

$$A_{ji} = \frac{4E_{nm}^3}{3\hbar^4 c^3}\,S_{ji} \tag{9.20}$$

The line strength is an energy maintained over a volume, and has the dimensions ML^5T^{-2}. Oscillator strengths are related to line strengths as follows:

$$f_{nm} = \frac{m_e c^3 \hbar^2 g_n}{2e^2 E_{mn}^2 g_m}\,A_{nm} = \frac{303 \cdot 76 S_{nm}}{g_m \lambda_0} \tag{9.21}$$

Furthermore,

$$A_{nm} = \frac{2 \cdot 026_0 \times 10^{18} S_{nm}}{g_n \lambda_0^3} \tag{9.22}$$

with λ_0 in Ångstrom units in both equations.

Line strengths for electric dipole transitions are often expressed in atomic units of

$$a_0 e^2 = 6 \cdot 459_6 \times 10^{-36}\ \text{cm}^2\ \text{(e.s.u.)}^2 \tag{9.23}$$

For the absorption of radiation accompanied by a transition (resonance absorption), the Doppler broadening due to thermal motion of the gas atoms must be taken into account. Under optically thin conditions, the maximum absorption cross-section $\sigma_{\max}$ is given by

$$\sigma_{\max} = \left(\frac{m}{2\pi kT}\right)^{1/2} \frac{\pi e^2 \lambda_0 f}{m_e c}\ \text{cm}^2 \tag{9.24}$$

Furthermore

$$\int_0^\infty \sigma(\lambda)\,\mathrm{d}\lambda = \frac{\pi e^2 \lambda_0^2 f}{m_e c^2} = 8 \cdot 85 \times 10^{-13} \lambda_0^2 f \qquad (9.25)$$

with λ_0 in Ångstrom units.

Optical resonance absorption cross-sections are several orders of magnitude larger than geometrical cross-sections of atoms.

This chapter is also concerned with transitions into continua. For an atom, the photoionization cross-section $_{\phi 0}\sigma_{1e}$ can be related to the oscillator strength for a transition into the first ionization continuum $f(\varepsilon)$. This quantity is a function of the kinetic energy ε of the ejected electron:

$$_{\phi 0}\sigma_{1e} = \frac{\pi e^2 h}{m_e c} \frac{\mathrm{d}f(\varepsilon)}{\mathrm{d}\varepsilon} \qquad (9.26)$$

To obtain the differential oscillator strength per electron, the cross-section in megabarns (10^{-18} cm^2) is divided by 111; and to obtain the photoionization cross-section, the differential oscillator strength per electron volt is multiplied by 10 975. One can write

$$\frac{\mathrm{d}f}{\mathrm{d}\varepsilon} = K\sigma(\varepsilon) \qquad (9.27)$$

with $K = 0 \cdot 12387 \times 10^{-19}$ cm^{-2} Rydberg^{-1}.

Resonance radiation, emitted in the transition from an excited state to the ground state, will not emerge from a mass of gas unattenuated by absorption followed by re-radiation. The progress of radiation through a gas by this means is known as radiative transfer; it was first discussed by Milne[6], but more recent developments[7, 8] should be studied.

In the theory of radiative transfer, it was at first assumed that the frequencies of the absorbed and re-emitted photons were identical, that is, that the scattering was 'coherent' in frequency. This assumption leads to predicted decay constants for radiation at the centre of a resonance line which are too small by two orders of magnitude. The discrepancy disappears when only incoherent scattering is considered[9]. The transport equation can be solved by variational techniques or by numerical methods[10]. Decay constants more recently measured[8] show good agreement with theory, with negligible contribution from diffusion of the excited atoms. Systems which are not in local thermodynamic equilibrium have been investigated[11].

In the foregoing discussion of emission and absorption, the absence of radiative transfer (optically thin conditions) has been assumed. Since these conditions are not always easy to fulfil in absorption spectroscopy, use is sometimes made of the back-scattering technique; in this case, re-radiation is observed in a backward direction, coming from the front layer of atoms; it is even possible that specular reflexion could be achieved by such a technique. This method is currently employed for the study of Zeeman level-crossing in hyperfine structure.

32*

9.2 DETERMINATION OF OSCILLATOR STRENGTHS

For many years, spectroscopists have accumulated values of oscillator strengths[12] (transition probabilities, line strengths) using a variety of experimental techniques, which are outlined in this section. Parallel with this activity, quantum theory calculations have been made for many transitions. This type of work must continue for many years until data are available for all important transitions. A knowledge of transition probabilities is of great importance in astrophysics, in space technology and in plasma diagnostics. Compilations of data[13-19] are available.

The experimental techniques, which have been reviewed[20, 21], fall into the following classes:

1. Absorption techniques,
2. Emission techniques,
3. Natural linewidth and similar techniques,
4. Anomalous dispersion,
5. Lifetime techniques,
6. Electron spectroscopy.

Absorption Techniques

Absorption techniques[22] make use of equation 9.14 to deduce absolute values of transition probabilities for levels combining with the ground state (or a state sufficiently close to it to be present in the gas under thermal equilibrium). One measures the absorption of radiation by a column of gas at known temperature at the line centre, or edges, or over the entire line; one must also know the atomic number density and path length of radiation through the gas.

Furnaces have been used to vaporize elements in these measurements. Atomic beams effusing from ovens have also been used[23, 24], and transition probabilities equivalent to linewidths as narrow as $0 \cdot 001 \times 10^{-14}$ cm can be measured. Beam intensity measurements have been made by chemical titration[25].

Emission Techniques

In emission techniques, a thermal equilibrium of atoms in different states of excitation is established, and the relative intensities of the different emission lines are measured.

The equation

$$I_{ji} = A_{ji} n_j h l \tag{9.28}$$

represents the intensity emitted (energy cm^{-2} sec^{-1} steradian^{-1}) from an optically thin homogeneous plasma layer of thickness l. The populations of different states are calculated from the Boltzmann equation

$$n_j = n_a \left(\frac{g_j}{U(T)} \right) \exp \left(\frac{-E_j}{kT} \right) \tag{9.29}$$

where n_a is the total number density of species a, and g_j is the statistical weight of state j. The atomic partition function

$$U(T) = \sum_{j=0}^{j*} g_j \exp\left(\frac{-E_j}{kT}\right) \qquad (9.30)$$

where the summation is cut off at an energy E_{j*}, determined by collective effects is the plasma. The equations are applied to determine the relative transition probabilities of different transitions from the same level. High-temperature arcs[26, 27] are utilized to establish a high temperature, and magnetically driven and conventional shock tubes[28] have also been used.

Photoabsorption experiments on positive ions also require a high-temperature plasma, for which a shock tube can be used. But the electric fields in a hot plasma can depress the ionization potential, and can also depress one-electron excited states of the neutral atom, and cause transitions from two-electron excited states into the depressed continuum.

Other techniques of high-temperature realization include flash pyrolysis, or heating by the discharge of a capacitor bank through a quartz spiral surrounding a quartz tube containing the material (in finely divided form, if a solid).

Linewidth Techniques

It will be seen from equations 9.4 and 9.5 that, if the natural linewidth of a level can be measured, the oscillator strength follows immediately. Although pressure and Doppler broadening effectively mask the natural linewidth by a large factor, there are methods whereby accurate linewidth measurement can be made[29, 30]. Autoionization linewidths are sufficiently large to dominate under nearly all circumstances.

An elegant oscillator strength measurement technique is the double-resonance method of Brossel and Kastler[31, 32]. This technique measures the natural linewidth by eliminating the Doppler broadening in the following way. Double resonance consists of simultaneous optical and radiofrequency transitions, involving respectively the electric and magnetic dipoles. Suppose that mercury vapour is contained in a steady magnetic field and illuminated with resonance radiation, 2537 Å, which raises $^1S_0$ to $^3P_1$, $m = 0$. Transitions to $m = \pm 1$ are induced by application of a radiofrequency magnetic field at the Larmor frequency, perpendicular to the steady magnetic field. These transitions are detected by modulation, both in intensity and in polarization, of the spontaneously emitted optical radiation. The natural linewidth is monitored by noting the change in modulation when the magnetic field is varied. Scattering of polarized light[33] has also been employed. There is a problem of providing sufficiently high radiofrequency power; alternating magnetic fields as high as 90 G are sometimes necessary, and they can cause breakdown in gases; the following effect, that of 'level-crossing', offers an alternative.

Magnetic Zeeman sub-levels shift in frequency v linearly with variation of magnetic field. Since the values of $\partial v/\partial H$ are different, there exist values of H at which two levels coincide, or cross. At zero H, this is known as the

Hanle effect[34, 35] and can be used for lifetime derivation. At non-zero H, one can observe a dispersion-shaped signal from transitions between sub-levels, and the lifetime of the level can be related to the field separation ΔH between the two peaks of this signal:

$$\tau = \frac{1}{\sqrt{3}\pi} \Delta H \frac{\partial v}{\partial H} \qquad (9.31)$$

Also, for nuclear spin zero and two levels differing in m by 2,

$$\tau = \frac{1}{\sqrt{3}\pi} \left(\frac{2g_J\mu_0}{h} \right) \Delta H \qquad (9.32)$$

where μ_0 is the Bohr magneton and g_J the Landé g factor of the atomic state.

In the 'magnetic depolarization' technique[36], polarized resonance radiation preferentially excites magnetic sub-levels, thereby affecting the polarization of scattered radiation. The interference arising from level-crossing is studied either by intensity variation as the magnetic field is varied, or by means of the anisotropy of polarization of the scattered radiation.

Errors in lifetime determination by level-crossing can be caused by coherence narrowing[37].

Anomalous Dispersion

The variation of refractive index $n(\lambda)$ of a gas in the vicinity of an absorption line of wavelength λ is related to the oscillator strength[38] by the equation

$$n-1 \simeq \frac{e^2 n_i f_{ik}}{4\pi m_e c^2} \left(\frac{\lambda_0^3}{\lambda - \lambda_0} \right) \qquad (9.33)$$

The most elegant technique for measuring the variation is due to Rozhdestwensky[39] and has been successfully revived and reviewed[40-42]. In this 'hook' method, patterned interference fringes are observed along the spectrum in an illuminated gas examined with a Jamin or Mach–Zender interferometer crossed with a stigmatic spectrograph. In addition to atomic spectra, molecular spectra have now been investigated.

Magnetorotation of the plane of polarization of radiation in the neighbourhood of a line (Faraday effect) may also be related to oscillator strengths[43, 44].

Lifetime Techniques

A variety of experiments[45-50] is directed at actual lifetime measurement. Levels are excited by radiation modulated with pulses or sinusoidally, and the delay or phase difference in the spontaneously emitted radiation s measured.

Fast Kerr cell modulation of radiation is now readily available, and ultrasonic standing waves have also been used for modulation. Electron excitation by pulsed electrons was pioneered some years ago[51] and is still in use[52]. Delayed coincidence counts are made of triangular 4×10^{-8} sec pulses of 30 eV electrons, and of the photon pulses of suitably filtered radiation produced therefrom. It is seen from equation 9.4 that, for the introduction of a variable delay T, the logarithm of the coincidence count rate is linearly proportional to $-T/\tau_{nm}$; the values of T must be greater than the resolving time of the circuit. Measurements of the lifetimes of a number of transitions in helium have been made, and more recently[53] spectrographic frequency resolution has been added for observation on neon. A particularly promising innovation is the incorporation of momentum analysis of the scattered electron.

High-energy ion beam measurements of lifetime are now becoming popular. One possible method[54] consists of measuring the intensity of the radiation which is emitted along an ion beam as it passes through a gas. The capture process

$$H_2^+ + He \rightarrow H'$$

($n = 3, 4, 5$) produces a uniform distribution of each state throughout the gas, but the subsequent decay processes retard the build-up of radiation by a time from which the lifetime can be calculated. The intensity of emission $j \rightarrow i$ after a time t is

$$I_{ji} \propto \frac{1 - [1 - \exp(-A_j \, \Delta t)] \exp(-A_j t)}{A_i \, \Delta t}$$

where Δt is the time of particle flight through the gas.

The other beam technique[55, 56], beam foil spectroscopy, is more difficult technically but has achieved much more success; it is already capable of high precision and versatility. The ion beam is accelerated to 1–5 MeV energy, which enables it to pass through a very thin metal or carbon foil (500–1000 Å) producing proportions of excited particles, both neutral and charged. The spatial variation of radiation intensity down the beam is exponential, yielding the appropriate lifetimes; a sum of exponentials is appropriate when there is cascading. The characteristic decay lengths are of order 5 mm, and the upper limit of lifetime which can be measured is 10^{-8} sec.

Lifetimes of metastable states may also determined by beam experiments. Electrons are crossed with an atomic beam, and the downstream attenuation of the detected metastables is measured[57].

Electron Spectroscopy

Another method of determining oscillator strengths is by 'electron spectroscopy'. The intensity of high-energy electrons scattered through zero angle with energy loss appropriate to excitation is, according to the Bethe–Born approximation, proportional to the oscillator strength. The experiments are discussed in Chapter 5 and in the literature[58].

Calculations

Calculations of oscillator strengths are also proceeding using various approximations. The line strength S_{nm} is related to the radial wave functions R_n and R_m as follows:

$$S_{nm} = \frac{\mathcal{S}_m \mathcal{S}_n}{4l^2 - 1} \left(\int_0^\infty R_n R_m r \, dr \right)^2 \tag{9.34}$$

where $\mathcal{S}$ are relative multiplet strengths, which have been tabulated[59, 60]. The simplest radial wave functions used for such calculations are of the self-consistent field type, but there is a superior Coulomb approximation which was used for the first time in the important and widely applicable calculations of Bates and Damgaard[61]. More accurate techniques include the 'screening approximation'[62, 63], in which use is made of wave functions expanded in powers of nuclear charge, and the 'nuclear charge expansion' method[64], in which inverse powers of the nuclear charge are used.

Correspondence between theory and experiment is in general satisfactory to within 5–20 per cent for combinations in the visible. For transitions in the ultra-violet, the divergences are often much greater.

9.3 OSCILLATOR STRENGTHS; SUM RULES AND MAGNITUDES

Quantum-mechanical arguments can be adduced for equations connecting the sum of oscillator strengths of energy levels of an atom with various atomic parameters[5]. These equations are known as the oscillator strength sum rules, but only the two most important will be discussed here.

The first, or Thomas–Reiche–Kuhn sum rule is

$$\sum_n f_{gn} + \int_0^\infty \frac{df}{d\varepsilon} \, d\varepsilon = N \tag{9.35}$$

where N is the total number of electrons in the atom. The subscript g represents the ground state.

The second rum rule is

$$\sum_n f_{gn}(E_g - E_n)^2 + \int_0^\infty E^2 \frac{df}{d\varepsilon} \, d\varepsilon = \frac{\alpha}{4} \tag{9.36}$$

where α is the atomic polarizibility.

The continuum oscillator strengths play an important part in these summations. Photoionization data permit an experimental test to be made. For neon, equation 9.35 yields $N = 10\cdot6 \pm 0\cdot4$, and equation 9.36 yields $\alpha = 0\cdot43 \pm 0\cdot02 \times 10^{-24}$ cm^3. The correct values would be 10 and $0\cdot398 \times 10^{-24}$ cm^3 respectively.

Equation 9.36 can be applied for oscillator strengths f_{mn} for excited states $m \neq g$, using the polarizibility α_m of the atom in state m.

The magnitudes of allowed electric dipole transition oscillator strengths are not easily predictable empirically. In general, higher frequency transitions have smaller oscillator strengths. The oscillator strengths of the one-electron system were calculated some years ago, and are proportional to v^{-3}. Since these frequencies decrease rapidly for transitions between adjacent levels with increasing principal quantum number n, the upper levels are comparatively long-lived. It is probable that such high n levels (Rydberg states) are long-lived in more complex systems. Discontinuities due to perturbations have been reported in the graph of $f(n)$ for atomic hydrogen.

9.4 PHOTOABSORPTION PROCESSES

Photoionization and molecular photodissociation are usually studied in absorption, partly because of the astrophysical and geophysical importance of the absorption of radiation from stars, and in particular from the sun. Some excellent books and articles on the subject exist[65, 66], the most recent by Marr[67], by Fano and Cooper[68] and by Biberman and Norman[69].

There is an additional reason why the measurement of photoionization functions is important: photoionization is the reverse process of radiative recombination yielding the ground state. Since this latter process is at present impossible to study experimentally because of the low intensity of the continuous radiation and the large probability of competing processes such as dissociative and collisional-radiative recombination, the principle of detailed balance may be successfully invoked, in the form:

$$\frac{e_1 \sigma_{\phi 0}}{\phi_0 \sigma_{e1}} = \frac{g_0 e}{2 g_1 m_e c^2} \frac{E_\phi^2}{E_e} \tag{9.37}$$

where E_e is the electron kinetic energy; g_0 and g_1 are statistical weights referring to the ground states of the atom and the ion respectively; the free electron spin has been taken into account in the factor 2.

It is important to 'get a feeling for' the energies E_ϕ of photons of frequencies v and wavelengths λ in comparison with the energies of the inelastic processes

$$E_\phi = hv = \frac{hc}{\lambda} \tag{9.38}$$

With wavelengths expressed in Ångstrom units, and their inverse in wave numbers $\bar{v}$, or numbers of waves per centimetre (cm^{-1}, or kayzers),

$$E_\phi = 1 \cdot 2396 \times 10^4 \bar{v} \tag{9.39}$$

with E_ϕ in electron volts. The energies corresponding to different $\bar{v}$ and different λ may be gauged approximately from *Figure 1.4*. For example, the alkali metal vapours can be ionized by photons in the near ultra-violet, whilst metal vapours, rare gases and molecular gases require photons in the vacuum ultra-violet, or even soft X-rays.

The processes of ionization and molecular dissociation can be brought about by photons of any energy above the threshold. For these processes, there is therefore an absorption continuum starting at the threshold; for ionization of atoms, the absorption cross-section usually falls as the photon energy increases, but most cross-sections are still sizable, perhaps 10 per cent of the maximum, for a range of several electron volts (that is, several hundred Ångstrom units) in the spectrum. Typical photo-ionization cross-section functions have maxima of the order of 500 cm^2 cm^{-3}atm.$^{-1}$ at s.t.p. (the units in which photon absorption coefficients are usually measured). The absorption equation and the definition of units was given in Chapter 1, where the more familiar units of cm^2 cm^{-3} torr^{-1} at 0°C were also introduced. A further unit of cross-section much used for photoabsorption data is the megabarn (1 Mbn = 10^{-18} cm^2).

One of the purposes of studying the photoionization of atoms and molecules is to establish their ionization potentials. Tabulation of recent ionization potential data is available[70], and some data are given in the Appendix. There often arise complications due to Rydberg states and to autoionizing states.

The photoabsorption function $\sigma(E)$ is directly proportional to the 'optical spectrum' $f(E)$, where f is the oscillator strength. Because of the form of the Bethe approximation[71], the function $E^{-1}f(E)$ is sometimes known as the 'excitation spectrum'. It is possible to measure not only the total photo-absorption function $\sigma(E)$ but also, using an ionization chamber, the photo-ionization function $\sigma_i(E)$. For atoms the two are identical, but for molecules the difference $\sigma(E) - \sigma_i(E)$ is equal to the photodissociation function $\sigma_d(E)$. The function $E^{-1}\sigma_d(E)$ is sometimes known as the 'superexcitation spectrum'[72]. For polyatomic molecules it is a smooth function, possibly owing to cooperative effects, but for diatomic molecules there is often structure. The energy region above the dissociation limit is one which can contain two-electron excited states, and these are buried in two continua, the ionization and the dissociation. Such states are supposedly more likely to decay by dissociation than by ionization, but the competition is affected by isotopic substitution.

Line and band absorptions ($\phi 0/0'$ and $\phi 00/00'$), corresponding to the conditions of equation 9.7, are often called resonance processes, because they are only likely when the photons are very close to the energy of excitation. Strictly speaking, the term 'resonance radiation' refers only to radiation arising from transitions to the ground state. Line absorptions have maximum values of the order of 5×10^5 cm^2 cm^{-3} atm.$^{-1}$ at the centre of the line, with natural line widths of the order of 5×10^{-3} Å obscured by Doppler, collision and other forms of broadening[73, 74]. The absorption continua are usually responsible for greater loss of continuous black-body radiation than line absorptions, although there may be a few regions of the spectrum and situations where line absorptions contribute considerably.

At wavelengths longer than the onset of ionization or molecular dissociation, there is a weak photoabsorption background due to Rayleigh scattering[75]. In this process, the radiant energy is taken from the beam and re-radiated as spherical waves, each gas molecule acting as an individual scattering centre. The differential scattering cross-section[76] is

$$\frac{d\sigma}{d\Omega}(\theta) = \frac{8\pi^4}{\lambda^4}\alpha^2 \left(\frac{6+6\varrho_n}{6-7\varrho_n}\right)\left[1+\left(\frac{1-\varrho_n}{1+\varrho_n}\right)\cos^2\theta\right] \qquad (9.40)$$

where α is the polarizability, and the depolarization ϱ_n is defined as the ratio of the intensity of the transverse scattered radiation polarized parallel to the plane containing the incident beam and the observed scattered beam to the intensity of the transverse scattered radiation polarized perpendicular to this plane. The total scattering cross-section

$$\sigma_{\text{tot}} = \frac{128\pi^5}{3\lambda^4} \alpha^2 \left(\frac{6+3\varrho_n}{6-7\varrho_n}\right) \tag{9.41}$$

Measurements have been made[77] at 1216 Å of scattering cross-sections, and for rare gases they agree with the above formula ($\sim 6\times10^{-24}$ cm^2 in argon).

9.5 EXPERIMENTAL PHOTOABSORPTION TECHNIQUES

The intensity conditions of photon absorption experiments are not nearly as stringent as those in crossed-beam experiments, such as those for the study of photodetachment; a spectrally monochromated beam is always used, with single-beam spectrophotometric technique. Such a beam may be obtained, as described in Section 3.9, by means of a discharge producing a suitable line spectrum: the appropriate line is selected by a monochromator, which usually possesses a movable diffraction grating.

In absorption measurements, the gas may be introduced before, during or after the spectrograph. The first position was used by Ditchburn and his colleagues[66]. For wavelengths below 2000 Å, the spectrograph must be maintained under vacuum to reduce background from scattered radiation. The second position was used in the absorption measurements of Weissler and his associates[78] in a region below the cut-off of lithium fluoride windows at 1050 Å; it has the advantage that greater intensities can be achieved with the source placed very close to the slit of the spectrograph, which is filled with the gas under investigation. For wavelengths ~ 1200 Å, no lenses are available and mirrors are inefficient. Descriptions of techniques in this wavelength region, the vacuum ultra-violet, are available[79]. The third position, which has been used by Watanabe and his group[80] who locate the absorption cell behind the spectrograph, has the advantage that there are fewer unwanted photons to produce photochemical complications in the gas.

Collision chambers closed off by windows of plastic Zapon, which is fairly transparent in the region of 100 Å, have been used in absorption measurements[81]. It is possible with these windows to use an X-ray tube as a light source[82].

Absorption experiments have been carried out on hot gases for the study of continua of atomic radicles such as nitrogen. For studies on metal vapours, these vapours must be in equilibrium with a sufficiently large area of metal at a measurable high temperature; the collision chamber is maintained at the same or at a higher temperature. Metal vapours can have harmful effects upon optical glass surfaces, but a technique of limiting the effusion by buffering the vapour between helium flows has proved effective (helium is virtually transparent over a wide waveband).

In ionization measurements, the earliest experiments[83] studied only the relative ionization produced by radiation at different wavelengths. This was supplemented[84] by comparison with total absorption data; in more recent experiments[85], the absorption is measured in the identical apparatus used for the study of ionization. In this apparatus, illustrated in *Figure 9.1*, the ionization chamber contains three separate collecting electrodes, spaced sufficiently far apart for the photon absorption between them to be appreciable. From the difference between the collected electron currents the absorption coefficient is deduced; from the absolute value of the electron currents the photoionization cross-sections is inferred. A uniform transverse electric field is applied in the electrode region by means of guard rings, so that the electrons are collected over a well-defined beam path length. The usual absorption and beam cross-section equation is applied, with absorption coefficient μ in units of cm^2 cm^{-3} atm.$^{-1}$

An absorption experiment must not only demonstrate Lambert's law (absorption coefficient independent of pathlength), in the form

$$\mu = \frac{T}{273}\,\frac{760}{pl}\,\ln\left(\frac{I_0}{I}\right) \qquad (9.42)$$

where radiation of intensity I_0 is reduced in a path l through gas at pressure p torr and temperature $T°$K to an intensity I. The experiment must also demonstrate Beer's law: μ so determined must be independent of total pressure p. In some experiments, when weakly bound molecules are present, it is found that

$$\mu_v = C + Ap \qquad (9.43)$$

at constant temperature. Under these conditions, the number of atoms is proportional to p and the number of molecules to p^2. The absorption C of the atoms may be determined by extrapolation.

The photoionization experiment of *Figure 9.1* has an important fall-out benefit; it is possible, with an atomic gas, to make an absolute calibration of a relative ultra-violet photon detector using the photoionization chamber. This is discussed in Section 3.15. Here it will only be reiterated that the

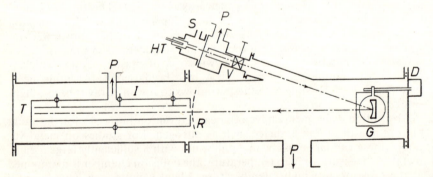

Fig. 9.1. Schematic diagram of photoionization apparatus: S, gas discharge source of radiation; HT, high-voltage electrode; V, vacuum valve; G, reflexion grating of radius 1 m; D, mechanical drive for rotation of grating for wavelength selection; R, Rowland focusing circle; P, to vacuum pumps; T, thermocouple radiation detector; I, collection system for ions and electrons

(From Weissler[65], by courtesy of Springer-Verlag)

principle involved is that a single photon produces a single ion, so an absolute measurement of the number of ions produced during a measured absorption enables the relative measurement of the number of photons absorbed to be made absolute. Various detectors of vacuum ultra-violet radiation have been calibrated in this way: thermocouples, multipliers and, probably the simplest and best, photoemission of electrons from a metal surface.

The mass-analysis of positive ions produced is a necessary part of photo-ionization studies of molecules. One must avoid mass-discrimination, by designing the collision chamber and mass-spectrometer so that the true proportions of each mass-number formed are registered. The same considerations are involved as in the dissociative ionization of molecules by electrons and by heavy-particle beams.

Photoabsorption studies of positive ions are made by passing the beam through plasma; no experiments with trapped ions have been reported.

Good optical resolution is necessary in photoabsorption studies. Particularly in the energy region just below the first and second ionization limits, absorption spectra are usually packed with sharp and diffuse bands and lines. Consequently absorption cross-sections may vary by many orders of magnitude depending on whether they are measured on or between the bands or lines. Unless the highest possible resolution is achieved, reasonably accurate cross-sections may be expected only in bands which are very diffuse because of predissociation. High resolution implies low intensity, therefore detection sensitivity is important. Deconvolution might prove valuable.

9.6 CALCULATION OF PHOTOABSORPTION AND PHOTOIONIZATION CROSS-SECTIONS

In the regions of the spectrum where absorption processes are not found, photons are still scattered by the Rayleigh and Raman processes. The former can be considered in terms of the radiation of oscillating dipoles induced by the incident light; the latter arises from similar causes, but involves a frequency change characteristic of the scattering molecule. Since the cross-sections are of the order of 10^{-8} of those for photoionization and photodissociation, they are not considered further here, but it should be noted that experimental investigations have been made even in the vacuum ultra-violet[86].

The photoionization cross-section at frequency v can be expressed in terms of the differential oscillator strength

$$\sigma = \frac{\pi e^2}{mc^2} \frac{df}{dv} = \frac{8\pi^3 e^2 v}{3cg_i h v'} |M_{if}|^2 \tag{9.44}$$

or

$$\sigma = \frac{4\pi^2 \alpha a_0^2}{3g_i} (E_i + k^2) |M_{if}|^2 \tag{9.45}$$

where $k^2 = hv'$ is the energy of the photon in excess of the ionization threshold, g_i is the statistical weight of the initial state of the atom, and α is the fine structure constant. The matrix element for the transition between initial

(i) and final (f) states is

$$|M_{if}|^2 = \sum_i \sum_f \left| \int \psi_f^* \sum_\mu r_\mu \psi_i \, d\tau \right|^2 \tag{9.46}$$

The first exact evaluation of a photoionization cross-section was made by Menzel and Pekeris[87], who arrived at an expression valid for a hydrogen-like system of nuclear charge Z:

$$\sigma(\nu) = G(n) \left(\frac{32\pi^2 e^6 R Z^4}{3^{3/2} h^3 \nu^3 n^5} \right) \tag{9.47}$$

where n is the principal quantum number of the initial state, and $G(n)$ is the Gaunt factor, which does not differ from unity by more than 20 per cent. With $G(n) = 1$, the formula reduces to that obtained by Kramers[88]. The analytic form of $G(n)$ is available[89].

For many-electron systems[90], this formula is inaccurate; it is usually replaced by a central field approximation, in which wave functions are taken to be linear combinations of determinates of spin-orbitals. The cross-section can be written

$$\sigma = A\mathcal{e} \left[C_{l-1} \left| \int_0^\infty R_i(n, l) (r) R_f(E_e, l-1) (r) r^3 \, dr \right|^2 \right.$$

$$\left. + C_{l-1} \left| \int_0^\infty R_i(n, l) (r) R_f(E_e, l+1) (r) r^3 \, dr \right|^2 \right] \tag{9.48}$$

where all R represent radial wave functions for principal and azimuthal quantum numbers n and l; all C are numerical coefficients tabulated in the original papers; A is a normalization factor, and $\mathcal{e}$ is a correcting factor approximated by a product over the electrons not taking part in the transition:

$$\mathcal{e} = \prod \left| \int_0^\infty R_i(n, l) (r) R_f(n, l) (r) r^2 \, dr \right|^2 \tag{9.49}$$

The cross-section is very sensitive to small changes in the form of the wave functions, and the different formulations of the matrix element lead to different cross-sections:

Dipole length matrix element, $\quad \int \psi_f^* \left(\sum_\mu r_\mu \right) \psi_i \, d\tau$;

Dipole velocity matrix element, $\quad \dfrac{\hbar^2}{m_e E} \int \psi_f^* \left(\sum_\mu \nabla_\mu \right) \psi_i \, d\tau$;

Dipole acceleration matrix element, $\quad \dfrac{\hbar^2}{m_e E^2} \int \psi_f^* \left(\sum_\mu \nabla_\mu V \right) \psi_i \, d\tau$;

where

$$\sum_\mu \nabla_\mu V = \sum_\mu \frac{Z}{r_\mu^2} \tag{9.50}$$

for nuclear charge Z. A number of dipole length, velocity and acceleration calculations have been carried out using Hartree–Fock methods[91].

A more general approach, capable of avoiding much laborious computation, is based on the 'quantum defect method'[89, 92, 93]. The differential equation satisfied by the radial wave function $R(\varepsilon, l | r)$ is

$$\left[\frac{d^2}{dr^2} - \frac{l(l+1)}{r^2} + \varepsilon - V(r) \right] R(\varepsilon, l | r) = 0 \qquad (9.51)$$

where ε (atomic units) is the energy above the photoionization threshold; $V(r) = -2Z/r$ for small r, and $V(r) = -2Z'/r$ for large r, with

$$Z' = Z - N + 1 \qquad (9.52)$$

for N outer shell electrons. For $\varepsilon < 0$, bounded solutions exist at eigenvalues

$$\varepsilon_{n, l} = -\frac{Z'^2}{n_l'^2} \qquad (9.53)$$

where n_l', the effective quantum number, differs from the principal quantum number n by the quantum defect

$$\delta_{n, l} = n - n_l' \qquad (9.54)$$

For $\varepsilon > 0$, one puts $\varepsilon = k^2$ and obtains a solution in the form

$$F(\varepsilon, l | r) = f(\varepsilon, l | r) + B(\varepsilon, l) g(\varepsilon, l | r) \qquad (9.55)$$

where f and g are linearly independent solutions of the radial equation for all finite ε, $(|\varepsilon| \neq 0)$. Since $B(\varepsilon_{n\ l}, l)$ is known from the energies of the bound states, $B(k^2, l)$ may be found. The method has been used with success to calculate oscillator strengths of bound states, as well as photoionization functions for He, Li, C, N, O, Na, Mg, K, O^+, Ne^+, Na^+, Mg^+, Si^+ and Ca^+.

Other techniques for calculation include the random-phase approximation[94], and the assumption of unrelaxed ionic core[95]; the latter is especially applicable to higher photon energies[96], where maxima occur in the functions, not only in alkali metals but even in rare gases. Theoretical treatment of rare gases has also been given[97, 98] in terms of collective interactions between many electrons in the atom.

Calculations of photoionization of molecules are confined to hydrogen[99]. But threshold laws for photoionization of diatomic molecules have been given by Geltman[100] and applied to photodetachment data.

Dissociation of molecules by photons is governed very much by the vibrational overlap integral; calculations have been made for H_2 [101] and O_2 [102], and for the photodissociation of H_2^+ [103].

9.7 PHOTOABSORPTION DATA

Molecular photoabsorption functions start with a photodissociation continuum at the longest wavelength, since dissociation energies are lower than ionization energies. For example, in oxygen the Schumann–Runge bands $(B^3\Sigma_u^- - X^3\Sigma_g^-)$ occupy the long wavelength region below 1760 Å, and the r associated dissociation continuum extends from 1670 Å to about 1350 Å (see *Figures 9.7* and *9.13a*).

The interpretation of photoabsorption data is complicated by many factors. The existence of energy levels just within the ionization or dissociation continua may result in preionization or predissociation. The photo-excitation of these levels is, however, a resonance process. The continuous absorption due to dissociation or ionization can be complicated not only by those processes, but by dissociative ionization into:

1. An excited ion and a ground state atom;
2. A ground state ion and an excited atom;
3. An excited ion and an excited atom;
4. Two ions of opposite sign (the positive ion may be in a ground or excited state).

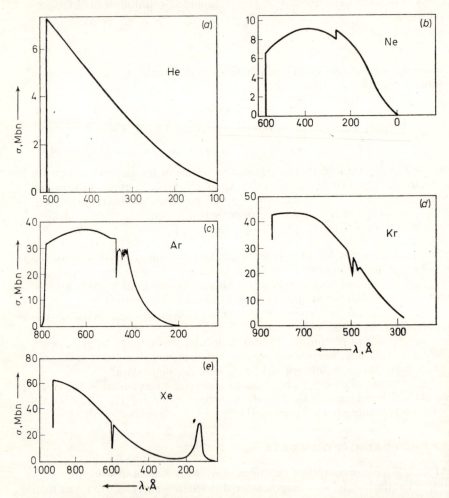

Fig. 9.2. Photoionization cross-sections for rare gases: (a) helium (measurements, Samson and Kelly); (b) neon (measurements, Samson and Kelly); (c) argon (measurements, Samson); (d) krypton (measurements, Samson and Kelly); (e) xenon (measurements, Samson and Kelly)

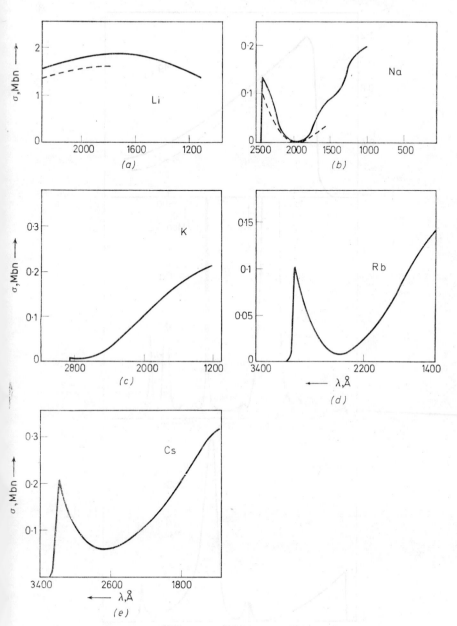

Fig. 9.3. Photoionization cross-sections for alkali metals: (a) lithium (measurements, Hudson and Carter; broken line, calculation of Burgess and Seaton); (b) sodium (measurements, Hudson and Carter; broken line, calculations of Burgess and Seaton); (c) potassium (measurements, Marr and Creek); (d) rubidium (measurements, Marr and Creek); (e) caesium (measurements, Marr and Creek)

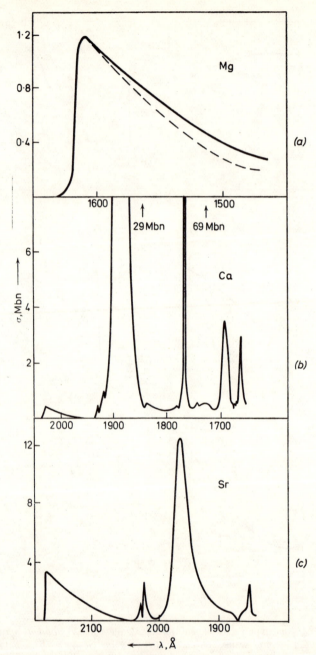

Fig. 9.4. Photoionization cross-sections for alkaline earths: (a) magnesium (measurements, Ditchburn and Marr; broken line, calculations of Peach); (b) calcium (measurements, Ditchburn and Hudson); (c) strontium (measurements, Hudson and Young)

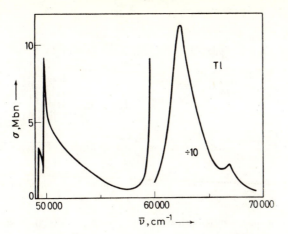

Fig. 9.5. Photoionization cross-section of thallium (Marr and Heppinstall), showing struc-ture arising from two Beutler–Fano profiles, centred around 50 000 cm⁻¹ and 60 000 cm⁻¹

Some photoabsorption and photoionization functions are shown in *Figures 9.2–9.8.* Data are also available for H, Be, B, C, N, O, F, Al, Zn, Ga, Cd, In, Ba, He^+, C^+, N^+, O^+, F^+, Ne^+, Na^+ and Ca^+, as well as N_2, O_2, H_2O, CO_2, NH_3, NO, N_2O, O_3, CH_4, C_2H_4, C_2H_6, K_2 and Na_2. Data are given in the literature[65–67], and include a bibliography[104].

Photoabsorption measurements on atomic hydrogen[106] show that the cross-section is 6·2 Mbn at threshold and falls off as the inverse third power of frequency. Atomic oxygen has also been investigated[107].

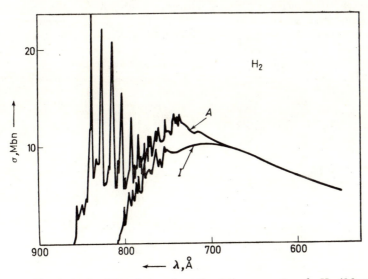

Fig. 9.6. Photoionization (I) and photoabsorption (A) cross-sections for H_2 (Metzger and Cook[105])

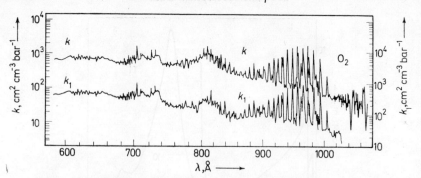

Fig. 9.7 Photoionization (k_1) and photoabsorption (k) cross-sections for O_2

The photoionization functions of atoms have sharp onsets and very slowly oscillate with increasing energy, as can be seen from *Figures 9.2* and *9.3* for alkali metals and rare gases[108, 109]. This general behaviour is in agreement with calculations. The continuum differential oscillator strength $df/d\varepsilon$, if extrapolated to energies below the continuum, should pass through the points $\frac{1}{2}n^*f(\varepsilon)$ calculated for excited states of the neutral atom, of effective quantum number n^* and oscillator strength f.

In addition to the continua, resonance transitions are observed above the ionization potential. Where atomic levels lie above the continuum, radiationless transition can take place with the emission of an electron, in the process known as autoionization. The f value is often greater than that for normal photoionization and must be taken into account in calculating total transition probabilities. Detailed investigations have been made by Garton and his co-workers[110] as well as by Beutler[111] and other investigators[112, 113]. A particularly fine measurement of absorption due to of autoionization absorptions in krypton, carried out by Huffman, Tanaka and Larrabee[114], is shown in *Figure 9.9*. The step at the $J = \frac{1}{2}$ onset arises from the high density of levels below and the absence of levels above.

Many of the levels in molecular continua can be regarded as Rydberg states. They converge to levels of the positive ion which can now be iden-

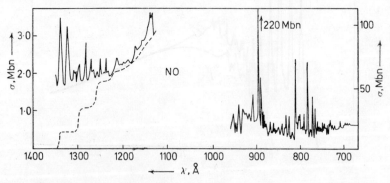

Fig. 9.8. Photoionization (broken line) and photoabsorption cross-sections for NO (Watanabe, Cook and Ching)

tified by photoelectron spectroscopy. Since there is a large volume of high-resolution data for the Rydberg states, the quantum defects can be calculated and the levels identified[115].

The development of X-ray and XUV techniques has made possible the study of photoabsorption continuously from the onset down to wavelengths

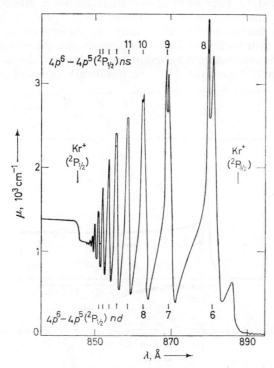

Fig. 9.9. Photoabsorption function for krypton near the ionization limit, showing two autoionization series[114]

of a few Ångstrom units. Three important features emerge from the atomic data:

1. The calculation of continuum and line oscillator strengths for inner shell transitions becomes possible.
2. High-energy maxima such as that in *Figure 9.2e* are available for comparison with the theories[96-98] proposed for their explanation. At present, the theory based on collective effects gives better agreement with experiment than the orthodox quantum theory. The maxima occur at the following wavelengths: neon, 2·6 Å; argon 2·2 Å; krypton 1·8 Å; and xenon, 2·1 Å and 1·6 Å.
3. There is autoionization structure around the inner shell ionization onsets (*Figure 9.10*), as well as above the first continuum.

Autoionization levels arise from the excitation of an electron in an inner sub-shell and often involve the excitation of more than one electron. The levels lie above the one or more ionization continua and are able to decay energetically into them. If the decay is an allowed transition, the life-time is in the range 10^{-12}–10^{-14} sec; if it is a forbidden transition, the life-time can be longer than 10^{-6} sec.

A continuum may be regarded as a set of closely spaced discrete levels that are singled out by particular boundary conditions at the walls of a large box surrounding the atom. Each of these discrete levels has a different configuration of wave function. An autoionization level 'repels' the continuum states of nearby energy with which it interacts, thereby creating a 'hole' in the spectral distribution of optical absorption by direct radiationless transition into the continuum. An overlap integral governs the energy width of this 'hole'. Fano[116] has treated the spectral distribution and arrived at a formula

$$\sigma(\varepsilon) = \sigma_a + \sigma_b \frac{(q+\varepsilon)^2}{1+\varepsilon^2} \tag{9.56}$$

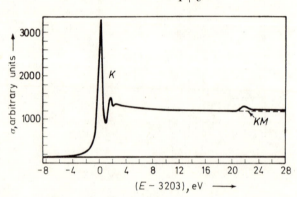

Fig. 9.10. *K absorption edge of argon, showing structure due to K excitation and KM excitation (Fano and Cooper)*

where

$$\varepsilon = \frac{E-E_0}{\frac{1}{2}\Gamma} \tag{9.57}$$

σ_a is the potential scattering cross-section, and σ_b is the resonance scattering cross-section, both at energy E; Γ is the line width, E_0 the energy of the level, and

$$q = -\mathrm{arc\ cot}\ \delta_r \tag{9.58}$$

is a quantity known as the 'line parameter', which is related to the resonant scattering phase shift δ_r; it can be positive when decay into two continua are possible, but is negative when decay into only one continuum is allowed. Some examples, such as cadmium, are known of series of autoionization levels during the course of which a second continuum channel of decay becomes open; the sign of the line parameter changes at this point. A graphical representation of functions $(q+\varepsilon)^2/(1+\varepsilon^2)$ is given in *Figure 4.8a*. It will be seen that the line shape possesses the characteristic 'Fano dip'; an absorption line displays a bright sideband. Such behaviour is characteristic of autoionization levels[112], and several series of two-electron excited levels have been discovered in helium, both by photoabsorption and by using electron and heavy-particle spectroscopy. A comparison of calculations of helium two-electron excited levels with measurements is given in *Figure 9.11*. The line shapes correspond well with equation 9.56. It is even possible (as with indium) for the line to be of such breadth as to extend below the ionization potential. A combination of very broad lines dominates the photoionization functions of thallium (*Figure 9.5*), and alkaline earths (*Figure 9.4*).

Autoionization levels often have oscillator strengths approaching unity, although the entire continuum oscillator strength is usually small. Molecules also display autoionization levels; the levels in H_2 are well known and identified[119, 120].

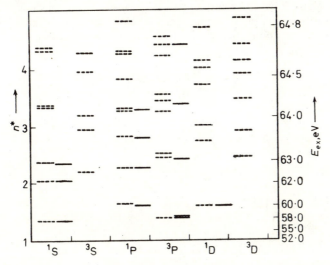

Fig. 9.11. *Two-electron excited levels of helium, with effective quantum number n* given by* $E_{ex} = 24{\cdot}59 + 13{\cdot}6\,(3 - n^{*-2})$ eV: *full lines, experiments; broken lines, calculations*[118]

Autoionization levels which are forbidden to decay into a continuum are, of course, not observed optically. Such levels can be classed as metastable. The $^4P_{5/2}$ $1s\,2s\,2p$ state of lithium has been identified[121] around 57 eV in an experiment using electron impact and delayed detection of the ions. The decay time proceeding by the spin–spin interaction is of order 10^{-6} sec; the equivalent line width, 10^{-9} eV, would render the excitation process invisible in absorption. The $^4F_{9/2}$ level is found in alkali metals but not in lithium. A search has been made for sextet nitrogen $1s^2\,2s\,2p^3\,3s$ around 17 eV. Selection rules for autoionization levels were first written by Shenstone and are given by Condon and Shortley[1].

Autoionization absorption measurements in Ca, Tl, Pb, Kr, Xe and Ar have been reported[122]; this paper also contains references to previous ultra-violet autoionization studies, in Ar, Kr, Xe, Na, K, Rb, Cs, Cu, Ag, Au, Mg, Ca, Sr, Ba, Zn, Cd, Hg, Al, Ga, In, Tl, Si, Pb, Mn, I and Y. Further references[123, 124] and reviews[125–127] are available.

Inner shell ionization edges can show structure (for example, the **krypton**

N I edge exhibits holes). The L edge in argon exhibits lines below the ionization edge[128], in which an inner shell electron is raised to a higher quantum number level, $1s^2\,2s^2\,2p^5\,3s^2\,3p^6\,nl$. Xenon M IV and V edges are similar. The K edge in argon is illustrated in *Figure 9.10*.

9.8 PHOTO-MASS-SPECTROMETRY AND PHOTOELECTRON-SPECTROSCOPY

The primary purpose of photo-mass-spectrometry and photoelectron-spectroscopy is to determine the energies of molecular 'inner ionization potentials'. These are the excited states of the positively ionized molecule; they are formed by the removal of one electron from the molecule; the energy of the state, and the equilibrium nuclear separation, enable deductions to be made about the orbital from which this electron comes.

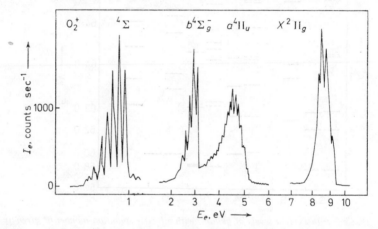

Fig. 9.12. Photoelectron spectrum of oxygen excited by helium resonance radiation (Turner and May)

The photoabsorption functions for molecules (see, for example, *Figures 9.6–9.8)* are too much complicated by levels in the dissociation and ionization continua to allow identification of the inner ionization potentials. The first simplification is achieved by incorporating mass-analysis of the ions produced, so that the photoionization functions show their characteristic structure without interference from dissociation functions and dissociative ionization functions. The characteristic photoionization function can approximate to a step-function; ideally a flight of steps for an atom containing a series of inner ionization potentials. For a molecule there is vibrational and also rotational sub-structure, so the function approximates to a series of flights of steps. However, levels in the continua still interfere with the identification of these functions. Data are to be found in the literature[119, 120] and in particular in the experiments of Berkowitz, Chupka and Inghram[129], which incorporate resolution superior to other experiments in that rotational transitions are resolved.

In order to study inner ionization potentials without interference from continuum levels, the more refined technique of photoelectron-spectroscopy has been developed by Turner and others[130, 131]. In this technique, resonant ultra-violet photons, usually from a helium discharge (584 Å, 21·21 eV), are directed into the gas, and the energy spectrum of the electrons ejected is measured by momentum analysis (in early experiments, by retardation technique). Since the resonance photons are effectively monoenergetic, the electron energy spectrum is a mirror of the continuum oscillator strength function. It is necessary for the ionization process to take place directly for registration in the photoelectron spectrum. Thus the registration of inner ionization potentials is unperturbed by continuum levels.

A typical photoelectron spectrum is shown in *Figure 9.12*, and dissociative ionization measurements in *Figure 9.13*.

The relative heights of the peaks for excitation of vibrational states of a given electronic level are proportional to the square of the overlap integrals, so a calculation can be made of the potential energy curve of the state of the positive ion from the photoelectron spectrum. However, autoionization processes can interfere with the overlap picture; and the use of a variety of different energies of photon source is necessary in order to avoid this.

The first photoelectron spectra were taken with retardation analysis using cylindrical grid and collector, rather similar to the Lozier experiment without the aligning cylinders. But for appreciable collision path length, the momentum resolution of this system is poor, and spherical retardation grids were introduced[133] with success. In more recent studies, the cylindrical momentum analyser has been used.

The angular distribution of photoelectrons is expected to be given by the equation

$$\frac{dN}{d\Omega} = \frac{\sin^2 \theta}{(1 - \beta \cos \theta)^4} \tag{9.59}$$

where β is the relativistic velocity of the photoelectron.

Techniques of calculation of Franck–Condon factors have received stimulus from photoelectron-spectroscopy[134, 135]. The accuracy of bond lengths of excited states inferred from these experiments and calculations is encouraging.

A review of photoelectron spectroscopy is available[136]. The important analogous studies using X-ray sources ('ESCA') have also been discussed[137]. Inner shell lines from atoms are observed, and these show characteristic chemical shifts, appropriate to the molecule under examination.

9.9 TWO-QUANTUM PROCESSES

Two-quantum processes are those in which radiation of frequency v is absorbed or emitted during a transition of energy $E = 2hv$. They are governed by different selection rules from single-photon processes; the parity does not change in electric dipole transitions involving two quanta.

Thus, the $2s$–$1s$ two-quantum transition in the hydrogen atom is actually more probable than other processes by which the $^2S_{1/2}$ state is destroyed (spontaneous emission to $2p \, ^2P_{1/2}$, and weak magnetic dipole radiation). The two-quantum transition probability has been shown[138] to be 8·2 sec^{-1},

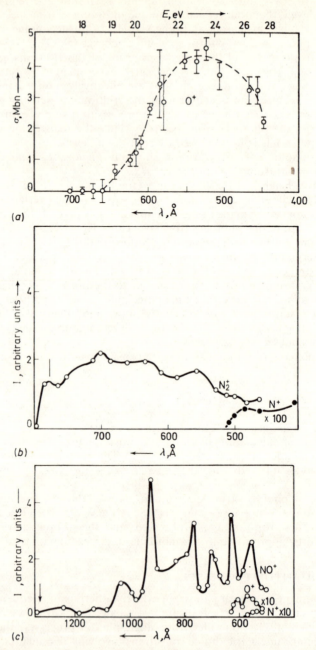

Fig. 9.13. Dissociative and non-dissociative ionization functions for: (a) O_2; (b) N_2; (c) NO (From Weissler et al.[132], by courtesy of the Optical Society of America)

and because of the arbitrary division of energy between the quanta, a continuum is emitted; this is observed from planetary nebulae[139]. Coincidence studies of the two photons have been reported, not only in hydrogen but in calcium[140].

Two-quantum processes are many orders of magnitude less probable than single-quantum processes. Thus they can only be observed in absorption[141] with very intense radiation such as may be obtained using laser beams. The pseudo-cross-section for an upward transition is given by the relation[142]:

$$\sigma = \left(\frac{e}{m_e c^2}\right)^2 \left(\frac{\lambda_r^2}{\varepsilon^2 \, \Delta v}\right) F \qquad (9.60)$$

where ε is the dielectric constant of the material, F is the incident photon flux, Δv is the width of the absorption band around $2v$, and λ_r is the wavelength of the radiation. Thus the number of excitations is proportional to the square of the radiation intensity.

The processes are most probable when Δv is smallest, that is, when the energy of the transition is closest to $2hv$. In the first solid state magnetic dipole transition for which a two-photon absorption was observed, $\Delta v = 5 \times 10^3$ cm^{-1}, but it is possible to conceive of processes which can be excited by line radiation with much smaller Δv values. For transitions into the continuum, ionization and detachment, the probabilities are smaller than for line processes[143].

In two-quantum absorption processes, there is usually a non-resonance transition to an intermediate state, excited by the first photon. This state is then excited by the remaining electromagnetic enegy, but there need be no conservation of energy $E = hv$ in the first excitation. When there is no possible stable intermediate state, as in two-quantum photodetachment, this mechanism is not possible.

Calculations of two-quantum photoionization of hydrogen have been reported[144].

Upward two-quantum transitions between two energy levels are comparatively rare in free-atom experiments, although well-known in solid state physics[145]. A transition in helium has been reported[146], and another in caesium[147]. It is not easy to find appopriate transitions which are accidentally resonant with laser frequencies.

Ionization of gases by pulsed laser light has been fairly extensively studied because of its application as a generator of plasma. For rare gases and atmospheric gases it is necessary that several photons combine in a single ionization process. The intensity of ionization is found to be proportional to a high power of the intensity of the radiation, but quantitative pseudo-cross-sections are difficult to obtain. Alkali metal vapours can be ionized by three ruby laser photons, and the experimental data[148] agree with calculations[149].

REFERENCES

1. CONDON, E. U. and SHORTLEY, G. H. *The Theory of Atomic Spectra*, 1935. London; Cambridge University Press.
2. KUHN, H. *Atomic Spectra*, 1962. London; Longmans, Green.

3. HERZBERG, G. *Atomic Spectra and Atomic Structure*, 1944. New York; Dover Publications.
4. HERZBERG, G. *Molecular Spectra and Molecular Structure*, 3rd edn, 1966. New York; Van Nostrand: Vol. 1, Diatomic Molecules; Vol. 2, Infrared and Raman Spectra; Vol. 3, Polyatomic Molecules.
5. BETHE, H. A. and SALPETER, E. E. *Quantum Mechanics of One- and Two-electron Atoms*, 1957. Berlin; Springer-Verlag.
6. MILNE, E. A. *Mon. Not. Roy. astr. Soc.* 85 (1924) 117; *J. Lond. math. Soc.* 1 (1926).
7. CHANDRASEKHAR, W. *Radiative Transfer*, 1950. Oxford; Oxford University Press; BARRAT, J. P. *J. Phys. Radium* 20 (1959) 541, 633, 657.
8. PHELPS, A. V. and McCOUBREY, A. O. *Phys. Rev.* 118 (1960) 1561; PHELPS, A. V. *Phys. Rev.* 110 (1958) 1362.
9. HOLSTEIN, T. *Phys. Rev.* 72 (1947) 1212; 83 (1951) 1159.
10. BIBERMAN, L. M. *J. exp. theor. Phys.* 17 (1947) 416.
11. HUMMER, D. G. and SEATON, M. J. *Mon. Not. Roy. astr. Soc.* 125 (1963) 437.
12. GARSTANG, R. H. *Symp. int. astr. Un.* 26 (1966) 57.
13. GLENNON, B. M. and WIESE, W. L. United States National Bureau of Standards Miscellaneous Publication 278, 1966; 'Bibliography of Atomic Transition Probabilities', Supplement 1968.
14. CARLISS, C. H. and BOZMAN, W. R. United States National Bureau of Standards Monograph 53, 1962.
15. GRIEM, H. R. *Plasma Spectroscopy*, 1964. New York; McGraw-Hill.
16. SLATER, J. C. *Quantum Theory of Atomic Structure, Vol. II*, 1960. New York; McGraw-Hill.
17. International Astronomical Union Commission Report 14a; *Trans. int. astr. Un.* 9 (1955) 214; 10 (1958) 220; 11 (1961).
18. ALLEN, C. S. *Astrophysical Quantities* 1955. London; University of London Press; *Mon. Not. Roy. astr. Soc.* 121 (1960) 299.
19. ROHRLICH, F. *Astrophys. J.* 129 (1959) 441, 449.
20. FOSTER, E. W. *Rep. Progr. Phys.* 27 (1964) 469.
21. WIESE, W. L. *Methods of Experimental Physics, Vol. 7B*, ed. W. L. Fite and B. Bederson, 1968. New York; Academic Press.
22. KING, A. S. *Astrophys. J.* 108 (1948) 429; KING, R. B. *Astrophys. J.* 95 (1942) 78; 105 (1947) 376; 108 (1948) 87; KING, R. B. and KING, A. S. *Astrophys. J.* 82 (1935) 377; 87 (1938) 24.
23. BELL, G. D., DAVIS, M. H., KING, R. B. and ROUTLY, P. M. *Astrophys. J.* 127 (1958) 775; 129 (1959) 437.
24. BELL, G. D. and KING, R. B. *Astrophys. J.* 133 (1962) 718.
25. FAIRCHILD, C. E. and CLARKE, K. C. *Bull. Amer. phys. Soc.* 11 (1962) 7, 458.
26. MASTRUP, F. and WIESE, W. *Z. Astrophys.* 44 (1958) 259; HEY, P. *Z. Phys.* 157 (1959) 79; WIESE, W. L. and SHUMAKER, J. B. *J. opt. Soc. Amer.* 51 (1961) 937.
27. ALLEN, C. W. and ASAAD, A. S. *Mon. Not. Roy. astr. Soc.* 45 (1955) 521; 117 (1957) 36.
28. NICHOLLS, R. W., PARKINSON, W. H. and REEVES, E. M. *Appl. Optics* 2 (1963) 919.
29. LAWRENCE, G. M. and SAVAGE, B. D. *Phys. Rev.* 141 (1966) 67.
30. PRAG, A. B., FAIRCHILD, C. E. and CLARK, K. C. *Phys. Rev.* 137 (1965) A1358.
31. BROSSEL, J. and KASTLER, A. *C.R. Acad. Sci., Paris* 229 (1949) 1213; BROSSEL, J. and BITTER, F. *Phys. Rev.* 86 (1952) 308; KASTLER, A. *J. Phys. Radium* 19 (1958) 797.
32. SERIES, G. W. *Rep. Progr. Phys.* 22 (1959) 280.
33. BYRON, F. W., McDERMOTT, M. N., NOVICK, R. *Phys. Rev.* 134 (1964) A615.
34. DE ZAFRA, R. L. and KIRK, W. *Amer. J. Phys.* 35 (1967) 573.
35. LURIO, A. and NOVICK, R. *Phys. Rev.* 134 (1964) A608.
36. LURIO, A., DE ZAFRA, R. L. and GOSHEN, R. J. *Phys. Rev.* 134 (1964) A1198.
37. BARRAT, J. P. *J. Phys. Radium* 20 (1959) 541, 633.
38. KORFF, S. A. and BREIT, G. *Rev. mod. Phys.* 4 (1932) 471.
39. ROZHDESTWENSKY, D. *Ann. Phys., Paris* 39 (1912) 307.

40. LADENBURG, R. and WOLFSOHN, G. *Z. Phys.* 63 (1930) 616.
41. OSTROVSKII, Y. I. and PENKIN, N. P. *Optika Spektrosk.* 9 (1960) 371; PENKIN, N. P. *J. quantve. Spectrosc. radiat. Transf.* 4 (1964) 41; ROZHDESTWENSKY, D. and PENKIN, N. P. *J. Phys., Moscow* 5 (1960) 1941.
42. MARLOW, W. C. *Appl. Optics* 6 (1967) 1715.
43. WEINGEROFF, M. *Z. Phys.* 67 (1931) 679.
44. STEPHENSON, G. *Proc. Roy. Soc.* A64 (1951) 458.
45. HULPKE, E., PAUL, E. and PAUL, W. *Z. Phys.* 177 (1964) 257.
46. OSBERGHAUS, O. and ZIOCK, K. *Z. Naturf.* 11a (1956) 762.
47. STROKE, H. H. *Phys. To-day* 19 (1966) 55.
48. BENNETT, W. R. *Appl. Optics*, Supplement 2 on Chemical Lasers, p. 23, 1965.
49. LAWRENCE, G. M. *J. quantve. Spectros. radiat. Transf.* 5 (1965) 359.
50. LAWRENCE, G. M. and SAVAGE, B. D. *Phys. Rev.* 141 (1966) 67.
51. HERON, S., MCWHIRTER, R. W. P. and RHODERICK, E. H. *Nature, Lond.* 174 (1954) 564; *Proc. Roy. Soc.* A234 (1956) 565.
52. KLOSE, J. *Z. Phys. Rev.* 141 (1966) 181.
53. BENNETT, R. G. and DALBY, F. W. *J. chem. Phys.* 31 (1959) 434; 32 (1960) 1111; 32 (1960) 1716.
54. ANKUDINOV, V. A., BOBASHEV, S. V. and ANDREEVE, E. P. *Soviet Phys. JETP* 21 (1965) 26.
55. KAY, L. *Proc. phys. Soc., Lond.* 85 (1965) 163.
56. BASHKIN, S. M., HEROUX, L. and SHAW, J. *Phys. Lett.* 13 (1964) 229; BASHKIN, S. M. *Nucl. Instrum. Meth.* 28 (1964) 88; *Beam Foil Spectroscopy*, 1968. London; Gordon and Breach.
57. READ, F. H. *Proc. phys. Soc., Lond.* 1 (1968) 893.
58. SKERBELE, A. M. and LASSETTRE, E. N. *J. chem. Phys.* 40 (1964) 1271.
59. ALLEN, C. W. *Astrophysical Quantities*, 1955. London; University of London Press; *Mon. Not. Roy. astr. Soc.* 121 (1960) 299.
60. ROHRLICH, F. *Astrophys. J.* 129 (1959) 441, 449.
61. BATES, D. R. and DAMGAARD, A. *Phil. Trans.* A242 (1949) 101.
62. LAYZER, D. *Ann. Phys.* 8 (1959) 271.
63. VARSAVSKY, C. M. *Astrophys. J. Suppl. Ser.* 6, No. 53 (1961) 75.
64. CROSSLEY, R. J. S. and DALGARNO, A. *Proc. Roy. Soc.* A286 (1965) 510.
65. WEISSLER, G. L. *Handb. Physik.* 21 (1956).
66. DITCHBURN, R. W. quoted in *Atomic and Molecular Processes*, Ed. D. R. Bates, 1962. New York; Academic Press; DITCHBURN, R. W., JUTSUM, P. J. and MARR, G. V. *Proc. Roy. Soc.* A219 (1953) 89; DITCHBURN, R. W. *J. quantve. Spectros. radiat. Transf.* 2 (1962) 361.
67. MARR, G. W. *Photoionization Processes in Gases*, 1967. Academic Press; New York.
68. FANO, U. and COOPER, J. W. *Rev. mod. Phys.* 40 (1968) 441.
69. BIBERMAN, L. M. and NORMAN, G. E. *Soviet Phys. Usp.* 10 (1967) 52.
70. WATANABE, K., NAKAYAMA, T. and MOTTL, J. R. *J. quantve. Spectros. radiat. Transf.* 2 (1962) 369.
71. SEATON, M. J. *Phys. Rev.* 113 (1958) 814.
72. PLATZMAN, R. L. *Vortex*, 23, No. 8 (1962) 20.
73. CHEN, S. and TAKEO, M. *Rev. mod. Phys.* 29 (1957) 20.
74. BREENE, R. G. *Rev. mod. Phys.* 29 (1957) 94.
75. RAYLEIGH, LORD *Phil. Mag.* 5, No. 47 (1899) 375; 6, No. 35 (1918) 373.
76. PENNDORF, R. *J. opt. Soc. Amer.* 47 (1957) 176.
77. GILL, P. and HEDDLE, D. W. O. *J. opt. Soc. Amer.* 53 (1963) 847.
78. WEISSLER, G. L., LEE, PO and MOHR, E. I. *J. opt. Soc. Amer.* 42 (1952) 84; LEE, PO *J. opt. Soc. Amer.* 45 (1955) 703.
79. DITCHBURN, R. W. *Proc. Roy. Soc.* A229 (1955) 44; PLATT, J. R. and KLEVENS, H. B. *Rev. mod. Phys.* 16 (1944) 182.
80. WATANABE, K., INN, E. C. Y. and ZELIKOFF, M. *J. chem. Phys.* 21 (1953) 1026.
81. LOWRY, J. F., TOMBOULIAN, D. H. and EDERER, D. *Phys. Rev.* 137 (1965) A1054.

82. HENKE, B. L. *Advanc. X-ray Analysis* 4 (1961) 224.
83. KINGDON, K. H. *Phys. Rev.* 21 (1923) 403.
84. MOHLER, F. L. and BOECKNER, C. *J. Res. nat. Bur. Stand.* 3 (1929) 303.
85. WAINFAN, N., WALKER, W. C. and WEISSLER, G. L. *J. appl. Phys.* 24 (1953) 1318; *Phys. Rev.* 99 (1955) 542.
86. HEDDLE, D. W. O. *J. quantve. Spectros. radiat. Transf.* 2 (1962) 349.
87. MENZEL, D. H. and PEKERIS, C. L. *Mon. Not. Roy. astr. Soc.* 96 (1935) 77; BURGESS, A. *Mon. Not. Roy. astr. Soc.* 118 (1958) 477.
88. KRAMERS, H. A. *Phil. Mag.* 46 (1923) 836.
89. PEACH, G. *Mon. Not. Roy. astr. Soc.* 124 (1962) 371.
90. BATES, D. R. *Mon. Not. Roy. astr. Soc.* 106 (1946) 423, 432; ARMSTRONG, B. H. *Proc. phys. Soc., Lond.* 74 (1959) 136; BATES, D. R. and SEATON, M. J. *Mon. Not. Roy. astr. Soc.* 109 (1949) 698; BATES, D. R., OPIK, U. and POOTS, G. *Proc. phys. Soc., Lond.* A66 (1953) 1113.
91. DALGARNO, A., HENRY, R. J. W., and STEWART, A. L. *Planet. Space Sci.* 12 (1964) 235; SEATON, M. J. *Proc. Roy. Soc.* A208 (1951) 408; A208 (1951) 418; STEWART, A. L. and WEBB, T. G. *Proc. phys. Soc., Lond.* 82 (1963) 532.
92. SEATON, M. J. *Mon. Not. Roy. astr. Soc.* 118 (1958) 504.
93. BURGESS, A. and SEATON, M. J. *Mon. Not. Roy. astr. Soc.* 120 (1960) 121.
94. ALTICK, P. L. Lawrence Radiation Laboratory, Berkeley, California, Report CCRL-10510 UC 34 Phys TID 4500.
95. COOPER, J. W. *Phys. Rev.* 128 (1962) 681; McGUIRE, E. J. Cornell University Physics Department, Technical Report 6 ARO (D), Project 2810p.
96. COOPER, J. W. *Phys. Rev.* 128 (1962) 681.
97. BRANDT, W. and LUNDQVIST, S. *J. quantve. Spectros. radiat. Transf.* 7 (1967) 411.
98. AMUSIA, M. YA. *Proc. 5th Int. Conf. Phys. electron. atom. Collisions*, p. 97, 1967. Leningrad; Akademii Nauk.
99. FLANNERY, M. R. and OPIK, U. *Proc. phys. Soc., Lond.* 86 (1965) 491.
100. GELTMAN, S. *Phys. Rev.* 112 (1958) 176.
101. DITCHBURN, R. W. and YOUNG, P. A. *J. atmos. terres. Phys.* 24 (1962) 127.
102. JARMAIN, W. and NICHOLLS, R. W. *Proc. phys. Soc., Lond.* 84 (1964) 417.
103. DUNN, G. H. Joint Institute of Laboratory Astrophysics Report 92, 1968.
104. KIEFFER, L. J. Joint Institute of Laboratory Astrophysics, Information Center Report 5, 1968.
105. METZGER, P. H. and COOK, G. R. *J. quantve. Spectros. radiat. Transf.* 4 (1964) 107.
106. BEYNON, J. D. E. and CAIRNS, R. B. *Proc. phys. Soc., Lond.* 86 (1965) 1343.
107. CAIRNS, R. B. and SAMSON, A. R. *Phys. Rev.* 139 (1965) A1403.
108. EDERER, D. L. and TOMBOULIAN, D. H. *Phys. Rev.* 113 (1964) A1525.
109. LUKIRSKII, A. P., BRYTOV, I. A. and ZIMKINA, T. H. *Optics Spectrosc., N.Y.* 17 (1964) 234.
110. GARTON, W. R. S. *Nature, Lond.* 166 (1950) 150; *Proc. phys. Soc., Lond.* A65 (1952) 268; GARTON, W. R. S. and CODLING, K. *Proc. phys. Soc., Lond.* 75 (1960) 87; GARTON, W. R. S. and RAJARATNAM, V. *Proc. phys. Soc., Lond.* A68 (1955) 1107; PERY-THORNE, A. and GARTON, W. R. S. *Proc. phys. Soc., Lond.* 76 (1960) 833.
111. BEUTLER, H. *Z. Phys.* 86 (1933) 495, 710; BEUTLER, H., DEUBNER, A. and JÜNGER, H. O. *Z. phys.* 98 (1936) 181.
112. DITCHBURN, R. W. and HUDSON, R. D. *Proc. Roy. Soc.* A256 (1960) 53; KAISER, T. R. *Proc. phys. Soc., Lond.* 75 (1960) 152.
113. MARR, G. V. *Proc. phys. Soc., Lond.* A224 (1954) 83.
114. HUFFMAN, R. E., TANAKA, Y. and LARRABEE, J. C. *J. chem. Phys.* 39 (1963) 902, 910; 40 (1964) 356.
115. LINDHOLM, E. *Ark. Fys.* 40 (1969) 97.
116. FANO, U. *Phys. Rev.* 124 (1961) 1866.
117. MADDEN, R. P. and CODLING, K. *Phys. Rev. Lett.* 10 (1963) 516; *J. opt. Soc. Amer.* 12 (1964) 106; *Astrophys. J.* 54 (1964) 268; SAMSON, J. A. R. *Phys. Rev.* 132 (1963) 2122.

118. FANO, U. *Atomic Physics*, p. 209, 1969. London; Plenum.
119. DIBELER, V. H., REESE, R. M. and KRAUSE, M. *J. chem. Phys.* 42 (1965) 2045.
120. COOK, G. R. and METZGER, P. H. *J. opt. Soc. Amer.* 54 (1964) 968.
121. FELDMAN, P. and NOVICK, R. *Proc. 3rd Int. Conf. Phys. electron. atom. Collisions*, p. 201, 1963. Amsterdam; North Holland Publishing Company.
122. HUFFMAN, R. E. United States Air Force, Physical Science Research Paper 66, 1964.
123. COMES, F. J. and LESSMAN, W. *Z. Naturf.* 16a (1961) 1396.
124. SCHONHEIT, E. *Z. Naturf.* 16a (1961) 1094.
125. VILESOV, F. I. *Soviet Phys. Usp.* 6 (1964) 888.
126. SAMSON, J. A. R. *Advanc. atomic molec. Phys.* 2 (1966) 178.
127. STEWART, A. L. *Advanc. atomic molec. Phys.* 3 (1967) 1.
128. ZIMKINA, T. *Proc. 5th Int. Conf. Phys. electron. atom. Collisions*, p. 132, 1967. Leningrad; Akademii Nauk.
129. BERKOWITZ, J., CHUPKA, W. A. and INGHRAM, M. G. *J. chem. Phys.* 26 (1957) 842; 27 (1957) 85, 87; 29 (1958) 653; 33 (1960) 533.
130. AL-JOBOURY, M. I. and TURNER, D. W. *J. chem. Soc.* 5 (1963) 5141.
131. DOOLITTLE, P. H. and SCHOEN, R. *Phys. Rev. Lett.* 14 (1965) 348.
132. WEISSLER, G. L., SAMSON, J. A. R., OGAWA, M. and COOK, G. R. *J. opt. Soc. Amer.* 49 (1959) 338.
133. FROST, D. C., MCDOWELL, C. A. and VROOM, D. A. *Phys. Rev. Lett.* 15 (1965) 612.
134. NICHOLLS, R. W. *J. quantve. Spectros. radiat. Transf.* 2 (1964) 433.
135. WACKS, M. R. *J. Res. nat. Bur. Stand.* 68A (1964) 631.
136. TURNER, D. W. *Physical Methods in Advanced Inorganic Chemistry*, Ed. M. A. O. Hill and P. Day, Chap. 3, 1968. London; Interscience; *Chem. in Brit.* 4 (1968) 435; *Molecular Photoelectron Spectroscopy*. 1970. New York; Wiley–Interscience.
137. SIEGBAHN, K. *et al. Nova Acta Soc. Sci. upsal.* 20 (1967).
138. SHAPIRO, J. and BREIT, G. *Phys. Rev.* 113 (1950) 179.
139. SEATON, M. J. *Mon. Not. Roy. astr. Soc.* 115 (1955) 279.
140. COMMINS, E. D. *Atomic Spectra and Radiation Processes Conf.* The Institute of Physics and The Physical Society 1965.
141. KAISER, W. and GARRETT, C. G. B. *Phys. Rev. Lett.* 7 (1961) 229.
142. GOEPPERT-MAYER, M. *Ann. Phys., Lpz.* 9 (1931) 273; KLEINMAN, D. Unpublished calculations.
143. HAMMERLING, P. *Lasers Symp.* Royal Society, 1962.
144. LIPELES, M., NOVICK, R. and TOLK, N. *Phys. Rev. Lett.* 15 (1965) 690.
145. MCMAHON, J., SORET, R. A, and FRANKLIN, A. R. *Phys. Rev. Lett.* 14 (1965) 1060.
146. DEWRI, R. P. Thesis, University of London, 1969.
147. ABELLA, I. D. *Phys. Rev. Lett.* 9 (1962) 453.
148. RIZZO, J. E. and KLEWE, R. C. *Brit. J. appl. Phys.* 17 (1966) 1137.
149. BEBB, H. B. *Phys. Rev.* 153 (1967) 23.

Chapter 10

ELASTIC COLLISIONS BETWEEN ATOMIC PARTICLES

10.1 INTRODUCTION

This chapter is concerned with elastic collisions between atomic species, both charged and uncharged. From measured total and differential cross-sections, information is obtained about the interaction energies between systems. Classical analysis of scattering is discussed in Chapter 2, but the quantum analysis discussed here enables both wave functions and inter-actions to be deduced. This is already to some extent possible; but, although there is a singular relationship between energy and wave function, the latter will not in its entirety be sensitive to the interaction. For sufficiently low-energy collisions, the interaction deduced from differential studies is adia-batic; however, for fast-particle scattering, the inelastic processes may be so strongly coupled that the interaction deduced from experiment is diabatic.

Studies of elastic scattering have gained momentum in recent years, and a number of review articles have appeared[1-4]. One of these[4] includes a cata-logue of neutral systems studied, types of measurement, and references complete to 1967.

It has been a conclusion of quantum theory for many years that the total scattering cross-section is finite if measured with sufficiently high angular resolution; classical analysis leads to an infinite cross-section. But experi-mental verification of this was only obtained in recent years, by Helbig and Pauly[2], who achieved an angular resolution of 5″ in the thermal energy scattering of potassium from various atomic targets.

10.2 SCATTERING OF FAST IONS AND ATOMS

The elastic scattering of fast ions and of fast atoms by atomic and molecular gases is important because of the light it throws on the interactions between the atomic systems at smaller nuclear separations than those investigated by near-thermal energy studies. The collisions differ from electron collisions in gases principally because the much smaller wavelengths of the projectiles diminish the diffraction effects.

Fast-atom scattering experiments were first carried out by Amdur and his colleagues[5], and ion beam scattering by Simons and his co-workers[6]. Other fast scattering measurements have been made by Berry[7], and by Everhart[8],

Fedorenko[9] and their colleagues. A determined attack upon the theoretical problems has been made by Mason and his collaborators[10, 11], and also by other workers[12]. Important new quantum effects have been found[13].

Experiments approximating sufficiently closely to ideal conditions are capable of yielding quantitative information about interatomic force fields $V(r)$ provided that there is already some approximate information about them. Absence of such information can lead to interpretation of accurate experiments by means of unrealistic functions; there is also a danger that the departure of the experiment from ideal conditions will lead to serious errors in the force fields. Much of the difficulty arises from the fact that there may be more than one impact parameter for which a particular value of projectile polar scattering angle θ results. Attractive and repulsive atomic force fields cannot be distinguished by the scattering they produce; therefore it is especially difficult to analyse the scattering in atomic fields which contain both components, dominating at different nuclear separations r.

Thus the formal inversion procedures[14] for deriving the interaction $V(r)$ from the differential scattering function $d\sigma_c/d\Omega_c(\theta_c)$ are not always used.

At the smallest scattering angles, the collision process cannot be described classically and no relationship between $V(r)$ and $d\sigma_c/d\Omega_c(\theta_c)$ can easily be deduced. For the study of $V(r)$, therefore, measurements at the smallest angles are not made. The critical angle θ_{cc} below which classical theory fails is given by the Massey–Mohr criterion[15]:

$$\theta_{cc} \sim \frac{\lambda}{2r_0} = \frac{\pi \hbar}{\mu v_1 r_0} \tag{10.1}$$

where λ is the projectile wavelength, r_0 is the internuclear distance at closest approach, v_1 is the impact velocity, and μ is the reduced mass of the system. Experiments now show that classical formulation holds to angles smaller than this criterion by almost an order of magnitude. The angle θ_{cc} is measured in the centre-of-mass system of coordinates, and subsequent discussion in this section is in this system unless specified.

In laboratory coordinates the Massey–Mohr critical scattering angle for He in He is $\sim 10°$ at thermal energies, $\sim \frac{1}{2}°$ at 10 eV and $\sim \frac{1}{20}°$ at 1 keV.

Another condition of validity of classical mechanics may be deduced from the momentum uncertainty[16]:

$$\frac{\hbar v_1}{2r_0 V(r_0)} \ll 1$$

The two types of scattering experiment to which analysis is normally applied are:

1. The measurement over a range of impact energies of total cross-section $\sigma(\theta_{c0})$ for scattering into all angles greater than the aperture θ_{c0} subtended by the beam collector;
2. The measurement of differential cross-sections at single impact energy.

Results of the first experiments were analysed[17] using the classical relationship

$$\sigma(\theta_{c0}) \propto \left(\frac{C}{\theta_{c0} \mathscr{E}} \right)^{2/s}$$

for an interaction

$$V(r) = \pm \frac{C}{r^s} \tag{10.2}$$

where θ_{co} is measured in centre-of-mass coordinates, whilst $\mathcal{E} = \frac{1}{2}\mu v_1^2$ represents the relative impact energy in this system. The aperture must be large enough for the scattering to be classical; at $\theta_{co} = 0$,

$$\sigma(0) \propto \left(\frac{C}{\hbar v_1}\right)^{2/(s-1)} \tag{10.3}$$

The measurement provides data from which the impact parameter function $b(\theta_c)$ can be calculated, using the equation

$$\theta_c(b, \mathcal{E}) = \pi - 2b \int_{r_0}^{\infty} \left[1 - \frac{b^2}{r^2} - \frac{V(r)}{\mathcal{E}}\right]^{-1/2} r^{-2} \, dr \tag{10.4}$$

where r_0, the closest distance of approach, is given by

$$1 - \frac{b^2}{r_0^2} - \frac{V(r_0)}{\mathcal{E}} = 0 \tag{10.5}$$

The contribution to the scattering from separations close to r_0 is by far the most important. Within this general equation there can be many functions $V(r)$ which reproduce the scattering data. Therefore the form of $V(r)$ must be assumed on the basis of previous knowledge.

The solution is greatly simplified when θ_c is small. For scattering of heavy particles by light ones ($m_1 \gg m_2$), this condition may apply even for large laboratory scattering angles. Elimination of b from equation 10.4 yields

$$\theta_c = \frac{r_0}{\mathcal{E}} \int_{r_0}^{\infty} [V(r_0) - V(r)] (r^2 - r_0^2)^{-3/2} r \, dr \tag{10.6}$$

The total cross-section can be written

$$\sigma(\theta_{co}) = \pi r_0^2 \left[1 - \frac{V(r_0)}{\mathcal{E}}\right] \simeq \pi r_0^2 \tag{10.7}$$

For the interaction

$$V(r) = \pm \frac{C}{r^s} \tag{10.8}$$

the solution has the form

$$\sigma(\theta_{co}) = \pi \left[\frac{C_s C}{\mathcal{E}\theta_{co}}\right]^{2/s} \left(1 \pm \frac{\theta_{co}}{C_s}\right) \tag{10.9}$$

with

$$C_s = \frac{\pi^{1/2} \Gamma(\tfrac{1}{2}s + \tfrac{1}{2})}{\Gamma(\tfrac{1}{2}s)} \qquad (10.10)$$

A plot of $\ln \sigma(\theta_{co})$ against $\ln \mathscr{E}$ is linear and of slope $-2/s$, and C may be calculated from the intercept.

The solution for the differential cross-section[18] is

$$\frac{d\sigma}{d\Omega}(\theta) = \frac{d\sigma_c}{d\Omega_c}(\theta_c) = \frac{1}{s}\left[\frac{(s-1)C_sC}{2\mathscr{E}}\right]^{2/s} \theta_c^{-(2s+2)/s} \qquad (10.11)$$

At small distances from the nucleus, an exponential (Born–Mayer) potential is appropriate:

$$V(r) = \pm A \exp\left(-\frac{r_s}{k^s}\right) \qquad (10.12)$$

with A, K and s constants. For this interaction, $\sigma(\theta_{co})^{s/2}$ is linear with $\ln \mathscr{E}$, as has been found, for example, for H_3^+ in He. Tabulations of some $\sigma(\theta_{co})(\mathscr{E})$ functions for Morse two-term exponential functions have been given[11].

A method of obtaining the potential function numerically, without the need of analytical assumptions, has been worked out by Firsov[12] and applied to high-energy scattering data[8] (*Figure 10.1*); since these data are governed by the force field at small nuclear distances, analytical assumptions would be particularly uncertain, although it is found that the Firsov function (equation 2.16) agrees fairly well with the numerical potentials. The He–He and Ar⁺–Ar repulsive interactions are illustrated in *Figure 10.2*. The Amdur equation referred to in *Figure 10.2b* is

$$V = 196 \exp(-4.21r)$$

V being in electron volts and r in Ångstrom units; this equation represents all the data within about 20 per cent.

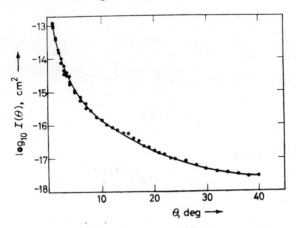

Fig. 10.1. Elastic scattering[8] of 2 keV Ar⁺ by Ar, in laboratory coordinates; a smooth line is drawn through the experimental points

(From Lane and Everhart[8], by courtesy of the American Physical Society)

34*

In the Firsov analysis, the differential cross-section is related to the impact parameter by the equation

$$2\pi \frac{d\sigma_c}{d\Omega_c}(\theta_c) \sin \theta_c \, d\theta_c = -2\pi b \, db \tag{10.13}$$

Therefore

$$b^2 = 2 \int_{\theta_c}^{\pi} \frac{d\sigma_c}{d\Omega_c}(\theta_c) \sin \theta_c \, d\theta_c \tag{10.14}$$

One may introduce equation 10.4 in the form

$$\theta_c = \pi - \int_{r_0}^{\infty} 2\frac{b}{r}\left[\left(1-\frac{V(r)}{\mathcal{E}}\right)r^2 - b^2\right]^{-1/2} dr \tag{10.15}$$

and define a new variable

$$\psi = \left[1-\frac{V(r)}{\mathcal{E}}\right]r^2 \tag{10.16}$$

Substitution into equation 10.15 leads eventually to

$$r(\psi) = \psi^{1/2} \exp\left[\frac{1}{\pi}\int_{\psi^{1/2}}^{\infty} \theta_c(b)\,(b^2-\psi)^{-1/2}\,db\right] \tag{10.17}$$

If the function $r(\psi)$ is known numerically for any value of ψ, then, from equations 10.16 and 10.17, $V(r)$ can be calculated numerically. For simplification, new variables are introduced:

$$u = \frac{\beta}{b} \tag{10.18}$$

and

$$s = \frac{\beta}{\psi^{1/2}} \tag{10.19}$$

where

$$\beta = \frac{Z_1 Z_2 e^2}{\mathcal{E}} \tag{10.20}$$

Equation 10.17 becomes

$$\frac{r}{\beta} = \frac{1}{s}\exp\left[\frac{I(s)}{\pi}\right] \tag{10.21}$$

with

$$I(s) = \int_0^s \frac{\theta_c(u)}{u}\left(1-\frac{u^2}{s^2}\right)^{-1/2} du \tag{10.22}$$

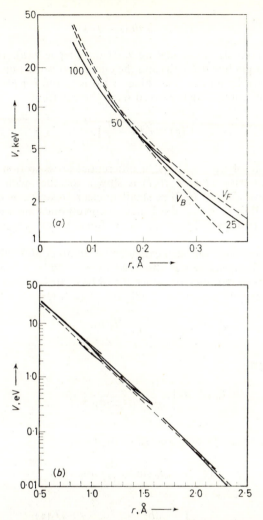

Fig. 10.2. Repulsive potential functions: (a) Ar⁺–Ar *at small nuclear separations (full lines, potentials derived from experimental data at impact energies 25* keV *and 100* keV; *broken line* V_B, *the Bohr function; broken line* V_F, *the Firsov function)*
   (From Lane and Everhart[8], by courtesy of the American Physical Society);
(b) comparison of beam-derived results for He–He *with theoretical calculations and with empirical potentials (based on high-temperature thermal conductivity data and on viscosity and virial coefficient data) (broken line, Amdur's equation)*

That is,

$$I(s) = \int_0^s \frac{\theta_c(s)\,\dfrac{u}{s}}{u\left(1 - \dfrac{u^2}{s^2}\right)^{1/2}}\,\mathrm{d}u - \int_0^s \frac{\left[\theta_c(s)\,\dfrac{u}{s} - \theta_c(u)\right]}{u\left(1 - \dfrac{u^2}{s^2}\right)^{1/2}}\,\mathrm{d}u \qquad (10.23)$$

or

$$I(s) = \tfrac{1}{2}\pi\theta_c(s) - I_1(s) \qquad (10.24)$$

The procedure consists of calculating $\theta_c(b)$ and thence $\theta_c(u)$. The difference between the straight line $\theta(s)(u/s)$ and the curve $\theta(u)$ is the numerator of the integrand of $I_1(s)$. From $I(s)$, the value of r/β is found for each value of s. The potential function is then derived from the equation

$$V(r) = \mathcal{E}\left(1 - \frac{sr}{\beta}\right)^{-2} \qquad (10.25)$$

As an example of this analysis, a differential cross-section for Ar^+–Ar high-energy scattering (*Figure 10.1*) is shown together with the Ar^+–Ar potential function (*Figure 10.2*) at small nuclear separations derived from these data. It will be seen that this function corresponds more closely to the Firsov function (equation 2.16) than to the Born–Mayer screened Coulomb function.

The Firsov procedure has been extended by Smith et al.[19] in the following way. For small-angle scattering, the quantity $\tau = \mathcal{E}\theta_c$ is a function almost solely of the impact parameter b, since it can be expressed as a power series in reciprocal energy in which the coefficient of each term is a function of b alone[20]:

$$\tau = \mathcal{E}\theta_c = \tau_0(b) + \frac{\tau_1(b)}{\mathcal{E}} + \dots \qquad (10.26)$$

with the first term

$$\tau_0 = -b \int_b^\infty \frac{dV}{dr} \frac{1}{(r^2 - b^2)^{1/2}} \, dr \qquad (10.27)$$

A scaling principle can be constructed of the form

$$\mathcal{F}(\tau, \eta) = \theta_c \sin \theta_c \frac{d\sigma_c}{d\Omega_c} (\theta_c, \mathcal{E}) \qquad (10.28)$$

or

$$\mathcal{F}(\tau, \eta) = \frac{1}{2} \left| \frac{db^2}{d(\ln \tau)} \right| = \mathcal{F}_0(\tau) + \frac{\mathcal{F}_1(\tau)}{\mathcal{E}} \qquad (10.29)$$

By means of such analysis, differential scattering experiments at different energies can be compared. Integrating the first term of equation 10.29 yields

$$b_0^2(\tau) = 2 \int_\tau^\infty \mathcal{F}_0(\tau) \, d(\ln \tau) \qquad (10.30)$$

whence $\tau_0(b)$ is obtained. The interaction potential is calculated by an integration

$$V(r) = \frac{2}{\pi} \int_r^\infty \frac{\tau_0(b)}{(b^2 - r^2)^{1/2}} \, db \qquad (10.31)$$

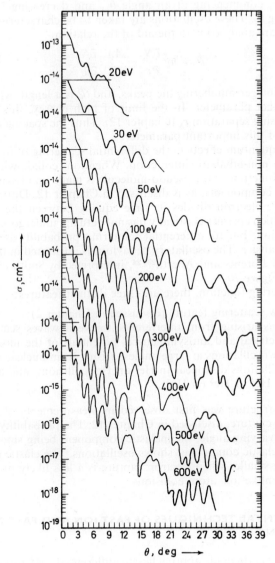

Fig. 10.3. Elastic differential scattering cross-sections for He⁺–He, *in laboratory coordi-nates*[23, 24]

A further feature of interference scattering[13, 22] arises in those collisions for which the distance of closest approach coincides with the nuclear separation of a pseudo-crossing of two interactions.

In certain differential scattering functions, sets of strong regular oscillations appear[21], commencing at an angle θ_{cx} and decreasing in amplitude with increasing θ_c. These oscillations are taken to be characteristic of curve-crossing, and are analysed with the aid of the relation

$$\mathcal{R} = hv\,\frac{\partial N}{\partial \tau} = \frac{4\pi}{k}\,\frac{\partial N}{\partial \theta_c} \qquad (10.32)$$

Here N is an integer numbering the peaks, and $\mathcal{R}$ is a length which is effectively the impact parameter. In the limit of high energy, this approaches the curve-crossing separation r_x (Chapter 12). Thus the spacing of the peaks is a measure of this important parameter.

Oscillatory quantum effects in the differential scattering of ions by atoms can arise from non-adiabatic interaction. When ions collide with their own gas, the wave function of the pseudo-molecule is made up of symmetric and antisymmetric components, as is discussed in Chapter 12. During the collision, the active electron divides its time equally between the two atoms. The fast ion which emerges elastically has not actually undergone a complete inelastic collision, but its differential scattering probability is affected by this electron transfer. The oscillatory behaviour is illustrated in the beautiful experiments of Lorents and Aberth[23, 24], interpreted by Smith *et al.*[12]; data are shown in *Figure 10.3*.

In this scattering function, there are three distinct features:

1. A rainbow scattering feature (discussed in Chapter 2);
2. Oscillations arising from interference between waves scattered from the symmetrical and antisymmetrical potentials of the molecular ion;
3. Secondary oscillations at large angles, arising from nuclear interchange (these oscillations disappear in He^3 on He^4 collisions, and appear only in He^4 on He^4 or He^3 on He^3).

Oscillatory structure was first observed in measurements of differential scattering with capture, discussed in Chapter 12. The probability of capture oscillates with varying angle, the inelastic component being smoothly varying, while the elastic component shows oscillations. The elastic component oscillations are smaller when electron capture is a less likely process, as in non-resonant unlike ion–atom collisions.

10.3 EXPERIMENTAL TECHNIQUES OF FAST ION AND FAST ATOM SCATTERING

Five different experimental approaches to differential and total scattering have been used by the following authors: Amdur and colleagues[5]; Simons and colleagues[6] and Ziegler[25]; Berry[7]; Everhart and colleagues[8] and Fedorenko and colleagues[9]; and Ramsauer and colleagues[26, 27].

The scattered ions must in differential scattering experiments be proved to have undergone only elastic collisions. Both collidants must finish in the same energy levels as those which they held before the collision. The best method of achieving this is by means of high-resolution electrostatic momentum analysis after the collision.

Elastic scattering experiments have been carried out with both ion and fast atom beams; the latter are usually produced by charge transfer neutralization. The principal unsolved problem is that of ensuring that the population of excited atoms is negligible.

The measurement of total cross-section for particles scattered through angles greater than a certain angle $\sigma(\theta_0)$ has been made[5] using a short collision chamber, and a detector separated from it. The collision path length is not very accurately defined by such an arrangement, since the effusion of gas, particularly from the exit orifice, must be taken into account. It is therefore good modern practice to conduct this experiment using crossed beams; in this way, the collision region and also the collision density are better defined. This also applies to differential scattering experiments, although the development of techniques for movable collision chamber exit slits, discussed in Section 3.3, has been timely. For $\theta \gtrsim 30°$, the simplest collision chamber is constructed with flexible bellows[8].

A fundamental problem of differential scattering experiments is to determine the angular resolution of the apparatus. From the dimensions of collimating slits, the angular distribution of the beams can be inferred, using geometrical optics. Analysis of the effects of finite slit width has long been available[28], but the effects of slit length are also important[23, 24].

Suppose the calculated pre-collision angular distribution function is $P(\alpha, \beta)$ and the calculated post-collision angular distribution function is $P'(\alpha', \beta')$ (both of these can be measured). If the experimental measured scattering function is $R(\theta_0, \alpha, \beta, \alpha', \beta')$ when the exit slit is set at laboratory polar angle θ_0, then the differential scattering cross-section $d\sigma/d\Omega(\theta)$ is given by:

$$R(\theta_0, \alpha, \beta, \alpha', \beta') = \int P(\alpha, \beta)\, P'(\alpha', \beta')\, \frac{d\sigma}{d\Omega}(\theta)\, d\alpha\, d\beta\, d\alpha'\, d\beta' \quad (10.33)$$

with

$$\theta = [(\theta_0 - \alpha' + \alpha)^2 + (\beta' - \beta)^2]^{1/2} \quad (10.34)$$

The angle variables α and α' are across the slits, whilst β and β' are along them. Evaluation of this equation (a deconvolution process) is not straightforward, but techniques are discussed in Chapter 1 and elsewhere[29].

Thermal motion of the target gas must be taken into account, and of course is simplified in crossed-beam experimentation.

10.4 INTERACTION ENERGIES DERIVED FROM SCATTERING EXPERIMENTS

In Table 10.1 are summarized some interpretations of atom–atom and ion–atom scattering experiments, following Mason and Vanderslice[11].

There are very few collision processes that have been studied under comparable conditions by different workers; for example, it is possible to make only limited comparisons between the work of Amdur and of Berry, since the ranges of interaction are different; at least it may be said that there is no serious disagreement between different measurements.

Table 10.1. SOME INTERPRETATIONS OF ELASTIC SCATTERING EXPERIMENTS

System	Potential function (eV)	Range (Å)	Reference
He-He	$3.47r^{-5.02}$	0.97-1.48	Amdur, Jordan and Colgate[5]
	$4.71r^{-5.94}$	1.27-1.59	Amdur and Harkness[5]
Ne-Ne	$312r^{-9.90}$	1.76-2.13	Amdur and Mason, 1955[5]
Ar-Ar	$849r^{-8.33}$	2.18-2.69	Amdur and Mason, 1954[5]
H-He	$2.34r^{-2.29}$	1.16-1.71	Amdur and Mason, 1956[5]
He-Ar	$62.1r^{-7.25}$	1.64-2.27	Amdur, Mason and Harkness[5]
H+-He	$1.90\left[\left(\dfrac{0.76}{r}\right)^4 - 2\left(\dfrac{0.76}{r^2}\right)^2\right]$	0.63-1.59	Simons, Muschlitz and Unger[6]
	$-6.72r^{-5}$	1.59-2.65	Mason and Vanderslice, 1957[10]
He+-He (attraction)	$2.16\left\{\exp\left[4.66\left(1-\dfrac{r}{1.080}\right)\right] - 2\exp\left[2.33\left(1-\dfrac{r}{1.080}\right)\right]\right\}$	0.9-3.8	Cramer and Simons[6] Mason and Vanderslice, 1958[10]
He+-He (repulsion)	$4.32\exp\left[2.33\left(1-\dfrac{r}{1.080}\right)\right]$	0.9-3.8	Cramer and Simons[6] Mason and Vanderslice, 1958[10]
H+-H₂	$2.7\left\{\exp\left[6\left(1-\dfrac{r}{1.5}\right)\right] - 2\exp\left[3\left(1-\dfrac{r}{1.5}\right)\right]\right\}$	1.5-3.7	Simons, Fontana, Muschlitz and Jackson[6] Mason and Vanderslice, 1959[10]
H⁻-He	$18.1\exp\left(-\dfrac{r}{0.491}\right)$	0.77-2.0	Bailey, May and Muschlitz[6] Mason and Vanderslice, 1958[10]

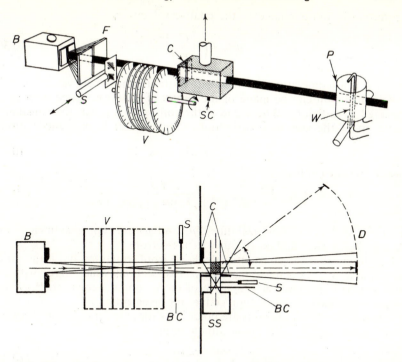

Fig. 10.5. *Schematic diagrams of two atomic beam scattering equipments for measuring differential cross-sections of velocity-selected beams in collision with gas and with another beam: B, primary beam source; F, fore slit; V, velocity selector; S, beam shutter; BC, beam chopper; C, collimating slits; SC, scattering chamber; W, hot-wire surface ionizer; P, positive ion collector; SS, secondary beam source; D, detector*

chamber in which the gas pressure is measured, and one or more detectors. More recently, velocity selectors have been inserted in the path of the atomic beam, and crossed-beam experiments have been carried out, using mechanical chopping and phase-sensitive detection. Intensity limitations common to the detection of thermal energy atoms have so far prevented the performance of experiments with two crossed velocity-selected beams. Typical experimental set-ups are illustrated in *Figure 10.5*.

In experiments designed to study the total collision cross-section for scattering of collision-free effusing beams by gases, correction must be made for the distribution of relative velocities of collision, using the methods of Chapter 1. The early experiments, which are most impressive for their period, are fully described in textbooks[36, 37] and original papers[38-41]; in the present section are described recent studies[42, 43], especially those due to Bernstein and his colleagues[44-46].

From the temperature variation of the total collision cross-section for a beam in a gas one may calculate, with the aid of Massey–Mohr theory[15], the long-range constant C in the isotropic potential function:

$$V(r) = -\frac{C}{r^6} \tag{10.35}$$

The cross-section is related to the relative velocity v_r:

$$\sigma = P\left(-\frac{C}{v_r}\right)^{2/5} \tag{10.36}$$

where P is a temperature-independent constant, whose best value[47] is 8.08.

A comparison can be made of the values of C so calculated with those expected from theory. Where one of the collidants is polar, the constant is made up of the sum of contributions (Section 2.3) from induced dipole

$$C_{ind} = \alpha_2\mu_1^2 + \alpha_1\mu_2^2 \tag{10.37}$$

and from dispersion

$$C_{disp} = \frac{3eh}{2\sqrt{m_e}} \frac{\alpha_1\alpha_2}{(\alpha_1/N_1)^{1/2} + (\alpha_2/N_2)^{1/2}} \tag{10.38}$$

where α_1 and α_2 are polarizabilities, μ_1 and μ_2 are dipole moments, and N_1 and N_2 are the numbers of outer shell electrons.

It is found that the constants calculated from equations 10.37 and 10.38 are capable of producing some eighty cross-sections which agree with experiment within a few per cent. Use of more complex equations[48] for the dispersion does not affect appreciable improvement.

When both of the collidants are polar, as is the case for collisions between alkali halide molecules, a dipole–dipole term is included:

$$C_{dd} = \frac{2\mu_1^2\mu_2^2}{3kT} \tag{10.39}$$

For different gas and beam temperatures T_1 and T_2, a substitution is made for T:

$$T = \frac{2T_1T_2}{T_1+T_2} \tag{10.40}$$

However, the agreement between experiments with certain molecular targets (NO, H_2S, CHF_3, CH_2F_2, *cis*-$C_2H_2Cl_2$, NH_3) is no better than semi-quantitative.

It is now known[49] that the apparent cross-sections and resulting values of C measured in these experiments suffered from McLeod gauge error (Section 3.2).

For the quantum analysis of scattering in terms of the interaction energies and phase shifts, the reduced notation given in Section 2.7 is adopted. The differential cross-section is related to the phase shifts η_l:

$$\frac{d\sigma_c}{d\Omega_c}(\theta_c) = \frac{1}{4k^2}\left[\left|\sum_l (2l+1)\sin 2\eta_l P_l(\cos\theta_c)\right|^2 \right.$$
$$\left. + \left|\sum_l (2l+1)(\cos 2\eta_l - 1)P_l(\cos\theta_c)\right|^2\right] \tag{10.41}$$

with $k = (2\mu E_c/\hbar^2)^{1/2}$. Often in the application of this equation the Legendre polynomials are replaced by asymptotic expressions, valid for large l; also, the summation is replaced by an integral.

The total scattering cross-section is given by

$$\sigma = \frac{4\pi}{k^2} \sum_{l=0}^{\infty} (2l+1) \sin^2 \eta_l \tag{10.42}$$

The radial wave equation possesses radial eigenfunction solutions R_l expressed in terms of reduced radius ϱ:

$$\frac{d^2 R_l(\varrho)}{d\varrho^2} + \left[A^2 - \frac{2\mu r_m^2}{\hbar^2} V(\varrho) - \frac{l(l+1)}{\varrho^2} \right] R_l(\varrho) = 0 \tag{10.43}$$

Solutions of this equation are obtained with the aid of various approximations which connect the phase shifts η_l with the interaction energy $V(r)$ or reduced interaction $U(\varrho)$: the Born approximation; the Jeffrey's WKB approximation; the high-energy approximation; and the Massey and Mohr approximation.

The Born approximation yields

$$\eta_l = -\frac{\pi\mu}{\hbar^2} \int_0^{\infty} [J_{(l+1/2)}(kr)]^2 V(r) \, r \, dr \tag{10.44}$$

where $J_{(l+1/2)}$ is the Bessel function.

The Jeffrey's WKB approximation, modified by Langer, yields in reduced notation:

$$\eta_l = A \left\{ \int_{\beta}^{\infty} \left(1 - \frac{\beta^2}{\varrho^2} \right)^{1/2} d\varrho - \int_{\varrho_0}^{\infty} \left[1 - \frac{U(\varrho)}{\mathscr{E}} - \frac{\beta^2}{\varrho^2} \right]^{1/2} d\varrho \right\} \tag{10.45}$$

Typical calculations are those for $\text{Li}^+\text{–He}$[50]; discussion of the technique is available[51].

The high-energy approximation, as used for nuclear interactions at very high energies, gives

$$\eta_l = \frac{1}{2\hbar g} \int_{-\infty}^{\infty} V(z) \, dz \tag{10.46}$$

with $r^2 = b^2 + z^2$.

Massey and Mohr[15] made the following approximation: for heavy particles with wavelengths very much smaller than the range of interaction, many partial waves must be used, with phase-shifts rapidly varying with l up to some value l_i; therefore the $\sin^2 \eta_l$ term sum can be replaced up to $l = l_i$ by $\frac{1}{2}$. For $l > l_i$ (large b), the interaction is sufficiently weak for phase shifts to be evaluated by Born or Jeffrey's WKB approximations, as above. Bernstein[52] has improved upon this, by computing η_l for many values of l.

The Massey–Mohr phase shift is

$$\eta_l = -\frac{\mu}{\hbar^2} \int_{r_0}^{\infty} \frac{V(r)}{[k^2 - (l+\frac{1}{2})^2/r^2]^{1/2}} \, dr \tag{10.47}$$

and the total scattering cross-section becomes

$$\frac{\sigma_{MM}}{\pi r_m^2} = \frac{2s-3}{s-2}\left[C\frac{f(s)A}{K}\right]^{2/(s-1)} \tag{10.48}$$

Landau and Lifschitz[54] have derived the total cross-section

$$\sigma_{LL} = \frac{2\pi(s-2)\sigma_{MM}}{(s-1)(2s-3)[2/(s-1)]!\sin[\pi/(s-1)]} \tag{10.49}$$

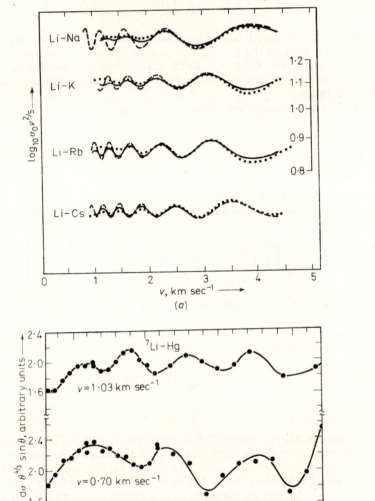

Fig. 10.6.(a) *Undulations in* σ_0 (v) *for lithium–alkali metal scattering plotted as* $v^{2/5}\log_{10}\sigma_0$ *versus* v *(broken lines are calculations for chosen* ε *and* r_m; *solid lines have velocity distribution folded in)* ; (b) *undulations in differential scattering cross-section for* [7]Li–Hg *plotted as* dσ $\theta^{4/3}\sin\theta$ *versus* θ *(full line is drawn smoothly through experimental points)*

Over a range of impact velocity, the Massey–Mohr cross-section σ_{MM} is a mean value, and the real cross-section oscillates about it. The number of maxima in $\sigma_0(v)$ is related to the number of quantum-mechanical bound states of the pseudo-molecule of collision[53]. Putting

$$\sigma = \sigma_{MM} + \Delta\sigma \tag{10.50}$$

the undulatory cross-section function is given by

$$\Delta\sigma(v) = -4\beta_0 \left(\frac{2\pi}{-\theta_0' A}\right)^{1/2} \cos\left(2\eta_0 - \frac{\pi}{4}\right) \tag{10.51}$$

The undulatory behaviour of $\sigma_0(v)$ (*Figure 10.6a*), which is well-known in experiment[34, 35], is smaller for heavier atoms. It occurs because there is a broad maximum in the function $\eta(l)$, which provides a significant number of non-random phases. In *Figure 10.6b* are shown the corresponding undulations in differential cross-section. From the extrema in $\sigma_0(v)$, the interaction parameters ε and r_m can be inferred[55]. For a pure van der Waals' interaction $s = 6$, the graph of $\ln \sigma_0$ against $\ln v$ has a slope of $-\frac{2}{5}$ (this is found to be correct within ± 2 per cent) on which the undulations are superposed. A number N is assigned to each maximum, and the slope of the graph of $N - \frac{3}{8}$ against v^{-1} gives the product εr_m in the interaction potential; the curved portion of this graph can yield $\varepsilon^2 r_m$, and also a third parameter[56, 57]. For a Lennard–Jones 6–12 potential

$$n - \frac{3}{8} = \frac{\varepsilon r_m}{\hbar v_n}\left[\frac{3n}{16(n-1)}\right]\left[\frac{5\pi n}{32(n-1)f(n)}\right]^{5/(n-6)} \tag{10.52}$$

where v_n is the velocity at which the nth extremum occurs. Also

$$C = \frac{n}{n-6}\varepsilon r_m^6 \tag{10.53}$$

and from equation 10.50

$$\frac{|\Delta\sigma_{max}|}{\pi r_m^2} = 4\beta_0\left(\frac{2\pi}{-\theta_0' A}\right)^{1/2} \tag{10.54}$$

For differential scattering, equation 10.11 is reproduced[34, 35, 58, 59] by

$$\left(\frac{d\sigma_c}{d\Omega_c}\right)\theta_c \propto \theta_c^{-4/3}\operatorname{cosec}\theta_c \propto \theta_c^{-7/3}$$

but there are undulations and rainbow scattering.

Differential cross-sections become nearly independent of angle at angles smaller than a critical value

$$\theta_c = \frac{\hbar}{2\mu v}\left(\frac{\pi}{\sigma_{tot}}\right)^{1/2} \tag{10.55}$$

This corresponds to the quantum uncertainty in the position of the scatterer during collision.

Table 10.2. RESULTS OF RESEARCH ON NEUTRAL ATOMS

Target	H	D	Li	Na	K	Rb	Cs	Ga	He	He*	Ne	Ne*	Ar	Kr	Xe	Ag	H_2	D_2	Li_2	K_2	N_2	O_2	I_2	KCl	KBr	CsF	CsCl	CsBr	TlF
Na	X		2	C	C	C	C																X		X				
K	H		2	C	C	C	C																	X	X				
Rb	H		2	C	C	C	C																						
Cs	.		2	C	C	C	C																						
He	X	2	2	X	2	X	C	2	C	C	C	C	C	C	C		X	2	X							C	C	C	C
Ne	H	2	X	X	2	X	C	C	C	C	H	C	C	C	C		2	C	C						X	C	C	C	C
Ar	H	2	2	2	2	C	C	H	H	H	C	C	C	C	C	X	2	2	2					C		C	C	C	C
Kr		2	2	2	2	X	2	H	H		C	C	C	C	C	X	2	2	X					C		X	C		
Xe		2	2	2	2	C	2	C	H		C		C	C	C	C	C	2			H			C			C	C	X
Zn				X	X																								
Cd				X	X	X																							
Hg	C	X	X	2	2	2	2	2	2		C	C	C	C	C	2	2	C	X			X	X			C	C	X	X
H_2	X	H	X	X	X	X	C	C	H	H	H			C		C	2	C	X	X	H					C	C		
D_2				X	X	X	C	C						C		C			X		X	X		C					
N_2				C	C	C	C	H					C	H		C			X		X	X							X
O_2				X	X	X	X																C						
Cl_2				C	C	X	C																						
Br_2	C			C	C	X	X																						
I_2	X			X	X	X	X																						
HCl																													
HBr			2	2	X	X	X																						
HI					2	2																		C		C			
DBr				2																									
CO				X	C		X				H		H																
NO				X	C		C																			C			
ICl				X	X	X	X																						
IBr				X	X	X	X																						

C, Data have yielded one potential constant, usually the van der Waals' constant C/r^6;
X, Some experimental data exist, but no potential parameters have been extracted;
2, Data have yielded two or more potential constants, usually ϵ and r_m;
H, Only high-energy data are available.

The rainbow scattering angle is independent of r_m, but sharply independent of E_c/ε, so the separate determination of both ε and r_m becomes possible[60]. The rainbow scattering angle θ_{cR} is also weakly dependent upon the repulsive interaction parameter α:

$$\theta_{cR} = -\frac{30\pi}{32k} \left[\frac{32f(n)(n-1)n}{30\pi} \right]^{-6/(n-6)} \tag{10.56}$$

and

$$\tilde{q} = \frac{1}{2} \frac{d^2\theta}{d\beta^2} = \frac{45\pi n}{32k} \left[\frac{32f(n)(n-1)n}{30\pi} \right]^{-8/(n-6)} \tag{10.57}$$

If these equations are used and a value of n assumed, ε and r_m may be determined from the variation of rainbow angle with impact velocity. Table 10.2 summarizes the status of research on neutral systems.

One very interesting feature of atomic interactions is the appearance of resonance states[61-63]. Calculations have been made of quasi-bound positive energy states of H_2, lying below the rotational energy barrier. These orbiting or tunnelling resonances[64] may be observed spectroscopically; only two have been observed, at $v = 14$, $J = 4$ and at $v = 14$, $J = 51$, and their widths have been obtained.

10.6 GAS DIFFUSION

Interaction energies are most successfully studied by scattering experiments, but additional information can be obtained from virial coefficients and also from the diffusion of gases[65-67], which will be discussed briefly.

The Chapman–Enskog formula[68, 69] for the diffusion coefficient in a binary gas mixture of atom densities n_1 and n_2 is

$$D_{12} = \frac{3}{16(n_1+n_2)} \left(\frac{2\pi kT}{\mu} \right)^{1/2} \left(\frac{1+\varepsilon_0}{\bar{\sigma}_d} \right) \tag{10.58}$$

with

$$\bar{\sigma}_d = \frac{1}{2} \int_0^\infty \tau^2 \sigma_d(\tau) \exp(-\tau) \, d\tau \tag{10.59}$$

The 'reduced temperature' τ is

$$\tau = \frac{\mu v_r^2}{kT} \tag{10.60}$$

for relative velocity v_r and reduced mass μ. The correction factor ε_0 is usually much less than 0.05[70]. If the momentum transfer cross-section σ_d is temperature independent, D_{12} is proportional to $(T/\mu)^{1/2}$.

Following the methods of Chapter 2, σ_d can be related to the scattering phase shifts η_l;

$$\sigma_d = \frac{4\pi}{k^2} \sum_{l=0}^\infty [(l+1) \sin^2(\eta_l - \eta_{l+1})] \tag{10.61}$$

with

$$k = \frac{\mu v_r}{\hbar} \tag{10.62}$$

35*

The Jeffrey's WBK approximation for the phase shift yields

$$\eta(b) = k \left\{ \int_{r_c}^{\infty} \left[1 - \frac{V(r)}{\varepsilon} \frac{b^2}{r^2} \right]^{1/2} dr - \int_{r_c}^{\infty} \left(1 - \frac{b^2}{r^2} \right) dr \right\} \qquad (10.63)$$

but in collisions between heavy particles a large number of phase shifts contribute, so the summation in equation 10.61 may be replaced by an integral. In this case:

$$\sigma_d = 4\pi \int_0^{\infty} b \cos^2 \left\{ \int_{r_c}^{\infty} \frac{dr}{r \left[\frac{r^2}{b^2} - 1 - \frac{r^2}{b^2} \frac{V(r)}{\varepsilon} \right]^{1/2}} \right\} db \qquad (10.64)$$

This is identical with the formula obtained by classical methods. A quantum description is only necessary when there is a multiplicity of interaction potentials; this is always the case for the passage of an excited atom through its own gas, and also for ground state atoms in their own gas, unless the electron shells are closed. The success of the classical theory is due to the factor $(1-\cos\theta)$ in the formula for σ_d, which suppresses the small-angle contribution, important only in quantum theory.

For an interaction potential $V(r) = Cr^{-n}$, σ_d is proportional to $T^{-2/n}$, and D_{12} at constant density to $T^{(2/n+1/2)}$. Therefore, for the r^{-6} long-range low-temperature interaction between ground state atoms, $D \propto T^{5/6}$, and, for the approximate r^{-12} shorter-range high-temperature interaction, D is approximately proportional to $T^{2/3}$. For a He–N_2 mixture, $T^{0.691}$ is found experimentally[71].

Equations 10.61–10.64 have been used in calculations of D_{12}, with the aid of the modified Buckingham potential[72], and of the Lennard–Jones 6–12 potential[65, 73–77]:

$$V(r) = 4A \left[\left(\frac{A}{r} \right)^{12} - \left(\frac{B}{r} \right)^6 \right] \qquad (10.65)$$

For pure gases, the parameters A and B may be derived from velocity data, and combined as follows:

$$A_{12} = (A_{11}A_{12})^{\frac{1}{2}} \qquad (10.66)$$

$$B_{12} = \tfrac{1}{2}(B_{11}+B_{12}) \qquad (10.67)$$

to give, typically,

$$D_{\text{He–Ar}} = 0.653 \text{ cm}^2 \text{ sec}^{-1} \text{ at } 273°\text{K}$$

for comparison with $D_{\text{exp}} = 0.641$ cm^2 sec^{-1}. Binary diffusion coefficients vary from about 0.1 cm^2 sec^{-1} for large atoms to about 1 cm^2 sec^{-1} for small atoms. The available experimental data include measurements[78–81] carried out on most of the common gases, by classical experimental techniques[82] at room temperature and below, and by a new point-source technique[83] at higher temperatures.

The diffusion coefficients of excited atoms will, in general, be smaller than those of ground state atoms, because of the strong long-range interaction potentials. The diffusion may be complicated by inelastic processes. For excited atoms in their own gas (as well as non-closed shell ground atoms in their own gas), the multiplicity of interaction potentials complicates the situation, as in the case of ionic mobility. Since two systems are equally likely to approach each other along either of the potentials, the phase shifts η_l are determined by the even configuration (for example, $H_2\,^1\Sigma_g$) when l is odd, and by the odd configuration ($H_2\,^3\Sigma_u$) when l is even. Although there is no relation between the classical and quantum expressions for σ_d, it may be shown that $\sigma_d \simeq 2\sigma_t$, where σ_t is the cross-section for the excitation exchange process $0'0/00'$ (Chapter 13). This normally results in large σ_d and small D_{12}; thus for self-diffusion in atomic hydrogen at s.t.p., $D_{11} \sim 0.3$ cm^2 sec^{-1}, which is much lower than the value expected on classical grounds. For Ne ^3P metastable atoms[84], $D_{12} \sim 0.2$ cm^2 sec^{-1}, whilst for ground state neon (closed shell) $D_{11} \sim 0.5$ cm^2 sec^{-1}. For He 2^3S, calculations[85] have shown that the symmetry effect is unusually small.

10.7 THE MOBILITY OF IONS IN GASES

Under the action of a static electric field, the motion of ions in gases contains both random and drift components, as does the motion of electrons (discussed in Chapter 4); the drift velocity v_d of the ions is found to be proportional to the field strength X. The constant of proportionality K is called the ionic mobility:

$$v_d = KX \qquad (10.68)$$

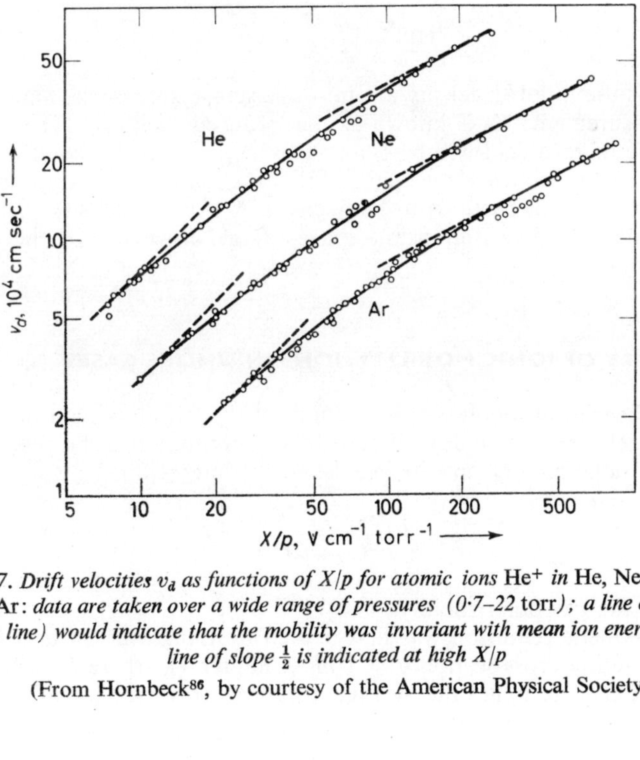

Fig. 10.7. *Drift velocities v_d as functions of X/p for atomic ions He$^+$ in He, Ne$^+$ in Ne and Ar$^+$ in Ar: data are taken over a wide range of pressures (0.7–22 torr); a line of slope unity (broken line) would indicate that the mobility was invariant with mean ion energy; a broken line of slope $\frac{1}{2}$ is indicated at high X/p*

(From Hornbeck[86], by courtesy of the American Physical Society)

This 'constant' is affected by the gas temperature and pressure, and by the energy acquired by the ions in the field. The parameter which determines the mean energy acquired by the ions in the field is X/p, because the velocity acquired in the mean free time t_f is $(eXt_f)/m_+$. Since t_f is inversely proportional to the pressure p, the energy obtained between collisions varies as $(X/p)^2$. A distinction must be made between conditions where the energy acquired from the field is smaller than thermal energy (low X/p, < 2 V cm^{-1} torr^{-1}) and the contrasting condition of high X/p, where the acquired energy is much greater than thermal. Typically, the dependence of drift velocity upon X/p is as shown in *Figure 10.7*. At low X/p the $v_d(X/p)$ function is linear (slope 1 on log–log plot), but at high X/p it is one-half power (slope $\frac{1}{2}$ on log–log plot). This effect is found also for H_2, D_2[87] and O_2[88]; it was predicted on the basis of a hard-sphere model by Wannier and Hornbeck[89].

Ionic mobility is related to ionic diffusion, and thereby to the diffusion cross-section. A knowledge of the variation of this cross-section with gas temperature is important in the study of ion–atom interaction energies at large nuclear separations which are inaccessible to ion beam scattering experiments; for this reason, ionic mobilities will be discussed in this chapter, even though the important inelastic contributions to them must be considered at the same time.

Ionic mobilities at pressures in the range of several torr are of the order of several hundred cm^2 V^{-1} sec^{-1} and are the same for positive and negative ions of the same species. Mobilities are found to be inversely proportional to number density and are usually reduced to s.t.p.:

$$K_0 = \frac{Kp}{760} \times \frac{273}{T} = \frac{n_0 K}{2 \cdot 69 \times 10^{19}} \qquad (10.69)$$

where n_0 is the number density at atmospheric pressure at the temperature of the measurement; K_0 is known as the 'reduced mobility'. The mobility appropriate to zero field will be denoted by K_{zero} or simply by K.

In mobility experiments it is assumed that the ion densities are sufficiently small for ion–ion interactions to be neglected[90]. An ion density of 10^8 ions cm^{-3} would introduce significant space-charge distortion of the electric field.

10.8 THEORY OF IONIC MOBILITY; IONS IN UNLIKE GASES

Classical theories of mobility take account of the two dominant potentials: the 'gas-kinetic' repulsion dominant at small internuclear distances; and the 'small-ion' polarization force giving rise to the interaction

$$V(r) = -\frac{\alpha e^2}{2r^4} \qquad (10.70)$$

where α is the gas polarizability.

The gas-kinetic repulsion was first treated by Langevin[91] in terms of the elastic scattering cross-section $\sigma = 1/nl_f$ (Chapter 1), where l_f is the mean free path. This determines the average distance $\bar{s}$ traversed by an ion in the field direction x between collisions:

$$\bar{s} = \frac{1}{2} a f^2 = \frac{Xe}{2m_+ \bar{v}^2} \frac{\int_0^\infty x^2 \sigma \exp(-\sigma x) \, dx}{\int_0^\infty \sigma \exp(-\sigma x) \, dx} \tag{10.71}$$

or

$$\bar{s} = \frac{Xe}{m_+ \bar{v}^2 \sigma^2} \tag{10.72}$$

The drift velocity v_d is equal to this distance divided by the mean free time $(l_f/\bar{v})$, so

$$v_d = \frac{Xe}{m_+ \bar{v} \sigma} \tag{10.73}$$

If the distributions of both the free paths and the velocities are considered as well as the difference in masses (m_+ and m_0) of the ion and neutral, then it follows[92] that the mobility

$$K = 0{\cdot}815 \frac{eL}{m_0 v_0} \left(1 + \frac{m_0}{m_+}\right)^{1/2} \tag{10.74}$$

where m_0 is the gas atom mass and v_0 its mean velocity, and $L = 1/n_0 \pi d^2$, d being the sum of the gas kinetic radii of ion and atom. This hard-sphere model is unrealistic in that the mobilities so predicted are about four times too high and depend upon ionic charge, on mean free path and on $T^{-1/2}$, none of which is found in experiment.

If the polarization force predominates in influence over the gas-kinetic repulsion, then the modified Langevin theory introduced by Hassé[93] yields

$$K = \frac{0{\cdot}5105}{[\varrho(\varepsilon - 1)]^{1/2}} \left(1 + \frac{m_0}{m_+}\right)^{1/2} \tag{10.75}$$

with gas density ϱ and dielectric constant ε. A parameter which is determined by $d^4/(\varepsilon - 1)$ can be introduced to govern the extent to which the mobility is dominated by one or the other equation. In many cases, the Hassé theory predicts mobilities in good agreement with experiment, but the assumed c/r^8 repulsive force is probably too soft, just as the impenetrable sphere force is too hard.

A more general approach which can be adapted to both classical and quantum theory follows from the Einstein relation

$$K = \frac{eD_+}{kT} \tag{10.76}$$

The positive ion diffusion coefficient D_+ is related to the diffusion cross-section σ_d by the Chapman–Enskog formula. One proceeds in a manner similar to equations 10.58 and 10.59:

$$D_+ = \frac{3\pi^{1/2}}{16(n_+ + n_0)} \left(\frac{2kT}{\mu}\right)^{1/2} \left(\frac{1 + \varepsilon_0}{P}\right) \tag{10.77}$$

for reduced mass μ, where ε_0 is a small correction factor ($\leqslant 0\cdot 13$) and

$$P = \tfrac{1}{2} \int_0^\infty \tau^2 \sigma_d(\tau) \exp(-\tau)\, d\tau \qquad (10.78)$$

with reduced temperature

$$\tau = \frac{\mu v_+^2}{2kT} \qquad (10.79)$$

The classical expression for σ_d at impact energy E is

$$\sigma_d = 4\pi \int_0^\infty b \cos^2 \alpha'\, db \qquad (10.80)$$

for impact parameter b and

$$\alpha' = \int_0^\infty \frac{dr}{r_0 \phi(r)} \qquad (10.81)$$

with

$$\phi(r) = \left[\frac{r^2}{b^2} - 1 - \frac{r^2}{b^2}\frac{V(r)}{E}\right]^{1/2} \qquad (10.82)$$

r_0 is the outermost zero of the function $\phi(R)$. The classical evaluation[91, 93] of σ_d leads to

$$\sigma_d = 2\cdot 210\pi q^2 \qquad (10.83)$$

where the diffusion radius

$$q = \left(-\frac{C}{E}\right)^{1/4} \qquad (10.84)$$

and

$$C = -\tfrac{1}{2}\alpha e^2 \qquad (10.85)$$

But Wannier[90, 94] has proposed that

$$\sigma_d = 2\pi q^2 \qquad (10.86)$$

is a better classical approximation. Quantum theory calculations have been shown by Dalgarno, McDowell and Williams[95] to lead to

$$\sigma_d = 2\cdot 120\pi q^2 \qquad (10.87)$$

There is thus little difference between the results of classical and quantum theory. As is the case for gas diffusion, this situation stems from the $1 - \cos\theta$ factor in the diffusion cross-section σ_d, which suppresses the small-angle scattering, important in quantum theory. The quantum calculations lead to a reduced mobility:

$$K_0 = \frac{35\cdot 9}{\sqrt{\alpha\mu}}\ \ \text{cm}^2\,\text{V}^{-1}\,\text{sec}^{-1} \qquad (10.88)$$

at gas density 2.69×10^{19} atoms cm^{-3}; α is in (a.u.)^3, and μ is the reduced mass in a.m.u. The mobility is temperature independent for a pure r^{-4} interaction, and it follows that, for a particular gas, $K_0 \sqrt{\mu}$ is independent of the nature of the ion and of its charge state. The temperature dependence that follows from the more complex polarization plus dispersion interaction

$$V(r) = -\left(\frac{a}{r^4} + \frac{b}{r^6}\right) \tag{10.89}$$

is not very sensitive:

$$K_0 \sqrt{\mu} \propto \left(1 + \frac{2b}{3a}\sqrt{\frac{kT}{a\pi}}\right)^{-1}$$

At higher temperatures, the mobility is increasingly dependent upon the short-range repulsive forces, and ultimately decreases steadily with increasing temperature. At sufficiently low temperatures, the reduced mobility becomes temperature invariant.

Since, according to equations 10.80–10.82, the diffusion cross-section is inversely dependent upon impact velocity, the mean free time t_f between collisions is independent of energy. This condition is known as the 'constant mean free time condition', and when it applies the intercollision velocity is Xet_f/m_+, and the intercollision energy is proportional to $(X/p)^2$. The energy contributed to the ions is negligible compared with their random energy when

$$\left(\frac{m_g}{m_+} + \frac{m_+}{m_g}\right)Xel_f \ll kT_g$$

For an interaction $V(r) \propto r^{-n}$, $\sigma_d \propto v^{-4/n}$, so

$$K_0 \propto T^{(2/n-1/2)}$$

Although the ion drift velocity is proportional to field strength X where this is sufficiently small, the mobility at higher field strengths can, according to the theory of Kihara[96], be expressed as a power series in X^2:

$$K = K_{zero} + K_2 X^2 + K_4 X^4 + \ldots \tag{10.90}$$

where K_n are complicated functions of collision integrals. Mason and Schamp[97] have shown that this expression should be valid up to $X/p \simeq 10$ V cm^{-1} torr^{-1}. It is possible with the aid of their analysis to derive the interaction energy from the zero-field mobility and its variation with X/p. For pure r^{-4} interaction, all K_n with $n > 0$ vanish, and, at temperatures corresponding to maxima or minima in the $K(T)$ function, K_2 vanishes.

It has already been mentioned that on a hard-sphere model v_d varies with $(X/p)^{1/2}$ and is independent of temperature at high X/p[98, 99].

10.9 THEORY OF IONIC MOBILITY; IONS IN THEIR PARENT GAS

There are two possible interactions between an atomic ion and its own gas atom, $V^+(r)$ and $V^-(r)$. If η_l^+ and η_l^- are the lth order phase shifts associated with elastic scattering by these potentials, then the diffusion cross-section

$$\sigma_d = 4\pi \int_0^\infty b \sin^2(\eta_l^+ + \eta_l^-)\, db \tag{10.91}$$

for impact parameter b[95]. It has been known since the pioneering calculations of Massey and co-workers[15, 100], that the cross-section σ_t for the 'symmetrical resonance' charge transfer process

$$X^+ + X \rightarrow X + X^+ \tag{10/01}$$

is exactly half this value, provided that the impact energy is not so small that effects arising from large-angle scattering need to be taken into account. In most cases the result will be valid above 400°K:

$$\sigma_t = \tfrac{1}{2}\sigma_d \tag{10.92}$$

Thus the symmetrical resonance charge transfer cross-sections available from beam experiments (Chapter 12) should be consistent with the mobility

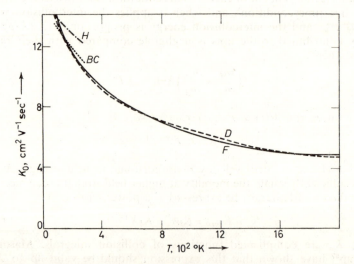

Fig. 10.8. *Comparisons of rare gas reduced mobilities with theory for* He$^+$–He: *full line F, experimental data of Frost*[104]; *broken line D, calculations of Dalgarno*[105]; *dotted line BC, experiments of Biondi and Chanin; chain line H, calculations of Holstein*[106]

measurements for ions in their parent gas[101]. The charge transfer cross-section σ_t varies with impact velocity as follows:

$$\sigma_t = (a \ln v - b)^2 \tag{10.93}$$

where a and b are constants[102]. It is possible to make use of this equation to check the consistency of the available data for σ_d and σ_t. When the average diffusion cross-section $\bar{\sigma}_d$ is established in units of πa_0^2, the temperature dependence of the mobility may be approximated by the relation

$$K_0 = \frac{2 \cdot 10 \times 10^4}{\bar{\sigma}_d \sqrt{\mu T}} \quad \text{cm}^2 \, \text{V}^{-1} \, \text{sec}^{-1} \tag{10.94}$$

In this equation,

$$\bar{\sigma}_d = \frac{1}{2} \int_0^\infty x^2 \sigma_d(x) \exp(-x) \, dx \tag{10.95}$$

and

$$x = \frac{\mu v^2}{kT}$$

Rare gas atomic ion cross-sections have been calculated using these equations, and are discussed further in Section 12.1.

The most accurate theoretical potentials for He$^+$–He have been set up by Dickinson[103] and are consistent with differential scattering experiments; the calculated charge transfer cross-section function is illustrated in *Figure 12.7* (page 618). The oscillations in the function are caused by orbiting and resonance phase shifts in the scattering, and may be expected to appear at energies as high as 0·37 eV. The total cross-section is averaged well by the Schiff–Landau–Lifschitz expression, the Massey–Mohr approximation being uniformly 7 per cent lower:

Experimental mobility data over a gas temperature range are shown in *Figure 10.8* for He$^+$–He, in *Figure 10.9* for He$_2^+$–He, and in *Figure 10.10* for Ne$^+$–Ne and Ar$^+$–Ar.

A classical treatment of mobility proceeding by symmetrical resonance charge transfer was given by Wannier[90, 94] and by Sena[111] in terms of the 'relay race' action by which the positive charge is passed from atom to atom. When random velocity is neglected in comparison with drift velocity, the post-collision ion velocity $v_p = 0$, so $v_d = \langle \frac{1}{2} v_a \rangle$, the ante-collision velocity being averaged over all intercollision path lengths. The ante-collision velocity is given by the relation

$$\tfrac{1}{2} m_+ v_a^2 = Xel \tag{10.96}$$

and the relative probability of an ion travelling a distance l without undergoing charge transfer is $l_c^{-1} \exp(-l/l_c)$, where $l_c^{-1} = n\sigma_t$ is the mean free

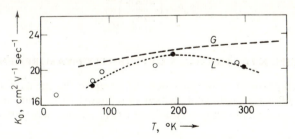

Fig. 10.9. Comparisons of rare gas reduced mobilities with theory for He$_2^+$–He: *closed circles, data of Biondi and Chanin[107-109]; open circles, Tyndall and Pearce[110]; L, calculations of Langevin; G, calculations of Geltman*

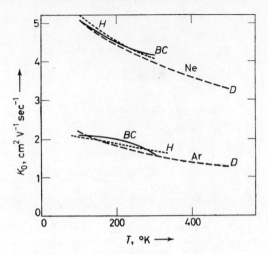

Fig. 10.10. Comparisons of rare gas reduced mobilities with theory for Ne⁺–Ne and Ar⁺–Ar: H, calculations of Holstein[106]; BC, experiments of Biondi and Chanin; D, calculations of Dalgarno[105]

path for charge transfer; therefore

$$v_d = \left(\frac{eX}{2m_+}\right)^{1/2} \frac{1}{l_c} \int_0^\infty l^{1/2} \exp\left(-\frac{l}{l_c}\right) dl \qquad (10.97)$$

or

$$v_d = \left(\frac{95\pi e}{m_+ n_L \sigma_t}\right)^{1/2} \left(\frac{X}{p}\right)^{1/2} \qquad (10.98)$$

Here n is the gas atom density, n_L is Loschmidt's number and the pressure p is in torr; σ_t is averaged over an appropriate energy distribution. This type of argument is valid only when the impact velocity variation of $\sigma_t|$ is sufficiently small to be neglected (constant mean free path conditions).

Wannier's formula is given in a rather different form:

$$v_d = \left(\frac{2eXl_c}{\pi m_+}\right)^{1/2} \qquad (10.99)$$

10.10 EXPERIMENTAL METHODS OF MOBILITY MEASUREMENT

With the exception of techniques depending upon the measurement of the ambipolar diffusion coefficient, which is related to the ionic mobility, all mobility measurements are essentially of drift velocity in a uniform electric field.

Historically, the earliest measurements were made by Rutherford, who observed the sharp diminution of drifting ion current on raising the frequency of an alternating field sufficiently high to prevent the ions reaching an electrode in one half cycle. The method was refined by Tyndall and Grindlay[112] and later by Bradbury[113]. Another early technique, known as the 'air blast technique' compared the velocities of ionization and of the gas molecules, but it was marred by turbulence effects. Use was also made of the magnetic deflexion technique well-known in electron drift velocity analysis[114, 115], but it was not until the development of the 'four-gauze' method by Tyndall, Starr and Powell[116] that mobility experiments were placed on a really sound footing.

The production of the ions in a drift tube may be achieved by one of the following methods:

1. A pulsed discharge may be struck between electrodes in the appropriate gas in the drift tube.
2. Photoelectrons from a metal surface or electrons from a thermionic filament may be accelerated in the gas in the drift tube.
3. A mass-analysed beam of ions may be passed through an orifice into the drift tube[117, 118].
4. X-rays and α-particles from radioactive sources may be used, as in the first mobility experiments.
5. Surface ionization sources producing alkali metal ions may be used in drift tubes.

In modern experiments, the determination of the ion drift velocity is achieved by the conventional methods of time measurement electronics.

Double-gate Method

Two gating pulses or a.c. signals are applied to electrodes, spaced a known distance apart, in the uniform electric field that is required in a drift velocity measurement. The uniform electric field is maintained by a series of closely spaced thick annular electrodes with potential difference between each. Only when there is the appropriate phase difference between the signals or delay between the pulses can ions pass both obstacles. Two types of electrode have been used:

1. The first derives from the Tyndall 'four-gauze' method and has been used by Hornbeck[86], Varney[119], and by Davies, Dutton and Llewellyn-Jones[120]. The signals or pulses are applied to two pairs of grids through which the ions drift to a collector. When the 'step' in the field is not removed by the pulse or the peak of the signal, the ions, being unable to pass it, must drift to the grid or radially outwards.
2. A special grid is constructed with alternate wires insulated from each other; there are no cross-wires. When an a.c. signal is applied between the sets of wires, the ions can only pass through them at the moments when the inter-wire potential difference is effectively zero; when an appreciable transverse field exists between the grid-wires, the ions are collected at the appropriate set. This type of filter grid was described by Bradbury and Nielson[121], and more recently it has been used by Crompton and Elford[122].

In both versions of the double-gate method, a steady ion current enters the region of the experiment, and a pulse of ions or a series of half sine-waves falls upon the collector when the delay or phase is correct. There is no reason why the collected current should not be integrated, although there is a loss of intensity dictated by the duty cycle.

The radial diffusion suffered by the ions during their passage through a drift tube may be calculated from the mean radial displacement $\bar{x}$ of a swarm of particles diffusing through a gas[123] after a time t:

$$\bar{x} = \left(\frac{4D_+ t}{\pi}\right)^{1/2} \tag{10.100}$$

Using equations 10.114 and 10.115

$$\frac{\bar{x}}{L} = 0 \cdot 172 V^{-1/2} \tag{10.101}$$

at temperature 300°K. Here L is the length of the drift tube and V is the potential difference in volts between the ends.

Pulsed Ion Source

An alternative drift velocity measurement technique consists of pulsing the source of ions and observing the collected current as a function of time on a calibrated oscilloscope. This is the method of Biondi and Chanin[107], who used a pulsed discharge as source, with a block of ions drifting out through the mesh which forms the discharge cathode, into a region of uniform field. The method can work down to very low values of X/p.

A similar technique has been used by Hornbeck and Molnar[124, 125] who use a simple arrangement of two electrodes in a gas. A block of photoelectrons is released from the cathode by ultra-violet light from a triggered spark source; the electrons pass through perforations in the anode, and, by their acceleration in a steady electric field, a Townsend avalanche is produced. A current of ions which is exponentially dependent upon time arrives at the cathode, and from its characteristic time the mobility is deduced.

In modern ion drift experiments, it is nearly always essential that the detected ions, whose velocity has been measured, be identified in mass-number. Experiments conducted without such identification can often measure the mobility of an ion which is identified incorrectly by inference (for example, by incorrect application of equation 10.88). Alkali metal ion fluxes from solid state sources are largely atomic, but molecular ions are never easy to identify by inference. Moreover, ions passing through drift tubes undergo inelastic processes in which their mass number is changed (for example, ion–atom interchange, charge transfer, dissociation, clustering), and it is these processes which arouse much of the interest in drift experiments.

Ion identification is usually achieved by terminating the drift tube with a metal plate in which there is an axially placed sampling orifice sufficiently small (0·05 mm diameter) to allow a high vacuum to be maintained behind it. A mass-spectrometer and particle multiplier record the ion flux. For many mass-spectrometers, acceleration of the ions is necessary, but for the quadrupole instrument this is not so; the instrument is especially suited to this type of sampling, one reason being that the resolution may be adjusted electronically so that tests for absence of mass-discrimination can be made during the experiment.

Mobility from Ambipolar Diffusion

Another method of determining mobility is to infer it from the measured ambipolar diffusion coefficient. The decay of electron density in the after-glow of a discharge is governed by diffusion as well as by electron–ion recom-bination and attachment processes. In the absence of negative ions, the diffusion is ambipolar (Section 1.4); the plasma condition of a macroscopic balance of charge

$$|n_+ - n_e| \ll n_e$$

is maintained at all times.

The microwave cavity time-dependent afterglow technique[126] (see also Chapter 7) has been used to determine the ambipolar diffusion coefficient D_a, which is related to the positive ion and electron mobilities K_+ and K_e and diffusion coefficients D_+ and D_e by the equation

$$D_a = \frac{D_+ K_e + D_e K_+}{K_+ + K_e} \tag{10.102}$$

The use of this diffusion coefficient enables the particle current densities J_e and J_+ to be equated ($J_+ = J_e$), whilst the electron diffusion equations for a space configuration of electrons $F(n_e)$ are still satisfied. The equation

$$J_e = -D_e \text{ grad } n_e - K_e F(n_e) \tag{10.103}$$

becomes

$$J = -D_a \text{ grad } n, \tag{10.104}$$

where $n \simeq n_e \simeq n_+$ and $J \simeq J_e \simeq J_+$; and in decaying plasma

$$\text{div } J + \frac{dn}{dt} = 0 \tag{10.105}$$

or

$$-D_a \nabla^2 n + \frac{dn}{dt} = 0 \tag{10.106}$$

The decay of space-averaged electron density, proceeding by diffusion fol-lowed by fast wall-recombination, is observed to be exponential with time, being defined by the equation

$$\frac{n_e}{n_{e0}} = \exp\left(\frac{-t}{\tau}\right) \tag{10.107}$$

Thus one may write

$$\nabla^2 n_e + \frac{n_e}{\tau D_a} = 0 \tag{10.108}$$

This equation is solved for the conditions of finite n_e and densities vanishing at the walls of the container. For example, in the simplest geometry (a sphere of radius R), equation 10.108 is written in terms of a radius variable r:

$$\frac{d^2(rn)}{dr^2}+\frac{rn}{D_a\tau}=0 \tag{10.109}$$

whose solution is

$$n=\frac{C}{r}\sin\frac{r}{\sqrt{D_a\tau}} \tag{10.110}$$

where

$$\frac{R}{\sqrt{D_a\tau}}=s\pi \tag{10.111}$$

and $s=1,2,\ldots$ is an integer defining the mode of diffusion. Thus the experimental determination of the diffusion time τ yields the ambipolar diffusion coefficient. For $s=1$, the fundamental mode,

$$\tau=\frac{R^2}{\pi^2 D_a} \tag{10.112}$$

Both the positive ion and electron diffusion coefficients are related to the appropriate mobilities by the Einstein relation[123]

$$\frac{D_+}{K_+}=\frac{D_e}{K_e}=\frac{kT}{e} \tag{10.113}$$

where e is the electronic charge and k the Boltzmann constant. Since $K_e\gg K_+$,

$$D_a\simeq 2D_+ \tag{10.114}$$

$$K_+\simeq\frac{eD_a}{2kT}=1\cdot17\times10^4\frac{D_a}{2T} \tag{10.115}$$

with D_a in cm^2 sec^{-1} and K in cm^2 V^{-1} sec^{-1}. In the absence of attachment processes, an exponential electron density decay may readily be related to the mobility; however, it has been shown in the analysis of Oskam[127] that the presence of higher diffusion modes is not necessarily disproved by the observation of a purely exponential decay of space-averaged electron density, nor is this electron density always directly proportional to the frequency shift of the microwave cavity resonance[128]. The numerical analysis of the combined diffusion and recombination equations is discussed in Chapter 7.

Ambipolar diffusion in the presence of more than two charged components has been discussed in Chapter 1. For a single positive ion species and electrons, there is only a single ambipolar diffusion coefficient, governing both the electron and the positive ion decay, and approximately equal to the diffusion coefficient for positive ions:

$$D_a\simeq D_+\left(1+\frac{T_e}{T_g}\right) \tag{10.116}$$

But the presence of a second positive ion component results in three ambipolar diffusion coefficients, one for each ion (1 and 2) and one for the electrons:

$$D_{ae} \simeq \frac{n_1}{n_e} D_{a1} + \frac{n_2}{n_e} D_{a2} \qquad (10.117)$$

$$D_{a1} \simeq D_{+1}\left(1 + \frac{T_e}{T_g}\right) \qquad (10.118)$$

$$D_{a2} \simeq D_{+2}\left(1 + \frac{T_e}{T_g}\right) \qquad (10.119)$$

Thus the diffusion of one ion is unaffected by the presence of the other, but if one ion can be converted into the other inelastically, then the ambipolar diffusion coefficient for the electrons will be time variant, producing a departure from exponential decay.

In an electro-negative gas, where a proportion β of negative ions is present, $\beta = n_-/n_e$ and

$$n_e + n_- \simeq n_+ \qquad (10.120)$$

There are again three ambipolar diffusion coefficients:

$$D_{ae} \simeq (1+\beta)D_+\left(1 + \frac{T_e}{T_g}\right) + \beta D_-\left(\frac{T_e}{T_g} - 1\right) \qquad (10.121)$$

$$D_{a-} \simeq 2(1+\beta)\left(\frac{D_e + D_-}{D_e}\right) - D_-\left(\frac{T_e}{T_g} - 1\right) \qquad (10.122)$$

$$D_{a+} \simeq D_+\left(1 + \frac{T_e}{T_g}\right) \qquad (10.123)$$

The positive ion diffusion is unaffected by the presence of negative ions, but both electron and negative ion diffusion are time invariant only when β is time invariant, or alternatively when $\beta \ll 1$. In oxygen afterglows, for example, a purely exponential electron decay is usually observed, with D_{ae} corresponding to a constant β, which may exceed unity by an order of magnitude; but the mechanism by which β is apparently maintained constant is not yet understood.

Modern afterglow studies of ambipolar diffusion must be accompanied by identification of the ionic species, and measurement of their time-dependence.

Electron density measurements in time-dependent afterglows have usually been made by means of microwave cavity measurements of plasma, but other electron density diagnostic techniques can be suitable, for example, metal probes[129].

A complication that arises in diffusion studies is the phenomenon of diffusion cooling[130, 131]. The product of ambipolar diffusion coefficient and pressure should be pressure independent: $pD_a = $ constant. However, with pressure decreasing below about 0·1 torr, pD_a is observed to diminish, which is consistent with the idea that the mean electron temperature in an afterglow actually falls below the gas temperature. Equation 10.114 should be written

$$D_a = D_+ \left(1 + \frac{|A\bar{v}_e}{\bar{v}_+}\right) \qquad (10.124)$$

where A is a parameter of the order of unity. For electron velocities much smaller than the ion velocities, $D_a \simeq D_+$, that is, one-half the value D_a achieves for equal electron and ion temperatures.

The mechanism of the electron temperature reduction is simply that the rate of energy loss due to faster diffusion of higher energy electrons exceeds the rate of energy gain from the neutral gas atoms. This applies only when the energy exchange ('thermal contact') between electrons and atoms is relatively weak, as would be the case for heavy atoms.

10.11 INTERPRETATION OF MOBILITY DATA

Mobility data may be shown to conform to theoretical expectations in the following ways. Drift velocities in weak fields are proportional to the ratio X/p, and, at sufficiently high fields, proportional to $(X/p)^{1/2}$, as is shown in *Figure 10.7* (page 539). The low-field drift velocity dependence implies a low-field constancy of mobility with increasing X/p. This constancy is realistic in the limit of low X/p but, from typical data shown in *Figure 10.11*, it appears that for ions in unlike gases the mobility usually increases with increasing X/p until a maximum value is reached; at higher values of X/p, it falls again.

Interpretation of the X/p variation of mobility suffers from the difficulty[90] that the distortion of the velocity distribution in the direction of the field at high X/p renders the Boltzmann transport calculations relating K and σ_d invalid. However it is possible to make a comparison of this variation with the theory of Kihara[96] by using the method devised by Mason and Schamp[97] (equation 10.129). Comparison of interactions for K^+ ions in rare gases so obtained with those deduced from gas temperature variation is made[136] in *Figure 10.12*.

A comparison of experimental data with equation 10.88 has been made by Dalgarno[105]. Some values of a parameter K_0' defined as

$$K_0' = K_0 \sqrt{\mu} \qquad (10.125)$$

are tabulated in Table 10.3. The value of K_0' should be constant at $35·9/\sqrt{\alpha}$ for each gas for pure r^{-4} interaction; xenon and krypton approximate most closely to this condition, but the numerical constant in equation 10.88 must apparently be multiplied by a numerical factor 1·07 to obtain the best agreement, as is shown by the tabulations of 'theoretical K_0' values' in the first two rows of Table 10.3. It was proposed that the experimental measurements of Tyndall's group are all subject to an error of 7 per cent in this

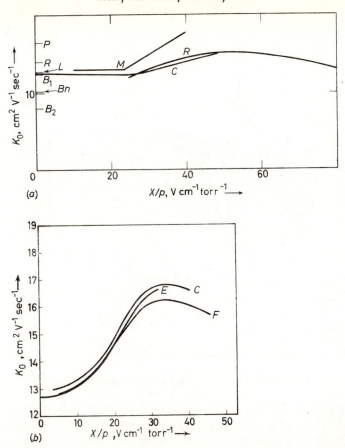

Fig. 10.11. *Reduced mobility as a function of* X/p: *(a) hydrogen ions (unidentified) in hydrogen (C, Chanin[132]; M, Mitchell and Ridler[133]; R, Rose[87]; P, Persson and Brown[134]; B_1 and B_2, Bradbury[113]; Bn, Bennet[135]; L, Lauer [125]); (b) K$^+$ ions in hydrogen (C, Creaser[136]; E, Elford; F, Fleming and Rees)*

direction, but according to the careful repetition of the work by Crompton and Elford[122], by Davies, Dutton and Llewellyn-Jones[120] and by Rees[137], no such error is to be found. The error arises from the neglect of inverse sixth power interaction in the theory.

A graphical representation of the mobilities of many ions in nitrogen is given in *Figure 10.13*; the conformity with equation 10.88 is well illustrated. Atomic ions in molecular gases may be expected to possess constant K_0' values, since it is possible to replace the angular-dependent interaction by spherically symmetrical interaction obtained by averaging over all molecular orientations; but K_0' values for different ions in H_2 and CO show rather large variations.

For interpretation of the temperature variation of atomic ion mobilities, it is necessary to apply trial interaction functions to bring the experimental data into harmony with theory. Thus Meyerott[139] and Dalgarno, McDowell

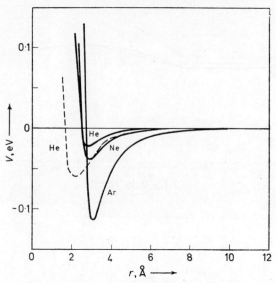

Fig 10.12. Interactions[136] for K+ in rare gases derived from mobility data: Ne, K+ *in neon;* Ar, K+ *in argon;* He, K+ *in helium (full lines, derived from X/p variation; broken line, derived from temperature variation of zero field mobility)*

and Williams[95] have applied interactions such as

$$V(r) = 74\cdot2 \exp(-2\cdot75r) - \frac{1\cdot39}{r^4} \quad \text{Rydberg units} \quad (10.126)$$

to the data for alkali metals in helium. For K+ ions in argon, Sida[140] has obtained reasonable agreement with the measurements of Hoselitz[141]. Neglect of the inverse sixth power term is an unfortunate feature of the analysis of Dalgarno, McDowell and Williams, but recent computations of McDowell have partly removed this limitation.

Table 10.3. SOME VALUES OF K_0' AT TEMPERATURES 290°–292°K

		Helium	Neon	Argon	Krypton	Xenon
Theoretical values	$35\cdot9/\sqrt\alpha$	30·5	21·9	10·8	8·9	6·9
	$38\cdot4/\sqrt\alpha$	32·8	23·5	11·6	9·6	7·4
Ions	Li+	38·6	30·4, 25·2	11·4	9·4	7·3
	Na+	41·9	26·8	11·5	9·3	7·5
	K+	41·0	27·4	11·7	9·6	7·4
	Rb+	39·3	27·2	11·7	9·5	7·4
	Cs+	36·3	25·5	11·5	9·5	7·4
	Ar+	39·9				7·7
	Hg+	26·5, 37·0				

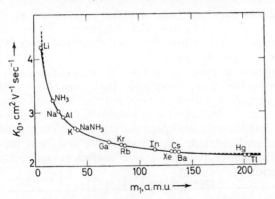

Fig. 10.13. Reduced mobilities of ions in N_2 *as a function of ion mass* m_1, *after Brata*[138]; *broken line represents normalized* $(1+m_2/m_1)^{1/2}$ *dependence*

Calculations have also been made for general interactions of the type:

$$V(x) = \frac{C}{r_{min}^4} \left[p \exp \left(\frac{-4(x-1)}{p} \right) - \frac{1}{x^4} \right] \qquad (10.127)$$

with

$$p = 1 - \frac{A r_{min}^4}{B} \qquad (10.128)$$

and $x = r/r_{min}$, where r_{min} is the nuclear separation at which the potential function passes through a minimum.

For interpretation of the X/p dependence of mobility on the theory of Kihara, Mason and Schamp[97] used the interaction

$$V(r) = \frac{A}{2} \left[1 + C \left(\frac{B}{r} \right)^{12} - 4C \left(\frac{B}{r} \right)^6 - 3(1-C) \left(\frac{B}{r} \right)^4 \right] \qquad (10.129)$$

They were able to reproduce most of the experimental data, taking A, B and C as empirical constants and making allowance for clustering (see *Figure 10.12*). Use of their potential leads to a dissociation energy 0·05–0·06 eV for LiHe$^+$.

The multiplicity of rare gas ion mobilities, as measured in afterglows and in drift tubes, caused some confusion in past years. Molecular, atomic and

Table 10.4. MOBILITIES (IN cm^2 V^{-1} sec^{-1}) OF RARE GAS IONS[131, 142]

Ions	Microwave		Drift tube	
	Atomic	*Molecular*	*Atomic*	*Molecular*
Helium	10·4	20·7	10·5	20·3
Neon	4	7·5	4	6·5
Argon	1·5	2·6	1·6	2·65

doubly-charged ions have now been identified mass-spectrometrically; their mobilities are tabulated in Table 10.4. Clustered ions are not detected unless the apparatus is at liquid nitrogen temperatures, when He_3^+ is found, as well as other rare gas clusters.

The interpretation of rare gas ion mobilities derived from microwave studies is only possible on the assumption that molecular ions dominate in the afterglow at pressures of several torr, but that atomic ions dominate at much lower pressures. That the He_2^+ ion dominates in high-pressure helium afterglows was first shown mass-spectrometrically by Phelps and Brown[143], after large measured recombination coefficients had been supposed[144] to arise from a dissociative process; a rate coefficient was derived for the three-body reaction

$$He^+ + 2\,He \rightarrow He_2^+ + He$$

which is further discussed in Chapter 14. The first microwave determination of helium mobility led to a value which differed from the early drift-tube measurements in which the ions were formed from α-particles. Moreover, the quantum theory calculations[15] are in better agreement with the drift tube measurements, and it was pointed out[139] that on theoretical grounds the molecular ion should have a larger mobility.

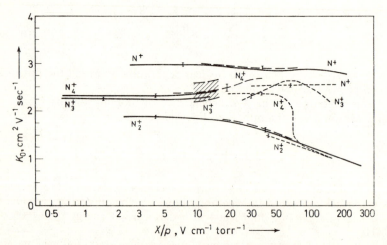

Fig. 10.14. Reduced mobilities of identified nitrogen ions in nitrogen at 300°K as functions of X/p: short broken lines, Saporoschenko: full lines, Moseley et al.; long broken lines, McKnight, McAfee and Edelson

The most interesting new discovery in this field[145] has been that in a drift tube two different mobilities for mass number 8 $\left(He_2^+ \text{ species}\right)$ are found. It is proposed[146] that one of them is the $^4\Sigma_u^+$ metastable ion, whereas the other is the ground state $X\,^2\Sigma_u^+$.

Extensive studies have been made during the past few years of the mobilities of the identified ionic species N^+, N_2^+, N_3^+ and N_4^+ in nitrogen[147-151], and H^+, H_2^+, H_3^+ and H_5^+ in hydrogen[152-155]. Nitrogen data are illustrated in *Figure 10.14*.

Mobility data for negative ions are not extensive; mass-spectrometric identification of ions has greatly benefited the understanding of them. Oxygen negative ion mobilities[156] have been studied[157] and the ions identified[158]; the mobilities are shown in *Figure 10.15*.

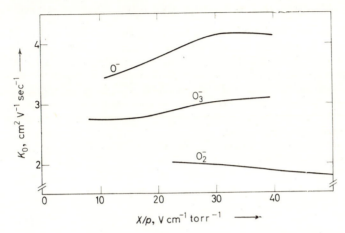

Fig. 10.15. *Reduced mobilities of negative ions in oxygen[156] as functions of X/p*

The Mobility of Ions in Gas Mixtures

The mobility of an ion is inversely proportional to the gas density:

$$K^{-1} = Gn_0 \qquad (10.130)$$

where G is a constant. If it is assumed that the nature of the ions does not change when two gases A and B are mixed, that X/p is small, and that ion–ion interactions and three-body collisions may be neglected, then

$$K^{-1}(A+B) = G(A)\,n(A) + G(B)\,n(B) = \frac{f(A)}{K(A)} + \frac{f(B)}{K(B)} \qquad (10.131)$$

where $f(A)$ and $f(B)$ are molar fractions, such that

$$f(A) + f(B) = 1$$

This relation[159] is known as Blanc's law. Much of the early data for alkali ions, and some more recent data[160], conform to this law; deviations from it may be due to the following causes:

1. It has been shown by Overhauser[161] that fixed and 'statistical' clustering (the increase in the local concentration of the more highly polarizable gas) leads to renormalization of the constants G.
2. The different velocity dependences of A^+A and A^+B mobilities may require the use of a more rigorous kinetic treatment, which could lead to curvature in the function $K(A+B)f(A)$. The effect is better known in the analysis of electron drift velocities in mixtures.

Blanc's law is found to be inapplicable to the analysis of the drift velocities of ions in isotopic mixtures, which determines in part the distribution of isotopes in discharges[162].

10.12 CLUSTERING, INTERCHANGE AND ATMOSPHERIC MOBILITIES

Since the earliest mobility studies, the phenomena of 'ageing' of ions and of anomalous apparent reduced masses in equation 10.88 have been well-known. It was proposed that ions could travel through gas with a loosely held cluster or sheath of atoms or molecules surrounding them. Especially likely to cluster are small ions, and molecules with permanent dipole moments. One must distinguish between clusters which are substantially stable and those in which a continual interchange of cluster molecules takes place during passage through a gas.

The formation of a cluster requires the dissipation of impact energy[163] by third-body stabilization.

Now that mass-spectrometric identification of ions in a gas can be achieved by monitoring through an orifice, detailed knowledge of clustering processes is becoming available.

For a pure polarization force surrounding hard spheres, the energy condition for stable clusters in a gas can be written:

$$\frac{\alpha e^2}{r^4} \geqslant 3kT$$

However, the likelihood of clustering is greatly reduced when the hard core is replaced by an inverse twelfth power repulsive potential[164]. This is in general agreement with the experimental data. Thus, Munson and Hoselitz[165] found that the only alkali ion to form clusters in rare gases was Li^+, which is favoured by its small radius. The $LiHe^+$ dissociation energy was shown to be ~ 0.07 eV.

The heavier rare gas atoms form clusters around Li^+ more easily than do light rare gas atoms. It is calculated that, at s.t.p., N_2 is not expected to cluster around Li^+, whereas two O_2 molecules and 28 CO_2 molecules can do so. Since there are probably about 50 vibrational levels in the typical cluster potential trough, classical calculations are unlikely to be much modified by quantum effects.

The first observation of the clustering of H_2O molecules around Li^+ was made by Munson and Tyndall[166]. Clustering of up to six H_2O molecules was found to be possible, and the relevance of this finding to the surrounding of ions in aqueous solution will be apparent. Nevertheless, the liquid water surroundings of an ion modify severely the energy conditions. The hydration in liquid is as much a matter of stability of orientation of the H_2O molecules as of exchange[167].

Studies of clustering in water vapour continue to be made[168, 169].

The dissociation of ion–atom cluster B^+A in a gas A at temperature T and zero electric field has been shown[170] to be governed by the following equilibrium equation

$$\frac{n(B^+A)}{n(B^+)} = 10.46 \times 10^{-4} \mu^{1/2} r^2 p \exp\left[\frac{(\alpha + \mu^2/3kT)e^2}{r^4}\right] \qquad (10.132)$$

where α is the polarizability of A, and μ its permanent dipole moment; the pressure p is in atmospheres, and the collisional radius r in Ångstrom units. The probability of clustering two A atoms is proportional to the square of the probability of clustering one.

The cluster atom interchange process

$$B^+A_1 + A_2 \rightarrow B^+A_2 + A_1$$

must affect the collisional momentum exchange and therefore the mobility. At high fields, it is possible[171] to relate the mobility to the interchange cross-section, on the assumption that the inelastic process dominates. The argument is an extension of the classical treatment of ion in its gas leading to equation 10.98. If rare gas molecular ion mobilities were dominated by this process, then cross-sections $\sim 10^{-15}$ cm^2 would be implied. In particular, the interchange process may be important in N_4^+ mobility in nitrogen. As is seen in *Figure 10.14*, a low mobility has been reported for this ion at values of X/p greater than a critical value.

Rare gas atomic ions are capable of combining not only with their own species but also with other rare gases: $NeAr^+$ has been reported in the mobility tube[172] and in a decaying plasma[173], and $HeNe^+$ has been studied spectroscopically[174-176]. These species are to be regarded as molecular ions rather than as clusters.

Atmospheric Ionic Mobility Studies

The earliest ionic mobility measurements were conducted in atmospheric air, but most of the studies reported in this chapter were made in pure gases in the pressure range $10^{-1} - 10$ torr. However, mobility studies of ions in atmospheric air are of importance in environmental research; they have been reviewed[177].

Ions of large molecular weight exist in atmospheric air. They are formed by cosmic and other ionizing radiation, followed by clustering, and can possess either positive or negative charge; those of negative charge have been thought to produce physiological effects. In corona discharges, which are used for the technological production of atmospheric negative ions, various long-lived excited species, chemical radicles and chemically active species are also formed; it may be that these are associated with some of the reported effects.

Since the mobilities of atmospheric ions are distributed over a range of four orders of magnitude, it is likely that there is a wide variation in their molecular weights; nevertheless, there is probably no single relation between the two parameters. Equation 10.88, in which the mobilities of heavy ions are approximately mass-independent, is a 'small-ion' equation. Multi-molecular ions do not necessarily exert simple inverse fourth power interactions. Furthermore, they are particularly subject to atom–atom interchange and other forms of inelastic process which seriously diminishes the mobility.

The 'mobility spectrum' or distribution of mobilities in an atmospheric specimen, is measured by means of an 'aspiration condenser', which consists of a flowing airstream in a tube fitted with anti-turbulence channels. At the wall of the tube, a source of ionization (for example, a radioactive source) is suddenly introduced, in such a way that the ions will be formed at the wall. Down the centre of the tube passes a concentric cable, and at a suitable distance downstream from the ionization source the cable is stripped to form a detector of ions at the inner wire. The outer wall of the concentric cable is maintained negative to the outer wall of the tube (for positive ion studies). Analysis of the current–time characteristic yields the distribution function of drift velocities (the mobility spectrum). The current is plotted against the inverse of drift velocity, which is proportional to the time.

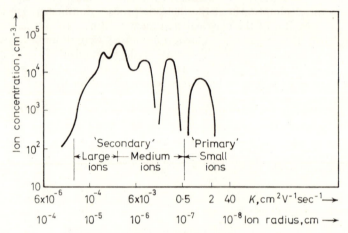

Fig. 10.16. Typical atmospheric mobility spectrum[178]

Ideally it consists of a piece-wise linear graph. Each of the linear sections is extrapolated back to zero v_d^{-1}, the distances between the points of intersection with the ordinate axis being proportional to the heights of the peaks in the mobility spectrum. Where it is necessary to study the mobility spectrum of ions already existing in an atmospheric specimen, an instantaneous introduction of the specimen into a flowing gas stream must be made. The flow velocity must be measured.

Typical atmospheric mobility spectra (*Figure 10.16*[178]) consist of three groups, corresponding to small primary ions, medium size secondary ions (possibly containing only one or two cluster sheaths), and, separated from these, and at mobilities 10^{-4} times smaller than those of the primary group, a group of large ions. These phenomena are related to the 'ageing' of ions as they pass through gases.

REFERENCES

1. PAULY, H. and TOENNIES, J. P. *Advanc. atomic. molec. Phys.* 1 (1965) 195.
2. PAULY, H. *Fortschr. Phys.* 9 (1961) 613.
3. BERNSTEIN, R. B. *Advances in Chemical Physics, Vol. 10: Molecular Beams*, Ed. J. Ross, p. 75, 1966. New York; Interscience.
4. BERNSTEIN, R. B. and MUCKERMAN, J. T. *Intermolecular Forces*, Ed. J. O. Hirschfelder, p. 389, 1967. New York; Interscience.

5. AMDUR, I. *J. chem. Phys.* 17 (1949) 844; 28 (1958) 987; *Science* 118 (1953) 567; AMDUR, I., DAVENPORT, D. E. and KELLS, M. C. *J. chem. Phys.* 18 (1950) 525; AMDUR, I., GLICK, C. F. and PEARLMAN, H. *Proc. Amer. Acad. Arts Sci.* 76 (1948) 101; AMDUR, I. and HARKNESS, A. L. *J. chem. Phys.* 22 (1954) 664; AMDUR, I., JORDAN, J. E. and COLGATE, S. O. *J. chem. Phys.* 34 (1961) 1525; AMDUR, I., KELLS, M. C. and DAVENPORT, D. E. *J. chem. Phys.* 18 (1950) 1676; AMDUR, I. and MASON, E. A. *J. chem. Phys.* 22 (1954) 670; 23 (1955) 415, 2268; 25 (1956) 624, 630, 632; *Phys. Fluids* 1 (1958) 370; AMDUR, I., MASON, E. A. and HARKNESS, A. L. *J. chem. Phys.* 22 (1954) 1071; AMDUR, I., MASON, E. A. and JORDAN, J. E. *J. chem. Phys.* 27 (1957) 527; AMDUR, I. and PEARLMAN, H. *J. chem. Phys.* 8 (1940) 7; AMDUR, I. and PEARLMAN, H. *J. chem. Phys.* 9 (1941) 503; AMDUR, I. and ROSS, J. *Combust. Flame,* 2 (1958) 412.
6. BAILEY, T. L., MAY, C. J. and MUSCHLITZ, E. E. *J. chem. Phys.* 26 (1957) 1446; CRAMER, W. H. *J. chem. Phys.* 28 (1958) 688; 30 (1959) 641; CRAMER, W. H. and MARCUS, A. B. *J. chem. Phys.* 32 (1960) 186; CRAMER, W. H. and SIMONS, J. H. *J. chem. Phys.* 26 (1957) 1272; MUSCHLITZ, E. E. *J. appl. Phys.* 28 (1957) 1414; MUSCHLITZ, E. E., BAILEY, T. L. and SIMONS, J. H. *J. chem. Phys.* 24 (1956) 1202; 26 (1957) 711; SIMONS, J. H. and CRAMER, W. J. *J. chem. Phys.* 18 (1950) 473; SIMONS, J. H., FONTANA, C. M., FRANCIS, H. T. and UNGER, L. G. *J. chem. Phys.* 11 (1943) 312; SIMONS, J. H., FONTANA, C. M., MUSCHLITZ, E. E. and JACKSON, S. R. *J. chem. Phys.* 11 (1943) 307; SIMONS, J. H., FRANCIS, H. T., FONTANA, C. M. and JACKSON, S. R. *Rev. sci. Instrum.* 13 (1942) 419; SIMONS, J. H., FRANCIS, H. T., MUSCHLITZ, E. E. and FRYBURG, G. C. *J. chem. Phys.* 11 (1943) 316; SIMONS, J. H. and FRYBURG, G. C. *J. chem. Phys.* 13 (1945) 216; SIMONS, J. H. and GARBER, C. S. *J. chem. Phys.* 21 (1953) 689; SIMONS, J. H. and McALLISTER, S. A. *J. chem. Phys.* 20 (1952) 1431; SIMONS, J. H. and UNGER, L. G. *J. chem. Phys.* 11 (1943) 322; 13 (1945) 221.
7. BERRY, H. W. *Phys. Rev.* 75 (1949) 913; 99 (1955) 553.
8. LANE, E. H. and EVERHART, E. *Phys. Rev.* 120 (1960) 2064; FULS, E. N., JONES, P. R., ZIEMBA, F. P. and EVERHART, E. *Phys. Rev.* 107 (1957) 704; EVERHART, E., STONE, G. and CARBONE, R. J. *Phys. Rev.* 99 (1955) 1287; JONES, P. R., ZIEMBA, F. P., MOSES, H. A. and EVERHART, E. *Phys. Rev.* 113 (1959) 182.
9. KAMINKER, D. M. and FEDORENKO, N. V. *J. tech. Phys., Moscow* 25 (1955) 2239; FEDORENKO, N. V., FILIPPENKO, L. G. and FLAKS, I. P. *J. tech. Phys., Moscow* 30 (1960) 49.
10. KREEVOY, M. M. and MASON, E. A. *J. Amer. chem. Soc.* 79 (1957) 4851; MASON, E. A. and KREEVOY, M. M. *J. Amer. chem. Soc.* 77 (1955) 5808; MASON, E. A., ROSS, J. and SCHATZ, P. N. *J. chem. Phys.* 25 (1956) 626; MASON, E. A. and VANDERSLICE, J. T. *J. chem. Phys.* 27 (1957) 917; 28 (1958) 253, 432, 1070; 29 (1958) 361; 30 (1959) 599 *Phys. Rev.* 108 (1957) 293; 114 (1959) 497; *Industr. Engng. Chem.* 50 (1958) 1033.
11. MASON, E. A. and VANDERSLICE, J. T. *Atomic Collisions,* Ed. D. R. Bates, 1962. New York; Academic Press.
12. FIRSOV, O. B. *J. exp. theor. Phys.* 24 (1953) 297; 33 (1957) 696; 34 (1958) 447; KELLER, J. B., KAY, T. and SCHMOYS, J. *Phys. Rev.* 102 (1956) 557.
13. SMITH, F. T., LORENTS, D. C., ABERTH, W. and MARCHI, R. P. *Phys. Rev. Lett.* 15 (1965) 742.
14. HYLLERAAS, E. A. University of Oslo Report 19, 1963.
15. MASSEY, H. S. W. and MOHR, C. B. O. *Proc. Roy. Soc.* A144 (1934) 188.
16. BOHM, D. *Quantum Theory,* 1951. New York; Prentice-Hall.
17. KENNARD, E. H. *Kinetic Theory of Gases,* 1938. New York; McGraw-Hill.
18. MOTT-SMITH, H. M. *Phys. Fluids* 3 (1950) 721.
19. SMITH, F. T., MARCHI, R. P., ABERTH, W., LORENTS, D. C. and HEINZ, O. *Phys. Rev.* 161 (1967) 31.
20. LEHMANN, C. and LEIBFRIED, G. *Z. Phys.* 172 (1962) 465.
21. SMITH, F. T. *Proc. 5th Int. Conf. Phys. electron. atom. Collisions,* 1967. Leningrad; Akademii Nauk.

22. AFROSIMOV, V. V., GORDEEV, YU. S., POLYANSKY, A. M. and SHERGIN, A. P. *Abstr. Proc. 5th Int. Conf. Phys. electron. atom. Collisions*, p. 475, 1967. Leningrad; Akademii Nauk.
23. LORENTS, D. C. and ABERTH, W. *Phys. Rev.* 139 (1965) 1017; 144 (1966) 109.
24. LORENTS, D. C., ABERTH, W. and HESTERMAN, V. W. *Phys. Rev.* 147 (1966) 849.
25. ZIEGLER, B. *Z. Phys.* 136 (1953) 108.
26. RAMSAUER, C., KOLLATH, R. and LILIENTHAL, D. *Ann. Phys., Lpz.* 8 (1930) 709.
27. RAMSAUER, C. and KOLLATH, R. *Ann. Phys., Lpz.* 16 (1933) 570.
28. FILIPPENKO, L. G. *Soviet Phys. tech. Phys.* 5 (1960) 52.
29. WITTKOWER, A. B. Thesis, University of London, 1967.
30. WEINBAUM, S. *J. chem. Phys.* 3 (1935) 547.
31. MOISEIWITSCH, B. L. *Proc. phys. Soc., Lond.* A69 (1956) 653.
32. CZAVINSZKY, P. *J. chem. Phys.* 31 (1959) 178.
33. WEBER, G. G., GORDON, N. H. and BERNSTEIN, R. B. *J. chem. Phys.* 44 (1966) 2814.
34. HOSTETTLER, H. U. and BERNSTEIN, R. B. *Phys. Rev. Lett.* 5 (1960) 318.
35. GROBLICKI, P. J. and BERNSTEIN, R. B. *J. chem. Phys.* 42 (1965) 2295.
36. RAMSEY, N. F. *Molecular Beams*, 1956. London; Oxford University Press.
37. MASSEY, H. S. W. and BURHOP, E. H. S. *Electronic and Ionic Impact Phenomena*, 1952. London; Oxford University Press.
38. MAIS, W. *Phys. Rev.* 45 (1934) 773.
39. ROSIN, S. and RABI, I. *Phys. Rev.* 48 (1935) 373.
40. ROSENBERG, P. *Phys. Rev.* 55 (1939) 1267.
41. RABI, I. *Rev. sci. Instrum.* 6. (1935) 251.
42. JAWTUSCH, W., SCHUSTER, G. and JAECKEL, R. *Z. Phys.* 141 (1955) 146.
43. KODERA, K. and TAMURA, T. *Bull. chem. Soc. Japan* 31 (1958) 206.
44. ROTHE, E. W. and BERNSTEIN, R. B. *J. chem. Phys.* 31 (1959) 1619.
45. SCHUMACHER, H., BERNSTEIN, R. B. and ROTHE, E. W. *J. chem. Phys.* 33 (1960) 584.
46. BERNSTEIN, R. B. *J. chem. Phys.* 34 (1961) 361; 33 (1960) 795.
47. BERNSTEIN, R. B. and KRAMER, K. H. *J. chem. Phys.* 38 (1963) 2507.
48. MASSEY, H. S. W. and BUCKINGHAM, R. A. *Nature, Lond.* 138 (1936) 77.
49. ROTHE, E. W. and NEYNABER, R. H. *J. chem. Phys.* 42 (1965) 3306.
50. WEBER, G. G. and BERNSTEIN, R. B. *J. chem. Phys.* 42 (1965) 2166.
51. BERNSTEIN, R. B. *J. chem. Phys.* 33 (1960) 795.
52. BERNSTEIN, R. B. *J. chem. Phys.* 33 (1960) 795; 34 (1961) 361; 36 (1962) 1403.
53. BERNSTEIN, R. B. *J. chem. Phys.* 38 (1963) 2599.
54. LANDAU and LIFSCHITZ, *Quantum Mechanics*, p. 146. 1955. London; Macmillan.
55. BERNSTEIN, R. B. and O'BRIEN, T. J. P. *Disc. Far. Soc.* 40 (1965) 35.
56. DUREN, R. and PAULY, H. *Z. Phys.* 175 (1967 227; 177 (1963) 146.
57. LEBEDEFF, S. A. *J. chem. Phys.* 40 (1964) 2716.
58. PAULY, H. *Z. Phys.* 157 (1959) 54.
59. HELBING, R. and PAULY, H. *Z. Phys.* 179 (1964) 16.
60. BERNSTEIN, R. B. *Science* 144 (1964) 141; ROTHE, E. W. *et al., Phys. Rev.* 130 (1963) 2333.
61. WAECH, T. G. and BERNSTEIN, R. C. University of Wisconsin Report, WIS-TCI-211, 1967.
62. BERNSTEIN, R. B. *et al.*, University of Wisconsin Report, WIS-TCI-133, 1965.
63. LEVINE, R. D., JOHNSON, J. T., MUCKERMAN, J. T. and BERNSTEIN, R. B. *Chem. Phys. Lett.* 1 (1968) 517.
64. FORD, K. W., HILL, D. L., NAKANO, M. and WHEELER, J. A. *Ann. Phys., New York* 7 (1959) 239.
65. HIRSCHFELDER, J. O., CURTISS, C. F. and BIRD, R. B. *Molecular Theory of Gases and Liquids*, 1954. Wiley; New York.
66. WESTENBERG, A. A. *Combust. Flame*, 1 (1957) 217.
67. DALGARNO, A. *Atomic and Molecular Processes*, Ed. D. R. Bates, 1962. New York; Academic Press.

68. CHAPMAN, S. and COWLING, T. G. *The Mathematical Theory of Non-Uniform Gases*, 1958. London; Cambridge University Press.
69. KIHARA, T. *Rev. mod. Phys.* 24 (1952) 45; 25 (1953) 844.
70. MASON, E. A. *J. chem. Phys.* 27 (1957) 782.
71. WALKER, R. E. and WESTENBERG, A. A. *J. chem. Phys.* 29 (1958) 1139, 1147.
72. MASON, E. A. *J. chem. Phys.* 22 (1954) 169.
73. KIHARA, T. and KOTANI, M. *Proc. phys.-math. Soc. Japan* 25 (1943) 60.
74. DE BOER, J. and VAN KRANENDONK, J. *Physica, Eindhoven*, 14 (1948) 442.
75. HIRSCHFELDER, J. O., BIRD, R. B. and SPOTZ, E. L. *J. chem. Phys.* 16 (1948) 968; 17 (1949) 1343.
76. ROWLINSON, J. S. *J. chem. Phys.* 17 (1949) 101.
77. WEISSMAN, S., SAXENA, S. C. and MASON, E. A. *Phys. Fluids* 3 (1960) 510.
78. WALKER, R. E. and WESTENBERG, A. A. *J. chem. Phys.* 31 (1959) 519; 32 (1960) 436, 1314.
79. AMDUR, I. and SCHATZKI, T. F. *J. chem. Phys.* 29 (1957) 1049.
80. SRIVASTAVA, B. N. and SRIVASTAVA, K. P. *J. chem. Phys.* 30 (1959) 984; SRIVASTAVA, K. P. and BARUA, A. K. *Indian J. Phys.* 33 (1959) 229.
81. WISE, H. *J. chem. Phys.* 31 (1960) 1414; WISE, H. and ABLOW, C. M. *J. chem. Phys.* 29 (1958) 634.
82. JOST, W. *Diffusion in Solids, Liquids and Gases*, 1952. New York; Academic Press.
83. WALKER, R. E. and WESTENBERG, A. A. *J. chem. Phys.* 29 (1958) 1139, 1147.
84. PHELPS, A. V. *Phys. Rev.* 114 (1959) 1011.
85. BUCKINGHAM, R. A. and DALGARNO, A. *Proc. Roy. Soc.* A213 (1952) 327, 506.
86. HORNBECK, J. A. *Phys. Rev.* 84 (1951) 615.
87. ROSE, D. J. *J. appl. Phys.* 31 (1960) 643.
88. VARNEY, R. N. *Phys. Rev.* 88 (1952) 362; 89 (1953) 708; *J. chem. Phys.* 31 (1959) 1314.
89. WANNIER, G. H. and HORNBECK, J. A. *Phys. Rev.* 82 (1951) 458.
90. WANNIER, G. H. *Bell Syst. tech. J.* 32 (1953) 170.
91. LANGEVIN, P. *Ann. Chim. (Phys.)*, 5 (1905) 245.
92. LOEB, L. B. *Kinetic Theory of Gases*, 2nd edn, 1934. New York; McGraw-Hill.
93. HASSÉ, H. R. *Phil. Mag.* 1 (1926) 139.
94. WANNIER, G. H. *Phys. Rev.* 83 (1951) 281; 87 (1952) 795.
95. DALGARNO, A., McDOWELL, M. R. C. and WILLIAMS, A. *Phil. Trans.* A250 (1958) 411.
96. KIHARA, T. *Rev. mod. Phys.* 15 (1953) 831.
97. MASON, E. A. and SCHAMP, H. W. *Ann. Phys., New York* 4 (1958) 233.
98. HORNBECK, J. A. *Phys. Rev.* 83 (1951) 374; 84 (1951) 621.
99. VARNEY, R. N. *Phys. Rev.* 88 (1952) 362; 89 (1953) 708.
100. MASSEY, H. S. W. and SMITH, R. A. *Proc. Roy. Soc.* A142 (1933) 142.
101. SHELDON, J. W. *Phys. Rev. Lett.* 8 (1962) 64.
102. DEMKOV, U. H. *Ann. Leningr. Univ.* 146 (1952) 74.
103. DICKINSON, A. S. *Proc. phys. Soc., Lond.* B1 (1968) 387, 395.
104. FROST, L. S. *Phys. Rev.* 105 (1957) 354.
105. DALGARNO, A. *Phil. Trans.* 250 (1958) 250.
106. HOLSTEIN, T. *J. phys. Chem.* 56 (1952) 832.
107. BIONDI, M. A. and CHANIN, L. M. *Phys. Rev.* 94 (1954) 910; *Phys. Rev.* 122 (1962) 843.
108. CHANIN, L. M. and BIONDI, M. A. *Phys. Rev.* 106 (1957) 473.
109. COURVILLE, G. E. and BIONDI, M. A. *J. chem. Phys.* 37 (1962) 616.
110. TYNDALL, A. M. and PEARCE, A. F. *Proc. Roy. Soc.* A149 (1935) 426.
111. SENA, L. A. *J. Phys., Moscow* 10 (1946) 179.
112. TYNDALL, A. M. and GRINDLAY, G. C. *Proc. Roy. Soc.* A110 (1926) 341, 358.
113. BRADBURY, N. E. *Phys. Rev.* 40 (1932) 508.
114. TOWNSEND, J. S. and TIZARD, H. *Proc. Roy. Soc.* A88 (1913) 336.
115. HERSHEY, A. V. *Phys. Rev.* 56 (1939) 909.
116. TYNDALL, A. M., STARR, L. H. and POWELL, C. F. *Proc. Roy. Soc.* A121 (1928) 172.

117. BLOOMFIELD, C. H. and HASTED, J. B. *Disc. Faraday Soc.* 37 (1964) 176; *Brit. J. appl. Phys.* 17 (1966) 449.
118. KANEKO, Y., MEGILL, L. R. and HASTED, J. B. *J. chem. Phys.* 45 (1966) 3741.
119. VARNEY, R. N. *Phys. Rev.* 89 (1953) 708.
120. DAVIES, D. E., DUTTON, J. and LLEWELLYN-JONES, F. *Proc. 5th Int. Conf. Ioniz. Phenom. Gases*, 1961. Amsterdam; North Holland Publishing Company.
121. BRADBURY, N. E. and NIELSEN, R. A. *Phys. Rev.* 49 (1936) 388.
122. CROMPTON, R. W. and ELFORD, M. T. *Proc. phys. Soc., Lond.* 74 (1959) 497.
123. EINSTEIN, A. *Ann. Phys., Lpz.* 17 (1905) 549.
124. HORNBECK, J. A. and MOLNAR, J. P. *Phys. Rev.* 84 (1951) 621.
125. LAUER, E. J. *Phys. Rev.* 83 (1951) 374; *J. appl. Phys.* 23 (1952) 300.
126. BIONDI, M. A. and BROWN, S. C. *Phys. Rev.* 75 (1949) 1700; 76 (1949) 302.
127. OSKAM, H. F. *Philips Res. Rep.* 13 (1958) 365.
128. PERSSON, K. *Phys. Rev.* 106 (1957) 191.
129. GOODALL, C. V. Thesis, University of Birmingham 1966.
130. BIONDI, M. A. *Phys. Rev.* 93 (1954) 1136.
131. MULCAHY, M. J. and LENNON, J. J. *Proc. phys. Soc., Lond.* 80 (1962) 626.
132. CHANIN, L. M. *Phys. Rev.* 123 (1961) 526.
133. MITCHELL, J. H. and RIDLER, K. E. *Proc. Roy. Soc.* A146 (1934) 911.
134. PERSSON, K. and BROWN, S. C. *Phys. Rev.* 100 (1955) 729.
135. BENNETT, W. H. *Phys. Rev.* 58 (1940) 992.
136. CREASER, R. P. Thesis, Australian National University, Canberra, 1969.
137. REES, Ph.D. Thesis, University of Wales, 1959.
138. BRATA, L. *Proc. Roy. Soc.* A141 (1933) 454.
139. MEYEROTT, R. *Phys. Rev.* 66 (1944) 242.
140. SIDA, D. W. *2nd Int. Conf. Phys. electron. atom. Collisions*, University of Colorado, 1961.
141. HOSELITZ, K. *Proc. Roy. Soc.* A177 (1941) 200.
142. BIONDI, M. A. and CHANIN, L. M. *Phys. Rev.* 106 (1957) 473; 94 (1954) 910.
143. PHELPS, A. V. and BROWN, S. C. *Phys. Rev.* 86 (1952) 102.
144. BATES, D. R. *Phys. Rev.* 77 (1950) 718.
145. MADSON, J. M., OSKAM, H. J. and CHANIN, L. M. *Phys. Rev. Lett.* 15 (1965) 1018.
146. BEATY, E. C., BROWNE, J. C. and DALGARNO, A. *Phys. Rev. Lett.* 16 (1966) 723.
147. WOO, S-B., *J. chem. Phys.* 42 (1965) 1251.
148. McKNIGHT, L. G., McAFEE, K. B. and SIPLER, D. P. *Phys. Rev.* 164 (1967) 62.
149. KELLER, G. E., MARTIN, D. W. and McDANIEL, E. W. *Phys. Rev.* 140 (1965) A1535.
150. MOSELEY, J. T., SNUGGS, R. M., MARTIN, D. W. and McDANIEL, E. W. *Phys. Rev.* 178 (1969) 240.
151. MOSELEY, J. T., SNUGGS, R. M., MARTIN, D. W. and McDANIEL, E. W. *Phys. Rev.* 178 (1969) 240.
152. SAPOROSCHENKO, M. *Phys. Rev.* 139 (1965) A349; *J. chem. Phys.* 42 (1965) 2760.
153. CHANIN, L. M. *Phys. Rev.* 123 (1961) 526.
154. MILLER, T. M., MOSELEY, J. T., MARTIN, D. W. and McDANIEL, E. W. *Phys. Rev.* 173 (1968) 115.
155. ALBRITTON, D. L., MILLER, T. M., MARTIN, D. W. and McDANIEL, E. W. *Phys. Rev.* 171 (1968) 94.
156. BURCH, D. S. and GEBALLE, R. *Phys. Rev.* 106 (1957) 183, 188.
157. REES, J. A. *Aust. J. Phys.* 18 (1965) 41.
158. McKNIGHT, L. G. *Gaseous Electronics Conf.* American Physical Society, 1966.
159. BLANC, A. *J. Phys.* 7 (1908) 825.
160. McDANIEL, E. W. and CRANE, H. R. *Rev. sci. Instrum.* 28 (1957) 684.
161. OVERHAUSER, A. W. *Phys. Rev.* 76 (1949) 250.
162. KAGAN, YU. and PEREL, V. I. *J. exp. theor. Phys.* 7 (1958) 87.
163. MAGEE, J. L. and FUNABASHI, K. *Radiat. Res.* 10 (1959) 622.
164. BLOOM, S. and MARGENAU, H. *Phys. Rev.* 85 (1952) 670.
165. MUNSON, R. J. and HOSELITZ, K. *Proc. Roy. Soc.* A172 (1939) 43.

166. MUNSON, R. J. and TYNDALL, A. M. *Proc. Roy. Soc.* A172 (1939) 28.
167. HASTED, J. B. *Prog. Dielect.* 3 (1961) 102.
168. DAWSON, P. H. and TICKNER, A. W. *J. chem. Phys.* 37 (1962) 672.
169. KEBARLE, P., ARSHADI, M. and SCARBOROUGH, J. *J. Amer. chem. Soc.* 89 (1967) 6393, 5753; 88 (1966) 28.
170. HIRSCHFELDER, J. O., CURTISS, C. F. and BIRD, R. B. *Molecular Theory of Gases and Liquids*, 1954. New York; Wiley.
171. HASTED, J. B. *Proc. phys. Soc., Lond.* 86 (1965) 795.
172. BLOOMFIELD, C. H. and HASTED, J. B. *Brit. J. appl. Phys.* 17 (1966) 449; *Disc. Faraday Soc.* 37 (1964) 176.
173. OSKAM, H. J. and MOSHARRAFA, M. A. *Gaseous Electronics Conf.* American Physical Society, 1964.
174. OSKAM, H. J. and JONGERIUS, H. M. *Physica, Eindhoven*, 24 (1958) 1092.
175. DRUYVESTEYN, N. J. *Nature, Lond.* 128 (1931) 1076.
176. HENDERSON, W. R., MATSEN, F. A. and ROBERTSON, W. W. *Gaseous Electronics Conf.* American Physical Society, 1964.
177. KNOLL, M., EICHMEIER, J. and STEIN, R. W. *Advanc. Electron.* 19 (1964) 177.
178. JUNGE, C. *Ann. Met., Hamburg* 5, suppl. 1 (1952); ISRAEL, H. *Atmos. elekt.* 1 (1957).

Chapter 11

IONIZATION AND EXCITATION
BY ATOMIC PARTICLES

11.1 INTRODUCTION; IONIZATION AT LOW IMPACT ENERGIES

Inelastic collisions between atomic systems include collisions involving positive ions, excited atoms, and also neutral atomic and molecular systems. The processes fall into two classes: those involving excitation, ionization and dissociation (Chapter 11); and those in which there is only exchange, either of energy or of an electron (Chapters 12 and 13). The two types of process are governed by different theory, but the experimental methods employed to study ionic collisions of either type are similar and will be jointly discussed. Collisions which result only in interchange among the atoms of molecular collidants are included in the discussion of chemical collisions (Chapter 14). All collisions of excited species are considered in Chapter 13.

A fast projectile particle, as it passes through a gas making ionizing collisions, loses relatively little energy and is scattered only through small angles; but the ions produced are scattered at nearly 90° to the projectile beam and normally receive only a small 'knock on' kinetic energy, less than 1 per cent of the projectile energy.

The cross-section functions for ionization and excitation by atomic particles behave in the conventional manner, rising with increasing impact energy until a maximum is reached, and eventually falling off at high energies. A number of features are incompletely understood, such as the behaviour at low impact energies, the role of potential energy curve pseudo-crossovers, the mechanism of energy transfer in multiple ionization, and the importance of the ionization with capture process $n0/(n-m)(m+p)e^p$ The broad features of some typical cross-section functions are shown in *Figures 11.1* and *11.2*.

Born approximation and other quantum calculations for ionization of simple atomic systems by protons, hydrogen atoms, etc., have been made, but the detailed calculation of heavy ion–atom cross-sections has not yet been achieved by quantum theory, although a statistical approach is possible.

There are two simple classical limits which may be applied to the ionization process. Consider the binary encounter between an incident particle of mass m_1 and a target atom orbital electron of mass m_e; when the mass ratio m_1/m_e is equal to the energy ratio E_p/E_i, the ionization process becomes energetically possible in a simple encounter. Since this is in the coordinate framework of the orbital electron, the exact limit for ionization is somewhat

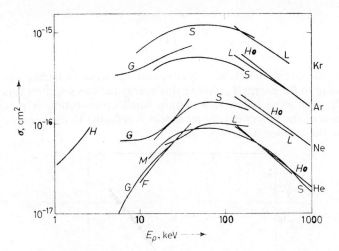

Fig. 11.1. Total ionization functions for protons in rare gases: S, Soloviev et al.[1]; Ho, Hooper et al.[2]; L, Gilbody et al.[3]; G, Gilbody and Hasted[3]; M, Mapleton [4]; H, argon data of Hasted[3]; F, Fedorenko et al.[5]

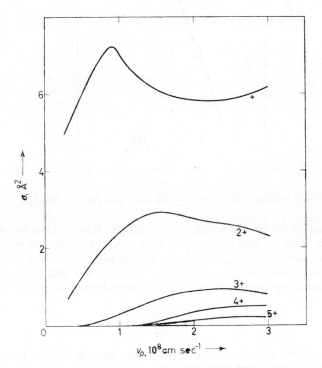

Fig. 11.2. Cross-section functions for formation of multiply-charged argon ions in He^+–Ar collisions, after Fedorenko and Afrosimov[5]

different:

$$E_p = \frac{m_1 E_i}{4m_e}\left(1 - \frac{v_p}{v_e}\right) \tag{11.1}$$

where v_e is the orbital electron velocity, and v_p is the projectile velocity corresponding to the energy E_p. Above this energy the cross-section function rises rapidly, but even below it a certain small cross-section is possible; the crudeness of the binary encounter model is reflected in the vagueness of this limit in the cross-section function.

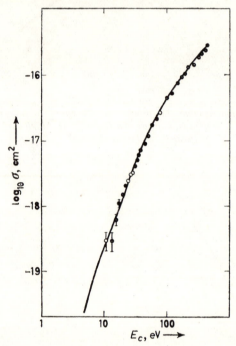

Fig. 11.3. Low-energy centre-of-mass system ionization function for N_2 in O_2 (smooth line) and O_2 in N_2 (open circles)

(From Utterback[8], by courtesy of the American Physical Society)

The second limit gives the actual impact velocity below which the process cannot take place by any mechanism. In a collision at kinetic energy E_p between two atomic particles of masses m_1 and m_2, the amount of kinetic energy actually available for conversion into internal energy is $(E_p m_2)/(m_1+m_2)$; therefore ionization cannot take place below the impact energy limit

$$E_{p0} = \frac{E_i(m_2+m_2)}{m_2} \tag{11.2}$$

The most sensitive measurements[6-8] show that this limit is very close to the lowest impact energy at which ionization cross-sections—as small as 10^{-23} cm²—can in fact be observed.

At the lowest energies, the cross-section function for ionization by atomic particles does not show the linear energy rise well-known in electron ionization but, instead, either the power law

$$\sigma \propto (E_p - E_{p0})^2$$

illustrated in *Figure 11.3* or an even faster rise (*Figure 11.4*). Historically, it proved difficult to determine the smallest impact energy at which ionization can take place; in successive experiments of greater sensitivity, ionization at

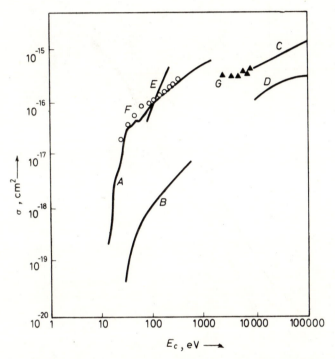

Fig. 11.4. Total ionization centre-of-mass system cross-section for He–He *(lower) and* Ar–Ar *(upper): A, Hayden and Amme; B, Hayden and Utterback; C, Afrosimov et al.; D, Solov'ev et al.; E, Calculations of Rosen; F, Rostagni; G, Sluyters et al.*

lower energies was detected. Some measurements[9, 10] are in serious disagreement with the general run of measurements[11-14]. The small cross-sections render the experiments particularly viable to stray electrons, and the first studies[6] which obtained the correct threshold of ionization (equation 11.2) relied on the registration of the slow ions formed collisionally.

In this discussion no distinction has been made between ionization by fast positive ions and by fast neutral atoms. The latter process is in general the more probable, especially at low energies where curve-crossing can be important. Low-energy ionization by neutrals is easier to measure than that by ions, since the only charged particles in the chamber are those arising from the collision process.

37*

Utterback's[8] experiments were carried out by passing fast neutral beams of N_2 and O_2 through gas; collection of electrons is carried out by parallel plate collectors, which also serve to apply a small transverse electric field, as in high-energy experiments. The great sensitivity (10^{-20} cm² cross-section threshold) arises from the fact that these are only neutral particles in the collision chamber until an ionizing collision takes place. The fast neutral particles are produced by charge transfer from an ion beam, and the possible

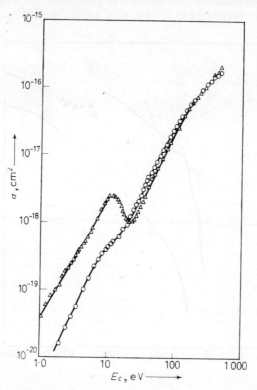

Fig. 11.5. Total centre-of-mass system cross-section for production of negative charge by N_2–N_2 *(triangles) and* CO–N_2 *(circles), after Amme and Utterback*

presence of long-lifetime excited states, electronic or vibrational, is investigated by applying to the data for beam A in gas B and for beam B in gas A a common centre-of-mass coordinate system. Since no distinction is made between electrons produced from A and from B, there should be a close correspondence between the two sets of data, which is in fact observed (*Figure 11.3*). In this coordinate system, the rate of rise appears approximately proportional to the square of the excess of kinetic over ionization energy, and the theoretical limit of onset is approached reasonably well. Studies have been made[15-17] of combinations of N_2, CO, CO_2, CH_4, C_2H_2, NO, He, Ne and Ar, and data are displayed in *Figures 11.3–11.5*. An important feature of these studies is the enhanced cross-sections (or even maxima) illustrated in *Figure 11.5*. Such features are similar to those in single electron capture processes by multiply-charged ions in which there is curve-crossing, and it seems reasonable to suppose that a similar crossing is involved in the ionization functions. One proposal is that the Coulomb states $A^- + B^+$ couple to $A + B$. The state A^- need not be stable for longer than the collision time, so the collision products can be $A + B^+ + e$, or $A^- + B^+$. Maximum cross-sections as high as 100 Å² have been reported at ~ 6 eV for Cs–Br₂ collisions.

The interchange process

$$N_2 + CO \rightarrow CN^- + NO^+$$

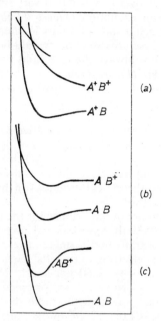

Fig. 11.6. Possible potential energy curves for ionization processes: (a) ionizing by a positive ion, proceeding via an autoionizing state; (b) ionization by neutral atom; (c) associative ionization

has been investigated both by collection of negative ions and by mass-analysis of the positive ions. The threshold for this process is lower than that appropriate to collisional ionization. Atom–atom interchange has been observed by isotope techniques in N₂–N₂ ionizing collisions.

The elimination of metastable atoms from the neutral beams used in low-energy ionization experiments is important. The beams are produced by symmetrical resonance charge transfer, and the test of identity of cross-sections for atom A through gas B and for atom B through gas A is usually decisive. For example, the structure in He through H₂ cross-section functions does not appear in those for H₂ through He, indicating the possible presence of helium metastables.

It was first suggested by Weizel[18] that at the lowest impact energies ionization processes take place through the excitation of two electrons. The two-electron excited state can possess sufficient internal energy for a re-arrangement to take place, with the liberation of one electron into the continuum. It is proposed that at small internuclear distance both the excitation and the rearrangement take place at pseudo-crossing of potential energy curves (*Figure 11.6*). The exact details of these processes are difficult to elucidate, since little is known as yet about many of the two-electron excited states.

For this and other reasons, the prediction of cross-sections for ionization by either neutrals or positive ions has hardly been attempted at low energies except in the simplest cases. For ionization by neutrals (00/01e), a potential energy curve pseudo-crossover may occur at very small nuclear separations, without any two-electron excitation (*Figure 11.6b*). At one time it was believed that, because such a state of affairs is impossible for ionization by positive ions (10/11e), these cross-sections would be smaller than those for neutral atoms. This is not always the case, and pseudo-crossing processes can be invoked for 10/11e (*Figure 11.6a*).

There is also the possibility of associative ionization

$$A + B \rightarrow AB^+ + e$$

(*Figure 11.6c*) which is the reverse of dissociative recombination.

Early data for ionization at low energies gave rise to the belief that two identical (K^+–Ar) or nearly identical (Ar^+–Ar) electronic systems would be most effective at producing free electrons in collision. Although there do not seem to be adequate theoretical grounds for this 'iso-electronic effect', experimental data support it[19].

Atomic ionization data are of importance for the interpretation of meteor ionization observations[20, 21]. It has been calculated[22] from radar echoes obtained from ionization trails that a single meteor atom has a one in five chance of producing an electron during its passage through the earth's atmosphere; its initial velocity is 60 km sec^{-1}. Massey and Sida[23] have calculated momentum loss cross-sections for calcium atoms in neon, and with the aid of these they show that the radar data indicate an ionization cross-section of the order of $0 \cdot 3$–$12 \ \pi a_0^2$ in the range 20–1000 eV. This unexpectedly large deduced cross-section may be associated with some curve-crossing. Laboratory measurements of similar processes have now been made[24]; it would appear that the ionization cross-section of the oxygen molecule and of the nitrogen molecule by potassium atoms are of this order.

11.2 CHARGE-CHANGING COLLISIONS

At higher energies of impact than those considered in the last section, beam experiments can be conducted in two complementary ways: the projectile particle species may be separated and measured; alternatively, the slow charged particles may be separated and measured. These methods derive historically from the studies of stopping and charge equilibration of α radiation, and from the studies of the ionization of gases by the same radiation. Both types of experiment have produced much valuable data.

The possibility of occurrence of multiple ionization processes of the type

$$X^+ + Y \rightarrow X^{m+} + Y^{n+} + (m+n-1)e$$

implies that $m+n$ pieces of data are provided if both particles and slow ions are charge-analysed; but the range of possible ionizing collisions is described by mn cross-sections (neglecting negative ion formation). Therefore the complete information, including all cross-sections $_{10}\sigma_{mn}$, could only be obtained with a coincidence counting experiment[25], in which both projectile particle and slow ion were charge-analysed and simultaneously recorded.

Collision Chambers at Higher Pressures

In passing through a collision chamber containing gas at a sufficiently low pressure, small fractions of the projectile particles (ions) are converted by ionizing processes to other charge states; the fractions correspond to collision cross-sections determined by equation 1.24. But as the pressure in the collision chamber is raised, converted projectile particles are re-converted by electron capture back to the original state of charge. At sufficiently high pressures an equilibrium between the original state and the converted state(s) is set up, with each particle undergoing many conversions and reconversions. When only two possible states are considered (1 and 0), the equilibrium fractions $F_{1\infty}$ and $1-F_{1\infty}$ of original and converted particles that traverse the gas are related to the cross-sections and the low-pressure fraction F_1 by the Wien equations (equations 1.31 and 1.32):

$$_{10}\sigma_{01} = \frac{1-F_{1\infty}}{\pi} \ln \left(\frac{1-F_{1\infty}}{F_1-F_{1\infty}} \right) \tag{11.3}$$

$$_{00}\sigma_{10} = \frac{F_1}{\pi} \ln \left(\frac{1-F_{1\infty}}{F_1-F_{1\infty}} \right) \tag{11.4}$$

They are derived as follows.

Consider a uniform beam current I_0 of singly charged ions directed through a path length l in gas at pressure p torr and temperature T. The number of atoms per cm^2 traversed by the beam is given by the target parameter

$$\pi = \frac{pl}{kT} \tag{11.5}$$

and, for $T = 20°C$,

$$\pi = 3\cdot30\times10^{16} \, pl \tag{11.6}$$

Under 'single collision conditions', the current of collision products I_p formed in a process for which the cross-section is σ, is given by

$$I_p = I_0\pi\sigma \tag{11.7}$$

When the pressure is raised by dp, the change dF_0 produced in the neutral fraction of the beam emerging from the chamber is, for a two-component system,

$$dF_0 = -dF_1 = -(F_0 \, _{00}\sigma_{01} + F_1 \, _{10}\sigma_{01}) \, d\pi \tag{11.8}$$

so that the fractions of neutrals and singly charged ions F_0, F_1 in the beam are given by

$$F_1 = F_{1\infty} + F_{0\infty} \exp \left[-\pi(_{00}\sigma_{10} + _{10}\sigma_{01}) \right] \tag{11.9}$$

and

$$F_0 = F_{0\infty}\{1 - \exp[-\pi({}_{00}\sigma_{10} + {}_{10}\sigma_{01})]\} \tag{11.10}$$

with

$$F_{0\infty} = \frac{{}_{10}\sigma_{01}}{{}_{00}\sigma_{10} + {}_{10}\sigma_{01}} \tag{11.11}$$

and

$$F_{1\infty} = \frac{{}_{00}\sigma_{10}}{{}_{00}\sigma_{10} + {}_{10}\sigma_{01}} \tag{11.12}$$

The fractions $F_{0\infty}$ and $F_{1\infty}$ are reached after passage through an infinite collision path. Equations 11.3 and 11.4 follow by rearrangement; a similar derivation may be made in the case where the initial beam is neutral.

Similar differential equations can be deduced[26, 27] for the case where the singly charged ion beam is part of a three-component system in which interconversion between the charge states 0, 1 and 2 is possible and is described by cross-sections σ_{10}, σ_{02}, etc.:

$$\frac{dF_0}{d\pi} = -F_0(\sigma_{01} + \sigma_{02} + \sigma_{20}) + F_1(\sigma_{10} - \sigma_{20}) + \sigma_{20} \tag{11.13}$$

$$\frac{dF_1}{d\pi} = -F_0(\sigma_{01} - \sigma_{21}) - F_1(\sigma_{10} + \sigma_{12} + \sigma_{21}) + \sigma_{21} \tag{11.14}$$

$$F_2 = 1 - (F_0 + F_1) \tag{11.15}$$

From which solutions can be obtained:

$$F_0 = F_{0\infty} + [P_0 \exp(\pi q) + N_0 \exp(-\pi q)] \exp\left[\tfrac{1}{2}\pi(a+g)\right] \tag{11.16}$$
$$F_1 = F_{1\infty} + [P_1 \exp(\pi q) + N_1 \exp(-\pi q)] \exp\left[\tfrac{1}{2}\pi(a+g)\right] \tag{11.17}$$
$$F_2 = F_{2\infty} + [P_2 \exp(\pi q) + N_2 \exp(-\pi q)] \exp\left[\tfrac{1}{2}\pi(a+g)\right] \tag{11.18}$$

with:

$$P_1 = -\frac{1}{2q}[(s-q)(1-F_{1\infty}) + bF_{0\infty}]$$

$$N_1 = 1 - F_{1\infty} - P_1$$

$$P_0 = \frac{P_1}{b}(s+q)$$

$$P_2 = \frac{-P_1}{b}(b+s+q)$$

$$N_0 = \frac{N_1}{b}(s-q)$$

$$N_2 = -\frac{N_1}{b}(b+s+q)$$

$$q = \tfrac{1}{2}[(g-a)^2 + 4bf]^{1/2}$$

$$a = -(\sigma_{10} + \sigma_{12} + \sigma_{21})$$

$$b = \sigma_{01} - \sigma_{21}$$

$$f = \sigma_{10} - \sigma_{20}$$

$$g = -(\sigma_{01} + \sigma_{02} + \sigma_{20})$$

The equilibrium fractions are given by the equations:

$$F_{0\infty} = \frac{f\sigma_{21} - a\sigma_{20}}{ag - bf} \qquad (11.19)$$

$$F_{1\infty} = \frac{b\sigma_{20} - g\sigma_{21}}{ag - bf} \qquad (11.20)$$

$$F_{2\infty} = \frac{\sigma_{20}(a-b) + g(a+\sigma_{21}) - f(b+\sigma_{21})}{ag - bf} \qquad (11.21)$$

In order that the equations may be solved to give all six cross-sections, the determination of the three equilibrium fractions must be supplemented by the determination of any three cross-sections under single collision conditions.

Four-component systems have been analysed, and it is found that the functions $F(\pi)$ can pass through maxima before the equilibrium values are reached. The five-component system has been set up on a computer[28].

It is relatively simple to solve the n-component problem[29], which is of importance at energies of several MeV, provided that the assumption is made that only one electron can be gained or lost by a projectile in a collision. It is not yet possible to estimate the quality of this assumption. The set of equations

$$\frac{dF_i}{d\pi} = \sum_{\substack{j \\ j \neq i}} \sigma_{ji} F_j - \sum_{\substack{j \\ j \neq i}} \sigma_{ij} F_i \qquad (11.22)$$

must be solved so that the cross-sections σ_{ij} for conversion of charge state i to charge state $j(|j-i| = 1)$ can be computed from the experimentally determined functions $F_i(\pi)$ and $F_j(\pi)$. When the one-electron gain or loss assumption is made, the equation

$$\frac{dF}{d\pi} = AF \qquad (11.23)$$

can be written, where F is a vector whose ith component is F_i, and A is a tridiagonal matrix:

$$A = \begin{bmatrix} -\sigma_{12} & \sigma_{21} & 0 & 0 & \cdots \\ \sigma_{12} & (-\sigma_{21} - \sigma_{23}) & \sigma_{32} & 0 & \cdots \\ 0 & \sigma_{23} & (-\sigma_{32} - \sigma_{31}) & \sigma_{43} & \cdots \\ \cdot & \cdot & \cdot & \cdot \\ \cdot & \cdot & \cdot & \cdot \\ \cdot & \cdot & \cdot & \cdot \end{bmatrix} \qquad (11.24)$$

The solution of equation 11.24 is

$$F(\pi) = F_0 \exp(A\pi) \qquad (11.25)$$

where F_0 is the initial charge distribution vector. It is necessary to write a computer programme to find the value of A for which $\sum_i (F_{i(exp)} - F_{i(calc)})^2$ is minimum, given F_0 and $F_{exp}(\pi)$.

Experiments on projectile and particle separation and measurement are conducted with the aid of electrostatic or magnetic fields imposed upon the collision chamber. A neutral particle beam passing straight through such a chamber is attenuated by the continual removal of the charged particles produced during its passage. A charged particle beam passes along a curved path (the circular path in a uniform magnetic field is the most suitable), and is attenuated by the continual removal of the more highly charged and the neutral particles produced during its passage.

The performance of two separation experiments of this type in the same collision chamber, one with charged particles and one with neutral particles, has been used as an alternative to equilibrium measurements.

Since a measurement made under single collision conditions is one of a small decrement in a large current, the projectile particle removal technique is more difficult than that of slow charge collection. However, the projectile particle detection and slow charge collection methods measure different cross-sections and are complementary in the sense that, at high energies, each produces a part of the total knowledge.

Among the errors peculiar to projectile-collecting experiments are: the change in calibration of a detector when gas is admitted to the collision chamber; the deflexion of beam particles away from the collector by elastic scattering collisions when gas is admitted; and the change in primary beam composition, on the admission of gas, due to charge-changing collisions in the pre-collision chamber background gas.

Charge-changing collision techniques are concerned basically with the relative efficiency of the processes of charge transfer 10/01 and neutral atom stripping 00/10e. In general, the latter is more efficient for molecular than for atomic targets. The pure two-component system is never found, because of the role played by metastable atoms and ions. Even the helium system has three components[30].

Charge-changing collisions are of special importance in the technology of tandem and multi-stage electrostatic generators. For the passage of protons through hydrogen, the equilibrium proportions are:

	30 keV	> 1 MeV
H^+	28%	99·9%
$H°$	70·6%	< 0·1%
H^-	1·4%	0%

Thus, to convert protons to H^- one uses 30 keV collisions, and to reconvert to protons, collisions in the MeV range of energy are used. Charge-changing in such applications is normally conducted in long narrow tubes, containing target gas, which can often be condensed rather than pumped. However, the condensation can block the tube. Multiple stripping processes are also important for electrostatic generators, since the generation of multiply-charged ions enhances the attainable energy.

A systematic study has been made[31] of the charge-changing collision cross-sections σ_{ij} for conversion of projectile ions ($2 \leqslant Z \leqslant 18$; 2 cm sec^{-1} $\leqslant v \leqslant 12 \times 10^8$ cm sec^{-1}) from charge states i to j, in targets which include rare gases and polymer films[32]. In general $\sigma_{i(i+1)}$ increases with increasing Z, but the effect is pronounced only when there is a small number of electrons in the outer shell of the ion. Furthermore, $\sigma_{i(i+1)} \propto \exp(-\mu i)$ at a given impact velocity v; μ is a constant which decreases as velocity increases. The individual loss cross-section function $\sigma_{i(i+1)}(v)$ passes through a maximum at a velocity v_m which is approximately given by

$$v_m = \gamma \sqrt{\frac{2E_i}{m_e}} \qquad (11.26)$$

where E_i is the appropriate ionization potential, and γ a constant varying from 1·3 for helium to 2 for krypton. The contrast with the adiabatic maximum rule should be noted.

A useful quantity sometimes measured is the total loss cross-section, defined as

$$\sigma_{it} = \frac{1}{q_i} \sum_{s=1}^{q} s\sigma_{i(i+s)} \qquad (11.27)$$

Many experiments have been reported on the charge distributions produced in fast projectile beams passing through thick gas or solid targets[33]. The equilibrium charge distribution function can generally be described by a Gaussian of halfwidth d:

$$F_{i\infty} = \left(\frac{1}{2\pi d^2}\right)^{1/2} \exp\left[-\frac{(i-\bar{i})^2}{2d^2}\right] \qquad (11.28)$$

Deviations from the Gaussian shape are more pronounced for ions with $Z_+ \geqslant 10$. This arises because there is often a point at which $F_{(i+1)}/F_i$ ceases to increase with increasing i. Shell effects may also contribute. The most probable charge $\bar{i}$ increases with increasing velocity, and at fixed velocity it increases with increasing Z; the halfwidth d increases as $Z^{1/2}$. Values of $\bar{i}$ are higher in solids than in the equivalent gas target. Possibly this is because of a reduction of electron binding energy by excitation, but there may be a small contribution from polarization of the solid by the fast ions.

11.3 TOTAL SLOW CHARGE COLLECTION

Total electron production measurements for ion beam collisions with gases have been carried out[2, 3, 5, 13, 14, 34, 35] over a wide energy range, 50 eV–2 MeV. In total charge collection experiments, the electrons are collected at a positively charged electrode which forms part of a 'condenser' system maintaining a uniform electric field transverse to the ion beam (*Figure 11.7*). The electric field is increased until the measured electron current is independent of it ('saturation conditions'); this condition can only be met when the energy with which the electrons are formed is small compared with the energy they acquire in the field. For incident ion energies $\sim$ 1 keV, the saturation condition can set in with transverse fields as low as a few V cm^{-1}.

In this system, the secondary electrons formed by the incidence of ions on the negatively charged electrode may be suppressed by a negative grid in front of it. A correction must be made for secondary electrons produced at the grid.

In a transverse electric field, without magnetic field, the slow charged particles follow parabolic paths to the metal plates. If the potential difference between the plates is comparable with the energy of formation of the slow

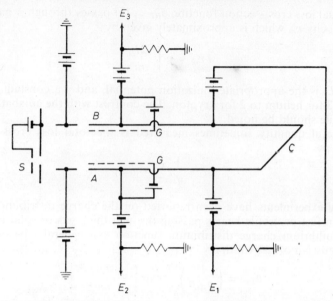

Fig. 11.7. 'Condenser' electrode system for collection of total charge: S, collision chamber entrance slit with electric field arranged to suppress secondary electrons; C, primary beam collector with electric field arranged to trap secondary particles; E_1, to primary beam electrometer; A, electron collector; B, positive ion collector; E_2, E_3, to electrometers; G, secondary electron suppression grids; guard electrodes are also included for the minimization of fringing field

charged particles, the latter will not be collected at points transversely corresponding to their points of formation; and under certain circumstances the path length of collection will not correspond to the length of the collecting plates. However, if the potential difference between the plates is sufficiently large, the electron paths will be almost exactly transverse. The achievement of 'saturation conditions'—invariance of collected current with varying electric field—is an indication that the collision path length is what it appears to be.

It is sometimes the practice[36] to measure the currents collected at several successive 'condenser' electrodes. This procedure does not offer very great advantages. The errors in physical measurement and electrical alignment of a single collecting electrode are far smaller than those arising from other sources, such as pressure measurement; certain of these other errors are likely to be increased by the use of several collecting electrodes. A statistical advantage may be achieved with a single collecting electrode by pressure variation. However, the use of several collecting electrodes provides a check on the invariance of the beam composition as it passes through the collision chamber, and also on the absence of interference from end effects and of pronounced forward or back scattering effects arising at the collision chamber entrance of the Faraday collection cage.

At higher incident ion energies, the study of ionization by measuring the electron flux is not easy, and unexplained discrepancies between different observers have appeared[2, 3, 5, 13, 14]. Discrepancies in charge transfer observations are smaller because slow positive ions and not electrons are being measured. Electrons formed in collisional ionization can have energies of hundreds of electron volts, and secondary electrons can arise from many surfaces; electrons are also produced by the incidence of metastable atoms upon surfaces. It is important to study collisional ionization by measuring not only the electrons but the slow positive ions produced.

The collision between a singly charged positive ion and an atom can give rise to ionizing collisions of the general type:

$$A^+ + B \rightarrow A^{m+} + B^{n+} + (m+n-1)e \qquad (10/mn)$$

These processes may be sub-classified as exchange collisions, in which the ion loses its charge (in general $10/0n$), and ionizing collisions, in which $m > 1$ and/or $n > 0$. The simplest exchange collision is charge transfer ($10/01$), and where $n > 1$ it is reasonable to suppose that the collision process $10/0n$, although ionizing, has partly exchange character; collisions of this type, such as $10/02$, are designated 'transfer ionization' or 'ionization with capture'. The charge transfer process is often referred to as 'electron capture' and as 'charge exchange', but the latter term may be misleading because of confusion with the electron exchange interaction or process. Ionizing collisions in which the electrons are torn from the projectile ion are referred to as 'stripping'.

The collection of the total 'slow' charge produced in these collisions involves various measurements.

Total Electron Current

Measurement of the total electron current produced is achieved by collecting the total negative current to the positive electrode of the system which maintains the transverse electric field. For the present, slow negative ions will be neglected. Secondary electrons arising from the impact of slow positive ions upon the negative collecting electrode must be eliminated or at least estimated. As has been seen, the most satisfactory method is a simple suppression technique in which a negatively charged grid of known transparency is placed in front of the positive ion collector. The cross-section calculated from electron current measurements is:

$$\sigma_e = \sum_m \sum_n (m+n-1) \, _{10}\sigma_{mn}$$

or

$$\sigma_e = \underbrace{\sum_n (n-1) \, _{10}\sigma_{0n}}_{\substack{\text{Charge transfer} \\ +\text{ionization with} \\ \text{capture}}} + \underbrace{\sum_n n \, _{10}\sigma_{1n}}_{\text{Ionization}} + \underbrace{\sum_{\substack{m \\ m>1}} \sum_n (m+n-1) \, _{10}\sigma_{mn}}_{\text{Stripping}} \qquad (11.29)$$

Total Slow Positive plus Negative Flux

In other experiments, measurements have been made of the total slow positive plus negative flux produced. This is achieved by collecting simultaneously and adding the currents produced at both the positive and negative electrodes of the system. As the ionization of the target particle produces equal amounts of positive and negative charge, which balance, this measurement is only of exchange processes plus, at sufficiently high energies, stripping processes. It was originally devised to measure charge transfer. At sufficiently high energies, stripping is more probable than charge transfer, so the measured total current becomes negative instead of positive. The symbol σ_{cs} is used to refer to the cross-section when the charge transfer exceeds the stripping, and σ_{cs} to the cross-section when the stripping exceeds the charge transfer:

$$\sigma_{cs} = \sum_m \sum_n (1-m)\,_{10}\sigma_{mn} = \sum_n \,_{10}\sigma_{0n} - \sum_m \sum_n (m-1)\,_{10}\sigma_{mn} \qquad (11.30)$$

$$\underbrace{\phantom{\sum_n \,_{10}\sigma_{0n}}}_{\substack{\text{Charge}\\\text{transfer}+\\\text{ionization}\\\text{with}\\\text{capture}}} \qquad \underbrace{}_{m>1 \text{ part of stripping}}$$

Total Slow Positive Ion Flux

Measurements have also been made of the total slow positive ion flux.

In earlier work it was considered easier to measure electron currents, because the measurement of total positive ions only gives the purely ionizing collision cross-section as a difference. However, more serious discrepancies appeared between the measurements of different workers than had arisen in charge transfer measurements, and it was thought that these might arise from stray electrons such as those produced by surface collisions of metastable atoms formed in collisions. Stray ions are very much rarer than stray electrons, so it was decided to make measurements of the cross-section for total positive ion production, σ_+:

$$\sigma_+ = \sum_m \sum_n n\,_{10}\sigma_{mn} = \sum_n n\,_{10}\sigma_{0n} + \sum_n n\,_{10}\sigma_{1n} + \sum_m \sum_n n\,_{10}\sigma_{mn} \qquad (11.31)$$

$$\underbrace{}_{\substack{\text{Charge}\\\text{transfer}+\\\text{ionization}\\\text{with capture}}} \quad \underbrace{}_{\text{Ionization}} \quad \underbrace{}_{\substack{m>1\\\text{remainder}\\\text{of}\\\text{stripping}}}$$

Therefore

$$\sigma_e = \sigma_+ - \sigma_{cs} = \sigma_+ + \sigma_{sc} \qquad (11.32)$$

The value of σ_+ was achieved in Leningrad by measuring the total positive current to the negative collecting electrode of the system maintaining the transverse electric field, the secondary electrons from both electrodes being suppressed by grids. This method has the disadvantage that the transparency of the grid in front of the negative collecting electrode must be known; use was therefore made in London and elsewhere of a magnetic field parallel to the incident beam, which not only suppresses the secondary electrons but also the electrons formed by ion–atom collisions, causing them to travel trochoidally along the incident beam. Under these conditions, the small number of higher energy electrons are more easily controlled than they are in the transverse electric field. It is significant that the discrepancies found in measurements of largely σ_e disappear with the newer technique; but the technique of measurement of σ_e is not to be discarded, since with sufficient precautions accurate data can be obtained.

Fast Neutral Detection

Because of the contribution from stripping to slow charge collection, high-energy charge transfer is best measured by detecting the fast neutrals. After the collision chamber, the projectile ion beam is deflected electrically or magnetically and the neutral beam is measured. There is a contribution from ionization with capture, and the cross-section σ_0 is:

$$\sigma_0 = \sum_n {}_{10}\sigma_{0n} \tag{11.33}$$

Charge transfer+
ionization with
capture

At high energies, all the cross-sections of theoretical interest do not emerge from these measurements. For a knowledge of the exchange cross-sections an independent measurement of stripping must be undertaken, and even then a separation of charge transfer from ionization with capture cannot be made. However, the total ionization cross-section, which can be compared with statistical ionization theory, is σ_e. It should be noted that σ_+ is not a purely ionization collision; also that the stripping process, as defined in equation 11.29, includes some stripping with target ionization. Parts of stripping and of ionization with capture are included in all three cross-sections.

Slow ions formed from the target gas may be extracted from a collision path of known length by means of a sufficiently strong transverse electric field, and may then be mass-analysed, for example, by means of a sector magnetic instrument. By contrast, post-collision projectile electrostatic analysis has been used for measurement of multiply-charged projectile ion fluxes, and in particular for the study of stripping-sections[35].

11.4 THEORY OF IONIZATION AND COMPARISON WITH DATA

Classical calculations of ionization by protons have been made along the lines of Section 2.6. The most complete are the Monte Carlo calculations of Abrines and Percival[37] for protons or atomic hydrogen, which are compared with experiment in *Figure 11.8*.

Quantum calculations have been made of ionization and excitation processes at high impact energies, using the first Born and the distortion approximations; a review is available[38].

The cross-section for excitation of a hydrogenic atom of charge $Z_a e$ from state p to state q, by the impact at velocity v_1 of a bare nucleus of charge $Z_b e$ has been expressed[39] in the first Born approximation as:

$$\sigma_{pq} = \left[\frac{8Z_b^2}{s^2} \int_{t_{\min}}^{\infty} |\mathcal{I}(p,q)|^2 \, t^{-3} \, dt \right] \pi a_0^2 \tag{11.34}$$

with

$$\mathcal{I}(p,q) = \int \chi_p^* \chi_q \exp{(itr_a)} \, d^3r_a \tag{11.35}$$

$$s^2 = \frac{m_e v_1^2}{2R} \tag{11.36}$$

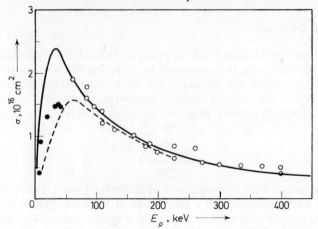

Fig. 11.8. Ionization cross-section function for H^+–H: *open circles, Ireland and Gilbody; closed circles, Fite et al.; full line, calculations of Bates and Griffing[39]; broken line, classical Monte Carlo calculations[37]*

and

$$t_{min} = \frac{E_{ex}}{2s}\left(1 + \frac{m_e E_{ex}}{4\mu s^2}\right) \tag{11.37}$$

where χ represent eigenfunctions normalized to unity, E_{ex} is the excitation energy, r_a the radius vector of the electron, t the time variable, and R the Rydberg constant.

In the impact parameter treatment, which has been shown to be mathematically equivalent[40]:

$$\sigma_{pq} = 2\pi \int_0^\infty b \mathcal{P}_{pq}\, db \tag{11.38}$$

where the probability of transition

$$\mathcal{P}_{pq} = \frac{1}{\hbar^2 v_1^2}\left| \int_{-\infty}^{\infty} \mathcal{O}_{pq}(r) \exp\left(\frac{-iE_{ex}Z}{\hbar v}\right) dZ \right|^2 \tag{11.39}$$

with

$$Z = v_1 t \tag{11.40}$$

and

$$\mathcal{O}_{pq}(r) = \int \chi_p^* V \chi_q\, d^3 r_a \tag{11.41}$$

$\mathcal{O}$ is the interaction potential due to the projectile and r the internuclear position vector. The transition probability becomes:

$$\mathcal{P}_{pq} = \frac{1}{\hbar^2 v_1^2}\left| \int_{-\infty}^{\infty} \mathcal{O}_{qp}(r) \exp\left[\frac{-i}{\hbar v_1}\gamma_{pq}(Z)\right] dZ \right|^2 \tag{11.42}$$

where

$$\gamma_{pq}(Z) = E_{ex}Z + \int_0^Z [\mathcal{O}_{pp}(r) - \mathcal{O}_{qq}(r)]\, dZ \tag{11.43}$$

At energies below the limit of validity of the Born approximation, the method of perturbed stationary states was at one time popular. Since it takes no account of the reluctance of the eigenfunctions to follow the rotation of the nuclear line during collision[41, 42], it has been superseded by the more modern distortion approximation[43, 44], in which full account is taken of the diagonal elements of the scattering matrix. These diagonal elements are thought to be more important than non-diagonal elements, of which no account is taken, because they are not multiplied by oscillating exponential factors in the formula for the cross-section. In general, the distortion calculations lead to smaller cross-sections than the Born at low energies.

A list of ionization and excitation cross-section calculations made by means of the Born approximation and other quantum theory methods appears in Table 11.1. Graphical comparisons are made between the distortion approximation and the first Born approximation in *Figure 11.9*, between H^+ and He^+ excitation (*Figure 11.10*), between theory and experiment for ionization (*Figures 11.8* and *11.11*), and between different types of transition (*Figure 11.12*). At sufficiently high energies the agreement with experiment is good, but not so complete as for fast electron collisions.

In addition to the total cross-sections, differential cross-sections have been calculated[56] for the angular distribution of electrons ejected in the ionization of hydrogen atoms by 1000 eV incident hydrogen atoms. It is found that for the more energetic electrons (> 1 eV) there is strong peaking in the forward direction, but for lower energy electrons the distribution is more nearly isotropic.

Table 11.1. QUANTUM THEORY CALCULATIONS OF EXCITATION AND IONIZATION BY ATOMIC PARTICLES (IONS AND ATOMS ARE TAKEN TO BE IN THE GROUND STATE UNLESS QUANTUM NUMBERS n AND l ARE SPECIFIED)

Process	Reference
H^+, He^{2+}, $Li^{3+}+H \rightarrow H^+$, He^{2+}, $Li^{3+}+H$ ($2s$, $2p$, $3s$, $3d$, continuum)	39
$H^+ + He \rightarrow H^+ + He$ ($1s2p$ ^1P)	45, 46
$H^+ + He \rightarrow H^+ + He$ ($1s3p$ ^1P)	46
$H^+ + Na \rightarrow H^+ + Na$ ($3p$)	47
$He^{2+} + He \rightarrow He^{2+} + He^+ + e$	48
$H^+ + He \rightarrow H^+ + He^+ + e$	49
$H^+ + Li \rightarrow H^+ + Li^+ + e$	50
$H^+ + He \rightarrow H^+ + He^+(2p, 3p, 4p, 3d, 4d) + e$	51
$H + H \rightarrow H(nl) + H(n'l')$ (either nl or $n'l'$ may assume any value including $1s$ and continuum)	39
$He + H \rightarrow He(1s2p$ $^1P) + H(1s, 2s, 2p, 3s, 3p, 3d)$	45
$He + H^+ \rightarrow He(1s2p$ $^1P) + H(1s, 2s, 2p, 3s, 3p, 3d)$	45
$H + He(2s$ $^3S) \rightarrow H(1s, 2s, 2p, 3s, 3d) + He(2p, 3p)$	52
$Ne + He(2s$ $^3S) \rightarrow Ne(nl) + He(2p, 3p)$	52
$He^+ + H \rightarrow He^{2+} + e + H(1s, 2s, 2p, 3s, 3p, 3d)$	53
$H + He \rightarrow H^+ + e + He$	54
$H^+ + H \rightarrow H^+ + H(2s, 2p)$	41
$H + H \rightarrow H + H(2s, 2p)$	41
$H^+ + Ne \rightarrow H^+ + Ne^+ + e$	55

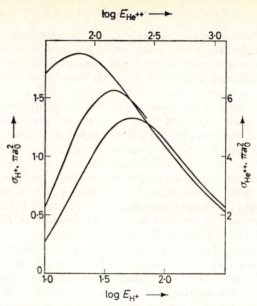

Fig. 11.9. *Scaled comparison of Born and distortion approximation calculations by Bates for 2p excitation of hydrogen by* H^+ *and* He^{2+} [43, 44]: *the two processes have the same cross-section function in the Born approximation (higher curve), but distortion depresses the* He^{2+} *cross-section below that of the* H^+

(From Bates[38], by courtesy of Academic Press)

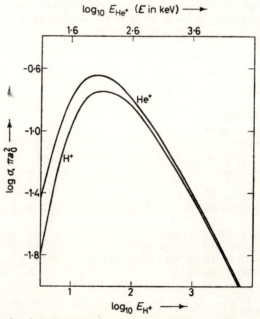

Fig. 11.10. *Calculated cross-section functions for excitation of helium to the level* $1s2p$ ^1P, *by* H^+ *and* He^+ [45]

(From Bates[38], by courtesy of Academic Press)

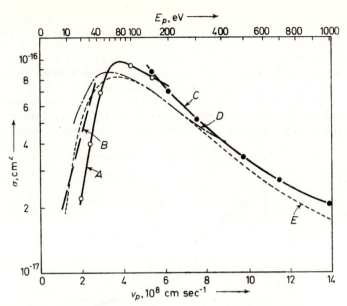

Fig. 11.11. Cross-section function for total ionization of helium by protons: A, Fedorenko et al.; B, Gilbody and Hasted; C, McDaniel et al.; D, Born approximation calculations of Mapleton; E, classical calculations of Gryzinski

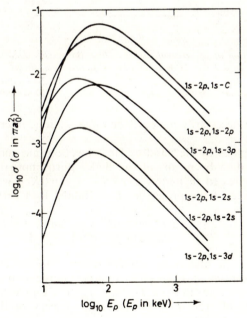

Fig. 11.12. Born approximation calculations for double-transition excitation in H–H collisions[39]

(From Bates[38], by courtesy of Academic Press)

Higher order Born approximations[47, 57] have also been employed (*Figure 11.13*). The second Born approximation may be regarded as allowing for the possibility that a transition $p \rightarrow q$ may take place through a $p \rightarrow n \rightarrow q$ series of virtual transitions; the distortion approximation, on the other hand, corresponds to the $p \rightarrow p \rightarrow q$ and $p \rightarrow q \rightarrow q$ sequences. The second Born approximation is especially important if there is a state n for which $p \rightarrow n$ and $n \rightarrow q$ are strong transitions, whilst $p \rightarrow q$ is a weak transition. The third Born approximation introduces a contribution from the $p \rightarrow q \rightarrow p \rightarrow q$ sequence, which may be considerable in strong transitions. These

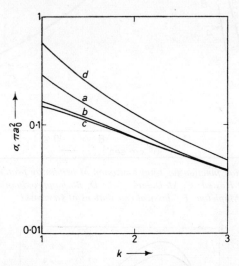

Fig. 11.13. *Comparison of approximations: (a) first Born; (b) distortion; (c) second Born including only* $1s \rightarrow 1s \rightarrow 2s$ *and* $1s \rightarrow 2s \rightarrow 2s$ *distortion effects; (d) second Born including also* $1s \rightarrow 2p \rightarrow 2s$ *polarization effect, for* $H \rightarrow H$ $2s$ *excitation by protons*[57]. *The abscissa unit* $k = (m_H v)/\hbar$, *therefore* $E_p = 29 \cdot 6k^2$ keV

sequences do not necessarily increase the cross-section; interference effects may diminish it.

Ionization of gases by He^+ and by He^{2+} has been measured[58] and compared with H^+ data by means of the Bethe–Born approximation[59]. The calculation is expressed in matrix elements of $\exp(iKz)$, where K is the momentum change of the projectile; K is limited to a certain value K_0, leading to

$$\sigma_{n,l} = \frac{2\pi Z'^2 e^4 c_{n,l} Z_{n,l}}{m_e v_0^2 |E_{n,l}|} \ln \left(\frac{2m_e v_0^2}{C_{n,l}} \right) \tag{11.44}$$

where n and l are the quantum numbers referring to the electron that is removed from the atom; Z' is the effective point charge on the ion, and $Z_{n,l}$ is the number of electrons in the n, l orbital; $c_{n,l}$ is a dipole matrix element, $C_{n,l}$ is an energy of order $E_{n,l}$, the ionization energy of the n, l orbital. Equation 11.44 can be written

$$\sigma_i = AZ'^2 \frac{m_+}{E_p} \ln \left(\frac{BE_p}{m_+} \right) \tag{11.45}$$

where A and B are constants characteristic of the target. Using an empirical effective Z' for He^+, proton ionization data can be employed to calculate the ionization by He^+.

Molecular target ionization processes have also been studied. The analogous processes for electron impact (dissociation, Chapter 5, and dissociative ionization, Chapter 6) are useful for comparison. Many of the same processes are found to occur: for example, the simultaneous production of H^- and H^+ from H_2 has been observed[60].

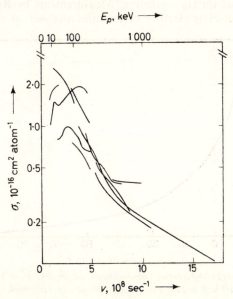

Fig. 11.14. *Measured cross-section functions for the sum of processes* $H_2^+ + H_2 \rightarrow H + H^+ + H_2$ *and* $H_2^+ + H_2 \rightarrow H^+ + H^+ + e + H_2$. *The key will be found in the data collection*[71] (From Barnett, Gauster and Ray[71], by courtesy of the Oak Ridge National Laboratory)

Dissociation studies of molecular ion beams H_2^+ and H_3^+ in gas collisions have been reported[61-66]; the application to injection into thermonuclear machines is important. After passage through a gas cell, the dissociated ions are separated by magnetic or electrostatic analysis. Unfortunately, discrepancies have arisen from loss of dissociated particles owing to scattering, and from inaccurate determination of gas cell thickness. There are also effects arising from variation in vibrational populations of the H_2^+ ions, which depend on the energies of the bombarding electrons in the ion source [67, 68]. Kinetic energy measurements of the products[69, 70] demonstrate the importance of vibrational transitions at impact energies $\geqslant 10$ keV. Data are illustrated in *Figure 11.14*.

11.5 HEAVY-PARTICLE SPECTROSCOPY

Heavy-particle spectroscopy is a technique for determining energy levels in the ionization continuum by means of structure in the ejected electron energy distribution function. Some important experiments are due to Rudd and his colleagues[72], but there are earlier measurements[73, 74]. Autoionizing states of He, Ne, Ar and O_2 have been studied.

The smooth (background) energy distribution of electrons from a heavy-particle ionizing collision is correctly predicted[75] to fall off, at its most rapid, in the region $\geqslant 100$ eV, as E_e^{-4}; the maximum energy that can be attained by an emitted electron is correctly given by equation 2.126. Distributions at lower energies, due to Rudd and Jorgensen[76], are shown in *Figure 11.15*. Born approximation calculations[39] for H ionization are in fair agreement with the experiments on H_2 ionization. Measurements by Rudd have been made with a parallel-plate electron momentum analyser, at polar scattering

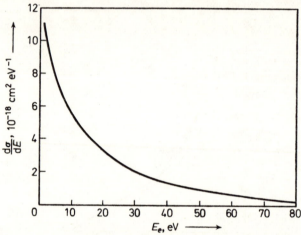

Fig. 11.15. *Electron energy differential cross-section (all angles) from* H^+–H_2 *collisions at 100* keV *impact energy, after Rudd and Jorgenson*

angle θ. A characteristic centre-of-mass electron energy peak E is observed at energy E', where

$$E = E' - 2\left(\frac{E'E_p m_e}{m}\right)^{1/2} \cos\theta + \frac{\bar{E}_g m_e}{m} \qquad (11.46)$$

for target gas of mean thermal energy $\bar{E}_g$, and m is the mass of the emitting ion or atom. In Ar^+–Ar collisions, two groups of lines are observed, one from each particle; one is Doppler shifted. Characteristic data are illustrated in *Figure 11.16*. Inner shell Auger transitions, involving the distribution of energy between an inner shell and one or more outer shell electrons, have characteristic electron energy spectra. An Auger energy diagram for neon is shown in *Figure 6.15* (page 401). Two-electron states need not decay into the first continuum; they may undergo a downward radiative transition.

The identification of two-electron states in helium (*Figure 11.17*) is assisted by comparison of the levels appearing in the proton and in the hydrogen atom excitation spectra of helium. Triplet states which are formed by electron exchange appear only in the hydrogen atom and not in the proton spectra. It can be claimed that at least one of the helium two-electron series was discovered by heavy-particle spectroscopy, as opposed to electron or photon experiments.

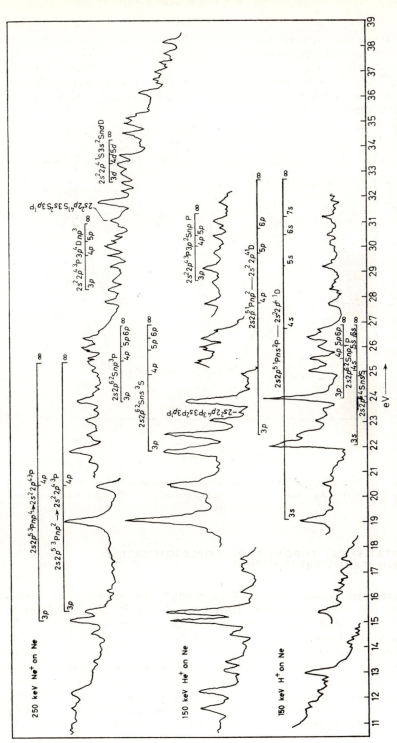

Fig. 11.16. Electron energy spectra, due to Rudd and Lang, of neon bombarded by 250 keV Ne⁺, 150 keV He⁺, and 150 keV H⁺

In the high impact energy collisions (10^4–10^5 eV) normally employed in heavy-particle spectroscopy, the atoms are able to separate before the electron is ejected. The energy calculated is appropriate to the fully separated atom. But for sufficiently slow collisions, such as have been studied by Berry[77], the atoms are not fully separated at the time of electron ejection, and the energy calculated is that of the pseudo-molecule potential energy curve at the time of ejection.

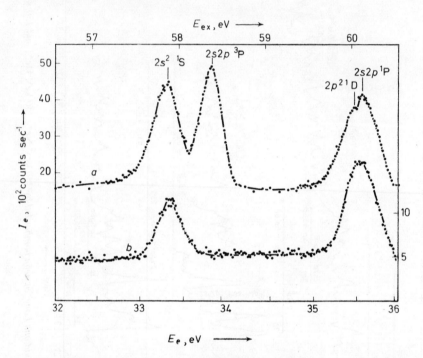

Fig. 11.17. Electron energy spectrum of helium under bombardment: (a) by 75 keV H_2^+; (b) by 75 keV H^+. Note the 2s2p ^3P level appearing only in the collision where exchange is possible

11.6 STATISTICAL THEORY OF MULTIPLE IONIZATION. THE 'TRANSFERRED ENERGY'

An important statistical theory of the multiple ionization of heavy atoms by atoms or ions was given by Russek and Thomas[78]. The theory can be applied successfully to the calculation of the angular distribution of ions scattered in different ionization states[79, 80], and it can go a long way towards yielding total ionization cross-sections.

The theory consists of a calculation of the statistical distribution among the orbital electrons of the energy E_T transferred from the kinetic to the internal degrees of freedom of the particles; this calculation can be tested on its own and is remarkably successful. It is rather more difficult to relate E_T to the atomic parameters, but its experimental determination is possible[81, 82].

In the statistical calculation it is assumed:

1. That E_T is statistically distributed only among the outer electrons; two closed shells of eight electrons will be considered. The energy scale is divided into cells of width ε, and these are all taken to have the same statistical weight.
2. That the ionization energy for each electron is identical and independent of how many other electrons have escaped. This implies that in the ionization process the orbital electrons are forced into 'positions' greatly different from their normal orbits; a sphere with eight equally spaced electrons on its surface has been imagined.

These assumptions are justified only because the collision process is not adiabatic. The collision time is sufficiently short for the atomic energy levels to be broadened by the uncertainty principle to a half-width $\Delta E = hv/a$; they are thus 'smeared' into a continuum which extends from the first excited state to the ionization energy. The broadening of the K and L energy levels is negligible by comparison, therefore these electrons can adjust adiabatically to the changing potential during the collision.

This rather striking picture of the collision process is justified by the fact that alteration of the assumptions seriously reduces the agreement of theory with experiment.

The ionization energy is taken as a convenient multiple of ε, namely 4ε; a subsequent calculation has shown that reduction of ε from this arbitrary value to a very small quantity has little effect upon the argument. The value of E_T is considered to be equal to $m\varepsilon$, where m is an integer; it is bounded by the limits $m\varepsilon \leqslant E_T \leqslant (m+1)\varepsilon$. The probability $P_n(m)$ that a neutral atom will be n times ionized if an energy $m\varepsilon$ is transferred to its internal motion is given by the number of ways in which the energy $m\varepsilon$ can be divided among the eight outer electrons in such a way that n and only n electrons receive $\geqslant 4\varepsilon$, divided by the total number of ways in which the energy $m\varepsilon$ can be distributed among the eight electrons:

$$P_n(m) = \binom{8}{n} \sum_{i=0}^{m-4n} \frac{K_n(i)\, Q_{(8-n)}\, (m-4n-i)}{K_8(m)} \tag{11.47}$$

where $\binom{8}{n}$ is the binomial coefficient,

$$K_n(m) = \prod_{i=1}^{n-1} \frac{m+i}{(n-1)!} \tag{11.48}$$

and

$$Q_{(n-1)}(m) = \sum_{r=0}^{3} Q_n(m-r) \tag{11.49}$$

For $0 \leqslant m \leqslant 3$,

$$Q_2(m) = m+1$$

for $4 \leqslant m \leqslant 6$,

$$Q_2(m) = 7-m$$

and, for $7 \geqslant m$,

$$Q_2(m) = 0$$

In collisions between a singly charged ion and a neutral atom there is assumed to be an even chance that upon separation this single electron deficiency will be associated with either atom; hence there is 50 per cent probability that the projectile will be singly charged even before electron evaporation is taken into consideration. The modified ionization probability $\tilde{P}_n$ is

$$\tilde{P}_n = \tfrac{1}{2}P_n(m) + \tfrac{1}{2}P_{n-1}(m) \tag{11.50}$$

Plots of $\tilde{P}_n(m)$ for $0 \leqslant n \leqslant 8$ as functions of E_p/ε are shown in *Figure 11.18a*. In *Figure 11.18b* are displayed measurements[83] of probability of ionization at polar scattering angles $\theta = 2°$ in Kr–Kr collisions as a function of impact energy. It is possible to make a quantitative comparison of these calculations with experimental data in the following way.

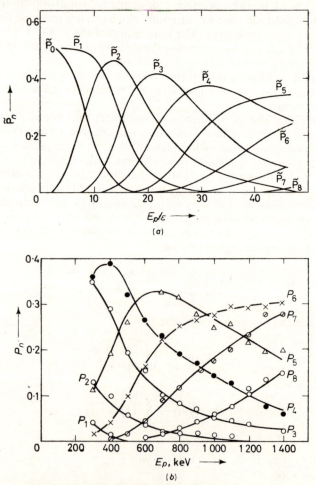

*Fig. 11.18. (a) $\tilde{P}_n$ as a function of E_p/ε for $0 \leqslant n \leqslant 8$
(From Russek and Thomas[78], by courtesy of the American Physical Society);
(b) measurements[83] of probability of ionization $\tilde{P}_n$ at $\theta = 2°$ for Kr$^+$–Kr*

Measurements are made[79, 80] of the relative proportions of charge states formed in the scattering of Ar^+ by Ar at a series of polar angles θ. As θ decreases, the transferred energy E_T increases and, successively, the cross-overs and maxima in the $\tilde{P}_n(E_T)$ functions make their appearance. From these measurements the single relation between θ and E_T cannot be determined, but it may be supposed that the crossovers and maxima occur in the correct order; in Table 11.2 they are compared with the values obtained from *Figure 11.18a*. The agreement is impressive at the three impact energies studied, and it acts as a stimulant to the experimental determinations of E_T.

Table 11.2. EXPERIMENTAL AND THEORETICAL VALUES OF IONIZATION CURVE INTERSECTIONS

Intersection or peak	Experimental values			Theoretical values
	25 keV	50 keV	100 keV	
$\tilde{P}_2 \times \tilde{P}_1$	0·42			0·42
$\tilde{P}_3 \times \tilde{P}_0$	0·09	0·07		0·09
$\tilde{P}_2$	0·42	0·51		0·46
$\tilde{P}_3 \times \tilde{P}_1$	0·24	0·22		0·25
$\tilde{P}_4 \times \tilde{P}_0$	0·07	0·05		0·02
$\tilde{P}_3 \times \tilde{P}_2$	0·34	0·33	0·36	0·39
$\tilde{P}_4 \times \tilde{P}_1$	0·15	0·13	0·13	0·11
$\tilde{P}_3$	0·41	0·38	0·36	0·42
$\tilde{P}_4 \times \tilde{P}_2$	0·23	0·24	0·24	0·25
$\tilde{P}_5 \times \tilde{P}_0$		0·03		0·00
$\tilde{P}_4 \times \tilde{P}_3$		0·33	0·32	0·35
$\tilde{P}_5 \times \tilde{P}_2$		0·13	0·14	0·13
$\tilde{P}_6 \times \tilde{P}_1$		0·02	0·02	0·01
$\tilde{P}_4$		0·38	0·37	0·37
$\tilde{P}_5 \times \tilde{P}_3$			0·24	0·24
$\tilde{P}_6 \times \tilde{P}_2$			0·06	0·07
$\tilde{P}_5 \times \tilde{P}_4$			0·32	0·32
$\tilde{P}_6 \times \tilde{P}_3$			0·14	0·14
$\tilde{P}_7 \times \tilde{P}_2$			0·04	0·03
$\tilde{P}_5$			0·35	0·35
$\tilde{P}_6 \times \tilde{P}_4$			0·24	0·24

The energy transferred in an ionizing collision may be determined from a measurement of target scattering angle ϕ and target ion kinetic energy E_2. It is found that, for ionization into a given charge state, a distribution of values is appropriate. These values correspond to a distribution of impact parameters, and may be expressed in terms of the distance of closest approach r_0, with the aid of an assumption concerning the force field. The Bohr screened-Coulomb function (equation 2.9)

$$V(r) = \frac{Z_1 Z_2 e^2}{r} \exp\left(-\frac{r}{\beta}\right) \qquad (11.51)$$

where

$$\beta = \frac{a_0}{Z_1^{2/3} + Z_2^{2/3}} \qquad (11.52)$$

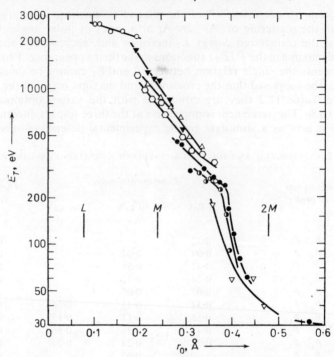

Fig. 11.19. $E_T(r_0)$ *functions for* Kr^+–Kr *collisions, showing structure arising from inner shells; impact energies vary from 200 keV (five open circles at top) to 6 keV (single closed circle at bottom)*

has been used to obtain $E_T(r_0)$ functions from the experimental data. Functions obtained with the coincidence technique described below are shown in *Figure 11.19*, from which it is easily seen that a step appears at about the distances appropriate to the L electron orbits.

A semi-statistical theory has also been proposed by Firsov[84], accounting for the form and magnitude of the total ionization functions as follows:

$$\sigma = \sigma_0\left[\left(\frac{v_p}{v_0}\right)^{1/5} - 1\right]^2 \qquad (11.53)$$

with

$$\sigma_0 = \frac{32 \cdot 7 \, \text{Å}^2}{(Z_1 + Z_2)^{2/3}} \qquad (11.54)$$

and

$$v_0 = \frac{23 \cdot 3 \, E_i \times 10^8 \, \text{cm sec}^{-1}}{(Z_1 + Z_2)^{2/3}} \qquad (11.55)$$

where Z_1 and Z_2 are the collidant nuclear charges, and E_i is the smaller of the two ionization potentials. Comparison with experiment[85] is encouraging; comparison with a single set of heavy-particle data[13, 14] is given in *Figure 11.20*. The deviations increasing with impact energy may be because the

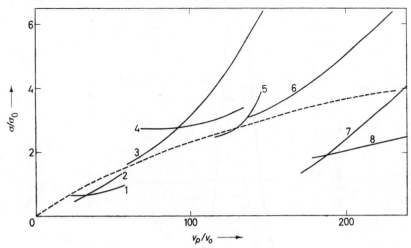

Fig. 11.20. *Comparison of some total ionization data*[13, 14] *with the calculated function of Firsov*[84]*: 1, Ne+–Ne; 2, Ar+–Ne; 3, Ar+–Ar; 4, Ne+–Ar; 5, Kr+–Ar; 6, Ar+–Kr; 7, Kr+–Kr; 8, Ne+–Kr; broken line represents the Firsov equations 11.53–11.55*

experiments weight the multiple ionization cross-sections too strongly, that is, $\sum_m \sum_n (m+n-1)_{10}\sigma_{mn}$ is measured; in the theory, only $\sum_m \sum_n {}_{10}\sigma_{mn}$ is calculated.

Comparisons[86] of ionization and stripping in collisions of fast alkali metal beams with rare gases show that the Firsov function is rather high in these cases. The underlying assumption that the ionization process arises from shell penetration is vindicated, in that the ionization cross-sections of the rare gases exceed the stripping cross-sections of the alkali metals. In these experiments, an iso-electronic effect is observed, namely an anomalously large cross-section when the collidants possess the same or nearly the same number of electrons. The effect is stronger for exactly iso-electronic collisions (rare-gas atoms with rare gases)[19].

The impact velocity variation of E_T in collisions between many-electron atomic systems is not sufficiently well understood at high energies. Ionization probabilities by the simplest atoms have been shown, by Born approximation calculations and by experiments, to fall off as $E_p^{-1} \ln E_p$ at the highest impact energies, and experiments bear this out (see *Figure 11.18*). However, the situation is more complicated for heavy ions; the probability of multiple ionization is found to increase with increasing ion velocity, and it is possible that, as the impact energy is increased, the depth of ionization increases until the nucleus can be entirely stripped.

Detailed experimental studies have been made of the angular distributions and velocity distributions of scattered target ions[87], of scattered projectile ions[79, 88], and of the electrons produced[73, 74]. Some typical data due to Fedorenko *et al.* are shown in *Figures 11.21* and *11.22*; multiply-charged product ions, both those arising from the projectile and those arising from the target, fall into two groups, corresponding to the 'hard' and 'soft' scattering of Section 2.6. The projectile ions are scattered into two groups, having different

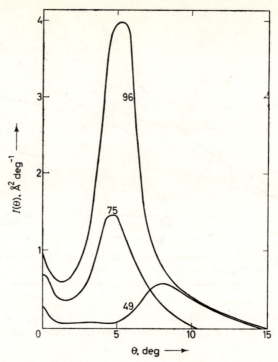

Fig. 11.21. *Differential scattering of* Ar$^+$ *by* Ar, *with stripping to form* Ar^{2+}; *impact energies are marked in keV, and distinct areas of hard and soft scattering are apparent*

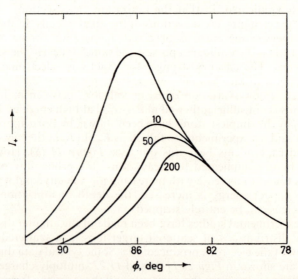

Fig. 11.22. *Differential scattering of argon target, triply ionized by 75 keV, Ne$^+$: although distinct areas of hard and soft scattering are not apparent, it will be seen that the slower ions (retarding potential marked in volts) are scattered closer to 90°*

ranges of scattering angle θ. The target ions do not fall into two groups as far as scattering angles ϕ are concerned, but they do fall into two energy groups, one of which is near-thermal, and the other extending to energies of tens or hundreds of electron-volts; the distribution of the less energetic particles cannot be easily measured. The geometrical features of differential scattering have been discussed by various workers[89-91]. Application of equations 2.124 and 2.125 leads to values of the energy transferred E_T which are consistent with statistical theory[79, 80]. A consequence of the presence of hard scattering is that maxima in the emitted electron energy distributions due to a given process $10/mn$ were expected[79], and have now been found.

For accurate measurements of E_T it is necessary to keep track of both the atomic constituents of a single collision. This requires the simultaneous detection of each of them, which has been achieved by Afrosimov and his colleagues[82] using coincidence counting technique. The incident positive ion was scattered through a small polar angle θ, whilst the target particle was scattered, in the plane of θ, through an angle ϕ. For a fixed θ detector, a second detector was scanned in ϕ and the number N of coincidences measured. For given θ and ϕ, the value of E_T was calculated from the equation

$$Q \equiv E_T = E_p\left[1 - \frac{\sin^2\phi + (m_+/m_0)\sin^2\theta}{\sin^2(\phi+\theta)}\right] \qquad (11.56)$$

It was found for Ar^+–Ar that the distribution function $N(E_T)$ was not monotonic but possessed three broad maxima (*Figure 11.23*). When the

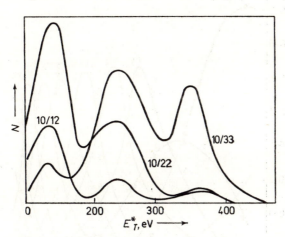

Fig. 11.23. E_T^* distribution functions of Ne[+]–Ar 75 keV ionizing collisions (10/12, 10/22 and 10/33); $r_0 = 0.25$ Å

appropriate ionization energies were subtracted from Q, the quantity R^* given by

$$R^* \equiv E_T^* = Q_{max} - \sum_{m, n} E_i(m, n) \qquad (11.57)$$

was found to be independent of the final states m and n of ionization, at a single impact energy E_p.

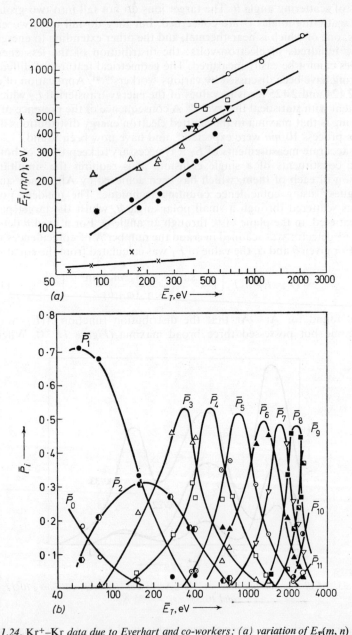

Fig. 11.24. Kr^+–Kr *data due to Everhart and co-workers: (a) variation of* $E_T(m, n)$ *with the inelastic energy loss* E_T *averaged over all states of ionization; since* $\bar{E}_T$ *is dependent upon impact energy,* $\bar{Q}_{m, n}$ *is also dependent upon it (crosses,* $m = 1$ *and* $n = 1$; *filled circles,* $m = 2$ *and* $n = 2$; *open triangles,* $m = 3$ *and* $n = 3$; *closed inverted triangles,* $m = 4$ *and* $n = 4$; *open squares,* $m = 5$ *and* $n = 5$; *open circles,* $m = 6$ *and* $n = 6$); *(b) probability* $\bar{P}_i$ *of ionization to a given state against* $\bar{E}_T$

A further series of experiments[92] showed that, over a range of impact energies, variation of R^* was possible; an analysis has been made of the average energy loss $\bar{Q}_{m,n}$ for ionization states m and n, which also shows variation with E_p (*Figure 11.24a*). However, there is a variation of $\bar{Q}_{m,n}$ with probability of ionization, which is characteristic of the statistical theory of ionization (*Figure 11.24b*). It is also possible to obtain from the data[81] the closest distance of nuclear approach for each collision, assuming suitable interaction energy (*Figure 11.19*).

The questions which must be answered are:

1. What is the nature of the characteristic excitation?
2. How does this characteristic energy become distributed amongst the many electrons which can be ejected?
3. Does this distribution take place during the collision process, or after the two atoms have separated?

The third of these questions can be answered by a statistical analysis of the correlation between ionization states of the two atomic systems, and the solution throws light upon the first two questions. The statistical theory of Russek seems to provide the best answer to the second question.

There have been various proposals as to the nature of the characteristic excitations. It is unlikely that they are inner shell Auger states, since an analysis[93] of the continuum of electron energies measured by Rudd[72] shows that there are far too many multiple ionization electrons for them to arise in this way, or indeed from the diabatic curve-crossing mechanism proposed by Fano and Lichten[94]. There seems to be a large background continuum of energy transfer.

Lichten[95] considered violent ion–atom collisions, in which the electron shells interpenetrate, in terms of the promotion of inner shell electrons predicted by molecular orbital theory. Molecular orbital calculations have been made of the H_2^+-like orbitals of the systems, $Ar+Ar$ and $Ne+Ne$, which are displayed in *Figure 11.25* together with orbitals for H_2^+ and He_2. Transitions between diabatic molecular orbitals at crossings can leave a 'promoted' electron stranded in a higher level after a collision. The idea of pseudo-crossing loses its validity when the interaction which causes avoidance is comparable to the largest separation of the levels, that is, when the promotion energy is smaller than the sub-shell splitting. There can also be effects due to configuration interaction. Examples of the interpretation of these diagrams are as follows.

In H^+–H collisions, the probability of $2p$ excitation or of charge transfer terminating in a $2p$ state can be calculated from the energy levels of *Figure 11.25a*. For 2 keV protons, the adiabatic criterion yields $\Delta E = 0 \cdot 14/b$, and the separation between $2p\sigma$ and $2p\pi$ energy levels is such that impact parameter $b = 0 \cdot 9$ a.u. A broad maximum in the excitation probability occurs in this region.

For He^+–He collisions, the lowest excited state is $(1s\sigma_g)^2 (2s\sigma_g)^2 \, ^2\Sigma_g^+$, and this excitation leads to an equal probability of capture or excitation into state He 2^3S. The transitions are

$$(2p\sigma_u)^2 \rightarrow (1\sigma_g)(2s\sigma_g)$$

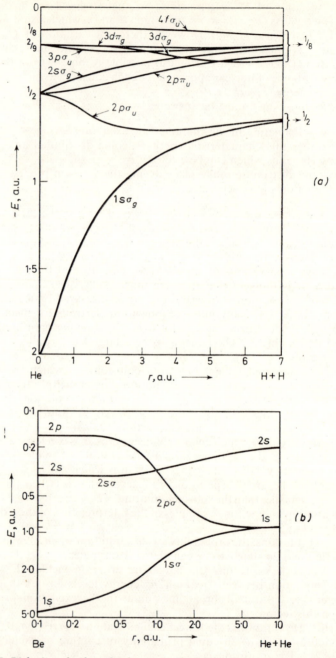

Fig. 11.25. Diabatic molecular orbitals proposed by Lichten[95]; (a) H+H[96]; (b) He+He;
(c) Ne+Ne; (d) Ar+Ar

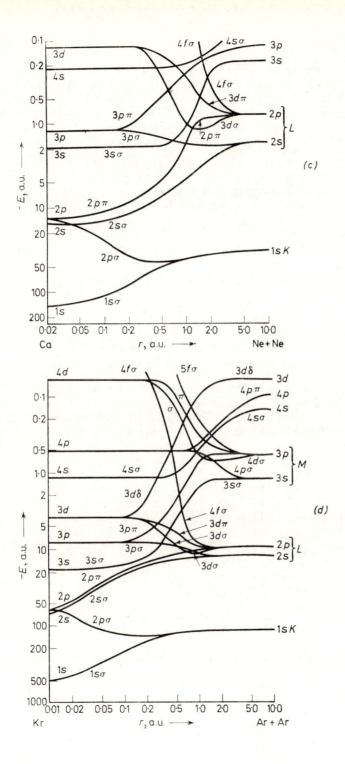

(c)

(d)

39*

with energy defect

$$\Delta E = \varepsilon(2s\sigma_g) + \varepsilon(1s\sigma_g) - 2\varepsilon(2p\sigma) \qquad (11.58)$$

which is zero at separation $r \sim 1 \cdot 9$ a.u.

For Ne^+–Ne collisions the differential scattering measurements of Jones *et al.*[97] show characteristic energy losses (excitations) at 18·9 eV and 48·4 eV. The first arises from $4f\sigma_u \rightarrow 3p\pi_u$ and the second from $3p\sigma_u \rightarrow 2p\pi_u$. At 750 eV, a characteristic ejected electron energy loss is observed in backward scattering. This arises from promotion of a K electron in a $2p\sigma$–$2p\pi$ crossing:

$$2\varepsilon(2p) - \varepsilon(1s) = \varepsilon(2p) + \varepsilon(2p) - \varepsilon(1s) = -97 + 852 = 755 \text{ eV} \qquad (11.59)$$

There is also an energy loss, deduced from the coincidence multiple ionization experiments, of 840 eV, which is interpreted as the promotion energy

$$\varepsilon(2p) - \varepsilon(1s) = 874 - 22 = 852 \text{ eV} \qquad (11.60)$$

For Ar^+–Ar collisions there are promotion energies

$$\varepsilon(3p) - \varepsilon(2p) = -16 + 244 = 228 \text{ eV} \qquad (11.61)$$

and

$$\varepsilon(3p) + [\varepsilon(3p) - \varepsilon(2p)] = -34 + 228 = 194 \text{ eV} \qquad (11.62)$$

Characteristic fast-electron peaks have been observed[98] at 190 ± 50 eV, and it has been proposed that the characteristic losses in the multiple ionizing collisions are 234 ± 24 eV and 289 ± 27 eV. The last of these is therefore unexplained.

Amusia has proposed[99] that there is a collective oscillation of M shell electrons in argon contributing an energy of the order of 200 eV. The excitation contributes in surface collisions[100] Ar^+–Cu in much the same way as it does in ionizing collisions. A characteristic energy loss at ~ 230 eV has been obtained[101, 102] using an electron spectrometer in which the zero-angle scattered electron energy loss is measured. Since no corresponding Auger process has been reported, the energy of excitation can only be distributed amongst several outer shell electrons. A 230 eV level in argon cannot be interpreted by involvement of L shell electrons; a collective oscillation of M shell electrons is a possible interpretation.

Theoretical treatment of collective oscillation has been given[103]. Such effects in metals are described in terms of particle–hole pairs, using the notation

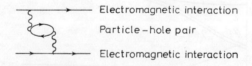

Electromagnetic interaction

Particle–hole pair

Electromagnetic interaction

11.7 EXCITATION BY ATOMIC AND MOLECULAR PARTICLES

Excitation is one of the important processes in the stopping of fast particles in gases, and it is also a process of geophysical importance.

Apart from early measurements without mass-analysis of the impacting ion beams, the first experiments were made with the auroral processes in mind[104]. It is important that complicating collision processes be minimized and it is therefore necessary to conduct the experiments under single collision conditions. The Amsterdam[105] and London[106] groups, Carleton and co-workers[107], and Fan and his colleagues[104, 108] pioneered such investigations. A mass-analysed ion beam is directed into a collision chamber containing gas; the beam is collected and measured, and electrodes are sometimes included for charge exchange and ionization measurement. A spectrometer, monochromator or detector sensitive to photons of a limited wavelength region receives light emergent through slits or ports in the collision chamber. By means of the Doppler shift of radiation emitted along the ion path, a method of distinguishing between photons emitted from projectiles and targets is available.

Excitation of either projectile or target or both is possible in a fast collision (processes 10/1'0, 10/10', 10/1'0'); simultaneous ionization is also possible (10/2'0e, 10'11'e); there remains a third process giving rise to radiation, namely electron capture or charge transfer into an excited state (10/0'1, 10/01', 10/0'1'). Cross-section functions for this type of process are qualitatively different from those for the excitation processes without transfer.

In measurements of excitation of gases by fast-ion and fast-atom beams, account must be taken of:

1. The change of ion beam composition as it passes through the collision chamber; this can be of great importance when the excitation cross-sections are much smaller than the charge transfer cross-sections.
2. The anisotropic distribution of radiation.
3. Resonance absorption or imprisonment of radiation.
4. Secondary effects of the photons and electrons produced in the collision chamber.
5. Cascade transitions contributing to radiation.

Consider a level j being excited by a beam of fast particles of intensity I particles cm^{-2} sec^{-1}, the process having a cross-section σ_{0j}. The rate of formation of j state atoms per unit volume is given by:

$$\frac{\mathrm{d}n_j}{\mathrm{d}t} = n_0 I \sigma_{0j} + \sum_i n_i A_{ij} - \sum_k n_j A_{jk} \qquad (11.63)$$

where $J_{jk} = n_j A_{jk}$ and $J_{ij} = n_i A_{ij}$ (for $i > j > k$), represent the total emission of quanta corresponding to the transition $j \to k$, and the A symbols represent the transition probabilities. It follows that, for maximum n_j,

$$\sigma_{0j} = \frac{\sum_k J_{jk} - \sum_i J_{ij}}{n_0 I} \qquad (11.64)$$

For the deduction of σ_{0j} from the 'emission cross-sections', all the relevant quanta emissions must be known.

Born approximation and distorted-wave calculations for the simplest excitation processes, by both fast ions and fast atoms, are discussed in Section 11.4; the Born cross-section functions show maxima at the energies to be expected on the basis of the adiabatic maximum rule, and fall off in proportion to $E_p^{-1} \ln E_p$ at high energies. The Lyman α emissions from the colli-

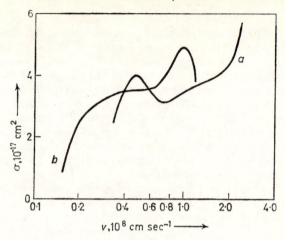

Fig. 11.26. Comparison of excitation[109] of H $1s \rightarrow 2p$: (a) by H^+; (b) by He^+

sions H^+–H and He^+–H have been measured[109] with iodine counters; the functions are shown in *Figure 11.26*.

Heavy-particle excitation cross-sections are in general smaller than those for ionization, but rather larger cross-sections have been observed for processes giving rise to Lyman α radiation (see *Figure 11.26*). The low-energy enhancement in the He^+–H_2 function is probably to be attributed to the charge exchange process:

$$He^+ + H_2 \rightarrow He + H^+ + H\ 2p$$

The emission functions for Balmer transitions[110] are of the same order of magnitude as those for the Lyman α radiation.

Calculations[111] of the excitation of the hydrogen atom from a level of energy E_2 to a level of energy E_1 lead to the expression

$$\sigma \simeq 2\pi \left(\frac{3\pi e^2 a_0}{hv} \right)^2 \left[\frac{4}{\pi^2} + \ln \left(\frac{\pi h v^2}{24 e^2 a_0 \omega} \right) \right] \qquad (11.65)$$

where

$$\hbar\omega = E_1 - E_2 = \Delta E \qquad (11.66)$$

For these collisions the maximum cross-section is reached at a velocity

$$v_{max} \simeq \left(\frac{6 e^2 a_0 \omega}{h} \right)^{1/2} \qquad (11.67)$$

which is proportional to $\Delta E^{1/2}$.

The Born and Jeffrey's WKB expressions for p–H excitation can be expressed in parametric form[112] so that regression analysis can be applied to the data to yield the parameters. The magnitudes of the cross-sections are proportional to n^{-3} (n being the principal quantum number)[113-115] both in the Born approximation and in experiment. Application of the Born approximation to p–H excitation shows that, for large impact parameter, excitation to the p states greatly exceeds those to the s and d states. The high impact energy limit given by the Bethe approximation is upheld. There should be a close correspondence between the excitation by protons and by electrons, but some unexplained differences have been found. The excitation functions H $1s \rightarrow 2p$ by protons and by He[116] (*Figure 11.26*) demonstrate the importance of coupling between different states of H_2^+ and H–He[+].

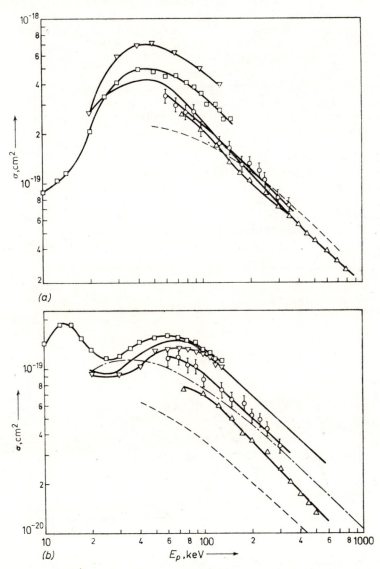

(a)

(b)

E_p, keV ⟶

Fig. 11.27. Excitation cross-sections: (a) for He 4^1S by protons (b) for He 4^1D by protons; open circles, Robinson ad Gilbody[118]; inverted triangles, Dodd and Hughes; squares, Van den Bos et al.; triangles, Thomas and Bent; full line, Denis et al.; broken line, calculations of Gaillard; chain dotted line, calculations of Bell and Kingston

The excitation of helium levels by protons has been studied in detail[117], and some functions are displayed in *Figure 11.27*[118]. The adiabatic maximum rule can be applied to the sum of the *s*, *p* and *d* excitation functions (for given multiplicity and principal quantum number), but not to the individual level functions. At maximum, the optically allowed transition functions are larger than those for forbidden transitions. At very large and at very small impact velocities, σ (nS) > σ (nP) > σ (nD). Another interesting finding is that the ^{3}P and ^{1}P functions are almost identical; the contribution of electron exchange collisions is therefore large. Unfortunately discrepancies remain between the different laboratories in which experimental studies are in progress. For ^{1}P levels, the Born approximation holds above 400 keV, but for other configurations the position is not clear[98, 106, 119].

In He$^+$–He collisions, radiation arises from both excitation[120] and charge transfer[121] processes. At energies $\gtrsim$ 10 000 eV, and with observation at 60° to the beam path, the two radiations are distinguishable by Doppler shift. For principal quantum number $3 \leqslant n \leqslant 6$, the lower energy cross-sections are almost identical for the two processes; intermediate molecular states play an important role.

The excitation of helium to the metastable level He 2^3S by 600 eV helium ions has been studied in differential scattering[122]. Oscillations arising from a curve-crossing are found.

Studies of excitation of molecules by ions are rather less developed than those of atoms. Much of the impetus has come from the need to understand the proton excitation of lines observed in the airglows of Earth and Venus[123].

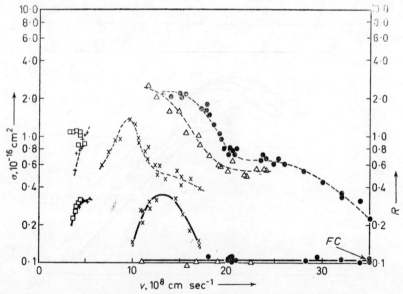

Fig. 11.28. Cross-section functions[107, 124] *(broken lines) for excitation of the 0, 0 Meinel band of* N$_2^+$ *in collision with* N$_2$: *vibrational enhancement ratios R are shown as full lines; closed circles, proton projectiles; triangles, deuteron projectiles; crosses, He$^+$ projectiles; squares, Ne$^+$ projectiles; plus signs, N$^+$ projectiles; FC, Franck–Condon calculation*

Sheridan[124] and Carleton[107] have reported studies of protons in nitrogen, which are illustrated in *Figure 11.28*. Although no rotational excitation was observed, some interesting enhancement of vibrational levels was reported. In the excitation (with ionization) of N_2^+ $B^2\Sigma_u^+$ from N_2 by protons, the ratio R of $v = 1$ to $v = 0$ agrees with the predicted Franck–Condon value, but in the excitation by He^+, Ne^+ and N^+ the ratio is enhanced.

Vibrational excitation of molecules has been treated in a many-state approximation, and also with perturbation theory, by Rapp and others[125]. The impulse approximation has been used to calculate rotational excitation[126], and experiments have been carried out[127] using a quadrupole state selector. Information about interaction energies can be obtained by applying the Born approximation[128].

REFERENCES

1. SOLOVIEV, E. S., ILYIN, R. N., OPARIN, V. A. and FEDORENKO, N. V. *J. exp. theor. Phys.* 42 (1962) 659.
2. HOOPER, J. W., MCDANIEL, E. W., MARTIN, D. W. and HARMER, D. S. *Phys. Rev.* 121 (1961) 1123.
3. GILBODY, H. B., HASTED, J. B., IRELAND, J. V., LEE, A. R., THOMAS, E. W. and WHITEMAN, A. S. *Proc. Roy. Soc.* A274 (1963) 40; GILBODY, H. B. and HASTED, J. B. *Proc. Roy. Soc.* A240 (1957) 382.
4. MAPLETON, R. A. *Phys. Rev.* 109 (1958) 1166; 122 (1961) 528.
5. AFROSIMOV, V. V., ILYIN, R. N. and FEDORENKO, N. V. *J. exp. theor. Phys.* 34 (1958) 1398; 36 (1959) 41; *J. tech. Phys., Moscow*, 28 (1958) 2266; FOGEL, Y. M., KRUPNIK, L. I. and SAFRONOV, I. G. *J. exp. theor. Phys.* 28 (1955) 589; FOGEL, Y. M., ANKUDINOV, V. A., PILIPENKO, D. V. and TOPOLIA, N. V. *J. exp. theor. Phys.* 34 (1958) 579; FEDORENKO, N. V., AFROSIMOV, V. V. and KAMINKER, D. M. *J. tech. Phys., Moscow* 26 (1956) 1926; FEDORENKO, N. V. and AFROSIMOV, V. V. *J. tech. Phys., Moscow* 26 (1956) 1941; AFROSIMOV, V. V., ILYIN, R. N., OPARIN, V. A., SOLOVIEV, E. S. and FEDORENKO, N. V. *J. exp. theor. Phys.* 41 (1961) 1048; FEDORENKO, N. V. *Usp. fiz. Nauk* 68 (1959) 481; ILYIN, R. N., AFROSIMOV, V. V. and FEDORENKO, N. V. *J. exp. theor. Phys.* 36 (1959) 41; FEDORENKO, N. V. *J. tech. Phys., Moscow* 24 (1954) 769, 784, 2113; KAMINKER, D. M. and FEDORENKO, N. V. *J. tech. Phys., Moscow* 25 (1955) 1843; AFROSIMOV, V. V. and FEDORENKO, N. V. *J. tech. Phys., Moscow* 27 (1957) 2557, 2573.
6. HORTON, F. and MILLEST, D. M. *Proc. Roy. Soc.* A185 (1946) 381.
7. MOE, D. *Phys. Rev.* 104 (1956) 694.
8. UTTERBACK, N. G. *Bull. Amer. phys. Soc.* 11, 7 (1962) 487.
9. BATHO, H. F. *Phys. Rev.* 42 (1932) 753; RUDNICK, P. *Phys. Rev.* 38 (1931) 1342.
10. BERRY, H. W., VARNEY, R. N. and NEWBERG, S. *Phys. Rev.* 61 (1942) 53; VARNEY, R. N. *Phys. Rev.* 50 (1936) 159.
11. ROSTAGNI, A. *Nuovo Cim.* 11 (1934) 99.
12. WOLF, F. *Ann. Phys., Lpz.* 23 (1935) 285, 627; 25 (1936) 527, 737; 27 (1936) 543; 29 (1937) 33; 30 (1937) 313.
13. GILBODY, H. B. and HASTED, J. B. *Proc. Roy. Soc.* A240 (1957) 382.
14. KEENE, J. P. *Phil. Mag.* (7) 40 (1949) 369.
15. UTTERBACK, N. G. and MILLER, G. H. *Phys. Rev.* 124 (1961) 1477.
16. HAYDEN, H. C. and UTTERBACK, N. G. *Phys. Rev.* 135 (1964) A1575.
17. UTTERBACK, N. G. *Phys. Rev. Lett.* 12 (1964) 295; *J. chem. Phys.* 44 (1966) 2540.
18. WEIZEL, W. *Z. Phys.* 76 (1932) 250.
19. FLAKS, I. P. *Zh. tech. Phys., Moscow* 31 (1961) 367.
20. ÖPIK, E. J. *Physics of Meteor Flight in the Atmosphere*, 1958. New York; Interscience.

21. EVANS, S. and HALL, J. E. *J. atmos. terr. Phys. Suppl.* 2 (1955) 18.
22. GREENHOW, J. S. and HAWKINS, G. S. *Nature, Lond.* 170 (1952) 355.
23. MASSEY, H. S. W. and SIDA, D. W. *Phil. Mag.* (7) 46 (1955) 190.
24. BYDIN, I. F. and BUKHTEEV, A. M. *Dokl. Akad. Nauk SSSR* 119 (1958) 1131.
25. HASTED, J. B. *Penetration of Charged Particles in Matter*, Ed. E. A. Uehling, Publication 752, 1958. United States National Academy of Sciences, National Research Council.
26. ALLISON, S. K. *Rev. mod. Phys.* 30 (1958) 1137; *Phys. Rev.* 110 (1958) 670; 109 (1958) 76.
27. ALLISON, S. K. and GARCIA-MUNOZ, M. *Atomic and Molecular Processes*, Ed. D. R. Bates, 1962. New York; Academic Press.
28. COLLINS, L. E. and STROUD, P. T. *Proc. phys. Soc., Lond.* 90 (1967) 641.
29. NIZAMUDDIN, S. Thesis, University of Manchester, 1968.
30. GILBODY, H B., BROWNING, R., LEVY, G., McINTOSH, A. I. and DUNN, K. F. *Proc. phys. Soc.* B1 (1968) 863.
31. AFROSIMOV, V. V., IL'IN, R. N., and SOLOV'EV, E. S. *Soviet Phys. tech. Phys.* 5 (1961) 661; ALLISON, S. J., CUEVAS, J. and GARCIA-MUNOZ, M. *Phys. Rev.* 120 (1960) 1266; NIKOLAEV, V. S., DMITRIEV, I. S., FATEEVA, L. N. and TEPLOVA, YA. A. *Soviet Phys. JEPT* 13 (1961) 695; 14 (1961) 67; DMITRIEV, I. S., NIKOLAEV, V. S., FATEEVA, L. N. and TEPLOVA, YA. A. *Soviet Phys. JETP* 15 (1962) 11; 16 (1963) 259; PIVOVAR, L. I., TUBAEV, I. and NOVIKOV *Soviet Phys. JETP* 14 (1961) 20; 15 (1962) 1035; PIVOVAR, L. I., NOVIKOV and TUBAEV, I. *Soviet Phys. JETP* 19 (1964) 318; 21 (1965) 681; PIVOVAR, L. I., NOVIKOV and DOLGOV *Soviet phys. JETP* 22 (1966) 508; KOZLOV, V. F. and BONDAR, S. A. *Soviet Phys. JETP* 23 (1966) 195; FLAKS, I. P. and SOLOV'EV, E. S. *Soviet Phys. tech. Phys.* 3 (1958) 564.
32. STEPHENS, K. G. and WALKER, D. *Phil. Mag.* 46 (1955) 563; REYNOLDS, WYLY and ZUCKER *Phys. Rev.* 98 (1955) 1825; HECKMAN, HUBBARD, E. L. and SIMON *Phys. Rev.* 129 (1963) 1240.
33. HUBBARD, E. L. and LAUER, E. J. *Phys. Rev.* 98 (1955) 1814; HECKMAN, HUBBARD, E. L. and SIMON, *Phys. Rev.* 129 (1963) 1240; NIKOLAEV, V. S., DMITRIEV, I. S., FATEEVA, L. N. and TEPLOVA, YA. A. *Soviet Phys. JETP* 6 (1958) 1019; 12 (1961) 627; REYNOLDS, WYLY and ZUCKER *Phys. Rev.* 98 (1955) 474; STEPHENS, K. G. and WALKER, D. *Proc. Roy. Soc* A229 (1955) 376; ALMQUIST, E., BROUDE, C., CLARK, M. A., KUEHNER, J. A. and LITHERLAND, A. E. *Canad. J. Phys.* 40 (1962) 954; BOOTH, W. and GRANT, I. S. *Nucl. Phys.* 63 (1965) 481; NORTHCLIFFE, L. C. *Phys. Rev.* 120 (1960) 1744; MARTIN, F. W. *Phys. Rev.* 140 (1965) A75; MOAK, C. S., LUTZ, H. O., BRIDWELL, L. B., NORTHCLIFFE, L. C. and DATZ, S. *Phys. Rev. Lett.* 18 (1967) 41; GRODZINS, L., KALISH, R., MURNICK, D., VAN DE GRAAFF, R. J., CHMARA, F. and ROSE, P. H. *Phys. Lett.* 24B (1967) 282.
34. SCHWIRZKE, F. *Z. Phys.* 157 (1960) 510.
35. SOLOV'EV, E. S., IL'IN, R. N., OPARIN, V. A. and FEDORENKO, N. V. *Soviet Phys. JETP* 15 (1962) 459.
36. DONAHUE, T. M. and HUSHFAR, F. *Phys. Rev. Lett.* 3 (1959) 470; *Nature, Lond.* 186 (1960) 1038; DONOHUE, T. M. and CURRAN, R. K. *Phys. Rev.* 118 (1960) 1233; CURRAN, R. K., DONAHUE, T. M. and KASNER, W. H. *Phys. Rev.* 114 (1959) 490.
37. ABRINES, R. and PERCIVAL, I. C. *Phys. Lett.* 13 (1966) 216; *Proc. phys. Soc., Lond.* 88 (1966) 861, 873, 885.
38. BATES, D. R. *Atomic and Molecular Processes*, 1962. New York; Academic Press.
39. BATES, D. R. and GRIFFING, G. W. *Proc. phys. Soc., Lond.* A66 (1953) 961; A67 (1954) 663; A68 (1955) 90.
40. ARTHURS, A. M. *Proc. Camb. phil. Soc.* 57 (1961) 904.
41. BATES, D. R. *Proc. Roy. Soc.* A245 (1958) 299.
42. BATES, D. R. *Proc. Roy. Soc.* A243 (1957) 15.
43. BATES, D. R. *Proc. phys. Soc., Lond.* 73 (1959) 227; 77 (1961) 59.
44. MITTLEMAN, M. H. *Phys. Rev.* 122 (1961) 499.
45. MOISEIWITSCH, B. L. and STEWART, A. L. *Proc. phys. Soc., Lond.* A67 (1954) 1069.

46. BELL, R. J. *Proc. phys. Soc., Lond.* 78 (1961) 903.
47. BELL, R. J. and SKINNER, B. G. *Proc. phys. Soc., Lond.* (1962).
48. ERSKINE, G. A. *Proc. Roy. Soc.* A224 (1954) 362.
49. MAPLETON, R. A. *Phys. Rev.* 109 (1958) 1166.
50. McDOWELL, M. R. C. and PEACH, G. *Phys. Rev.* 121 (1961) 1383.
51. DALGARNO, A. and McDOWELL, M. R. C. *The Airglow and the Aurora*, Ed. A. Dalgarno and E. B. Armstrong, 1955. New York; Pergamon Press.
52. ADLER, J. and MOISEIWITSCH, B. L. *Proc. phys. Soc., Lond.* A70 (1957) 117.
53. BOYD, T. J. M., MOISEIWITSCH, B. L. and WILLIAMS, A. L. *Proc. phys. Soc., Lond.* A70 (1957) 110.
54. BATES, D. R. and WILLIAMS, A. *Proc. phys. Soc., Lond.* A70 (1957) 306.
55. BATES, D. R., McDOWELL, M. R. C. and OMHOLT, A. *J. atmos. terr. Phys.* 10 (1957) 51.
56. DALGARNO, A. and GRIFFING, G. W. *Proc. Roy. Soc.* A248 (1958) 415.
57. KINGSTON, A. E., MOISEIWITSCH, B. L. and SKINNER, B. G. *Proc. Roy. Soc.* A258 (1960) 237.
58. McDANIEL, E. W. *et al. Phys. Rev.* 136 (1964) A379, A385.
59. BETHE, H. A. *Ann. Phys., Lpz.* 5 (1930) 325.
60. DUKELSKII, V. M. *Soviet Phys. JETP* 6 (1958) 657.
61. WIND, H. *Atomic Collision Processes in Plasma Symp.* p. 135. Culham Laboratory, Harwell, 1964.
62. RIVIERE, A. C. and SWEETMAN, D. R. *Proc. phys. Soc., Lond.* 78 (1961) 1215.
63. BARNETT, C. F and RAY, J. A. *Atomic Collision Processes*, p. 743, Amsterdam; North Holland Publishing Company; IL'IN, R. N. *et al. Soviet Phys. JETP* 19 (1964) 817.
64. BERKNER, K. T. *et al. Phys. Rev.* 146 (1966) 9.
65. WILLIAMS, J. F. and DUNBAR, D. N. F. *Phys. Rev.* 149 (1966) 62.
66. GUIDINI, J. *C. R. Acad. Sci. Paris* 253 (1961) 829.
67. TUNITSKII, N. N. *Dokl. Akad. Nauk USSR.* 191 (1955) 903.
68. McGOWAN, N. W. and KERWIN, L. *Canad. J. Phys.* 42 (1964) 972.
69. CAUDANO, R., DELFOSSE, J. M. and STEYAERT, J. *Ann. Soc. sci. Brux.* 76 (1963) 127.
70. VALCKX, F. P. G. and VERVEER, P. *Abstr. 4th Int. Conf. Phys. electron. atom. Collisions*, p. 333, 1965. Hastings-on-Hudson, New York; Science Brookcrafters.
71. BARNETT, C. F., GAUSTER, W. B. and RAY, J. A. Oak Ridge National Laboratory Report 3113, 1961.
72. RUDD, M. E., JORGENSON, T. and VOLZ, D. J. *Phys. Rev. Lett.* 16 (1966) 929; EDWARDS, A. K. Thesis, University of Nebraska, 1967; RUDD, M. E., SAUTTER, C. A. and BAILEY, C. L. *Phys. Rev.* 151 (1966) 20; RUDD, M. E. *Phys. Rev. Lett.* 15 (1965) 580; 13 (1964) 503; RUDD, M. E. and JORGENSEN, T. *Phys. Rev.* 131 (1963) 666; KUYATT, C. and JORGENSEN, T. *Phys. Rev.* 130 (1963) 1444; RUDD, M. E. and LANG, D. V. *Abstr. 4th Int. Conf. Phys. electron. atom. Collisions*, p. 153, 1965. Hastings-on-Hudson, New York; Science Bookcrafters; EDWARDS, A. K. and RUDD, M. E. *Phys. Rev.* 170 (1968) 140.
73. BLAUTH, E. *Z. Phys.* 147 (1957) 228.
74. KUYATT, C. Ph.D. Thesis, University of Nebraska, 1961.
75. FERMI, E. *Z. Phys.* 29 (1924) 315.
76. RUDD, M. E. and JORGENSEN, T. *Phys. Rev. Lett.* 16 (1966) 929.
77. BERRY, H. W. *Phys. Rev.* 121 (1961) 1714; 127 (1962) 1634.
78. RUSSEK, A. and THOMAS, M. T. *Phys. Rev.* 109 (1955) 2015
79. FULS, E. N., JONES, P. R., ZIEMBA, F. P. and EVERHART, E. *Phys. Rev.* 107 (1957) 704; JONES, P. R., ZIEMBA, F. P., MOSES, H. A. and EVERHART, E. *Phys. Rev.* 113 (1959) 182
80. EVERHART, E., STONE, G. and CARBONE, R. J. *Phys. Rev.* 98 (1955) 1045, 1287; CARBONE, R. J., FULS, E. N. and EVERHART, E. *Phys. Rev.* 102 (1956) 1524; MORGAN, G. H. and EVERHART, E. *Phys. Rev.* 128 (1962) 667
81. HASTED, J. B. *Proc. phys. Soc., Lond.* 77 (1961) 269

82. AFROSIMOV, V. V., GORDEEV, YU. S., PANOV, M. N. and FEDORENKO, N. V. *J. tech. Phys., Moscow* 34 (1964) 1614, 1624, 1637
83. PIVOVAR, L. I., NOVIKOV, M. T. and TUBAYEV, V. M. *J. exp. theor. Phys.* 46 (1964) 471
84. FIRSOV, O. B. *Soviet Phys. JETP* 36 (1959) 1076
85. FEDORENKO, N. V. *Usp. Fiz. Nauk*, 68 (1959) 481
86. KIKIANI, B. I., OGURTSOV, G. N., FEDORENKO, N. V. and FLAKS, I. P. *Soviet Phys. JETP* 22 (1966) 264
87. FEDORENKO, N. V. and AFROSIMOV, V. V. *J. tech. Phys., Moscow* 26 (1957) 1827; *J. exp. theor. Phys.* 34 (1958) 1398
88. KAMINKER, D. M. and FEDORENKO, N. V. *J. tech. Phys., Moscow* 25 (1956) 1843, 2239.
89. JORDAN, E. B. and BRODE, E. B. *Phys. Rev.* 43 (1933) 112.
90. KAMINKER, D. M. and FEDORENKO, N. V. *J. tech. Phys., Moscow* 25 (1955) 2239.
91. SKALSKAYA, I. P. *J. exp. theor. Phys.* 24 (1954) 1912.
92. EVERHART, E. and KESSEL, Q. C. *Phys. Rev. Lett.* 14 (1965) 247; *Phys. Rev.* 146 (1966) 27; KESSEL, Q. C., RUSSEK, A. and EVERHART, E. *Phys. Rev. Lett.* 14 (1965) 484; KESSEL, Q. C. and EVERHART, E. *Phys. Rev.* 146 (1966) 16.
93. WEG, W. F., SNOEK, C., BIERMAN, D. J. and KISTEMAKER, J. *Proc. 5th Int. Conf. Phys. electron. atom. Collisions*, p. 95, 1967. Leningrad; Akademii Nauk.
94. FANO, U. and LICHTEN, W. *Phys. Rev. Lett.* 14 (1965) 627.
95. LICHTEN, W. *Phys. Rev.* 164 (1967) 131; 131, (1963) 229.
96. BATES, D. R., LEDSHAM, K. and STEWART, A. L. *Phil. Trans. Roy. Soc.* A246 (1953) 215.
97. JONES, P. R., COSTIGAN, P. and VAN DYK, G. *Phys. Rev.* 129 (1963) 63; JONES, P. R., BATRA, T. L. and RANGA, H. A. *Proc. 5th Int. Conf. Phys. electron. atom. Collisions*, p. 470, 1967. Leningrad; Akademii Nauk.
98. VAN DEN BOS, J., WINTER, G. J. and DE HEER, F. J. *Physica, Eindhoven* (1967).
99. AMUSIA, M. YA. *Phys. Lett.* 14 (1965) 36.
100. SNOEK, C., WEG, S. F., GEBALLE, R. and ROL, P. K. *Proc. 7th Int. Conf. Ioniz. Phenom. Gases*, vol. 1, p. 9, Belgrade, 1966.
101. HALE, G. C. and HASTED, J. B. *Nature, Lond.* 217 (1968) 945; BONESS, M. J. W., HALE, G. C., HASTED, J. B. and LARKIN, I. W. *Proc. 5th Int. Conf. Phys. electron. atom. Collisions*, p. 577, 1967. Leningrad; Akademii Nauk.
102. AFROSIMOV, V. V., GORDEEV, YU. S., LAVROV, V. M. and SHCHEMELININ, S. G. *Proc. 5th. Int. Conf. Phys. electron. atom. Collisions*, p. 127, 1967. Leningrad; Akademii Nauk.
103. BRANDT, W. *J. quantve Spectros. radiat. Transf.* 7 (1967) 411.
104. FAN, C. B. and MEINEL, A. B. *Astrophys. J.* 113 (1951) 50; 115 (1952) 330; 116 (1953) 205
105. SLUYTERS, T. J. M. and DE HAAS, E. *Rev. sci. Instrum.* 29 (1958) 597; SLUYTERS, T. J. M. and KISTEMAKER, J. *Physica, Eindhoven*, 25 (1959) 182; 28 (1962) 1184; *Proc. 5th Int. Conf. Ioniz. Phenom. Gases*, p. 54, 1961. Amsterdam; North Holland Publishing Company.
106. ROBINSON, J. M. and GILBODY, H. B. *Proc. phys. Soc., Lond.* 92 (1967) 589; THOMAS, E. W. and BENT, G. *Phys. Rev.* (1967).
107. CARLETON, N. P. and LAWRENCE, T. R. *Phys. Rev.* 109 (1958) 1159; 107 (1957) 110; SHERIDAN, W. F., OLDENBERG, O. and CARLETON, N. P. *2nd Int. Conf. Phys. electron. atom. Collisions*, University of Colorado, 1961.
108. HUGHES, R. H., WARING, R. C. and FAN, C. Y. *Phys. Rev.* 122 (1961) 525.
109. DUNN, G., GEBALLE, R. and PRETZER, D. *2nd Int. Conf. Phys. electron. atom. Collisions*, University of Colorado, 1961.
110. DIETERICH, E. J. *Phys. Rev.* 103 (1956) 632.
111. KRALL, N. A. and GERJUOY, E. *Phys. Rev.* 120 (1960) 143; 119 (1960) 705; *2nd Int. Conf. Phys. electron. atom. Collisions*, University of Colorado, 1961.
112. VAN DEN BOS, J. and DE HEER, F. J. *Physica, Eindhoven*, 34 (1967) 333; VAN DEN BOS, J. Thesis, University of Amsterdam, 1967.

113. HASTED, J. B. *Symp. Atomic Collision Processes in Plasmas*, p. 155, Culham Laboratory, Harwell, 1964.

114. OCHKUR, V. I. and PETRUNKIN, A. M. *Optika Spektrosk.* 14 (1963) 245.

115. VAN ECK, J., DE HEER, F. J. and KISTEMAKER, J. *Physica, Eindhoven*, 30 (1964) 1171.

116. YOUNG, R. A., STEBBINGS, R. F. and McGOWAN, J. W. Gulf General Atomic Report G. A. 7609, 1967.

117. VAN ECK, J., DE HEER, F. J. and KISTEMAKER, J. *Physica, Eindhoven*, 30 (1964) 1171.

118. ROBINSON, J. M. and GILBODY, H. B. *Proc. phys. Soc., Lond.* 92 (1967) 589.

119. DODD, J. G. and HUGHES, R. H. *Phys. Rev.* 135 (1964) A618.

120. DE HEER, F. J. and VAN DEN BOS, J. *Physica, Eindhoven*, 31 (1965) 365.

121. DE HEER, F. J., MULLER, L. W. and GEBALLE, R. *Physica, Eindhoven* 31 (1965) 1745.

122. LORENTZ, D. C., ABERTH, W. and HESTERMAN, V. W. *Phys. Rev. Lett.* 17 (1966) 849.

123. POLYAKOVA, G. N., FOGEL, YA. M. and ZATS, A. V. *J. exp. theor. Phys.* 52 (1967) 1495.

124. SHERIDAN, J. R. and CLARK, K. C. *Abstr. 4th Int. Conf. Phys. electron. atom. Collisions*, p. 276, 1965. Hastings-on-Hudson, New York; Science Bookcrafters.

125. SHARP, T. E. and RAPP, D. *Abstr. 4th Int. Conf. Phys. electron. atom. Collisions*, p. 188, 1965. Hastings-on-Hudson, New York; Science Bookcrafters; RAPP, D. *J. chem. Phys.* 32 (1960) 735; 40 (1964) 2813; KELLY, J. D. and WOLFSBERG, M. *Abstr. 4th Int. Conf. Phys. electron. atom. Collisions*, p. 186, 1965. Hastings-on-Hudson, New York; Science Bookcrafters.

126. BERNSTEIN, R. B. and KRAMER, K. H. University of Wisconsin, Theoretical Chemistry Institute Report WIS-TCI-91, 1965.

127. TOENNIES, J. P. *Z. Phys.* 182 (1965) 257; 193 (1966) 76.

128. PAULY, H. and TOENNIES, J. P. *Advanc. atomic molec. Phys.* 1 (1965) 195.

Chapter 12

CHARGE TRANSFER PROCESSES

12.1 INTRODUCTION; THEORY OF SYMMETRICAL RESONANCE PROCESSES

In the single charge transfer process[1], often termed 'charge exchange' or 'electron capture', one electron, and very little kinetic energy, is transferred from the target atom to the projectile ion;

$$X^+ + Y \rightarrow X + Y^+ + \Delta E \qquad \qquad 10/01$$

In general, the collision is a glancing one, capable of taking place at comparatively large impact parameters, except at very high energies. The product ion is scattered nearly perpendicular to the impact momentum vector, whilst the projectile scattering angle is small. The energy defect ΔE of the process is equal to the difference between the two atomic ionization potentials. The process

$$X^+ + X \rightarrow X + X^+$$

possessing zero energy defect, is known as 'symmetrical resonance charge transfer'; its cross-section may be as large as, or larger than, gas kinetic, and falls with increasing impact velocity v as follows:

$$\sigma^{1/2} = a - b \ln v \qquad \qquad (12.1)$$

with a and b constants[2]. Calculations of the cross-sections in terms of transitions between the symmetrical and anti-symmetrical configurations of the pseudo-molecular ion were first given by Massey and Smith[3], who used the method of perturbed stationary states. More recent discussions are given by Bates et al.[4, 5], and experiments have been reviewed[6].

In the velocity region[7] below 10^8 cm sec^{-1}, but above $10^6 \mu^{-1/2}$ cm sec^{-1}, (reduced mass μ in a.m.u), it is possible to apply the Jeffrey's WKB approximation[8] and to show that the result is equivalent to the semi-classical impact parameter approximation[9], which proceeds along the following lines. The projectile is taken as travelling along the x-axis, with impact parameter b, and constant impact velocity v; scattering is neglected. The nuclear separation r is then

$$r^2 = x^2 + b^2 \qquad \qquad (12.2)$$

The effects of all states except the specified initial and final ones p and q are neglected. The amplitude coefficients $c(x, b)$, whose initial boundary conditions are

$$|c_p(-\infty, b)|^2 = 1 \tag{12.3}$$

and

$$c_p(-\infty, b) = 0 \tag{12.4}$$

are given by

$$i\frac{\partial c_p}{\partial x} = \frac{c_q}{v} U_{pq} \exp\left[i\alpha x - \frac{1}{v}\int_0^x (U_{pp} - U_{qq})\,dx\right] \tag{12.5}$$

$$i\frac{\partial c_q}{\partial x} = \frac{c_p}{v} U_{qp} \exp\left[-i\alpha x - \frac{1}{v}\int_0^x (U_{pp} - U_{qq})\,dx\right] \tag{12.6}$$

with $\alpha = \Delta E/v$; the symbol U is used to indicate the matrix elements[10] of the interaction potential V:

$$U_{pq} = \frac{2m_e}{\hbar^2} \iint V(r, r_a, r_b)\psi_p\psi_q^*\,dr_a\,dr_b \tag{12.7}$$

where r is the nuclear separation, and r_a and r_b are the electron coordinates of atoms a and b.

The probability that the systems are finally in state q is

$$P(b) = |c_q(\infty, b)|^2 \tag{12.8}$$

and the cross-section

$$\sigma = 2\pi \int_0^\infty bP(b)\,db \tag{12.9}$$

It is possible to neglect the difference between the diagonal matrix elements, U_{pp} and U_{qq}. The Rosen–Zener formula[11] for the solution of equations 12.5–12.8 is

$$P = \frac{\Lambda^2}{\Omega^2} \sin^2 \frac{\Omega}{v} \tag{12.10}$$

with

$$\int_{-\infty}^\infty U_{pq}\,dx = \int_{-\infty}^\infty U_{qp}\,dx = \Omega(b) \tag{12.11}$$

and

$$\int_{-\infty}^\infty U_{pq} \exp(ci\alpha x)\,dx = \Lambda(b, v) \tag{12.12}$$

The solution is valid for charge transfer, and for excitation transfer 0′0/00′ when there is an exact energy balance, or when the interaction is weak enough for the Born approximation to be valid.

A simple approximation was introduced by Firsov[12]. From equation 12.10 it follows that, as Ω/v varies, P oscillates between zero and Λ^2/Ω^2; thus Ω/v can be replaced by $\frac{1}{2}\Lambda^2/\Omega^2$, so that

$$\sigma = 2\pi \int_0^{b*} \frac{\Lambda^2}{2\Omega^2} b\,db \tag{12.13}$$

with b^* the greatest root of the equation

$$P(b^*) = \frac{1}{\pi} \qquad (12.14)$$

If the ion and atom are taken as point centres of positive charge, the atom possessing the active electron, semi-empirical[7] orbitals may be used to express the stationary state energies and hence P in terms of the ionization potential E_i:

$$P(b, v) \propto \sin^2 \left\{ \int\limits_{-\infty}^{\infty} \frac{E_i r}{\hbar v a_0} \exp\left[-\left(\frac{E_i}{R}\right)^{1/2} \frac{r}{a_0} \right] \mathrm{d}x \right\} \qquad (12.15)$$

where R is the Rydberg. Equations 12.13 and 12.15 imply a probability of charge transfer varying periodically with impact parameter as is shown in *Figure 12.1*. At small impact parameters, the probability oscillates rapidly between zero and unity. Failure of $P(b)$ extrema to reach these limits is attributable to coupling to other channels, either open or closed, the latter having

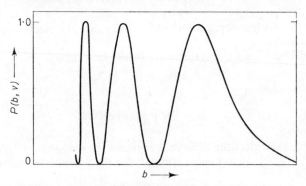

Fig. 12.1. *Periodically varying probability of symmetrical resonance charge transfer* $P(b, v)$ *as a function of impact parameter* b

an imaginary wave number and an energy above that permitted in conservation. The Firsov approximation is equivalent to replacing this varying function $P(b)$ by $\frac{1}{2}$ between zero impact parameter and an impact parameter b^*, so that

$$\sigma = \tfrac{1}{2}\pi b^{*2} + 2\pi \int\limits_{b^*}^{\infty} bP(b, v)\, \mathrm{d}b \qquad (12.16)$$

The calculation of b^* from equations 12.14–12.16 may be made by means of an equation given by Rapp and Francis[7, 11]:

$$b^{*3/2}\left(\frac{2\pi}{\gamma a_0}\right)^{1/2} \frac{E_i}{\hbar v}\left(1 + \frac{a_0}{\gamma b^*}\right)\exp\left(\frac{-\gamma b^*}{a_0}\right) = \frac{\pi}{6} \qquad (12.17)$$

with

$$\gamma = \left(\frac{E_i}{R}\right)^{1/2} \qquad (12.18)$$

Equation 12.17 is most easily solved by substituting an average value for the pre-exponential b^* term. This leads to a cross-section function of the form

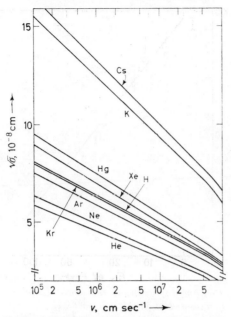

Fig. 12.2. Variation of symmetrical resonance charge transfer $\sqrt{\sigma}$ with v, calculated by Rapp and Francis[7]

of equation 12.1 over more than a decade of impact velocities, but curvature sets in when the impact velocity is larger, and also when it is so large that Jeffrey's WKB approximation becomes unusable.

Symmetrical resonance charge transfer cross-sections derived with the aid of these approximations are strongly dependent on the ionization potential; this will be apparent from *Figures 12.2* and *12.3*, where some calculations due to Rapp and Francis[7], and some experimental data at an arbitrary impact velocity, are represented graphically.

A similar equation for b^* was given by Firsov:

$$\alpha b^* - \left(2\beta - \frac{1}{2}\right) \ln \alpha b^* - \frac{\beta - \frac{1}{8}}{\alpha b^*} = \ln A + \ln \left[\frac{(2\pi^3)^{1/2}\alpha\hbar}{(2\gamma)! \, m_e v}\right] \quad (12.19)$$

with

$$\beta = \left(\frac{m_e e^4}{2\hbar^2 E_i}\right)^{1/2} \quad (12.20)$$

if the electron is in a Coulomb field, but $\beta = 0$ if it is not (negative ion charge transfer $\bar{1}0/0\bar{1}$). The parameter α is given by

$$\alpha = \left(\frac{2m_e E_i}{\hbar^2}\right)^{1/2} \quad (12.21)$$

and $\ln A$ is an optional normalization constant.

Solutions of these equations are among those shown in the compilation of data displayed in *Figures 12.4* and *12.5*. Rearrangement of the Rapp–Francis and Firsov expressions enables them to be compared[6]; this comparsion is illustrated in *Figure 12.6*, which is in terms of a collision length ϱ_1 given by

$$\sigma = \frac{1}{2}\pi\varrho_1^2 \quad (12.22)$$

40

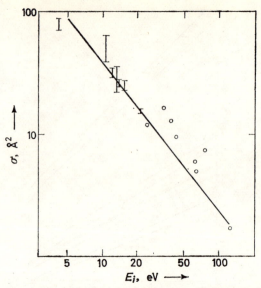

Fig. 12.3. *Variation of measured symmetrical resonance charge transfer processes with ionization potential, all measured at identical impact velocity* $v = 10^{-7}$cm sec^{-1}: *two-electron and three-electron processes are included, and a straight line is drawn arbitrarily through the data*

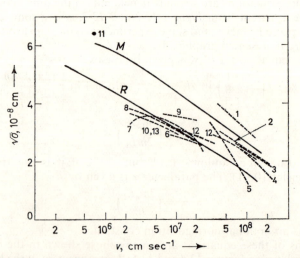

Fig. 12.4. He$^+$–He *symmetrical resonance charge transfer function. Broken lines indicate experimental data, and full lines indicate calculations:* 1^{13}, 2^{14}, 3^{15}, 4^{16}, 5^{17}, 6^{18}, 7^{19}, 8^{20}, 9^{21}, 10^{14}, 11^{22}, 12^{14}, 13^{15}, M^{23}, R^7

(From **Rapp and Francis**[7], by courtesy of Lockheed Missiles and Space Company)

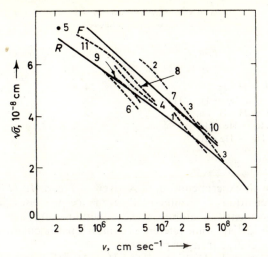

Fig. 12.5. Ar⁺–Ar symmetrical resonance charge transfer function. Broken lines indicate experimental data, and full lines indicate calculations: 1^{13}, 2^{14}, 3^{15}, 4^{14}, 5^{22}, 6^{24}, 7^{25}, 8^{26}, 9^{27}, 10^{28}, 11^{29}, R^7, F^{12}

(From Rapp and Francis[7], by courtesy of Lockheed Missiles and Space Company)

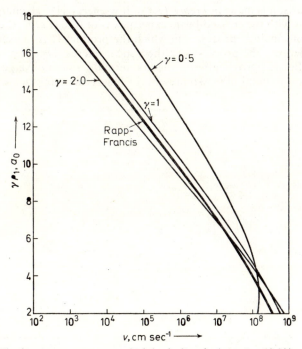

Fig. 12.6. The function $\gamma\varrho_1$ versus v, which leads directly (equation 12.22) to symmetrical resonance charge transfer cross-sections in both Firsov (thin lines) and Rapp–Francis (thick line) formulations; in the Firsov formulation the function is dependent on γ, while in the Rapp–Francis formulation it is not

Table 12.1. CALCULATIONS OF SYMMETRICAL RESONANCE CHARGE
TRANSFER PROCESSES

Process	References
$H^+ + H \rightarrow H + H^+$	34–37
$H^+ + H (2s, 2p) \rightarrow H^+ + H (2s, 2p)$	38
$H^- + H \rightarrow H + H^-$	39
$He^+ + He \rightarrow He + He^+$	2, 3, 7, 12, 40
$He^{2+} + He \rightarrow He + He^{2+}$	41

Comparison with experimental data is given in *Figure 12.11* (see page 621).

Further calculations of symmetrical resonance processes, using nodeless spherically symmetrical atomic wave functions, are available[2, 12, 30–33]. Listed in Table 12.1 are some references to individual calculations of symmetrical resonance charge transfer processes.

Calculations made by Dickinson[42] for He^+–He at energies in the very low impact velocity region show that oscillations in the cross-section function are to be expected (*Figure 12.7*). These diminish with increasing energy, and are difficult to observe experimentally. Such oscillations are discussed further in Section 12.7.

A comparison is made in *Figure 12.8* of low-energy H^+–H charge transfer calculations with the crossed-beam experiments of Fite *et al.*[43]; the measured cross-sections include processes in which the product H is in excited states. At high energies, the process is of theoretical interest because the series of which the Born approximation is the first term does not converge; however, the Born approximation may represent the first term of another series. Com-

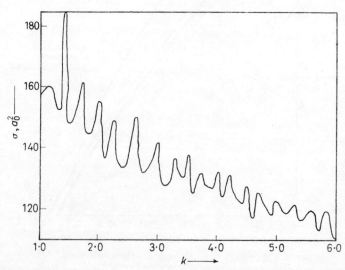

Fig. 12.7. Oscillating total He^+–He charge transfer function calculated by Dickinson[42] (momentum variable $k = \mu v/\hbar$)

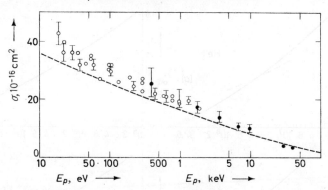

Fig. 12.8. H⁺–H *symmetrical resonance charge transfer functions: low-energy experimental data[43], compared with the calculations of Dalgarno and Yadav[35] (broken line)*

parison of theory and experiment is complicated not only by the effects of excited state products but also by relativistic effects. A fuller discussion is given in Section 12.8.

Classical calculations of symmetrical resonance charge transfer cross-sections have been reported[44]; a purely classical treatment yields a cross-section independent of energy, but if a quantum tunnel effect contribution is included, the form of equation 12.1 is obtained. Comparison of the calculation for Cs⁺–Cs with experiment (*Figure 12.9*) yields better order of magnitude correspondence than equations 12.16 and 12.17. The observed oscillations in this function (see *Figure 12.25*, page 642) are not reproduced.

The symmetrical resonance two-electron capture process has been treated quantum mechanically[45] by a variational method of calculating the differences between the symmetric and antisymmetric interactions. Owing to an inadequate trial function, the results were in poor agreement with rare-gas ion

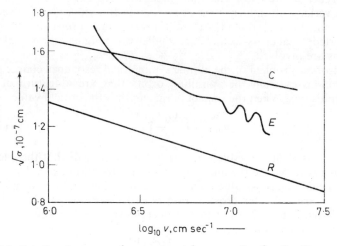

Fig. 12.9. Cs⁺–Cs *charge transfer experimental cross-section function E, compared with classical (C)[44] and quantum (R) calculations*

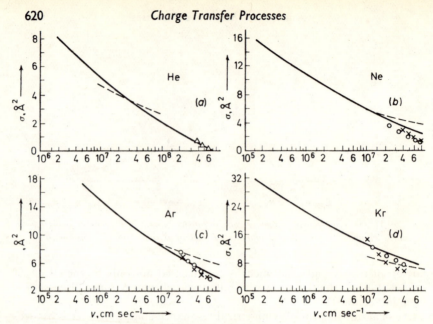

Fig. 12.10. Symmetrical resonance two-electron capture cross-sections: (a) helium; (b) neon; (c) argon; (d) krypton. Measurements of Allison[48], Islam et al.[46], Flaks and Solov'ev[25]; broken lines, calculations of Fetisov and Firsov[45]; full lines, calculations of Ianev[47]

experiments[46], as was shown graphically in the first edition of this book. More recent calculations[47] are much more successful, and the comparison with experiment is made in *Figure 12.10*.

Exact symmetrical resonance charge transfer between diatomic molecular ions and their molecules would only be possible if the equilibrium nuclear separations were identical in molecule and ion, and if the vibrational and rotational energies were to correspond. Since this is not the case, the Franck–Condon overlap will influence the cross-section function. Qualitatively the cross-section functions resemble those for the atomic process[49]. But it has been reported[50] that cross-sections are smaller when higher vibrational states of the ions are involved.

An important problem of the atomic symmetrical resonance charge transfer function arises from the spiralling orbits that are a feature of low-velocity collisions. When the impact parameter has the critical value

$$b_{orb} = \left(\frac{4\alpha e^2}{\mu v^2} \right)^{1/4} \tag{12.23}$$

the system passes into a stable orbit. It has been argued[7] that the effect of this orbiting is to reduce the minimum nuclear separation equivalent to impact parameter b_{orb} to a value equivalent to b^*. Therefore the symmetrical resonance charge transfer cross-section is better approximated by $\frac{1}{2}\pi b_{orb}^2$ than $\frac{1}{2}\pi b^{*2}$ (equation 12.16) at impact energies sufficiently small for b_{orb} to exceed b^*; in this region

$$\sigma_c = \frac{\pi}{2v} \left(\frac{4\alpha e^2}{\mu} \right)^{1/2} \tag{12.24}$$

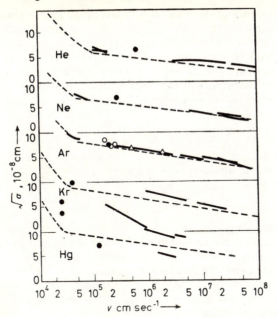

Fig. 12.11. Symmetrical resonance charge transfer and mobility data displayed as $\sqrt{\sigma}$ against v; broken lines represent calculations from the Rapp–Francis formulation (equation 12.16) and from equation 12.24; experimental data are represented as full lines and points

Refinements have been proposed by Wolf and Turner[51]. A graphical representation of $\sqrt{\sigma}$ against v calculated from equations 12.16 and 12.24 is given in *Figure 12.11*, together with experimental data, some of which are deduced from ionic mobilities[6]. The experimental data show very little orbiting effect. It is only at the lowest energies ($T \lesssim 300°K$) that orbiting can become important, except for systems with large polarizability. The Cs^+–Cs measurements of Bullis[52] are of special interest as the lowest energy heavy-particle beam experiments so far reported; other very low energy beam experiments have been described[53, 54].

12.2 CHARGE TRANSFER BETWEEN UNLIKE IONS AND ATOMS. THE ADIABATIC CRITERION

The atomic charge transfer process between unlike A^+ ions and B atoms differs from the symmetric resonance process in that it involves an electronic transition and also an internal energy defect

$$\Delta E = E_i(A) - E_i(B)$$

The cross-section functions are small at the lowest energies, rise to a maximum, and fall off in the manner of the symmetrical resonance process. The adiabatic criterion proposed by Massey[55] is important in these collisions. There is a region of low impact energy where the relative motion of the atoms is so slow that the electronic motion of transition can adjust itself to small changes of internuclear distance; this adiabatic adjustment makes the transition an unlikely event. However, if the impact energy falls outside this 'adiabatic region' and the time of transition is comparable to the time of collision, the likelihood of transition is no longer small. The time of collision is taken as a/v, where v is the impact velocity and a is known as the 'adiabatic parameter', a distance of the order of atomic dimensions over which the charge transfer transition is significant. The time of transition is $h/|\Delta E|$. Thus the condition

$$v \ll \frac{a\,|\Delta E|}{h}$$

characterizes the adiabatic region of energy, and

$$v = \frac{a\,|\Delta E|}{h}$$

marks the termination.

The criterion was applied by Hasted[56] to the charge transfer cross-section functions in a manner which minimizes the danger of false deduction from inaccurate experiment; the criterion was examined at the laboratory impact energy E_m at which the cross-section is maximum, and the assumption was made that the criterion holds as an equality at this energy. The adiabatic parameter is then given by

$$a = \frac{0 \cdot 57 E_m^{1/2}}{(m_1/e)^{1/2}\,|\Delta E|} \tag{12.25}$$

with a in Ångstrom units, m_1/e in a.m.u, and ΔE and E_m in electron volts; m_1 is the projectile ion mass.

The analysis of a large volume of experimental data leads to a surprisingly small probable error (10–20 per cent), with a value $am = 7$ Å common to different types of reaction, involving the transfer of m electrons. The 'adiabatic maximum rule' (equation 12.25) is thus a convenient method of estimating the energy of a cross-section function maximum, apparently valid over a large range of impact energy. It is important to remember that, in all but the most refined experiments, the projectile ion population in different energy levels of long lifetime must be taken into account. In the energy region below 1 keV, it is usually possible by application of the rule to be certain of the dominance of a particular energy level; over a wide energy range there is sometimes more than one cross-section function maximum, corresponding to more than one process. Interpretations of this kind are greatly strengthened by observations[57] such as that of *Figure 12.12* made with ion beams produced either by electron impact or by surface ionization; the latter procedure eliminates excited states and with them the appropriate maxima. However, the collision products can be formed in more than one energy level, and the complications arising from this are considered in Section 12.5. It will be seen from the two-state approximation discussed below that the cross-sections are governed by the energy defects, the ionization potentials and by the statistical weights. The adiabatic maximum rule is entirely consistent with the results of this two-state approximation.

The adiabatic maximum rule has been found applicable[58] to ionization and excitation collisions of protons (and similar light projectiles) with atoms. The

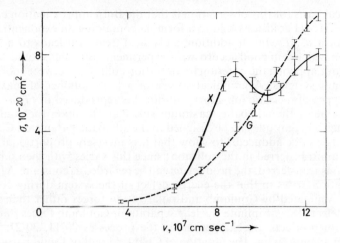

Fig. 12.12. *Two-electron capture cross-section functions*[54] *for* Li^+–Ar: *broken line G, ground state ions produced by surface ionization; full line X, ions produced by electron impact*

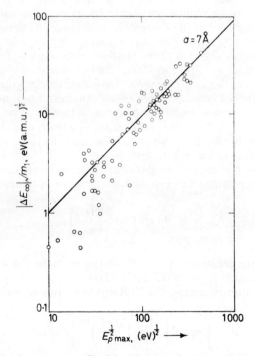

Fig. 12.13. *Graphical representation*[58] *of the adiabatic maximum rule for a large number of different one-electron inelastic ion–atom collisions: the straight line represents equation 12.25 with a = 7 Å; no correction has been made for potential interaction* $(|\Delta E|$ *is taken as* $|\Delta E_{\infty}|)$

theoretical basis for this observation is that the Born approximation calculations for H^+–H collisions lead to a formula containing an exponential term of the form $a|\Delta E|/\hbar v$; in addition, a classical treatment leads to a similar condition which is in good accord with experiment. But one should be cautious about applying the criterion to ionization collisions between more complex atomic systems. A graphical survey of a large number of maxima in several types of inelastic ion–atom collision[58] is reproduced in *Figure 12.13*.

The value of the adiabatic maximum rule depends upon the confidence with which a magnitude can be assigned to the adiabatic parameter. Classical arguments can be adduced[59] to show that it is inversely proportional to the momentum transferred in the collision; since this varies with the number m of electrons transferred, the product am can be regarded as constant. A further complication arises in that the energy defect of the system during collision may be dominated by Coulomb and polarization forces rather than by the energy defect ΔE_∞ at infinite nuclear separation. Coulomb forces dominate when both products are charged, as in the processes 20/11, 30/21, 30/12, $10/\bar{1}2$, $00/\bar{1}1$, and so on. The notation of Chapter 1 will be found particularly helpful in discussing charge transfer processes, which are of the general type $n0/(n-m)m$. Polarization forces dominate where the products include one neutral atom, as is the case in the simple charge transfer 10/01. A discussion of procedure suitable for arriving at a mean energy defect averaged over the collision path has been given[60]. The use of averaged energy defects in equation 12.25 leads to reduced probable errors in the calculations of the adiabatic parameters, or of the energies E_m of maximum cross-section, using the mean value $a = (7/m)$ Å. But it is arguable that the use of ΔE_∞ rather than averaged energy defect is theoretically more sound.

At impact energies which are moderate, but greater than that of the maximum cross-section, the non-resonance ion–atom charge transfer function follows the form of the symmetrical resonance process. For ion–atom pairs of widely different ionization energies, the cross-section function is not readily found by interpolation between the functions of *Figure 12.2*. In the extreme case of alkali ions in rare gases[61], the cross-sections increase with increasing alkali ion Z, despite increasing endothermicity.

The low energy rise of non-resonance cross-section functions is complex. Fairly close to the maximum an exponential rise is a realistic approximation[62]:

$$\sigma = A \exp\left(-\frac{B|\Delta E|}{hv}\right) \tag{12.26}$$

where A and B are constants. This is similar in form to the Landau–Teller formula for excitation transfer (Chapter 13).

The two-state approximation due to Rapp and Francis[7] yields

$$\sigma = \int \operatorname{sech}^2\left[\frac{|\Delta E|}{hv}\left(\frac{a'\pi b}{2\gamma}\right)^{1/2}\right]2\pi b \, db \tag{12.27}$$

where

$$a' = \frac{\hbar^2}{m_e e^2} \tag{12.28}$$

and

$$\gamma = \left(\frac{E_i}{R}\right)^{1/2} \tag{12.29}$$

This is compared with experiment in *Figures 12.14* and *12.15*. The results are consistent with the adiabatic maximum rule. Algebraic simplification leads to expressions including tabulated integrals from which the cross-section function can quickly be determined[63].

Although equation 12.27 is not unsuccessful, the assumptions on which it is based have been shown[64] to be mathematically incorrect: similar two-state calculations which avoid the incorrect assumptions have been given[65].

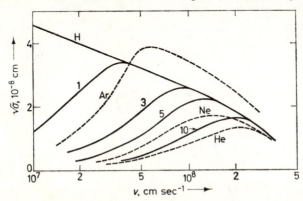

Fig. 12.14. Comparison of calculations using equation 12.27 with experimental data for proton charge transfer in rare gases: full lines represent calculations of Rapp and Francis for H^+–H, *and for* $E = 1$ eV, 3 eV, 5 eV *and* 10 eV *as marked; broken lines represent data of Hasted and Stedeford[14] for* H^+–Ar ($|\Delta E| = 2{\cdot}160$ eV), H^+–Ne ($|\Delta E| = 7{\cdot}964$ eV), *and* H^+–He ($E = 10{\cdot}986$ eV)

(From Rapp and Francis[7], by courtesy of Lockheed Missiles and Space Company)

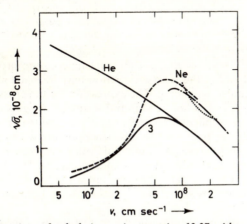

Fig. 12.15. Comparison of calculations using equation 12.27 with experimental data for He^+–Ne *charge transfer* ($|\Delta E| = 3{\cdot}022$ eV): *full lines show* He^+–He *resonant process, and process with* $|\Delta E| = 3$ eV; *broken lines represent data of Hasted and Stedeford[14]*

(From Rapp and Francis[7], by courtesy of Lockheed Missiles and Space Company)

Equation 12.27 reduces, at very low impact velocities, to

$$\sigma \simeq \frac{10 \cdot 8}{\pi} \left(\frac{\gamma}{a'}\right)^2 \left(\frac{v\hbar}{\Delta E}\right)^4 \qquad (12.30)$$

But the available data do not conform to this v^4 dependence, being often orders of magnitude higher. An extreme example is the Ne^+–Ar cross-section[66] shown in *Figure 12.16*. These unexpectedly large values arise from coupling to levels other than the ground states. It is now possible to make calculations involving three and four states, as is discussed in Section 12.6.

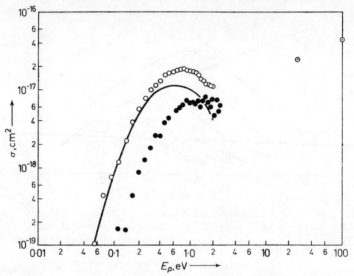

Fig. 12.16. Total cross-section (Bohme et al.) for Ne^+–Ar charge transfer: open circles indicate experiments carried out with ion beams containing excited states; closed circles indicate experiments carried out with ground state ions; full line represents the difference; circles with centres represent earlier data of Hasted and Stedeford[14]

A comparison of cross-sections for an inelastic collision process proceeding in the forward and in the reverse direction may be made with the aid of the principle of detailed balancing. The expression for the cross-section ratio contains a ratio f_2/f_1 of the final and initial statistical weights of the system. Not only is this ratio applicable to forward and reverse reactions, but it also provides a comparison between the cross-sections of two processes which possess different statistical weights but would otherwise have identical cross-sections (for example, accidental resonance processes H^+–H and O^+–H, whose energy defects are virtually the same but whose statistical weight ratio is 3/2). These collision cross-sections functions have been measured[66] and are displayed in *Figure 12.17*.

In two colliding atomic systems, the number of states of the pseudo-molecule (deduced from the Wigner–Witner rules) exceeds that of the two isolated atomic systems. The final states of the atomic systems could be formed from several pseudo-molecule levels. There will always be a correspondence

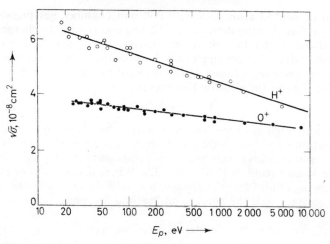

Fig. 12.17. H$^+$–H *and* O$^+$–H *charge transfer cross-section functions*[67]

of identity between at least one of the A^+B and AB^+ pseudo-molecule states, and the transition will occur via one of the corresponding pair of states. Thus the ratio of the probability of forward to reverse reactions is f_2/f_1, the final and initial numbers of pseudo-molecule states. In this argument, the time variation of orientation of the wave functions during the collision is neglected.

At present, the number of processes for which data can be adduced to test these ideas is rather small. In addition to the H$^+$–H and O$^+$–H comparison, there are two rare gas pairs[68] with a 1 : 1 ratio and three with a 3:1 ratio. In all cases the comparison is satisfactory at the same relative velocity for forward and reverse processes.

12.3 EXPERIMENTAL METHODS OF CHARGE TRANSFER MEASUREMENT

The study of charge transfer cross-sections in beams may be carried out either by detection of the slow ions produced in the collision or by the measurement of the almost undeflected neutralized projectiles. It is also possible to measure the radiation that is produced in those processes in which the products are formed in excited levels.

At thermal energies, the symmetrical resonance charge transfer process may be distinguished experimentally by the contribution which it makes to ionic mobility. The non-resonance processes are likely to have comparatively small rate constants unless very close to resonance, but may be studied by the monitoring of ion densities in time-dependent or flowing afterglows, or by the drift-tube technique.

Measurements of the neutralized projectile flux formed when an ion beam passes through a gas has been successfully achieved at energies in the kilovolt region[16]. The fraction of the ion beam which retains its charge after passing through the collision chamber is deflected away from the neutral atom detector by an electric or magnetic field; the cross-section is calculated from the variation of neutral atom flux with collision chamber gas pressure. It is necessary either that the detector should respond equally to charged and to neutral atoms, or that the relative sensitivities should be known. The experiments are conducted under single collision conditions, but when the collision chamber gas pressure is raised a proportion of the neutrals are re-converted to positive ions by the electron loss process

$$X+Y \rightarrow X^+ +Y+e$$

At sufficiently high gas pressure, the emergent beam contains an equilibrated proportion of positive ions and neutrals, and from this proportion the ratio of capture (charge transfer) to loss cross-sections may be determined, using the Wien equations (1.31 and 1.32). Where dissociation of molecular ions, stripping or formation of negative ions is likely, more complicated analysis is necessary, and the equilibrium method loses much of its value (see Chapter 11).

Dissociative charge transfer processes of the type

$$XY^+ +Z \rightarrow X+Y+Z^+$$

can only be studied if the neutral detector is able to distinguish between $X+Y$ and XY. At sufficiently high energies, single particle counting can be used[69] for this purpose.

The majority of charge transfer measurements have been made by collecting the slow ions produced in collisions, as was first done by Goldmann[70]. This may be effected either by making use of the large scattering angle of the slow products[19] or by controlling the slow products in electric and magnetic fields[18]. The second method has greater possibilities, since it makes it easier to define with precision the path length from which the collision products are collected. Uniform electric fields perpendicular to the ion beam are increased until no variation of collected ion current is observed with varying field; the length of the collecting plate is then taken as the collection path length. Since ionization processes compete, it is necessary to collect the sum of positive and negative charge, so that the effects of the ionization are neutralized. Even so, the stripping cross-section is still included in the measurement at high impact energies (see Chapter 11).

At impact energies below the kilovolt region, it is found that the tightest control can be kept upon the slow ions by means of parallel electric and magnetic fields, perpendicular to the primary ion beam; the slow ions travel to the collecting plates in helical paths; the helix can be tightened by increasing the magnetic field. Advantages accrue from performing the entire experiment in a magnetic field[14], which serves to control a Nier ion source, to mass-analyse the ions with 180° focusing, and to control the slow ions formed in the collision chamber; secondary electrons produced at collimating slits are also tightly controlled. However, it is necessary to take great care that electric fields do not penetrate unduly into the collision chamber through the entrance and exit orifices.

Total charge collection techniques are inadequate for the study of molecular charge transfer and for multiple processes, such as

$$X^+ +YZ \rightarrow X+YZ^+ \qquad X^+ +YZ \rightarrow X+Y^+ +Z$$
$$X^+ +YZ \rightarrow X+Y+Z^+ \qquad X^+ +Y \rightarrow X^- +Y^{2+}$$

and

$$X^{2+} + Y \rightarrow X + Y^{2+} \qquad\qquad X^2 + Y \rightarrow X^+ + Y^+$$

One method of studying these processes is to mass-analyse the slow ions extracted from the collision chamber with, for example, a sector[71] or with a quadrupole mass-spectrometer. An image of the collision region is formed upon the entrance slit of the mass-spectrometer. The length of the object defines the path length appropriate to the calculation of the cross-section. When the sector mass-spectrometer entrance and exit slits are widened, the 'peaks' in mass observed with electrostatic or magnetic scanning should assume a 'flat-topped' appearance. When this is the case, there is no mass-discrimination; it is a minimal condition to be observed in order that the flux of slow ions collected be proportional to the number produced over a defined length of beam. If this proportionality can be obtained, a measurement of the relative proportions of all the mass peaks, together with a single total charge measurement, would be sufficient to determine all the possible target ion cross-sections. With the quadrupole mass-spectrometer, the resolution can be reduced electronically until flat-topped peak conditions are achieved and the mass-discrimination reduced to negligible proportions.

For a discussion of collision chamber extraction technique, one should consult recent publications[72]. For collision studies in the impact energy region 10^4–10^5 eV, atomic ions formed from atoms require extraction potentials of the order of 50 V; but for atomic ions formed from molecules, even higher potentials are sometimes necessary. Transverse potentials bend the primary beam out of course and this effect can be counteracted by added electrodes carrying reversed potentials. The ion optics preceding the mass-spectrometer must be relatively free from spherical aberration, which can be achieved by making the apertures several times wider than the diameter of the extracted beam at its widest point.

In some experiments, the primary beam after collision is mass-analysed. Fast neutral particles may be separated by the application of magnetic or electric fields, and may be counted individually with suitable detectors.

No great error is involved in neglecting ionization in very low energy studies of molecular charge transfer processes[71], but at higher energies where charge transfer, ionization and ionization with capture are all probable, neither mass-analysis of the slow ions nor of the primary beam after collisions is sufficient to separate all the cross-sections. As was seen in Chapter 11, in the general collision process $10/mn$

$$X^+ + Y \rightarrow X^{m+} + Y^{n+} + (m+n-1)e$$

there are mn cross-sections and only $m+n+1$ pieces of data unless coincidence counting is used, in which case all cross-sections can, in principle, be obtained. For many purposes it would be possible to neglect some of the cross-sections in comparison with others.

The relative magnitudes of the charge-changing cross-sections for H in H_2 may be compared by studying *Figure 12.18*.

Crossed-beam methods are used for studies of charge transfer in atomic hydrogen, oxygen and nitrogen; a schematic diagram of apparatus for such measurement is shown in *Figure 12.19*. The advantages of crossed-beam technique, principally the precision in defining the collision region, might

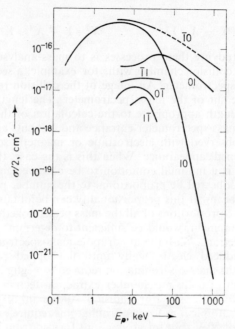

Fig. 12.18. *Charge-changing cross-section functions for* H *in* H_2[71]: *only the projectile initial and final charge states are specified*
(From Hasted[58], by courtesy of Academic Press)

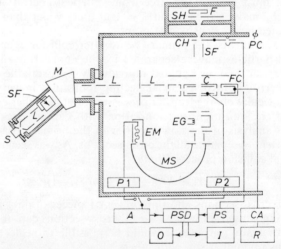

Fig. 12.19. *Schematic diagram of apparatus for crossed-beam studies of charge transfer: F, furnace; SH, shutter; CH, rotating chopper; φ, PC, light source and photoelectric cell; SF, sweeping field for removal of unwanted charged particles; S, ion source; ML, ion lenses; C, collision region; FC, Faraday cage; EG, electron gun; MS, mass-spectrometer; EM, particle multiplier; P1, P2, pre-amplifiers; A, narrow-band amplifier; PSD, phase-sensitive detector; PS, phase shifter; CA, electrometer amplifier; O, oscilloscope; I, integrator; R, recorder*

with advantage be applied to other charge transfer cross-section measurements.

Charge transfer experiments are particularly sensitive to the presence of long-lifetime excited ions in the beam. Ground state ion sources are discussed in Chapter 3, and it is necessary to stress the importance of making sure that the ion beam population is what it is imagined to be.

In the collection of collision products arising either from the projectile particles or from the target gas, the dependence of the flux I_c upon gas pressure is only linear in the low-pressure limit (single collision conditions). At rather higher pressures, an additional square-law pressure term appears. Where only one inelastic process is possible, this may be regarded as arising from the second term in the expansion of an exponential:

$$\frac{I_c}{I_0} \simeq \pi\sigma - \frac{1}{2}\pi^2\sigma^2 \dots \tag{12.31}$$

for target parameter π. Physically the transmitted current is reduced because at the latter end of the collision chamber the primary beam responsible for producing the collision product current has been reduced by an amount proportional to gas pressure.

However, a positive π^2 term can arise because a primary beam particle, having undergone an inelastic collision, may be more capable of producing the collision product than it was in its initial state. It is not often possible to make use of this π^2 term, but its analysis has been considered[73, 74] in the study of the charge degradation of multiply-charged ions, by charge transfer processes $n0/(n-1)1$. It is of course possible to analyse the π term in the presence of the π^2 term, and indeed most single collision experiments do precisely this. An example is the measurement of two-electron capture and loss[13, 75]. In an H^+ beam, it is possible to collect and measure the fast H^- component as a function of gas pressure. Neglecting terms of third order and higher, the pressure dependence is

$$\frac{I_-}{I_+} = \pi\sigma_{1\bar{1}} + \frac{1}{2}\pi^2(\sigma_{10}\sigma_{0\bar{1}} + \sigma_{1\bar{1}}\sigma_{10} + \sigma_{1\bar{1}}^2 - \sigma_{1\bar{1}}\sigma_{\bar{1}0} - \sigma_{1\bar{1}}\sigma_{\bar{1}1}) \tag{12.32}$$

from which the $\pi\sigma_{1\bar{1}}$ component can be separated by least-squares analysis; thus the method is applicable to the study of two-electron capture.

Aston banding techniques[76, 77] have been used to study molecular ion collisions. Such peaks or bands were first observed and discussed in the pioneer mass-spectrometer studies of Aston. They arise because an ion, dissociating in collision with a gas molecule after acceleration in a mass-spectrometer, retains a velocity appropriate to the energy of the undissociated ion. In conventional mass-spectrometers, these peaks are broadened because the region in which collisions can occur is not equipotential and because of the release of kinetic energy. But by separate pumping of the source and analysis chambers of a sector instrument, the two being connected only by a very fine slit and the source pressure being kept low, it may be arranged that collisions occur only in the analysis chamber, free of electric field, but before the beam undergoes magnetic deflexion. Under these conditions, the apparent ratio m^*/e of an Aston peak is related to the primary molecular ion mass m_p and charge e_p and the fragment ion mass m_f and charge e_f, by the equation:

$$\frac{m^*}{e^*} = \frac{m_f^2 e_p}{m_p e_f^2}$$ (12.33)

At impact energies too low for beam techniques to operate, flowing afterglow experiments have been employed for charge transfer measurement; these is also the drift-tube method, as well as merged-beam technique.

A further technique which can be applied to low-energy charge transfer measurement is that of ion cyclotron resonance broadening[78].

The drift-tube method[79] developed from the theoretical and experimental work done on the analysis of ions drifting in gases under the action of uniform electric fields[80]. Under 'constant mean free path' conditions, the mean impact energy of these ions is simply related to their drift velocity v_d:

$$E = \tfrac{1}{2}m_+ v_d^2 + \tfrac{1}{2}m_g v_d^2 + \tfrac{3}{2}kT_g$$ (12.34)

where T_g is the gas temperature.

The drift velocity can be measured by well-established electric shutter techniques, and is a function only of X/p, the ratio of field strength to pressure. To set up the collision events, this parameter is maintained constant. Buffer gas pressures of the order of 1 torr are maintained in the collision chamber, in order to minimize the radial and axial diffusion of the ions. The latter produces negligible effects, and the former introduces a correction factor. An ion beam is injected into the collision chamber, which contains a buffer gas chosen for its inactivity to low-energy inelastic processes of ground state ions. Along the axis of the chamber a uniform electric field is maintained, and at the exit there is a sampling orifice, followed by a mass-spectrometer and detector. Measured proportions of reactant gas are introduced, and if these are small they will not affect the drift process appreciably, but only the sampled ion currents. The cross-section for conversion of ion A^+ to B^+ is related to the sample currents I_A and I_B:

$$\sigma = \frac{v_d}{v_r} \frac{1}{n_0 l} \ln\left(\frac{I_B}{I_A} + 1\right)$$ (12.35)

where v_d and v_r are respectively the drift and random velocities of A^+, l is the length of the drift space, and n_0 is the density of reactant gas. Tests are made to ensure that the injected beam thermalizes to its normal drift conditions before the first drift velocity measurement grid is reached. For very small cross-sections, it is possible to use the reactant gas as its own buffer. Cross-sections as small as 10^{-20} cm² can be measured.

12.4 ACCIDENTAL RESONANCE CONDITIONS

Suppose that a charge transfer process between an unlike ion and an atom is, by chance, exactly resonant, having zero energy defect. This will only be the case at infinite nuclear separation; as the particles approach, the interaction energies of initial and final states are not identical. The charge transfer process is not similar to a symmetrical resonance process, in that an electronic transition takes place. At sufficiently small velocities, the probability of transition never reaches $\frac{1}{2}$, even at the smallest impact parameters. Bates and Lynn[81] have discussed such collisions in detail.

The mechanism whereby the charge transfer takes place is fundamentally different from the mechanism in the symmetrical resonance process. When the nuclear separation is infinite, there is a pair of degenerate eigenfunctions of the quasi-molecule; but here the charge transfer which involves an electronic transition between them has a low probability because the overlap is small when the separation is large. As the nuclei approach, the overlap increases but the difference between the associated eigenenergies also increases, so the transition is inhibited at low velocities. The exact cancellation which prevents the symmetrical resonance cross-section from being affected by distortion does not occur for accidental resonance. The collision will still be adiabatic at the lowest impact velocities, because the mean energy defects averaged over the collision are non-zero even though $\Delta E_\infty = 0$.

An important discussion of the dominance of exchange in the near-resonance collision was given by Demkov[82]. A useful expression for the probability P of transition can be written in terms of the nuclear separation r_{ex} at which the exchange contribution U_{pq} is equal to the energy defect $U_{pp} - U_{qq}$:

$$P = \left| \frac{1}{2} \operatorname{sech} \left[\frac{\pi \, \Delta E}{2(2mE_i)^{1/2}} \left(\frac{dr}{dt} \right)_{r_{ex}}^{-1} \right] \right|^2 \tag{12.36}$$

The collision $O^+ - H$ is an exact accidental resonance, but should, at sufficiently low impact energies, cease to exhibit the large cross-section shown in *Figure 12.17*. These energies have not yet been reached in experiment. The collision $I^+ - Hg$ is also very close to resonance, but has been found to fall off at low energies[83].

The 'recombination energies' of atomic ions which are accurately known have been used[84] to infer inner ionization energies of molecules. Charge transfer measurements are made with a large number of atomic ions, and those for which there is an accidental resonance with an inner ionization energy will show large cross-sections. These studies have largely been superseded by superior methods of investigation, in particular photoelectron spectroscopy. It is important in making inferences from dissociative charge transfer data to be certain of the kinetic energies of the ions produced. The use of onset energies of charge transfer has also proved successful[85].

An application of the conditions of accidental resonance has been made to the calculation of molecular charge transfer rates[86] at near-thermal energies. Inward-spiralling orbiting occurs, and the time of collision is lengthened by a factor of between 2 and 10, depending on impact parameter; this implies an increase in the apparent adiabatic parameter. An exothermic charge transfer process is supposed to take place with the molecular products vibrationally and rotationally excited into the level or levels which are closest to resonance. It had previously been assumed that exothermic molecular processes behaved as perfect accidental resonance processes because a suitably placed level could always be found. With a larger adiabatic parameter this need not necessarily be the case, bearing in mind that rotational levels are not spaced equally but as $J(J+1)$. Detailed calculations have been made using a two-state approximation in the Landau–Teller form

41 *

of equation 12.26, with adiabatic parameter $a = 100$ Å. A comparison of the available room-temperature data for charge transfer, ion–atom interchange and atom–atom interchange with these calculations is given in *Figure 12.20*. The method has been extended with the aid of the Demkov expression (equation 12.36), and in this form can account for the energy

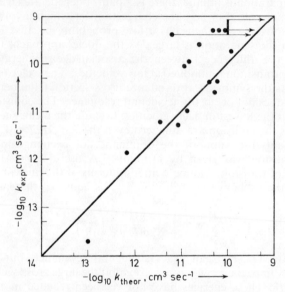

Fig. 12.20. Comparison[86] of thermal energy charge transfer, ion–atom interchange and atom–atom interchange rates with nearest resonance calculations

variation of cross-sections for atomic ions with diatomic molecules; this variation includes a minimum in the energy range 0·05–5 eV, as well as the characteristic maximum at **a** rather higher energy.

12.5 THE ROLE OF EXCITED STATES. EXPERIMENTS WITH EXCITED AND GROUND STATE IONS

Several experimental investigations have been carried out with atomic ions produced in ion sources capable of eliminating most of the long-lived excited ions in the beam[87, 88]. Most ion sources (for example, conventional electron impact, oscillating electron, and radiofrequency discharge sources) produce a proportion of excited ions whose lifetimes are sufficiently long for the excitation to persist right up to the moment of collision. In general, the larger the internal energy of the excited ion, the lower the proportions in which it is likely to be produced. It is sometimes possible to make reliable deductions of these proportions from analysis of the operation of the source; indirect inferences may be made from the experimentally determined charge transfer cross-section functions, provided that comparison can be made between conventional ion source data and 'ground state ion source' data. Beams can be produced with ions entirely in the ground state by means of:

1. Surface ionization;
2. Monochromated ultra-violet radiation;
3. Controlled-energy electron fluxes.

Surface ionization sources are limited to atoms of very low ionization potential, that is, to the alkali atoms. The first experimental comparison of inelastic cross-sections obtained with ground state positive ions and with conventional ion sources was made[89] using Li^+ produced by surface ionization.

The production of monochromated photon fluxes of sufficiently high energy to ionize gases requires the use of vacuum ultra-violet techniques and monochromation. For this reason, these ion sources are unlikely to be as widely accessible as those using electrons of controlled energy. Ionization functions by electrons are zero at threshold, whereas those for ionization by photons are finite at threshold. When ionization yielding a number of excited levels of the ion is possible, one might expect a piecewise linear function for electron impact and a step function for photons. Thus electrons are inferior tools for the production of ground state ion fluxes.

A compromise must be made between the maximization of ion flux for the collision experiment and the optimization of the resolution of the electron energy selection. Where the lowest long-lifetime electronic excited state of the ion lies several electron volts above the ground state, thermionic electron energy distributions are sufficiently narrow to allow effective elimination of the excited states; a conventional or specially adapted electron impact ion source is adequate[90-92] and will naturally produce a greater ion flux than a source using momentum-analysed electrons. The latter[93] is necessary when the lowest long-lifetime state of the ion lies within about 1 eV of the ground state.

For example, it is found that the dependence of the charge transfer cross-section O^+-N_2 upon the ion source electron energy[94] is marked. The experiment yields a large cross-section for the production of N_2^+ by excited O^+ ^2D, while the cross-section for the ground state ion is small although ion–atom interchange can occur. Elimination of excited ions from the beam can be achieved by passing it through nitrogen gas. The higher the ion beam energy, the less serious the effect of elastic scattering in this 'filtering' technique. The technique is indispensable when the threshold ionization function is such that ground state ions cannot easily be produced without momentum-analysed electrons. However, a Nier–Bleakney electron impact source with capillary gas feed[86, 95] is adequate for the production of ground state O^+ beams. *Figure 12.21* shows the dependence of O^+ intensity upon electron energy for ion source gas CO; the curve follows closely the mass-spectrometric investigations[96], and allows the electron energy scale to be corrected to the sharp break at O^+ ^2D appearance. The O^+ ^2D–Ar charge transfer is 1·17 eV exothermic, while the ground state process O^+ ^4S–Ar is 2·15 eV endothermic; at 0·8 eV impact energy, the latter cannot take place. *Figure 12.21* also shows the electron energy dependence of the Ar^+ intensity produced in 0·8 eV collisions. Under these circumstances, it is possible to extract reliable ground state O^+ collision cross-sections from data obtained several electron volts below O^+ ^2D appearance.

Singly charged heavy rare-gas ions produced by collisions with electron

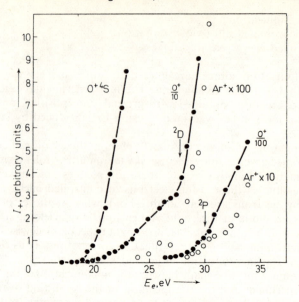

Fig. 12.21. Ground state ion source appearance potential functions of O^+ *ions from* CO *by electron impact (360 μA electron current; pressure* $< 1 \times 10^{-5}$ *torr): open circles represent production of* Ar^+ *ions by inelastic* O^+ ^2D*–Ar collisions at 0·8 eV impact energy; the yields of ions below* O^+ ^2D *threshold can be regarded as background*

beams energy-selected in analysers have been used[93] for symmetrical resonance charge transfer collisions. The lowest states of these ions can exist with inner quantum number $J = \frac{3}{2}$ and $J = \frac{1}{2}$ (the $J = \frac{1}{2}$ being, for example, the higher by 0·67 eV in krypton. Thus the possibility arises of collision processes

$$J = \tfrac{3}{2} \longrightarrow J = \tfrac{3}{2}$$
$$\times$$
$$J = \tfrac{1}{2} \longrightarrow J = \tfrac{1}{2}$$

At sufficiently low impact energies, the processes involving energy defects should be unlikely, whilst the symmetrical processes would be expected to have large cross-sections. In the initial experiments on Kr^+–Kr at 250 eV impact energy[97], the ratio of total $J = \frac{1}{2}$ to $J = \frac{3}{2}$ cross-sections, presumably dominated by symmetrical resonance processes, was found to be $\leqslant 0·1$. In subsequent experiments[93] using a finer resolution of electron energy, the ratio was found to increase with increasing impact energy and pass through a maximum value of 0·6 at 700 eV. Quantum calculations imply identical symmetrical resonance cross-sections for $J = \frac{3}{2}$ and $J = \frac{1}{2}$ at all except the lowest impact velocities. But recent experimental investigations have reported no difference between the two cross-sections, as would be expected on theoretical grounds.

12.6 PROCESSES TERMINATING IN EXCITED SPECIES

Charge transfer processes terminating in excited states, whether of the ionized or neutral species or both, may be investigated experimentally by monitoring the radiation emitted from the collision region, provided that the states in question are not metastable. When they are, the experimental techniques are more difficult, but in cases such as H $2s$ it is possible to quench the

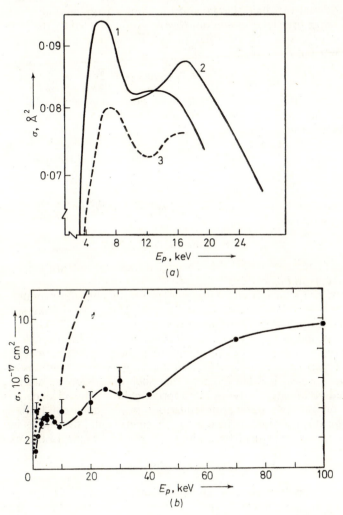

Fig. 12.22. Total cross-sections for production of H $2s$: *(a) in proton collisions with neon (1, experimental data of Jaecks et al.; 2, experimental data of Ankudinov et al.; 3, calculations of Poluektev and Presnyakov[102]); (b) in proton collisions with hydrogen atom (points with error bars, experiments of Stebbings et al.; broken line, calculations of Bates and Williams; dotted line, calculations of Bell and Skinner; full line with points, calculations of Wilets and Gallaher[103])*

state by an electric field, so that the Lyman α radiation from H $2p$ can be monitored.

The experimental data for the processes

$$H^+ + Ne \rightarrow H\ 2s + Ne^+$$

and

$$H^+ + H \rightarrow H\ 2s + H^+$$

are displayed in *Figures 12.22*. The neon cross-section function exhibits two maxima, of which one falls at the same energy as that of the H$^+$–Ne total charge transfer cross-section[98] which is dominated by the H $1s$ product.

Such processes are suitable for the investigation of coupling effects. The contribution of the path

$$H^+ + Ne \rightarrow H\ 1s + Ne^+ \rightarrow H\ 2s + Ne^+$$

is significant. Rare-gas studies[99, 100] have served to emphasize its significance.

Following the work of Lovell and McElroy[101], various workers have attempted atomic multi-state approximation calculations. Poluektev and Presnyakov[102] take the overall probability $\bar{P}$ of transition from initial state (0) to final state (2), including the contribution of intermediate state (1), as

$$\bar{P} = P_{02} + \tfrac{1}{2}P_{01}P_{12} \qquad (12.37)$$

The results of their calculations are displayed in *Figure 12.22a*. Calculations of cross-sections for the process

$$H^+ + Ne \rightarrow H\ 2s + Ne^+$$

have been made[103] with Sturmian eigenfunctions and are compared with experiments in *Figure 12.22b*.

Experiments have been conducted[104] with the collision products exposed to electric fields sufficient to cause Stark separation of the sub-states of the H $3s$ level, which decay producing H α and Lyman β radiation, the observed relative proportions of which are dependent on the electric field.

The possibility must be considered that coupling to intermediate atomic states contributes considerably to total capture cross-sections in the far adiabatic region. If often happens that such cross-sections are larger than expected on an atomic two-state approximation. The v^4 dependence predicted on the two-state approximation[7] is not apparent from the data, although this may in part be due to the experimental difficulties inherent in measuring very small cross-sections. A good example is the exothermic collision

$$Ne^+ + Ar \rightarrow Ne + Ar^+$$

for which the cross-section (shown above in *Figure 12.16*) is as large as 10^{-17} cm^2 at 4 eV impact energy[86]. The Ne$^+$ ions used in these experiments were produced by controlled energy electron impact, so that they could include only the ground state $\left(J = \tfrac{3}{2}, \tfrac{1}{2}\right)$. But there is a possibility of coupling to excited states (for example, to Ne$^+$ $3s$ ^{4}P) which can undergo the accidentally resonant process

$$Ne^+\ 3s\ ^4P + Ar\ 3p^6\ ^1S \rightarrow Ne\ 3s\left[1\tfrac{1}{2}\right] + Ar^+\ 3d\ ^4D$$

Experiments at low impact energies using optical techniques for the detection of excited species[105] have shown that certain rare-gas charge transfer cross-section functions terminating in specific states are not negligible in the adiabatic region and, furthermore, show considerable structure[106]. The failure of the two-state model to explain this leads one to consider non-adiabatic behaviour[107]. If there is a pseudo-crossing of two inelastic channels such that the two amplitudes were efficiently mixed, coherent interference would influence the final populations. The critical phase would be the phase difference between the two inelastic amplitudes at crossover, and would therefore be approximately proportional to the time taken by the system to travel from its minimum separation to crossover. This time is comparatively independent of impact parameter, so that oscillations may appear in the total cross-section function.

12.7 DIFFERENTIAL SCATTERING WITH CAPTURE. OSCILLATORY STRUCTURE

On the two-state impact parameter approximation, the probability P_0 of symmetrical resonance charge transfer, considered as a function either of b or of v, oscillates between zero and unity[108]. The oscillation is observable both in collisions at fixed polar scattering angle and varying impact velocity, and in collisions at fixed impact velocity and varying polar scattering angle. Both types of experiment were first carried out by Everhart and his colleagues[109]. Some data are illustrated in *Figure 12.23*. For collisions between unlike ions and atoms, the maxima of the oscillations are not constant at unity, but decrease with increasing b (*Figure 12.24*). Even for symmetrical resonance collisions, the probabilities do not reach zero or unity. This is because of wave effects[110] and because of coupling to other states[111].

The oscillations in the differential scattering functions are governed by the quantity Ea, which is calculated from the interaction energies $E(s)$ along the path variable s, the closest distance of approach r_0 being inferred from the experimental scattering angle[10, 11]:

$$Ea = \int_{-\infty}^{\infty} E(s)\,\mathrm{d}s \tag{12.38}$$

and

$$s^2 \simeq b^2 + r_0^2 \tag{12.39}$$

For the H_2^+ ion, the wave functions and energies of the appropriate states are available:

$$E(r) = E(1s\sigma_g) - E(2p\sigma_u) \tag{12.40}$$

and a comparison of experimental data and calculated data for Ea for the H^+–H collision is given in Table 12.2. However, the experimental data cannot be correctly reproduced without a phase-constant β. This constant should be zero, and the constants K_1 and K_2 should be zero and unity respectively in the reproduction of the experimental data by the equation:

$$P_0 = K_1\left(\frac{1}{v}\right) + K_2\left(\frac{1}{v}\right) \sin^2\left(\frac{\pi Ea}{hv} - \beta\right) \tag{12.41}$$

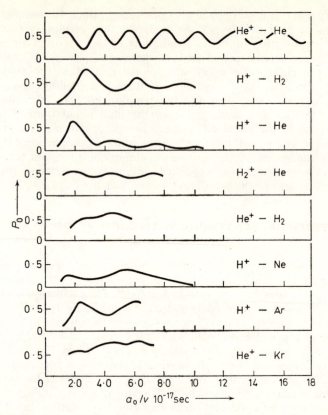

Fig. 12.23. Differential scattering with capture: probabilities of capture as functions of collision time (after Everhart and co-workers[109])

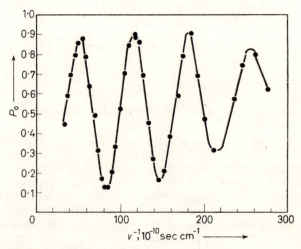

Fig. 12.24. Differential scattering with capture for H^+–H at $\theta = 3°$: probability as function of reciprocal velocity, showing equally spaced oscillations

Table 12.2. COMPARISON OF EXPERIMENTAL AND CALCULATED
VALUES OF Ea

n	Ea_{exp} (eV Å)	Ea_{calc} (eV Å)	r_0 (Å)
1	63·7	70·0	0·041
2	63·6	69·9	0·073
3	61·8	68·9	0·114
4	60·0	67·8	0·162
5	54·5	66·0	0·219

But in order to explain the measurements, K_1 and K_2 must be assumed to vary slowly with $1/v$. Some values of Ea and β at high energies are tabulated in Table 12.3. Differential scattering measurements have been reported[112] for H–H, H–H$_2$ and H$^-$–H.

Table 12.3. HIGH-ENERGY VALUES OF Ea AND β

Collision	Ea (eV Å)	$\dfrac{\beta}{\pi}$
H$^+$–H	63·7±1	0·28±0·01
H$^+$–He	84·6±1	0·26±0·02
He$^+$–He	102±3	0·23±0·08

Everhart's experiments on differential scattering with capture involve a measurement, at the same polar scattering angle, of both charged and neutral components. However, it is possible to obtain information about charge transfer (except in the adiabatic region) from measurements of only the differential scattering of the charged particles[113]. This is discussed in Chapter 10. The structure in the scattering functions arises from rainbow effects, from nuclear interchange, and from interference between waves scattered from the symmetric and antisymmetric potentials of the quasi-molecular ion.

Under certain circumstances, oscillating behaviour is found in total charge transfer collision functions, both symmetric and asymmetric[114]. Typical data are shown in *Figure 12.25*. In the two-state approximation, the resonant total charge transfer cross-section is of the form

$$\sigma = 2\pi \int_0^\infty b \sin^2 (\eta_g - \eta_u) \, \mathrm{d}b \qquad (12.42)$$

with the term $\sin^2 (\eta_g - \eta_u)$ replaced by $\frac{1}{2}$ out to the impact parameter at which it falls to zero. When the phase difference $\eta_g - \eta_u$ is velocity invariant at a certain impact parameter b, the simple theory is inaccurate because, at energies where the stationary value of $\eta_g - \eta_u$ is close to an odd multiple of $\pi/2$, the cross-section passes through a maximum. This effect 'shows through' the integration over impact parameter. The stationary phase may be due either

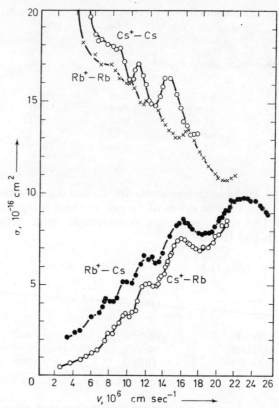

Fig. 12.25. Oscillating total charge transfer functions for alkali metals (after Perel and co-workers[114])

to a maximum in the interaction energy difference, or to the potential core. Recent calculations[115] show that oscillations should be present in all charge transfer cross-section functions, albeit to a varying extent, and should be sharpest at the lowest impact velocities.

12.8 BEHAVIOUR OF INELASTIC ION–ATOM COLLISION CROSS-SECTIONS AT HIGH ENERGIES

The contrasting high-energy fall-off of ionization and charge transfer cross-section functions[116–122] is evidence of a very marked difference between the two processes. Since complexities such as the participation of inner-shell electrons complicate the issue when many-electron atomic systems are studied, only the simplest collisions (for instance H^+–H, H^+–H_2, H^+–He, He^+–He and between H^+ and rare gas atoms) are considered here.

The Born approximation calculations predict that the charge transfer cross-section falls off very rapidly with impact velocity ($\sigma \propto v^{-12}$). In the Bethe–Born approximation, the ionization, loss and optically allowed excitation cross-sections fall off much less rapidly ($\sigma \propto v^{-2} \ln v^2$), and the optically forbidden excitation cross-sections fall off as $\sigma \propto v^{-2}$. The charge transfer data[123] for H^+–H_2 and H^+–He certainly show the rapid fall-off ($\sigma \propto v^{-10}$), but H^+–Ar only varies as $v^{-3 \cdot 7}$. In the same experiments, the loss cross-section H–H_2 follows v^{-2}, and H–N_2 and H–Ar follow v^{-1}. The capture of electrons by α-particles[124] in gases is best represented by $v^{-5 \cdot 6}$. Typical H_2^+ cross-section functions[69] are shown in *Figure 12.26*, and data extending over a wider range of energies are now available[125]. Ionization data of Hooper *et al.*[126] for protons in rare gases fall as $\sigma \propto v^{-c}$ with c having the following values: He, $1 \cdot 52$; Ne, $1 \cdot 38$; Ar, $1 \cdot 42$; Hg, $1 \cdot 72$; N_2, $1 \cdot 40$; O_2, $1 \cdot 50$; CO, $1 \cdot 46$. This dependence holds over the energy region $0 \cdot 15$–$1 \cdot 1$ MeV and presumably at higher energies. The predicted log v term is discernible in the ionization data of Gilbody *et al.*[68].

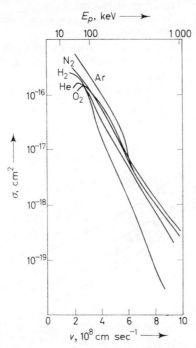

Fig. 12.26. *High-energy charge transfer cross-sections for* H_2^+ *in various gases*[69]

An important question arises as to whether the Born approximation is correctly formulated in the calculations of charge transfer, particularly the collisions H^+–H[127, 128]. At sufficiently high velocities, only vanishingly small impact parameter collisions contribute to the charge transfer process. In the Born approximation, the differential cross-section for charge transfer accompanied by scattering of the incident proton[10] from direction n_i to n_f in centre-of-mass coordinates is:

$$d\sigma = \left(\frac{\mu}{2\pi\hbar^2} \right)^2 |\psi_f| V_{1e} + V_{12} |\psi_i|^2 \qquad (12.43)$$

where μ is reduced mass, ψ_i and ψ_f are the initial and final wave functions,

V_{1e} and V_{12} are the interaction energies respectively of the proton and electron, and of the two protons; these are combined into the matrix element of
the interaction prior to transfer. In the early calculations of Brinkmann and
Kramers[117] using an impact parameter method, V_{12} was neglected[129, 130],
but later authors[118-120] include it. It can be shown[131] that, at sufficiently
high energies, V_{12} is only of order m_e/m_+, as was originally believed by
Brinkmann and Kramers. The series of which the Born approximation is
the first term does not converge in this process, but the approximation may
represent the first term in another series. Relativistic effects are also important in the MeV region.

It might be supposed that, at sufficiently high impact energies, charge
transfer data[123] for molecular hydrogen H^+-H_2 would provide a good test
of H^+-H calculations. However, it has been shown[132] that for the purposes
of this collision process molecular hydrogen is not the equivalent of two
hydrogen atoms even at 10^5 eV energies.

Comparison of H^+-H calculations with experiment is difficult because
of the contribution of processes which terminate in excited states of the
hydrogen atom. Some structure in the experimental data of *Figure 12.27*
may indicate such a contribution. Recent calculations[131, 134] include a

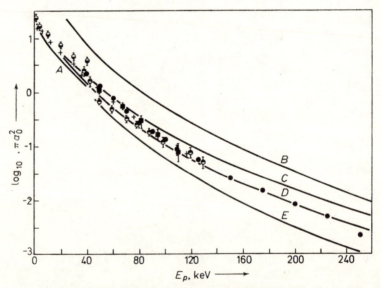

Fig. 12.27. *High-energy* H^+-H *charge transfer cross-sections*[133]: *closed circles, Wittkower
et al.; open-top circles, Fite et al.; closed-top circles, Gilbody and Ryding; crosses, McClure;
square, calculations of Abrines and Percival; A, calculation of Ferguson; B, calculation of
Brinkmann and Kramers; C, calculation of McElroy; D, calculations of Jackson and Schiff,
and of Bates and Dalgarno; E, calculation of Coleman and McDowell*

form of impulse approximation in which the wave function takes account
of the interaction between the bound electron and the incident ion. New
calculations have also been made using the Fadeev equations. Classical
Monte Carlo calculations[135] are highly successful and are displayed in
Figure 12.27. The earliest treatment of high-energy charge transfer[136] was

in fact classical, and yields a v^{-10} dependence; it is rendered invalid by uncertainty relations.

The cross-section functions for capture into excited states have been calculated for the H^+–H collisions[37, 117, 118, 137, 138]. After passing through maxima in the 10–20 keV region, they decrease rapidly; for fixed azimuthal quantum numbers l they decrease with increasing principal quantum number n, and for fixed n they usually decrease with increasing l. At very high energies, the cross-sections are proportional to n^{-3}.

Theories of high-energy charge transfer by many-electron atoms[139-141] have been based on the Thomas–Fermi atom. Bohr's treatment[139] yields:

$$\sigma_c = 4\pi a_0^2 Z_+^5 Z_0^{1/3} \left(\frac{v_0}{v}\right)^6 \qquad (12.44)$$

where the charge Z^+ on the nucleus of the ion is assumed to be much smaller than Z_0, the charge on the nucleus of the target atom. It is also assumed that v greatly exceeds $v_0 = e^2/\hbar$. The theory of Bohr and Lindhard[141] yields

$$\sigma_c = \pi a_0^2 Z_+^2 Z_0^{1/3} \left(\frac{v_0}{v}\right)^3 \qquad (12.45)$$

but for hydrogen or helium targets

$$\sigma_c = \pi a_0^2 Z_+^3 Z_0^3 \left(\frac{v_0}{v}\right)^7 \qquad (12.46)$$

Other general theories have been given[142, 143].

12.9 IONIZATION WITH CAPTURE

Ionization with capture is a collision process of the type

$$A^+ + B \rightarrow A + B^{2+} + e \qquad \text{10/02e}$$

One must consider how far such collision cross-sections differ from, and perhaps are more likely than those for multiple ionization in which no capture takes place

$$A^+ + B \rightarrow A^+ + B^{2+} + 2e \qquad \text{10/12}$$

The question is of particular importance because the statistical theory of Russek and Thomas[144] is reasonably satisfactory in interpreting multiple ionization collisions, but takes no account of any inclusion of capture.

The two collision processes given above can only be distinguished experimentally if the charge state of the primary beam is determined simultaneously with that of the slow ionized product, possibly by detection with coincidence. This was pointed out more than ten years ago[145], but the first measurements directed at measuring the cross-section are much more recent[146]. The coincidence technique was developed at Leningrad for the study of the conversion of energy from kinetic to internal in the multiple ionization process, but for studies of ionization with capture the technique was combined with extraction by means of a transverse electric field of slow ions from the collision chamber. Transferred momentum, which might seriously complicate this technique, was minimized in the first experiments by the use of the proton as projectile and rare gases as targets.

It was found that the cross-sections for ionization with capture (10/02 and 10/03) are indeed much larger than the corresponding ionization cross-sections $10/n2$ and $10/n3$ where $n = 1, 2, \ldots$ The cross-section functions maximize at much lower energies, as would be expected on the basis of the adiabatic maximum rule. No detailed comparison of a wide variety of processes is yet possible, but it is of interest that the adiabatic parameter for process $10/02e$ is smaller than that for process $10/03e^2$. It has been proposed that the ionization with capture process is dominated by capture into auto-ionizing states and so may be treated by theory similar to that for pure capture processes.

12.10 PSEUDO-CROSSING POTENTIAL ENERGY CURVES

When a charge transfer collision is considered in terms of the two adiabatic potential energy curves for the initial and final states of the system, situations can occur where these curves would pass across each other. For adiabatic crossing it is possible to calculate two sets of eigenenergies for the system at nuclear separations near the crossover. The adiabatic curves do not cross, but have the form of the 'pseudo-crossing' shown in the inset of *Figure 12.28* (see page 648). Transitions can occur by exchange in the region of the pseudo-crossing between the distorted energy curves; the coupled equations which describe the system were solved by Stueckelberg[147], and the probabilities of transition at crossover calculated by Landau[148] and by Zener[149]. Calculations of inelastic collision cross-sections have been made, with the aid of the Landau–Zener formula, by Bates and his colleagues[150–153]. Although these are of value in the interpretation of experimental data[14], it will be seen that allowance should be made for transitions occurring over a wider range of nuclear separations[154, 155].

The system approaching along the energy E_m may pass over to the energy E_n, but may again pass over to E_m on the return journey. Thus, if P is the probability of a transition, the probability $\mathcal{P}$ of the system not finishing in the same state as before the collision is

$$\mathcal{P} = 2P(1-P) \tag{12.47}$$

To determine the probability of transition at pseudo-crossing, it is necessary to calculate the interaction matrix elements

$$U_{mn} = \int \phi_m^*\left(\frac{r_e}{r}\right) \mathcal{H}\phi_n\left(\frac{r_e}{r}\right) \, \mathrm{d}r_e \tag{12.48}$$

where $\mathcal{H}$ represents the Hamiltonian operator, and ϕ_m and ϕ_n are orthogonal linear combinations of the wave functions ψ_m and ψ_n. The variable r_e is the position vector of the active electron with respect to the midpoint of r, the nuclear separation; this takes the value r_x at crossover. Application of time-dependent perturbation theory yields the Landau–Zener formula for the probability of transition:

$$P = \exp(-\omega) \tag{12.49}$$

and

$$\omega = \frac{2\pi U_{nm}^2/\hbar}{v(U_{nn}' - U_{mm}')} \tag{12.50}$$

The matrix elements can be taken as corresponding to the separation of the energy curves; the separation at crossover, corresponding to U_{nm}, will be denoted by ΔU_{r_x}. The cross-section can be expressed in the form[150-153]

$$\sigma = 4\pi r_x^2 I(\eta)P' \tag{12.51}$$

where

$$I(\eta) = \int_1^\infty \exp(-\eta x)[1 - \exp(-\eta x)]x^{-3}\,dx \tag{12.52}$$

and

$$\eta = 247\sqrt{\mu n}\,\frac{\sqrt{\Delta U_{r_x}}}{\Delta E_\infty E_p^{1/2}} \tag{12.53}$$

These equations are applicable to collisions in which an electron is transferred from an $(n+1)$ charged ion to a neutral atom; P' is the probability that the systems approach along the specified potential energy curves, and μ is the reduced mass. This particular type of collision process $(n+1)0/n1$ is selected because the long-range Coulomb interaction between the products distorts the potential energy curves in such a way that the pseudo-crossover usually occurs at a suitable calculable nuclear separation. Many of the other collisions in which pseudo-crossovers play a part are much more complicated, but in the simplest systems distorted by Coulomb forces the matrix elements can be calculated, as has been done[150] for the mutual neutralization process

$$H^+ + H^- \rightarrow H(ns) + H(ns, np)$$

discussed in Chapter 7.

For the charge transfer processes

$$X^{(n+1)} + Y \rightarrow X^{n+} + Y^+$$

where $n = 1, 2, \ldots$, the experimental data have been compared with the predictions of the Landau–Zener formula along the following lines[73, 74]. Moiseiwitsch[156] has calculated the integral $I(\eta)$ which is displayed in *Figure 12.28*; it will be seen that it increases with increasing η, until the maximum value ~ 0.1 is reached when $\eta = 0.414$, there being a decrease at higher values of η. Cross-section function maxima are used to calculate ΔU_{r_x}, and r_x is calculated as follows. Neglecting polarization interaction between the reactants

$$r_x \simeq \frac{ne^2}{\Delta E_\infty} = \frac{27.2n}{\Delta E_\infty}\ \text{a.u.} \tag{12.54}$$

with the energy defect ΔE_∞ in electron volts. Equation 12.51 with $I(\eta) \sim 0.1$ fixes the maximum value of the cross-section, which is a useful check upon

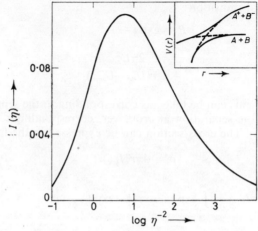

Fig. 12.28. Calculation of the integral I (η) ; inset shows form of the pseudo-crossing potential energy curves for the process $A^+ + B^- \rightleftharpoons A + B$

(From Hasted[58], by courtesy of Academic Press)

the validity of the interpretation. Although those multiply-charged ion collisions of importance for pseudo-crossing theory have been studied experimentally in the low-energy region, the presence of excited states can complicate either one or both sides of the reaction equation. Nevertheless, it is likely that the experimental cross-section function is dominated by the ground state process.

Despite the numerous simplifications, the available data[73, 74] are found to be consistent with the Landau–Zener formula, with $\log_{10} \Delta U_{r_x}$ a smoothly varying function[157] of r_x^{-1}. In the summary of data shown in Tables 12.4 and

Table 12.4. QUANTUM CALCULATIONS OF CURVE-CROSSING PROCESSES

	Collision		Reference
1	Li^{2+}–H	(Li$^+$ ^{1}S)	151
2	Li^{2+}–H	(Li$^+$ ^{3}S)	151
3	B^{2+}–H	(B$^+$ ^{1}P)	151
4	Al^{3+}–H		151
5	Al^{3+}–H	(Al^{2+} ^{2}P)	151
6	Be^{2+}–H		150
7	Mg^{2+}–H		150
8	Si^{2+}–H		150
9	H$^-$–Li$^+$	(Li 2s)	153
10	Be^{3+}–He	(Be^{2+} 2^{3}S)	150
11	Be^{3+}–He	(Be^{2+} 2^{3}P)	150
12	Li^{3+}–He	(Li^{2+} 2s or 2p)	150
13	Be^{3+}–He	(Be^{2+} 2^{1}P)	150
14	Al^{3+}–He	(Al^{2+} 3s ^{2}S)	150
15	Mg^{3+}–He	(Mg^{2+} 3s ^{3}P)	150

Table 12.5. EXPERIMENTAL STUDIES OF CURVE-CROSSING PROCESSES

	Collision		ΔE_∞ (eV)	r_z^{-1} (a.u.)$^{-1}$	$E_{p\,max}$ (keV)	ΔU_{r_z} (eV)	$\Delta U_{r_z}(calc)$ (eV)	Reference
16	N^{2+}–He		5.02	0·18	610	0·77	0·27	74, 177
17	Ar^{2+}–He		3·02	0·11	470	0·42	0·0032	74, 177
18	Kr^{3+}–He		11·1	0·20	1 900	1·65	1·17	73, 177
19	Kr^{4+}–He		19·0	0·23	7 000	2·52	—	73
20	N^{2+}–Ne		8·04	0·29	950	1·32	—	74
21	Ar^{2+}–Ne		6·04	0·22	3 500	0·57	—	74
22	Kr^{2+}–Ne		2·7	0·1	1 400	0·35	0·03	73, 177
22a	Kr^{2+}–Ne	$(Kr^{2+}\ ^1D)$	4·5	0·17	1 400	0·59	—	73
22b	Kr^{2+}–Ne	$(Kr^{2+}\ ^1S)$	6·8	0·25	1 400	1·00	—	73
23	Kr^{3+}–Ne		14·0	0·26	2 800	1·59	0·76	73, 177
23a	Kr^{3+}–Ne	$(Kr^{2+}\ ^1S)$	9·9	0·18	2 800	1·13	—	73
23b	Kr^{3+}–Ne	$(Kr^{2+}\ ^1D)$	12·2	0·22	2 800	1·38	—	73
24	Kr^{4+}–Ne		22	0·27	400	1·02	—	73
24a	Kr^{4+}–Ne	$(Kr^{3+}\ ^4P)$	7	0·09	400	0·11	—	73
25	Ar^{2+}–Ar		11·86	0·30	3 500	1·40	1·75	177
26	Kr^{2+}–Kr		10·56	0·36	3 500	1·10	1.60	177
27	Xe^{2+}–Xe		9·07	0·23	4 500	0·94	0·93	177
28	Xe^{3+}–Ne		10·5	0·19	—	1·60	0·86, 1·3	177, 178
29	Xe^{4+}–Ne		23·9	0·26	—	5·40	1.17, 5·0	177, 178
30	B^{3+}–He	$(He^+\ 2s)$	13·34	0·24	14 000	2·3	2·28	179

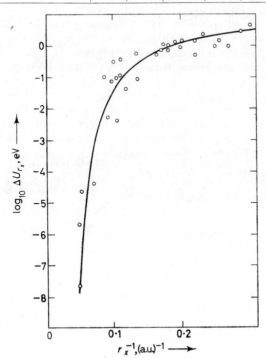

Fig. 12.29. Function ΔU_{r_z} versus r_x^{-1} appropriate to the data of Tables 12.4 and 12.5

12.5 and in *Figure 12.29*, the matrix element calculations of ΔU_{r_x} are compared with values deduced from the application of Equations 12.51–12.53 to the experimental data.

As was noted in Section 12.6, structured cross-section functions can arise from curve-crossing at energies as low as several electron volts, resulting in the formation of excited ions[158], for example,

$$He^+ + Ar \rightarrow He + Ar^{+\prime}$$

It has been assumed in the Landau–Zener treatment that the fast oscillating terms in the probabilities of transition cancel out except at the classical crossover. However, the range of separations over which transitions are likely is given[154, 155] by:

$$\Delta r = \frac{4\pi v \hbar s(r - r_x)}{U_{nn} - U_{mm}} \simeq 0 \cdot 5\,|\Delta U_{r_x}|\,r_x^2 \text{ a.u.} \qquad (12.55)$$

In order to calculate the probabilities of transition, it is necessary to integrate over the entire range of the variable $Z = vt$:

$$P = 2\left| \frac{1}{hv} \int_{-\infty}^{\infty} U_{nm} \exp\left[\frac{-i}{hv} \int_0^Z (U_{nn} - U_{mm})\,dZ \right] dZ \right| \qquad (12.56)$$

Calculations have been made along these lines, and it is found that the cross-section function maximum occurs at a higher energy than that calculated in the Landau–Zener approximation; a second maximum can be observed at higher impact energies, as is the case[6] for Kr^{3+}–Ne, and for C^{4+} in helium, neon and argon[159]. The 'oscillating terms' in the transition probabilities appear clearly in differential scattering experiments[160] and, at sufficiently low energies, in total cross-section functions. Semi-classical criticisms[161] and further theoretical treatment[162, 163] have been given.

In certain collision processes, such as $1\bar{1}/0'0'$, and molecular ion collisions, there can be a large number of different reactant and product excited states. These result in many potential curves, and many pseudo-crossovers[164]. Supposing that the probability of transition at each crossover is p_t, and that there are $2N$ crossovers, the probability of emergence in the final state is

$$P_N = \frac{2p_t(1 - p_t)}{1 - (1 - p_t)^2}\,[1 - (1 - p_t)^{2N}] \qquad (12.57)$$

For large N and small p_t one has the limit:

$$P_N \simeq 1 - \exp(-2Np_t) \qquad (12.58)$$

and, as $2Np_t$ becomes large, P_N approaches unity. For N moderately large and $p_t > \frac{1}{2}$, P_N approaches

$$P_N \rightarrow 2p_t(1 - p_t) + 2p_t(1 - p_t)^3$$

As p_t approaches unity this expression can be regarded as proportional to $1 - p_t$, so that practically all the reaction occurs at the first crossover. The cross-section is related to the probability by an integral of the type

$$\sigma(v) = 2\pi \int_0^{\infty} P_N(r)\left(1 + \frac{e^2}{\mu} v^2 r\right) r\,dr \qquad (12.59)$$

No detailed calculations of this type have been carried out, but a similar treatment has been applied to the charge transfer between protons and hydrocarbon molecules[165].

12.11 RADIATIVE CHARGE TRANSFER

The process

$$A^+ + B \rightarrow A + B^+ + h\nu$$

in which electromagnetic radiation carries away the excess energy, can under some circumstances be more likely than the similar process in which no radiation is produced. This radiative charge transfer only arises if the quasi-molecule of the collision is formed in a state which can radiate to another state, which in turn can dissociate in the required manner.

Daly and Powell[166] report studies in helium using the trapped-ion technique[167]. They observed the process

$$He^{2+} + He \rightarrow He^+ + He^+ (+h\nu)$$

at thermal energies. The non-radiative process would be exothermic by $5 \cdot 2$ eV. Even allowing for pseudo-crossing and for the possibility of the collision products being in excited states, it is likely to have a much smaller cross-section than that calculated for the radiative process[168]. Daly did not attempt to detect radiation.

12.12 TWO-ELECTRON CAPTURE PROCESSES*

Two-electron capture $(10/1\bar{2})$ by singly charged ions has been observed for some years in beam experiments, principally those conducted by Fogel and his colleagues[169]. In some cases, the negative ions in these processes (for example, N^-) are formed in metastable states. A small proportion of fast positive ions passing through a gas is converted into negative ions, which are observed as such provided that their lifetimes are sufficiently long for them to live until mass-analysed and collected. The lifetimes of the negative ion resonances observed in the elastic scattering of electrons by atoms[170] are far too short for this condition to be satisfied. Metastable negative ion states of microsecond lifetimes must be involved.

Some of the importance of two-electron capture processes lies in their use in double electrostatic generators[171]; positive ions are accelerated, converted into negative ions, and further accelerated in the same electric field so that the energy achieved is twice that available in the conventional electrostatic generator.

The 'growth rate' method of measuring two-electron capture cross-sections has been applied. It is outlined above with reference to equation 12.32.

Typical cross-section functions[169] for two-electron capture processes are shown in *Figure 12.30*. Over a range of one and a half orders of magnitude, the cross-section[172] is proportional to $\exp(-1/v)$. However, the lower energy rate of rise is less rapid. The metastable He^- ion is now known to

* The symmetrical resonance two-electron capture process is discussed in Section 12.1.

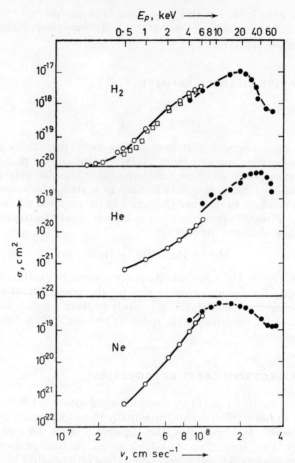

Fig. 12.30. Cross-sections for two-electron capture by protons in H_2, He *and* Ne *(after Kozlov and Bondar); exponential threshold is demonstrated approximately*

be formed from He^+ in two successive one-electron processes rather than a single two-electron process. This may also be true of other metastable negative ions.

12.13 NEGATIVE ION CHARGE TRANSFER

The principles of the charge transfer process apply equally well to negative and to positive ions. For negative ion charge transfer, the electron moves in the field of two atoms rather than that of two ions, and in this weaker field it moves more easily, so negative ion charge transfer cross-sections are in general larger than those of positive ions. For positive ions, the ionization energy (or mean ionization energy) plays a large part in determining the cross-section magnitude; for negative ions the electron affinity is the

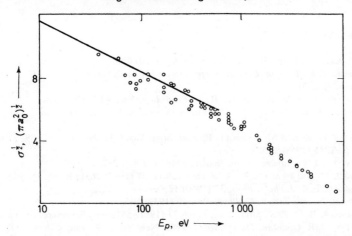

Fig. 12.31. H$^-$–H *charge transfer cross-sections: solid line represents perturbed stationary state calculations[174], and open circles represent experimental data[175]*

corresponding parameter, and since this is smaller than the ionization energy, the cross-section is larger.

In *Figure 12.31* the symmetrical resonance H$^-$–H data are compared with theory[173, 174]. Non-resonant processes, apparently behaving as expected, are illustrated in *Figure 12.32*. Alkali metal negative ion charge transfer processes, both resonant and non-resonant, have been measured and compared with theory for the purpose of inferring electron affinities.

Where the neutral atom in collision with the negative ion is itself incapable of forming a stable negative ion, then charge transfer is impossible and only collisional detachment, or a similar process, can take place. A fuller discussion is given in Chapter 8.

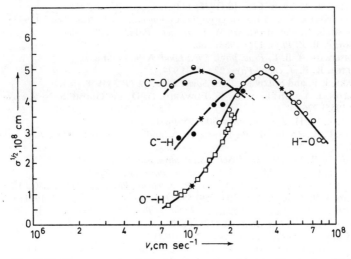

Fig. 12.32. Non-resonant negative ion charge transfer data (due to Snow[176])

REFERENCES

1. KALLMANN, H. and ROSEN, B. *Z phys.* 64 (1930) 806.
2. DEMKOV, U. H. *Ann. Leningr. Univ.* 146 (1952) 74; SENA, L. A. *J. exp. theor. Phys.* 9 (1939) 1320.
3. MASSEY, H. S. W. and SMITH, R. A. *Proc. Roy. Soc.* A142 (1933) 142.
4. BATES, D. R., MASSEY, H. S. W. and STEWART, A. L. *Proc. Roy. Soc.* A216 (1953) 437.
5. BATES, D. R. and McCARROLL, R. *Phil. Mag. Suppl.* 11, No. 4 (1962) 39; *Proc. Roy. Soc.* A245 (1958) 175.
6. HASTED, J. B. *Advanc. atomic molec. Phys.* 4 (1968) 237.
7. RAPP, D. and FRANCIS, W. E. *J. chem. Phys.* 37 (1962) 2631; RAPP, D. and ORTEN-BURGER, I. B. *J. chem. Phys.* 33 (1960) 1230.
8. ROSEN, N. and ZENER, C. *Phys. Rev.* 40 (1932) 502.
9. SKINNER, B. G. *Proc. phys. Soc., Lond.* 77 (1961) 551; see also references 11 and 15.
10. BATES, D. R. *Quantum Theory, vol.* 1, 1961. New York; Academic Press; *Atomic and Molecular Processes,* 1963. New York; Academic Press.
11. ROSEN, N. and ZENER, C. *Phys. Rev.* 40 (1932) 402.
12. FIRSOV, O. B. *J. exp. theor. Phys.* 21 (1951) 1001.
13. KEENE, P. J. *Phil. Mag.* 40 (1949) 369.
14. HASTED, J. B. *Proc. Roy. Soc.* A205 (1951) 421; A212 (1952) 235; A222 (1954) 74; HASTED, J. B. and STEDEFORD, J. B. H. *Proc. Roy. Soc.* A277 (1955) 466; HASTED, J. B. and SMITH, R. A. *Proc. Roy. Soc.* A235 (1956) 349; GILBODY, H. B. and HASTED, J. B. *Proc. Roy. Soc.* A238 (1956) 334; HASTED, J. B. and CHONG, A. Y. J. *Proc. phys. Soc., Lond.* 80 (1962) 893.
15. FEDORENKO, N. V., AFROSIMOV, V. V. and KAMINKER, D. M. *J. tech. Phys., Moscow* 26 (1957) 1861.
16. STIER, R. M. and BARNETT, C. F. *Phys. Rev.* 109 (1958) 385; 103 (1956) 896; MON-TAGUE, J. H. *Phys. Rev.* 81 (1951) 1026; RIBE, F. *Phys. Rev.* 83 (1951) 217; WHITTIER, A. C. *Canad. J. Phys.* 32 (1954) 275.
17. SMITH, R. A. *Proc. Camb. phil. Soc.* 30 (1934) 21.
18. WOLF, F. *Ann. Phys., Lpz.* 23 (1936) 185, 627; 25 (1936) 527, 737; 27 (1937) 543; 28 (1937) 361; 29 (1938) 33; 30 (1938) 313; 34 (1939) 341.
19. ROSTAGNI, A. *Nuovo Cim.* 12 (1935) 134.
20. DALLAPORTA, N. and BONFIGLIONI, G. *Comment. pontif. Acad. Sci.* 7 (1943) 141.
21. GHOSH, S. N. and SHERIDAN, W. F. *J. chem. Phys.* 26 (1957) 480.
22. ZIEGLER, B. *Z. Phys.* 136 (1953) 108.
23. MOISEIWITSCH, B. L. *Proc. phys. Soc., Lond.* A69 (1956) 653.
24. POTTER, R. F. *J. chem. Phys.* 22 (1954) 974.
25. FLAKS, I. P. and SOLOV'EV, E. S. *Soviet Phys. JETP* 3 (1958) 564.
26. DILLON, J. A., SHERIDAN, W. F., EDWARDS, H. D. and GHOSH, S. N. *J. chem. Phys.* 23 (1955) 776.
27. CRAMER, W. H. *J. chem. Phys.* 30 (1959) 641.
28. JONES, P. R., ZIEMBA, F. P., MOSES, H. A. and EVERHART, E. *Phys. Rev.* 113 (1959) 182.
29. KUSHNIR, R. M. *Bull. Acad. Sci. USSR, phys. Ser.* 23 (1959) 995.
30. GURNEE, E. F. and MAGEE, J. L. *J. chem. Phys.* 26 (1957) 1237.
31. KARMOHAPATRO, S. B. and DAS, T. P. *J. chem. Phys.* 29 (1958) 240; KARMOHA-PATRO, S. B. *J. chem. Phys.* 30 (1959) 538; *Proc. phys. Soc., Lond.* 77 (1961) 416.
32. IOVITSU, I. P. and IONESCU-PALLAS, N. *Soviet Phys. JETP* 4 (1960) 781.
33. HOLSTEIN, T. *J. phys. chem.* 56 (1952) 832.
34. McCARROLL, R. *Proc. Roy. Soc.* A246 (1961) 547.
35. DALGARNO, A. and YADAV, H. N. *Proc. phys. Soc., Lond.* A66 (1953) 173.
36. FERGUSON, A. F. *Proc. Roy. Soc.* A246 (1961) 540.

37. COLEMAN, J. P. and McDOWELL, M. R. C. *Proc. phys. Soc., Lond.* 85 (1965) 1097.
38. BOYD, T. J. M. and DALGARNO, A. *Proc. phys. Soc., Lond.* 72 (1958) 694.
39. DALGARNO, A. and McCARROLL, R. *Proc. phys. Soc., Lond.* A69 (1956) 615.
40. MOISEIWITSCH, B. L. *Proc. phys. Soc., Lond.* A69 (1956) 653.
41. FERGUSON, A. F. and MOISEIWITSCH, B. L. *Proc. phys. Soc., Lond.* 74 (1959) 457.
42. DICKINSON, A. S. *Proc. phys. Soc., Lond.* 31 (1968) 395.
43. FITE, W. L., STEBBINGS, R. F., HUMMER, D. G. and BRACKMANN, R. T. *Phys. Rev.* 119 (1960) 663.
44. BATES, D. R. and MAPLETON, R. A. *Proc. phys. Soc., Lond.* 87 (1966) 657; MAPLETON, R. A. *Proc. phys. Soc., Lond.* 87 (1966) 219.
45. FETISOV, I. K. and FIRSOV, O. B. *J. exp. theor. Phys.* 21 (1959–60), 67, 95.
46. ISLAM, M., HASTED, J. B., GILBODY, H. B. and IRELAND, J. V. *Proc. phys. Soc., Lond.* 79 (1962) 1118.
47. JANEV, R. K. *Abstr. 5th Int. Conf. Phys. electron. atom. Collisions,* p. 82, 1967. Leningrad; Akademii Nauk; KAMAROV, I. V. and JANEV, R. K. *Soviet Phys. JETP* 24 (1967) 1159; JANEV, R. K. and PESIC, S. S. *Proc. phys. Soc., Lond.* 92 (1967) 94.
48. ALLISON, S. K. and GARCIA-MUNOZ, M. *Atomic and Molecular Processes,* Chap. 19, Ed. D. R. Bates, 1962. New York; Academic Press.
49. STEBBINGS, R. F. and TURNER, B. R. Gulf General Atomic, San Diego, Report GA 2768, 1967.
50. AMME, R. C. and UTTERBACK, N. G. *Proc. 3rd Int. Conf. Phys. electron. atom. Collisions,* p. 847, 1963. Amsterdam; North Holland Publishing Company.
51. WOLF, F. A. and TURNER, B. R. Gulf General Atomic Report 7919, 1967.
52. BULLIS, R. H. *Abstr. 4th Int. Conf. Phys. electron. atom. Collisions,* p. 263, 1965. Hastings-on-Hudson, New York; Science Bookcrafters.
53. NICHOLS, B. J. and WITTEBORN, F. C. National Aeronautics and Space Administration, TN D3265, 1966.
54. MAHADEVAN, P. and MAGNUSON, G. D. *Proc. 5th Int. Conf. Phys. electron. atom. Collisions.* p. 405, 1967. Leningrad; Akademii Nauk.
55. MASSEY, H. S. W. *Rep. Progr. Phys.* 12 (1949) 248.
56. HASTED, J. B. *Proc. Roy. Soc.* A205 (1951) 421; A212 (1952) 235.
57. FOGEL, I. M., KOZLOV, V. F., KALYMKOV, A. A. and MURATOV, I. M. *J. exp. theor. Phys.* 36 (1959) 929.
58. HASTED, J. B. *Advanc. Electronics Electron Phys.* 13 (1960) 1.
59. DRUKHAREV, G. F. *Soviet Phys. JETP* 10 (1960) 603.
60. HASTED, J. B. and LEE, A. R. *Proc. phys. Soc., Lond.* 79 (1962) 702.
61. LAYTON, J. K. *Proc. 5th Int. Conf. Phys. electron. atom. Collisions,* p. 412, 1967. Leningrad; Akademii Nauk.
62. HASTED, J. B. *J. appl. Phys.* 30 (1959) 25.
63. LEE, A. R. and HASTED, J. B. *Proc. phys. Soc.* 85 (1965) 673.
64. SKINNER, B. J. *Proc. phys. Soc, Lond.* 77 (1961) 551.
65. BATES, D. R. *Disc. Faraday soc.* 33 (1962) 7.
66. ALAM, G. D., BOHME, D. K., HASTED, J. B. and ONG, P. P. *Proc. 5th Int. Conf Phys. electron. atom. Collisions.* p. 403, 1967. Leningrad; Akademii Nauk.
67. FITE, W. L., STEBBINGS, R. F., HUMMER, D. G. and BRACKMANN, R. T. *Phys. Rev.* 119 (1960) 663; Gulf General Atomic, San Diego, Report 2783, 1962.
68. GILBODY, H. B., HASTED, J. B., IRELAND, J. V., LEE, A. R., THOMAS, E. W. and WHITEMAN, A. S. *Proc. Roy. Soc.* A274 (1963) 40.
69. SWEETMAN, D. R. *Phys. Rev. Lett.* 3 (1959) 425; *Proc. Roy. Soc.* A256 (1960) 416; unpublished data.
70. GOLDMANN, F. *Ann. Phys., Lpz.* (5), 10 (1931) 460.
71. FEDORENKO, N. V. and AFROSIMOV, V. V. *J. tech. Phys., Moscow* 26 (1957) 1872; *J. exp. theor. Phys.* 34 (1958) 1398.
72. BROWNING, R. and GILBODY, H. B. *Proc. phys. Soc., Lond.* 1 (1968) 1149.
73. HASTED, J. B. and CHONG, A. Y. J. *Proc. phys. Soc., Lond.* 80 (1962) 893.
74. HASTED, J. B. and SMITH, R. A. *Proc. Roy. Soc.* A235 (1956) 349.

75. FOGEL, I. M., MITIN, R. V. and KOVAL, A. G. *J. exp. theor. Phys.* 31 (1956) 397; FOGEL, I. M., ANKUDINOV, V. A. and SLABOSPITSKII, R. E. *J. exp. theor. Phys.* 32 (1957) 453; FOGEL, I. M. and MITIN, R. V. *Soviet Phys. JETP* 3 (1965) 334.
76. MELTON, C. E. and WELLS, G. F. *J. chem. Phys.* 27 (1957) 1132.
77. McGOWAN, W. and KERWIN, L. *Canad. J. Phys.* 38 (1962) 642.
78. WOBSCHALL, D., FLUEGGE, R. and GRAHAM, J. R. *Proc. 19th Gaseous Electronics Conf.* American Physical Society, 1966.
79. KANEKO, Y., MEGILL, L. R. and HASTED, J. B. *J. chem. Phys.* 45 (1966) 3741.
80. WANNIER, G. H. *Phys. Rev.* 83 (1951) 281; DALGARNO, A., McDOWELL, M. R. C. and WILLIAMS, A. *Phil. Trans.* A250 (1958) 411.
81. BATES, D. R. and LYNN, N. *Proc. phys. Soc., Lond.* A253 (1959) 141.
82. DEMKOV, N. YU. *Proc. 3rd Int. Conf Phys. electron. atom. Collisions*, p. 831, 1963. Amsterdam; North Holland Publishing Company.
83. EDMONDS, P. H. and HASTED, J. B. *Proc. phys. Soc., Lond.* 84 (1964) 99.
84. GUSTAFSSON, E. and LINDHOLM, E. *Ark. Fysik* 18 (1960) 219; LINDHOLM, E. and KOCH, H. *Ark. Fysik* 19 (1961) 11; LINDHOLM, E. *Ark. Fysik* 8 (1954) 257, 433; *Proc. phys. Soc., Lond.* A66 (1953) 1068.
85. GIESE, C. F. and MAIER, W. B. *J. chem. Phys.* 39 (1963) 197.
86. BOHME, D. K., HASTED, J. B. and ONG, P. P. *Proc. phys. Soc., Lond.* 81 (1968) 879.
87. HASTED, J. B. *Proc. Roy. Soc.* A222 (1954) 74.
88. AMME, R. C. and UTTERBACK, N. G. *Proc. 3rd Int. Conf. Phys. electron. atom. Collisions.* p. 847, 1963. Amsterdam; North Holland Publishing Company.
89. FOGEL, YA. M., KOZLOV, V. F., KALYMKOV, A. A. and MURATOV, V. I. *J. exp. theor. Phys.* 36 (1959) 929.
90. STEBBINGS, R. F., TURNER, B. R. and RUTHERFORD, J. A. *J. geophys. Res.* 71 (1966) 771.
91. BOHME, D. K., NAKSHBANDI, M. M., ONG, P. P. and HASTED, J. B. *Proc. 7th Int. Conf. Ioniz. Phenom. Gases*, p. 16, 1967. Amsterdam; North Holland Publishing Company.
92. BAEDE, A. P. M., MOUTINHO, A. M. C., DE VRIES, A. E. and LOS, J. *Chem. phys. Lett.* 3 (1969) 530.
93. HUSSAIN, M. and KERWIN, L. *Abstr. 4th Int. Conf. Phys. electron. atom. Collisions*, p. 296, 1965. Hastings-on-Hudson, New York; Science Bookcrafters.
94. STEBBINGS, F. R. *Advanc. atomic molec. Phys.* 4 (1968) 299.
95. EDMONDS, P. H. and HASTED, J. B. *Proc. phys. Soc., Lond.* 84 (1964) 99.
96. CUTHBERT, J., FARREN, J., PRAHALLADA RAO, S. B. and PREECE, E. R. *Proc. phys. Soc., Lond.* 88 (1966) 91.
97. SCOTT, J. T. and HASTED, J. B. *Spectrometry Symp.* p. 1, Institute of Petroleum and American Society for Testing and Materials, 1964.
98. STEDEFORD, J. B. H. and HASTED, J. B. *Proc. Roy. Soc.* A227 (1955) 466.
99. JAECKS, D., VAN ZYL, B. and GEBALLE, R. *Phys. Rev.* A137 (1965) 230.
100. ANKUDINOV, V. A., BOBASHEV, S. V. and ANDREEV, E. P. *J. exp. theor. Phys.* 48 (1965) 40, 1.
101. LOVELL, S. E and McELROY, M. B. *Proc. Roy. Soc.* A283 (1965) 100.
102. POLUEKTEV, I. A. and PRESNYAKOV, L. P. *Abstr. 5th Int. Conf. Phys. electron. atom. Collisions*, p. 71, 1967. Leningrad; Akademii Nauk.
103. WILETS, L. and GALLAHER, D. F. *Phys. Rev.* 147 (1966) 13.
104. ANKUDINOV, V. A., BOBAHSEV, S. V. and ANDREEV, E. P. *Abstr 5th Int. Conf. Phys. electron. atom. Collisions*, p. 304, 1967. Leningrad; Akademii Nauk.
105. LIPELES, M., NOVICK, R. and TOLK, N. *Phys. Rev. Lett.* 15 (1965) 815; LORENTS, D. C, ABERTH, W. and HESTERMAN, V. W. *Phys. Rev. Lett.* 17 (1966) 849.
106. DWORETSKY, S. H., NOVICK, R. and TOLK, N. *Proc. 6th Int. Conf. Phys. electron. atom. Collisions*, p. 294, 1969. Cambridge, Mass.; Massachusetts Institute of Technology Press.
107. ROSENTHAL, H. *Proc. 6th. Int. Conf. Phys. electron. atom. Collisions*, p. 302, 1969. Cambridge, Mass.; Massachusetts Institute of Technology Press.

108. BATES, D. R. and McCARROLL, R. *Advanc. Phys.* 11 (1962) 39.
109. ZIEMBA, F. P., LOCKWOOD, G. J., MORGAN, G. H. and EVERHART, E. *Phys. Rev.* 118 (1960) 1522; EVERHART, E., HELBIG, H. F. and LOCKWOOD, G. J. *Proc. 3rd Int. Conf. Phys. electron. atom. Collisions*, p. 865, 1963. Amsterdam; North Holland Publishing Company.
110. MASSEY, H. S. W. and SMITH, R. A. *Proc. Roy. Soc.* A142 (1933) 142; SMITH, F. J. *Proc. phys. Soc., Lond.* 84 (1964) 889.
111. BATES, D. R. and WILLIAMS, D. A. *Proc. phys. Soc., Lond.* 83 (1964) 425.
112. KEEVES, W. C., LOCKWOOD, G. J., HELBIG, H. F. and EVERHART, E. *Phys. Rev.* 166 (1968) 68.
113. LORENTS, D. C. and ABERTH, W. *Proc. 4th Int. Conf. Phys. electron. atom. Collisions*, p. 269, 1965. Hastings-on-Hudson, New York; Science Bookcrafters; MARCHI, R. P. and SMITH, F. T. *Proc. 4th Int. Conf. Phys. electron. atom. Collisions*, p. 269, 1965. Hastings-on-Hudson, New York; Science Bookcrafters; SMITH, F. T., MARCHI, R. P., ABERTH, W., LORENTS, D. C. and HEINZ, O. *Phys. Rev.* 161 (1967) 31.
114. PEREL, J., VERNON, R. H. and DALEY, H. L. *Phys. Rev.* 138 (1965) A837, 336; SMITH, F. J. *Phys. Lett.* 20 (1966) 271.
115. SMITH, F. J. *Proc. 5th Int. Conf. Phys. electron. atom. Collisions*, p. 1068, 1969. Cambridge, Mass.; Massachusetts Institute of Technology Press.
116. OPPENHEIMER, J. R. *Phys. Rev.* 32 (1928) 361; 31 (1928) 349.
117. BRINKMANN, H. C. and KRAMERS, H. A. *Proc. Acad. Sci. Amst.* 33 (1930) 973.
118. BATES, D. R. and DALGARNO, A. *Proc. phys. Soc., Lond.* A66 (1953) 972.
119. JACKSON, J. D. and SCHIFF, H. *Phys. Rev.* 89 (1953) 359; JACKSON, J. D. *Canad. J. Phys.* 32 (1954) 60; SCHIFF, H. *Canad. J. Phys.* 32 (1954) 393.
120. MAPLETON, R. A. *Phys. Rev.* 122 (1961) 528.
121. BRANSDEN, B. H., DALGARNO, A., JOHN, T. L. and SEATON, M. J. *Proc. phys. Soc., Lond.* 71 (1958) 877.
122. BRANSDEN, B. H., DALGARNO, A. and KING, N. M. *Proc. phys. Soc., Lond.* A67 (1954) 1075.
123. STIER, P. M. and BARNETT, C. F. *Phys. Rev.* 109 (1958) 385; 103 (1956) 896.
124. RUTHERFORD, J. J. *Phil. Mag* (6) 47 (1924) 277.
125. BERKNER, K. H., KAPLAN, S. N., PYLE, R. V. and WELSH, L. M. *Proc. 5th Int. Conf. Phys. electron. atom. Collisions*, p. 418, 1967. Leningrad; Akademii Nauk.
126. HOOPER, J. W., McDANIEL, E. W., MARTIN, D. W. and HARMER, D. S. *Phys. Rev.* 121 (1961) 1123.
127. GERJUOY, E. *Rev. mod. Phys.* 33 (1961) 544.
128. BATES, D. R. and McCARROLL, R. *Advanc. Phys.* 11 (1962) 1.
129. SAHA, M. N. and BASU, D. *Indian J. Phys.* 19 (1945) 121.
130. TAKAYANAGI, K. *Sci. Rep. Saitama Univ.* 1 (1952) 9.
131. McCARROLL, R. *Proc. Roy. Soc.* A264 (1961) 547.
132. TUAN, T. F. and GERJUOY, E. *Phys. Rev.* 117 (1960) 756.
133. WITTKOWER, A. B., RYDING, G. and GILBODY, H. B. *Proc. phys. Soc., Lond.* 89 (1966) 541.
134. CHESHIRE, I. M. *Proc. 3rd Int. Conf. Phys. electron. atom. Collisions*, p. 757, 1963. Amsterdam; North Holland Publishing Company; *Proc. Heavy Particle Collisions Conf.* p. 15, The Institute of Physics and The Physical Society, 1968.
135. ABRINES, R. and PERCIVAL, I. C. *Proc. phys. Soc., Lond.* 88 (1966) 873.
136. THOMAS, L. H. *Proc. Roy. Soc.* A114 (1927) 561.
137. SAHA, M. N. and BASU, D. *Indian J. Phys.* 19 (1945) 121.
138. OPPENHEIMER, J. R. *Phys. Rev.* 31 (1928) 349.
139. BOHR, N. *Dan. Matt. Fys. Medd.* 18, No. 8 (1948).
140. BELL, G. I. *Phys. Rev.* 90 (1953) 548.
141. BOHR, N. and LINDHARD, J. *Dan. Matt. Fys. Medd.* 28, No. 7 (1957).
142. GLUCKSTERN, R. L. *Phys. Rev.* 98 (1955) 1817.
143. BRUNINGS, G., KNIPP, J. K. and TELLER, E. *Phys. Rev.* 60 (1941) 657.
144. RUSSEK, A. and THOMAS, M. T. *Phys. Rev.* 109 (1958) 2015.

145. HASTED, J. B. National Academy of Science, National Research Council Report 29, 1960.
146. AFROSIMOV, V. V., MANAEV, YU. A., PANOV, M. N. and FEDORENKO, N. V. *Abstr. 5th Int. Conf. Phys. electron. atom. Collisions*, p. 210, 1967. Leningrad; Akademii Nauk.
147. STUECKELBERG, E. C. G. *Helv. phys. acta*, 5 (1932) 370.
148. LANDAU, L. *Phys. Z. Sowjet*, 2 (1932) 46.
149. ZENER, C. *Proc. Roy. Soc.* A137 (1932) 696.
150. BATES, D. R. and MOISEIWITSCH, B. L. *Proc. phys. Soc., Lond.* A67 (1954) 805.
151. DALGARNO, A. *Proc. phys. Soc., Lond.* A67 (1954) 1010.
152. BATES, D. R. and LEWIS, J. T. *Proc. phys. Soc., Lond.* A68 (1955) 173.
153. BATES, D. R. and BOYD, T. J. M. *Proc. phys. Soc., Lond.* A69 (1956) 910; *Proc. phys. Soc., Lond.* A70 (1957) 809.
154. BATES, D. R. *Proc. Roy. Soc.* A258 (1960) 22.
155. MORDVINOV, YU. P. and FIRSOV, O. B. *J. exp. theor. Phys.* 39 (1960) 437.
156. MOISEIWITSCH, B. L. *J. atmos. terr. Phys.* 23, Spec. Suppl. vol. 2 (1955) 23.
157. MAGEE, J. L. *J. chem. Phys.* 8 (1940) 687.
158. LIPELES, N., NOVICK, R. and TOLK, N. *Phys. Rev. Lett.* 15 (1965) 815; DWORETSKY, S., NOVICK, R., SMITH, W. W. and TOLK, N. *Proc. 5th Int. Conf. Phys. electron. atom. Collisions*, p. 280, 1967. Leningrad; Academii Nauk.
159. ZWALLY, H. J. and KOOPMAN, D. W. *Proc. 6th Int. Conf. Phys. electron. atom. Collisions*, p. 1025, 1969. Cambridge, Mass.; Massachusetts Institute of Technology Press.
160. ALAM, G. D., BOHME, D. K., HASTED, J. B. and ONG, P. P. *Proc. 5th Int. Conf. Phys. electron. atom. Collisions*, p. 403, 1967. Leningrad; Akademii Nauk; HASTED, J. B., IQBAL, S. M. and YOUSAF, M. M. *J. Phys.* B4 (1971) 1.
161. LEBEDEFF, S. A. New York University, Courant Institute of Mathematics and Science Report, 1966.
162. COULSON, C. A. and ZALEWSKI, K. *Proc. Roy. Soc.* A268 (1965) 437.
163. BYKHOVSKY, V. K. and NIKITIN, E. E. *Soviet Phys. JETP* 20 (1965) 500.
164. MAGEE, J. L. *Disc. Faraday Soc.* 12 (1952) 33.
165. MAGEE, J. L. *Mass Spectrometry Conf.* American Society for Testing and Materials, 1958.
166. DALY, N. R. and POWELL, R. E. *Proc. phys. Soc., Lond.* 89 (1966) 281.
167. BAKER, F. A. and HASTED, J. B. *Phil. Trans.* A261 (1966) 33.
168. ALLISON, D. C. S. and DALGARNO, A. *Proc. phys. Soc., Lond.* 85 (1965) 845.
169. KOZLOV, V. F. and BONDAR, S. A. *J. exp. theor. Phys.* 50 (1966) 297.
170. BURKE, P. G. and SCHEY, H. M. *Phys. Rev.* 126 (1962) 147; SCHULZ, G. J. *Proc. 3rd Int. Conf. Phys. electron. atom. Collisions*, p. 124, 1963. Amsterdam; North Holland Publishing Company.
171. JORGENSEN, T. *Phys. Rev.* 140 (1965) A1481.
172. HASTED, J. B. *J. appl. Phys.* 30 (1959) 25.
173. HUMMER, D., FITE, W. L. and STEBBINGS, R. *Phys. Rev.* 119 (1960) 668.
174. DALGARNO, A. and MCDOWELL, M. R. C. *Proc. phys. Soc., Lond.* A69 (1956) 615.
175. HUMMER, D., FITE, W. L. and STEBBINGS, R. *Phys. Rev.* 119 (1960) 668.
176. SNOW, W. R. Thesis, University of Washington, Seattle, 1966.
177. HASTED, J. B., LEE, A. R. and HUSSAIN, M. *Proc. 3rd Int. Conf. Phys. electron. atom. Collisions* p. 802, 1964. Amsterdam; North Holland Publishing Company.
178. OGURTSOV, G. N. and FLAKS, I. P. *Soviet Phys. JETP* 15 (1962) 502.
179. ZWALLY, H. J. and CABLE, P. G. *Phys. Rev. Lett.* (1971)

Chapter 13

COLLISIONS OF EXCITED SPECIES

13.1 LINE BROADENING

The processes which lead to the broadening of spectral lines emitted in the presence of atoms, molecules, ions and electrons, may logically be considered under the heading of collisions of excited atoms and molecules. The subject is so extensive that only the principal theories will be enumerated here, together with a few important results obtained with their aid. Reviews of the subject are available[1-5].

The 'natural' width of a spectral line in emission or absorption is governed by the radiation damping. For emission from a state of natural lifetime τ, the width in frequency is $\Delta \nu \simeq \tau^{-1}$. The width can be increased by Doppler effect, and also by number-density-dependent effects arising from interaction of the emitting or absorbing atom with neutral atoms, with ions and with electrons.

An atom whose velocity component in the line of sight of the observer is v_0, emits radiation of wavelength λ given approximately by

$$\lambda \simeq \lambda_0 \left(1 - \frac{v_0}{c} \right) \tag{13.1}$$

where λ_0 is the wavelength emitted by a stationary atom. The thermal motion of atoms of mass m at temperature $T°K$ gives rise to a frequency distribution of the radiation in all directions. The observed intensity I varies with frequency ν as follows:

$$I = \exp \left[-\frac{m\lambda_0^2(\nu - \nu_0)^2}{kT} \right] \tag{13.2}$$

The whole width $(\Delta \nu)_{1/2}$ of the line at the intensity which is one half of the maximum intensity (fullwidth at half maximum or FWHM), is then given by

$$(\Delta \nu)_{1/2} = \frac{2}{\lambda_0} \left(\frac{kT}{m} \ln 2 \right)^{1/2} \tag{13.3}$$

The Doppler broadening is temperature-dependent rather than pressure-dependent; lines may be sharpened by cooling the gas with liquid nitrogen. The absorption of a Doppler-broadened line by a gas at a different temperature from that of the emitting atoms is an important problem in the diagnostics of excited state densities[6].

Line broadening due to interaction with atomic particles may be considered as arising from the following cause. Elastic scattering of the radiating atoms both before, during and after the radiation process is to be expected; but the interaction potentials of the pre-radiation and post-radiation atoms with their neighbours are different. The radiative process takes place in a time short compared with the scattering process, so, as with the Franck–Condon principle in diatomic molecules, the radiative transition is nearly 'vertical'. Since the two interaction potentials are not equidistant over their path, a distribution of vertical transition energies will be found; this implies a distribution of radiated frequencies.

The first definitive attempt to treat pressure-dependent line broadening was made by Lorentz[7]. It was be shown classically[8] that the half-width of the line was the same in emission as absorption and was directly proportional to atom density n_0:

$$(\Delta v)_{1/2} = 4\pi^2 \varrho^2 \bar{v} n_0 \qquad (13.4)$$

where $\bar{v}$ is the mean velocity of the atoms, and ϱ is the 'optical collision diameter', which is related to the form of the interaction potentials. It will be seen from Table 13.1, which gives typical values of ϱ, that they exceed gas-kinetic collision diameters, presumably because excited states are involved. Direct proportionality of $(\Delta v)_{1/2}$ to gas density is observed in experiment at sufficiently small densities.

Table 13.1. TYPICAL VALUES OF OPTICAL
COLLISION DIAMETER

Gas atom	Line	ϱ (Å)
Ar	Na 5896 Å	11·2
H_2	Na 5896 Å	7·16
N_2	Cs 3876 Å	18·4
Ar	Cs 3876 Å	26·2

Comparison of the half-widths with interaction potentials is much easier for molecules than for atoms, since the forces are more exactly known[9]. Classical Fourier integral theory[10] has made it possible, by considering the lack of coherence between radiation before and after the collision, to relate $(\Delta v)_{1/2}$ to the force constants C for inverse sixth power interaction ($p = 6$). If ΔC is the difference between the force constants before and after radiation,

$$(\Delta v)_{1/2} = 25 \cdot 8\pi \left(\frac{kT}{2m}\right)^{3/10} \langle |\Delta C|^{2/5}\rangle n_0 \qquad (13.5)$$

The brackets $\langle \ \rangle$ imply an average over statistical weight probabilities.

Associated with the broadening is a shift of the frequency of maximum intensity of the line, towards either the red or the violet. This is related to the interaction potentials[11]:

$$\frac{(\Delta v)_{1/2}}{(\Delta v)_m} = 2 \cot \left(\frac{\pi}{p-1}\right) \tag{13.6}$$

$$(\Delta v)_m = v_m - v_0 \tag{13.7}$$

where v_m is the frequency of maximum intensity of the shifted and broadened line.

It has been shown[8] that classical Fourier integral theory does not yield the observed pressure effects. Quantum-mechanical Fourier integral theory has been developed to rectify this, but at pressures much larger than those considered in these theories (several atmospheres pressure and consequently many collisions during a transition), it is necessary to adopt a statistical approach. Classical statistical theory[12] leads to a red shift:

$$(\Delta v)_m = -2\pi^4 \left(\tfrac{2}{3}\right)^3 |\Delta C| n_0^2 \tag{13.8}$$

and

$$(\Delta v)_{1/2} = 1 \cdot 644\pi^4 |\Delta C| n_0^2 \tag{13.9}$$

whilst for large Δv,

$$I |v - v_m| \propto |v - v_m|^{-3/2} \tag{13.10}$$

It is recognized that classical statistical theory is not entirely adequate at high pressures, but a quantum-mechanical statistical theory[13] leads to similar results. The most difficult region to treat is that of intermediate pressures, that is, the region where the atom is likely to undergo a small number of collisions during the emission.

Resonance transitions, in which there exists an inverse third power interaction potential between the excited atom and the similar atoms which are its neighbours, undergo a special type of 'resonance broadening'[14]; it is found that

$$(\Delta v)_{1/2} = \frac{e^2 f_{on} n_0}{m v_0} \tag{13.11}$$

where f_{on} is the oscillator strength for the transition and m is the magnetic quantum number of the state. As is instanced by the difference between He 2^1P and He 2^3P emissions, resonance lines terminating in the ground state are much broader than others.

Since the early days of study of spectral lines in emission from excited atoms in gas discharges, attention has been paid to the effects of neighbouring electric charges[15]. Holtsmark calculated the Stark effect of these charges upon the lines. For the effects of positive ions:

$$(\Delta v)_{1/2} = 3 \cdot 25 \, a n_+^{2/3} \tag{13.12}$$

in which a is the constant of proportionality appropriate to the linear Stark splitting in a macroscopic field X:

$$2(\Delta v)_{Stark} = aX \tag{13.13}$$

There is a similar broadening to be expected from molecules which have permanent dipole moments μ, or quadrupole moments Q:

$$(\Delta v)_{1/2,\ \text{dipole}} = 4\cdot54a\mu n_\mu \qquad\qquad (13.14)$$

or

$$(\Delta v)_{1/2,\ \text{quadrupole}} = 5\cdot53aQn_Q^{4/3} \qquad\qquad (13.15)$$

In the simple Holtsmark theory the finite radii of the electric charges, dipoles and quadrupoles were not considered. It has remained for more recent investigators to fit the Holtsmark and Lorentz theories into one coherent theory[16, 17]. The Holtsmark broadening is too small to explain observed broad lines emitted from plasmas. The importance of line broadening in plasmas arises from the need to diagnose electron density from the observed width of emitted lines; this is possible in the range $n_e = 10^{15}$–10^{17} cm^{-3}.

A general theory was proposed by Baranger[18], and certain differences which make the calculations very much easier but also less complete were worked out by Griem[19]. A simple approximate formula[3] derived from the general theory is

$$(\Delta v)_{1/2} = 3\cdot86\times10^{-5}n_e T_e^{-1/2}Z^{4/5}(\alpha_i-\alpha_j)^{2/5} \qquad\qquad (13.16)$$

where an atom of nuclear charge Z is emitting from level i to level j, which possess polarizabilities α_i and α_j. The Baranger treatment is unique in including the important effects of inelastic scattering during the radiative process. It enables broadening to be considered without the adiabatic assumption being made. The transitions between the Stark sub-levels, which are induced by electron impact, are taken into account.

The broadening of lines emitted from gas lasers is of importance for the reason that, whereas ordinary pressure-broadening is often obscured by Doppler broadening, in the laser line this is not so, and a chance to make comparison of pressure-broadening with theory appears. The normal shape of a laser line is a broad bell shape, displaying a downward dip in the top (the 'Lamb dip')[20]. The width of the dip in the lowest mode of operation is typically ~ 10 Hz; the dip is unaffected by temperature, but pressure-broadening fills in the gap unsymmetrically. From detailed analysis of its shape are obtained the inverse sixth power interaction constants between excited states, which are two orders of magnitude larger than those between ground states of atoms.

The broadening of lines emitted by states of very high principal quantum number ($18 \leqslant n \leqslant 50$) is of special character[21], owing to the very large atomic radii which enable the emitting atom to contain more than one ground state atom within its outermost electron orbital, even at moderate pressures. In this situation the line width is dominated by the elastic electron scattering cross-section σ at the energy of the outermost bound electron ($< 0\cdot04$ eV), and is given by the expression $(2\hbar/m_e)(\pi\sigma)^{1/2}n_0$, plus a small polarization term. At present this represents the only method of studying electron scattering in the sub-thermal energy region. The cross-section for caesium rises to a value $\sim 1\cdot4\times10^{-13}$ cm^2 as the electron energy decreases; the very large value may indicate that the negative ion possesses an excited state close to the continuum.

13.2 COLLISION-INDUCED RADIATION

The changes induced in radiation processes by collisions are responsible not only for the broadening of spectral lines emitted from gases and plasmas but for certain special types of emission. The Lorentz broadening of a spectral line may be considered[22] as the quantum analogue of a change in the frequency of vibration of an oscillator atom caused by the approach of the perturbing atom, so that the phases of the unperturbed vibration before and after the collision no longer correspond. Although, in general, relatively small frequency changes are involved in collision broadening, there are exceptional cases where broad diffuse maxima appear in emission. These were first observed on the short-wave side of the mercury 2537 Å line by Kuhn and Oldenberg[23] when heavy rare gases were added to the vapour. The presence of two intensity maxima is consistent with the idea of quasi-molecule formation in either of two vibrational states, depending on the direction of the relative momenta at the time of collision; the occurrence of two maxima is among the first pieces of evidence for space quantization upon impact. This phenomenon is known as collision-induced radiation.

If radiation from quasi-molecular states is allowed, then it follows that atomic states from which dipole radiation is not possible may radiate during a collision with another atom, provided that a suitable state of the quasi-molecule can be found. This phenomenon also comes under the heading of collision-induced radiation. Its significance as a process influencing the destruction of He 2^1S metastables in collisions was first proposed by Nikerson[24], in order to explain the presence of a helium band of continuous radiation near 600 Å. The presence of a sharp band edge 385 cm^{-1} on the red side of the 602 Å forbidden line was interpreted as indicating the existence of a maximum in the difference of the $2^1\Sigma_g$ and $1^1\Sigma_g$ He$_2$ potential energy curves. Such a maximum was also predicted by Buckingham and Dalgarno[25].

A quantum theory formulation of collision-induced radiation has been given by Kramers and Ter Haar[26] and applied to the destruction of He 2^1S atoms by Burhop and Marriott[27]. The mean cross-section at thermal energies is given by

$$\bar{\sigma} = \frac{4\pi\mu}{3c^3h^4k^2T_{\text{gas}}} \int_0^\infty r^2[\mu(r)]^2\, I(r)[V_m(r)-V(r)]^3\, dr \qquad (13.17)$$

with

$$I(r) = \int_{v_{r\,\text{min}}}^\infty v_r \exp\left(-\frac{\mu v_r^2}{2kT}\right) \left\{\left[v_r^2 - \frac{2V_m(r)}{\mu}\right]^{1/2}\right.$$

$$\left. -\left[v_r^2 - \frac{2V_m(r)}{\mu} - \frac{v_r^2 b_{\text{max}}^2}{r^2}\right]^{1/2}\right\} dv_r \qquad (13.18)$$

Here $\mu(r)$ is the dipole moment corresponding to the appropriate molecular transition when the nuclear separation is r; b_{max} is the maximum value of impact parameter consistent with the chosen values of relative velocity v_r,

43

and $V(r)$ and $V_m(r)$ are the force fields respectively for two ground state atoms and for metastable and ground state atoms; the reduced mass of the system is μ.

For He 2^1S in collision with ground state helium atoms, the appropriate quasi-molecule states are $2^1\Sigma_u$ and $2^1\Sigma_g$, of which the latter can radiatively pass to $1^1\Sigma_g$. Using the 2^1S–1^1S potental energy curves calculated by Buckingham and Dalgarno[25], a mean cross-section $\sigma = 0\cdot9 \times 10^{-25}$ cm^2 at 300°K was calculated. The very low value is due to the barrier which appears in these potential energy curves, but even if this were absent the mean cross-section would only be 5×10^{-22} cm^2.

Quasi-molecule bands have been reported[28] in caesium mixed with xenon, but on the long wavelength side of the Cs I lines, which contradicts some predictions[12, 14] but supports others[29].

13.3 EXCITATION EXCHANGE (SENSITIZED FLUORESCENCE) AND QUENCHING

Excited atoms in collision with their own species can readily exchange the excitation, allowing the radiation to diffuse through the gas. In collision with a different species, the process can occur frequently when the energy defect is sufficiently small, but for larger energy defect the probability of its occurrence is slight. In collisions of excited atoms with molecules the likelihood of occurrence of a small energy defect is often large, the molecule being left with vibrational–rotational excitation; hence the radiation from excited atoms is readily quenched by molecular impurities. Similar considerations apply to collisions of metastable atoms; deactivation of the metastables by excitation exchange with molecules will often occur.

The excitation exchange process is known as sensitized fluorescence. Excitation exchange between unlike gases is an important process in the optical pumping of two-gas lasers.

Information about these processes is mostly obtained from experiments at thermal energies. The collision process $0'0/00'$, in which an excited atom in collision with a ground state atom falls to the ground state without radiation, whilst raising the second atom to an excited state, was first proposed by Franck[30]. It is not essential that either atom should pass into or out of its ground state, but this is the most common situation.

When the process is exothermic, of energy defect ΔE, internal energy is converted into kinetic energy and distributed amongst the collidants, so their velocities would be

$$\frac{1}{2} m_1 v_1^2 = \frac{\Delta E m_2}{m_1 + m_2} \tag{13.19}$$

and

$$\frac{1}{2} m_2 v_2^2 = \frac{\Delta E m_1}{m_1 + m_2} \tag{13.20}$$

assuming zero ante-collision velocity. The velocities of the radiating atoms give rise to Doppler frequency shifts $\Delta \nu$:

$$\frac{\Delta \nu}{\nu} = \frac{v}{c} \cos \phi \tag{13.21}$$

where ϕ is the angle between the line of observation of the radiation and the velocity vector v. For thermally agitated atoms, an average value of cos ϕ is appropriate. The Doppler shifts are usually too small to measure, but Rasetti[31] first measured the $\Delta\lambda = 0.17$ Å shift of Na D_1 and D_2 when excited by Hg 6^3P; the energy defect is as large as 2.8 eV.

It is not necessary that an excitation exchange collision should be exothermic, provided that sufficient kinetic energy of collision is available to make up the energy defect of an endothermic process; for example, the endothermic excitation of Cd 6^3S_1 by mercury states has been observed at high temperatures.

The first experiments upon excitation exchange (sensitized fluorescence) were conducted by Cario and Franck[32, 33], who irradiated mixed vapours of mercury and thallium with a mercury resonance lamp and measured the emission of the thallium lines. It was shown by Beutler and Josephy[34] that, other things being equal, the excitation exchange process is most probable for the smallest energy defect. The excitation of various sodium levels by Hg 6^3P_1 and Hg 6^3P_0 is shown in *Figure 13.1* in the form of intensity I of

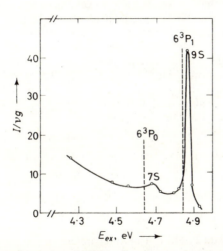

Fig. 13.1. *Instensity of emission, divided by frequency and statistical weight, in arbitrary units, for levels of sodium excited by* Hg *and* 6^3P_1 *and* 6^3P_0, *as a function of energy of the sodium level[34]; energies of the appropriate levels are indicated*

(From Mitchell and Zemansky[35], by courtesy of Cambridge University Press)

sodium emission divided by frequency ν and statistical weight g of the upper sodium state. When this quantity is compared with the energy of the sodium level, in its relation to the mercury levels, the 'accidental resonance' nature of the collision becomes apparent. Measurements of this type, coupled with the similar comparisons of charge transfer data in the 10^2–10^3 eV impact energy region[36] stimulated the eventual formulation of the adiabatic criterion.

For excitation transfer between alkali metal states, Franck formulated the empirical rule $\sigma \propto \Delta E^{-\alpha}$, for energy defect ΔE, and α a constant. This is of much the same form as the theory given by Stueckelberg[37] for the case of non-crossing energy levels of exchange interaction

$$V_{12} = \beta r^{-s} \tag{13.22}$$

The expression he deduced was

$$\sigma = \pi r_0^2 k_s x^{s-1} \exp\left(-x\right) \tag{13.23}$$

where

$$x = \frac{\Delta E r_0}{\hbar v}$$

$$k_s = \frac{8\pi^2}{4^s \Gamma(\tfrac{1}{2}s)^2}$$

$$\Delta E = V_{22}(\infty) - V_{11}(\infty)$$

and r_0 is the nuclear separation at which $V_{12} = \Delta E$. The Stueckelberg expression has recently been improved[38], but has been shown[39] to contain sources of error which cause serious discrepancies between calculation and recent experiments.

There is evidence that, in an excitation exchange collision, of all the possible transfers of energy with essentially identical ΔE, the most likely one will be that in which the total resultant spin of the two-atom system remains unchanged ('Wigner's rule'). This prediction[40] was substantiated in an experiment of Beutler and Eisenschimmel[41], who investigated the relative probabilities of excitation of Hg $^1D_2$ and $^3D_{1, 2, 3}$, all of which have nearly the same energy and lie fairly close to Kr $^3P_2$. The radiation from Hg $^3D_{1, 2, 3}$ (at that time unresolvable) was found to be more strongly sensitized by Kr $^3P_2$ than was that from Hg $^1D_2$. Thus

$$\text{Kr } ^3P_2 + \text{Hg } 6\,^1S_0 \rightarrow \text{Kr } ^1S_0 + \text{Hg } 8\,^3D_{1, 2, 3}$$
$$S = 1 \quad S = 0 \quad\quad S = 0 \quad\quad S = 1$$

goes strongly, whilst

$$\text{Kr } ^3P_2 + \text{Hg } 6\,^1S_0 \rightarrow \text{Kr } ^1S_0 + \text{Hg } 8\,^1D_2$$
$$S = 1 \quad S = 0 \quad\quad S = 0 \quad\quad S = 0$$

goes weakly. This result could not be interpreted on the basis that the state multiplicity of the pseudo-molecule affected the detailed balance since it was not possible to separate the $^3D_1$, $^3D_2$ and $^3D_3$ emission lines.

Some similar experiments due to Maurer and Wolf yielded effective thermal energy excitation transfer cross-sections in helium (listed in Table 13.2), from which it may be seen that the 'accidental resonance effect' (small ΔE, large cross-section) is apparent, although the Wigner rule is obeyed weakly.

The conservation of total spin can only be applied weakly as a selection rule governing exchange collisions. Evidence was first reported by Lees and Skinner[42], from the observation of anomalous apparent electron excitation functions for helium radiation of wavelengths 4472 Å, 3820 Å and 3705 Å, that the cross-sections for processes such as

$$\text{He}(n\,^1P) + \text{He}(1\,^1S) \rightarrow \text{He}(1\,^1S) + \text{He}(n\,^3D)$$

may be larger than gas kinetic.

Table 13.2. THERMAL ENERGY EXCITATION TRANSFER
CROSS-SECTIONS σ

Process in helium	ΔE (eV)	σ (10^{-15} cm²)
$3^1P \rightarrow 3^3P$	0·079	2·1
$3^1P \rightarrow 3^3D$	0·013	11·9
$4^1P \rightarrow 4^1D$	0·006	67
$4^1P \rightarrow 4^3D$	0·006	15
$4^1P \rightarrow 4^1S$	0·068	3
$4^1P \rightarrow 4^3S$	0·148	1·5
$5^1P \rightarrow 5^1D$	0·003	51
$5^1P \rightarrow 5^3D$	0·003	27
$5^1P \rightarrow 5^1S$	0·034	1·5
$5^1P \rightarrow 5^3S$	0·071	0·9

At low pressures the electron excitation functions for He n^3D lines have their own characteristic shape, but as the helium pressure is raised, the functions take on the appearance of allowed transitions such as ^1P. It has been proposed[43] that many ^1P states, including those with large (4–15) principal quantum number n, transfer excitation energy to neighbouring F states, having closely corresponding principal quantum numbers; these states, which possess both singlet and triplet qualities, in turn populate low-level triplet states by radiative transitions.

Dissociative excitation transfer

$$M' + AB \rightarrow M + A' + B \qquad 0'(00)/00'0$$

or

$$M' + AB \rightarrow M + A + B' \qquad 0'(00)/000'$$

has been studied in connection with population inversion[44, 45]. The transition occurs to repulsive states AB', and the cross-section, although difficult to calculate, can be as large as 10^{-15} cm² for an energy defect of 1–2 eV. When the process is non-dissociative, the molecular state AB' will contain rotational and vibrational excitation as well as electronic; in either case, population inversion and light amplication can occur following the excitation transfer collision process[46].

Quenching collisions are those which destroy the resonance radiation from discharges ('resonance lamps') when small quantities of another gas are added[35]. The state in which the second gas finds itself after the collision is not specified. Resonance states are destroyed by excitation exchange collisions of the type

$$X' + YZ \rightarrow X + YZ^v \qquad 0'0/0\,00^v$$

which even at thermal energies compete with spontaneous emission. Resonance radiation can hardly be quenched at all by the addition of atomic gases, since conversion of internal energy to kinetic energy is a very inefficient process in thermal energy collisions between atomic systems.

The quenching Q is defined as the ratio of the intensities I and I_0 of resonance radiation, respectively with and without the foreign gas; from this quantity the collision cross-section is calculated. In the absence of radiation diffusion and collision broadening, the Stern–Volmer formula[47] is derived as follows.

Equating the rates of formation ε and destruction of resonance states: without the foreign gas

$$\varepsilon = \frac{n_*}{\tau}$$

and with the foreign gas

$$\varepsilon = \frac{n_*}{\tau} + Zn_*$$

The emitted radiation energy is written as $Fn_* h\nu/\tau$, so

$$I_0 = Fh\nu\varepsilon \tag{13.24}$$

and

$$I = \frac{Fh\nu\varepsilon}{1 + \tau Z} \tag{13.25}$$

where F is a geometrical factor, and n_* is the density of resonance states. There follows directly the Stern–Volmer formula

$$Q = \frac{1}{1 + \tau Z} \tag{13.26}$$

where τ is the lifetime of the state; Z is the number of collisions per unit time and volume, and for a mixture of two species is related[48] to the cross-section σ as follows:

$$Z = \frac{2n_1 n_2 \sigma}{\pi} \left[2\pi kT \left(\frac{1}{m_1} + \frac{1}{m_2} \right) \right]^{1/2} \tag{13.27}$$

A plot of Q^{-1} against p is known as a Stern–Volmer diagram.

At the time when resonance radiation quenching experiments were first conducted, collisions were described not in terms of a cross-section, but a collision radius which was also denoted by the Greek sigma; if this is written ς, we have $\sigma = \pi\varsigma^2$. The cross-sections given by Mitchell and Zemansky[35] are thus in units of $\pi\text{Å}^2$.

Studies of quenching of radiation from atoms moving with velocities greater than thermal were made with the aid of a discovery by Terenin[49] that the NaI molecule dissociates into excited Na 3^2P atoms when irradiated in the ultra-violet region. It is possible to avoid radiation diffusion entirely, since at the beginning of the experiment there are no ground state sodium atoms. Various workers were able to study the velocity dependence of the quenching cross-section with I_2 and Br_2 molecules. The excess energy from the photodissociation is distributed between the two atoms of the NaI molecule, resulting in sodium atom velocities

$$v_1^2 = \frac{2\Delta E m_2}{m_1(m_1 + m_2)} \tag{13.28}$$

which may be derived from measurements of the Doppler shifts of the sodium emission. A cross-section function deduced from experiments with radiation of different wavelengths is shown in *Figure 13.2*.

Attempts to correlate the energy defect of the quenching processes with cross-sections are only partially successful[35].

Quenching studies have been conducted in flames[50, 51], which offer the advantage that a greater variety of metals and also of reactive gases can be studied. Flames consist of mixtures of gases, but if measurements are made

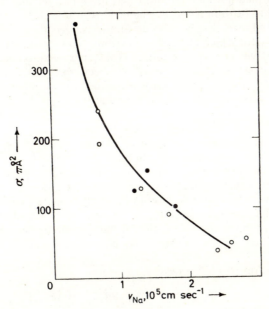

Fig. 13.2. Quenching cross-section for I_2 and Br_2 radiation by NaI as a function of velocity of sodium atoms

(From Mitchell and Zemansky[35], by courtesy of Cambridge University Press)

in a set of isothermal flames differing widely in composition, then pure gas quenching rates can be deduced using the equation

$$\frac{P_f}{\Delta P} = \frac{\alpha W}{4\pi} \frac{A}{A + \sum_X k[X]}$$

(13.29)

When the flame is illuminated with resonance radiation, the absorbed power is ΔP. The fluorescence power P_f is emitted into a known solid angle W. A factor α is introduced to allow for re-absorption of fluorescent power. The term $k[X]$ is the quenching rate for species X, and A is the appropriate Einstein coefficient. Rates have been reported for all alkali metals and thallium in H_2, N_2, O_2, CO, CO_2, H_2O, Ar and He, and the cross-sections for the molecules vary between 0·03 Å² and 32·5 Å². Cross-sections for quenching by atoms are too small for it to be possible to do more than give upper limits.

Quenching of molecular radiation from $N_2 C\,^3\Pi_u$ and $N_2^+ B\,^2\Sigma_u^+$ has been studied using excitation of the nitrogen by 50 keV electrons[52].

Flowing afterglow systems have been used for nitrogen quenching studies[53], and studies of the excited OH radicle have also been reported. Quenching cross-sections for Hg 6^3P_1 have been reported[54] for H_2, N_2, H_2O, NH_3, CO, CO_2, CH_4, CF_3H and C_2H_6. These cross-sections are all less than 10^{-15} cm^2 and are presumed to be dominated by vibrational and rotational excitation of the molecule, but the cross-sections for NO, O_2, N_2O, H_2S, CS_2, HCN, higher alkanes, olefines and C_2H_2 are all larger than 10^{-15} cm^2 and are presumed to be dominated by chemical reaction.

It has been shown that quenching and sensitized fluorescence collisions between atoms, for which there is no possibility of vibrational excitation, are only probable when a very small internal energy difference is involved. Sensitized fluorescence cross-sections have been observed by Krause and his colleagues[55] for upward and downward transitions between the sodium D line states in collision with thermal argon atoms

$$\text{Na } 3^2P_{3/2} + \text{Ar} \rightarrow \text{Na } 3^2P_{1/2} + \text{Ar}$$

the internal energy defect being converted into kinetic energy. In these experiments the sodium vapour is illuminated with only $3^2P_{3/2}$ resonance radiation, and in the presence of the rare gas the population of $3^2P_{1/2}$ states is diagnosed by observation of emitted radiation. The studies are made possible by the availability of high-resolution interference filters by means of which the alkali doublets (except lithium) can be separated. Experiments are normally conducted in the pressure region of half an atmosphere. Kinetic equations[55] are written for the collision frequencies ν in terms of the intensities of emission.

Consider the reversible process

$$^2P_{1/2} + {}^2S_{1/2} \underset{\sigma_2}{\overset{\sigma_1}{\rightleftharpoons}} {}^2S_{1/2} + {}^2P_{3/2}$$

Irradiation with $^2P_{1/2}$ frequency of intensity I_{i1} produces fluorescence at the $^2P_{3/2}$ frequency, having intensity I_{f1}. A ratio $\eta_1 = I_{f1}/I_{i1}$ can be defined, and it follows that

$$n_0\bar{v}_r\sigma_1 = v_1 = \frac{1+\eta_1}{\tau_1(-\eta_1+\eta_2^{-1})} \tag{13.30}$$

where τ_1 is the natural lifetime of $^2P_{1/2}$, and η_2 refers to the ratio of intensities observed when the irradiation is with $^2P_{3/2}$ frequency. Similarly

$$n_0\bar{v}_r\sigma_2 = v_2 = \frac{1+\eta_2}{\tau_2(-\eta_2+\eta_1^{-1})} \tag{13.31}$$

Non-resonant collision cross-sections are found to be dependent upon the energy defect; in approximate terms one may apply Franck's rule $\sigma \propto \Delta E^{-\alpha}$, where α is a constant specific to a particular alkali metal. Selection rules for these collisions are:

1. Conservation of total angular momentum along collision axis z ($\sum J_z = $ constant);
2. Orientation of a $^2S_{1/2}$ atom must be unchanged by the collision.

From experiments with the rare gas replaced by the vapour of another alkali metal, excitation exchange cross-sections are obtained which allow the detailed balance to be checked. The balance is satisfactory except in cases where three-body effects become important. Satellite bands due to alkali metal–rare gas molecules have been reported[56], and sensitized fluorescence data are described in further references[57, 58].

The alkali metal sensitized fluorescence processes of the type

$$^2P_{1/2} + Ar \; ^1S_0 \rightleftharpoons \; ^2P_{3/2} + Ar \; ^1S_0 + \varDelta E$$

have been treated in terms of the van der Waals' interaction arising from the dipole induced in the rare gas by the quadrupole moment of the excited alkali metal. A semi-classical theory due to Zener[59] leads to an expression for the collision probability

$$P \simeq \left[\frac{2\pi W \tau_0}{h} (1 + 2\beta) \exp(-\beta) \right]^2 \tag{13.32}$$

with

$$\beta = \frac{\pi^2 \varDelta E \tau_0}{2h} \tag{13.33}$$

W is the matrix element of the interaction energy, τ_0 is the time of inter-action ($\propto v_r^{-1}$). The theory has also been applied[60] to the satellite bands observed close to the resonance line in Cs–Ar mixtures.

Another approach to sensitized fluorescence in alkali metal–rare gas mixtures stems from the original proposal of Fermi[61] to treat electrons bound in atomic excited states as though they were free electrons of kinetic energy equal to the classical orbit value (see also Section 13.1). The fluorescence cross-sections are comparable to the appropriate elastic electron–rare gas scattering cross-sections at the calculated energies. The major factor influencing the conversion rate is the energy defect $\varDelta E$. For collisions of the type

$$^2P_{1/2} + Ar \; ^1S_0 \rightleftharpoons \; ^2P_{3/2} + Ar \; ^1S_0 + \varDelta E$$

with identical $\varDelta E$ but different rare gases, the best guide to the magnitude of the rate is the elastic scattering cross-section of the active electron at its orbital velocity.

Yet another theoretical approach to these collisions can be made from the standpoint of pseudo-crossing of the energy levels[62] of the quasi-molecule of collision. Calculations have been made of the repulsive $^2\varPi$ state of the Li–He which can cross with an attractive $^2\varSigma$ state; these two states diverge to different sublevels of lithium, $J = \frac{3}{2}$ and $J = \frac{1}{2}$. A similar approach[63] is through transitions to negative and positive ion states at pseudo-cross-overs.

Effects similar to sensitized fluorescence and quenching have been noticed in studies of the enhancement of 'step-wise excitation'. In the irradiation of gases it is known that most higher energy states are in part excited in step-wise processes involving lower energy states. The lifetimes of atoms in the lower excited state influence the production of the higher excited states, the metastable Hg 6^3P_0, for example, being more efficient than the short-lifetime 6^3P_1. The addition of nitrogen[64] enhances the step-wise radiation of mercury and cadmium by processes of the type:

$$\text{Hg } 6^3P_1 + N_2 \rightarrow \text{Hg } 6^3P_0 + N_2^v \qquad\qquad 0'0/0^m0^v$$

This has been verified by studying the line reversal of second step radiation, and also by destroying the Hg 6^3P_0 by the addition of hydrogen[65]. The enhancement of Hg–Tl sensitized fluorescence by molecular nitrogen has also been attributed[66] to exchange collisions of the type $0'0/0^m0^v$.

13.4 METASTABLE AND LONG-LIFETIME EXCITED SPECIES

This section discusses some properties of metastable species whose lifetimes exceed 10^{-3} sec and may even exceed 1 sec. Selection rules determining which states are unable to pass spontaneously to a lower state by electric dipole transitions are given in the Appendix. Methods of preparing and of detecting fluxes of metastable atoms are described in Chapter 3. A list of the common metastable species of atoms and molecules, and their energies, appears in Table 13.3.

In addition, the following classes of excited species are sufficiently long-lived to enable them to undergo collision processes in beams and plasmas:

1. High principal quantum number states, not only of the hydrogen atom, but also of other atoms and ions[67];
2. Vibrationally excited non-polar molecules.

The principal feature of excited species of atom, whether metastable or not, is their very large radii in comparison with ground states. Naturally this feature greatly influences the cross-sections, for example, for the Hornbeck–Molnar process discussed below. The following approximate expressions for the radii r^* may be used[68]:

$$r^* = \frac{n^*(2n^*+1)}{3(Z-s)} a_0 \qquad\qquad (13.34)$$

An elastic collision diameter for a metastable in a gas may be approximated by a sum of ground state and excited state radii, plus an empirical correction (1·8 Å). For effective quantum number n^* and screening constant s, it is sufficient to use Slater's rules[69], as follows. The radial wave function has the form

$$R(r) = r^{n^*-1} \exp\left(-\frac{Z-s}{n^*r}\right) \qquad\qquad (13.35)$$

and the total energy of an electron is $-[(Z-s)/n^*]^2$. For principal quantum number $n = 1, 2, 3, 4, 5$ and 6, the value of n^* is respectively 1, 2, 3, 3·7, 4·0 and 4·2. For the calculation of s, the electrons are considered in the following groups: $1s$; $2s, 2p$; $3s, 3p$; $3d$; $4s, 4p$; $4d$; $4f$; etc. The screening constant s is calculated as a sum of contributions from every other electron in the atom, as follows: zero from any electron in a shell outside that of the electron under consideration; 0·35 from any electron in the same shell (0·3 for $1s$); 0·85 from each electron with a principal quantum number smaller

Table 13.3. METASTABLE SPECIES OF ATOMS AND MOLECULES AND IONS

Species	Configuration	Energy (eV)
H	$2^2S_{\frac{1}{2}}$	10·196
He	2^3S_1	19·82
	2^1S_0	20·61
Li	$^4P_{\frac{5}{2}}$	56
Be	2^3P	2·72
B	2^4P	3·57
C	2^1D	1·26
	2^1S	2·68
	2^5S	4·18
N	2^2D^o	2·38
	2^2P^o	3·58
O	2^1D	1·97
	2^1S	4·19
	2^5S^o	9·14
F	$2^4P_{\frac{5}{2},\frac{3}{2},\frac{1}{2}}$	12·69, 12·72, 12·74
Ne	$2^3P_{2,0}$	16·62, 16·71
Ar	$3^3P_{2,0}$	11·55, 11·72
Kr	$4^3P_{2,0}$	9·91, 10·3
Xe	$5^3P_{2,0}$	8·31, 9·44
Zn	$4^3P_{0,2}$	4·00, 4·08
Cd	$5^3P_{0,2}$	
Hg	$6^3P_{0,2}$	4·64, 5·43
	6^3D_3	9·05
Tl	$6^2P^o_{\frac{3}{2}}$	0·96
H_2	$C^3\Pi_u$	11·86
NO	a^4	4·4
CO_2	$?\ ^3\Pi$	
CH	$^2\Delta$	
C_2	$^1\Pi_u$	
He_2	$2^3\Sigma$	
CO	$a^3\Pi$	6·01
	$b^3\Sigma^+$	10·5
O_2	$^1\Delta_g$	0·98
	$^1\Sigma_g^+$	1·5
	$^5\Pi$	12
N_2	$A^3\Sigma_u^+$	6·17
	$a^1\Pi_g$	8·54
	$a'^1\Pi_g$ (vib. ex)	10·2–11·4
		11·4–11·6
He^+	$2^2S_{\frac{1}{2}}$	40·80
Li^{2+}	$2^2S_{\frac{1}{2}}$	91·81
Li^+	2^3S	59·01
	2^1S	74·54
B^+	2^3P	4·63
C^{2+}	2^3P	6·49
C^+	2^4P	5·33
N^{2+}	2^4P	7·10
N^+	2^1D_2	1·90

Table 13.3 — *continued*

Species	Configuration	Energy (eV)
N$^+$ *(cont.)*	2^1S_0	4·05
	2^5S_2	5·85
O^{2+}	2^1D_2	2·51
	2^1S_0	5·35
	2^5S_2	7·48
O$^+$	2^2D	3·32
	2^2P	5·02
F$^+$	2^1D_2	2·59
	2^1S_0	5·57
	2^5S_2	21·90
Ne^{2+}	2^1D_2	3·20
	2^1S_0	6·91
	2^5S_2	38·94
Na$^+$	2^3P	32·9
Al$^+$	3^3P	4·64
Si$^+$	3^4P	5·48
P$^+$	3^1D_2	1·10
	3^1S_0	2·67
S$^+$	3^2D	1·84
	3^2P	3·04
Cl$^+$	3^1D_2	1·44
	3^1S_0	3·46
	3^5S_2	13·37
Ar^{2+}	3^1D_2	1·74
	3^1S_0	4·12
	3^5S_2	21·61
Ar$^+$	$3^4D_{\frac{7}{2}}$	16·40
Hg$^+$	$6^2D_{\frac{5}{2}}$	4·40
	$6^2D_{\frac{3}{2}}$	6.27
Tl$^+$	6^3P_0	6·14
	6^3P_2	7·65
Br$^+$	4^3P_1	0·39
	4^3P_0	0·48
	4^1D_2	1·41
	4^5S_2	11·64
Kr^{2+}	4^3P_1	0·56
	4^3P_0	0·66
	4^1D_2	1·82
	4^5S_2	18·06
Kr$^+$	$4^4D_{\frac{7}{2}}$	14·90
	$4^4F_{\frac{9}{2},\frac{7}{2}}$	15·62
	$4^2F_{\frac{7}{2}}$	16·32
Xe^{2+}	5^3P_1	0·59
	5^3P_0	0·01
	5^1D_2	2·12
	5^5S_2	15·06
Xe$^+$	5^5D_2	13·86
	$5^4D_{\frac{7}{2}}$	11·83

Table **13.3**–*continued*

Species	Configuration	Energy (eV)
Xe$^+$(*cont.*)	$5^4F_{\frac{7}{2}}$	12·25
	$5^4F_{\frac{9}{2}}$	12·32
	$5^2F_{\frac{7}{2}}$	14·24
Cs$^+$	5^3P_2	13·31
	5^3P_1	13·37
	5^3P_0	15·17

by 1; and 1·0 from each electron still further in (provided that the shell of the electron under consideration is s or p; if it is d or f, then 1·0 from every electron inside this shell).

Certain metastable states of neutral atoms are of sufficiently high energy to lie in the continuum. For metastability of such states, downward transitions must be forbidden not only to states of the neutral atom, but into the continuum. Even if forbidden by selection rules, the latter transitions can still take place in about 10^{-6} sec, but this is a sufficiently long period for the species to be detected[70]. In lithium the state $1s\ 2s\ 2p\ ^4P_{5/2,\ 3/2,\ 1/2}$ has been identified around 60 eV. In potassium and rubidium there are levels $p^5s\,d^4F_{9/2}$, and in caesium this level exists, and also another, not well identified; all are around 20 eV. Decay of the quartet states take place by the spin–spin operator $\mu_1 r_{12}\mu_2 r_{12}/r_{12}^5$, which is approximately equal to $\alpha^2 e^2/r_{12}$. Therefore the lifetime is a factor α^4 greater than that for an allowed transition into the continuum; α is the fine structure constant, and the lifetime of Li ^4P is 5·6 μsec. The electron excitation function for this level shows structure due to cascading; the state was detected in a crossed-beam experiment by the delayed production of an electron. A search for N $1s^2\ 2s\ 2p^3\ 3s\ ^6S$ is being made[71] around 17 eV.

13.5 SCATTERING AND DEACTIVATION OF METASTABLES

Since the symmetrical resonance excitation transfer collision of a metastable atom in its own gas is indistinguishable from elastic scattering except in angle and polarization, both processes are included in the scattering cross-section and in the diffusion cross-section; the total and diffusion cross-sections are as usual

$$\sigma = 2\pi \int_0^\pi \frac{d\sigma}{d\Omega}(\theta)\sin\theta\,d\theta \qquad (13.36)$$

and

$$\sigma_d = 2\pi \int_0^\pi \frac{d\sigma}{d\Omega}(\theta)(1-\cos\theta)\sin\theta\,d\theta \qquad (13.37)$$

The diffusion of metastables is similar to the diffusion of ions in their own gases (Chapter 10); the inelastic process results in scattering through a large polar scattering angle ($\theta \sim \pi/2$), whilst in the elastic process θ is small. The diffusion coefficient D_m is related to σ_d as follows:

$$D_m \simeq \frac{3\pi^{1/2}}{16} \frac{(2\mu kT)^{7/2}}{n_m n_0 P_m} \tag{13.38}$$

with

$$P_m = \int_0^\infty v^5 \sigma_d(v) \exp\left(-\frac{\frac{1}{2}\mu v^2}{kT}\right) dv \tag{13.39}$$

In a gas containing a proportion of its own metastable atoms, the time variation of the space-averaged metastable density n_m is given by the equation

$$\frac{dn_m}{dt} = D_m \nabla^2 n_m - k_{2m} n_m n_0 - k_e n_e n_m \tag{13.40}$$

with solution

$$n_m(t) = n_{0m} \exp\left(-\frac{t}{\tau}\right) \tag{13.41}$$

where

$$\frac{1}{\tau} = \frac{D_m}{\Lambda^2} + k_{2m} n_0 + k_e n_e \tag{13.42}$$

Here Λ is the diffusion length, and k_{2m} and k_e represent rate constants for volume deactivation processes involving two-body collisions with neutral gas atoms and with electrons respectively.

Helium possesses two important atomic metastable states, of which the 2^1S is higher than the 2^3S by 0·79 eV. Phelps and his colleagues[72] performed experiments in which the densities of both of these states in the afterglow of a pulsed helium discharge were monitored by the measurement of the absorption of radiation of wavelength suitable for raising the metastable atoms to a higher energy level. Radiation of wavelength $\lambda = 3889$ Å is suitable for 2^3S atoms and 5016 Å for 2^1S. The $2^3\Sigma$ molecule is formed in a binary He 2^3S–2^1S collision, and may be detected by the absorption of 4650 Å radiation. Since the original experiments, high-frequency sources of helium resonance radiation[73] have made possible the application of fast time-dependent electronics to the optical diagnosis of the helium afterglow.

It is usual for diffusion coefficients such as D_m to be inversely proportional to pressure; since $k_{2m} n_0$ is directly proportional to pressure, the separation of these two terms in equation 13.42 can readily be achieved. However, the electron density term $k_e n_e$ can only be separated with the aid of a simultaneous experiment measuring electron density. This can be achieved by means of a microwave cavity. Flowing afterglow technique can also be applied.

Phelps was able to separate three diffusion coefficients in an elegant series of experiments; his results are given in Table 13.4. The agreement between the experimental diffusion coefficient and that calculated for He 2^3S using Jeffreys' approximation by Buckingham and Dalgarno[25] (370 cm² sec⁻¹) is fair, and early measurements of Ebbinghaus[74] (400 cm² sec⁻¹) also compare well. Biondi[72], using a purely microwave method, reports 520 cm² sec⁻¹ at

Table 13.4. HELIUM METASTABLE DIFFUSION
COEFFICIENTS OBTAINED AT 300° K AND 1 torr

	D_m (cm^2 sec^{-1})
He 2^{1}S	440 ± 50
He 2^{3}S	470 ± 25
He$_2$ 2^{3}S	310 ± 50

an unspecified temperature. An approximately linear temperature variation of the diffusion coefficient from 160 cm^2 sec^{-1} at 100°K to 1200 cm^2 sec^{-1} at 1000°K is derived from the quantum theory calculations. Diffusion data for rare gas metastables are summarized in Table 13.5.

The early experiments of Meissner and Graffunder[78] and Zemansky[79], using a technique similar in principle, yielded diffusion cross-sections for other rare gas metastables. Their σ_d for He 2^{3}S were 34·8×10^{-16} cm^2 and 54·4×10^{-16} cm^2, corresponding to different thicknesses of absorption tube, whilst the calculation of Buckingham and Dalgarno yields 24×10^{-16} cm^2. For Ne ^{3}P their experiments gave $\sigma_d = 7·66×10^{-16}$ cm^2; and for Ar ^{3}P, $\sigma_d = 33·9×10^{-16}$ cm^2 at 300°K and 50×10^{-16} cm^2 at 80°K, the variation with temperature being in the expected direction. For Hg 6^{3}P$_0$ metastables, Samson[80] and Webb and Messenger[81] found $\sigma_d = 15·5\,\pi$Å^2 at

Table 13.5. DIFFUSION DATA FOR SOME METASTABLES

Collision	D_{mp} cm^2 sec^{-1} torr^{-1}	Reference
He^m–He	440, 470	72
	520 ± 20	72
	400	74
Ne^m–Ne	200 ± 20	72
	120 ± 10	75
	150 ± 30	72
Ar^m–Ar	45 ± 4	75
	67·5	76
Xe^m–Xe	13 ± 1	75
N$_2^m$–Ar	157	75
Hg$_2^m$–Hg	17	77

$300°K$, $17\cdot7\,\pi Å^2$ at $374°K$ and $18\cdot4\,\pi Å^2$ at $486°K$, all in mixed nitrogen and mercury vapour; the temperature variation is unexpected. For Hg 6^3P_0 in mercury, Coulliette[82] found $\sigma_d = 48 \times 10^{-16}$ cm², using a variant of Webb's[83] method.

A study of mercury metastable diffusion in mercury was carried out by Webb[83], who measured the resonance frequency when two alternating signals are applied: (*i*) to the acceleration of bombarding electrons, and (*ii*) to a collecting grid designed to accept the electrons liberated by the arrival of the metastable atoms at a metal plate after diffusion through the gas. Only at a certain resonance frequency will the metastable diffusion time correspond to a phase difference of $2n\pi$, producing a signal at the detector. At an unspecified temperature, probably $\sim 300°K$, it was found that $\sigma_d = 20\cdot0\,\pi Å^2$.

Total collision cross-sections for He 2^3S and for neon thermal metastable atoms in rare gases have been measured, using beam techniques, by Stebbings[84] and others[85-87]. The experiment is carried out in absorption, with a collision chamber placed between the metastable source and a surface electron ejection detector. The angular resolution necessary for the measurement of a true total collision cross-section must be very fine ($\sim 1°$); it is necessary to experiment with different resolutions in order to verify that the resolution is adequate. It is also necessary to identify correctly the metastable atoms emerging from the source.

A further complication arises from the fact that the beam and gas have comparable velocity distributions. Unless the appropriate beam–gas mathematical analysis is made (Chapter 1), the gas densities appropriate to the collision region will be incorrectly estimated[88]. Table 13.6 gives corrected data; taking the temperature as $300°K$, the agreement with Buckingham and Dalgarno's[25] calculated value for He 2^3S in helium is satisfactory.

Table 13.6. TOTAL CROSS-SECTIONS
FOR METASTABLE ATOMS

Collision	σ (πa_0^2)
He^m–He	124
He^m–Ne	132
He^m–Ar	214
Ne^m–He	105
Ne^m–Ne	130
Ne^m–Ar	265

More recent experiments[86] have included magnetic separation of He 2^3S and 2^1S and have provided differential scattering data, which agree with the interaction potential[25]. The velocity dependence of total cross-sections has also been measured[87] using a selector and post-collision magnetic separation of the helium states. The excitation transfer contribution is only $\sim 10\%$ of the total cross-section. Analysis is made of the cross-section function along the lines of Chapter 10, using a Lennard–Jones 6–12 potential. The van der Waals' constants are consistent with a polarizability α(He 2^3S) = $46\cdot4 \times 10^{-24}$ cm³.

Total collision cross-sections in rare gases have been reported for thermally excited Ga $^2P_{3/2}$, in different Zeeman substates $m_J = +\frac{3}{2}, -\frac{1}{2}$; they differ by $\sim 1\%$.

Deactivation is possible in collisions of metastables with gases, but two-body metastable deactivation collision of the second kind are inefficient unless the internal energy is small. Mercury metastable transitions were studied in collisions with nitrogen by Samson[80] in afterglow experiments. The transition $6^3P_0 \rightarrow 6^3P_1$ can proceed in either direction, with exchange of internal energy for thermal kinetic energy of the nitrogen molecules. The cross-sections are as follows:

$6^3P_0 \rightarrow 6^3P_1$ $0 \cdot 21 \times 10^{-16}$ cm^2

$6^3P_1 \rightarrow 6^3P_0$ $0 \cdot 97 \times 10^{-16}$ cm^2, $0 \cdot 86 \times 10^{-16}$ cm^2

$6^3P_1 \rightarrow 6^1S_0$ $0 \cdot 069 \times 10^{-16}$ cm^2

$6^3P_0 \rightarrow 6^1S_0$ $0 \cdot 052 \times 10^{-16}$ cm^2

For He 2^1S Phelps[72] has obtained data indicating a binary deactivation collision with ground state helium atoms, with cross-section 3×10^{-20} cm^2. This is much larger than would be expected on the basis of a three-body deactivation process, such as has been invoked[89] to explain the high-pressure pseudo-cross-section, $1 \cdot 5 \times 10^{-20}$ cm^2 at 850 torr, obtained by Jesse and Sadauskis[90] from studies of the energy necessary to produce an ion pair when α-particles pass through helium. The question has been raised[27] as to whether the process is one of collision-induced radiation. A cross-section of $9 \cdot 6 \times 10^{-21}$ cm^2 was found by Biondi[72] for He 2^1S deactivation; also $8 \cdot 9 \times 10^{-20}$ cm^2 for Ne ^3P in neon and, for Ar ^3P in argon, $2 \cdot 1 \times 10^{-20}$ cm^2. These values are more or less consistent with experiments on electron multiplication in gas mixtures[91], from which may be deduced the ratio of deactivation to ionization cross-sections. There seems little doubt that a number of deactivation cross-sections of the order of 10^{-20} cm^2 have been observed, but the processes which bring them about are not understood. A more recent experiment[92] yields $(5 \cdot 4 \pm 1 \cdot 0) \times 10^{-16}$ cm^2 for Ne ^3P in N$_2$.

In modern experiments with fast He 2^1S and 2^3S beams obtained by charge transfer of He$^+$ with caesium, very much larger deactivation cross-sections are observed in helium gas[93]. The observation is made of $3^1P \rightarrow 2^1P$ and $3^3P \rightarrow 2^3S$ light emitted along the beam path: these states are excited collisionally by the gas atoms. The cross-sections are ~ 14 Å^2 (2^1S) and ~ 9 Å^2 (2^3S) at 200 eV, and fall off with increasing impact energy; there is oscillatory structure in the cross-section function, which does not agree with theoretical predictions[94]. The low thermal energy cross-section is attributable to the maxima in the adiabatic interaction energy functions[25]; the oscillatory structure is attributable to pseudo-crossings between the adiabatic symmetrical and antisymmetrical interactions.

The appearance of forbidden lines, particularly in the auroral emission from the earth's upper atmosphere, but also in certain discharges, facilitates the study of deactivation. Forbidden lines appear from a region containing excited atoms when the lifetimes are not sufficiently long for diffusion of the excited atoms out of the region to be complete or for deactivation processes to become effective. The forbidden O^1S radiation occurs in pure oxygen and oxygen mixture discharges, and its emission under steady state conditions in cylindrical tubes was studied by Kvifte and Vegard[95]. With the aid of various assumptions, they obtained deactivation cross-sections 3.6×10^{-6} and 2.9×10^{-7} of gas-kinetic, for deactivation by O_2 and Ne respectively. Experiments in neon were conducted by Phelps, and Twiddy has extended the technique to argon.

A three-body process similar in principle to collision-induced radiation is partly responsible for the deactivation of metastables in gases. For the collision of two atoms a 'sticky' combination process is rare, and collision-induced radiation is relatively unlikely; but in the presence of a third atom (or molecule), the collision is more feasible. The process has already been invoked[89] to explain high-pressure deactivation of He 2^1S in helium. The three-body association process for He 2^3S in collision with two ground state helium atoms was reported by Phelps[72], who observed the resulting $He_2 \, ^3\Sigma_u^+$. The molecule has a lifetime of at least 0.05 sec, and can be monitored by its absorption of 4650 Å radiation. The three-body rate coefficient for the processes was found, by the substitution of a term $k_{3m}n_0^2 n_m$ for $k_{2m}n_0 n_m$ in equation 13.40, to be 2.5×10^{-34} cm^6 sec^{-1} at 300°K.

Some interesting chemical deactivation collisions of $O_2 \, ^1\Delta_g$ have been reported[96, 97].

13.6 ELECTRON AND PHOTON COLLISIONS WITH METASTABLES

Excitation of excited atoms by electrons has formed the subject of various calculations[98, 99]. The cross-sections can achieve enormous peak values ($\sim 10^5 \pi a_0^2$), but the peaks are extremely sharp, and even the thermal velocity distributions of thermionic electrons are sufficiently broad for the excitation rates to be much reduced.

Collisions of the second kind between electrons and metastables are also of importance. These are processes resulting in downward transitions in which the internal energy is converted into kinetic energy of the electron. The conversion

$$He \, 2^1S + e \rightarrow He \, 2^3S + e + 0.79 \, eV$$

was first observed by Phelps[72], who correlated the rate of conversion of singlet to triplet states with the electron density in the afterglow, independently measured by microwave technique. The mean cross-section for 300°K electrons was found to have the extremely large value 3.0×10^{-14} cm^2. This transition involves electron exchange, so the principal contribution arises from electrons with very small angular momentum; the maximum cross-section permissible in theory is 4×10^{-14} cm^2. Unfortunately quantum theory calculation of the cross-section for this process[100] leads to a value of the order of 5×10^{-15} cm^2; the discrepancy may possibly be due to the use of a poor He 2^1S wave function, or to coupling with the 2^1P and 2^3P states.

Ionization of metastables by electron collisions is of importance in ionized gases, since the formation of a proportion of the ions can take place in a two-stage process:

$$e + X \rightarrow X^m + e$$

$$e + X^m \rightarrow X^+ + 2e$$

Mean cross-sections of the order of Å^2 in magnitude are reported[101], and calculations have also been made[102]. The process of metastable ionization by electrons has been used, close to threshold, as a metastable detection device.

As regards photon absorption by metastables, it has already been noted that the absorption of radiation of suitable wavelengths is one of the methods by which metastable atom densities can be monitored. The calculations of Bates and Damgaard[103] may be consulted to obtain the oscillator strengths.

Mention must be made of one of the most important series of experiments in modern physics, the Lamb–Retherford experiments to determine the separation in energy of the $2^2S_{1/2}$ and $2^2P_{1/2}$ states of H, and similar one-electron systems[104]. According to the Dirac theory these states should be completely degenerate, but spectroscopic evidence on the point proved inconclusive. In the Lamb–Retherford experiment, an atomic hydrogen beam effuses from a furnace and is crossed with an electron beam, after which it contains a proportion of H $2s\,^2S_{1/2}$ metastable atoms; these atoms have a lifetime sufficiently long for them to pass through the apparatus until detected by electron ejection from a metal surface. In static electric fields the metastability is much reduced by Stark effect mixing of S and P states, causing a reduction in the electron current, since the $2^2\,P_{1/2}$ level can radiate speedily. Electromagnetic radiation of the correct frequency will also raise the metastables to the $2^2P_{1/2}$ level, but since only fixed-frequency oscillators in the microwave region were readily available, Zeeman splitting in a steady magnetic field of variable magnitude was added. In this way it was shown that the separation of these levels is 1062 ± 5 MHz ($\nu' = 0{\cdot}033$ cm^{-1}).

The separation arises from interaction of the atomic electron with the radiation field, and quantitative treatment has since been actively pursued with the aid of field theory. The separation between the two states of the He$^+$ ion, $2^2S_{1/2}$ and $2^2P_{1/2}$, has been measured as $14\,020\pm100$ MHz; since the metastable ions could not easily be distinguished from ground state ions by surface effects[105], the $2^2P_{1/2}$ state population was monitored by detecting the ultra-violet photons emitted. The Lamb shifts are strongly dependent on nuclear charge and may be observed spectroscopically for one-electron systems with $Z \geqslant 3$.

13.7 PRODUCTION OF IONS IN COLLISIONS OF METASTABLES WITH ATOMS AND MOLECULES

A process was first proposed by Kruithof, Penning and Druyvesteyn[91] to account for the X/p variation of the Townsend ionization coefficient for a swarm of electrons drifting through a gas containing a small proportion of a gas of lower ionization potential than the energy of the metastable state and the principal gas. It is known as 'Penning ionization', and being an exothermic process it can take place at thermal energies:

$$\text{He } 2^1\text{S} + \text{Ar} \rightarrow \text{He} + \text{Ar}^+ + e + \Delta E$$

In afterglows of pulsed discharges in pure helium, Biondi[72] observed that the electron density actually increases with increasing time in the first hundred microseconds. This is due to a process of the type

$$He\ 2^1S + He\ 2^1S \rightarrow He + He^+ + e$$

which gives rise to a variation of electron density:

$$\frac{dn_e}{dt} = \frac{dn_+}{dt} = k_{mm}n_m^2 \tag{13.43}$$

The Penning ionization of a second gas gives rise to a variation of electron density:

$$\frac{dn_e}{dt} = \frac{dn_+}{dt} = k_p n_m n \tag{13.44}$$

where n represents the number density of the atoms of the second gas. Neglecting the collisional deactivation of the metastables,

$$\frac{dn_m}{dt} = D_m \nabla^2 n_m - k_p n_m n - k_{mm}n_m^2 \tag{13.45}$$

$$n_m = n_{0m} \exp\left(-\frac{t}{\tau_m}\right) \tag{13.46}$$

and

$$\frac{1}{\tau_m} = \frac{D_m}{\Lambda^2} + k_p n_m n + k_{mm}n_m^2 \tag{13.47}$$

Allowing for decay by ambipolar diffusion, the electron density variation is given by

$$n_e = A \exp\left(\frac{-t}{\tau_e}\right) - B \exp\left(\frac{-2t}{\tau_m}\right) \tag{13.48}$$

with

$$A = n_{0e} + B \tag{13.49}$$

and

$$B = \frac{k_p n_{0m}^2}{2/\tau_m - 1/\tau_e} \tag{13.50}$$

Variation of n and total gas pressure permits separation of the diffusion and volume processes. The ionization cross-sections deduced from these experiments are included, together with those from beam experiments, in Table 13.7.

In the experiments of Muschlitz and Sholette[107] a helium jet was crossed with an electron beam of 25–35 eV energy, and the charged particles removed. Over this region of electron energy the ratio of the 2^1S and 2^3S excitation cross-sections changes considerably, so provided their ratio is accurately known, the proportions of the two states present in the beam can be controlled and known. The metastable beam so formed is directed through a collision chamber containing gas, and the positive ions are drawn out electrically and collected. An estimate is made of the current due to elastically scattered metastables emitting electrons from the metal surfaces. It is difficult

Table 13.7. PENNING IONIZATION CROSS-SECTIONS

Collision	Cross-section σ_p (Å²)	Reference
He 2^3S–Ne	0·28	106
He 2^1S–Ne	4·1	106
He 2^3S–Ar	6·6	106
	7·6	107
He 2^1S–Ar	7·6	107
	12	72
He 2^3S–Kr	10·3	106
	16·5	90
	9·0	107
He 2^1S–Kr	9·0	107
He 2^3S–Xe	13·9	106
	20·2	90
	12·0	107
He 2^1S–Xe	12·0	107
He 2^3S–H$_2$	6·1	106
	1·7	90
	2·6	107
	2·4	108
He 2^1S–H$_2$	1·7	107
He 2^3S–N$_2$	6·4	106
	6·1	90
	7·0	107
He 2^1S–N$_2$	7·0	107
He 2^3S–O$_2$	14	107
He 2^1S–O$_2$	14	107
He 2^3S–CO	7	107
He 2^1S–CO	7	107
He 2^3S–He 2^3S	100	72
He 2^3S–Hg	140	72
He 2^1S–Hg	140	72
Ne ^{3}P–Ar	2·6	72
	7·6	108
	0·00075	109
Ne ^{3}P–H$_2$	4·2	108

to calculate the error in path length that derives from the possibility of elastically scattered particles producing ionization. The Penning ionization cross-section is calculated from the pressure dependence of the ion current, under single collision conditions. Similar experiments have recently been reported[174].

No difference was found between the cross-sections for ionization by 2^1S and 2^3S metastables, except in collisions with hydrogen; this was confirmed by separation of the two beams in a non-uniform magnetic field. It will be noticed that in Table 13.7 there are discrepancies between these cross-sections σ_p and some of those obtained in earlier experiments, such as those of Jesse and Sadauskis[90]. Some data obtained by the Texas group[106] have also been included in this table; the technique involves the monitoring of metastable densities in afterglows containing proportions of second gas. Agreement between the data obtained with widely different techniques is encouraging.

Comparison may be made between the Penning ionization cross-section σ_p and the metastable momentum transfer cross-section σ_d, usually unknown experimentally, but calculable with classical orbit theory. It is found that $\sigma_p/\sigma_d \sim 0.2$, a factor which is rather smaller than the electron ejection coefficient for metastables incident upon metal surfaces. Orders of magnitude of σ_p may be estimated in this way. Energy analysis of the emitted electrons supports this comparison[110].

An electron capture model of Penning ionization has been proposed[111] and found to be reasonably successful; there are also other calculations[112]. Penning ionization of N_2 by He^m conforms to the Franck–Condon principle.

In the dissociative Penning process

$$X^m + YZ \rightarrow X + Y^+ + Z + e$$

transitions to an anti-bonding state occur, and isotope effects are to be expected, arising from the competition between downward transitions and separation of the atoms.

Detection of ions and electrons from the Penning process has been used by Cermak[113] both for bracketing the unknown internal energies of metastable and other long-lifetime states of molecules, and for determining inner ionization potentials of molecules.

The bracketing technique is carried out by introducing a series of gases of known ionization potential into a collision chamber fed with a flux of thermal energy excited particles. The lowest ionization potential gas from which no positive ions are produced, and the highest ionization potential gas from which positive ions are produced, mark the limits which can be placed upon the metastable internal energy.

The determination of inner ionization potentials using Penning ionization[114, 115] is a technique basically similar to that of photoelectron spectroscopy (Chapter 9), but utilizing a metastable atom source instead of vacuum ultra-violet photons. The energy spectrum of the emitted electrons is measured, and subtraction of a characteristic electron energy from the known metastable atom energy yields the inner ionization potentials of the molecule. The differences between Penning ionization electron spectroscopy and photoelectron spectroscopy arise from the metastables exerting van der Waals' forces on the molecule, and also from the possibility of electron exchange. Resonance effects have been observed.

Associative ionization of an atom by a metastable is also possible:

$$X^m + Y \rightarrow XY^+ + e$$

it is generally known as the 'Hornbeck–Molnar process', although the first observations, including that of He_2^+ formed at 19.8 eV appearance potential, were made some years earlier[116, 117].

Modern experiments on the Hornbeck–Molnar process are usually conducted in electron-impact mass-spectrometer sources containing sufficient gas (pressure $\sim 10^{-4}$ torr) for single collisions of excited atoms with gas. The appearance potential of the molecular ion should correspond to the energy of the excited atom. Other states besides the lowest lying metastables can participate in these processes[118], and because the electron impact thresholds do not provide the most sensitive method of identifying these states, photoexcitation technique has been used[119]; about two dozen levels able to participate in Hornbeck–Molnar processes were identified for argon, krypton and xenon. The lifetimes of some of these levels have been shown by pulse techniques[118] to be as large as 0.5 μsec. Studies of the process are accumulating[119–128].

Higher energy collisions of excited species with atoms can result in the formation of positive ions[129]. High principal quantum number states of hydrogen are ionized in collision with gases[130]. High principal quantum number states of atoms and molecules can be collisionally ionized even at thermal energies. This process, known as 'collision-induced-ionization' has been observed in studies over a range of gas pressures of the ionization produced in photon absorption just below the ionization potential[131–133]. At the lowest gas pressures there is for polyatomic species still a finite ionization probability, corresponding to spontaneous transitions from vibrationally excited molecules. However, there is a range over which the probability of ionization is directly proportional to pressure; at the higher pressures, deactivation processes predominate.

13.8 COLLISION PROCESSES OF VIBRATIONALLY AND ROTATIONALLY EXCITED MOLECULES[134]

Vibrationally excited molecules[135] can undergo deactivating collisions in which they revert to their ground state, of vibrational quantum number $v = 0$. These can either be of the exchange type in which the target particle undergoes an upward transition, vibrational or electronic; alternatively the entire vibrational energy may be converted into kinetic energy. The latter process is the reverse of vibrational excitation by collision with a molecule, so the two cross-sections can be related by the principle of detailed balancing. Using the notation of Chapter 1 (with molecules denoted by 0), such processes might be written $0^v0/00$, $00/00^v$, $0'0/00^v$, $0^v0/00'$, $0^v0/00^v$, $0^v0/0^v0^v$. All these collisions except the symmetrical resonance collisions involve an interchange between vibrational and kinetic energy.

The process $0'0/00^v$ and its reverse can proceed via potential energy hypersurface crossings[136], involving an intermediate state and interconverting electronic and translational energy; it is known that certain near-resonance processes of this type are extremely probable. The exchange processes between vibrational states of a molecule,

$$XY^v + XY \rightarrow XY^v + XY^v \qquad\qquad v0/(v-1)1$$
$$v \quad\; 0 \quad\; v-1 \quad 1$$

are usually near-resonance, and are also probable. It is also possible that vibrational deactivation could occur by means of an interchange collision; for example, with the helium molecular ion

$$\mathrm{He} + \mathrm{He}_2^+(v) \rightarrow \mathrm{He}_2^+(v' \neq v) + \mathrm{He}$$

A similar process for molecular oxygen has been proposed[137].

Experimental studies of vibrational deactivation nearly all refer to the comparatively improbable collision $0^v0/00$, in which a $v = 1$ molecule converts its vibrational into translational energy. This process governs the time of equilibration of vibrational and kinetic temperatures; the vibrational energy levels are sufficiently separated for most experiments to be conducted in gases with only the $v = 0$ and $v = 1$ states appreciably populated; if higher states are present they quickly pass to $v = 1$ by $v0/(v-1)1$ cascade.

For a total collision frequency per second

$$v = 4\pi_0 \varsigma_0^2 \left(\frac{\pi k T}{m} \right)^{1/2} \tag{13.51}$$

with gas kinetic radius ς_0, the time of deactivation is

$$\tau_d^{-1} = P_{10} v \tag{13.52}$$

where P_{10} is the probability of vibrational deactivation per collision. This applies to a small number of molecules in state 1, with no possibility of the reverse process re-populating the state; but in a Boltzmann distribution the deactivation is governed by a relaxation time

$$\tau_r^{-1} = v P_{10} \left[1 + \exp \left(-\frac{\Delta E}{kT} \right) \right] \tag{13.53}$$

where ΔE is the energy of deactivation.

Because the specific heat of a gas at constant volume C_v depends upon the vibrational state of the molecules, the speed of sound v_s in the gas will also show such dependence, since

$$v_s^2 = \frac{RT}{m} \left(1 + \frac{R}{C_v} \right) \tag{13.54}$$

However, for sound of high frequency, the gas is unable to follow the pressure fluctuations, so the specific heat C_v no longer possesses its low frequency value C_{v0} but instead, a smaller value $C_{v\infty}$. For angular frequency ω, the speed of sound is:

$$v_s^2(\omega) = \frac{RT}{m} \left[1 + \frac{R(C_{v0} + C_{v\infty}\omega^2\tau_r^2)}{C_{v0}^2 + C_{v\infty}^2\omega^2\tau_r^2} \right] \tag{13.55}$$

where R is the gas constant. There is also an absorption of the sound, with coefficient

$$\mu_s = \frac{2\pi\omega C_v \tau_r}{C_{v0}} \left\{ \frac{v_s^2(\omega)/v_s^2(0) - 1}{[v_s^2(\omega)/v_s^2(0)] \left[(C_{v\infty}\tau_r\omega/C_{v0})^2 + 1 \right]} \right\} \tag{13.56}$$

The relaxation time, and hence P_{10}, is determined by speed of sound measurement over a suitable temperature range, at a suitable fixed frequency. An inverse variation of τ_r with gas pressure should be obtained, and virial corrections must be applied. Sound speeds in the frequency range from 50 kHz to 10 MHz are measured at 300°K or below by means of an acoustic interferometer, with piezo-electric transducer source.

Other methods of studying vibrational deactivation processes depend upon measurement of time variation of the kinetic and vibrational temperatures when energy is supplied to a gas. The rapid adiabatic compression behind a shock wave results in an increase of temperature and therefore a decrease of density at constant pressure. After a period of activation, part of the energy is converted into vibrational degrees of freedom, resulting in a fall in temperature and a rise in density[138]; its time dependence, measured optically as a 'density profile', is therefore related to P_{10}. Line reversal technique can be used to measure kinetic temperatures, and infra-red emission[139] has been used to study vibrational temperatures in shock tubes. Similar compressions and rarefactions, with the accompanying temperature changes and vibrational deactivation, are found to take place during the aerodynamic flow of gas past obstacles[140].

Vibrational energy exchange between different molecules has been studied by infra-red fluorescence.

The study of vibrational deactivation by means of calorimetric detection of excitation has been achieved in a flowing afterglow system[141]. Vibrationally excited molecules are deactivated by a sufficient number of surface collisions inside a calorimeter.

Where it is possible to excite a vibrational level by pulsed infra-red radiation (as is the case with a polar molecule), sound is generated in the gas, with an intensity depending upon the relation between the pulse repetition frequency and the relaxation time[142]. Unfortunately experimental difficulties have hitherto limited this technique to semi-quantitative results[143].

The first expression for the probability of deactivation by transition between two non-crossing potential energy curves was deduced from classical arguments by Landau and Teller[144]. A similar result can be obtained by quantum theory[134]:

$$P_{10} = K \exp 3\left(\frac{-2\pi^4 \mu v^2}{\alpha^2 kT}\right)^{1/3} \tag{13.57}$$

or

$$P_{10} = K \exp\left(-3y\right) \tag{13.58}$$

where α is the parameter governing the exponential repulsion between the colliding molecules:

$$V(r) = V_0 \exp\left(-\alpha r\right) \tag{13.59}$$

An explicit value of the constant K has been derived by Cottrell and Ream[145]:

$$K = \frac{4\pi}{\sqrt{3\nu}} n_0 \varsigma_0^2 B v^* \tag{13.60}$$

with

$$B = \left(\frac{\mu}{\mu_{\text{int}}}\right)\left(\frac{\mu v}{h\alpha^2}\right) 32\mu^4 \tag{13.61}$$

The critical velocity for energy transfer is

$$v^* = \left(\frac{4\pi^2 kT\nu}{\alpha\mu}\right)^{1/3} \tag{13.62}$$

The reduced mass appropriate to a pair of molecules is represented by μ, and the reduced mass appropriate to the atoms in the molecule is written as μ_{int}.

Herzfeld and his colleagues[146] have shown that the change of critical velocity during the energy transfer and the change due to the attractive potential introduce corrections:

$$P_{10} = \frac{4\pi}{\sqrt{3v}} n_0 \varsigma_0^2 B v^* \exp\left(-3y + \frac{h}{2kT} + \frac{D}{kT}\right) \qquad (13.63)$$

where D is the depth of the potential minimum. For interactions which are not dominated by the exponential repulsion, some additional terms must be added to this equation[147]; even the above correction terms must be regarded as uncertain.

Even if the repulsive potential constant α is not previously known, the temperature dependence of the relaxation times provides a good test of these equations. A critical survey of the data has been made by McCoubrey *et al.*[148] and it is shown that the temperature dependence is as predicted; the correction terms in equation 13.63 bring the values of α down to the level of those obtained from gas viscosity measurements. Repulsion constants are tabulated in Table 13.8. The experimental data from which this comparison with theory is made are taken from a large number of authors[149]. The values of P_{10} are not listed, but typically they are of the order of 10^{-6} for a diatomic molecule at room temperature, rising to about 10^{-3} at $1000°K$. In certain gases, anomalies such as sticky collisions lead to high values[150]. Polyatomic molecules have been treated according to a theory worked out by Tanczos[151]. Vibrationally excited polar molecules possess larger P_{10} than non-polar molecules[152].

The experimental data are much less profuse for collision processes $0^v0/00^v$ in which vibrational energy is exchanged between molecules, with only a small part converted into kinetic energy. Denoting these molecules by their vibrational quantum number, it has been seen that the most-probable such collisions could be written $30/21$, $20/11$, etc. For identical molecules some shock wave tube measurements of probabilities per collision have been reported[153, 154].

Table 13.8. COMPARISON OF RANGE PARAMETERS

Molecule	α (experimental) (Å^{-1})	α (viscosity) (Å^{-1})
O_2	6·5	5·5
N_2	7·1	5·1
Cl_2	4·9, 6·7	4·9
Br_2	6·3	5·0
N_2O	7·5	5·1
COS	6·6	4·9
CS_2	5·9	4·8
CH_4	6·5	5·0
CF_4	7·6	4·2
C_2H_4	6·0	4·7
C_3H_6	8·2	4·0

There is also some spectroscopic evidence of vibrational excitation exchange between unlike molecules[155]; in particular, NO $A^2\Sigma$ can exchange with $N_2\,X^1\Sigma$, CO $X^1\Sigma$ and NO $X^2\Pi$. The probabilities are proportional to a function of the type exp $(|\Delta E|/kT)$; that is, they are larger the more closely the process approaches accidental resonance (*Figure 13.3*). The process

$$O_2^v + O_2^v \rightarrow O_2{}^1\Delta_g + O_2$$

has been reported.

Evidence of exchange between unlike molecules may be obtained from ultrasonic dispersion in binary mixtures[156]. It is not easy to extract this

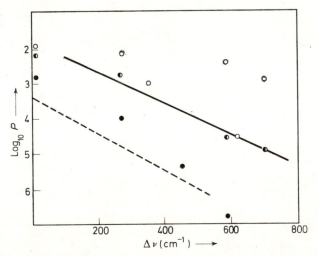

Fig. 13.3. Collision numbers for vibrational energy exchange as functions of frequency difference: broken line, calculations for two non-hydride diatomic molecules; closed circles, data for non-hydride diatomic molecules; full line, calculation for one diatomic molecule and one polyatomic hydride; divided circles, data for one diatomic molecule and one polyatomic hydride; open circles, data for one diatomic molecule and one triatomic hydride

evidence, since there are four 0ᵛ0/00 processes, which give rise to two relaxation times:

$$\tau_1^{-1} = \frac{1-x}{\tau_{AA}} + \frac{x}{\tau_{AB}} \tag{13.64}$$

$$\tau_2^{-1} = \frac{x}{\tau_{BB}} + \frac{1-x}{\tau_{BA}} \tag{13.65}$$

for molar fraction x and collisions between molecules *AA*, *BB*, *AB* and *BA*. Where two relaxation times are observed, their variation with molar fraction is consistent with these equations, so the fast equilibrium between A^v and B^v has usually relatively little influence.

The collision process in which an excited atom is deactivated by a molecule, which is thus excited vibrationally, was first reported by Zemansky[157].

The emission of the atomic radiation is quenched by such processes as:

$$\text{Hg }^3P_1 + N_2 \rightarrow \text{Hg }^3P_0 + N_2^v$$
$$\text{Hg }^3P_1 + CO \rightarrow \text{Hg }^3P_0 + CO^v$$
$$\text{Hg }^3P_1 + NO \rightarrow \text{Hg }^3P_0 + NO^v$$
$$\text{Na }^2P + N_2 \rightarrow \text{Na }^2S + N_2^v$$
$$\text{Na}^2 P + CO \rightarrow \text{Na }^2S + CO^v$$
$$\text{Na }^2P + NO \rightarrow \text{Na }^2S + NO^v$$

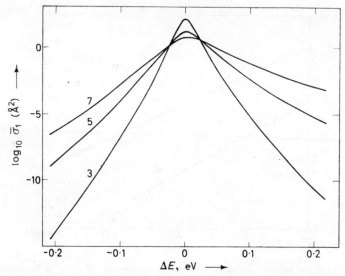

Fig. 13.4. $\text{Log}_{10}\,\bar{\sigma}_1$ *(300°K) where* $\sigma_1 = \sigma/u_e^2 u_{\text{vib}}^2$, *for effective quenching cross-sections of mercury resonance radiation, shown as a function of energy defect* ΔE; *values of* α *in units of* 10^8 cm^{-1} *are indicated;* u_e^2 *and* u_{vib}^2 *may be regarded as scaling factors, of which the former is constant for invariant molecular mass, and the latter is inversely proportional to the frequency of vibration*

(From Dickens, Linnett and Sover[158], by courtesy of the Faraday Society)

The Born approximation, and also the method of distorted waves, have been applied to calculating the cross-sections for such collisions[158]. It is found that the process is extremely improbable except where the energy defect is very small or where there is a pseudo-crossing[159]; it is also very sensitive to the exponential repulsive interaction potential parameter α. *Figure 13.4* gives the results of these calculations for quenching of the mercury line, at 300°K, for different values of α: it is profitable to insert $\alpha \sim 5 \times 10^8$ cm^{-1} obtained from ultrasonic relaxation data[153]. Reasonable semi-quantitative accord with experiment is found, but the possibility of pseudo-crossing of potential energy curves must be taken into account for a comprehensive treatment. It is not clear whether this collision process will always behave as an accidental resonance process, that is, be probable only for small $|\Delta E|$.

The process of vibrational deactivation of high quantum number levels has been studied[160] in atmospheres of inert gas, using infra-red emission to measure the vibrational populations[161], and flash photolysis and exothermic chemical reactions[162] to produce the high quantum number populations. It is found that the most efficient deactivators are free radicles, which show incipient chemical activity, and molecules with nearly identical vibrational energies.

Where a single molecular species is highly vibrationally excited, it approaches a vibrational Boltzmann equilibrium long before it approaches a kinetic equilibrium.

Rotational Relaxation

Since rotational quanta are much smaller than vibrational quanta, it is expected and found that rotational relaxation times are very small ($\sim 10^{-9}$ sec). The times are very difficult to measure; for hydrogen it has been achieved[163] by means of ultrasonic dispersion. It was found that $P_{exp}^{-1} = 350$, which agrees well with the values $P_{20}^{-1} = 329$ and $P_{31}^{-1} = 338$, calculated[164] by a distorted-wave technique. For heavier homonuclear molecules, it has been shown[165] that the mean deactivation number is given by

$$\langle P^{-1} \rangle = 2 \left(\frac{d_0}{r_0} \right)^{-2} \tag{13.66}$$

where d_0 is the molecular internuclear distance and r_0 the kinetic collision diameter; but this expression leads to values which are four times larger than the experimental values for O_2, N_2 and HCl, for which $\langle P^{-1} \rangle$ is 4·1, 5·3 and 7 respectively[166-168]. For polyatomic molecules the values are even closer to unity.

The study of rotational relaxation has been stimulated by the use of the electrostatic quadrupole selector for rotational states[169]. In measurements using the polar molecule TlF, cross-sections with $\Delta J = 1$ were found to be ten times larger than those with $\Delta J = 2$. The cross-sections with $J = 2$ and 3 were found to vary from 2 $Å^2$ for helium up to 500 $Å^2$ for NH_3, in which dipole alignment is possible during the collision.

Collision-induced transitions between rotational levels have been studied using microwave double resonance technique; the selection rules for these transitions were verified experimentally[170].

The exchange of rotational and electronic energy of excitation has been studied[171] for the alkali doublet resonance states. The conversion of one sublevel into the other by collisions with molecules which thereby become rotationally excited is a probable process, with cross-sections dependent upon the energy defect.

Some recent observations suggest the possibility of efficient interconversion of rotational and vibrational energy[172, 173].

REFERENCES

1. CH'EN, S. and TAKEO, M. *Rev. mod. Phys.* 29 (1957) 20.
2. BREENE, R. G. *Rev. mod. Phys.* 29 (1957) 94.
3. VAN REGEMORTEUR, H. *Ann. Rev. Astrophys.* 3 (1965) 71.

4. BREENE, R. G. *Handbuch der Physik*, *Vol. 27*, 1964. Berlin; Springer-Verlag.
5. GRIEM, H. R. *Plasma Spectroscopy*, 1964. New York; McGraw-Hill.
6. HAMBERGER, S. *Proc. 6th Int. Conf. Ioniz. Phenom. in Gases*, p. 1919, 1961. Amsterdam; North Holland Publishing Company.
7. LORENTZ, H. A. *Proc. Acad. Sci. Amst.* 8 (1906) 591.
8. VAN VLECK, J. H. and WEISSKOPF, V. F. *Rev. mod. Phys.* 17 (1945) 227.
9. HOLSTEIN, T. *Phys. Rev.* 79 (1950) 744.
10. WEISSKOPF, V. *Phys. Z.* 34 (1933) 1.
11. FOLEY, H. M. *Phys. Rev.* 69 (1946) 616.
12. MARGENAU, H. *Phys. Rev.* 82 (1951) 156.
13. JABLONSKI, A. *Phys. Rev.* 68 (1945) 78.
14. MARGENAU, H. and WATSON, W. W. *Rev. mod. Phys.* 11 (1939) 1.
15. HOLTSMARK, J. *Ann. Phys., Lpz.* 58 (1919) 577; *Z. Phys.* 20 (1919) 162.
16. KIVEL, B., BLOOM, S. and MARGENAU, H. *Phys. Rev.* 98 (1955) 495.
17. BARANGER, M. *Atomic and Molecular Processes*, Ed. D. R. Bates, 1961. New York; Academic Press.
18. BARANGER, M. *Phys. Rev.* 111 (1958) 481, 494; 112 (1958) 855.
19. GRIEM, H. R., BARANGER, M., KOLB, A. C. and OERTEL, G. *Phys. Rev.* 125 (1962) 177; GRIEM, H. R., KOLB, A. C. and SHEN, K. *Phys. Rev.* 116 (1959) 4.
20. LAMB, W. *Phys. Rev.* 134 (1965) A1429.
21. MAZING, M. A. and VRUBLEVSKAYA, N. A. *J. exp. theor. Phys.* 50 (1966) 343.
22. WEISSKOPF, V. *Phys. Z.* 34 (1933) 1.
23. KUHN, H. and OLDENBERG, O. *Phys. Rev.* 41 (1932) 72.
24. NIKERSON, J. L. *Phys. Rev.* 47 (1935) 707.
25. BUCKINGHAM, R. A. and DALGARNO, A. *Proc. Roy. Soc.* A213 (1952) 327, 506.
26. KRAMERS, H. A. and TER HAAR, D. *Bull. astr. Insts Netherlds* 10 (1946) 137.
27. BURHOP, E. H. S. and MARRIOTT, R. *Proc. phys. Soc., Lond.* A69 (1956) 271.
28. HERMAN, R. and HERMAN, L. *J. quantve Spectros. radiat. Transf.* 4 (1964) 487.
29. LINDHOLM, E. *Ark. Mat. Astr. Fys.* 32A (1945) 17.
30. FRANCK, J. *Z. Phys.* 9 (1922) 259.
31. RASETTI, F. *Nature, Lond.* 118 (1926) 47.
32. CARIO, G. *Z. Phys.* 10 (1922) 185.
33. CARIO, G. and FRANCK, J. *Z. Phys.* 11 (1922) 161; 17 (1923) 202.
34. BEUTLER, H. and JOSEPHY, B. *Z. Phys.* 53 (1929) 747.
35. A review of collision processes relevant to this chapter was published in 1934 by A. C. G. MITCHELL and M. W. ZEMANSKY, *Resonance Radiation and Excited Atoms*; reprinted in 1961 by the Cambridge University Press.
36. WOLF, F. *Ann. Phys., Lpz.* 23 (1935) 285, 627; 25 (1936) 527, 737; 27 (1936) 543; 29 (1937) 33; 30 (1937) 313.
37. STUECKELBERG, E. C. G. *Helv. phys. acta* 5 (1932) 370.
38. NIKITIN, E. E. *Chem. phys. Lett.* 2 (1968) 402.
39. GALLAGHER, A. *Phys. Rev.* 179 (1969) 105.
40. WIGNER, E. *Gott. Nachr.* (1927) 375.
41. BEUTLER, H. and EISENSCHIMMEL, W. *Z. phys. Chem.* B10 (1930) 89; *Z. Electrochem.* 37 (1931) 582.
42. LEES, J. H. and SKINNER, H. W. B. *Proc. Roy. Soc.* A137 (1932) 186.
43. ST JOHN, R. M. and FOWLER, R. G. *Phys. Rev.* 122 (1961) 1813; 128 (1962) 1749; *J. opt. Soc. Amer.* 50 (1960) 28; *Rev. sci. Instrum.* 33 (1962) 1089; *Ann. Phys. (New York)* 15 (1961) 461.
44. SHULER, K. E., CARRINGTON, T., LIGHT, J. C. *Appl. Opt. Suppl. Chem. Lasers*, 8 (1965).
45. BENNETT, W. R. *Appl. Opt. Suppl. Chem. Lasers*, 8 (1965).
46. PATEL, C. K. *Lasers (Advances)*, 2 (1967) 1.
47. STERN, O. and VOLMER, M. *Phys. Z.* 20 (1919) 183.
48. SAMSON, E. W. *Phys. Rev.* 40 (1932) 940.
49. TERENIN, A. *Z. Phys.* 37 (1926) 98.

50. JENKINS, D. R. *Proc. Roy. Soc.* A293 (1966) 493.
51. JENKINS, D. R. *Proc. Roy. Soc.* A293 (1966) 493.
52. BROCKLEHURST, B. *Trans. Faraday Soc.* 60 (1964) 2151.
53. YOUNG, R. A. and BLACK, G. *J. chem. Phys.* 47 (1967). 2311.
54. CVETANOVIC *Prog. Reaction Kinet.* 2 (1964) 39.
55. CHAPMAN, G. D., KRAUSE, L. and BROCKMAN, I. H. *Canad. J. Phys.* 42 (1964) 535.
56. JEFIMENKO, O. *Proc. 3rd Int. Conf. Phys. electron. atom. Collisions*, p. 1065, 1953. Amsterdam; North Holland Publishing Company.
57. CHAPMAN, G. D. and KRAUSE, L. *Canad. J. Phys.* 43 (1965) 563.
58. CZAJKOWSKI, M. and KRAUSE, L. *Canad. J. Phys.* 43 (1965) 1259.
59. ZENER, C. *Phys. Rev.* 38 (1931) 277.
60. JEFIMENKO, O. *J. chem. Phys.* 39 (1963) 2457.
61. FERMI, E. *Nuovo Cim.* 11 (1934) 157.
62. BEAHN, T. J., CONDELL, W. J. and MANDELBERG, H. *Phys. Rev.* 141 (1966) 83.
63. MORI, M., WATANABE, T. and KATSUURA, K. *J. phys. Soc., Japan*, 19 (1964) 380, 1504; WATANABE, T. *Abstr. 4th Int. Conf. Phys. electron. atom. Collisions*, p. 67, 1965. Hastings-on-Hudson, New York; Science Bookcrafters.
64. WOOD, R. W. *Proc. Roy. Soc.* 106 (1924) 679; *Phil. Mag.* 50 (1925) 775; 4 (1927) 466.
65. KLUMB, H. and PRINGSHEIM, P. *Z. Phys.* 52 (1928) 610.
66. DONAT, K. *Z. Phys.* 29 (1924) 345; LORIA, S. *Phys. Rev.* 26 (1925) 573.
67. LATYPOV, Z. Z., KUPRIYANOV, S. E. and TUNITSKII, N. N. *J. exp. theor. Phys.* 46 (1964) 833.
68. HIRSCHFELDER, J. O. and ELIASON, M. A. *Ann. N. Y. Acad. Sci.* 67 (1957) 451.
69. SLATER, J. C. *Phys. Rev.* 36 (1930) 57.
70. FELDMAN, P. and NOVICK, R. *Phys. Lett.* 11 (1963) 278; *Proc. 3rd Int. Conf. Phys. electron. atom. Collisions*, p. 201, 1963. Amsterdam; North Holland Publishing Company.
71. INNES, F. R. and OLDENBERG, O. *J. chem. Phys.* 38 (1963) 2306.
72. PHELPS, A. V. *Phys. Rev.* 99 (1955) 1307; BIONDI, M. A. *Phys. Rev.* 82 (1951) 543; 88 (1952) 660; PHELPS, A. V. and MOLNAR, J. P. *Phys. Rev.* 89 (1953) 1203; PHELPS, A. V. and PACK, J. L. *Rev. sci. Instrum.* 26 (1955) 45; PHELPS, A. V. Westinghouse Scientific Paper 6-94439-6-P3, 1957.
73. HAMBERGER, S. *Proc. 6th Int. Conf. Ioniz. Phenom. in Gases*, p. 1919, 1961. Amsterdam; North Holland Publishing Company.
74. EBBINGHAUS, E. *Ann. Phys., Lpz.* 7 (1930) 267.
75. MOLNAR, J. P. *Phys. Rev.* 83 (1951) 933.
76. ENGSTROM, R. W. and HUXFORD, W. S. *Phys. Rev.* 58 (1940) 67.
77. McCOUBREY, A. O. and MAITLAND, C. G. *Phys. Rev.* 101 (1956) 603.
78. MEISSNER, K. W. and GRAFFUNDER, W. *Ann. Phys., Lpz.* 84 (1927) 1009.
79. ZEMANSKY, M. W. *Phys. Rev.* 34 (1929) 213.
80. SAMSON, E. W. *Phys. Rev.* 40 (1932) 940.
81. WEBB, H. W. and MESSENGER, H. A. *Phys. Rev.* 40 (1932) 466; 33 (1929) 319.
82. COULIETTE, J. H. *Phys. Rev.* 32 (1928) 636.
83. WEBB, H. W. *Phys. Rev.* 24 (1924) 113.
84. STEBBINGS, R. F. *Proc. Roy. Soc.* A241 (1957) 270.
85. HASTED, J. B. and MAHADEVAN, P. *Proc. Roy. Soc.* A249 (1958) 42.
86. RICHARDS, H. L. and MUSCHLITZ, E. E. *J. chem. Phys.* 41 (1964) 559.
87. ROTHE, E. W., NEYNABER, R. H. and TRUIJILLO, S. M. *J. chem. Phys.* 42 (1965) 3310.
88. ROTHE, E. Private communication, 1962.
89. BURHOP, E. H. S. *Proc. phys. Soc., Lond.* A67 (1954) 276.
90. JESSE, W. P. and SADAUSKIS, J. *Phys. Rev.* 88 (1952) 417.
91. KRUITHOF, A. A. and DRUYVESTEYN, M. J. *Physica, Eindhoven*, 4 (1937) 450; KRUITHOF, A. A. and PENNING, F. M. *Physica, Eindhoven*, 4 (1937) 430; GLOTOV, I. I. *Phys. Z. Sowjet.* 12 (1937) 256.
92. KASNER, W. H. and BIONDI, M. A. Westinghouse Report 64-928-113-P7, 1964.

93. HOLLSTEIN, M., SHERIDAN, J. R., PETERSON, N. R. and LORENTS, D. C. *Proc. 6th Int. Conf. Phys. electron. atom. Collisions*, p. 967, 1969. Cambridge, Mass.; Massachusetts Institute of Technology Press.
94. EVANS, S. A. and LANE, N. F. *Bull. Am. Phys. Soc.* 14 (1968) 262; *Proc. 6th Int. Conf. Phys. electron. atom. Collisions*, p. 983, 1969. Cambridge, Mass.; Massachusetts Institute of Technology Press.
95. KVIFTE, G. and VEGARD, L. *Geofys. Publ.* 17 (1947) 1.
96. HARTECK, P. and REEVES, R. R. *Disc. Faraday Soc.* 1964.
97. ARNOLD, S. G., FINLAYSON, N., OGRYZLO, E. A. *J. chem. Phys.* 44 (1966) 2529.
98. VRIENS, L., BONSEN, T. F. M., WALLINGA, H. and FLUIT, J. M. *Proc. 5th Int. Conf. Phys. electron. atom. Collisions*, p. 58, 1967. Leningrad; Akademii Nauk.
99. SOBELMAN, I. I. *J. exp. theor. Phys.* 48 (1965) 965.
100. MARRIOTT, R. *Proc. phys. Soc., Lond.* A70 (1957) 288.
101. RADCLIFFE, S. W., McLAREN, T. I. and HOBSON, R. M. *New Experimental Techniques in Propulsion and Energetics Problems*, 1967. Munich; AGARD.,
102. WATANABE, T. and KATSUURA, K. *Gaseous Electronics Conf.*, American Physical Society, 1966.
103. BATES, D. R. and DAMGAARD, A. *Phil. Trans.* A242 (1949) 101.
104. LAMB, W. E. and RETHERFORD, R. C. *Phys. Rev.* 79 (1950) 549; 81 (1951) 222.
105. HAGSTRUM, H. D. *J. appl. Phys.* 31 (1960) 897.
106. BENTON, E. E., FERGUSON, E. E., MATSEN, F. A. and ROBERTSON, W. W. *Phys. Rev.* 128 (1962) 206; FERGUSON, E. E. *Phys. Rev.* 128 (1962) 210.
107. MUSCHLITZ, E. E. and SHOLETTE, W. P. University of Florida Report G-5967, 1960.
108. SCHUT, T. G. and SMIT, J. A. *Physica, Eindhoven*, 10 (1943) 440.
109. HOFFMAN, A. *Z. Phys.* 119 (1942) 223.
110. HOTOP, H. and NIEHAUS, A. *Proc. 6th Int. Conf. Phys. electron. atom. Collisions*, p. 882, 1969. Cambridge, Mass.; Massachusetts Institute of Technology Press.
111. BELL, R. J., DALGARNO, A. and KINGSTON, A. E. *Proc. phys. Soc.* B1 (1968) 18.
112. LUDLAM, *et al. J. chem. Phys.* 46 (1967) 127.
113. CERMAK, V. *J. chem. Phys.* 44 (1966) 1318, 3774, 3781; CERMAK, V. and HERMAN, Z. *Coll. Czech. chem. Commun.* 29 (1964) 953; 30 (1965) 169; CERMAK, V. *13th American Society for Testing and Materials Conf.* 1965; OLMSTED, *et al. J. chem. Phys.* 42 (1965) 2321.
114. CERMAK, V. *J. chem. Phys.* 44 (1966) 1318, 3774, 3781.
115. HOTOP, H. and NIEHAUS, A. *Proc. Heavy Particle Collisions Conf.* p. 194, The Institute of Physics and The Physical Society, 1968.
116. ARNOT, F. L. and McEWAN, M. B. *Proc. Roy. Soc.* A166 (1938) 543; A171 (1939) 106; ARNOT, F. L. and MILLIGAN, J. C. *Proc. Roy. Soc.* A153 (1936) 359; A165 (1938) 133.
117. MOHLER, F. L. *J. Res. nat. Bur. Stand.* 5 (1930) 51, 399.
118. MUNSON, R. J., FRANKLIN, J. L. and FIELD, J. H. *J. chem. Phys.* 36 (1962) 3332; 37 (1962) 1790; 38 (1963) 1542; BECKER, P. M. and LAMPE, F. W. *J. chem. Phys.* 42 (1965) 3857.
119. HUFFMAN, R. E. and KATAYAMA, D. H. *J. chem. Phys.* 45 (1966) 138; TETER, M. P., NILES, F. E. and ROBERTSON, W. W. *Gaseous Electronics Conf.* American Physical Society, 1965.
120. HERMAN, Z. and CERMAK, V. *Proc. 5th Int. Conf. Phys. electron. atom. Collisions*, p. 602, 1967. Leningrad; Akademii Nauk.
121. HENGLEIN, A. and MUCCINI, J. A. *Z. angew. Chem.* 72 (1960) 630; *Z. Naturf.* 15a (1960) 584.
122. KAUL, W. and TAUBERT, R. *Z. Naturf.* 17a (1962) 88; 18a (1963) 432.
123. COMES, F. J. *Z. Naturf.* 18a (1963) 539.
124. HERMAN, Z. and CERMAK, V. *Nature, Lond.* 199 (1963) 588.
125. PAHL, M. *Z. Naturf.* 18a (1963) 1276.
126. FONTIJN, A. and BAUGHMAN, G. L. *J. chem. Phys.* 38 (1963) 1784.
127. KOYANO, I. and TANAKA, I. *J. chem. Phys.* 40 (1964) 2734.

128. CERMAK, V. *J. chem. Phys.* 43 (1965) 4527.
129. DONNALLY, B., CLAPP, T., SAWYER, W. and SCHULTZ, M. *Phys. Rev. Lett.* 42 (1964) 502.
130. RIVIERE, A. C. and SWEETMAN, D. R. *Proc. 3rd Int. Conf. Phys. electron. atom. Collisions*, p. 734, 1963. Amsterdam; North Holland Publishing Company; BUTLER, S. T. and MAY, R. M. University of Sydney School of Physics Report T. 12, 1964.
131. PERSON, J. C. *Proc. 6th Int. Conf. Phys. electron. atom. Collisions*, p. 990, 1969. Cambridge Mass.; Massachusetts Institute of Technology Press.
132. HOTOP, H. and NIEHAUS, A. *J. chem. Phys.* 47 (1967) 2506; *Z. Phys.* 215 (1968) 215, 395.
133. MELTON, C. E. and HAMILL, W. H. *J. chem. Phys.* 41 (1964) 1469.
134. CALLEAR, A. B. *Photochemistry and Reaction Kinetics*, ed. P. G. Ashmore, F. S. Dainton and T. M. Sugden, p. 133, 1967. Cambridge; Cambridge University Press.
135. B. STEVENS, *Collisional Activation*, 1967. Oxford; Pergamon Press.
136. MAGEE, J. L. and RI, T. *J. chem. Phys.* 9 (1941) 638; LAIDLER, K. J. *J. chem. Phys.* 10 (1942) 34, 43.
137. BATES, D. R. *J. atmos. terr. Phys.* 6 (1955) 171; BATES, D. R. and MOISEIWITSCH, B. L. *J. atmos. terr. Phys.* 8 (1956) 305.
138. BETHE, H. A. and TELLER, E. Aberdeen Proving Ground Report X, 1941, p.117.
139. WINDSOR, M. W., DAVIDSON, N. and TAYLOR, R. *7th Combustion Symp.* London, 1958.
140. HUBER, P. W. and KANTROWITZ, A. *J. chem. Phys.* 15 (1947) 275.
141. MORGAN, J. E. and SCHIFF, H. I. *Canad. J. Chem.* 41 (1963) 903.
142. SLOBODSKAYA, P. V. *Izv. Akad. Nauk SSSR, Ser. Fiz.* 12 (1948) 656.
143. JACOX, M. E. and BAUER, S. H. *J. phys. Chem.* (1957) 833.
144. LANDAU, P. and TELLER, E. *Phys. Z. Sowjet.* 10 (1936) 34.
145. COTTRELL, T. L. and REAM, N. *Trans. Faraday Soc.* 51 (1955) 159, 1453.
146. SCHWARTZ, R. N., SLAWSKY, Z. I. and HERZFELD, K. F. *J. chem. Phys.* 10 (1952) 1591; SCHWARTZ, R. N. and HERZFELD, K. F. *J. chem. Phys.* 22 (1954) 767.
147. WIDOM, B. *Disc. Faraday Soc.* 33 (1962) 37.
148. McCOUBREY, J. C., MILWARD, R. C. and UBBELOHDE, A. R. *Trans. Faraday Soc.* 57 (1961) 1472.
149. BLACKMAN, V. H. *J. Fluid. Mech.* 1 (1956) 61; GAYDON, A. G., CLOUSTON, J. G. and GLASS, I. I. *Proc. Roy. Soc.* A248 (1958) 429; LUKASIK, S. J. and YOUNG, J. E. *J. chem. Phys.* 27 (1957) 1149; HUBER, P. W. and KANTROWITZ, A. *J. chem. Phys.* 15 (1947) 275; KNOETZEL, H. and KNOETZEL, L. *Ann. Phys., Lpz.* 2 (1948) 393; SMILEY, E. F. and WINKLER, E. H. *J. chem. Phys.* 22 (1954) 2018; EUCKEN, A. and BECKER, R. *Z. phys. Chem.* B20 (1933) 467; SHIELDS, F. D. *J. acoust. Soc. Amer.* 32 (1960) 180; RICHARDSON, E. G. *J. acoust. Soc. Amer.* 31 (1959) 152; EUCKEN, A. and AYBAR, S. *Z. phys. Chem.* B46 (1940) 195; COTTRELL, T. L. and MARTIN, P. E. *Trans. Faraday Soc.* 53 (1957) 1157; EDMONDS, P. D. and LAMB, J. *Proc. phys. Soc., Lond.* 72 (1958) 940; EUCKEN, A. and NIIMANN, N. *Z. phys. Chem.* B36 (1937) 163; BUSCHMANN, K. F. and SCHAFER, K. *Z. Phys. Chem.* B50 (1941) 73; EUCKEN, A. and JAACKS, H. *Z. phys. Chem.* B30 (1935) 85; CORRAN, P., LAMBERT, J. D., SLATER, R. and WARBURTON, B. *Proc. Roy. Soc.* A244 (1958) 212; McGRATH, W. D. and UBBELOHDE, A. R. *Proc. Roy. Soc.* A227 (1954) 1; LAMBERT, J. D. and ROWLINSON, J. S. *Proc. Roy. Soc.* A204 (1950) 424; FRICKE, E. F. *J. acoust. Soc. Amer.* 12 (1940) 245; AMME, R. and LEGVOLD, S. *J. chem. Phys.* 30 (1959) 163; GRAVITT, J. C. *J. Acoust. Soc. Amer.* 32 (1960) 560; ANGONA, F. C. *J. acoust. Soc. Amer.* 25 (1953) 1116; RICHARDS, W. T. and REID, J. A. *J. chem. Phys.* 2 (1934) 193.
150. BAUER, H. J., KNESER, H. O. and SITTIG, E. *J. chem. Phys.* 30 (1959) 1119.
151. TANCZOS, F. I. *J. chem. Phys.* 25 (1956) 439.
152. BASCO, N., CALLEAR, A. B. and NORRISH, R. G. W. *Proc. Roy. Soc.* A260 (1960) 459.
153. HERZFELD, K. F. and LITOWITZ, T. A. *Absorption and Dispersion of Ultrasonic Waves*, 1960. New York; Academic Press.

154. COTTRELL, T. L. and McCOURBEY, J. C. *Molecular Energy Transfer in Gases*, 1961. London; Butterworth.
155. CALLEAR, A. B. *Disc. Faraday Soc.* 33 (1962) 28.
156. LAMBERT, J. D., EDWARDS, A. J., PEMBERTON, D. and STRETTON, J. L. *Disc. Faraday Soc.* 33 (1962) 61.
157. ZEMANSKY, M. W. *Phys. Rev.* 36 (1930) 919.
158. DICKENS, P. G., LINNETT, J. W. and SOVER, O. *Disc. Faraday Soc.* 33 (1962) 52; WITTEMAN, W. J. *J. chem. Phys.* 35 (1961) 1.
159. POLANYI, J. C. *J. quantve. Spectros. radiat. Transf.* 3 (1963) 471.
160. LIPSCOMB, F. J., NORRISH, R. G. W. and THRUSH, B. A. *Proc. Roy. Soc.* A233 (1956) 455.
161. BROIDA, H. P. and CARRINGTON, T. *J. chem. Phys.* 38 (1963) 136.
162. GARVIN, D., BROIDA, H. P. and KASTOWSKI, H. J. *J. chem. Phys.* 23 (1960) 880.
163. STEWART, J. L. and STEWART, E. S. *J. acoust. Soc. Amer.* 24 (1952) 194.
164. BROUT, R. *J. chem. Phys.* 22 (1954) 934.
165. BROUT, R. *J. chem. Phys.* 22 (1954) 1189.
166. GREENSPAN, M. *J. acoust. Soc. Amer.* 31 (1959) 155.
167. ANDERSON, W. H. and HORNIG, D. F. *Molec. Phys.* 2 (1959) 49.
168. BREAZEALE, M. A. and KNESER, H. O. *J. acoust. Soc. Amer.* 32 (1960) 885.
169. TOENNIES, J. P. *Disc. Faraday Soc.* 33 (1962) 96.
170. OKA, T. *J. chem. Phys.* 45 (1966) 754; 47 (1967) 13; RONN, R. M. and WILSON, E. B. *J. chem. Phys.* 46 (1962) 3262.
171. KRAUSE, L. *Appl. Opt.* 5 (1966) 1375.
172. MILLIKAN, R. and OSBURG, C. A. *J. chem. Phys.* 41 (1964) 2196.
173. COTTRELL, T. L. and MATHESON, N. J. *Trans. Faraday Soc.* 58 (1962) 2336; 59 (1963) 824.
174. DUNNING, F. B. Thesis, University of London, 1970.

Chapter 14

CHEMICAL INTERCHANGE COLLISIONS

In the first edition of this book the final chapter was devoted to the ion–atom interchange process[1-6], whilst a short discussion of chemical reactive scattering was included in Chapter 10. There has been much recent interest in the study of the chemical interchange collision using scattering techniques

$$A + BC \rightarrow AB + C$$

These subjects will now be grouped together, and a brief study of the collisional aspects of the chemical interchange process will be attempted—introduced by way of ion–atom interchange but without a comprehensive discussion of chemical gas reactions.

14.1 ION-MOLECULE REACTION CLASSIFICATION. TECHNIQUES OF TOTAL CROSS-SECTION MEASUREMENT

The importance of ion–molecule reactions arises from their roles in the chemistry of discharges, in gaseous radiation chemistry, flames, mass-spectrometry, and planetary aeronomy. They can be classified as follows:

1. Charge transfer

$$A^+ + BC \rightarrow A + BC^+$$

2. Dissociative charge transfer

$$A^+ + BC \rightarrow A + B^+ + C$$
$$A^+ + BC \rightarrow A + C^+ + B$$

3. Ion interchange

$$AB^+ + C \rightarrow A + BC^+$$

4. Atom interchange

$$A^+ + BC \rightarrow AB^+ + C$$

In polyatomic systems many processes might be taken to be either ion interchange or atom interchange, but a distinction can still be made between the two classifications.

5. Addition (the rare 'sticky reaction' which can only take place via stabilization, possibly through inverse predissociation)

$$AB^+ + C \rightarrow ABC^+$$

6. Condensation—a term applied to such processes as

$$CH_3^+ + CH_4 \rightarrow C_2H_5^+ + H + H$$
$$CH_3^+ + CH_4 \rightarrow C_2H_5^+ + H_2$$

7. Proton transfer—processes such as

$$H_2S^+ + H_2O \rightarrow HS + H_3O^+$$

8. Hydrogen atom transfer—processes such as

$$C_2H_4^+ + C_2H_4 \rightarrow C_2H_5^+ + C_2H_3$$

9. Hydride ion (H^-) transfer

$$R_1^+ + R_2H \rightarrow R_1H + R_2^+$$

where R is a radicle.

Three-body processes of all these types are also possible.

Experimental techniques for ion–molecule reaction study include (*i*) mass-spectrometer sources, (*ii*) beam–gas, crossed-beam and merged-beam experiments, (*iii*) drift tubes, and (*iv*) afterglows. Ion cyclotron resonance experiments will not be discussed at this stage, although their promise is great.

Mass-spectrometer sources were used in pioneer ion–atom interchange studies[1-3]. The technique is inferior to crossed-beam, drift and afterglow methods, but is of value for survey studies, especially where complex molecules are involved. The production of the primary ions by radiation processes is an important advance[7] in this technique. An electron impact ion source is shown in *Figure 3.24* (page 190). The confinement of the electron beam by a magnetic field ensures that only 'primary' ions are formed by direct electron–molecule collisions in a narrow region. However, in the operation of the source, other 'secondary' ion currents detected in the mass-spectrometer are produced by ion–molecule collisions; this is supported by the following observations:

1. The secondary ion appearance potential corresponds not to the energy that would be appropriate to direct formation, but to the known appearance potential of another ion;
2. The secondary ion abundance is proportional to the square of the gas pressure in the source;
3. Many of the secondary ions are of such composition that a direct formation from the gas molecule would not be possible.

An 'apparent cross-section' Q of the experiment may be deduced, assuming single collision conditions, and knowing the distance l between the electron–molecule collision region and the source exit slit. For the process

$$A^+ + BC \rightarrow AB^+ + C$$

$$\frac{I_{AB^+}}{I_{A^+}} = Qln_0 \tag{14.1}$$

The impact energy distributin of the primary ions in their passage through the gas is wide, since they are being constantly accelerated. The apparent cross-section must be taken as an averaged value $\bar{\sigma}(\bar{E})$ and may be very different from $\sigma(E)$.

An assumed form of the cross-section function is combined with an assumed ion mean energy to yield a rate constant. It will be recalled that an ion moving in an inverse fourth power polarization field of force will pass into inward spiralling orbits when the impact parameter is less than a value given by

$$b_{orb}^4 = \frac{4\alpha e^2}{\mu v^2} \tag{14.2}$$

for molecule polarizability α. If all such collisions result in an (exothermic) ion–atom interchange process, then the cross-section[8] will be

$$\sigma = \pi b_{orb}^2 = \frac{\pi}{v} \left(\frac{4\alpha e^2}{\mu} \right)^{1/2} \tag{14.3}$$

and the crude rate constant σv will be energy independent.

The rate constant is deduced from the apparent cross-section Q (equation 14.1) as follows. The rate of formation of secondary ions is written:

$$\frac{dN_{AB+}}{dt} = kN_{A+}n_{BC} \tag{14.4}$$

the symbols N represent total numbers, but, as always, n represents a density. Substituting this into equation 14.1 gives:

$$\frac{dN_{AB+}}{dt} = \frac{kN_{A+}I_{AB+}}{I_{A+}lQ} \tag{14.5}$$

Since

$$I_{AB+} = e\,\frac{dN_{AB+}}{dt} \tag{14.6}$$

and

$$I_{A+} = e\,\frac{N_{A+}\bar{v}_1}{l} \tag{14.7}$$

where $\bar{v}_1$ is the (distance) average velocity of the A^+ ions, one has

$$\bar{v}_1 Q = k \tag{14.8}$$

In the electric field X produced by the application of potential between the repeller electrode, and the extraction orifice the final velocity achieved by the A^+ ions is given by

$$\tfrac{1}{2}m_1\bar{v}_1^2 = eXl \tag{14.9}$$

If the ions start with zero velocity, then

$$2\bar{v}_1 = v_1 \tag{14.10}$$

so

$$\bar{v}_1 = \left(\frac{eXl}{2m_1} \right)^{1/2} \tag{14.11}$$

Thus the rate constant is related to Q:

$$k = \left(\frac{Xel}{2m_1}\right)^{1/2} Q \tag{14.12}$$

The use of this equation to calculate rate constants from apparent cross-sections is only valid when the dependence $Q \propto X^{-1/2}$ is observed experimentally.

The use of equation 14.12 implies that the averaging of charges to obtain currents is carried out over distance; in a more rigid treatment it would be necessary to average over time, as has been done by Field, Franklin and Lampe[3].

In most mass-spectrometer source experiments, the gas pressure is maintained sufficiently low (say 10^{-5} torr) for single collision conditions to prevail. However, it is possible[9] to operate at higher pressures ($> 10^{-2}$ torr,) so that swarm analysis can be applied, the repeller electrode is adjusted in such a way that the ratio of field strength to pressure (X/p) achieves a value which can be related to the mean energy (estimated from independent ion mobility experiments). This is a crude version of the drift tube technique discussed below.

Mass-spectrometer source experiments can be refined in a number of ways. The primary ion identity can be made more certain by the use of suitably monochromated vacuum ultra-violet photons to produce the ionization. It is even more satisfactory to inject a mass-analysed ion beam into the source chamber, instead of the electron beam; but the ion energy cannot easily be lowered beyond ~ 3 eV.

A criticism that is made of conventional mass-spectrometer source experiments is that square law pressure dependence of a (secondary) ion current could also refer to ions formed by ionization of neutral products of chemical reactions, and also to products of chemi-ionization by excited neutrals.

A modification of the mass-spectrometer source technique which allows room temperature rates to be measured is due to Talrose and Frankevich[10]. Through a low-pressure mass-spectrometer source containing gas, a $t_i = 1$ μsec pulse of 50 eV electrons is directed. There elapses a delay time t_d, of up to 8 μsec, after which a pulse of potential is applied for time $t_s = 2\cdot5$ μsec to the repeller electrode so as to sweep the positive ions through a slit into the mass-spectrometer. During the delay there is no appreciable electric field in the source, and the primary ions, whose density is n_1^+, are free to react at near-thermal energy with the gas in the source. (Atomic ions formed with appreciable kinetic energy will not equilibrate kinetically, so they cannot be studied by means of this technique.) The rate of formation of product ions by interchange collisions at thermal energy is given by

$$\frac{dn_2^+}{dt} = k_i n_0 n_1^+ \tag{14.13}$$

or

$$n_2^+ = k_i n_1^+ n_0 t \tag{14.14}$$

where n_0 is the appropriate neutral atom density. The experiment consists of a comparison of the ratio of ion intensities I_2^+/I_1^+ at different delay times $t = t_d$. It is assumed that the currents are proportional to the source ion densities, which will only be the case if the extraction efficiency is mass-independent, if $t_i + t_d$ is sufficiently short for thermal diffusion to the walls to be negligible, and if t_s is sufficiently long for the extraction field to sweep ions even from the further side of the chamber. The linear variation of I_2^+/I_1^+, not only with n_0 but also with t_d, demonstrates the internal consistency of the method. However, the limitations on t_d are severe, and there can only be one order of magnitude between t_d and t_s. A pulsed photon source has been used in similar experiments[11]. Attempts have been made[12] to adapt the pulse technique for supra-thermal energies by the addition of a steady repeller potential, but this has the disadvantage that, as with the original mass-spectrometer source technique, the collisions take place over a wide range of energies. The potential can be applied to the repeller in the form of a pulse[13, 14], sufficiently short for the ions to achieve a uniform energy in a distance short compared with the collision path.

Beam–gas and (preferably) crossed-beam techniques are the most refined methods of total and differential cross-section measurement, but the lower limit of energy is probably about 1 eV. The principal problem is that of collection, with identification, of the ion species. The secondary beam possesses a considerable fraction of the primary beam's forward momentum, and the secondary ion momentum vectors are distributed over a laboratory solid angle which can be wide. One of the best collection techniques is by means of a wide-angle lens which focuses the collision products onto the entrance orifice of a mass-spectrometer[15–18]. The crossed-beam is to be preferred to beam–gas technique because of the superior definition of the collision region and the difficulty of arranging wide-angle collection in a gas collision chamber.

Experiments have been carried out with collision products extracted by electric fields both perpendicular to, and parallel to, the impacting beam. Since there is considerable momentum transfer in the interchange collision, neither method is satisfactory at the lowest energies, and it is preferable to allow the collision products to pass out of the collision region without application of extraction potential. For preference, a collecting system including mass-analysis is made traversable in scattering angle in the plane of the crossed beams; differential cross-sections are measured and summed to give the total cross-section. Such an apparatus has been used successfully by Bailey and his colleagues[19]; it also incorporates post-collision momentum analysis.

The drift tube technique[20] for ion–molecule reactions covers the energy range between the thermal energies (77°–450°K) of the flowing afterglow and the lower limit (~ 1 eV) of beam experiments. A flux of ions is caused to drift under the action of an electric field in a buffer gas at pressure ~ 1 torr. The reactant gas is maintained at a much smaller pressure (unless the probability of reaction is extremely small); the proportions of primary and secondary ion (I_1, I_2) are monitored, with mass-analysis, at an exit orifice; the cross-section

$$\sigma = \frac{v_d}{n l v_r} \ln \left(1 + \frac{I_2}{I_1}\right) \tag{14.15}$$

where v_d is the drift velocity of the primary ion, $v_r = \sqrt{2T/m_+}$ is its most probable random velocity, and l is the length of the drift tube containing reactant gas of density n. Thus it is necessary to measure the drift velocity by one of the techniques outlined in Chapter 10. The mean energy of collision in the laboratory frame is taken to be

$$\bar{E} = \tfrac{1}{2}(m_+ + m_g)v_d^2 + \tfrac{3}{2}kT \qquad (14.16)$$

where m_g is the mass of the buffer gas atom, which is chosen for its unreactivity. Radial diffusion corrections must be applied[20].

A variety of techniques can be used to provide the drift tube with primary ions. The most flexible is to inject a mass-separated ion beam of some hundreds of electron volts energy through an orifice. Thermalization should occur within ~ 1 cm, but tests must be made by means of drift velocity measurement. For the same pumping speed, intensity is greater for injection by way of an array of 0·025 mm capillaries of length 0·2 mm, rather than through a single orifice. A sophisticated 'injected ion drift tube' is illustrated in *Figure 14.1*; positive ions can be injected simultaneously with negative ions, or with fast neutral atoms, for study of special reactions. The drift tube is of variable length, for separation of three-body effects from two-body, and for study of the thermalization of the injected beams.

Drift tube measurements have also been made[21, 22] with electron or photon excitation of the gas in the tube, which serves as buffer, source of ions, and reactant. The range of processes which can be studied in this way is not insignificant. However, it is necessary for all these experiments to incorporate mass-spectrometric identification of the ions arriving at the rear of the drift tube. Early attempts were made to dispense with this identification and infer the identify of ions from their mobility[23], but this is dangerous and now unnecessary. The lower energy limit of the drift tube technique is governed by considerations of intensity.

The measurement of time-dependence of ion densities in the afterglow of a discharge has proved a very powerful technique for measurement of ion–atom interchange rate constants at thermal energies. A short discussion of afterglows is given in Chapter 3. The ions are allowed to thermalize in the gas, but some variation of mean impact energy can be achieved by heating the gas[24]. The initial experiments were conducted in a purely time-dependent mode[25, 26] following an initial experiment of Dickinson and Sayers[27]; however, great advantages of flexibility are achieved by operating in a fast-flowing gas in a tube of 5–20 cm diameter. In particular, the flowing afterglow enables the excitation of the reactant gas by the discharge to be avoided by introducing it downstream. In a flowing system[28, 29] with known axial flow velocity, the time variable becomes a distance variable, so a measurement of ion density at a certain distance downstream corresponds to a measurement at a certain time in the afterglow. However, it is usual to operate the experiment with pulsed excitation, so as to avoid the propagation of secondary effects (for example, those due to Penning ionization by metastables) down the tube. The decay of an ion species is governed by recombination, diffusion and ion–atom interchange processes. It is desirable to wait sufficiently long in the afterglow for recombination to be neglected without loss of accuracy. Then the kinetic equation for the space-averaged ion density n_+ in a time-dependent system is:

$$\frac{dn_+}{dt} = D_a \nabla^2 n_+ - \nu n_+ \qquad (14.17)$$

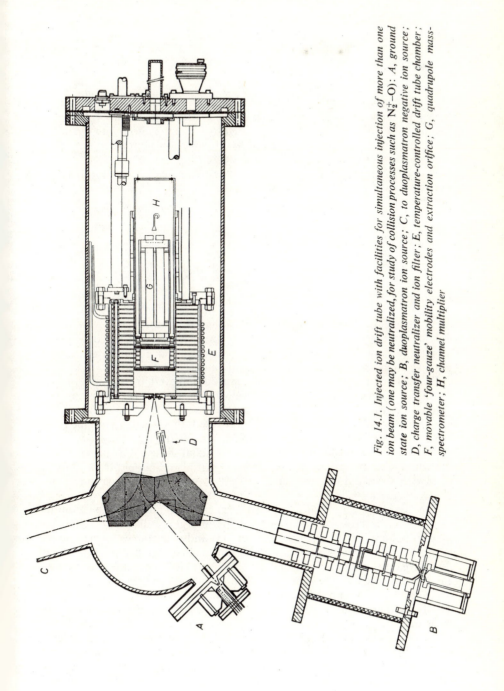

Fig. 14.1. *Injected ion drift tube with facilities for simultaneous injection of more than one ion beam (one may be neutralized, for study of collision processes such as N_2^+–O): A, ground state ion source; B, duoplasmatron ion source; C, to duoplasmatron negative ion source; D, charge transfer neutralizer and ion filter; E, temperature-controlled drift tube chamber; F, movable 'four-gauze' mobility electrodes and extraction orifice; G, quadrupole mass-spectrometer; H, channel multiplier*

where $$v = \bar{\sigma} n_0 \bar{v} \qquad (14.18)$$

$$n_+ = n_{+0} \exp\left(-\frac{t}{\tau}\right) \qquad (14.19)$$

$$\frac{p}{\tau} = \frac{p D_a}{\Lambda^2} + vp \qquad (14.20)$$

D_a is the ambipolar diffusion coefficient, Λ is the diffusion length and τ is the characteristic decay period. Since pD_a is constant, equation 14.20 enables the diffusion contribution to be separated from the ion–atom interchange by variation of total gas pressure p. A mean cross-section $\bar{\sigma}$ appropriate to mean collisional (thermal) velocity $\bar{v}$ can be deduced.

In a flowing system it is more usual to deduce the cross-section from the measured exponential decay of primary ions A^+, logarithmically dependent upon the rate Q_R of addition of reactant gas:

$$\ln\left(\frac{[A^+]}{[A^+]_0}\right) = -\frac{kQ_R l}{A\bar{v}_f^2} \qquad (14.21)$$

where $[A^+]$ is the concentration of A^+ a distance l downstream from the point of introduction of the reactant gas, and $[A^+]_0$ is the concentration at this point when no reactant gas is introduced; A is the cross-sectional area of the tube, and $\bar{v}_f$ the mean velocity of flow. Nevertheless, there are three separate measurement variables applicable to the ion density in pulsed flowing afterglow: (*i*) dependence on time after the pulse, (*ii*) dependence on distance down the tube, and (*iii*) dependence on concentration of reactant gas. All three of these can yield the same mean cross-section.

Excitation of a flowing afterglow can be achieved by passing the gas through a hollow cathode, with a central wire anode to which high tension is applied. A graphite cathode appears to introduce the least noise and sputtering. Microwave excitation has also been used and is usually applied to a resonant cavity surrounding a relatively narrow quartz tube ($\sim$ 1 cm diameter); this can result in a relatively slow radial diffusion of plasma to fill the whole afterglow tube (diameter 8–30 cm); also, high electron temperatures can persist down the tube, possibly owing to propagation of the electromagnetic wave.

The 'carrier' gas (pressure, say, 0·3 torr) can with advantage be admitted over the entire cross-sectional area of the afterglow tube, using a porous plug. If helium is used, then the helium ions and metastables formed by the excitation pulse will be able to produce ions from a specified gas introduced downstream through a fine capillary or a 'sprinkler' type tube; the reactant gas is introduced in the same way, still further downstream. A nose-cone orifice enclosing a pumped mass-spectrometer (often quadrupole) is mounted axially, and in some experiments is movable axially. Flow velocities are measured by pulse techniques, using photomultipliers or metal probes. Flow rate measurements are necessary but not sufficient, since in order to deduce the flow velocity from them, laminar flow must be assumed; this assumption is not necessarily valid, but correspondence between the two techniques is an indication that the flow is in fact laminar. A schematic diagram of a flowing afterglow system is given in *Figure 3.10* (page 153).

Merged beam technique has been applied to ion–atom interchange studies and is discussed in Section 3.2. It can be operated close to thermal energies of impact. Since there is considerable momentum transfer in an ion–atom interchange collision, the ionized products are emitted in a forward cone which can have quite a large angle. In a beam–gas experiment, it is not easy to collect and mass-analyse all the products simultaneously; in a crossed ion and molecular beam experiment, differential scattering measurements[30] must be made in order that the total cross-sections can be deduced by summing over all angles. But the merged-beam technique possesses the unique advantage that the scattering cone is narrowed in the laboratory frame until it is approximately the same as the beam divergence cone. This is because the scattering event takes place while both particles are travelling fast forward in the laboratory. The collision products are readily removed from the projectile ions by a demerging magnet. Measurements have been attempted for the collision[31]

$$Na^+ + O_2 \rightarrow NaO^+ + O$$

which is unobservable in the flowing afterglow, since the ion (isoelectronic with NeO) is presumably dissociated by thermal energy gas collisions, except at temperatures so low that the forward process is impossible. Although the merged-beam technique is costly and difficult, it is quite possible that it will supersede the high-pressure techniques for two-body processes; however, it cannot be applied to three-body processes, which are of great importance at thermal energies.

Other techniques for ion–atom interchange studies include time of flight mass-spectrometry and ion cyclotron resonance.

14.2 ANALYSIS OF ION–ATOM INTERCHANGE DATA

The first theoretical approach to ion–atom interchange was essentially classical[2, 32, 33]. It was assumed that, in any ion–molecule 'complex' formed in spiralling orbits, the time of collision would be sufficiently long for the total energy to be distributed amongst all possible degrees of freedom. Of all the possible channels in which the species can emerge, that with lowest internal energy is supposed to be greatly favoured; in the simplest case where only one channel is exothermic, this one will be formed in the great majority of spiralling orbit collisions. The orbiting cross-section (equation 14.3) is taken to be equal to the ion–atom interchange cross-section; it will be noticed that it is inversely proportional to impact velocity when the interaction energy is purely polarization, inverse fourth power. Actually the classical interaction is more complicated than this, and the additional inverse sixth power dispersion forces must be considered. Collisions between ions and polar molecules have been treated[34] from the point of view of the additional interaction:

$$V = -\frac{\mu e}{r^2} \cos \theta \qquad (14.22)$$

for dipole moment μ and angle θ between negative end of the dipole and the impact velocity vector. This interaction gives rise to an additional cross-section

$$\sigma_{\text{dipole}} = \frac{\pi \mu e}{E_c} \qquad (14.23)$$

where E_c is the impact energy in the centre-of-mass frame. This treatment assumes that $\cos \theta = 1$ (that is, that there is complete rotational adjustment during orbiting). The effects of this unrealistic assumption may be assessed by computing the degree of adjustment using quantum-mechanical expressions originally applied to rotational Stark effect. It would seem that in the low impact velocity limit the adjustment is perfect, but at high impact velocities it is most imperfect. The effect of this is that the cross-section should be smaller than the sum $\sigma_{\text{polarization}} + \sigma_{\text{dipole}}$ and that its energy dependence should be less rapid than $E^{-1/2}$. However, some polar molecule cross-sections have shown to be E^{-1} in experiment.

A classical phase-space technique of calculating the cross-sections has been worked out by Light and his colleagues[35]. The probability of formation of any given product in a strong coupling collision is taken as proportional to the ratio of the phase-space available to that product divided by the total phase-space available, with conservation of energy and of angular momentum. By a strong coupling collision is meant a collision for which the only good quantum numbers for the three-or-more-body system are those corresponding to the total energy and total angular momentum[33]. In the low impact energy limit, the cross-section for an exothermic process approaches the orbiting limit (equation 14.3). As the impact energy increases, additional channels become open to the decay of an orbiting complex; vibrational excitation of molecular products must be taken into account. Therefore the cross-section for a particular ion–atom interchange process becomes smaller than the orbiting limit as the impact energy increases. Endothermic processes possess cross-section functions which increase gradually above threshold energy, as an increasing amount of phase-space becomes available in these channels. At higher energies still, they in their turn become less probable. For endothermic processes the phase-space theory yields in the centre-of-mass frame a threshold law

$$\sigma \propto E_c - E_{\text{threshold}}$$

provided that $E_c^{1/2} \gg \mu / \alpha^{1/2}$. This is an equality for $E_c = 0.07 \text{ eV}$ with reduced mass $\mu = 2 \text{ a.m.u.}$ and $\alpha = 1 \text{ Å}^3$. The phase-space theory predicts a large cross-section for systems in which the reduced mass of the final system is large, and for those in which the polarizability of the final neutral molecule is large.

The high-energy fall-off of both experimental cross-sections and those predicted by the phase-space theory is very much more rapid than the $E^{-1/2}$ predicted as the orbiting limit. Other classical treatments yield a similar result, and an impulse approximation has been applied[36], to yield an $E^{-5.5}$ dependence, except for the case $m(A) \simeq m(B)$ in the process $A^+ + BC \rightarrow B + A^+C$, when an E^{-1} dependence might be expected. The treatment assumes that C suffers a binary collision with part or all of A^+ and then suffers another with part or all of B; its energy of motion relative to A^+ is thereby reduced until it is insufficient for the two to separate.

Consideration has not yet been given to the questions of whether the classical approximation is adequate for these collisions, whether the spiralling orbit collisions are completely efficient in producing interchange, and whether interchange can occur in non-spiralling orbit collisions. There are two key experimental results which bear upon these questions. First, there are a fairly large number of the simplest tri-nuclear ion–atom interchange processes whose cross-sections at the lowest impact energies are orders of magnitude smaller than the orbiting cross-section. A cross-section function of this sort is illustrated in *Figure 14.2*, and it will be seen that it approaches the orbiting limit only at higher impact energies. Second, there is evidence from the experiments of Henglein and his colleagues[39], as well as other workers, that non-spiralling collisions can give rise to interchange, rather in the manner of the 'spectator-stripping' collisions, discussed in Section 14.3. This evidence[30, 40, 41] derives from Wien filter measurements of post-collision velocities, but retarding and electrostatic deflexion velocity analyses have also been used. It appears that at impact energies of a few electron volts the largest proportion of interchange reactions take place in stripping collisions rather than in spiralling orbits. But at the lowest impact energies the spiralling orbit collisions can come into their own, and are predominant in the interchange process. There is obviously great value in the experimental

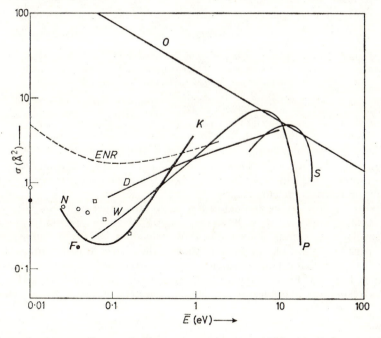

Fig 14.2 Cross-section function for the processes $O^+ + N_2 \rightarrow NO^+ + N$: *S, beam measurements of Stebbings et al.; P, unpublished mass-spectrometer source measurements of Paulson; D, drift tube data of Kaneko, Megill and Hasted[20]; W, photoionization mass-spectrometer source measurement of Warneck[7]; F, flowing afterglow measurements of Ferguson et al.; N, time-dependent afterglow measurements of Nakshbandi and Hasted[37]; O, spiralling orbit limit; ENR broken line, calculations using the nearest resonance method with inclusion of the exchange term[38]*

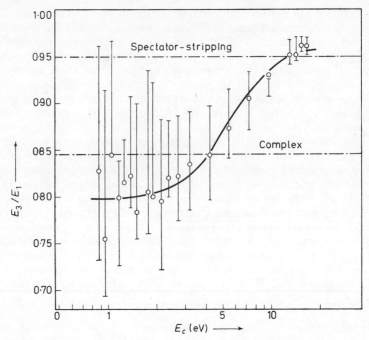

Fig. 14.3. Kinematic analysis of an ion–atom interchange[44] process in terms of ratio of collision energies E_3/E_1, showing proportion of stripping and spiralling orbit collisions as a function of impact energy E_c in centre of mass frame

method of determining the relative proportions of stripping and spiralling, by kinematic analysis of measured post-collision velocities and scattering angles; classical calculations[42, 43] are less powerful because they are extremely sensitive to the interactions assumed. In *Figure 14.3* are displayed the results of a typical Henglein experiment in which the transition from spiralling to stripping regimes appears clearly.

Returning now to the anomalously small thermal energy cross-sections for tri-nuclear processes; these have been supposed[45] to arise from the adiabaticity of the transitions. Since the time of collision a/v is for spiralling orbits an order of magnitude larger than that of the charge transfer collisions to which the adiabatic maximum rule and two-state impact parameter treatment (Chapter 12) were originally applied, the time of transition $h/|\varDelta E|$ must be correspondingly longer for the adiabatic criterion to hold as an equality

$$\frac{a}{v_{\max}} \simeq \frac{h}{|\varDelta E|}$$

It will be recalled that the impact velocity $v_{\max}$ at which the cross-section function for single electron capture passes through a maximum is approximately given by this equality, with adiabatic parameter a taking the value 7 Å. The enhancement of the collision time by spiralling orbits or more complicated trajectories[42, 43] implies a larger adiabatic parameter. The rotation of the interparticle axis in the spiralling orbit is sufficiently slow for rotational adjustment of the electron cloud to be possible adiabatically. It might be supposed that the process is dominated by orbiting collisions in which the final products are in vibrational and rotational excited states such that the energy defect of the process is as small as possible (accidental resonance conditions). But with the longer collision time appropriate to complex formation, the energy defect must be correspondingly smaller for equality of adiabatic criterion. For an artifically large adiabatic parameter $a = 100$ Å, only a few wavenumbers are necessary for the collision to be adiabatic and the cross-section small; since rotational levels of quantum number J are spaced as $J(J+1)$, the spacing for large J in collisions involving only three nuclei is sufficient for the collisions to be adiabatic; for small J this is not the case. It turns out that detailed calculations for all the suitable available thermal energy data (300°–3000°K) (some twenty processes), taking into account rotational, vibrational and kinetic temperatures, and all possible channels, are reasonably successful in reproducing the order of magnitude of cross-sections and their dependence on impact energy and vibrational temperature. This is illustrated in *Figure 12.20* (page 634).

These calculations made use of essentially the Landau–Teller expression, in the form

$$\sigma = \sigma_m \exp\left(1 - \frac{a|\Delta E|}{hv}\right) \tag{14.24}$$

with adiabatic parameter $a = 100$ Å, and impact velocity v. Subsequent calculations[38] made use of Demkov's exchange theory of charge transfer[46] and succeeded in reproducing the minima in the cross-section functions typified by *Figure 14.2*.

For collisions in which more than three nuclei are involved, the vibrational and rotational modes give rise to more closely spaced levels, so the chance of the collisions being principally adiabatic is correspondingly smaller. Exothermic collisions which have small cross-sections at low energies are much less common among polyatomic systems; exothermic polyatomic ion–atom interchange cross-sections are in general within a factor of two of the spiralling orbit limit. Endothermic processes do not achieve such magnitudes until energies are appreciably above threshold, owing to shortage of available phase-space.

The energy transferred from kinetic to internal modes during the collision is an important parameter, and can be inferred from studies of post-collision scattering in crossed beams[47]. It appears that collisions are on the whole near-resonance, but not universally so. Endothermic channels have been found which could indicate curve-crossing.

The classical paths are a good deal more complicated than has been suggested, the interaction having azimuthal dependence; impeded rotation of the impacting molecule is possible as well as actual oscillations of the ions with respect to the molecule[42, 43]. If charge transfer is likely during the collision, the classical interaction will be strongly affected; it has been known for some years that departures of the order of a factor of two from the spiralling orbit limit are to be expected even in polyatomic collisions.

Two-body ion–atom interchange processes involving negative ions are well known, and apparently governed by much the same considerations as apply to those involving positive ions.

Tri-nuclear negative ion processes are sometimes affected by changes of geometrical configuration which arise from application of the Walsh rules.

Three-body ion–atom interchange processes, both positive and negative, are also well known, but whereas some hundreds of apparent two-body processes have been recorded, only some dozens of three-body rates are known.

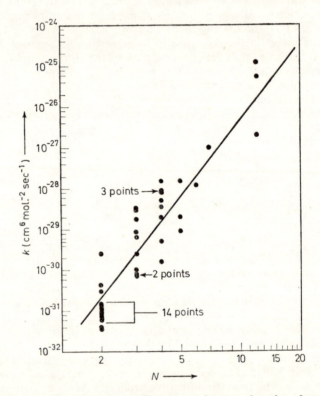

Fig. 14.4. Measured three-body rate coefficients as a function of number of atoms N in the collision complex: the full line represents the empirical relation $k = 10^{-32 \cdot 5 \pm 1} N^{6 \cdot 24}$

High-pressure techniques such as drift and afterglow are particularly suitable for three-body studies, but it is necessary to be quite certain that the variation of pressure in the experiment is sufficient to separate two-body and three-body affects, and this has not always proved possible in early studies. A three-body rate is a measure of the collision time[48], since the longer the collision time the greater the chance of arrival of a third body to play its part. The enormous differences between three-body thermal energy rates lends support to this view; the lifetime of the collision complex may be determined by statistical mechanics; the three-body rates are found to be exponentially proportional to the number of atoms in the complex (see *Figure*

14.4). RRK theory yields an order of magnitude estimate of complex lifetime

$$\tau = 10^{-13} \left(\frac{E - E^*}{E} \right)^{1-s} \tag{14.25}$$

where E is the total energy available, E^* the well-depth, and s the number of active vibrational modes.

The sharp maximum in the rate

$$He^+ + He + He \rightarrow He_2^+ + He$$

is unexpected[49], and may be associated with the role of resonance states of He_2^+.

14.3 CHEMICAL INTERCHANGE PROCESSES BETWEEN NEUTRAL SPECIES

Insofar as the chemical two-body atom–atom interchange process

$$A + BC \rightarrow AB + C$$

and the two-body 'recombination' process

$$A + B \rightarrow AB'$$

have their counterparts in ionic collisions, some consideration of their study may not be out of place in this chapter: total rate measurement and its theoretical interpretation is discussed briefly, together with the measurement of differential scattering with reaction, known as 'reactive scattering'.

Total rate constant measurement for volume processes is normally achieved in a flowing system essentially similar to the flowing afterglow described above. There have been other types of reaction chamber (particularly large spherical vessels into which constituents are introduced at the centre), but flowing systems are most strongly favoured. Chemical methods of analysing the collision products have been used in many experiments, but without doubt the most powerful detection techniques are spectroscopic and mass-spectroscopic. The former can be used either (*i*) in the absorption of resonance radiation, (*ii*) in the measurement of emitted radiation for the detection of excited states, or (*iii*) by exciting the species by irradiation and observing radiation emitted at a different frequency. The mass-spectroscopic detection technique requires ionization of the neutral species, usually by electron impact; the species is allowed to effuse from an orifice in the reaction zone into a high-vacuum system containing ionizer, mass-spectrometer and ion detector.

Theoretical treatment of the chemical collision had for many years been in a rudimentary state; only the effective energy threshold of the reaction, or 'activation energy' was considered as an important parameter, and not the actual cross-section. The reason for this is that gas reactions were studied under thermodynamic equilibrium, the temperature being such that only the high-energy tail of impact energies exceeded the activation energy; since the Maxwell–Boltzmann distribution implies an exponentially falling high-energy tail, only the activation energy could be deduced with any accuracy from the temperature variation of the rate.

Nevertheless the stimulus of the reactive scattering techniques discussed below, together with the ability to monitor the excited state populations following a reaction, have changed all this. Not only reactive scattering, but also infra-red spectroscopy ('chemiluminescence'), has been used[50] to study the excitation of collision products. Intensive theoretical studies have been made using classical theory[51–57]. One most important experimentally available parameter is the fraction f_T of the impact energy which is transferred into vibrational and rotational energy. This proportion, or distribution

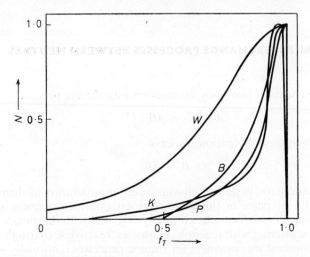

Fig. 14.5. Distribution of transferred energies $N(f_T)$ for the reaction $K + Br_2 \rightarrow KBr + Br$. The maximum possible value corresponds to a 'horizontal' reaction which is equivalent to accidental resonance: W, Warnock et al.[62], experiments; B, Birely and Herschbach[61], experiments; K, Karplus[58], theory; P, Polanyi[60], theory

of proportions $N(f_T)$, is predicted to be different by different types of theoretical approach. For example the 'nearest resonance technique' for ion–atom interchange discussed in the previous section is based on the extreme assumption that the proportion in all collisions is unity ('horizontal reaction'). The predictions of Raff and Karplus[58], Smith[59], Polanyi and colleagues[60] and the reactive scattering experiments of Birely and Herschbach[61] and of Warnock *et al.*[62] lead to less extreme situations; for alkali metal reactions of the type

$$K + Br_2 \rightarrow KBr + Br$$

experimental results and calculations are shown in *Figure 14.5*. However, these reactions are 'harpoon' in character and therefore not typical. The strongly ionic character of KBr is responsible for a Coulomb attraction between $K^{(+)}$ and $Br^{(-)}$, with the other Br acting only as a spectator ('spectator-stripping'):

$$K + Br_2 \rightarrow K^+ + Br_2^- \rightarrow K^+ Br^- + Br$$

Because of the discovery[63] of the discrimination in detecting molecular beams of K and KBr by surface ionization (see Chapter 3), the initial researches into reactive scattering have concentrated on such processes. The extension into 'covalent' reactions[63], involving species which must be detected by ionization, mass-analysis, acceleration and particle counting, produces much more severe problems of background noise than do the alkali metal processes. The chemical cross-sections are relatively small (4 Å^2 for reactive scattering, contrasted with 850 Å^2 for elastic scattering).

Reactive scattering[64] takes place principally at large angles; it is investigated by studying the in-plane differential scattering of collidants and products scattered by a beam crossed with the collidant beam. Velocity selection is incorporated after the collision. Classical analysis proceeds along the following lines.

The energy equation for a reaction between ground state particles $1 + 2 \rightarrow 3 + 4$ may be written

$$\Delta D + E_{12} = E_{34} + E_T \tag{14.26}$$

where ΔD is the difference between dissociation energies of products and reactants; E_{12} and E_{34} are relative impact energies before and after the collisions:

$$\left(\frac{2E_{12}}{\mu_{12}}\right)^{1/2} = v_1 - v_2 \tag{14.27}$$

$$\left(\frac{2E_{34}}{\mu_{34}}\right)^{1/2} = v_3 - v_4 \tag{14.28}$$

The velocity of the centre of mass of the system is

$$v_c = \frac{m_1 v_1 + m_2 v_2}{m_1 + m_2} \tag{14.29}$$

All subscripts 1 and 2 refer to projectile and target, and 3 and 4 refer to the collision products. Momentum conservation requires that:

$$v_1 - v_c = m_2(v_1 - v_2) \tag{14.30}$$

$$v_2 - v_c = \frac{-m_1(v_1 - v_2)}{m_1 + m_2} \tag{14.31}$$

$$v_3 - v_c = \frac{m_4(v_3 - v_4)}{m_3 + m_4} \tag{14.32}$$

$$v_4 - v_c = \frac{-m_3(v_3 - v_4)}{m_3 + m_4} \tag{14.33}$$

With sufficient information it is possible to construct a vector diagram, sometimes called a 'newton diagram', as in *Figure 14.6*. For each value of E_{34} the spectrum of recoil vectors for m_3 can range over a sphere of radius $m_4(v_3 - v_4)/(m_3 + m_4)$ about the tip of the v_c vector. The angle between $v_1 - v_2$

and $v_3 - v_4$ describes the angular distribution; it has azimuthal isotropy about $v_1 - v_2$. It will be seen that the laboratory distribution of recoil vectors of scattering angle θ depends upon the relative energy E_{12}, upon E_T and

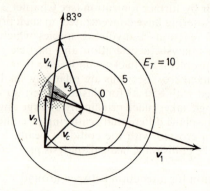

Fig. 14.6. Newton diagram for chemical scattering of potassium atoms by CH_3I, for which the data are shown in Fig. 14.7: 50% of the KI recoil vectors appear in the heavily shaped region, 90% within the lightly shaded region; E_T calibration spheres (drawn as circles) are in kcal mole^{-1} *The reaction is nearly always exothermic*

(From Kwei, Norris and Herschbach[65], by courtesy of W. A. Benjamin, Inc.)

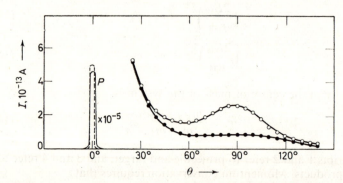

Fig. 14.7. Reactive scattering detector currents for the process $K + CH_3I \rightarrow CH_3 + KI$ *for orthogonal beams of potassium and CH_3I: solid circles represent normalized data obtained with a Pt–W detector more than 50 times as sensitive to K as to KI; open circles represent data with a tungsten detector equally sensitive to K and KI; the parent potassium beam $0.7°$ wide, shown diminished by factor of 10^5, is attenuated 7% by CH_3I beam*

(From Kwei, Norris and Herschbach[65], by courtesy of W. A. Benjamin, Inc.)

upon the differential cross-section $d\sigma/d\Omega(\theta)$. For the reaction

$$K + CH_3I \rightarrow KI + CH_3$$

the distribution $d\sigma/d\Omega(\theta)$, which is shown in *Figure 14.7* leads to a value $E_T \simeq 2.1$ kcal/mole.

Effects of the distribution of velocity in the beams have been considered[66, 67] and descriptions of experiments are available[68–70]. In some reactions there is forward peaking, indicating spectator-stripping, and in others there is backward peaking (rebound). In most cases a large fraction of the reaction exothermicity is converted into product excitation—that is, the processes are usually reasonably 'horizontal', as is shown in *Figure 14.5* for the process

$$K + Br_2 \rightarrow KBr + Br$$

The classical theory of the chemical collision has received considerable attention. It has been supposed that the interchange reaction proceeds by the frictionless motion of a particle upon a model of the potential energy surface for the collision. In general, the excitation of products is small when the moment of inertia of the product molecule is small, and when the reactant atom is light. The degree of excitation also depends upon the form of the saddle in the potential energy surface. It will be recalled from *Figure 2.3* (page 75) that the two troughs in the surface approach and meet over a saddle. However, the two troughs do not always butt symmetrically onto each other, as in a picture-frame: it is possible for the reactant trough to butt onto the side of the product trough; and on the classical[71] picture of motion of a particle on this surface, it will, on surmounting the saddle, pass into vibrational motion along the product trough. The height of this saddle is in fact the 'activation energy' of the reaction.

In the classical approximation the relative impact velocity v_r along a line connecting the centres of the reactants must exceed a certain value for the reaction to occur. The simplest expression[72] is:

$$\sigma(v_r) = \pi b^{*2}\left(1 - \frac{E_{act}^*}{\frac{1}{2}\mu v_r^2}\right) \tag{14.34}$$

where b^* is the largest impact parameter and E_{act}^* the minimum activation energy for which reaction occurs. This expression neglects all coupling of overall rotation of collision complex with the internal degrees of freedom.

Potential energy surfaces for tri-nuclear systems have been calculated using the London–Eyring–Polanyi–Sato (LESP) semi-empirical method, and the activation energies so deduced are in reasonable agreement with experiment. For more complicated systems, hypersurfaces are appropriate. When the surfaces are known, classical Monte Carlo calculations of the collision process can be carried out[73].

Much of the stimulus to understand reactions involving excited species comes from the need to understand chemical effects of irradiating gases with nuclear radiation (radiation chemistry of gases)[74]. A source of strength I curies emitting characteristic radiation at E eV will provide energy at a rate

$$Q \text{ (eV sec}^{-1}) = 3{\cdot}7 \times 10^{10} IE \tag{14.35}$$

The energy efficiency of a reaction is expressed as a parameter known as the G value, which is defined as the number of product molecules produced for each 100 eV of absorbed energy. But many processes for which G values have been determined are in fact more complex than one single reaction.

REFERENCES

1. TALROSE, V. L. and LYUBIMOVA, A. K. *Dokl. Akad. Nauk SSSR*, 86 (1952) 909.
2. STEVENSON, D. P. and SCHISSLER, D. O. *J. chem. Phys.* 23 (1955) 1353; 24 (1956) 926; 29 (1958) 282, 294.
3. FIELD, F. H., FRANKLIN, J. L. and LAMPE, F. W. *J. Amer. chem. Soc.* 78 (1957) 5967; 79 (1957) 2419, 2665, 6132; FIELD, F. H. and LAMPE, F. W. *J. Amer. chem. Soc.* 79 (1957) 4244; 80 (1958) 5587.
4. BIONDI, M. A. *Advanc. Electronics Electron Phys.* 16 (1963).
5. MCDANIEL, E. W. *et al. Ion–Molecule Reactions*, 1970. London; Wiley.
6. HASTED, J. B. *Advanc. Electronics Electron Phys.* 13 (1960) 1.
7. WARNECK, P. *J. chem. Phys.* 46 (1967) 502, 513.
8. GIOUMOUSIS, G. and STEVENSON, D. P. *J. chem. Phys.* 29 (1958) 294.
9. SAPOROSCHENKO, M. *Phys. Rev.* 111 (1958) 1550.
10. TALROSE, V. L. and FRANKEVICH, E. L. *J. phys. Chem., Moscow*, 34 (1960) 1275.
11. WARNECK, P., Geophysical Corporation of America Report 66-13-N, 1966; POSCHEN-REIDER, W. and WARNECK, P. *J. appl. Phys.* 37 (1966) 2812.
12. RYAN, K. R. and FUTRELL, J. H. *J. chem. Phys.* 42 (1965) 824.
13. HENCHMAN, M., OPAUSZKY, I. and MATUS, L. Report, University of Leeds, 1967.
14. KARACHEVTSEV, G. V., MARKIN, M. I. and TALROZE, V. L. *Kinet. Catal.* 5 (1964) 331.
15. GIESE, C. F. and MAIER, W. B. *J. chem. Phys.* 39 (1963) 739.
16. TURNER, B. R., FINEMAN, M. A. and STEBBINGS, R. F. *J. chem. Phys.* 42 (1965) 4088; *Planet. Space Sci.* 13 (1965) 1125.
17. HERMAN, Z., KERSLETTER, J. D., ROSE, T. L. and WOLFGANG, R. Report, University of Colorado, 1967.
18. LU, C. and CARR, H. E. *Rev. sci. Instrum.* 33 (1962) 823.
19. CHAMPION, R. L., DOVERSPIKE, L. D. and BAILEY, T. L. *J. chem. Phys.* 45 (1966) 4377, 4385.
20. KANEKO, Y., MEGILL, L. R. and HASTED, J. B. *J. chem. Phys.* 45 (1966) 3741.
21. GOLDEN, D. E., SINNOTT, G. and VARNEY, R. N. Lockheed Palo Alto Research Laboratory Report, 1967.
22. MILLER, T. M., MOSELEY, J. T., MARTIN, D. W. and MCDANIEL, E. W. *Phys. Rev.* 173 (1968) 115.
23. BLOOMFIELD, C. H. and HASTED, J. B. *Disc. Faraday Soc.* 37 (1964) 176; *Brit. J. appl. Phys.* 17 (1966) 449.
24. DUNKIN, D. B., FEHSENFELD, F. C., SCHMELTEKOPF, A. L. and FERGUSON, E. E. *J. chem. Phys.* 49 (1968) 1365.
25. FITE, W. L., RUTHERFORD, J. A., SNOW, W. R. and VAN LINT, V. *Disc. Faraday Soc.* 33 (1962) 264.
26. LANGSTROTH, G. F. O. and HASTED, J. B. *Disc. Faraday Soc.* 33 (1962) 257.
27. DICKINSON, P. G. and SAYERS, J. *Proc. phys. Soc., Lond.* 76 (1960) 137.
28. FEHSENFELD, F. C., FERGUSON, E. E. and SCHMELTEKOPF, A. L., *J. chem. Phys.* 44 (1966) 3022, 4087, 4095, 4537.
29. FEHSENFELD, F. C., FERGUSON, E. E. and SCHMELTEKOPF, A. L. *J. chem. Phys.* 25 (1966) 404; 46 (1967) 2019, 2802; FERGUSON, E. E., FEHSENFELD, F. C., DUNKIN, D. B., SCHMELTEKOPF, A. L. and SCHIFF, H. I. *Planet. Space Sci.* 12 (1964) 1169; FEHSENFELD, F. C., SCHMELTEKOPF, A. L., FERGUSON, E. E. and GOLDAN, P. D. *Planet. Space Sci.* 13 (1965) 219, 579, 823; 14 (1966) 969; FEHSENFELD, F. C., GOLDAN, P. D. and SCHMELTEKOPF, A. L. *J. geophys. Res.* 70 (1965) 4323; SCHOFIELD, K. *Planet. Space Sci.* 15 (1967) 643.
30. CHAMPION, R. L., DOVERSPIKE, L. D. and BAILEY, T. L. *J. chem. Phys.* 45 (1966) 4377, 4385.
31. FITE, W. L. unpublished data.
32. EYRING, H., HIRSCHFELDER, J. O. and TAYLOR, H. S. *J. chem. Phys.* 4 (1936) 479.
33. FIRSOV, O. B. *Soviet Phys. JETP.* 15 (1962) 906.

34. DUGAN, J. V. and MAGEE, J. L. NASA report TND 3229, 1966; *Gaseous Electronics Conf.* American Physical Society, 1964; THEARD, L. P. and HAMILL, W. H. *J. Amer. chem. Soc.* 84 (1962) 1134; MORAN, T. F. and HAMILL, W. H. *J. chem. Phys.* 39 (1963) 1413.
35. LIGHT, J. C. *J. chem. Phys.* 40 (1964) 3221; PECHUKAS, P. and LIGHT, J. C. *J. chem. Phys.* 42 (1965) 3281; WOLF, F. A. *J. chem. Phys.* 44 (1966) 1619.
36. BATES, D. R., COOK, C. J. and SMITH, F. J. *Proc. phys. Soc., Lond.* 83 (1964) 49; THOMAS, L. H. *Proc. Roy. Soc.* A114 (1927) 561.
37. NAKSHBANDI, M. M. and HASTED, J. B. *Planet. Space Sci.* 15 (1967) 1781.
38. HASTED, J. B. and MOORE, L. *Abstr. 6th Int. Conf. Phys. electron. atom. Collisions,* p. 328, 1969. Cambridge, Mass.; Massachusetts Institute of Technology Press.
39. DING, A., LACMANN, K. and HENGLEIN, A. *Berichte Bun. Phys. Chem.* 71 (1967) 596; DING. A. *Z. Naturf.* 24a (1969) 856; HENGLEIN, A., LACMANN, K. and KNOLL, B. *J. chem. Phys.* 43 (1965) 1048; HENGLEIN, A. and MUCCINI, G. A. *Z. Naturf.* 17a (1962) 452; 18a (1963) 753; HENGLEIN, A., LACMANN, K. and JACOBS, G. *Berichte Bun. Phys. Chem.* 69 (1965) 279; LACMANN, K. and HENGLEIN, A. *Berichte Bun. Phys. Chem.* 69 (1965) 286, 292.
40. TURNER, B. R., FINEMAN, M. A. and STEBBINGS, R. F. *J. chem. Phys.* 42 (1965) 4088.
41. DURUP, M. and DURUP, J. *Advanc. Mass Spectrosc.* 4 (1967) 677.
42. DUGAN, J. V. and MAGEE, J. L. *J. chem. Phys.* 47 (1967) 3103.
43. DUGAN, J. V., PALMER, R. W. and MAGEE, J. L. *Abstr. 6th Int. Conf. Phys. electron. atom. Collisions,* p. 333, 1969. Cambridge, Mass.; Massachusetts Institute of Technology Press.
44. DING, A. *Z. Naturf.* 24a (1969) 856.
45. BOHME, D. K., HASTED, J. B. and ONG, P. P. *Proc. phys. Soc.* B1 (1968) 879.
46. DEMKOV, YU. N. *Proc. 3rd Int. Conf. Phys. electron. atom. Collisions,* p. 831, 1963. Amsterdam; North Holland Publishing Company; DING, A. *Z. Naturf.* 24a (1969) 856.
47. HERMAN, Z., HIERL, P., LEE, A. and WOLFGANG, R. *Abstr. 6th Int. Conf. Phys. electron. atom. Collisions,* p. 78, 1969. Cambridge, Mass.; Massachusetts Institute of Technology Press.
48. BOHME, D. K. *Gaseous Electronics Conf.* American Physical Society, 1968.
49. ONG, P. P. and HASTED, J. B. *Proc. phys. Soc.* B2 (1969) 91.
50. POLANYI, J. C. *J. quantve. Spectros. radiat. Transf.* 3 (1963) 471.
51. MONTROLL, E. W. and SHULER, K. E. *Advanc. chem. Phys.* 1 (1958) 361.
52. KECK, J. *Disc. Faraday Soc.* 33 (1962) 173.
53. BAK, T. A. and LEBOWITZ, J. L. *Disc. Faraday Soc.* 33 (1962) 189.
54. KECK, J. *J. chem. Phys.* 29 (1958) 410; 32 (1960) 1035.
55. WALL, F. T., HILLER, L. A. and MAZUR, J. *J. chem. Phys.* 35 (1961) 255, 1284.
56. BUNKER, D. L. and BLAIS, N. C. *J. chem. Phys.* 37 (1962) 393, 2713; 39 (1963) 315.
57. ELIASON, M. A. and HIRSCHFELDER, J. O. *J. chem. Phys.* 30 (1959) 1426; ELIASON, M. A., STOGRYN, D. E. and HIRSCHFELDER, J. O. *Proc. nat. Acad. Sci., Wash.* 42 (1956) 546.
58. RAFF, L. M. and KARPLUS, M. *J. chem. Phys.* 41 (1964) 1267.
59. SMITH, F. T. *J. chem. Phys.* 31 (1959) 1352.
60. KUNTZ, P. J., NEMETH, E. M., POLANYI. J. C., ROSNER, S. D. and YOUNG, C. E. *J. chem. Phys.* 44 (1966) 1168.
61. BIRELY, J. H. and HERSCHBACH, D. R. *J. chem. Phys.* 44 (1966) 1690.
62. WARNOCK, T. T., BERNSTEIN, R. B. and GROSSER, A. E. *J. chem. Phys.* 46 (1967) 1685.
63. DATZ, S. and TAYLOR, E. H. *J. chem. Phys.* 39 (1963) 1896.
64. HERSCHBACH, D. R. *Advanc. chem. Phys.* 8 (1965).
65. HERSCHBACH, D. R. *J. chem. Phys.* 33 (1960) 1870; *Disc. Faraday Soc.* 33 (1962) 149; KWEI, G. H., NORRIS, J. A. and HERSCHBACH, D. R. *Bull. Amer. phys. Soc.* 5 (1960) 503; *J. chem. Phys.* 34 (1961) 1842; KINSEY, J. L., KWEI, G. H. and HERSCHBACH, D. R. *Bull. Amer. phys. Soc.* 6 (1961) 152.
66. DATZ, S., HERSCHBACH, D. R. and TAYLOR, E. H. *J. chem. Phys.* 35 (1961) 1549.

67. HERSCHBACH, D. R. *Disc. Faraday Soc.* 33 (1962) 149.
68. GROSSER, A. E. and BERNSTEIN, R. B. *J. chem. Phys.* 42 (1964) 1268; 43 (1965) 1140.
69. DATZ, S. and MINTURN, R. E. *J. chem. Phys.* 41 (1964) 1153.
70. WILSON, K. R., KWEI, G. H., NORRIS, J. A., HERM, R. R., BIRELY, J. H. and HERSCHBACH, D. R. *J. chem. Phys.* 41 (1964) 1154.
71. GLASSTONE, S., LAIDLER, K. J. and EYRING, H. *The Theory of Rate Processes*, 1951. New York; McGraw-Hill.
72. ELIASON, M. A. and HIRSCHFELDER, J. O. *J. chem. Phys.* 30 (1959) 1426.
73. BUNKER, D. L. and BLAIS, N. C. *J. chem. Phys.* 41 (1964) 2377.
74. LIND, S. C. *Radiation Chemistry of Gases*, 1961. New York; Reinhold.

APPENDIX

A.1 ATOMIC CONSTANTS AND CONVERSION FACTORS FOR SYSTEMS OF UNITS[1]

Units of Energy

$$1 \text{ eV} = 1.6020 \times 10^{-12} \text{ erg particle}^{-1}$$
$$= 1.1610 \times 10^{4} \, k \text{ °K}$$
$$= 23.053 \text{ kcal mole}^{-1}$$
$$= 8067.5 \text{ cm}^{-1}$$
$$1 \text{ cm}^{-1} = 1.23981 \times 10^{-4} \text{ eV/hc}$$
$$= 2.858 \times 10^{-3} \text{ kcal mole}^{-1}$$
$$1 \text{ kcal mole} = 7 \times 10^{-14} \text{ erg}$$
$$\tfrac{3}{2} \, kT \text{ at } 0°C = 5.657 \times 10^{-14} \text{ erg}$$
$$1 \text{ Rydberg (R)} = 13.595 \text{ eV} = 109\,778 \text{ cm}^{-1}$$

Atomic Units

$$\hbar = e = m_e = 1$$
$$1 \text{ a.u. of energy} = \mu e^{4}/\hbar^{2} = 2\text{R} = 27.2 \text{ eV}$$
$$(\mu \simeq 1.66 \times 10^{-24} \text{ g})$$
$$1 \text{ a.u. of length} = a_0 = \hbar^{2}/\mu e^{2} = 0.5292 \text{ Å}$$
$$\pi a_0^{2} = 8.806 \times 10^{-17} \text{ cm}^2$$
$$a_0^{2} = 2.803 \times 10^{-17} \text{ cm}^2$$
$$1 \text{ a.u. of velocity} = e^{2}/\hbar = 2.18 \times 10^{8} \text{ cm sec}^{-1}$$
$$1 \text{ a.u. of time} = \hbar^{3}/\mu e^{4} = 2.42 \times 10^{-17} \text{ sec}$$

Rationalized MKS Units

$$\mu_0 = 1 \cdot 257 \times 10^{-6} \text{ henry metre}^{-1} \text{ (H m}^{-1})$$

$$= 4\pi \times 10^{-7} \text{ newton second}^2 \text{ coulomb}^{-2} \text{ (N s}^2 \text{ C}^{-2})$$

$$\varepsilon_0 = 8 \cdot 854 \times 10^{-12} \text{ farad metre}^{-1} \text{ (F m}^{-1})$$

$$= (4\pi \times 9 \times 10^9)^{-1} \text{ coulomb}^2 \text{ newton}^{-2} \text{ metre}^{-2} \text{ (C}^2 \text{ N}^{-2} \text{ m}^{-2})$$

$$1 \text{ Tesla} = 1 \text{ weber metre}^{-1} \text{ (W m}^{-1}) = 10^4 \text{ gauss (G)}$$

Charge and Mass of particles

Electron charge $\qquad\qquad e = 1 \cdot 6030 \times 10^{-19}$ C

$$= 4 \cdot 8029 \times 10^{-10} \text{ statcoulomb}$$

$1 \text{ C} = 6 \cdot 2422 \times 10^{18} e$

Electron mass $\qquad\qquad m_e = 9 \cdot 1084 \times 10^{-28}$ g

Electron charge–mass ratio $e/m_e = 5 \cdot 2731 \times 10^{17}$ statcoulomb g^{-1}

$$= 1 \cdot 7592 \times 10^{11} \text{ C kg}^{-1}$$

$e^2/m_e c^2 = 2 \cdot 8 \times 10^{-13}$ cm

Proton mass $\qquad\qquad m_p = 1 \cdot 6598 \times 10^{-24}$ g

$m_p/m_e = 1836 \cdot 13$

Electron velocity $\quad v_e \simeq 5 \cdot 9 \times 10^7 \ V^{1/2}$ cm sec^{-1} (V in volts)

Proton velocity $\quad v_p \simeq 4 \cdot 7 \times 10^7 \ V^{1/2}$ cm sec^{-1} (V in kilovolts)

Physical Constants

Avogadro's number $\qquad N_0 = 6 \cdot 0248 \times 10^{23}$ molecule mole^{-1} at s.t.p.

Loschmidt's number $\qquad n_0 = 2 \cdot 6871 \times 10^{19}$ molecule cm^{-3} at s.t.p.

Velocity of light $\qquad\quad c = 2 \cdot 9979 \times 10^{10}$ cm sec^{-1}

Planck's constant $\qquad\quad h = 6 \cdot 6253 \times 10^{-27}$ erg sec

$$\hbar = 1 \cdot 0545 \times 10^{-27} \text{ erg sec}$$

Gas constant $\qquad\qquad R = 8 \cdot 3167 \times 10^7$ erg mole^{-1} ($^\circ$K)$^{-1}$

Boltzmann constant $\qquad k = 1 \cdot 38041 \times 10^{-16}$ erg ($^\circ$K)$^{-1}$

Fine structure constant $\quad \alpha^{-1} = 137 \cdot 0$

$$1 \text{ cm}^2 \text{ cm}^{-3} \text{ at 1 torr} = 2 \cdot 81 \times 10^{-17} \text{ cm}^2$$

A.2 STEPWISE IONIZATION POTENTIALS[2] OF ATOMS (eV)

Z	Element	1	2	3	4	5	6	7	8
1	H	13·595							
2	He	24·581	54·403						
3	Li	5·390	75·619	122·419					
4	Be	9·320	18·206	153·850	217·657				
5	B	8·296	25·149	37·920	259·298	340·127			
6	C	11·256	24·376	47·871	64·476	391·986	489·84		
7	N	14·53	29·593	47·426	77·450	97·863	551·925	666·83	
8	O	13·614	35·108	54·886	77·394	113·873	138·080	739·114	871·12
9	F	17·418	34·98	62·646	87·14	114·214	157·117	185·139	953·60
10	Ne	21·559	41·07	63·5	97·02	126·3	157·91		
11	Na	5·138	47·29	71·65	98·88	138·37	172·09	208·444	264·155
12	Mg	7·644	15·031	80·12	109·29	141·23	186·49	224·90	265·957
13	Al	5·984	18·823	28·44	119·96	153·77	190·42	241·38	284·53
14	Si	8·149	16·34	33·46	45·13	166·73	205·11	246·41	303·07
15	P	10·484	19·72	30·156	51·354	65·007	220·414	263·31	309·26
16	S	10·357	23·4	35·0	47·29	72·5	88·029	280·99	328·80
17	Cl	13·01	23·80	39·90	53·5	67·80	96·7	114·27	348·3
18	Ar	15·755	27·62	40·90	59·79	75·0	91·3		
19	K	4·339	31·81	46	60·90	82·6	99·7	128	143
20	Ca	6·111	11·868	51·21	67	84·39	109	139	159
21	Sc	6·54	12·80	24·75	73·9	92	111	141	172
22	Ti	6·82	13·57	27·47	43·24	99·8	120	151	174
23	V	6·74	14·65	29·31	48	65	129	161	185
24	Cr	6·764	16·49	30·95	50	73	91		

Z	Element	1	2	3	4	5	6	7	8
25	Mn	7·432	15·636	33·69		76		119	196
26	Fe	7·87	16·18	30·643					151
27	Co	7·86	17·05	33·49					
28	Ni	7·633	18·15	35·16					
29	Cu	7·724	20·29	36·83					
30	Zn	9·391	17·96	39·70					
31	Ga	6·00	20·51	30·70	64·2	93·4	127·5		
32	Ge	7·88	15·93	34·21	45·7	62·6	82		
33	As	9·81	18·63	28·34	50·1	68	88·6		
34	Se	9·75	21·5	32	43	59·7	$132·5 \pm 20$		
35	Br	11·84	21·6	35·9	47·3	$69·9 \pm 12$		155	193
36	Kr	13·996	24·56	36·9	$70·5 \pm 2·5$			103	
37	Rb	4·176	27·5	40					
38	Sr	5·692	11·027		57				
39	Y	6·38	12·23	20·5		77			
40	Zr	6·84	13·13	22·98	34·33		99		
41	Nb	6·88	14·32	25·04	38·3	50	103		
42	Mo	7·10	16·15	27·13	46·4	61·2	68		
43	Tc	7·28	15·26						
44	Ru	7·364	16·76	28·46				125	
45	Rh	7·46	18·07	31·05				126	153
46	Pd	8·33	19·42	32·92					
47	Ag	7·574	21·48	34·82					
48	Cd	8·991	16·904	37·47					
49	In	5·785	18·86	28·03	54·4	72·3			
50	Sn	7·342	14·628	30·49	40·72	56	108		
51	Sb	8·639	16·5	25·3	44·1	60	72	137	
52	Te	9·01	18·6	31	38				

Z	Element	1	2	3	4	5	6	7	8
53	I	10·454	19·09	32·1	45·5	76	85 ± 5	135 ± 20	170
54	Xe	12·127	21·2						
55	Cs	3·893	25·1						
56	Ba	5·210	10·001						
57	La	5·61	11·43	19·17					
72	Hf	7	14·9						
73	Ta	7·88	16·2						
74	W	7·98	17·7						
75	Re	7·87	16·6						
76	Os	8·7	17						
77	Ir	9							
78	Pt	9·0	18·56						
79	Au	9·22	20·5						
80	Hg	10·43	18·751						
81	Tl	6·106	20·42	34·2	50·7				
82	Pb	7·415	15·028	29·8	42·31	68·8			
83	Bi	7·287	16·68	31·93	45·3	56·0	88·3		
84	Po	8·43		25·56					
85	At								
86	Rn	10·746							
76	Fr	5·277	10·144						
88	Ra	6·9	12·1						
89	Ac			20 ?					

A.3 FIRST IONIZATION POTENTIALS OF SOME MOLECULES AND RADICLES

Molecule	E_i (eV)	Molecule	E_i (eV)
CH_4	12·99	HCOOH	11·05
C_2H_6	11·65	F_2	16·5
C_2H_4	10·507	HF	17·0
C_2H_2	11·41	Cl_2	11·32
C_6H_6	9·247	HCl	12·90
		Br_2	12·8
O_2	12·063	HBr	12·09
O_3	12·80	I_2	9·41
H_2O	12·618	HI	10·39
CO	14·013	N_2	15·576
CO_2	13·769	NO	9·266
CH_3OH	10·85	NO_2	9·78
C_2H_5OH	10·48	N_2O	12·893
HCHO	10·87	BF_3	15·7
CH_3CHO	10·21	BCl_3	12·0
NH_3	10·154	SiH_4	12·2
HCN	13·91	SiO_2	11·7
C_2N_2	13·57	$SiCl_4$	11·6
S_2	10·8	PH_3	10·0
CS_2	10·079	PCl_3	12·3
SO_2	12·34	H_2Se	10·46
H_2	15·427	H_2S	10·472
LiI	8·55	S_2	10·7
NaI	8·8	COS	11·25

Radicle	E_i (eV)
CH	11·13
CH_2	11·9
CH_3	9·95
C_2H_5	8·72
C_6H_5	7·76
OH	13·18
HO_2	11·53
CH_3CO	7·92
NH	13·1
CN	14·0
HS	11·1
CS	11·8
COOH	10·88

A.4 SOME IMPORTANT ATOMIC ENERGY LEVELS

The energies of atomic levels, deduced from atomic spectra, are of paramount importance in the study of atomic collisions; the most complete tabulation of the available data for atoms and positive ions is that of Charlotte Moore[2]. Supplements are compiled from time to time, and are published by the United States National Bureau of Standards. A short list of the more important lowest energy levels of the more important neutral atoms appears below.

Atom	Configuration	Wavenumber (cm^{-1})
H	$1s\ ^2S$	0
	$2p\ ^2P_{1/2}$	82258·907
	$2s\ ^2S_{1/2}$	82258·942
	$2p\ ^2P^o_{1/2}$	82259·272
	$3p\ ^2P^o_{1/2}$	97492·198
	$4p\ ^2P^o_{1/2}$	102823·835
He	$1s^2\ ^1S_0$	0
	$2s\ ^3S_1$	159850·318
	$2s\ ^1S_0$	166271·70
	$2p\ ^3P^o_{2,1,0}$	169081·0
	$2p\ ^1P^o_1$	171129·148
	$3s\ ^3S_1$	183231·08
	$3s\ ^1S_0$	184859·06
	$3p\ ^3P^o_{2,1,0}$	185559·0
Li	$^2S_{1/2}$	0
	$^2P^o_{1/2}$	14904
	$^2S_{1/2}$	27206
C	$^3P_0$	0
	$^3P_1$	16·4
	$^3P_2$	43·5
	$^1D_2$	10193·70
	$^1S_0$	21648·4
	$^5S^o_2$	33735·2
	$^3P^o_{0,1,2}$	60333·8
N	$^4S^o$	0
	$^2D^o$	19223·0
	$^2D^o$	19231·0
	$^2P_{3/2,1/2}$	28840
O	$^3P_2$	0
	$^3P_1$	158·5
	$^3P_0$	226·5
	$^1D_2$	15867·7
	$^1S_0$	33792·4
F	$^2P^o_{3/2}$	0
	$^2P^o_{1/2}$	404·0
	$^4P_{5/2}$	102406·50
	$^4P_{3/2}$	102681·24
	$^4P_{1/2}$	102841·20
Ne	$^1S_0$	0
	$3s\ [1\frac{1}{2}]^o_{2,1}$	134043·790

Atom	Configuration	Wavenumber (cm $^{-1}$)
Ne (cont.)	$3s\ [1\frac{1}{2}]^{\circ}_{0,\ 1}$	134820·591
	$3p\ [\frac{1}{2}]_1$	148259·746
Na	$^2S_{1/2}$	0
	$^2P_{1/2}$	16956·183
	$^2P_{3/2}$	16973·379
	$^2S_{1/2}$	25739·86
S	$^3P_2$	0
	$^3P_1$	396·8
	$^3P_0$	573·6
	$^1D_2$	9239·0
	$^1S_0$	22181·4
Cl	$^2P^{\circ}$	0
	$^2P^{\circ}_{1/2}$	881
	^4P	71954·00
	^4P	72484·20
	$^4P_{1/2}$	72822·64
Ar	$^1S_0$	0
	$4s\ [1\frac{1}{2}]^{\circ}_2$	93143·800
	$4s\ [1\frac{1}{2}]^{\circ}_1$	93750·639
	$4s'\ [\frac{1}{2}]^{\circ}_0$	94553·707
	$4s'\ [\frac{1}{2}]^{\circ}_1$	95399·870
	$4p\ [\frac{1}{2}]1$	104102·144
K	$^2S_{1/2}$	0
	$^2P^{\circ}_{1/2}$	12985·17
	$^2P^{\circ}_{3/2}$	13042·89
	$^2S_{1/2}$	21026·8
Br	$^2P^{\circ}_{3/2}$	0
	$^2P^{\circ}_{1/2}$	3685
	$^2P^{\circ}$	63429·82
	^4P	64900·50
	$^4P_{1/2}$	66877·16
Kr	^1S	0
	$5s\ [1\frac{1}{2}]^{\circ}_2$	79972·535
	$5s\ [1\frac{1}{2}]^{\circ}_1$	80917·561
	$5s'\ [\frac{1}{2}]^{\circ}_0$	85192·414
	$5s'\ [\frac{1}{2}]^{\circ}_1$	85847·501
	$5p[\frac{1}{2}]1$	91169·313
I	$^2P^{\circ}_{3/2}$	0
	$^2P^{\circ}_{1/2}$	7603·15
	$^4P_{5/2}$	54633·46
	$^4P_{3/2}$	61819·81
	$^4P_{1/2}$	60896·27
Xe	$^1S_0$	0
	$6s[1\frac{1}{2}]^{\circ}_2$	67068·047
	$6s[1\frac{1}{2}]^{\circ}_1$	68045·663
	$6s'[\frac{1}{2}]^{\circ}_0$	76197·292
	$6s'[\frac{1}{2}]^{\circ}_1$	77185·560
	$6p[\frac{1}{2}]1$	77269·649

Atom	Configuration	Wavenumber (cm)$^{-1}$
W	$^5D_0$	0
	$^5D_1$	1670·30
	$^5D_2$	3325·53
	$^5D_3$	4830·00
	$^5D_4$	6219·33
Hg	^1S	0
	$^3P_0^o$	37645·080
	$^3P_1^o$	39412·300
	$^3P_2^o$	44042·977
	$^1P_1^o$	54068·781
	$^3S_1$	62350·456

A.5 ATOMIC SELECTION RULES FOR ELECTRIC DIPOLE RADIATION[3,4]

Notation is as follows. The principal quantum number is denoted by n, and its value is often written in front of the term. The azimuthal quantum number of the ith electron is l, and the parity $(-1)^{\Sigma l_i}$ is written as a superscript o for **odd**, but without any superscript for even terms. In tables of energy levels the wave number is written in italics when the term is odd. The orbital angular momentum is denoted by L, and when

$$L = 0 \text{ the term is denoted by } S$$
$$L = 1 \text{ the term is denoted by } P$$
$$L = 2 \text{ the term is denoted by } D$$
$$L = 3 \text{ the term is denoted by } F$$
$$L = 4 \text{ the term is denoted by } G$$
$$\text{etc.} \qquad\qquad \text{etc.}$$

The spin angular momentum is denoted by S, and the multiplicity $(2S+1)$ is written as an anterior superscript. The total angular momentum (inner quantum number) is denoted by J, and its value is written as a subscript. For terms of odd multiplicity J takes the values $0, 1, 2, 3, \ldots$; for terms of even multiplicity J takes the values $\frac{1}{2}, \frac{3}{2}, \frac{5}{2}, \ldots$. Thus for the term $6^3P_0^o$, $n = 6$, $L = 1$, $S = 1$, $J = 0$ and the term is odd. The magnetic quantum number is denoted by M.

The rigorous selection rules for electric dipole radiation are:

1. $\Delta J = 0, \pm 1$, but $0 \to 0$ is forbidden.
2. $\Delta M = 0, \pm 1$.
3. The parity must change.

There are three further approximate selection rules:

4. Only one electron must jump, changing l by 1; $\Delta l = \pm 1$.
5. $\Delta L = 0, \pm 1$, but $0 \to 0$ is forbidden. This rule can be weakly violated, the transitions becoming progressively weaker as ΔL increases.
6. $\Delta S = 0$, unless (as is the case for heavy atoms) spin–orbit coupling is not negligible.

Spin–orbit coupling, which is the energy of the spin magnetic moment of the electron in the magnetic field produced by its own orbital motion, and which determines the fine structure in atomic spectra, enables singlet–triplet transitions to take place weakly.

Rules 1–6 may be violated by nuclear perturbations, external perturbations, or in two-quantum transitions. Magnetic dipole and electric quadrupole selection rules have been given[4].

The selection rules for transition from an autoionizing state of an atom into the state of the positive ion are as follows[5]:

If neither of the states is an S state, then the transition is not forbidden.

If one of the states is an S state, then (*i*) if the parities are the same, the other state must be S, D, G..., or (*ii*) if the parities are opposite, the other state must be P, F, H....

The multiplicities $(2S+1)$ of the states must differ by unity.

The restrictions are further relaxed when the autoionizing state is excited by impact with an electron or an atom, since electron exchange is possible. In this case: for atoms subject to LS coupling, the multiplicity can change at most by a number equal to twice the number of electrons in the impacting system; for atoms not subject to LS coupling, or atoms formed from the dissociation of molecules, there is no restriction on the change of spin except that the total spin must be conserved.

A.6 SIMPLIFIED DIATOMIC MOLECULAR SELECTION RULES FOR ELECTRIC DIPOLE RADIATION[4,6]

1. The quantum number Λ for the orbital angular momentum of the electrons is signified by the Greek capital:

$$\Sigma \text{ for } \Lambda = 0$$
$$\Pi \text{ for } \Lambda = 1$$
$$\Delta \text{ for } \Lambda = 2$$

   There is a selection rule $\Delta\Lambda = 0, \pm 1$.
2. The multiplicity $2S+1$ is written as an anterior superscript, and there is a spin selection rule $\Delta S = 0$.
3. The quantum number Ω denotes total electronic angular momentum about the internuclear axis, and is written as a subscript. It is made up of Λ and the spin component quantum number

$$\Sigma = S, S-1, S-2, \ldots, -S$$

   Now $\Omega = |\Lambda + \Sigma|$ (Hund's case *a*). There is a selection rule $\Delta\Sigma = 0$, so that $\Delta\Omega = 0, \pm 1$.
4. The total angular momentum of the atom is designated as J, with $J = \Omega$, $\Omega + 1, \Omega + 2, \ldots$. These are the first rotational levels of an electronic state. There is a general selection rule that $\Delta J = 0, \pm 1$, with $0 \rightarrow 0$ disallowed.
5. However for $\Omega = 0 \rightarrow \Omega = 0$, there is a rule $\Delta J \neq 0$.
6. The coupling of different momenta is such that Ω is not always defined.

In such cases (Hund's case b) it is usual to define a quantum number K, indicating the total angular momentum apart from the spin:

$$K = \Lambda, \Lambda+1, \Lambda+2, \ldots$$

and

$$J = (K+S), (K+S-1), \ldots, |K-S|$$

There is a selection rule that $\Delta K = 0, \pm 1$, and another:

7. $\Delta K \neq 0$ for $\Sigma \rightarrow \Sigma$ transitions.

8. A rotational level is termed positive or negative according to whether the eigenfunction remains unchanged or changes sign for reflexion at the origin. Positive terms combine only with negative.

9. For identical nuclei, a term is symmetric (s) or anti-symmetric (a) according to whether the total eigenfunction remains unchanged or changes sign for an exchange of the nuclei.
   Symmetric terms combine only with symmetric and anti-symmetric with anti-symmetric.

10. The Σ state is called Σ^+ or Σ^- according to whether its electronic eigenfunction remains unchanged or changes sign upon reflexion in any plane passing through the internuclear axis. It can be shown that $\Sigma^+ \nleftrightarrow \Sigma^-$. This rule is valid only if the spin–orbit interaction and the interaction of rotation and electronic motion are negligible.

12. An electronic state is even (gerade, g) or odd (ungerade, u) according to whether the electronic eigenfunction remains unchanged or changes sign for a reflexion at the centre of symmetry. Even states combine only with odd

$$g \nleftrightarrow g, g \leftrightarrow u, u \nleftrightarrow u$$

for nuclei of equal charge but not necessarily equal mass.

13. Of transitions involving change of vibrational quantum number v, $\Delta v = \pm 1$ is by far the strongest, but larger values are allowed; $\Delta v = 0$ is allowed, giving rise to a pure rotation spectrum. Dipole radiation from a purely vibrational transition, involving no rotational transition, is only possible in a polar molecule.

By custom the ground state of a diatomic molecule is designated X, and the first and (usually) successively higher excited states of the same multiplicity are designated A, B, C. Excited states of multiplicity different from X are designated a, b, c. No extra selection rules arise from these designations, which are entirely a matter of custom, in order to distinguish between different states of the same type. Occasionally asterisks are added to the term as an alternative method of distinction.

Garstang[4] adds the following: for all singlet ($S = 0$) states the distinction between Hund's cases a and b disappears, so that singlet states can be considered as belonging to either. All Σ states ($\Lambda = 0$) belong to case b. For Σ^+ states, rotational levels with even K (but any J) have $+$ symmetry, and vice versa. For Π and Δ states the levels for each J are doubly degenerate (Λ type doubling); one component has $+$ symmetry and the other $-$ symmetry. For identical nuclei the g states have $+$ levels with symmetry s and $-$ levels with symmetry a; whilst for u states the $+$ levels have symmetry a and the $-$ levels symmetry s.

A.7 WIGNER–WITMER RULES[6]

These are rules for deriving molecular quantum numbers from those of the atoms which make them up. For two atoms the rules are as follows:

$$
\begin{array}{lll}
S_g + S_g, & S_u + S_u, & \to \Sigma^+ \\
S_g + S_u, & & \to \Sigma^- \\
S_g + P_g, & S_u + P_u, & \to \Sigma^-, \Pi \\
S_g + P_u, & S_u + P_g, & \to \Sigma^+, \Pi \\
S_g + D_g, & S_u + D_u, & \to \Sigma^+, \Pi, \Delta \\
S_g + D_u, & S_u + D_g, & \to \Sigma^-, \Pi, \Delta \\
P_g + P_g, & P_u + P_u, & \to \Sigma^+(2), \Sigma^-, \Pi(2), \Delta \\
P_g + P_u, & & \to \Sigma^+, \Sigma^-(2), \Pi(2), \Delta \\
P_g + D_g, & P_u + D_u, & \to \Sigma^+, \Sigma^-(2), \Pi(3), \Delta(2), \Phi \\
P_g + D_u, & P_u + D_g, & \to \Sigma^+(2), \Sigma^-, \Pi(3), \Delta(2), \Phi \\
D_g + D_g, & D_u + D_u, & \to \Sigma^+(3), \Sigma^-(2), \Pi(4), \Delta(3), \Phi(2), \Gamma \\
D_g + D_u, & & \to \Sigma^+(2), \Sigma^-(3), \Pi(4), \Delta(3), \Phi(2), \Gamma \\
\end{array}
$$

Parities conserve throughout, inner quantum numbers are ignored. The even and odd atomic states are here designated g and u. The multiplicites $(2S+1)$ combine as follows:

$$
\begin{array}{lll}
\text{singlet} & +\text{singlet} & \to \text{singlet} \\
\text{singlet} & +\text{doublet} & \to \text{doublet} \\
\text{singlet} & +\text{triplet} & \to \text{triplet} \\
\text{singlet} & +\text{quartet} & \to \text{quartet} \\
\text{doublet} & +\text{doublet} & \to \text{singlet and triplet} \\
\text{doublet} & +\text{triplet} & \to \text{doublet and quartet} \\
\text{doublet} & +\text{quartet} & \to \text{triplet and quintet} \\
\text{triplet} & +\text{triplet} & \to \text{singlet, triplet and quintet} \\
\text{triplet} & +\text{quartet} & \to \text{doublet, quartet and sextet} \\
\text{quartet} & +\text{quartet} & \to \text{singlet, triplet, quintet, septet.} \\
\end{array}
$$

However, if the nuclei are rather far apart, one may expect a situation where Ω is well defined and related to J, but where Λ and S lack meaning (Hund's case c). In such a case, the possible Ω values are as follows. For unlike atoms with $J_1 \geqslant J_2$, values are:

$$
\begin{array}{l}
J_1 + J_2, \; J_1 + J_2 - 1, \; \ldots, \; \tfrac{1}{2} \text{ or } 0^+ \\
 J_1 + J_2 - 1, \; \ldots, \; \tfrac{1}{2} \text{ or } 0^- \\
\cdots\cdots\cdots\cdots\cdots\cdots\cdots\cdots\cdots \\
J_1 - J_2, \; \ldots, \; \tfrac{1}{2} \text{ or } 0^+ \text{ or } 0^-
\end{array}
$$

If $J_1 + J_2$ is half-integral, the smallest value of Ω is $\frac{1}{2}$. If $J_1 + J_2$ are both half-integral, the smallest value of Ω is 0, and there are equal numbers of 0^+ and 0^-, alternatively 0^+ and 0^- in each line.

If J_1 and J_2 are both integral, there is an odd number of 0^+ and 0^-, and the odd one, in the last line, is 0^+ or 0^- according as to whether the sum $J + J_2 + \sum l_1 (> \sum l_2)$ is even or odd (l_1 and l_2 are the azimuthal quantum numbers of the atomic electrons).

A molecule formed from identical atoms in different states has Ω values as above, but each term occurs twice, once as even (g) once as odd (u). A molecule formed from identical atoms in identical states has Ω values:

$$(2J)_g, (2J-1)_g, \ldots, 0_g^+ \text{ for integral } J$$

$$(2J-1)_u, \ldots, 0_u^-$$

$$\cdots\cdots\cdots$$

$$0_g^+$$

$$(2J)_u, (2J-1)_u, \ldots, 0_u^- \text{ for half-integral } J$$

$$(2J-1)_g, \ldots, 0_g^+$$

$$\cdots\cdots\cdots$$

$$0_g^+$$

Wigner–Witmer rules for diatomic molecules combining with atoms to form linear triatomic molecules[7] are as follows:

$$S_g + \Sigma^+, \quad S_u + \Sigma^- \rightarrow \Sigma^+$$

$$S_g + \Sigma^-, \quad S_u + \Sigma^+ \rightarrow \Sigma^-$$

$$S_g + \Pi, \quad S_u + \Pi \rightarrow \Pi$$

$$S_g + \Delta, \quad S_u + \Delta \rightarrow \Delta$$

$$P_g + \Sigma^+, \quad P_u + \Sigma^- \rightarrow \Sigma^-, \Pi$$

$$P_g + \Sigma^-, \quad P_u + \Sigma^+ \rightarrow \Sigma^+, \Pi$$

$$P_g + \Pi, \quad P_u + \Pi \rightarrow \Sigma^+, \Sigma^-, \Pi, \Delta$$

$$P_g + \Delta, \quad P_u + \Delta \rightarrow \Pi, \Delta, \Phi$$

$$D_g + \Sigma^+, \quad D_u + \Sigma^- \rightarrow \Sigma^+, \Pi, \Delta$$

$$D_g + \Sigma^-, \quad D_u + \Sigma^- \rightarrow \Sigma^+, \Pi, \Delta$$

$$D_g + \Pi, \quad D_u + \Pi \rightarrow \Sigma^+, \Sigma^-, \Pi, \Delta, \Phi$$

$$D_g + \Delta, \quad D_u + \Delta \rightarrow \Sigma^+, \Sigma^-, \Pi, \Delta, \Phi, \Gamma$$

A.8 DISSOCIATION ENERGIES D OF SOME MOLECULES AND RADICLES

Molecule	D (eV)		Molecule	D (eV)	
C_2	4.9 ± 0.3		NO_2	3.114_9	(N–O)
CN	8.1 ± 0.3		NO_2	4.505_6	(NO_2)
CO	11.11		HI	3.06 ± 0.01	
Cl_2	2.476		Hg_2	0.060 ± 0.003	
F_2	1.6 ± 0.35		N_2	9.762	
H_2	4.4776		NO	6.49 ± 0.05	
H_2^+	0.649		Na_2	0.75 ± 0.03	
HCl	4.431		NaCl	4.24 ± 0.05	
HBr	3.75 ± 0.02		O_2	5.084	
HF	5.8 ± 0.2		OH	4.45 ± 0.2	
H_2O	5.113_6		S_2	4.4 ± 0.1	
H_2S	3.26		SH	3.85 ± 0.2	
HCN	9.69	(C–N)	O_3	1.0_4	
HCN	5.6_5	(C–N)	SO_2	5.61_3	
N_2O	4.930_3	(N–N)	NH_3	4.3_8	
N_2O	1.6771_1	(N–O)	CH_4	4.40_6	
CO_2	5.453		C_2H_4	$7.2_6 \pm 0.3$	(C–C)
OCS	3.71		N_2O_4	0.5937	(N–N)

A.9 DIPOLE MOMENTS AND SCALAR POLARIZABILITIES[8–11]

Atom or molecule	Dipole moment μ (Debye units) (10^{-18} e.s.u.)	Polarizability α (Å^3)	Atom or molecule	Dipole moment μ (Debye units) (10^{-18} e.s.u.)	Polarizability α (Å^3)
He	0	0.206	HCl	1.07	2.58
Ne	0	0.408	HBr	0.79	3.52
Ar	0	1.64	HI	0.38	5.23
Kr	0	2.49	DCl	1.12	2.58
Xe	0	4.02	ICl	0.65	7.5
H_2	0	0.806	CO_2	0	$2.59, 2.93$
D_2	0	0.796	H_2O	1.84	1.45
N_2	0	1.74	H_2S	1.02	3.61
O_2	0	1.57	SO_2	1.62	3.78
CO	0.13	1.94	CCl_4	0	10.24
NO	0.16	1.70	CH_4	0	2.56
HF	1.91	0.79	NH_3	1.47	2.16
			C_2H_6	0	4.43

Quadrupole moments are: H_2 and D_2, 0.63×10^{-26} e.s.u.; CO, $\pm 10^{-26}$ e.s.u.; CO_2, -3×10^{-26} e.s.u.; H_2O, 2×10^{-26} e.s.u.

A.10 POTENTIAL ENERGY DIAGRAMS FOR SOME DIATOMIC MOLECULES

Potential energy diagrams for the following diatomic molecules are given in Figures A.1–A.8: H_2, N_2, NO, O_2, CO, HF, Br_2 and He_2^m.

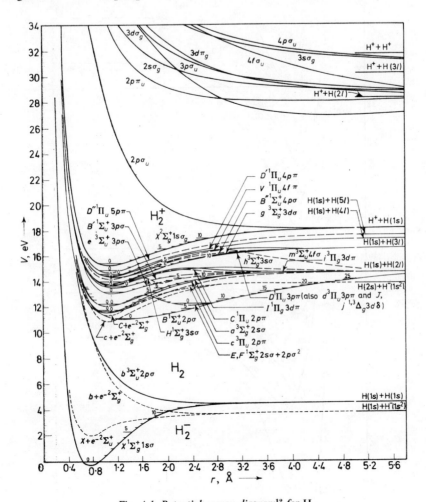

Fig. A.1. Potential energy diagram[12] for H_2

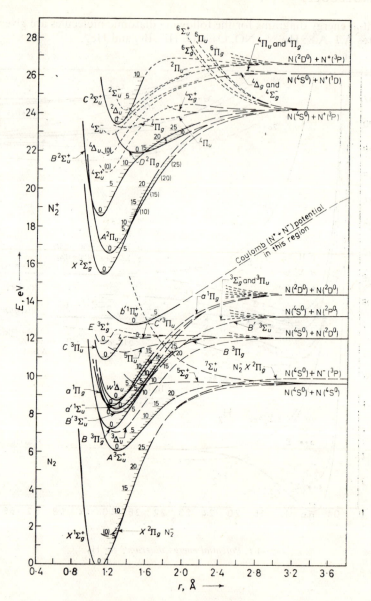

Fig. A.2. Potential energy diagram[13] for N_2

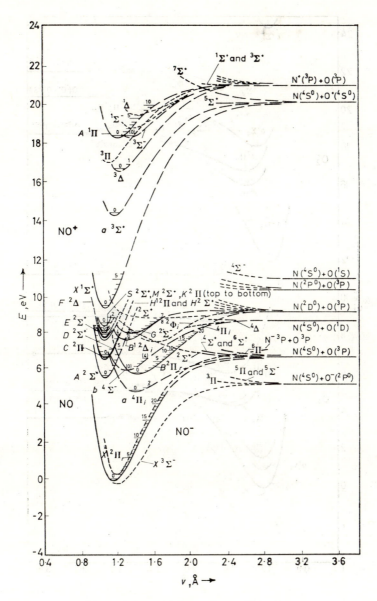

Fig. A.3. Potential energy diagram[13] for NO

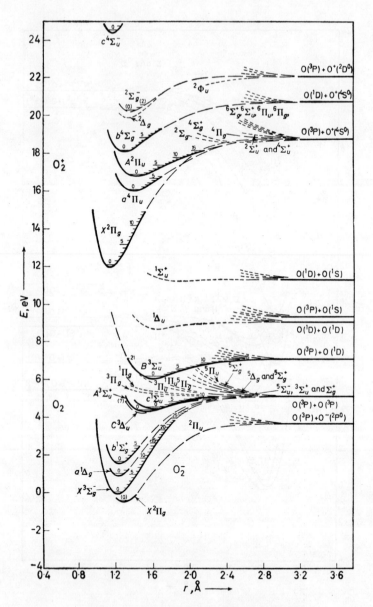

Fig. A.4. Potential energy diagram[13] for O_2

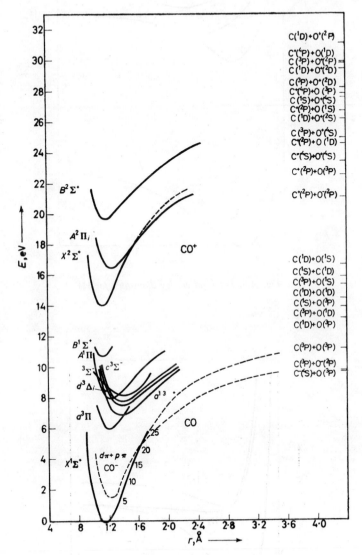

Fig. A.5. Potential energy diagram[14] for CO

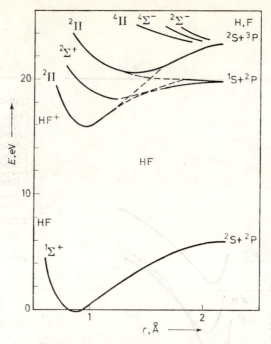

Fig. A.6. Potential energy diagram[15] for HF

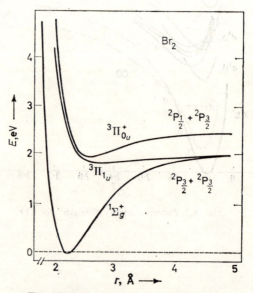

Fig. A.7. Potential energy diagram[6] for Br_2

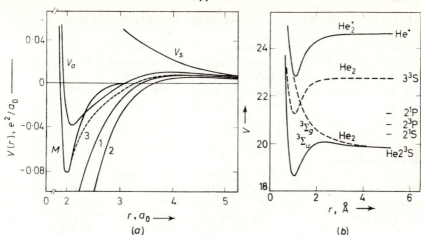

Fig. A.8. (a) Potential energy diagram of He ^{1}S *and* He 2^3S *(V_a and V_s are based on first order perturbation calculations; M is based on Morse interactions; 1 and 2 represent respectively V_a–10/r^6 and V_a–20/r^6; 3 is a possible energy variation linking M and 1)*

(From Buckingham and Dalgarno[16], by courtesy of The Royal Society);
(b) potential energy diagram for He[7]
(From Anderson[17] by courtesy of General Electric Company)

A.11 SOME IMPORTANT MEETINGS ON ATOMIC COLLISIONS AND RELATED FIELDS

International Conferences on the Physics of Electronic and Atomic Collisions (ICPEAC)

1st, 1959, New York: Unpublished.
2nd, 1961, Boulder, Colorado: Abstracts, Benjamin Inc., New York.
3rd, 1963, London, England: Proceedings, entitled 'Atomic Collision Processes', Ed. M. R. C. McDowell, North Holland Publishing Company, Amsterdam.
4th, 1965, Quebec: Abstracts, Science Bookcrafters Inc., Hastings-on-Hudson, New York.
5th, 1967, Leningrad: Proceedings, Nauk Publishing House, Leningrad.
6th, 1969, Boston, Mass: Abstracts, Massachusetts Institute of Technology Press, Cambridge, Mass.

International Conferences on Ionization Phenomena in Gases (Situated in Europe)

1st, 1953, Oxford, England: Unpublished.
2nd, 1955, Delft, Holland: Unpublished.
3rd, 1957, Venice, Italy: Proceedings, North Holland Publishing Company, Amsterdam.
4th, 1959, Uppsala, Sweden: Proceedings, Vols. I and II, North Holland Publishing Company, Amsterdam.

5th, 1961, Munich, Germany: Proceedings, North Holland Publishing Company, Amsterdam.
6th, 1963, Paris, France: Proceedings, SERMA, Paris.
7th, 1965, Belgrade, Yugoslavia: Proceedings, Gradevinska Knjiga Publishing House, Belgrade.
8th, 1967, Vienna, Austria: Proceedings, Springer-Verlag, Berlin.
9th, 1969, Bucharest, Romania: Proceedings, Academy of Socialist Republic of Romania, Bucharest.

British Conferences

1968, Belfast, *Conference on Heavy Particle Collisions:* Proceedings, The Institute of Physics and The Physical Society, London.
1969, Manchester, *1st National Congress on Atomic and Molecular Physics:* Abstracts, The Institute of Physics and The Physical Society, London.
1970, London, *2nd National Congress on Atomic and Molecular Physics:* Abstracts, The Institute of Physics and The Physical Society, London.

REFERENCES

1. *American Institute of Physics Handbook*, 1957. New York; McGraw-Hill.
2. CHARLOTTE E. MOORE *Atomic Energy Levels* Vol. 1, 1949; Vol. 2. 1952; Vol. 3, 1957. Washington; National Bureau of Standards.
3. HERZBERG, G. *Atomic Spectra and Atomic Structure*, 2nd Edn, 1944. New York; Dover; CONDON, E. U. and SHORTLEY, G. H. *The Theory of Atomic Spectra*, 1935. London; Cambridge University Press; SLATER, J. C. *Quantum Theory of Atomic Structure*, 1960. New York; McGraw-Hill; LAYZER, D. *Ann. Phys.* 8 (1959) 271.
4. GARSTANG, R. H. *Atomic and Molecular Processes*, Ed. D. R. Bates, 1962. New York; Academic Press.
5. RUDD, M. E. and SMITH, K. *Phys. Rev.* 169 (1968) 79.
6. HERZBERG, G. *Molecular Spectra and Molecular Structure, Vol. I: The Spectra of Diatomic Molecules*, 2nd edn, 1950. New York; Van Nostrand.
7. HERZBERG, G. *Molecular Spectra and Molecular Structure, Vol. III: Polyatomic Molecules*, 1968. New York; Van Nostrand.
8. ROTHE, E. W. and BERNSTEIN, R. B. *J. chem. Phys.* 31 (1959) 1619.
9. DALGARNO, A. *Planetary Aeronomy, X: Atomic Polarizabilities and Shielding Factors*, Geophysics Corporation of America, contract NAS w-395, 1963.
10. BUCKINGHAM, R. A. *Quart. Rev. chem. Soc., Lond.* 13 (1959).
11. SALOP, A., POLLACK, E. and BEDERSON, B. *Phys. Rev.* 124 (1961) 1431; CHAMBERLAIN, G. E. and ZORN, J. C. *Phys. Rev.* 129 (1963) 677.
12. SHARP, T. E. Report LMSC 5-10-69-9, Lockheed Palo Alto Research Laboratory, Palo Alto, California, 1970.
13. GILMORE, F. R. *J. quantve. Spectros. radiat. Transf.* 5 (1965) 369.
14. KRUPENIE, P. H. and WEISSMAN, S. *J. chem. Phys.* (1965) 1529.
15. PRICE, W. G. University of London, private communication.
16. BUCKINGHAM, R. A. and DALGARNO, A. *Proc. Roy. Soc.* A213 (1952) 327.
17. ANDERSON, J. M. General Electric Report 61-RL 2817G, Schenectady, 1961.

AUTHOR INDEX

Numbers in italic type indicate pages where full reference details to work by the appropriate author are given; other numbers indicate text pages on which these references are quoted

Abella, I. D., 513, *517*
Aberth, W., 61, *108;* 119, *288;* 120, 204, *296;* 524, 526, *561;* 527, 606, *611;* 639, *656;* 641, *657*
Abgrall, H., 470, *482*
Ablow, C. M., 113, *288;* 538, *563*
Abrines, R., 84, *110;* 581, 582, *608;* 644, *657*
Adamczyk, B., 241, 243, *299*
Adams, J., 267, *302*
Adirovich, E. I., 30, *55*
Adler, J., 583, *609*
Afanaseva, E. A., 435, 440, *449*
Afrosimov, V. V., 146, *290;* 204, *296;* 224, *297;* 243, *299;* 264, *301;* 269, *302;* 453, *480;* 526, *562;* 567, 577, 579, *607;* 576, *608;* 590, *610;* 595, 602, 616, 617, *654;* 629, 630, *655;* 645, *658*
Akishin, A. I., 265, *301*
Al Joboury, M. I., 511, *517*
Alam, G. D., 228, 237, *298;* 626, *655;* 650, *658*
Albritton, D. L., 556, *564*
Alder, K., 348, *380;* 387, *419*
Allen, A. J., 180, *292;* 188, *294*
Allen, C. W., 348, *380;* 490, *514;* 491, 494, *515*
Allen, J. E., 283, *303*
Allen, J. S., 132, 136, *289*
Allis, W. P., 37, *55;* 323, 328, 331, *341*
Allison, D. C. S., 651, *658*
Allison, J. K., 198, *295*
Allison, S. K., 15, *55;* 574, 576, *608;* 620, *655*
Almquist, E., 577, *608*
Altick, P. L., 501, *516*
Altshuler, S., 313, *341*
Amdur, I., 162, *291;* 264, *301;* 526, 528, 529, *561;* 538, *563*
Amme, R., 620, *655;* 634, *656;* 688, *695*
Amusia, M. Ya., 501, 507, *516;* 602, *610*
Anderson, J. B., 154, 159, *290;* 160, *291*
Anderson, J. M., 337, 338, *342;* 429, *448;* 438, *449;* 739, *740*
Anderson, N., 221, *297*
Anderson, R. J., 359, *381*
Anderson, W. H., 619, *696*
Andreev, E. P., 493, *515;* 638, *656*
Andres, R. P., 154, 159, 160, *290*

Andrick, D., 227, *298*
Angel, D. W., 257, *300;* 259, *301*
Angona, F. C., 688, *695*
Ankudinov, V. A., 493, *515;* 567, 577, *607;* 631, 638, *656*
Ansdell, D. A., 204, *296;* 452, *480*
Armstrong, B. H., 500, *516*
Arnold, S. G., 680, *694*
Arnot, F. L., 684, *694*
Arshadi, M., 558, *565*
Arthurs, A. M., 582, *608*
Asaad, A. S., 491, *514*
Ash, E. A., 146, *290*
Asundi, R. K., 389, 395, 396, 397, *419;* 465, *466;* 393, 464, *482*
Atkinson, W. R., 425, *447;* 443, *450*
Aubrey, B. B., 312, *340;* 437, *449*
Awan, A. M., 227, *298;* 322, 341, 376, *383*
Aybar, 688, *695*

Bacon, F., 417, *421*
Baede, A. P. M., 635, *656*
Bagot, C. H., 323, 339, *341;* 458, *481*
Bahcall, D., 453, *480*
Bailey, C. L., 588, *609*
Bailey, D. S., 284, *303*
Bailey, T. L., 193, 195, *295;* 476, *483;* 526, 528, *561;* 701, 705, 707, *716*
Bailey, V. A., 458, *481*
Bailey, V. I., 306, 323, 333, *340*
Bak, T. A., 712, *717*
Baker, F. A., 126, *289;* 291, *294;* 372, 378, *383;* 392, 408, *421;* 651, *658*
Baldwin, G. C., 180, 231, *292*
Balloffet, G., 186, *289*
Ballu, Y., 189, 227, *294*
Bandel, H. W., 310, *340*
Bannerberg, J. G., 125, *289*
Bannon, J., 125, *289*
Baranger, M., 662, *692*
Barber, M., 123, *288*
Barber, M. W., 286, *304;* 418, *422*
Barbiere, D., 307, *340*
Bardsley, N., 316, *341;* 407, *420;* 427, *448;* 440, *450;* 463, *481;* 479, *483*
Barker, M., 286, *304*

Barnard, G. B., 191, *294;* 238, *298*
Barnes, W. S., 50, *56*
Barnett, C. F., 264, 266, *301;* 270, *302;* 477, 478, *483;* 587, *609;* 616, 617, 627, *654;* 643, 644, *657*
Barrat, J. P., 489, 492, *514*
Barth, C. A., 147, *290*
Barua, A. K., 538, *563*
Basco, N., 212, *296;* 688, *695*
Bashkin, S. M., 493, *515*
Bastide, R. P., 196, 200, *295*
Basu, D., 644, 645, *657*
Bates, B., 259, *301*
Bates, D. R., 57, 100, *108;* 73, *110;* 96, 106, *111;* 424, 425, 426, *447;* 427, *448;* 429, 430, 439, *449;* 443, 444, *450;* 446, 462, *481;* 473, *483;* 475, 479, 494, *515;* 500, *516;* 556, *564;* 581, 582, 583, 584, 585, 588, *608;* 583, *609;* 600, *610;* 612, *654;* 613, 619, 625, *655;* 633, *656;* 639, 642, 643, 644, 645, *657;* 646, 648, 649, *658;* 681, *694;* 685, *695;* 706, *717*
Batho, H. F., 569, *607*
Bauer, E., 312, *341;* 427, *448;* 440, *450*
Bauer, H J., 688, *695*
Bauer, S. H., 687, *695*
Baughman G. L., 685, *694*
Beahn T. J., 671, *693*
Beaty, E. C., 556, *564*
Beauchamp, R. K., 204, *296*
Bebb H., B., 513, *517*
Becker, E. W., 155, 160, 161, *291*
Becker, G., 157, *291*
Becker, P. M., 685, *694*
Becker, R., 688, *695*
Bederson, B., 43, 44, *56;* 115, *288;* 217, *297;* 261, *300;* 312, *340;* 313, *341;* 369, *382;* 732, *740*
Bekefi, G., 332, 335, 338, *342;* 336, 339, *343;* 307, *340*
Bell, G. D., 490, *514*
Bell, G., I., 645, *657*
Bell, R. J., 583, 586, *609;* 684, *694*
Bellamy, E. H., 275, *302*
Belyaev, V. A., 113, 115, *288*
Bennett, R. G., 492, *515*

741

CHEMICAL FORMULA INDEX

SUBJECT INDEX

Beams *(Con.)*
  polarized, 369
  profiles of, 115, 116
Bellows, collision chamber, 119
Beta radiation, 179
Bethe approximation, 98, 345, 398, 405,
  586
  plot of, 398
Beutler–Fano profile, 317
Blanc's law, 557
Born approximation, 96, 98, 345, 360,
  379, 385, 533, 583, 607
  validity of, 97
Born–Oppenheimer approximation, 102,
  363
Breit–Wigner formula, 320
Bremsstrahlung, stimulated, 4
Broadening. *See* Line broadening
Buckingham–Corner potential, 68

Calibration
  of multipliers, 268
  of photon detectors, 188, 256, 350
Calorimetric detection, 264, 280, 284
Capacitance manometer, 124
Capillary source, atom beams, 156
Capture, electron. *See* Charge transfer
Cathodes, 168
Cavity resonators, 151
Centimetre band, 32
Central force fields, 85, *see also* Interaction energy
Centre-of-mass coordinates, 76
Centrifugal potential, 87
Channel multiplier, 257
Chaperone, 2, 21
Chapman–Enskog formula, 537, 541
Charge transfer, 5, 106, 114, 167, 205, 612,
  697
  dissociative, 697
  high-energy, 642
  measurement of, 627
  multiply-charged ion, 648, 651
  negative ion, 456, 652
  partial, 5
  radiative, 651
  symmetrical resonance, 106, 161, 612
  unlike ions and atoms, 621
  yielding excited species, 637
Charge-changing collisions, 13, 572
  collection, 120, 577
Chemical detection of atomic beams, 280
Chemical interchange, 697, 711
Chemical source, 167
Chopping. *See* Beams, crossed

Classical theory
  elastic scattering, 82
  inelastic collisions, 83
  ionization, 386
Classification of collisions, 2
Clausius–Clapeyron equation, 128
Clustering, 5, 558
  kinetics of, 558
  statistical, 557
Coincidence experiments, 411
Collective states, oscillations, 602
Collision chambers, 118, 573
Collision frequency, 10
  electron, 2, 10, 315, 333
Collision path length, 119
Collision time, 621
Collisional detachment, 6, 456, 475
Collisional radiative decay, 429
Collisional radiative recombination, 429,
  439
Collisions
  charge-changing. *See* Charge-changing collisions
  chemical, 711
  classification of, 2
  elastic, 2, 305
  electron–atom and electron–molecule,
    305
  excited species, 569
  experimental study of, 112
  inelastic, 2
  kinetics of, 23, 27
  notation, 3
  orbiting, 90, 479, 620, 705
  quantum theory of, 100
  superelastic, 4
  theoretical background of, 57
  three-body, 2, 21, *see also* Attachment,
    electron; Recombination
  two-body, definition, 1
  see also Cross-sections; Excitation; Ionization
Compound states. *See* Resonances
Computer control of experiments, 287
Concentrations, equilibrium, 26
Condensation, 698
Conferences, 739
Configuration interaction, 63
Control of experiments, 286
Convolution, 39, 41
Coordinates
  centre-of-mass, 76
  laboratory, 18, 76
  spherical polar, 93
Coronal regime, 54
Coster–Kronig transition, 379, 401